TABLE OF ATOMIC WEIGHTS 1971*

Scaled to the relative atomic mass, $A_r(^{12}C) = 12$

The values of $A_r(E)$ given here apply to elements as they exist in materials of terrestrial origin and to certain artificial elements.

Name	Symbol	Atomic Number	Atomic Weight	Name	Symbol	Atomic Number	Atomic Weight
Actinium	Ac	89	—				200.5_9
Aluminium	Al	13	26.98154				95.9_4
Americium	Am	95		Neodymium			144.2_4
Antimony	Sb	51	121.7_5	Neon	Ne	10	20.17_9
Argon	Ar	18	39.94_8	Neptunium	Np	93	237.0482
Arsenic	As	33	74.9216	Nickel	Ni	28	58.7_1
Astatine	At	85	—	Niobium	Nb	41	92.9064
Barium	Ba	56	137.3_4	Nitrogen	N	7	14.0067
Berkelium	Bk	97	—	Nobelium	No	102	—
Beryllium	Be	4	9.01218	Osmium	Os	76	190.2
Bismuth	Bi	83	208.9804	Oxygen	O	8	15.999_4
Boron	B	5	10.81	Palladium	Pd	46	106.4
Bromine	Br	35	79.904	Phosphorus	P	15	30.97376
Cadmium	Cd	48	112.40	Platinum	Pt	78	195.0_9
Caesium	Cs	55	132.9054	Plutonium	Pu	94	—
Calcium	Ca	20	40.08	Polonium	Po	84	—
Californium	Cf	98	—	Potassium	K	19	39.09_8
Carbon	C	6	12.011	Praseodymium	Pr	59	140.9077
Cerium	Ce	58	140.12	Promethium	Pm	61	—
Chlorine	Cl	17	35.453	Protactinium	Pa	91	231.0359
Chromium	Cr	24	51.996	Radium	Ra	88	226.0254
Cobalt	Co	27	58.9332	Radon	Rn	86	—
Copper	Cu	29	63.54_6	Rhenium	Re	75	186.2
Curium	Cm	96	—	Rhodium	Rh	45	102.9055
Dysprosium	Dy	66	162.5_0	Rubidium	Rb	37	85.467_8
Einsteinium	Es	99	—	Ruthenium	Ru	44	101.0_7
Erbium	Er	68	167.2_6	Samarium	Sm	62	150.4
Europium	Eu	63	151.96	Scandium	Sc	21	44.9559
Fermium	Fm	100	—	Selenium	Se	34	78.9_6
Fluorine	F	9	18.99840	Silicon	Si	14	28.08_6
Francium	Fr	87	—	Silver	Ag	47	107.868
Gadolinium	Gd	64	157.2_5	Sodium	Na	11	22.98977
Gallium	Ga	31	69.72	Strontium	Sr	38	87.62
Germanium	Ge	32	72.5_9	Sulfur	S	16	32.06
Gold	Au	79	196.9665	Tantalum	Ta	73	180.947_9
Hafnium	Hf	72	178.4_9	Technetium	Tc	43	—
Helium	He	2	4.00260	Tellurium	Te	52	127.6_0
Holmium	Ho	67	164.9304	Terbium	Tb	65	158.9254
Hydrogen	H	1	1.0079	Thallium	Tl	81	204.3_7
Indium	In	49	114.82	Thorium	Th	90	232.0381
Iodine	I	53	126.9045	Thulium	Tm	69	168.9342
Iridium	Ir	77	192.2_2	Tin	Sn	50	118.6_9
Iron	Fe	26	55.84_7	Titanium	Ti	22	47.9_0
Krypton	Kr	36	83.80	Tungsten	W	74	183.8_5
Lanthanum	La	57	138.905_5	Uranium	U	92	238.029
Lawrencium	Lr	103	—	Vanadium	V	23	50.941_4
Lead	Pb	82	207.2	Wolfram	W	74	183.8_5
Lithium	Li	3	6.94_1	Xenon	Xe	54	131.30
Lutetium	Lu	71	174.97	Ytterbium	Yb	70	173.0_4
Magnesium	Mg	12	24.305	Yttrium	Y	39	88.9059
Manganese	Mn	25	54.9380	Zinc	Zn	30	65.38
Mendelevium	Md	101	—	Zirconium	Zr	40	91.22

* Pure and Applied Chemistry, **30**, 637 (1972).

PHYSICAL CHEMISTRY

PHYSICAL CHEMISTRY

FOURTH EDITION

FARRINGTON DANIELS

ROBERT A. ALBERTY

Professor of Chemistry,
Dean of Science
Massachusetts Institute
of Technology

John Wiley & Sons, Inc.
New York London Sydney Toronto

Cover photograph courtesy CIBA-GEIGY Corp.

Library of Congress Cataloging in Publication Data:

Daniels, Farrington, 1889–1972.
  Physical Chemistry.

  Includes bibliographies.
  1. Chemistry, Physical and theoretical.
I. Alberty, Robert A., joint author. II. Title.

QD453.2.D36 1975 541 74-13604
ISBN 0-471-19480-8

Printed in the United States of America

10 9 8 7 6 5 4 3 2 1

PREFACE

This book is intended for a comprehensive first course in physical chemistry. It emphasizes the fundamentals that provide a basis for the understanding of chemistry.

The basic structure of this fourth edition is the same as the third edition: Part I, Thermodynamics; Part II, Dynamics; Part III, Quantum Chemistry; and Part IV, Structure. The number of chapters is the same as in the previous edition, but there are three new chapters, and three topics to which full chapters were devoted in the third edition have now been integrated into other parts of the text. The new chapters are Chapter 7, "Ionic Equilibria and Biochemical Reactions"; Chapter 16, "Magnetic Resonance Spectroscopy"; and Chapter 20, "Macromolecules." These are not new topics, but they are topics that receive increased emphasis in the present edition. Two of them reflect more attention to biological applications of physical chemistry, and the chapter on magnetic resonance reflects the increased importance of nuclear magnetic resonance and electron spin resonance. The topics in the third edition that no longer appear as chapters are "Gases," "Other Structural Methods," and "Nuclear and Radiation Chemistry." There is not sufficient space to mention all of the major changes in various chapters, but there is greater coverage of quantum theory, molecular electronic structure, photochemistry, and solid-state chemistry.

In this edition a major step has been taken toward the use of SI (International System of Units) units. SI is a system of carefully worked-out units that is suitable for representing all physical quantities and is used internationally. Although it is desirable to replace the thermochemical calorie with the joule (which is the SI unit of energy), more complete implementation of this change in physical chemistry instruction will have to wait for the availability of standard reference tables in joules. Except for using the calorie, I have tried to follow SI recommendations of base units, symbols, and abbreviations. Three other non-SI units that are retained are the atmosphere (exactly 101,325 Pa or newtons per square meter), the Torr (1/760 atm), and the Ångström (exactly 10^{-10} m). There are other non-SI units that the International Organization for Standardization (ISO) has recommended may be retained because of their practical importance or because of their use in specialized fields. These include the liter (10^{-3} m^3) and the electron volt (eV). The dyne, erg, and esu, which are cgs (centimeter-gram-second) units with special names, are not used in this edition.

Since the number of credits in physical chemistry courses, and therefore the need for more advanced material, varies at different universities, more topics have been included in this edition than can be covered in some courses. Some of the more advanced material has been set in smaller type to indicate that it might be skipped in a first course.

This edition contains 229 new problems. Different types of problems are offered at the end of each chapter to meet the needs of students with varying backgrounds and interests. There are three parallel sets to give ample choice. The answers are given for the first set of problems, and then the student is on his own. In each chapter several typical problems are worked out as *Examples*.

Outlines of Theoretical Chemistry, as it was then entitled, was first written in 1913 by Dr. Frederick H. Getman, who carried it through 1927 in four editions. The next four editions were written by Dr. Farrington Daniels. In 1955 I joined Dr. Daniels in the first edition of this book. The present edition, therefore, traces its origin back 62 years.

Dr. Daniels was looking forward to working actively on the fourth edition and, before he died on June 23, 1972, made a number of suggestions as to how the third edition could be improved. His good judgment and wise counsel have been greatly missed. Dr. Daniels was involved with this book for 45 years and during that time introduced many innovations in the teaching of physical chemistry.

Numerous individuals made useful suggestions in the preparation of previous editions. Quite a few of them were very kind to read drafts of chapters of the current edition and to make detailed recommendations. I especially acknowledge the suggestions provided by P. Bender, Joan B. Berkowitz, G. Blytas, M. J. Buerger, C. D. Cornwell, J. M. Deutch, W. H. Eberhardt, G. G. Hammes, W. Kauzmann, S. H. Kim, E. L. King, J. L. Kinsey, R. C. Lord, W. G. Miller, I. Oppenheim, J. Th. G. Overbeek, M. A. Paul, J. Ross, P. R. Schimmel, R. J. Silbey, J. I. Steinfeld, J. S. Waugh, and M. S. Wrighton.

I am also indebted to H. DeVoe, J. Edwards, R. T. Grimley, H. P. Gregor, N. R. Kestner, H. Kimmel, P. A. Lyons, R. S. Scott, P. Smith, J. E. Stuehr, and M. A. Wartell for reviews of the manuscript for this edition.

Valuable help in checking calculations and problems was provided by Curt Covey and Barry Nelson also assisted with the proofreading. I especially thank Lillian Alberty for the difficult job of typing the manuscript and for encouraging me in the preparation of this new edition.

Cambridge, Massachusetts, 1974   Robert A. Alberty

A NOTE TO THE STUDENT

A problems book consisting of selected problems from Problems Set A, including worked-out solutions, is available as a companion to this text. Please ask for *Physical Chemistry Problems and Solutions* by Robert A. Alberty.

R. A. A.

CONTENTS

SI UNITS

In the past physical chemists have used cgs (centimeter-gram-second) units and certain defined units like the calorie. More recently international usage has moved toward the four-unit system based on the meter, kilogram, second, and ampere. The name "Systeme International d'Unites" (International System of Units), with the abbreviation SI, was adopted by the 11th Conference Generale des Poids et Mesures in 1960. The SI system is founded on the seven base units listed in the following table.

Physical Quantity	Symbol for Quantity	Name of SI Unit	Symbol for SI Unit
Length	l	meter	m
Mass	m	kilogram	kg
Time	t	second	s
Electric current	I	ampere	A
Thermodynamic temperature	T	kelvin	K
Amount of substance	n	mole	mol
Luminous intensity	I_v	candela	cd

The definitions of the SI units are given in the Appendix. All quantities may be expressed in these units or in terms of derived units obtained algebraically by multiplication and division. The principal derived units used in physical chemistry are given in the following table.

Quantity	Unit	Symbol	Definition
Force	newton	N	$kg\ m\ s^{-2}$
Work, energy, quantity of heat	joule	J	$N\ m$
Power	watt	W	$J\ s^{-1}$
Pressure	pascal	Pa	$N\ m^{-2}$
Electric charge	coulomb	C	$A\ s$
Electric potential difference	volt	V	$kg\ m^2\ s^{-3}\ A^{-1}(= J\ A^{-1}\ s^{-1} = J C^{-1})$
Electric resistance	ohm	Ω	$kg\ m^2\ s^{-3}\ A^{-2}(= V\ A^{-1})$
Frequency	hertz	Hz	s^{-1} (cycle per second)
Magnetic flux density	tesla	T	$kg\ s^{-2}\ A^{-1}(= V\ m^{-2}\ s)$

Decimal multiples and fractions of these units are designated by means of the prefixes in the following table.

Fraction	Prefix	Symbol	Multiple	Prefix	Symbol
10^{-1}	deci	d	10	deka	da
10^{-2}	centi	c	10^2	hecto	h
10^{-3}	milli	m	10^3	kilo	k
10^{-6}	micro	μ	10^6	mega	M
10^{-9}	nano	n	10^9	giga	G
10^{-12}	pico	p	10^{12}	tera	T
10^{-15}	femto	f			
10^{-18}	atto	a			

PART
ONE

THERMODYNAMICS

Thermodynamics deals with the properties of systems at equilibrium. It has nothing to do with time. It provides exact relationships between various measurements and an answer to the question, "How far will this particular reaction go before equilibrium is reached?" It also provides the basis for reliable predictions of the effects of temperature, pressure and concentration on chemical equilibria. Thermodynamics is independent of any assumptions about molecular structure or of mechanisms by which equilibrium is reached. In short, thermodynamics is concerned only with initial and final states. Even so, it is one of the most powerful tools of physical chemistry, and because of its importance the first part of the book is devoted to it. Fortunately, thermodynamics can be developed completely without difficult mathematics, and so a nearly complete treatment can be given at the level of this book. We will begin to consider the applications of thermodynamics to chemistry with the zeroth, first, second, and third laws of thermodynamics. These principles will then be applied to chemical equilibria, electromotive force, phase equilibria, and surface phenomena.

Equilibrium conditions are independent of mechanism, and it is the great strength (and weakness) of thermodynamics that it is not concerned with mechanisms or models (as, for example, molecules) or with time. Later, in Part Three, we will see how various thermodynamic quantities may be calculated from information about individual molecules using statistical mechanics. The relations between the various thermodynamic quantities derived classically also apply in statistical mechanics. Statistical mechanics provides insight into thermodynamics, but it is difficult to apply statistical mechanics to liquids and highly interactive systems.

CHAPTER 1

FIRST LAW
OF THERMODYNAMICS

The quantitative concepts of temperature, work, internal energy, and heat play an important role in the understanding of chemical phenomena. These concepts will be developed in this chapter along with the relationship between heat and work as forms of energy. The chapter opens with a discussion of the scientific concept of temperature. The principle involved in defining temperature was not recognized until after the establishment of the first and second laws of thermodynamics, and therefore it is referred to as the "zeroth" law.

The first law expresses the concept of conservation of energy. This concept first appeared in mechanics and was later extended to include electrostatics and electrodynamics. Joule performed experiments in 1840–1845 that showed how heat could also be included in the conservation of energy. The first law leads to the definition of the internal energy U. One of the important applications of the first law in chemistry is interpreting the heat effects of chemical reactions. If heat capacities of reactants and products are known, the heat of reaction may be calculated at other temperatures after it has been measured at one.

1.1 SYSTEM, SURROUNDINGS,
STATE OF A SYSTEM, AND
STATE VARIABLES

A thermodynamic system is a part of the physical universe that is under consideration. It is separated from its surroundings by a boundary. If the boundary prevents any interaction with the surroundings, the system is called an isolated system. If matter can pass across the boundary, we have an open system. If it cannot we have a closed system. Heat may enter or leave a closed system.

A system may be taken through a series of changes in which work and heat pass through the boundary so that there is a change in the surroundings as well as the system. If the boundary does not permit the flow of heat, any process occurring in the system is said to be adiabatic, and the boundary is called an adiabatic wall.

When a system is at equilibrium under a given set of conditions it is said to be in a particular state. The state of a system may be identified from the fact that when it is in a definite state each of its properties has a definite value. It is found that, for a fixed amount of a fluid (gas or liquid), the state is completely defined by any two of the three variables—pressure, volume, and temperature. Such variables are referred to as state variables.

Thermodynamics is concerned with transformations from an initial state to a final state. In such a transformation heat may pass through the boundary of the system and work may be done on the system or on the surroundings.

1.2 THE ZEROTH LAW OF THERMODYNAMICS

In order to define temperature we must first consider the concept of thermal equilibrium. If two closed systems are brought together so that they are in thermal contact, changes take place in the properties of both. Eventually a state is reached in which there is no further change, and this is the state of thermal equilibrium. Thus we can readily determine whether two systems are at the same temperature by bringing them into contact and seeing whether observable changes take place in the properties of either system. If no change occurs, they are at the same temperature.

Now let us consider three systems A, B, and C. It is an experimental fact that if body A is in thermal equilibrium with body C, and body B is also in thermal equilibrium with body C, then A and B are in thermal equilibrium with each other. It is not obvious that this should necessarily be true, and so this empirical fact is referred to as the *zeroth law of thermodynamics*.

This law puts the concept of temperature on a firm basis in the following way: if two systems are in thermal equilibrium they have the same temperature; if they are not in thermal equilibrium they have different temperatures. Now, how can a temperature scale be set up?

To establish a temperature scale we start with body B in a state defined by volume V_B and pressure P_B. The values of V_A and P_A of a body of fluid A that is in equilibrium with B are determined experimentally. There are many combinations of P_A and V_A for which there is equilibrium and these pairs of values may be plotted on a graph of P_A versus V_A such as that in Fig. 1.1. According to the zeroth law of thermodynamics, this curve at constant temperature (an isotherm) is independent of the nature of body B since the same result would be obtained by using instead of B any other body in equilibrium with it. If the thermal state of B is changed and

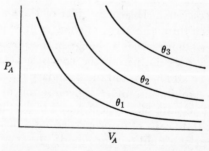

Fig. 1.1 Isotherms for fluid A. This plot, which is for a hypothetical fluid A, might look quite different for some other fluid.

the experiment repeated, another isotherm for fluid A will be obtained. Each isotherm obtained in this way may be assigned a temperature θ, and in this way a temperature scale may be set up. All bodies having the same temperature θ will remain unchanged in properties when they are brought into thermal contact with each other through a wall that allows the two bodies to have different pressures and different chemical compositions.

Many different temperature scales may be defined, but the simplest and most useful is that based on the behavior of ideal gases, as obtained by the extrapolation of the behaviour of real gases to zero pressure. This scale is identical with one based on the second law of thermodynamics, which is independent of the properties of any particular substance (see Section 2.4). In Chapter 17 the ideal gas temperature scale will be identified with that which arises in statistical mechanics.

The pairs of the variables P and V that correspond to the same temperature may be determined (for example, the curves in Fig. 1.1) and represented by means of a function.

$$f(P, V) = \theta \qquad (1.1)$$

where θ is the temperature. Such an equation is called the *equation of state* for the fluid. Real fluids all have different equations of state. According to this equation there exists a function of the state of a fluid, called the temperature, that has the same value for fluids that are in thermal equilibrium with each other. There are many temperature scales that may be defined on the basis of equation 1.1; indeed there are as many scales as fluids—more, if the right side of equation 1.1 is replaced with a function of θ. In order to use the behavior of gases at low pressures we start with Boyle's law (1662). As the pressure of a fixed amount of a gas is lowered, it follows the equation

$$PV = k \qquad \text{(at constant temp.)} \qquad (1.2)$$

more and more closely. The PV products of gases extrapolated to zero pressure are proportional to the number n of moles of gas and a function of temperature θ.

$$\lim_{P \to 0} (PV)_\theta = nf(\theta) \qquad (1.3)$$

It is convenient to take this function as RT, where R is the ideal gas constant and T is the ideal gas absolute temperature.*

$$\lim_{P \to 0} (PV)_T = nRT \qquad (1.4)$$

All that is needed to complete the definition of the temperature scale is to specify T at some standard condition so that the gas constant R may be calculated. The size of the unit of thermodynamic temperature, the Kelvin K in SI units,† is set

---

* It is not obvious that this absolute temperature corresponds with our ideas of hotter and colder. However, the connection can be made; A. B. Pippard, *The Elements of Classical Thermodynamics*, Cambridge University Press, 1960.

† We will use the SI (International System of Units) unit Kelvin, represented by K, for thermodynamic temperature. The K unit does not have a degree sign, but the degree sign is retained in the unit for the Celsius scale (°C).

by assigning the origin at absolute zero and 273.1600 K to the triple point of water (the temperature and pressure at which ice, liquid, and vapor are in equilibrium with each other in the absence of air).

The ice point (the temperature at which ice and water are in equilibrium in the presence of air at one atmosphere) sets the zero on the Celsius scale. The ice point is now defined as the temperature 0.0100° C below that of the triple point of water, that is, 273.1500 K. Thus the Celius temperature (t) is defined in terms of the thermodynamic temperature (T) by $t = T - 273.1500$.

1.3   UNITS OF THE IDEAL GAS CONSTANT R

Very careful experiments with molecular oxygen show that as the pressure is reduced indefinitely the PV product for one mole of oxygen (31.9988 g) approaches 22.41383 liter atm at 0° C (273.1500 K). We may calculate the gas constant using equation 1.4.

$$R = \frac{\lim_{P \to 0}(PV)_T}{nT} = \frac{22.41383 \text{ liter atm}}{(1 \text{ mol})(273.150 \text{ K})}$$

$$= 0.0820569 \text{ liter atm K}^{-1} \text{ mol}^{-1} \tag{1.5a}$$

A standard atmosphere is equal to the pressure required to support 76 cm of mercury at 0° at a point on the earth where the acceleration of gravity g is 9.80665 m s^{-2}. The density of mercury at 0° is 13.5951 g cm^{-3}.

The unit of pressure in the SI system is the pascal Pa, which is the pressure produced by the force of a newton on an area of a square meter. Thus the pressure of a standard atmosphere in pascals may be calculated as follows:

$$P = (0.76 \text{ m})(13.5951 \times 10^3 \text{ kg m}^{-3})(9.80665 \text{ m s}^{-2})$$

$$= 101,325 \text{ Pa} = 101,325 \text{ N m}^{-2}$$

where the density of mercury at 0° is 13.5951×10^3 kg m^{-3} or 13.5951 g cm^{-3}. Since pressure is force per unit area, the product of pressure and volume has the dimensions of force times distance, which is work or energy. Thus the gas constant may be expressed in the SI unit of energy, the joule J, by expressing the pressure in pascals.

$$R = \frac{PV}{nT} = \frac{(101,325 \text{ N m}^{-2})(22.41383 \times 10^{-3} \text{ m}^3)}{(1 \text{ mol})(273.150 \text{ K})}$$

$$= 8.31441 \text{ J K}^{-1} \text{ mol}^{-1} \tag{1.5b}$$

The thermochemical calorie is defined as 4.184 J. This definition was chosen to make the calorie approximately equal to the amount of heat required to raise the temperature of a gram of water one degree Celsius in the neighborhood of 15°. Most chemical literature uses the calorie, and so we will use the calorie in this book. However, there is an international plan to shift to SI units, and so it is important to be able to make calculations in SI units as well.

The value of R in cal K^{-1} mol^{-1} is

$$R = \frac{(8.31441 \text{ J K}^{-1} \text{mol}^{-1})}{(4.184 \text{ J cal}^{-1})} = 1.98719 \text{ cal K}^{-1} \text{mol}^{-1} \qquad (1.5c)$$

In calculations on the P-V-T relations of gases it is usually convenient to express R in liter atmospheres; in electrochemical problems involving volts and coulombs, R is best expressed in joules; and in thermochemical problems, R is expressed in calories or in joules.

Although the atmosphere is not an SI unit we will find it convenient to use the atmosphere for measuring gas pressures. Another convenient unit is the Torr, which is defined as $\frac{1}{760}$ of a standard atmosphere and corresponds with the pressure exerted by a millimeter of mercury at $0°$ C where the acceleration is 9.80665 m s^{-2}.

We will also find it convenient to apply the ideal gas law to a mole of gas and write it in the form

$$P\bar{V} = RT \qquad (1.6)$$

where $\bar{V}$ is the volume per mole. Throughout the book thermodynamic quantities per mole will be designated by a bar.

1.4   WORK

In thermodynamics heat and work are algebraic quantities that can be positive or negative. Force is a vector quantity; that is, it has direction as well as magnitude. We will use boldface type for vectors.

Force is defined by

$$\boldsymbol{F} = m\boldsymbol{a} \qquad (1.7)$$

where $\boldsymbol{F}$ is the force that will give a mass m an acceleration $\boldsymbol{a}$.

Work (w) is a scalar quantity defined by

$$w = \boldsymbol{F} \cdot \boldsymbol{l} \qquad (1.8)$$

where $\boldsymbol{F}$ is the vector force, $\boldsymbol{l}$ is the vector length of path, and the dot indicates a scalar product (that is, the product is taken of the magnitude of one vector by the projection of the second vector along the direction of the first). If the force vector of magnitude F and the vector length of magnitude l are separated by the angle θ, the work is given by $Fl \cos \theta$. In the SI system, the unit of work is the joule J; 1 J $= 1$ N m.

Work may be expressed as the product of two factors: an intensity factor and a capacity factor. Examples are given in Table 1.1. The differential quantity of work done by a force f operating over a distance dl is $f\,dl$. Since pressure P is force per unit area, the force on a piston is PA, where A is the surface area perpendicular to the direction of the motion of the piston. Thus the differential quantity of work done by an expanding gas that causes the piston to move

Table 1.1   Intensity and Capacity Factors for Various Types of Work

Type of Work	Intensity Factor	Capacity Factor
Mechanical (J)	Force (N)	Change in distance (m)
Volume expansion (J)	Pressure (N m^{-2})	Change in volume (m^3)
Surface increase (J)	Surface tension (N m^{-1})	Change in area (m^2)
Electrical (J)	Potential difference (V)	Quantity of electricity (C = A × s)
Gravitational (J)	Gravitational potential (height × acceleration) (m^2 s^{-2})	Mass (kg)

distance dl is $PA\,dl$. But $A\,dl = dV$, the increase in gas volume, and so the differential quantity of work is $P\,dV$.

Work is often conveniently measured by the lifting of weights. The work required to lift a mass m in the earth's gravitational field, which has an acceleration g, is mgh, where h is the height through which the weight is lifted.

The work w required to lift a kilogram $\frac{1}{10}$ m is

$$w = mgh = (1 \text{ kg})(9.807 \text{ m s}^{-2})(0.1 \text{ m}) = 0.9807 \text{ J} \qquad (1.9)$$

Work is an algebraic quantity, and so it is important to adopt a sign convention. We will take a positive value of w to indicate that *work is done by an external force on the system*. A negative value of w indicates that a system does work on its own surroundings, for example, when a gas expands against a piston. Earlier editions of this book, and a number of textbooks of thermodynamics, use the opposite sign convention for w.

1.5   JOULE'S EXPERIMENTS

Joule showed that under adiabatic conditions a given amount of work would heat the water in a calorimeter a certain number of degrees, independent of whether the work was used to turn a paddle wheel or was dissipated by an electrical current flowing through a resistance or by the friction of rubbing two objects together. Since a given change in state of the water in the calorimeter can be accomplished in different ways involving the same amount of work, or by different sequences of steps, the change in state is independent of the path and is dependent only on the total amount of work. This makes it possible to express the change in state of a system in an adiabatic process in terms of the work required, without stating the type of work or the sequence of steps used. The property of the system whose change is calculated in this way is called the *internal energy U*. Since the internal energy U of a system may be increased by doing work on it, we may calculate the increase in internal energy from the work w done on a system to change it from one state to another in an adiabatic process.

$$\Delta U = w \qquad \text{(in an adiabatic process)} \qquad (1.10)$$

The symbol Δ indicates the value of the quantity in the final state minus the value of the quantity in the initial state; $\Delta U = U_2 - U_1$ where U_1 is the internal energy in the initial state and the U_2 is the internal energy in the final state. *If the system does work on its surroundings, w is negative, and ΔU is negative if the process is adiabatic.*

1.6 HEAT

A given change in state of a system can be accomplished in ways other than by the performance of work under adiabatic conditions. A change equivalent to that in the Joule experiment described in the preceding section may be obtained by immersing a hot object in the water. When this is done, heat flows from the so-called heat reservoir (the hot object) to the water. We should not say, however, that the water now has more "heat" any more than we would say it has more "work" after it has been heated with moving paddle wheels. After the experiment the temperature of the water is higher, and it has a greater internal energy U.

Since the same change in state (as determined by measuring such properties as temperature, pressure, and volume) may be produced by doing work on the system or by allowing heat to flow in, the amount of heat q may be expressed in mechanical units. In experiments like that of Joule's, it is found that the expenditure of 4.184 J of work produces the same change of state as the transfer of 1 cal of heat.

Heat is an algebraic quantity, and so it is important to adopt a sign convention. We will take a positive value of q to indicate that heat is absorbed by the system from its surroundings. A negative value of q means that the system gives up heat to its surroundings. The change in internal energy U produced by the transfer of heat q to a system when no work is done is given by

$$\Delta U = q \qquad \text{(no work done)} \tag{1.11}$$

In words, *the heat absorbed by a closed system in a process in which no work is done is equal to the increase in internal energy of the system.* Or, put another way, if no work is done, the heat evolved is equal to the decrease in the internal energy of the system.

1.7 THE FIRST LAW OF THERMODYNAMICS

Now let us consider a change from state 1 to state 2 in which a closed system does work or has work done on it and gains or loses heat by being brought in contact with a heat reservoir. Applying equation 1.10 to the system and heat reservoir together, we see that

$$(U_2 - U_1) + (U_2' - U_1') = w \tag{1.12}$$

where $U_2 - U_1$ is the change in internal energy of the system and $U_2' - U_1'$ is the change in internal energy of the heat reservoir. The heat q gained by the system is equal to the negative of the heat gained by the heat reservoir so, by equation 1.11, $(U_2' - U_1') = -q$. Thus equation 1.12 becomes

$$U_2 - U_1 = \Delta U = q + w \tag{1.13}*$$

Equation 1.13 expresses the *first law of thermodynamics* for closed systems. The first law is not just equation 1.13 but includes the statement that the thermodynamic function U defined by this equation is a function of the state of the system only. It should be noted that the first law provides a means for determining changes in internal energy but not the absolute value of internal energy.

If there is no change of internal energy (as in the isothermal expansion of an ideal gas, Section 1.10), the work done must be the negative of the heat absorbed.

If ΔU is negative we may say that the system loses energy and that this energy is dissipated in heat that is evolved and work that is done by the system. The first law has nothing to say about how much heat is evolved and how much work is done except that equation 1.13 is obeyed. In other words, the entire decrease in internal energy could show up as work ($q = 0$). Another possibility is that even more than this amount of work would be done and heat would be absorbed ($q = +$) so that equation 1.13 is obeyed.

As an example of the application of equation 1.13, a muscle does work and evolves heat. The sum of the work done by the muscle and the heat evolved is equal to $-\Delta U$ for the change in state due to the chemical reactions that occur. Although the first law has nothing to say about the relative amounts of heat and work, the second law does.

The first law is frequently stated in the form that energy may be transformed from one form to another, but it cannot be created or destroyed, and the total energy of an isolated system is constant.†

---

* In earlier editions the first law was written $\Delta U = q - w$, and work w was defined as the work done by the system. In equation 1.13, the opposite sign convention has been used for work in order to emphasize the fact that a certain change in internal energy makes it possible to obtain a certain amount of heat and work.

† It is sometimes incorrectly stated that mass can be converted into energy according to Einstein's relation $E = mc^2$, where c is the speed of light. The mass m in this equation is the relativistic mass that is related to the rest mass m_0 by

$$m = m_0 \, (1 - v^2/c^2)^{-\frac{1}{2}}$$

where v is the speed of the object. The correct interpretation of $E = mc^2$ is that energy E and mass m are related through the necessarily positive proportionality constant c^2. A given mass m is therefore equivalent to a certain energy. Since, according to the first law of thermodynamics, energy is conserved, Einstein's relation requires that mass be included in this conservation principle. R. P. Bauman, *J. Chem. Ed.*, **43**, 366 (1966), gives a clear illustration of the meaning of relativistic mass and its relation to energy. If a ball of mass m_0 is kicked, its energy and mass are increased at the expense of the energy and mass of the kicker. As the ball bounces to rest it gives up energy and mass to the earth. In nuclear fission the rest mass of the fragments is less than the rest mass of the original atom, but the mass of the surroundings is increased through collision with the fission fragments.

1.8 EXACT AND INEXACT DIFFERENTIALS

The internal energy U is a state function, like V, because it depends only on the state of the system. The integral of the differential of a state function along any arbitrary path is simply the difference between values of the function at two limits. For example, if a system goes from state a to state b we can write

$$\int_a^b dU = U_b - U_a \qquad (1.14)$$

Since the integral is path independent, the differential of a state function is called an *exact differential*.

The quantities q and w are not state functions. The integrals of their differentials in going from state a to state b *depend on the path chosen*. Therefore their differentials are called *inexact differentials*. We will use $đ$ rather than d to indicate inexact differentials. In going from state a to state b the work w done is represented by

$$\int_a^b đw = w \qquad (1.15)$$

Note that the result of the integration is not written $w_b - w_a$ because the amount of work done depends on the particular path that is followed between state a and state b. For example, when a gas is allowed to expand, the amount of work obtained may vary from zero (if the gas is allowed to expand into a vacuum) to a maximum value that is obtained if the expansion is carried out reversibly as described in the next section.

If an infinitesimal quantity of heat $đq$ is absorbed by a system, and an infinitesimal amount of work $đw$ is done on the system, the infinitesimal change in the internal energy is given by

$$dU = đq + đw. \qquad (1.16)$$

It is interesting to note that the sum of two inexact differentials can be an exact differential. To further illustrate this point we consider the following:

The differential $dz = y\,dx$ is not an exact differential

$$\int_a^b y\,dx = \text{area I} \qquad (1.17)$$

because this area depends on the path between a and b, as may be seen from Fig. 1.2.

The differential $dz = y\,dx + x\,dy$ is an exact differential. Since $dz = d(xy)$

$$\int_a^b dz = \int_a^b d(xy) = x_b y_b - x_a y_a \qquad (1.18)$$

The reason $dz = y\,dx + x\,dy$ is an exact differential may be seen from Fig. 1.2.

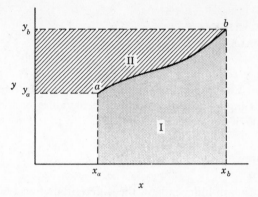

Fig. 1.2   Path of a system in going from state a to state b.

Since $\int_a^b y\,dx$ is area I, and

$$\int_a^b x\,dy = \text{area II} \tag{1.19}$$

then

$$\int_a^b dz = \int_a^b y\,dx + \int_a^b x\,dy = \text{area I} + \text{area II} \tag{1.20}$$

The sum of these areas is independent of the shape of the curve (path) between a and b. There is a simple test to see whether a differential is exact (see Section 2.13).

A cyclic process is a process in which a system is carried through a series of steps which eventually bring the system back to its *initial* conditions. The change in internal energy for a cyclic process is zero since dU is exact, and therefore from equation 1.14

$$\int_a^a dU = \oint dU = 0 \tag{1.21}$$

where the circle indicates integration around a cycle. The cyclic integrals of q and w are not in general equal to zero, and their values depend on the path followed.

1.9   REVERSIBLE PROCESSES

A *reversible* process is a process that may be reversed at any moment by changing an independent variable by an infinitesimal amount. Thus in the reversible expansion of a gas the expansion can be stopped at any point by increasing the pressure exerted by the piston by an infinitesimal amount. A reversible process is often spoken of as one that consists of a series of successive equilibria. Such processes are idealizations that may be closely approached, but not actually reached, in the laboratory. To carry out a finite process reversibly would require an infinite time. Reversible processes are of great conceptual importance because they yield the maximum amount of work which may be obtained from a given net change. This amount of work is just sufficient to return the system to its original conditions. When a process is carried out *irreversibly*, less work is obtained than would be required to return the system to its initial state.

Imagine a gas enclosed in a cylinder fitted with a frictionless and weightless piston and maintained at a constant temperature. The external pressure on the piston is decreased by an infinitesimal amount, dP, and the gas expands by an amount dV. In this expansion the pressure of the gas in the cylinder decreases until it becomes equal to the external pressure, and then the piston ceases to move. A second infinitesimal decrease in pressure produces a second expansion dV; as the pressure is decreased in successive amounts, the volume undergoes a series of increases. During each little expansion the pressure throughout the gas is constant (within an infinitesimal amount). In each little expansion the work done is the external pressure multiplied by $-dV$, and the total work obtainable in expanding the gas reversibly from the *initial* volume V_1 to the *final* volume V_2 is equal to the integral of the pressure times the differential of the volume.

$$w_{\text{rev}} = -\int_{V_1}^{V_2} P \, dV \tag{1.22}$$

Since the gas is at its equilibrium pressure (within an infinitesimal amount) at each stage in the expansion, we may substitute an expression for the dependence of pressure on volume obtained from equilibrium measurements. If the gas was allowed to expand rapidly, the pressure and temperature would not be uniform throughout the volume of the gas, and so such a substitution could not be made. *Only if the expansion is carried out reversibly at constant temperature can sufficient energy be obtained to reverse the process, compressing the gas to its original conditions.*

1.10   REVERSIBLE ISOTHERMAL
EXPANSION OF A GAS

The maximum work that can be obtained from the *isothermal* expansion of an ideal gas may readily be calculated. If the expansion is carried out reversibly at constant temperature, the pressure is always given by $P = nRT/V$. Substituting in equation 1.22 we obtain

$$w_{\text{rev}} = -\int_{V_1}^{V_2} \frac{nRT}{V} \, dV \tag{1.23}$$

and since the temperature and number of moles of gas are constant

$$w_{\text{rev}} = -nRT \int_{V_1}^{V_2} \frac{dV}{V} = -nRT \ln \frac{V_2}{V_1} = -2.303 \, nRT \log \frac{V_2}{V_1} \tag{1.24}$$

where ln represents natural logarithms and log represents base 10 logarithms so that $\ln x = 2.303 \log x$.

In integration the lower limit always refers to the initial state and the upper limit to the final state. If the gas is compressed, the final volume is smaller and w_{rev} is positive. The positive value means that work is done on the gas.

Example 1.1 What is the work of reversible expansion of a mole of ideal gas at $0°$ from 2.24 to 22.4 liters?

$$w_{rev} = -2.303 \, RT \log (\bar{V}_2/\bar{V}_1)$$
$$= -2.303(1.987 \text{ cal K}^{-1} \text{ mol}^{-1})(273.15 \text{ K}) \log 10$$
$$= -1250 \text{ cal mol}^{-1}$$
$$= -2.303(8.314 \text{ J K}^{-1} \text{ mol}^{-1})(273.15 \text{ K}) \log 10$$
$$= -5230 \text{ J mol}^{-1}$$

The work obtainable from the reversible isothermal expansion of 1 mole of gas may be expressed in terms of pressures instead of volumes, since at constant temperature $\bar{V}_2/\bar{V}_1 = P_1/P_2$. Then

$$w_{rev} = -RT \ln \frac{P_1}{P_2} = 2.303 \, RT \log \frac{P_2}{P_1} \qquad (1.25)$$

A number of processes may be carried out essentially reversibly in the laboratory. A liquid may be vaporized reversibly as described in the next section. In the reversible discharge of an electrochemical cell, the applied voltage from the external source is kept within an infinitesimal voltage of the electromotive force of the electrochemical cell.

1.11 REVERSIBLE VAPORIZATION OF A LIQUID

Imagine that a liquid is placed in a cylinder provided with a weightless, frictionless piston and that the cylinder is set into a large heat reservoir at the boiling temperature of the liquid. (The fact that no machine can be built with a weightless, frictionless piston in no way affects the conclusions drawn from this idealized process.) The vapor pressure of the liquid at this temperature is exactly equal to the pressure exerted by the piston, and the whole system is in a state of equilibrium. Now, if the temperature of the reservoir is raised by an infinitesimal amount, the vapor pressure of the liquid will be slightly greater, and the piston will be pushed back against the atmospheric pressure. As the volume increases, more liquid evaporates and the pressure in the cylinder is thus maintained constant; heat flows in from the heat reservoir to supply energy for vaporization and maintain the temperature constant.

This process of absorbing heat and doing external work is reversible, because at any time the vaporization can be stopped by decreasing the temperature by an infinitesimal amount or by increasing the pressure by an infinitesimal amount, thus making internal and external pressures exactly equal. Increasing the pressure still further by an infinitesimal amount causes the vapor to condense and give back the heat of vaporization to the heat reservoir.

If the liquid in the cylinder is water and the pressure is 1 atm, the temperature of vaporization will be 100°; and when one mole has been evaporated, the increase in volume can be calculated on the assumption that water vapor behaves as an ideal gas and that the volume of the liquid (0.018 liter mol^{-1}) is negligible:

$$w = -P\Delta\bar{V} = -(1 \text{ atm})(22.41 \text{ liter mol}^{-1})\frac{373.15 \text{ K}}{273.15 \text{ K}} = -30.6 \text{ liter atm mol}^{-1}$$

Since it is assumed that ideal gas law is obeyed, and we are considering the vaporization of 1 mol of water, RT may be substituted for $P\,\Delta\bar{V}$. Then,

$$w = -RT = -(0.08205 \text{ liter atm K}^{-1} \text{ mol}^{-1})(373.15 \text{ K})$$
$$= -30.6 \text{ liter atm mol}^{-1}$$

or

$$w = -RT = -(1.987 \text{ cal K}^{-1} \text{ mol}^{-1})(373.15 \text{ K}) = -741.4 \text{ cal mol}^{-1}$$

The work done by the system in vaporizing 1 mol to give an ideal gas depends only on the temperature and is independent of the pressure and volume. If the pressure is doubled, the volume change is halved, and the product $P\,\Delta\bar{V}$ is the same. For exact computations it is not accurate to consider a vapor at its boiling point to be an ideal gas; the volume change must be measured experimentally or calculated with a more exact equation of state.

The energy required to do this pressure-volume work comes from heat absorbed from the heat reservoir by the evaporating liquid. A great deal more energy than that required to do the pressure-volume work must be absorbed from the heat reservoir, however, in order to separate the molecules from their neighboring molecules in the liquid. To vaporize water at 373.15 K and atmospheric pressure, 539.7 cal g^{-1} is required. Thus

$$q = (18.02 \text{ g mol}^{-1})(539.7 \text{ cal g}^{-1}) = 9725 \text{ cal mol}^{-1}$$

By use of equation 1.13 the change in internal energy of a mole of water upon vaporization is given by

$$\Delta\bar{U} = q + w = 9725 - 741 = 8984 \text{ cal mol}^{-1}$$

The bar in $\Delta\bar{U}$ indicates that one mole is being considered.

1.12   ENTHALPY*

Constant-pressure processes are more common in chemistry than constant-volume processes because most operations are carried out in open vessels. If only

---

* Pronounced enthal'py to distinguish it from en'tropy (Section 2.4).

pressure-volume work is done and the pressure is constant, equation 1.13 may be written

$$\Delta U = q - P\,\Delta V \tag{1.26}$$

If the initial state is designated by 1 and the final state by 2, then

$$U_2 - U_1 = q - P(V_2 - V_1) \tag{1.27}$$

so that the heat absorbed is given by

$$q = (U_2 + PV_2) - (U_1 + PV_1) \tag{1.28}$$

Since the heat absorbed is given by the difference of two quantities that are functions of the state of the system, it is convenient to introduce a new state function, the *enthalpy H* which is defined by

$$H = U + PV \tag{1.29}$$

Thus equation 1.28 may be written $q = H_2 - H_1 = \Delta H$. In words, *the heat absorbed in a process at constant pressure is equal to the change in enthalpy if the only work done is pressure-volume work.*

When pressure-volume work is the only kind of work (electrical and other kinds being excluded), it is easy to visualize ΔU and ΔH; in a constant-volume calorimeter the evolution of heat is a measure of the decrease in internal energy U, and in a constant-pressure calorimeter the evolution of heat is a measure of the decrease in enthalpy H.

1.13 HEAT CAPACITY

The heat capacity is the amount of heat required to raise the temperature of a system a specified amount, or, the ratio of the heat absorbed to the temperature change, $q/\Delta T$. The heat capacity C at any particular temperature is the limit of this ratio as ΔT is made smaller and smaller.

For a chemically inert system of fixed mass the internal energy is a function of temperature and volume. Since U is a state function the differential dU is given by

$$dU = \left(\frac{\partial U}{\partial T}\right)_V dT + \left(\frac{\partial U}{\partial V}\right)_T dV \tag{1.30}$$

The first term is the change in internal energy due to the temperature change alone, and the second term is the change in internal energy due to the volume change alone. Since only pressure-volume work is involved,

$$dq = dU + P\,dV = \left(\frac{\partial U}{\partial T}\right)_V dT + \left[P + \left(\frac{\partial U}{\partial V}\right)_T\right] dV \tag{1.31}$$

where the second form has been obtained by introducing equation 1.30. Thus the amount of heat absorbed depends upon the volume change, as well as the temperature change.

If the volume of the system is held constant, then

$$dq_V = \left(\frac{\partial U}{\partial T}\right)_V dT \tag{1.32}$$

The quantity dq_V/dT may be measured experimentally at constant volume and is known as C_V, the heat capacity of the system at constant volume.

$$C_V = \frac{dq_V}{dT} = \left(\frac{\partial U}{\partial T}\right)_V \tag{1.33}$$

The second form has been obtained from equation 1.32.

If one mole of pure substance is considered, the heat capacity becomes a molar quantity.

$$\bar{C}_V = \left(\frac{\partial \bar{U}}{\partial T}\right)_V \tag{1.34}$$

The change in internal energy of one mole of substance heated from T_1 to T_2 at constant volume is

$$\Delta \bar{U} = \int_{T_1}^{T_2} \bar{C}_V \, dT \tag{1.35}$$

For a chemically inert system of fixed mass, the enthalpy H is conveniently taken to be a function of temperature and pressure. Thus the differential dH of the enthalpy caused by a change of temperature dT and a change of pressure dP is given by

$$dH = \left(\frac{\partial H}{\partial T}\right)_P dT + \left(\frac{\partial H}{\partial P}\right)_T dP \tag{1.36}$$

If the pressure is constant, then $dH = dq_P$, and equation 1.36 becomes

$$dq_P = \left(\frac{\partial H}{\partial T}\right)_P dT \tag{1.37}$$

The ratio dq_P/dT is the heat capacity at constant pressure C_P. Thus

$$C_P = \frac{dq_P}{dT} = \left(\frac{\partial H}{\partial T}\right)_P \tag{1.38}$$

For 1 mol of pure substance,

$$\bar{C}_P = \left(\frac{\partial \bar{H}}{\partial T}\right)_P \tag{1.39}$$

The heat capacity at constant pressure C_P is always larger than the heat capacity at constant volume C_V because pressure-volume work is done when a substance is heated at constant pressure. The equation for the difference $C_P - C_V$

may be derived by dividing equation 1.31 by dT at constant pressure

$$\frac{dq_P}{dT} = \left(\frac{\partial U}{\partial T}\right)_V + \left[P + \left(\frac{\partial U}{\partial V}\right)_T\right]\left(\frac{\partial V}{\partial T}\right)_P \tag{1.40}$$

$$C_P = C_V + \left[P + \left(\frac{\partial U}{\partial V}\right)_T\right]\left(\frac{\partial V}{\partial T}\right)_P \tag{1.41}$$

where equation 1.41 has been obtained by inserting equation 1.33 and 1.38. For one mole

$$\bar{C}_P - \bar{C}_V = \left[P + \left(\frac{\partial \bar{U}}{\partial \bar{V}}\right)_T\right]\left(\frac{\partial \bar{V}}{\partial T}\right)_P \tag{1.42}$$

The quantity $(\partial U/\partial V)_T$ may be measured in principle in an experiment devised by Joule. Imagine two gas bottles connected with a valve and immersed in a stirred liquid in a thermally isolated container. The two bottles constitute the system under consideration. The first bottle is filled with a gas under pressure, and the second is evacuated. When the valve is opened, gas rushes from the first bottle into the second. Joule found that there was no discernible change in the temperature of the stirred liquid as a result of this expansion, and so $dq = 0$. No work is done in this expansion and so $dw = 0$ and $dU = dq + dw = 0$. Since the temperature is constant, equation 1.30 becomes

$$dU = \left(\frac{\partial U}{\partial V}\right)_T dV = 0 \tag{1.43}$$

Since $dV \neq 0$,

$$\left(\frac{\partial U}{\partial V}\right)_T = 0 \tag{1.44}$$

In words, the internal energy of the gas is independent of volume at constant temperature. Joule's experiment is not very sensitive because of the large heat capacity of the stirred liquid and the small heat capacity of the gas. The quantity $(\partial U/\partial V)_T$, though usually not large, is different from zero for real gases.

Equation 1.44 applies to an ideal gas, and at constant temperature the internal energy of an ideal gas is also independent of pressure

$$\left(\frac{\partial U}{\partial P}\right)_T = 0 \tag{1.45}$$

There is no interaction between the molecules of an ideal gas, and so the energy does not change with the distance between molecules. Equations 1.44 and 1.45 for an ideal gas are not really separate from the ideal gas law, since they may be derived from the second law of thermodynamics by application of the ideal gas law (see Section 2.14).

Since for an ideal gas $(\partial \bar{U}/\partial \bar{V})_T = 0$ and $(\partial \bar{V}/\partial T)_P = R/P$, then equation 1.42 becomes

$$\bar{C}_P - \bar{C}_V = R \tag{1.46}$$

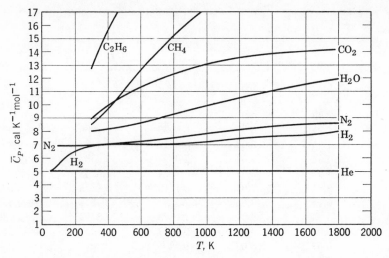

Fig. 1.3   Influence of temperature on the molar heat capacities of gases at constant pressure.

This relationship may be visualized quite easily for an ideal gas. When a mole of ideal gas is heated at constant pressure, the work done in pushing back a piston is $P \, \Delta \bar{V} = R \, \Delta T$. For a 1 K change in temperature the amount of work done is R, and this is just the extra energy required to heat a mole of ideal gas at constant pressure over that required to heat it 1 K at constant volume.

The general expression for $\bar{C}_P - \bar{C}_V$ for gases is given in the next chapter because its derivation uses the second law of thermodynamics (Problem 2.16).

In general, the more complex the molecule, the greater its molar heat capacity, and the greater the temperature effect. The dependence of $\bar{C}_P$ on temperature is shown for a number of gases in Fig. 1.3. The molar heat capacities of simple gases can be calculated on the basis of spectroscopic data more accurately than they can be determined directly, especially at high temperatures.

To calculate the heat absorbed when a gas is heated from one temperature to another, it is convenient to express $\bar{C}_P$ as a function of temperature. The parameters for the empirical equation

$$\bar{C}_P = a + bT + cT^2 \tag{1.47}$$

that are applicable in the range 300–1500 K are given in Table 1.2.

To calculate the heat absorbed per mole at constant pressure when the temperature of a substance is raised, the equation $d\bar{H} = \bar{C}_P \, dT$ is integrated between the desired temperature limits.

$$\Delta \bar{H} = \bar{H}_{T_2} - \bar{H}_{T_1} = \int_{T_1}^{T_2} \bar{C}_P \, dT = \int_{T_1}^{T_2} (a + bT + cT^2) \, dT \tag{1.48}$$

$$= a(T_2 - T_1) + \frac{b}{2}(T_2^2 - T_1^2) + \frac{c}{3}(T_2^3 - T_1^3) \tag{1.49}$$

Table 1.2 Molar Heat Capacities of Gases at Constant Volume and Pressure in cal K^{-1} mol^{-1}

Gas	Values at 25°		Parameters in Equation 1.47		
	$\bar{C}_V$	$\bar{C}_P$	a	$b \times 10^3$	$c \times 10^7$
He	2.98	4.97	4.98	0	0
H_2	4.91	6.90	6.9469	−0.1999	4.808
N_2	4.95	6.94	6.4492	1.4125	−0.807
O_2	5.05	7.05	6.0954	3.2533	−10.171
CO	4.97	6.97	6.3424	1.8363	−2.801
HCl	5.01	7.05	6.7319	0.4325	3.697
Cl_2	6.14	8.25	7.5755	2.4244	−9.650
H_2O	5.93	7.91	7.1873	2.3733	2.084
CO_2	6.92	8.96	6.3957	10.1933	−35.333
NH_3	6.57	8.63	6.189	7.887	−7.28
CH_4	6.59	8.60	3.422	17.845	−41.65
C_2H_6	10.65	12.71	1.375	41.852	−138.27

1.14 HEAT CAPACITY OF SOLIDS

The simplest type of solid we can consider is a crystal of an element with an atom at each lattice point. If we assume that each atom behaves like a simple harmonic oscillator with three degrees of freedom (corresponding with the x, y, and z directions) then we would expect $C_V = 3R$, where R is the contribution for each degree of freedom since each degree of freedom for an atom in a crystal has both kinetic and potential energy associated with it. (See Section 9.5.) This is in approximate agreement with the values of C_V at room temperature for elements heavier than potassium. This is the basis of Dulong and Petit's law (1819): at room temperature the product of the specific heat at constant pressure and the atomic weight of the solid elements is a constant, $\bar{C}_P$, equal to approximately 6.4 cal K^{-1} mol^{-1}. This law played an important part in the early determination of atomic weights because it could be used to select the multiple of the known equivalent weight (obtained from quantitative analysis) that is required to give the atomic weight.

However, solid elements at the beginning of the periodic table have lower heat capacities, and the heat capacities of all solids approach zero as the temperature is reduced toward absolute zero, as illustrated in Fig. 1.4. This effect could not be understood in terms of classical physics and was only explained in terms of quantum mechanics.

In 1907 Einstein gave a quantum mechanical treatment of the heat capacity of an idealized crystal. On the basis of the assumption that all the individual oscillators had the same frequency ν, he was able to show that the heat capacity should vary with temperature in approximately the manner illustrated in Fig. 1.4.

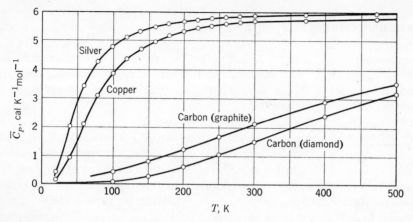

Fig. 1.4 Molar heat capacities of solid elements as a function of temperature.

According to the Einstein theory, plots of $\bar{C}_V$ versus T for different elements should be made superimposable by dividing T by a factor θ that could be different for each atomic crystal. The quantity θ, called the characteristic temperature of a substance, is usually of the order of 100–400 K. For substances that approach a heat capacity of 6 cal K^{-1} mol^{-1} only at quite high temperatures, the value of θ is larger; for diamond it is 1860 K. This behavior is characteristic of solids with strong interatomic forces, like diamond and graphite. For such solids the frequency spectrum extends to higher frequencies, and these higher frequencies are not completely excited at room temperature. For metals like copper and silver, which are soft and malleable and have low melting points, θ is smaller (315 and 215 K, respectively). In such solids with weak interatomic forces the lattice vibrations have low frequencies that are fully excited below room temperature.

The Einstein theory did not give an exact representation of the variation of heat capacity with temperature. An improved theory was derived by Debye who allowed for a range of vibration frequencies rather than a single one.

Debye showed that at sufficiently low temperatures the heat capacity is proportional to the cube of the absolute temperature, as given by

$$\bar{C}_V = \frac{12}{5}\,\pi^4 R\left(\frac{T}{\Theta}\right)^3 \tag{1.50}$$

where Θ is the Debye characteristic temperature. This equation is useful for extrapolating heat capacity data below 15 K, where it is difficult to obtain experimental values.

The vibration frequencies of diatomic molecules are much higher than those of atomic crystals. Thus the vibrations of most atomic crystals are completely activated at temperatures far below those required to activate vibrations of diatomic molecules.

1.15 THERMOCHEMISTRY

Thermochemistry deals with the heat absorbed or evolved in a chemical reaction, in a phase change, or dilution of a solution. Exothermic reactions evolve heat and have negative values of ΔH or ΔU, and endothermic reactions absorb heat and have positive values of ΔH or ΔU.

The change in enthalpy ΔH or internal energy ΔU for a chemical reaction depends on the states of the reactants and products. For example, the heat of combustion of graphite is different from that of diamond, and the heat of solution of $HCl(g)$ to form 1 M solution is different from that for a 0.1 M solution. To facilitate the tabulation of thermodynamic data certain standard states are adopted and thermodynamic properties are tabulated for these standard states. The standard state of a gas is the ideal gas at 1 atm at the temperature concerned; for a solid substance it is a specified crystalline state at 1 atm at the temperature concerned. The standard form of carbon is graphite, and the standard form of sulfur is rhombic sulfur. The standard state of a solute is the concentration required to give unit activity (Section 4.10). The temperature of the standard state needs to be specified. Thermodynamic functions are often tabulated at 25°, but it should be remembered that the term standard state does not in general imply 25°.

When the reactants in their standard states are converted into products in their standard states the changes in thermodynamic quantities are given a superscript zero.

The heat absorbed at constant pressure in reactions where pressure-volume work is the only work done is equal to ΔH, the enthalpy of the products minus the enthalpy of the reactants. In chemical reactions $\Delta H = H_2 - H_1$, where H_2 is the enthalpy of the final state of the number of moles of products written at the right-hand side of the equality sign, and H_1 is the enthalpy of the initial state of the number of moles of reactants written at the left. The absolute values of the enthalpy are unknown, but changes in enthalpy can be determined.

1.16 CALORIMETRIC
MEASUREMENTS

The heat evolved or absorbed by a chemical reaction is measured with a calorimeter. In the most common type, the reaction is allowed to take place in a reaction chamber surrounded by a weighed quantity of water in an insulated vessel, and the rise in temperature is measured with a sensitive thermometer. The product of the rise in temperature and the total heat capacity of the water and calorimeter is equal to the heat evolved. The heat capacity of the surrounding water is obtained by weighing the water and multiplying by its known specific heat; the heat capacity of the calorimeter is determined by carrying out a reaction of known heat evolution in the calorimeter or by introducing a known quantity of heat with an electric heater.

Since only rapid, complete reactions are suitable for thermochemical measurement, these reactions have provided the most important data of thermochemistry. To be sure that combustions will be complete, the material is ignited electrically in a heavy steel bomb containing oxygen under a pressure of 25 atm. Under these conditions all hydrocarbons are burned to water and carbon dioxide. Some reactions present difficulties since they do not go rapidly and completely to definite products. For example, the heat of combustion of a substance like ethyl chloride is not known with high accuracy because of the uncertain mixture of products obtained.

To facilitate the comparison of various reactions, experimental thermochemical data are usually corrected to give the *standard enthalpy change $\Delta H°$ for a reaction* at 25°. The standard enthalpy of combustion of a hydrocarbon is the standard enthalpy of reaction for complete oxidation of 1 mol of the substance to $CO_2(g)$ and $H_2O(l)$.

There are two types of calorimetric experiments: constant volume and constant pressure. In a constant-volume calorimeter no work is done and so the heat absorbed is equal to the increase in internal energy. However, for a constant-pressure calorimeter the first law is

$$\Delta U = q - P\,\Delta V \tag{1.51}$$

The difference in heat absorbed in the two types of experiments is illustrated in the following example.

Example 1.2 The heat of combustion of CO in a constant volume calorimeter is -67.370 kcal mol^{-1}. Calculate the heat of combustion in a constant pressure calorimeter assuming that the gases are ideal.

$$CO(g) + \tfrac{1}{2}O_2(g) = CO_2(g)$$

The work done against the atmosphere in the constant pressure calorimeter is $(n_2 - n_1)RT$, where n_2 is the number of moles of gaseous products and n_1 is the number of moles of gaseous reactants.

$$\Delta H = \Delta U + (n_2 - n_1)RT$$
$$= -67.370 - \frac{(0.5 \text{ mol})(1.987 \text{ cal K}^{-1} \text{ mol}^{-1})(298 \text{ K})}{1000 \text{ cal kcal}^{-1}}$$
$$= -67.636 \text{ kcal mol}^{-1}$$

In other types of calorimetric experiments very little heat is evolved, and very sensitive differential methods involving thermistors (semiconductors with an exponential dependence of resistance on temperature) are used. Some special calorimeters have a sensitivity of a microcalorie. Such calorimeters may be used to measure the heat evolved in reactions involving proteins and polynucleotides.

Heats of combustion are useful in calculating other thermochemical data. Also they have practical as well as theoretical importance. The purchaser of coal is interested in its heat of combustion per gram. The dietician must know, among other factors, the number of calories obtainable from the combustion of various foods. In nutrition the term "calorie" refers to kilocalorie.

1.17 APPLICATION OF FIRST LAW TO THERMOCHEMISTRY

Lavoisier and Laplace recognized in 1780 that the heat absorbed in decomposing a compound must be equal to the heat evolved in its formation under the same conditions. Thus, if the reverse of a chemical reaction is written, the sign of ΔH is changed. Hess pointed out in 1840 that the overall heat of a chemical reaction at constant pressure is the same, regardless of the intermediate steps involved. These principles are both corollaries of the first law of thermodynamics and are a consequence of the fact that the enthalpy is a state function. This makes it possible to calculate the enthalpy changes for reactions that cannot be studied directly. For example, it is not practical to measure the heat evolved when carbon burns to carbon monoxide in a limited amount of oxygen because the product will be an uncertain mixture of carbon monoxide and carbon dioxide. However, carbon may be burned completely to carbon dioxide in an excess of oxygen and the heat of reaction measured. Thus, for graphite at 25°

$$C(s) + O_2(g) = CO_2(g) \qquad \Delta H° = -94.0518 \text{ kcal}$$

The heat evolved when carbon monoxide burns to carbon dioxide can be readily measured also:

$$CO(g) + \tfrac{1}{2}O_2(g) = CO_2(g) \qquad \Delta H° = -67.6361 \text{ kcal}$$

Writing these equations in such a way as to obtain the desired reaction, adding, and canceling, we have:

$$
\begin{aligned}
C(s) + O_2(g) &= CO_2(g) & \Delta H° &= -94.0518 \text{ kcal} \\
CO_2(g) &= CO(g) + \tfrac{1}{2}O_2(g) & \Delta H° &= 67.6361 \text{ kcal} \\
\hline
C(s) + \tfrac{1}{2}O_2(g) &= CO(g) & \Delta H° &= -26.4157 \text{ kcal}
\end{aligned}
$$

It will be noticed that, since the second reaction has been reversed, the sign of $\Delta H°$ is changed from minus to plus. This indicates that 67.6361 kcal would be absorbed if the reaction $CO_2(g) = CO(g) + \tfrac{1}{2}O_2(g)$ occurred.

These data may be represented in the form of an enthalpy level diagram, as shown in Fig. 1.5.

1.18 ENTHALPY OF FORMATION

A chemical reaction with two reactants A_1 and A_2 and two products A_3 and A_4 may be represented as follows:

$$\nu_1 A_1 + \nu_2 A_2 = \nu_3 A_3 + \nu_4 A_4 \tag{1.52}$$

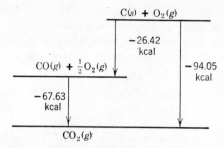

Fig. 1.5 Enthalpy level diagram for the combustion of $C(s)$ and $CO(g)$. The differences in level are standard enthalpy changes at 25°.

where the ν's are stoichiometric coefficients. We will adopt the convention that the stoichiometric coefficients are positive for products and negative for reactants so that a general chemical reaction may be written

$$0 = \sum \nu_i A_i \tag{1.53}$$

This symbolism is convenient because it allows the reactants and products to be numbered.

The enthalpy change for a reaction is equal to the sum of the enthalpies of the products minus the sum of the enthalpies of the reactants at the temperature concerned.

$$\Delta H = \sum \nu_i \bar{H}_i \tag{1.54}$$

Absolute enthalpies of substances are not known but, fortunately, thermochemistry is concerned only with differences in enthalpies of products and reactants. Thus the enthalpy of a substance relative to those of other substances is important, not the absolute value of its enthalpy. *Because of this we may introduce an arbitrary convention that the enthalpy of an element is zero at all temperatures.* As a consequence the enthalpy of any substance in its standard state is equal to ΔH_f°, the enthalpy change of the reaction in which a mole of the substance is formed from elements, each in its standard state. For example, the enthalpy of formation of CO_2 is the enthalpy change for the following reaction:

$$C(\text{graphite}) + O_2(g) = CO_2(g) \qquad \Delta \bar{H}^\circ = -94.0518 \text{ kcal mol}^{-1}$$

and so

$$\Delta \bar{H}^\circ_{f,298} = -94.0518 \text{ kcal mol}^{-1}$$

Since the enthalpy of combustion of hydrogen to form liquid water at 298 K is -68.3174 kcal mol^{-1}, the enthalpy of formation of $H_2O(l)$ is -68.3174 kcal mol^{-1}. The enthalpy of formation of gaseous water is less negative by the molar heat of vaporization of H_2O at 25°, which is 10.5195 kcal mol^{-1}. Therefore the enthalpy of formation of $H_2O(g)$ at 25° is -57.7979 kcal mol^{-1}.

When a substance cannot be formed directly in a rapid reaction from its elements, the enthalpy changes for a series of suitable reactions may be utilized in calculating the enthalpy of formation.

Example 1.3 Calculate the enthalpy of formation of $H_2SO_4(l)$ from the enthalpy change for the combustion of sulfur to SO_2, the oxidation of SO_2 to SO_3 using a platinum catalyst, and the heat of solution of SO_3 in H_2O to give $H_2SO_4(l)$ at 25°. The desired reaction is the sum of the following reactions, and the sum of the enthalpy changes is the enthalpy of formation of $H_2SO_4(l)$.

$$
\begin{array}{ll}
S(s) + O_2(g) = SO_2(g) & \Delta H° = -70.96 \text{ kcal} \\
SO_2(g) + \tfrac{1}{2}O_2(g) = SO_3(g) & \Delta H° = -23.49 \text{ kcal} \\
SO_3(g) + H_2O(l) = H_2SO_4(l) & \Delta H° = -31.14 \text{ kcal} \\
H_2(g) + \tfrac{1}{2}O_2(g) = H_2O(l) & \Delta H° = -68.32 \text{ kcal} \\
\hline
S(s) + 2O_2(g) + H_2(g) = H_2SO_4(l) & \Delta \overline{H}_f° = -193.91 \text{ kcal}
\end{array}
$$

Standard enthalpy changes for reactions may be calculated using enthalpies of formation as follows:

$$\Delta H° = \sum v_i \Delta \overline{H}_f° \tag{1.55}$$

That this yields the same result as equation 1.54 may be shown by considering the reaction

$$CO + \tfrac{1}{2}O_2 = CO_2$$

$$
\begin{aligned}
\Delta H° &= \Delta \overline{H}_{f,CO_2}° - \Delta \overline{H}_{f,CO}° \\
&= (\overline{H}_{CO_2}° - \overline{H}_C° - \overline{H}_{O_2}°) - (\overline{H}_{CO}° - \overline{H}_C° - \tfrac{1}{2}\overline{H}_{O_2}°) \\
&= \overline{H}_{CO_2}° - \overline{H}_{CO}° - \tfrac{1}{2}\overline{H}_{O_2}°
\end{aligned}
$$

This latter form may be obtained directly from equation 1.54.

The enthalpies of formation of many compounds, ions, and atoms are accurately known, and a few selected values from *Circular 500* and *Circular 461* of the National Bureau of Standards are given in Table 1.3.* In obtaining these accurate data great precision in calorimetry and great care in purification of the substances were required. From these enthalpies of formation the enthalpy changes for many reactions may be calculated by use of equation 1.55.

The standard state of a solute in aqueous solution is taken as the hypothetical ideal state of unit molality, in which the molar enthalpy of the solute is the same as in the infinitely dilute solution.

$-3109.9 + 55.07 + 2792.26$
-44.4

1.19 HEAT OF SOLUTION

When a solute is dissolved in a solvent, heat may be absorbed or evolved; in general, the heat of solution depends on the concentration of the final solution. The *integral heat of solution* is the enthalpy change for the solution of 1 mol of solute in n moles of solvent. The solution process may be represented by a chemical equation such as

$$HCl(g) + 5H_2O(l) = HCl \text{ in } 5H_2O \qquad \Delta H_{298}° = -15.31 \text{ kcal} \tag{1.56}$$

---

* *Circular 500* is being replaced by a series of revised tables. Data for the first 34 elements are given in NBS Technical Note 270-3 (1968).

Table 1.3[1] Enthalpy of Formation at 25° (ΔH_f° in kcal mol^{-1})

Elements and Inorganic Compounds

$O_3(g)$	34.0	$CO(g)$	−26.4157
$H_2O(g)$	−57.7979	$CO_2(g)$	−94.0518
$H_2O(l)$	−68.3174	$PbO(s)$	−52.5
$HCl(g)$	−22.063	$PbO_2(s)$	−66.12
$Br_2(g)$	7.34	$PbSO_4(s)$	−219.50
$HBr(g)$	−8.66	$Hg(g)$	14.54
$HI(g)$	6.20	$Ag_2O(s)$	−7.306
$S(\text{monoclinic})$	0.071	$AgCl(s)$	−30.362
$SO_2(g)$	−70.96	$Fe_2O_3(s)$	−196.5
$SO_3(g)$	−94.45	$Fe_3O_4(s)$	−267.0
$H_2S(g)$	−4.815	$Al_2O_3(s)$	−399.09
$H_2SO_4(l)$	−193.91	$UF_6(g)$	−505
$NO(g)$	21.600	$UF_6(s)$	−517
$NO_2(g)$	8.091	$CaO(s)$	−151.9
$NH_3(g)$	−11.04	$CaCO_3(s)$	−288.45
$HNO_3(l)$	−41.404	$NaF(s)$	−136.0
$P(g)$	75.18	$NaCl(s)$	−98.232
$PCl_3(g)$	−73.22	$KF(s)$	−134.46
$PCl_5(g)$	−95.35	$KCl(s)$	−104.175
$C(s, \text{diamond})$	0.4532		

Organic Compounds

Methane, $CH_4(g)$	−17.889	Propylene, $C_3H_6(g)$	4.879
Ethane, $C_2H_6(g)$	−20.236	1-Butene, $C_4H_8(g)$	0.280
Propane, $C_3H_8(g)$	−24.820	Acetylene, $C_2H_2(g)$	54.194
n-Butane, $C_4H_{10}(g)$	−29.812	Formaldehyde, $CH_2O(g)$	−27.7
Isobutane, $C_4H_{10}(g)$	−31.452	Acetaldehyde, $CH_3CHO(g)$	−39.76
n-Pentane, $C_5H_{12}(g)$	−35.00	Methanol, $CH_3OH(l)$	−57.02
n-Hexane, $C_6H_{14}(g)$	−39.96	Ethanol, $C_2H_5OH(l)$	−66.356
n-Heptane, $C_7H_{16}(g)$	−44.89	Formic acid, $HCOOH(l)$	−97.8
n-Octane, $C_8H_{18}(g)$	−49.82	Acetic acid, $CH_3COOH(l)$	−116.4
Benzene, $C_6H_6(g)$	19.820	Oxalic acid, $(CO_2H)_2(s)$	−197.6
Benzene, $C_6H_6(l)$	11.718	Carbon tetrachloride, $CCl_4(l)$	−33.3
Ethylene, $C_2H_4(g)$	12.496	Glycine, $H_2NCH_2CO_2H(s)$	−126.33

Ions in Water

H^+	0.000	SO_4^{2-}	−216.90	Cu^{2+}	15.39
OH^-	−54.957	HS^-	−4.22	Ag^+	25.31

[1] These data have been obtained from F. D. Rossini, D. D. Wagman, W. H. Evans, S. Levine, and I. Jaffe, "Selected Values of Chemical Thermodynamics Properties," *Natl. Bur. Standards Circ. 500*, U.S. Government Printing Office, Washington, D.C., 1952, and F. D. Rossini, K. S. Pitzer, W. J. Taylor, J. P. Ebert, J. E. Kilpatrick, C. W. Beckett, M. G. Williams, and H. G. Werner, "Selected Values of Properties of Hydrocarbons," *Natl. Bur. Standards Circ. C 461*, U.S. Government Printing Office, Washington, D.C., 1947.

Table 1.3   (*Continued*)

Ions in Water (Continued)

F^-	-78.66	NO_3^-	-49.372	Mg^{2+}	-110.41
Cl^-	-40.023	NH_4^+	-31.74	Ca^{2+}	-129.77
ClO_4^-	-31.41	PO_4^{3-}	-306.9	Li^+	-66.554
Br^-	-28.90	CO_3^{2-}	-161.63	Na^+	-57.279
I^-	-13.37	Zn^{2+}	-36.43	K^+	-60.04
S_2^-	10.0	Cd^{2+}	-17.30		

Gaseous Atoms

H	52.089	Br	26.71	N	85.565
F	18.3	I	25.482	C	171.698
Cl	29.012				

Solutes and Solutions

$NaOH(s)$	-101.99	$NaC_2H_3O_2(s)$	-169.8
in 100 H_2O	-112.108	in 100 H_2O	-173.827
in 200 H_2O	-112.1	in 200 H_2O	-173.890
in ∞ H_2O	-112.236	in ∞ H_2O	-174.122
$NaCl(s)$	-98.232	$HC_2H_3O_2(l)$	-116.4
in 100 H_2O	-97.250	in 100 H_2O	-116.705
in 200 H_2O	-97.216	in 200 H_2O	-116.724
in ∞ H_2O	-97.302	in ∞ H_2O	-116.743
$NaNO_3(s)$	-111.54	$HCl(g)$	-22.063
in 100 H_2O	-106.83	in 100 H_2O	-39.713
in 200 H_2O	-106.70	in 200 H_2O	-39.798
in ∞ H_2O	-106.651	in ∞ H_2O	-40.023

where "HCl in $5H_2O$" represents a solution of 1 mol of HCl in 5 mol of H_2O.

The integral heats of solution of HCl, NaOH, and NaCl are plotted versus the number of moles of water per mole of solute in Fig. 1.6. The symbol *aq* is used to represent an aqueous solution which is so dilute that additional dilution produces no thermal effect. As an illustration

$$HCl(g) + aq = HCl(aq) \qquad \Delta H^\circ_{298} = -17.96 \text{ kcal} \qquad (1.57)$$

When a solute is dissolved in a solvent which is chemically quite similar and there are no complications of ionization or solvation, the heat of solution may be nearly equal to the heat of fusion of the solute. It might be expected that heat would always be absorbed in overcoming the attraction between the molecules or ions of the solid solute when the solute is dissolved. Another process which commonly occurs, however, is a strong interaction with the solvent, referred to as solvation, which evolves heat. In the case of water the solvation is called hydration.

The importance of this attraction of the solvent for the solute in the process of solution is illustrated by the dissolving of sodium chloride in water. In the crystal

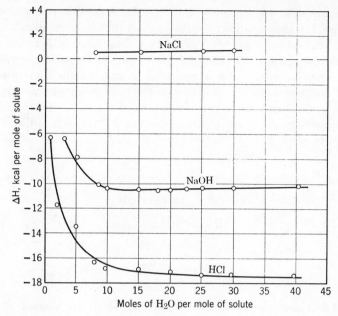

Fig. 1.6   Integral heats of solution at 25°.

lattice of sodium chloride, positive sodium ions and negative chloride ions attract each other strongly. The energy required to separate them is so great that nonpolar solvents like benzene and carbon tetrachloride do not dissolve sodium chloride; but a solvent like water, which has a high dielectric constant and a large dipole moment, has a strong attraction for the sodium and chloride ions and solvates them with a large decrease in the energy of the system. When the energy required to separate the ions from the crystal is about the same as the solvation energy as it is for dissolving NaCl in water, ΔH for the net process is close to zero. When NaCl is dissolved in water at 25° there is only a small cooling effect; q is positive. When Na_2SO_4 is dissolved in water at 25° there is an evolution of heat because the energy of hydration of the ions is greater than the energy required to separate the ions from the crystal.

The integral heat of dilution between two molalities m_1 and m_2 is the heat accompanying the dilution of an amount of solution of concentration m_1 containing 1 mol of solute with pure solvent to make a solution of concentration m_2.

The *differential heat of solution* is the heat of solution 1 mol of solute in a volume of solution so large that the addition of one more mole of solute does not change the concentration appreciably. The differential heat of solution depends on the concentration of the solution. It is impractical to measure the differential heat of solution directly, but it can be calculated from data on the integral heat of solution. The heat effect for the formation of a solution containing m moles of solute and 1000 g of solvent is $m\,\Delta H$, where ΔH is the integral heat of solution per mole of solute. If this quantity is plotted against the number of moles of solute m, the slope of the graph at a given concentration is the heat effect per mole of solute or the differential heat of solution, $d(m\,\Delta H)/dm$, at that concentration.

Integral heats of solution, heats of dilution, and heats of reaction in solution may be calculated from tabulated values of heats of formation in solution. The enthalpy of formation of water is neglected in calculations if there is the same number of moles of water on both sides of the balanced chemical equation. Also the enthalpy of formation of pure water is used for water in aqueous solution. This is done arbitrarily, much as the enthalpies of formation of the elements are taken equal to zero.

Example 1.4 Calculate the integral heat of solution of one mole of $HCl(g)$ in $200H_2O(l)$.

$$HCl(g) + 200H_2O(l) = HCl \text{ in } 200H_2O$$

$$\Delta H_{298}^{\circ} = \Delta \bar{H}_{f,\,HCl \text{ in } 200H_2O}^{\circ} - \Delta \bar{H}_{f,HCl(g)}^{\circ}$$

$$= -39.798 - (-22.063)$$

$$= -17.735 \text{ kcal}$$

Example 1.5 Calculate the integral heat of dilution for the addition of 195 mol of H_2O to 1 mol of HCl in 5 mol of H_2O. The enthalpy of formation of HCl in 5 mol of H_2O is -37.37 kcal and the enthalpy of formation in 200 mol of H_2O is -39.798 kcal.

$$HCl \text{ in } 5H_2O + 195H_2O(l) = HCl \text{ in } 200H_2O$$

$$\Delta H_{298}^{\circ} = \Delta \bar{H}_{f,HCl \text{ in } 200H_2O}^{\circ} - \Delta \bar{H}_{f,HCl \text{ in } 5H_2O}^{\circ}$$

$$= -39.798 - (-37.37)$$

$$= -2.43 \text{ kcal}$$

Example 1.6 Calculate ΔH_{298}° for the reaction

$$HCl \text{ in } 100H_2O + NaOH \text{ in } 100H_2O = NaCl \text{ in } 200H_2O + H_2O(l)$$

$$\Delta H_{298}^{\circ} = -97.216 - 68.3174 + 39.713 + 112.108$$

$$= -13.712 \text{ kcal}$$

For dilute solutions it is found that the heat of reaction of strong bases, like NaOH and KOH, with strong acids, like HCl and HNO_3, is independent of the nature of the acid or base. This constancy of the heat of neutralization is a result of the complete ionization of strong acids and bases and the salts formed by neutralization. Thus, when a dilute solution of a strong acid is added to a dilute solution of a strong base, the only chemical reaction is

$$OH^- + H^+ = H_2O \qquad \Delta H_{298}^{\circ} = -13.360 \text{ kcal mol}^{-1}$$

When a dilute solution of a weak acid or base is neutralized, the heat of neutralization may be somewhat less because of the absorption of heat in the dissociation of the weak acid or base. The neutralization of weak acid with a strong base in dilute solution may be considered to be the resultant of two successive steps.

$$HA(aq) = H^+(aq) + A^-(aq)$$

$$\frac{H^+(aq) + OH^-(aq) = H_2O(aq)}{HA(aq) + OH^-(aq) = H_2O(aq) + A^-(aq)}$$

so that the enthalpy change is

$$\Delta H^\circ_{neut} = \Delta H^\circ_{ioniz} - 13.360 \text{ kcal mol}^{-1}$$

For example, the heat of neutralization of HCN by NaOH is only -2.6 kcal mol^{-1} because the heat of ionization ΔH_{ioniz} of HCN is 10.8 kcal mol^{-1}.

1.20 ENTHALPIES OF FORMATION OF IONS

Since for strong electrolytes in dilute solution the thermal properties of the ions are essentially independent of the accompanying ions, it is convenient to use relative enthalpies of formation of individual ions. The sum of the enthalpies of formation of H$^+$ and OH$^-$ ions may be calculated from

$$H_2O(l) = H^+(aq) + OH^-(aq) \qquad \Delta H^\circ = 13.360 \text{ kcal}$$

$$\frac{H_2(g) + \tfrac{1}{2}O_2(g) = H_2O(l) \qquad\qquad \Delta H^\circ = -68.317 \text{ kcal}}{H_2(g) + \tfrac{1}{2}O_2(g) = H^+(aq) + OH^-(aq) \qquad \Delta H^\circ = -54.957 \text{ kcal}}$$

The separate enthalpies of formation of H$^+$ and OH$^-$ cannot be calculated, but if the enthalpy of formation of H$^+(aq)$ is arbitrarily assigned the value zero, it is possible to calculate *relative* enthalpies of formation for other ions. Denoting the electron by e, the conventions is that

$$\tfrac{1}{2}H_2(g) + aq = H^+(aq) + e \qquad \Delta H^\circ = 0$$

Therefore the enthalpy of formation of OH$^-$ is given by

$$\tfrac{1}{2}H_2(g) + \tfrac{1}{2}O_2(g) + aq + e = OH^-(aq) \qquad \Delta H^\circ = -54.957 \text{ kcal mol}^{-1}$$

On the basis of these values for the enthalpies of formation of H$^+$ and OH$^-$, the enthalpies of formation of other ions of strong electrolytes may be calculated.

From the enthalpy of formation of HCl(aq) it is possible to calculate the enthalpy of formation of Cl$^-(aq)$.

$$\tfrac{1}{2}H_2(g) + \tfrac{1}{2}Cl_2(g) + aq = H^+(aq) + Cl^-(aq) \qquad \Delta H^\circ = -40.023 \text{ kcal mol}^{-1}$$

$$\tfrac{1}{2}Cl_2(g) + aq + e = Cl^-(aq) \qquad\qquad \Delta H^\circ_f = -40.023 \text{ kcal mol}^{-1}$$

These and other enthalpies of formation of ions are given in Table 1.3.

1.21 DEPENDENCE OF THE HEAT OF REACTION ON TEMPERATURE

Let us consider the simple process $A \rightarrow B$, which may be carried out either at temperature T_1 or at temperature T_2. If we know the enthalpy of reaction at temperature T_1, we can calculate the enthalpy of reaction at temperature T_2 by means of the following two paths, provided that the heat capacities of A and B are known:

$$
\begin{array}{ccc}
 & A \xrightarrow[\;T_2\;]{\Delta H_2} B & \\
\int_{T_2}^{T_1}\bar{C}_{P,A}\,dT \Bigg\downarrow & & \Bigg\uparrow \int_{T_1}^{T_2}\bar{C}_{P,B}\,dT \\
 & A \xrightarrow[\;T_1\;]{\Delta H_1} B &
\end{array}
\qquad (1.58)
$$

The enthalpy change of reaction at T_2 is equal to the sum of three terms:

$$
\Delta H_2 = \int_{T_2}^{T_1}\bar{C}_{P,A}\,dT + \Delta H_1 + \int_{T_1}^{T_2}\bar{C}_{P,B}\,dT \qquad (1.59)
$$

since the enthalpy change is independent of the path chosen between two states. It is convenient to write this equation

$$
\Delta H_2 = \Delta H_1 + \int_{T_1}^{T_2}\Delta C_P\,dT \qquad (1.60)
$$

where $\Delta C_P = \bar{C}_{P,B} - \bar{C}_{P,A}$. If the heat capacities of reactant and product are independent of temperature between T_1 and T_2, ΔC_P may be placed in front of the integration sign and

$$
\Delta H_2 = \Delta H_1 + \Delta C_P(T_1 - T_2) \qquad (1.61)
$$

Example 1.7 Calculate the heat evolved in the freezing of water at constant pressure and a temperature of $-10°$.

$$
H_2O(l) = H_2O(s)
$$

Given: $\Delta H_{273} = -79.7$ cal g^{-1}; $C_{P,H_2O(l)} = 1.00$ cal K^{-1} g^{-1}; $C_{P,H_2O(s)} = 0.49$ cal K^{-1} g^{-1}. The following calculation is carried out for 1 gram of H_2O rather than 1 mole.

$$
\Delta H_{263} = \Delta H_{273} + [C_{P,H_2O(s)} - C_{P,H_2O(l)}](263 - 273)
$$
$$
\Delta H_{263} = \Delta H_{273} + (-0.51)(-10)
$$
$$
= -79.7 + 5.1 = -74.6 \text{ cal g}^{-1}
$$

1.22 GENERAL TREATMENT OF THE EFFECT OF TEMPERATURE ON THE ENTHALPY OF REACTION

For a general chemical reaction given in equation 1.52 the enthalpy change is given by equation 1.54. The rate of change of ΔH with temperature is obtained by differentiating equation 1.54 with respect to temperature at constant pressure.

$$
\left[\frac{d(\Delta H)}{dT}\right]_P = \sum \nu_i \left(\frac{d\bar{H}_i}{dT}\right)_P \qquad (1.62)
$$

Remembering that $(d\bar{H}/dT)_P = \bar{C}_P$, we see that

$$\left[\frac{d(\Delta H)}{dT}\right]_P = \sum \nu_i \bar{C}_{P,i} = \Delta C_P \tag{1.63}$$

This equation may be stated in words as follows: the change in enthalpy of reaction at constant pressure per degree rise in temperature is equal to the change in heat capacity at constant pressure of the system as the result of the reaction.

Equation 1.63 may be integrated between two temperatures T_1 and T_2 to obtain the relation between the enthalpy changes at these two temperatures:

$$\int_{\Delta H_1}^{\Delta H_2} d(\Delta H) = \Delta H_2 - \Delta H_1 = \int_{T_1}^{T_2} \Delta C_P \, dT \tag{1.64}$$

By use of this equation it is possible to calculate ΔH for a reaction at another temperature if it is known at one temperature and if the values of $\bar{C}_P$ for the reactants and products are known in the intervening temperature range. The empirical equations such as those given in Table 1.2 for expressing C_P as a function of temperature are very useful for this purpose.

Example 1.8 Calculate the heat of combustion of hydrogen at 1500 K.

$$2H_2(g) + O_2(g) = 2H_2O(g) \qquad \Delta H_{298}^\circ = -115{,}595.8 \text{ cal}$$

The molar heat capacities of Table 1.2 are substituted into the following equation:

$$\Delta H_{1500}^\circ = \Delta H_{298}^\circ + \int_{298}^{1500} (2\bar{C}_{P,H_2O} - \bar{C}_{P,O_2} - 2\bar{C}_{P,H_2}) \, dT$$

to obtain

$$\Delta H_{1500}^\circ = -115{,}595.8 + \int_{298}^{1500} (-5.6146 + 1.8931 \times 10^{-3} T + 4.723 \times 10^{-7} T^2) \, dT$$

$$= -119{,}767 \text{ cal}$$

Equation 1.64 is applicable only if there are no changes in phase in going from T_1 to T_2; additional terms must be introduced for the enthalpy changes accompanying phase transformations such as melting or vaporization.

1.23 BOND ENERGIES*

The standard bond dissociation energy DH° is the enthalpy change of the chemical reaction in which one mole of a specified bond is broken. The reactant and product are assumed to be in their standard states of hypothetical ideal gas at one atmosphere pressure and 25°. The bond dissociation energy $DH^\circ(A - B)$ of the bond between parts A and B of the molecule is the enthalpy change for the reaction

$$A - B(g) = A(g) + B(g) \tag{1.65}$$

and may be expressed in terms of enthalpies of formation as follows:

$$DH^\circ(A - B) = \Delta H_{f,A}^\circ + \Delta H_{f,B}^\circ - \Delta H_{f,AB}^\circ \tag{1.66}$$

---

* S. W. Benson, *J. Chem. Ed.*, **42**, 502 (1965).

Table 1.4[1]   Bond Dissociation Energies of Diatomic Molecules
and Radicals at 25° (kcal mol^{-1})

	$DH°$		$DH°$
H_2	104.2	HO	102.4
D_2	106.0	HF	135.8
O_2	119.2	HCl	103.0
N_2	226	HBr	87.5
C_2	143	HI	71.3
Pb_2	13	HC	81
F_2	38	HN	86
Cl_2	58	HS	85
Br_2	46.0	NO	151.0
I_2	36.1	CO	256.9
Na_2	18.7	CF	116

[1] From S. W. Benson, *J. Chem. Ed.*, **42,** 502 (1965).

Bond dissociation energies may be obtained from (1) equilibrium measurements by measuring the equilibrium constant over a range of temperature, (2) spectroscopic studies of vibrational energy levels, and indirectly from (3) chemical kinetics, and (4) mass spectroscopy. The bond dissociation energies of a number of diatomic molecules and radicals are given in Table 1.4 and bond dissociation energies for single and multiple bonds are given in Tables 1.5 and 1.6. Table 1.5 illustrates the fact that bond dissociation energies depend on the structure of the whole molecule and not just on the bond being broken.

Table 1.5[1]   Bond Dissociation Energies for Single Bonds at 25° (kcal mol^{-1})

	$DH°$		$DH°$
HO—H	119	CH_3—H	104
CH_3CO_2—H	112	CH_3CH_2—H	98
CH_3O—H	102	$(CH_3)_2CH$—H	94.5
HO_2—H	90	$(CH_3)_3C$—H	91
C_6H_5O—H	85	C_6H_5—H	103
		CH_2CH—H	103
		HCC—H	125
CH_3—CH_3	88	H_2N—NH_2	58
$(CH_3)_3C$—CH_3	80	HO—OH	51
$(CH_3)_3C$—$C(CH_3)_3$	67.5	CH_3—Cl	84
$(C_6H_5)_3C$—$C(C_6H_5)_3$	15	H_2N—H	103
HCC—CCH	150		
NC—CN	144	CH_3—NH_2	79
CH_3—CN	122	CH_3—OH	91
C_6H_5—C_6H_5	100	CH_3—F	108
CH_2CH—CH_3	29	CH_3—I	56

[1] From S. W. Benson, *J. Chem. Ed.*, **42,** 502 (1965).

Table 1.6[1] Bond Dissociation Energies for Multiple Bonds in Some Isoelectronic Sequences at 25° (kcal mol^{-1})

	$DH°$		$DH°$
N≡N	226	CH$_2$=CH$_2$	163
HC≡CH	230	CH$_2$=O	175
HC≡N	224	O=O	119
C≡O	257	HN=O	115
		HN=NH	109

[1] From S. W. Benson, *J. Chem. Ed.*, **42**, 502 (1965).

The data in Table 1.6 are for two groups of molecules that are isoelectronic; that is, they have the same number of electrons. It is striking that the triple-bond energies lie in a rather narrow range, but the double-bond energies do not.

References

F. Daniels, J. W. Williams, P. Bender, R. A. Alberty, C. D. Cornwell, and J. E. Harriman, *Experimental Physical Chemistry*, McGraw-Hill Book Co., New York, 1970.

K. G. Denbigh, *The Principles of Chemical Equilibrium*, Cambridge University Press, Cambridge, 1971.

R. E. Dickerson, *Molecular Thermodynamics*, W. A. Benjamin Inc., New York, 1969.

I. M. Klotz and R. M. Rosenberg, *Chemical Thermodynamics*, W. A. Benjamin Inc., New York, 1972.

G. N. Lewis, M. Randall, revised by K. S. Pitzer, and L. Brewer, *Thermodynamics*, McGraw-Hill Book Co., New York, 1961.

A. B. Pippard, *Elements of Classical Thermodynamics*, Cambridge University Press, Cambridge, 1960.

P. A. Rock, *Chemical Thermodynamics*, The MacMillan Co., London, 1969.

F. D. Rossini, K. S. Pitzer, W. J. Taylor, J. P. Ebert, J. E. Kilpatrick, C. W. Beckett, M. G. Williams, and H. G. Werner, "Selected Values of Properties of Hydrocarbons," *Natl. Bur. Standards Circ. C461*, U.S. Government Printing Office, Washington D.C., 1947.

F. D. Rossini, D. D. Wagman, W. H. Evans, S. Levine, and I. Jaffe, "Selected Values of Chemical Thermodynamic Properties," *Natl. Bur. Standards Circ. 500*, U.S. Government Printing Office, Washington D.C., 1952.

H. J. V. Tyrrell and A. E. Beezer, *Thermometric Titrimetry*, Chapman and Hall, London, 1968.

F. T. Wall, *Chemical Thermodynamics*, W. H. Freeman and Co., San Francisco, 1965.

Problems

1.1 How much work is done when a man weighing 75 kg (165 lb) climbs the Washington monument, 555 ft high? How many kilocalories must be supplied to do this muscular work, assuming that 25% of the energy produced by the oxidation of food in the body can be converted into muscular mechanical work?   *Ans.* 119 kcal.

1.2 Thirty-five liters of hydrogen is produced at a total pressure of 1 atm by the action of acid on a metal. Calculate the work done by the gas in pushing back the atmosphere in (a) liter-atmospheres, (b) calories and, (c) joules.

Ans. (a) 35 liter atm, (b) 847 cal, (c) 3540 J.

1.3 How many degrees will a 500 W electric heater raise the temperature of 10 liters of water in 1 hr if no heat is lost? The specific heat of water may be taken as 1 cal K^{-1} g^{-1}, and the density of water as 1 g cm^{-3} independent of temperature. *Ans.* 43.1°.

1.4 One hundred grams of benzene is vaporized at its boiling point of 80.2° at 760 Torr. The heat of vaporization is 94.4 cal g^{-1}. Calculate (a) w_{rev}, (b) q, (c) ΔH, (d) ΔU.

Ans. (a) -898, (b) 9440, (c) 9440, (d) 8542 cal.

1.5 One hundred grams of nitrogen at 25° and 760 Torr is expanded reversibly and isothermally to a pressure of 100 Torr (a) What is the value of w? (b) What is the value of w if the temperature is 100°? *Ans.* (a) -4290, (b) -5370 cal.

1.6 (a) Calculate the work in calories done when a mole of sulfur dioxide gas expands isothermally and reversibly at 27° from 2.46 to 24.6 liters, assuming that the gas is ideal. (b) Would an attractive force between molecules of the gas tend to make the work done larger or smaller? *Ans.* (a) -1370 cal mol^{-1}, (b) Smaller.

1.7 In an adiabatic expansion of a gas work is done on the surroundings at the expense of the internal energy of the gas and the temperature drops because $q = 0$. For a mole of ideal gas the first law is

$$d\bar{U} = -P\,d\bar{V} = \bar{C}_V\,dT$$

Show that for the reversible adiabatic expansion of an ideal gas,

$$\bar{C}_V \ln \frac{T_2}{T_1} = -R \ln \frac{\bar{V}_2}{\bar{V}_1}$$

1.8 Calculate the temperature increase and final pressure of helium if a mole is compressed adiabatically and reversibly from 44.8 liters at 0° to 22.4 liters. The molar heat capacity $\bar{C}_V$ of helium is constant and equal to 3.00 cal K^{-1} mol^{-1}. *Ans.* 159.1 K. 1.583 atm.

1.9 The equation for the molar heat capacity of *n*-butane is

$$\bar{C}_P = 4.64 + 0.0558T$$

Calculate the heat necessary to raise the temperature of 1 mole from 25 to 300° at constant pressure. *Ans.* 7961 cal.

1.10 How much heat is required to raise the temperature of 10 grams of argon (a monoatomic gas) through 10° (a) at constant volume, (b) at constant pressure? *Ans.* (a) 7.47, (b) 12.5 cal.

1.11 In an adiabatic calorimeter, oxidation of 0.4362 gram of naphthalene caused a temperature rise of 1.707°. The heat capacity of the calorimeter and water was 2460 cal K^{-1}. If corrections for oxidation of the wire and residual nitrogen are neglected, what is the enthalpy of combustion of naphthalene per mole? *Ans.* -1234 kcal mol^{-1}.

1.12 Calculate the enthalpy change for the transition of one mole of monoclinic sulfur to rhombic sulfur at room temperature. This transition is too slow for direct calorimeter measurement. The enthalpy of combustion of monoclinic sulfur is -71.03 kcal and for rhombic sulfur is -70.96 kcal. *Ans.* -0.07 kcal.

1.13 The following reactions might be used to power rockets:

(1) $H_2(g) + \frac{1}{2}O_2(g) = H_2O(g)$

(2) $CH_3OH(l) + 1\frac{1}{2}O_2(g) = CO_2(g) + 2H_2O(g)$

(3) $H_2(g) + F_2(g) = 2HF(g)$

(a) Calculate the enthalpy changes at 25° for each of these reactions per kilogram of reactants. (For HF(g), $\Delta \bar{H}_f^{\circ} = -64.2$ kcal.) (b) Since the thrust is greater when the molecular weight of the exhaust gas is lower, divide the heat per kilogram by the molecular weight of the product (or the average molecular weight in the case of reaction 2) and arrange the above reactions in order of effectiveness on the basis of thrust.

Ans. (a) $-3210, -1908, -3210$ kcal kg^{-1}. (b) (1) > (3) > (2).

1.14 Calculate the enthalpy of formation of $PCl_5(s)$, given the heats of the following reactions at 25°:

$$2P(s) + 3Cl_2(g) = 2PCl_3(l) \qquad \Delta H^{\circ} = -151,800 \text{ cal}$$

$$PCl_3(l) + Cl_2(g) = PCl_5(s) \qquad \Delta H^{\circ} = -32,810 \text{ cal}$$

Ans. -108.71 kcal mol^{-1}.

1.15 Using the data in Table 1.3 on enthalpies of formation, calculate the enthalpies of combustion at 25° of the following substances to $H_2O(l)$ and $CO_2(g)$: (a) *n*-butane, (b) methanol, (c) acetic acid. *Ans.* (a) -687.982, (b) -173.66, (c) -208.3 kcal mol^{-1}.

1.16 For acetone, $(CH_3)_2CO$, $\Delta \bar{H}_f^{\circ}$ is -61.4 kcal mol^{-1} at 25°. (a) Calculate the heat of combustion of $(CH_3)_2CO$ at constant pressure. (b) Calculate the heat evolved when 2 grams of $(CH_3)_2CO$ are burned under pressure in a closed bomb at 25°.

Ans. (a) -425.7 kcal mol^{-1}. (b) 14.68 kcal.

1.17 Calculate ΔH_{298}° for the reaction

(1) $CdSO_4(s) + H_2O(g) = CdSO_4 \cdot H_2O(s)$

using the information that

(2) $CdSO_4(s) + 400H_2O(l) = CdSO_4$ in $400H_2O$	$\Delta H^{\circ} = -10,977$ cal
(3) $CdSO_4 \cdot H_2O(s) + 399H_2O(l) = CdSO_4$ in $400H_2O$	$\Delta H^{\circ} = -6,095$ cal
(4) $H_2O(g) = H_2O(l)$	$\Delta H^{\circ} = -9,717$ cal

By means of heat-capacity measurements down to the neighborhood of absolute zero and measurements of equilibrium water-vapor pressures, M. N. Papadopoulos and W. F. Giauque [*J. Am. Chem. Soc.*, **77**, 2740 (1955)] have obtained $-15,451$ cal mol^{-1} for ΔH_{298} of reaction 1. *Ans.* -14.599 kcal mol^{-1}.

1.18 Calculate the enthalpies of reaction at 25° for the following reactions in dilute aqueous solutions:

(a) $HCl(aq) + NaBr(aq) = HBr(aq) + NaCl(aq)$

(b) $CaCl_2(aq) + Na_2CO_3(aq) = CaCO_3(s) + 2NaCl(aq)$

(c) $Li(s) + \frac{1}{2}Cl_2(g) + aq = Li^+(aq) + Cl^-(aq)$

Ans. (1) 0, (2) $+2.95$, (3) -106.577 kcal.

1.19 Calculate ΔH° at 1000 K for the reaction

$$CH_4(g) + 2O_2(g) = CO_2(g) + 2H_2O(g)$$

Ans. -191.358 kcal.

1.20 Calculate how high a 70-kg man could climb if he could convert the heat of combustion of one ounce of chocolate (150 kcal) completely into work of vertical displacement.

1.21 The surface tension of water is 71.97×10^{-3} N m^{-1} or 71.97×10^{-3} J m^{-2} at 25°; calculate the surface energy in calories of 1 mole of water dispersed as a mist containing droplets 1 μm (10^{-4} cm) in radius. The density of water may be taken as 1.00 g cm^{-3}.

1.22 What current must be passed through a 100-ohm heater to heat a 10-liter water thermostat 0.1 K min^{-1}? The specific heat of water may be taken as 1 cal K^{-1} g^{-1}, and the density as 1 g cm^{-3}.

1.23   A mole of ammonia gas is condensed at its standard boiling point of $-33.4°$ C by the application of a pressure infinitesimally greater than 1 atm. To evaporate a gram of ammonia at its boiling point requires the absorption of 327 cal. Calculate (a) w_{rev}, (b) q, (c) $\Delta \bar{H}$ (d) $\Delta \bar{U}$.

1.24   One mole of ideal gas at 25° and 100 atm is allowed to expand reversibly and isothermally to 5 atm. Calculate (a) the work done on the gas in liter-atmospheres, (b) the heat absorbed in calories, (c) $\Delta \bar{U}$ and (d) $\Delta \bar{H}$.

1.25   Calculate the maximum work obtained in the isothermal expansion of 10 grams of helium from 10 to 50 liters at 25°, assuming ideal gas behavior. Express the answer in (a) calories, (b) liter atmospheres and (c) joules.

1.26   One mole of methane (considered to be an ideal gas) initially at 25° and 1 atm pressure is heated at constant pressure until the volume has doubled. The variation of the molar heat capacity with absolute temperature is given by

$$\bar{C}_P = 5.34 + 11.5 \times 10^{-3}T$$

Calculate (a) $\Delta \bar{H}$, and (b) $\Delta \bar{U}$.

1.27   One hundred liters of helium at 0° and 1 atm is heated in a closed vessel to 800°. (a) Calculate the change in internal energy in kilocalories. (b) How much more heat would be required if the gas were heated at a constant pressure of 1 atm?

1.28   One mole of hydrogen at 25° and 1 atm is compressed adiabatically and reversibly into a volume of 5 liters. Assuming ideality, calculate (a) the final temperature, (b) the final pressure, and (c) the work done on the gas. (See problem 1.7.)

1.29   How many grams of cane sugar, $C_{12}H_{22}O_{11}$, must be oxidized to give the same number of calories of *heat* as the number of calories of work done by a 160-lb (72.7-kg) man in climbing a mountain 1 mile (1.609 km) high? The heat of combustion of $C_{12}H_{22}O_{11}$ is 1349.7 kcal mol^{-1}. It is found empirically that only about 25% of the heat value of food can be converted into useful work by men or animals, and accordingly the calculated grams of $C_{12}H_{22}O_{11}$ should be multiplied by 4 to give approximately the amount which would actually be oxidized.

1.30   The combustion of oxalic acid in a bomb calorimeter yields 673 cal g^{-1} at 25°. Calculate (a) $\Delta \bar{U}°$ and (b) $\Delta \bar{H}°$ for the combustion of 1 mole of oxalic acid ($M = 90.0$).

1.31   A solution of hemoglobin containing 5 g of the protein ($M = 64,000$) in 100 cm^3 of solution is completely oxygenated. Each mole of hemoglobin binds four moles of oxygen The temperature of the solution rises 0.031°. What is the enthalpy of reaction per mole of oxygen bound? The heat capacity of the solution may be assumed to be 1 cal K^{-1} cm^{-3}.

1.32   The enthalpy change for the combustion of toluene to $H_2O(l)$ and $CO_2(g)$ is -934.50 kcal mol^{-1} at 25°. Calculate the enthalpy of formation of toluene.

1.33   Construct an energy level diagram like Fig. 1.5 from the following data at 298°:

$$CH_4(g) + 2O_2(g) = CO_2(g) + 2H_2O(l) \qquad \Delta H° = -212.8 \text{ kcal}$$
$$CH_4(g) = C(s) + 2H_2(g) \qquad \Delta H° = +17.9 \text{ kcal}$$
$$C(s) + O_2(g) = CO_2(g) \qquad \Delta H° = -94.05 \text{ kcal}$$
$$H_2(g) + \tfrac{1}{2}O_2(g) = H_2O(l) \qquad \Delta H° = -68.32 \text{ kcal}$$

1.34   For 25° C the following results are given:

$$Be_3N_2(s) + 3Cl_2(g) \rightarrow 3BeCl_2(s) + N_2(g) \qquad \Delta H° = -214.4 \pm 0.1 \text{ kcal}$$
$$Be(s) + Cl_2(g) \rightarrow BeCl_2(s) \qquad \Delta H° = -118.1 \pm 0.6 \text{ kcal}$$
$$3Be(s) + 2NH_3(g) \rightarrow Be_3N_2(s) + 3H_2(g) \qquad \Delta H° = -118.4 \pm 0.4 \text{ kcal}$$

$$\tfrac{1}{2}N_2 + \tfrac{3}{2}H_2 \rightarrow 1 NH_3 \qquad \Delta H° = -11.04$$

Given the standard enthalpy of formation of $NH_3(g)$ at $25°$ as -11.04 kcal, compare the two values for the standard enthalpy of formation of Be_3N_2 which can be obtained from the foregoing data.

1.35 The following heats of solution are obtained at $18°$ when a large excess of water is used:

$$CaCl_2(s) + aq = CaCl_2(aq) \qquad \Delta H° = -18.0 \text{ kcal}$$
$$CaCl_2 \cdot 6H_2O(s) + aq = CaCl_2(aq) \qquad \Delta H° = +4.5 \text{ kcal}$$

Calculate the heat of hydration of $CaCl_2$ to give $CaCl_2 \cdot 6H_2O$ by (a) $H_2O(l)$ and (b) $H_2O(g)$. The heat of vaporization of water at this temperature is 586 cal g^{-1}.

1.36 Using these data for the integral heat of solution of m moles of cadmium nitrate in 1000 g of water, plot $\Delta H°$ against the molality m, and determine the differential heat of solution of cadmium nitrate in a 4-molal solution:

m (molality)	1.063	1.799	2.821	4.251	6.372	9.949
$\Delta H°$, kJ	-34.2	-56.7	-88.3	-126.0	-174.4	-228.5

1.37 Using the enthalpies of formation of $NaOH(aq)$ and $NaNO_3(aq)$ in infinitely dilute solution (Table 1.3) and the value for $\Delta \bar{H}_{f,OH^-}°$, show how the value for $\Delta \bar{H}_f°$ for NO_3^- in Table 1.3 is obtained.

1.38 (a) Estimate from enthalpies of formation and bond energies the heat of dissociation of $HCl(g)$ into atoms. (b) Estimate the heats of the following reactions at constant pressure:

$$C_2H_4(g) + Cl_2(g) = C_2H_4Cl_2(g)$$
$$C_2H_6(g) + 2Cl_2(g) = C_2H_4Cl_2(g) + 2HCl(g)$$

1.39 Given the enthalpy of formation of $H_2O_2(g)$ at $298°$ as -38.8 kcal and the enthalpy of formation of $OH(g)$ at $298°$ as 8 kcal, calculate ΔH for the reaction

$$H_2O_2(g) = 2OH(g)$$

1.40 Calculate the quantity of heat required to raise the temperature of one mole of oxygen from 300 to 1000 K at constant pressure.

1.41 Calculate the molar heat of combustion of carbon monoxide at constant pressure and $1327°$.

1.42 Calculate the heat of vaporization of water at $25°$. The specific heat of water may be taken as 1 cal $K^{-1} g^{-1}$. The heat capacity of water vapor at constant pressure in this temperature range is 8.0 cal K^{-1} mol^{-1}, and the heat of vaporization of water at $100°$ is 539.7 cal g^{-1}.

1.43 By calorimetric determination of the heat evolved in the cooling of a sample of mercuric bromide from the indicated temperature of 300 K, the results tabulated below were obtained. Calculate values for the heat of fusion of mercuric bromide, and the molar heat capacity C_P for liquid mercuric bromide at 525 K. The melting point is to be taken as 511.3 K:

T, K	$(\bar{H}_T° - \bar{H}_{300}°)$, cal mol^{-1}	T, K	$(\bar{H}_T° - \bar{H}_{300}°)$ cal mol^{-1}
544.1	9313	513.9	6251
533.4	9083	507.4	4114
526.8	8729	502.5	3923
522.3	8672	496.3	3974
516.6	8695	487.4	3656

1.44 . (a) Calculate the change in internal energy of oxygen when it is heated from $0°$ to $100°$ at constant volume. (b) Calculate the change in enthalpy for this process.

1.45   A mole of oxygen at 5 atm and in a volume of 4 liters is allowed to expand reversibly and adiabatically to a final pressure of 1 atm. What are the final (a) volume and (b) temperature? (See problem 1.7.)

1.46   From the following data calculate the value of $(\bar{H}_{298}^{\circ} - \bar{H}_0^{\circ})$ for $Al_2O_3(s)$, assuming the validity of the Debye T^3 law for the heat capacity below 10 K.

T, K	C_P°, J K^{-1} mol^{-1}	T	C_P°	T	C_P°	T	C_P°
10	0.009	90	9.69	180	43.79	270	72.37
20	0.076	100	12.84	190	47.53	280	74.84
30	0.263	110	16.32	200	51.14	290	77.19
40	0.691	120	20.06	210	54.60	298.16	79.01
50	1.492	130	23.96	220	57.92	273.16	73.16
60	2.779	140	27.96	230	61.10		
70	4.582	150	31.98	240	64.13		
80	6.895	160	35.99	250	67.01		
		170	39.94	260	69.76		

1.47   Calculate the enthalpy of formation of acetylene from the fact that the enthalpy change for combustion is -310.615 kcal mol^{-1}.

1.48   The enthalpy of formation of nitric oxide from nitrogen and oxygen has been calculated from spectroscopic data. A direct calorimetric determination is desirable, however. It has been found that phosphorus will burn completely to P_2O_5 in NO, leaving nitrogen, if the phosphorus is thoroughly ignited with a hot arc. Calculate the enthalpy of formation of NO from the following data.

Phosphorus was burned in a calorimeter in a stream of NO for 12.00 min and produced 1.508 grams of H_3PO_4. The calorimeter was surrounded by 1386 grams of water. The observed temperature rise was $2.222°$, and the cooling correction amounted to $0.037°$. The correction for the heat of stirring was 11.1 cal evolved per min. The heat capacity of the calorimeter, as determined with an electric heater, was 244 cal K^{-1}.

In a second experiment the NO was replaced by a mixture of half nitrogen and half oxygen. The amount of P_2O_5 produced in 10.00 min was equivalent to 2.123 grams of H_3PO_4. The observed temperature rise was $2.398°$, and the cooling correction was $0.032°$. The correction for heat of stirring was 12.2 cal evolved per min. The weight of water was the same as in the first experiment. What is the enthalpy of formation of NO?

1.49   Calculate the quantity of heat required to raise 1 mol of CO_2 from $0°$ to $300°$ (a) at constant pressure and (b) at constant volume. $\bar{C}_P$ in calories per degree per mole is given by

$$\bar{C}_P = 6.40 + 10.2 \times 10^{-3}T - 35 \times 10^{-7}T^2$$

1.50   One hundred grams of iron is dissolved in dilute acid at $25°$, giving a ferrous salt. Will more heat be evolved when the reaction is carried out in an open beaker or in a closed bomb? How much more?

1.51   From Table 1.3 of the enthalpies of formation calculate the heats of combustion at constant pressure at $25°$ of (a) CO, (b) H_2, (c) C_2H_6, and (d) C_2H_5OH.

1.52   Calculate the heat of hydration of $Na_2SO_4(s)$ from the integral heats of solution of $Na_2SO_4(s)$ and $Na_2SO_4 \cdot 10H_2O(s)$ in infinite amounts of H_2O, which are -0.56 kcal mol ^{1}

and $+18.85$ kcal mol^{-1}, respectively. Enthalpies of hydration cannot be measured directly because of the slowness of the phase transformation.

1.53 The integral heat of solution of 1 mole of H_2SO_4 in n moles of water is given in calories by the equation

$$\Delta H^\circ = \frac{-18,070n}{n + 1.798}$$

Calculate ΔH° for the following reactions: (a) solution of 1 mole of H_2SO_4 in 5 moles of water; (b) solution of 1 mole of H_2SO_4 in 10 moles of water; (c) solution of 1 mole of H_2SO_4 in a large excess of water, 100,000 moles for example; (d) addition of a large excess of water to a solution containing 1 mole of H_2SO_4 in 10 moles of water; (e) addition of 5 moles of water to a solution containing 1 mole of H_2SO_4 in 5 moles of water.

1.54 Calculate the enthalpies of reaction at 25° for the following reactions in dilute aqueous solutions:

(1) $AgNO_3(aq) + KCl(aq) = AgCl(s) + KNO_3(aq)$

(2) $Zn(s) + 2HCl(aq) = ZnCl_2(aq) + H_2(g)$

1.55 Calculate the heat evolved in the reaction $H_2(g) + Cl_2(g) = 2HCl(g)$ at 1023° and constant pressure.

1.56 Using data in this chapter and $\bar{C}_P$ for graphite

$$\bar{C}_P = 2.673 + 0.00261\,T - \frac{1.169 \times 10^5}{T^2}$$

calculate ΔH° for the following reaction at 600 K:

$$H_2O(g) + C(s) = CO(g) + H_2(g)$$

1.57 Calculate the theoretical flame temperature for the burning of ethane with a stoichiometric amount of oxygen. The heat evolved in the reaction is used to heat the products to the flame temperature, it being assumed that the reaction goes to completion and no heat is lost by radiation. The flame temperature obtained experimentally would be less than this theoretical value because at high temperature there will be other substances than CO_2 and H_2O in the products and losses due to radiation.

1.58 Calculate the enthalpy change for the dissociation of $N_2(g)$ at 25°, using data from Table 1.3. Explain why this result differs from that given in Table 1.6.

1.59 Calculate the enthalpy change at 25° for the reaction of a dilute aqueous solution of $AgNO_3$ with a dilute aqueous solution of NaCl.

CHAPTER 2

SECOND AND THIRD LAWS OF THERMODYNAMICS

The first law of thermodynamics states that when one form of energy is converted into another the total energy is conserved. It does not indicate any other restriction on this process. However, we know that many processes have a natural direction, and it is with this question that the second law is concerned. For example, a gas expands into a vacuum, but, although it would not violate the first law, the reverse never occurs. For a bar at uniform temperature to become hot at one end and cold at the other would not be a violation of the first law, yet we know it never occurs. The second law establishes a criterion for predicting whether a process can occur spontaneously, and so it is of great importance to chemistry.

The quantity that tells us whether a chemical reaction or a physical change can occur spontaneously in an isolated system is the entropy S. The entropy is a function of the state of the system, as is the internal energy U. Another way of saying this is that dS is an exact differential (Section 1.8).

The introduction of S completes the necessary set of thermodynamic quantities, but it is useful to define certain further quantities, especially the Gibbs free energy G and the chemical potential μ, in terms of other thermodynamic quantities. The quantities G and μ are especially useful in discussing phase equilibria (Chapters 3 and 4) and chemical equilibria (Chapters 5, 6, and 7). This chapter closes with the third law of thermodynamics that allows us to obtain the absolute value of the entropy of a substance.

2.1 INTENSIVE AND EXTENSIVE PROPERTIES

The macroscopic properties of matter, which are the concern of thermodynamics, are of two different kinds: intensive and extensive. Intensive properties do not depend on the amount of material in the system. Temperature T, pressure P, and density ρ are intensive properties. Extensive properties are directly proportional to the mass of the system if intensive properties are held constant. Volume V and internal energy U are extensive properties. In this chapter we will introduce new extensive properties: the entropy S, the Helmholtz free energy A, the Gibbs free energy G, and new intensive properties, the chemical potential μ, and the partial derivatives $\bar{V}_i$, $\bar{U}_i$, $\bar{H}_i$, $\bar{S}_i$, $\bar{A}_i$, and $\bar{G}_i$ of the corresponding extensive properties with respect to amount of substance.

2.2   SPONTANEOUS AND NONSPONTANEOUS CHANGES

We are familiar with the fact that many changes occur spontaneously; that is, when systems are simply left to themselves. Water runs downhill, chemical reactions proceed to equilibrium, and heat flows from hotter bodies to cooler bodies. For any spontaneous change it is possible to devise, in principle at least, a means for getting useful work. Thus, falling water can turn a turbine, a chemical reaction may be harnessed in a battery, and hot and cold reservoirs may be used to run a heat engine. Since work can be obtained from a spontaneous change, it is evident that in the occurrence of a spontaneous change the system loses capacity to do work.

It is a matter of experience that spontaneous changes do not reverse themselves; that is, water does not run uphill. The term *nonspontaneous* is applied to the reverse of a spontaneous change, for example, water flowing uphill. Nonspontaneous changes can be made to occur only by supplying energy from outside the system. For example, energy is required to pump water uphill, to recharge a battery, and to transfer heat from a cold reservoir to a hot one, as in a refrigerator. Since the energy required can be supplied only by some other spontaneous change, it is apparent that a spontaneous change may be reversed only by harnessing, in some way, energy from another spontaneous change.

2.3   THE SECOND LAW OF THERMODYNAMICS

We will regard the second law as a postulate that is to be judged on the basis of the results calculated with it. We know that the second law is true for macroscopic systems because there are no known violations of the relations derived from it.

There are various ways of stating the second law that sound quite different, but lead to the same results when applied. Kelvin stated the second law as follows: It is impossible to use a cyclic process† to transfer heat from a heat reservoir and convert it into work without at the same time transferring a certain amount of heat from a hotter to a colder body. Clausius stated the second law thus: it is impossible to use a cyclic process to transfer heat from a colder to a hotter body without at the same time converting a certain amount of work into heat. Historically the development of the second law of thermodynamics was closely connected with the studies of heat engines, as these statements indicate. However, this is pretty far removed from chemistry, and so we will discuss the second law in terms that are more directly applicable to chemical reactions. This is the approach used by Gibbs* to whom we are indebted so much for concepts of thermodynamics and statistical mechanics.

---

† A cyclic process is a series of steps that brings a system back to its initial condition.

* J. W. Gibbs, *Trans. Conn. Acad. Sci.*, **3**, 228 (1876). J. W. Gibbs, *The Collected Works of J. Willard Gibbs*, Yale University Press, New Haven, 1948.

Since ΔU and ΔH do not tell us whether a chemical reaction or a physical change will be spontaneous, we need an additional thermodynamic quantity for this purpose. This quantity is the entropy S.

2.4   ENTROPY

The entropy S is a function of the state of a system. That is, like the internal energy U, it has a certain value when the system is in a certain state. Its differential dS is defined as follows:

$$\text{For an infinitesimal reversible change: } dS \equiv dq_{\mathrm{rev}}/T \qquad (2.1)$$

where T is a temperature defined by this relation. The use of dq rather than dq reminds us that q is not a state function but depends on the path used in making the change. However, for a reversible process dq/T is an exact differential. We will find later (Section 2.14) that the temperature in equation 2.1 may be identified with the ideal gas temperature scale, and so we will use the same symbol. Because of 2.1 we may say that T is the integrating factor for heat since dq_{rev}/T is an exact differential (Section 1.8). The entropy change for a given change in state is *independent of path*. The entropy change for a finite change from state 1 to state 2 is given by the integral of equation 2.1 along a reversible path.

$$\Delta S = S_2 - S_1 = \int_1^2 \frac{dq}{T} \qquad (2.2)$$

Any reversible path between state 1 and state 2 will give the same value of the integral. The entropy is an extensive property, and so like the volume V and the internal energy U it depends on the mass of the system considered.

2.5   THE SECOND LAW IN TERMS OF ENTROPY

The second law may conveniently be stated in terms of the entropy. The differential of the entropy dS is either greater than or equal to dq/T for any infinitesimal process.

$$dS \geq dq/T \qquad (2.3)$$

Thus the second law distinguishes between two types of changes: those for which the differential of the entropy is greater than dq/T, which are irreversible changes, and those for which the differential of the entropy is equal to dq/T, which are reversible changes. Examples of reversible changes are the transfer of an infinitesimal quantity of heat dq from a reservoir of temperature T to a reservoir with an infinitesimally lower temperature or the conversion of an infinitesimal amount of a reactant to a product in an equilibrium-reaction mixture.

The second law provides a criterion for determining whether or not a chemical reaction can occur in the statement that for an irreversible process $dS > dq/T$. Thus if we can show that this inequality applies to a certain change or chemical reaction we will know that that change or chemical reaction can occur spontaneously.

It is often convenient to apply equation 2.3 to an isolated system, that is, one of constant internal energy U and volume V with no heat, work, or material flowing in or out. For an isolated system $dq = 0$ and so equation 2.1 and 2.3 become

$$\text{For an infinitesimal reversible change:} \quad dS = 0 \tag{2.4}$$

$$\text{For an infinitesimal irreversible change:} \quad dS > 0 \tag{2.5}$$

For a finite reversible change in an isolated system $\Delta S = 0$, and for a finite irreversible change $\Delta S > 0$. Thus when an irreversible change occurs in an isolated system the entropy increases. When all possibilities for increasing the entropy in spontaneous changes have been exhausted, the entropy will have a maximum value. For any infinitesimal change at equilibrium in an isolated system, $dS = 0$.

We will now consider some simple processes for which entropy changes are readily calculated. Three types of processes that may be carried out reversibly are phase transitions (e.g., the evaporation of a liquid into saturated vapor), heating a substance, and expansion of an ideal gas.

2.6 CALCULATION OF ENTROPY CHANGES

The transfer of heat from one body to another at an infinitesimally lower temperature is a reversible change, since the direction of heat flow can be reversed by an infinitesimal change in the temperature of one of the bodies. The fusion of a solid at its melting point and the evaporation of a liquid at a constant partial pressure of the substance equal to its vapor pressure are examples of isothermal transformations that can be reversed by an infinitesimal change in temperature. For these changes the entropy change is easily calculated. Since T is constant, performing the integration of equation 2.1 yields

$$S_2 - S_1 = \Delta S = \frac{q_{\text{rev}}}{T} \tag{2.6}$$

where q_{rev} represents the heat absorbed in the reversible change.

For a mole of substance at constant temperature and pressure,

$$\bar{S}_2 - \bar{S}_1 = \Delta \bar{S} = \frac{\Delta \bar{H}}{T} \tag{2.7}$$

This equation may also be used to calculate the entropy of sublimation, the entropy of fusion, or the entropy change for a transition between two forms of a solid. Since the heat gained by the system is equal to that lost by the surroundings, the entropy change for the surroundings is the negative of the entropy change for the system; for both the system and surroundings taken together, ΔS is zero if the transfer of heat is carried out reversibly, as required by equation 2.4.

Example 2.1 n-Hexane boils at $68.7°$, and the heat of vaporization at constant pressure is 6896 cal mol^{-1} at this temperature. If liquid is vaporized into the saturated vapor at this temperature, the process is reversible and the entropy change per mole is given by

$$\Delta \bar{S} = \frac{\Delta \bar{H}}{T} = \frac{6896 \text{ cal mol}^{-1}}{341.8 \text{ K}} = 20.18 \text{ cal K}^{-1} \text{ mol}^{-1} = 84.41 \text{ J K}^{-1} \text{ mol}^{-1}$$

The molar entropy of a vapor is always greater than that of the liquid with which it is in equilibrium, and the molar entropy of the liquid is always greater than that of the solid at the melting point. According to the disorder concept of entropy to be discussed later in Section 2.9, in which the entropy is a measure of the disorder of the system, the molecules of the gas are more disordered than those of the liquid, and the molecules of the liquid are more disordered than those of the solid.

The increase in entropy of a system due to an increase in temperature can be calculated since the temperature change can be carried out in a reversible manner. If the heating is carried out at constant pressure, the heat absorbed in each infinitesimal step is equal to the heat capacity C_P multiplied by the differential increase in temperature dT, and so

$$dS = \frac{C_P \, dT}{T} \tag{2.8}$$

Integrating equation 2.8 between the limits T_1 and T_2 gives

$$\int_{S_1}^{S_2} dS = \int_{T_1}^{T_2} \frac{C_P \, dT}{T} \tag{2.9}$$

If C_P is independent of temperature,

$$S_2 - S_1 = C_P(\ln T_2 - \ln T_1) = C_P \ln \frac{T_2}{T_1} = 2.303 C_P \log \frac{T_2}{T_1} \tag{2.10}$$

The fact that the entropy is always larger at the higher temperature agrees with the increased disorder of the motion of the molecules at the higher temperature. If the heating is carried out at constant volume, C_V is used instead of C_P.

If the heat capacities change with temperature, an empirical equation like that introduced in equation 1.47 may be inserted into equation 2.9 before integration.

Example 2.2   Calculate the increase in entropy of gaseous oxygen when a mole is heated at constant pressure from $25°$ to $600°$, using data from Table 1.2.

$$\Delta \bar{S} = \int_{298}^{873} \frac{(6.0954 + 3.2533 \times 10^{-3} T - 10.171 \times 10^{-7} T^2) \, dT}{T}$$

$$= (6.0954)(2.303) \log \tfrac{873}{298} + 3.2533 \times 10^{-3}(873 - 298)$$

$$\quad - \frac{10.171 \times 10^{-7}}{2}(873^2 - 298^2)$$

$$= 6.56 + 1.87 - 0.36$$

$$= 8.07 \text{ cal K}^{-1} \text{ mol}^{-1}$$

It is often inconvenient to fit heat capacity data to an empirical equation, and so equation 2.9 is integrated graphically or by use of a computer.

The change in entropy may be obtained by plotting $\bar{C}/T$ versus T or $\bar{C}$ versus log T and determining the area under the curve from temperature T_1 to temperature T_2.

$$S_2 - S_1 = \int_{T_1}^{T_2} C \, d \ln T = 2.203 \int_{T_1}^{T_2} C \, d \log T \qquad (2.11)$$

If the temperature and pressure are both changed reversibly, then for one mole of an ideal gas

$$dq = d\bar{U} + P \, d\bar{V} = \bar{C}_V \, dT + P \, d\bar{V} \qquad (2.12)$$

$$dq = \bar{C}_V \, dT + \frac{RT \, d\bar{V}}{\bar{V}} \qquad (2.13)$$

The infinitesimal entropy change per mole accompanying an infinitesimal change in temperature dT and in volume $d\bar{V}$ is given by

$$d\bar{S} = \frac{dq}{T} = \bar{C}_V \frac{dT}{T} + R \frac{d\bar{V}}{\bar{V}} \qquad (2.14)$$

Assuming that $\bar{C}_V$ is independent of temperature and volume, integration yields

$$\Delta \bar{S} = \bar{S}_2 - \bar{S}_1 = \bar{C}_V \ln \frac{T_2}{T_1} + R \ln \frac{\bar{V}_2}{\bar{V}_1} \qquad (2.15)$$

This equation applies to any change in the state of an ideal gas.

If the temperature is held constant,

$$\Delta \bar{S} = R \ln \frac{\bar{V}_2}{\bar{V}_1} \qquad (2.16)$$

This equation shows that for a tenfold increase in volume of an ideal gas $\Delta \bar{S} = (1.987)(2.303) = 4.59$ cal K^{-1} mol^{-1}. The entropy of the gas will change by this amount whether the isothermal expansion is carried out reversibly or irreversibly because $\Delta \bar{S}$ depends only on the entropies of the initial and final states.

2.7 ENTROPY CHANGES FOR IRREVERSIBLE CHANGES

The entropy change for an irreversible change may be calculated by considering a path by which the process can be carried out in a series of *reversible* steps. This is illustrated for the freezing of water below its freezing point.

The freezing of a mole of supercooled water at $-10°$ is an irreversible change, but it can be carried out reversibly by means of the following three steps for which the entropy changes are indicated:

$$H_2O(l) \text{ at } -10° \to H_2O(l) \text{ at } 0° \qquad \Delta\bar{S} = \int_{263}^{273} \bar{C}_{liq}\,\frac{dT}{T}$$

$$H_2O(l) \text{ at } 0° \to H_2O(s) \text{ at } 0° \qquad \Delta\bar{S} = \frac{q_{rev}}{T}$$

$$H_2O(s) \text{ at } 0° \to H_2O(s) \text{ at } -10° \qquad \Delta\bar{S} = \int_{273}^{263} \bar{C}_{ice}\,\frac{dT}{T}$$

For the crystallization of liquid water at 0°, $q_{rev} = -79.7$ cal g^{-1}. The specific heat of water may be taken to be 1.0 cal K^{-1} g^{-1}, and that of ice may be taken to be 0.49 cal K^{-1} g^{-1} over this range. Then the total entropy change of the water when 1 mole of liquid water at $-10°$ changes to ice at $-10°$ is simply the sum of the foregoing entropy changes:

$$\begin{aligned}
\Delta\bar{S} = &\,(18 \text{ g mol}^{-1})(1.0 \text{ cal K}^{-1} \text{ g}^{-1})(2.303) \log \tfrac{273}{263} \\
&+ \frac{(18 \text{ g mol}^{-1})(-79.7 \text{ cal g}^{-1})}{273 \text{ K}} \\
&+ (18 \text{ g mol}^{-1})(0.49 \text{ cal K}^{-1} \text{ g}^{-1})(2.303) \log \tfrac{263}{273} \\
= &\, 0.67 - 5.26 - 0.33 = -4.92 \text{ cal K}^{-1} \text{ mol}^{-1}
\end{aligned}$$

The decrease in entropy corresponds to the increase in structural order when water freezes.

The statement that the entropy of an isolated system increases in a spontaneous process may be illustrated by considering supercooled water at $-10°$ in contact with a large heat reservoir at this temperature. The entropy change for the isolated system upon freezing includes the entropy change of the reservoir as well as the entropy change of the water. If the heat reservoir is large, the heat evolved by the water upon freezing is absorbed by the reservoir with only an infinitesimal change in temperature. Since the heat of fusion of water at $-10°$ is 74.6 cal g^{-1} (Section 1.21), the entropy change of the reservoir in this reversible process is

$$\Delta S = \frac{(18 \text{ g mol}^{-1})(74.6 \text{ cal g}^{-1})}{263 \text{ K}}$$

$$= 5.10 \text{ cal K}^{-1} \text{ mol}^{-1}$$

The transfer of heat to a reservoir at the same temperature is a reversible process. The entropy change of the water is -4.92 cal K^{-1} mol^{-1} and the total entropy change of the system water plus reservoir is

$$\Delta S = 5.10 - 4.92 = 0.18 \text{ cal K}^{-1} \text{ mol}^{-1}$$

Thus the total entropy of the isolated system, including water and reservoir, increases, as required by inequality 2.5.

2.8 ENTROPY OF MIXING IDEAL GASES

If two gases at the same pressure and temperature are brought into contact, they will spontaneously diffuse into each other until the gas phase is macroscopically homogeneous. Since there is no interaction between molecules of ideal gases, there is no change in energy on mixing. The mixing occurs spontaneously only due to a change in entropy. The entropy change is the same as the entropy change that results when the gases are each allowed to expand isothermally from their initial volume to the final total volume of the mixture. Equation 2.16 may be used to calculate the entropy change for each gas. The entropy change for gas 1 is

$$\Delta S_1 = n_1 R \ln \frac{V_1 + V_2}{V_1} \tag{2.17}$$

where n_1 is the number of moles of gas 1, V_1 represents the initial volume of gas 1, and V_2 represents the initial volume of gas 2 so that $V_1 + V_2$ is the final volume available to the gas. The sum of the entropies of expansion of the two gases is

$$\Delta S_{\text{mix}} = n_1 R \ln \frac{V_1 + V_2}{V_1} + n_2 R \ln \frac{V_1 + V_2}{V_2} \tag{2.18}$$

where n_2 is the number of moles of gas 2. This equation is valid only for isothermal mixing. Since the mole fractions X_1 and X_2 of the two ideal gases are $X_1 \equiv n_1/(n_1 + n_2) = V_1/(V_1 + V_2)$ and $X_2 \equiv n_2/(n_1 + n_2) = V_2/(V_1 + V_2)$

$$\Delta S_{\text{mix}} = -R(n_1 \ln X_1 + n_2 \ln X_2) \tag{2.19}$$

If $n_1 + n_2 = 1$, the entropy change is a molar quantity, and equation 2.19 may be written

$$\Delta \bar{S}_{\text{mix}} = -R(X_1 \ln X_1 + X_2 \ln X_2) \tag{2.20}$$

The entropy of mixing is positive since $X_1 < 1$ and $X_2 < 1$. Therefore, according to equation 2.2 the diffusion of one gas into another is a spontaneous process in an isolated system even though there is no decrease in energy.

Equation 2.20 may be generalized to cover the mixing of a total of one mole of i components to form an ideal solution.

$$\Delta \bar{S}_{\text{mix}} = -R \sum_i X_i \ln X_i \tag{2.21}$$

Example 2.3 Calculate the entropy change when 1 mol of hydrogen at 1 atm pressure is mixed with 1 mol of nitrogen at 1 atm and constant temperature.

Assuming that the gases are ideal,

$$\begin{aligned}
\Delta S &= -2.303R(n_1 \log X_1 + n_2 \log X_2) \\
&= -(2.303)(8.314)(\log 0.5 + \log 0.5) \\
&= 11.52 \text{ J K}^{-1}
\end{aligned}$$

2.9   THE STATISTICAL INTER-
PRETATION OF THE ENTROPY
OF MIXING

As stated earlier, thermodynamics is not concerned with molecules or particular models of systems. To develop an intuitive feeling for thermodynamic functions, however, it is very helpful to think in terms of molecules. The calculation of thermodynamic properties from information about molecules is discussed in the chapter on statistical mechanics (Chapter 17). As an introduction to the ideas of statistical mechanics we consider here the mixing of two idealized crystals from a simple statistical point of view. For diffusion in an idealized crystal we can count the possible arrangements of molecules, and we will arrive at the conclusion that the mixed state is more probable than the unmixed state. Boltzmann recognized that this is the reason the mixed state is observed at equilibrium. The following derivation is designed to clarify this point.

Since the entropy is an extensive property, the entropy of a system consisting of two parts having the same intensive variables is simply the sum of the entropies of the two parts of the system: $S = S_1 + S_2$. If the number of equally probable arrangements for one part of the system is Ω_1 and for the other is Ω_2, the number of equally probable arrangements for the whole system is $\Omega_1\Omega_2$, since any arrangement of the first system can be combined with any arrangement of the second to specify an arrangement for the whole system. If we assume that the entropy is given by a function $f(\Omega)$ of the number of equally probable arrangements, then the entropy of the system is given by

$$S = S_1 + S_2 \tag{2.22}$$

$$f(\Omega_1\Omega_2) = f(\Omega_1) + f(\Omega_2) \tag{2.23}$$

For this to be true there must be a logarithmic relation between S and Ω. Boltzmann postulated that

$$S = k \ln \Omega \tag{2.24}$$

where k is the Boltzmann constant, that is, the molecular gas constant R/N_A where N_A is the Avogadro constant. The general quantitative expression for Ω is given in Chapter 17 after a discussion of the possible quantum states of a system, but we can illustrate the use of equation 2.24 now by applying it to the mixing of two idealized crystals.

Imagine that two small crystals are brought into contact so that atoms can diffuse from one to the other. The crystals are assumed to be sufficiently alike so that this can occur without a change in energy; in other words we assume that the lattice structures are the same and the various atom-atom interactions are identical so that the mixed crystals formed are of the ideal solid-solution type (Section 4.17). As illustrated in Fig. 2.1, there are initially four A atoms in crystal A and four B atoms in crystal B. Initially there is just this one possible arrangement of the atoms in this system

$$\Omega_{4:0} = 1$$

where $\Omega_{4:0}$ represents the number of arrangements with four A atoms to the left of the dividing plane and no A atoms to the right.

We are interested in the number of different arrangements of atoms in the various possible mixed crystals. Suppose one atom of A diffuses to the right of the plane and one atom of B diffuses to the left of the plane. Since the atom A can be placed on any one of the four

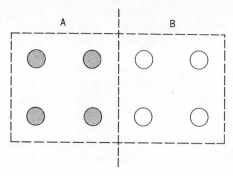

Fig. 2.1 Mixing of two idealized crystals. The black circles represent A atoms, and the open circles represent B atoms.

lattice points on the right and the atom B can be placed on any one of the four lattice points on the left, there are 16 distinguishable arrangements.

$$\Omega_{3:1} = 16$$

where $\Omega_{3:1}$ represents the number of arrangements with three A atoms to the left of the dividing plane and one to the right.

Suppose that two atoms of A diffuse to the right of the dividing plane and two atoms of B diffuse to the left. The first atom of A can occupy any one of the four sites and the second can occupy any one of the three remaining sites. This gives rise to 4×3 arrangements, but only $4 \times 3/2!$, where $2!$ is factorial 2, of these are distinguishable since the two A atoms are assumed to be identical. Each of the $4 \times 3/2!$ arrangements of A atoms on the right side can be combined with any one of the $4 \times 3/2!$ arrangements of B atoms on the left side to give 36 different arrangements

$$\Omega_{2:2} = \frac{4 \times 3}{2!} \cdot \frac{4 \times 3}{2!} = 36$$

Suppose that three atoms of A diffuse to the right of the dividing plane and three atoms of B diffuse to the left. Since the atom of A remaining on the left can be placed on any one of the four lattice points to the left and the atom of B can be placed on any one of the four lattice points to the right, there are again sixteen different arrangements.

$$\Omega_{1:3} = 16$$

If all four atoms of A diffuse to the right there is just one arrangement

$$\Omega_{0:4} = 1$$

Thus after the two crystals have been in contact for awhile, the atoms will be found to be in one of the 70 possible different arrangements.

$$\Omega = \Omega_{4:0} + \Omega_{3:1} + \Omega_{2:2} + \Omega_{1:3} + \Omega_{0:4} \tag{2.25}$$
$$= 1 + 16 + 36 + 16 + 1 = 70$$

Since the energy is the same for each of these arrangements, there is no reason to believe that one is more probable than another. In fact, it is a basic postulate of statistical mechanics that all states of a system with the same total energy and volume are equally probable. Thus at some later time the probability of finding two A atoms to the right of the plane and two B atoms to the left is $\frac{36}{70}$, whereas the probability of finding the crystals in their initial states is $\frac{1}{70}$. The equally mixed state is more probable than any other because there are more different ways in which the atoms can be arranged in this state than any other. If we had considered two crystals with larger numbers of atoms, this effect would have been much more striking. This is why spontaneous processes occur at fixed energy; a system goes spontaneously to a state with a larger number of possible arrangements because it spends equal time in each of the arrangements available to it, and there are so many more of the mixed ones.

The change in entropy for the interdiffusion of these two idealized crystals is readily calculated with the Boltzmann relation. Initially $S_1 = k \ln 1$ and at equilibrium, $S_2 = k \ln 70$. Thus the change in entropy

$$\Delta S = S_2 - S_1 = k \ln \frac{70}{1}$$

is positive as required for a spontaneous change in an isolated system.

Now let us generalize on this simple example by considering N_1 molecules of component 1 and N_2 molecules of component 2. We wish to calculate the number of ways, Ω, in which the molecules may be distributed among the sites. There are $N_1 + N_2$ choices of sites for the first molecule, $N_1 + N_2 - 1$ for the second, $N_1 + N_2 - 2$ for the third, etc., and so the total number of possibilities is $(N_1 + N_2)(N_1 + N_2 - 1)(N_1 + N_2 - 2) \cdots = (N_1 + N_2)!$. However, since molecules of type 1 are not distinguishable from each other, we must correct the total number of possibilities for the number of ways in which molecules of type 1 may be interchanged with each other. Since these molecules occupy N_1 lattice sites, the first may be placed on any one of N_1, the second on any one of $N_1 - 1$, etc., so that there are

$$N_1(N_1 - 1)(N_1 - 2) \cdots = N_1!$$

possibilities.

We must divide the number of ways of arranging N_1 molecules of 1 and N_2 molecules of 2, that is, $(N_1 + N_2)!$ by $N_1!$ to correct for the indistinguishability of molecules of type 1 and by $N_2!$ to correct for the indistinguishability of molecules of type 2 to get the number Ω_{mixed} of different mixed states.

$$\Omega_{\text{mixed}} = \frac{(N_1 + N_2)!}{N_1! N_2!} \tag{2.26}$$

By substituting $N_1 = 4$ and $N_2 = 4$, we obtain 70, just as in equation 2.25.

The number of distinguishable arrangements of the molecules of the pure components before mixing is

$$\Omega_1 = \frac{N_1!}{N_1!} = 1 \qquad \Omega_2 = \frac{N_2!}{N_2!} = 1$$

so that the entropy of the system before mixing is $S = k \ln 1 + k \ln 1 = 0$. Thus the entropy change on mixing is given by

$$\Delta S_{\text{mix}} = k \ln \frac{(N_1 + N_2)!}{N_1! N_2!} \tag{2.27}$$

When N_1 and N_2 are very large numbers, as is usually the case for molecular systems, Stirling's approximation may be used to eliminate the factorials.

$$\ln N! = N \ln N - N \tag{2.28}$$

This leads to

$$\Delta S_{mix} = k\{(N_1 + N_2) \ln (N_1 + N_2) - (N_1 + N_2) - [N_1 \ln N_1 - N_1] - [N_2 \ln N_2 - N_2]\}$$

$$= -k\left[N_1 \ln \frac{N_1}{N_1 + N_2} + N_2 \ln \frac{N_2}{N_1 + N_2}\right]$$

$$= -R[n_1 \ln X_1 + n_2 \ln X_2] \tag{2.29}$$

where $n_1 = N_1/N_A$ and $n_2 = N_2/N_A$ are numbers of moles and X_1 and X_2 are mole fractions. This equation is identical with equation 2.19 derived for the mixing of ideal gases.

These same ideas about the number of possible microscopic arrangements of a system may also be used to understand the expansion of a gas. Suppose we have an ideal gas in a bulb that is connected with an evacuated bulb. When the stopcock is opened the number of arrangements available to the system is vastly increased because of the increase in volume. The number of arrangements now available includes all those for the initial system (that is, gas all in one bulb) as well as a much greater number of new arrangements. Over a long enough period of time all of the possible arrangements will actually occur. Therefore, there is a chance that at a later time all of the gas molecules will be back in the first bulb. For a macroscopic amount of gas, however, the number of arrangements is so large that the probability of observing this occurrence is negligibly small. Furthermore, we can never predict when a large spontaneous fluctuation will occur.

Why the gas expands is sufficiently important to bear repeating once more. When the stopcock is opened the system can move over all of the possible arrangements. Each of these arrangements is equally probable. Since there are so many more arrangements which correspond to the mixed states, the chances are overwhelming that when we observe the system at a later time, the gas density will be found to be uniform in the two bulbs. The increase in entropy is determined by the increase in the number of arrangements.

Thermodynamics and equilibrium statistical mechanics do not deal with the rate of approach to equilibrium but only with the equilibrium state. Some time is required even for a gas to expand into another container, and for some chemical reactions the rate of approach to equilibrium is very slow.

2.10 CRITERIA OF CHEMICAL EQUILIBRIUM

Early attempts by Berthelot to discover a thermodynamic criterion of spontaneous chemical reactions led him in 1879 to the false conclusion that reactions which evolve heat are spontaneous. The discovery of spontaneous reactions which absorb heat proved that this idea was wrong. According to the second law a process will be spontaneous if its occurrence in an isolated system will lead to an increase in entropy of the system. In this section we will see how this statement leads to even more useful criteria of spontaneous chemical reactions.

It is of considerable practical importance to know whether a system is in equilibrium or just in a metastable state. By equilibrium we mean that a system is in such a state that it can undergo no spontaneous change under the given conditions. Thus at equilibrium any infinitesimal change which might take place in the system must be *reversible*, since any *irreversible* change would result in a displacement of the original equilibrium.

Let us consider a system in contact with a reservoir at temperature T in which an infinitesimal *irreversible* process occurs and the only work done is pressure-volume work. The quantity of heat dq is exchanged with the reservoir, and since the process is irreversible the entropy change dS for the system is greater than dq/T:

$$dS > \frac{dq}{T} \tag{2.30}$$

Since $T\,dS$ is greater than dq, $dq - T\,dS$ is negative:

$$dq - T\,dS < 0 \tag{2.31}$$

Because the only work done is pressure-volume work, $dq = dU + P\,dV$. Substituting this into equation 2.31, we have

$$dU + P\,dV - T\,dS < 0 \tag{2.32}$$

This inequality is always applicable if a spontaneous change occurs and the only work involved is pressure-volume work. If the volume and entropy of the system are held constant, then

$$(dU)_{V,S} < 0 \tag{2.33}$$

Thus, for any irreversible process in a system of constant volume that does not change its entropy, the internal energy decreases. This is the familiar condition for a conservative mechanical system that the stable state is the one of lowest energy.

The volume and internal energy of a system may be kept constant by isolating the system. For an isolated system, equation 2.30 becomes

$$dS > 0 \tag{2.34}$$

and so the entropy must increase in such an irreversible process. It must be remembered that inequalities 2.32, 2.33, and 2.34 apply only to systems where the only work done is of the pressure-volume type.

If the system is not isolated, there are entropy changes in the adjacent systems that must also be considered. If the volume is constant during the infinitesimal irreversible process, inequality 2.32 becomes

$$(dU - T\,dS)_V < 0 \tag{2.35}$$

which may also be written

$$d(U - TS)_{T,V} < 0 \tag{2.36}$$

The quantity $U - TS$ is referred to as the Helmholtz free energy and is represented by A.

$$A = U - TS \tag{2.37}$$

Thus from equation 2.36

$$(dA)_{T,V} < 0 \tag{2.38}$$

for a spontaneous process. Thus, in an irreversible process at constant T and V, the Helmholtz free energy A *decreases*.

Physical processes or chemical reactions are usually carried out in the laboratory at constant pressure and temperature. When P and T are constant, inequality 2.32

may be written

$$d(U + PV - TS)_{T,P} < 0 \qquad (2.39)$$

The quantity $U + PV - TS$ is referred to as the Gibbs free energy* and is represented by the symbol G.

$$G = U + PV - TS = H - TS \qquad (2.40)$$

from equation 2.39

$$(dG)_{T,P} < 0 \qquad (2.41)$$

Thus, in an irreversible process at constant T and P in which only pressure-volume work is done, the Gibbs free energy decreases.

If the processes just discussed were reversible, the inequalities would all be replaced with equal signs because of equation 2.4. The conditions for irreversibility and reversibility for processes involving only pressure-volume work are summarized in Table 2.1. Each line of this table represents a mathematical way of stating the

Table 2.1   Criteria for Irreversibility and Reversibility for Processes Involving No Work or Only Pressure-Volume Work

For Irreversible Processes	For Reversible Processes
$(dS)_{V,U} > 0$	$(dS)_{V,U} = 0$
$(dU)_{V,S} < 0$	$(dU)_{V,S} = 0$
$(dA)_{T,V} < 0$	$(dA)_{T,V} = 0$
$(dG)_{T,P} < 0$	$(dG)_{T,P} = 0$

second law of thermodynamics. Since the Gibbs free energy decreases in an irreversible process at constant T and P, it becomes a minimum at the final equilibrium state, where $dG = 0$ for any infinitesimal change. We can *imagine* a process occurring at equilibrium; for example, we may imagine the evaporation of an infinitesimal amount of water from the liquid into a vapor phase which is saturated with water vapor at constant temperature and pressure. For such a process $dG = 0$.

These same relations may be applied to finite changes as well as infinitesimal changes, replacing the d's by Δ's. It must be remembered, however, that spontaneous changes always go to the minimum (as in the case of the Gibbs free energy at constant T and P) or to the maximum (as in the case of the entropy of an isolated system) and not to some other condition, even though the change to some other condition satisfies the required inequality.

Although these criteria show whether a certain change is spontaneous, it does not necessarily follow that the change will take place with an appreciable speed. Thus,

---

* This is sometimes referred to as the free energy or the Gibbs energy. The Gibbs free energy G is named in honor of Prof. J. Willard Gibbs of Yale University, whose many important generalizations in thermodynamics have given him a position as one of the great geniuses of science.

a mixture of 1 mol of carbon and 1 mol of oxygen at 1 atm pressure and 25° has a Gibbs free energy greater than that of 1 mol of carbon dioxide at 1 atm and 25°, and so it is possible for the carbon and oxygen to combine to form carbon dioxide at this constant temperature and pressure. Although carbon may exist for a very long time in contact with oxygen, the reaction is theoretically possible. The reverse of a thermodynamically spontaneous change is of course a nonspontaneous change. Thus the decomposition of carbon dioxide to carbon and oxygen at room temperature, which involves an increase in Gibbs free energy, is nonspontaneous. It can occur only with the aid of an outside agency.

This discussion has been restricted to systems that do not have the capability of doing work other than pressure-volume work. If the system contained an electrochemical cell, then electrical work could be done and the criteria for equilibrium would be altered.

The infinitesimal Gibbs free energy change is in general

$$dG = dU + P\,dV + V\,dP - T\,dS - S\,dT \tag{2.42}$$

At constant pressure and temperature

$$dG = dU + P\,dV - T\,dS \tag{2.43}$$

Substituting $dU = dq + dw$, we have

$$dG = dq + dw + P\,dV - T\,dS \tag{2.44}$$

If a change is carried out by a reversible process and heat dq is transferred from a reservoir at the same temperature T as the system, then $dq = T\,dS$ and $dw = dw_{\text{rev}}$. For this case equation 2.44 can be written

$$-dG = -dw_{\text{rev}} - P\,dV \tag{2.45}$$

Thus, for a reversible process at constant temperature and pressure, the decrease in Gibbs free energy is equal to the maximum work that can be done by the system in excess of the pressure-volume work.

If a change is carried out by an irreversible process and heat dq is transferred from a reservoir at the same temperature T as the system, then $T\,dS > dq$. Substitution into equation 2.44 yields

$$-dG > -dw_{\text{irrev}} - P\,dV \tag{2.46}$$

For a finite change

$$-\Delta G > -w_{\text{irrev}} - P\,\Delta V \tag{2.47}$$

Thus, for an irreversible process at constant temperature and pressure, the decrease in Gibbs free energy is greater than the maximum work done by the system in excess of the pressure-volume work.

Since $\Delta G = \Delta A + P\,\Delta V$ at constant pressure, equation 2.45 shows that

$$-\Delta A = -w_{\text{rev}} \tag{2.48}$$

for a process at constant temperature and pressure. Thus the decrease in Helmholtz free energy is equal to the maximum amount of work which can be done by the system in an isothermal process.

2.11 THE GIBBS FREE ENERGY AS A CRITERION OF EQUILIBRIUM AT CONSTANT TEMPERATURE AND PRESSURE

At constant temperature, equation 2.40 becomes

$$\Delta G = \Delta H - T\,\Delta S \qquad (2.49)$$

Whether or not a process is spontaneous at constant temperature and pressure depends upon two terms, ΔH and $T\,\Delta S$. A change is favored if ΔH is negative or ΔS is positive; these changes correspond with a decrease in energy and an increase in disorder, respectively. If ΔH is large in magnitude and $T\,\Delta S$ is small in magnitude, the sign of ΔG will be determined solely by the sign of ΔH. If $T\,\Delta S$ is large in magnitude, and ΔH is small in magnitude, the sign of ΔG will be determined solely by the sign of ΔS. This always occurs at very high temperatures. In other cases both terms may be important. In any case the spontaneous process leads to the minimum possible value of $H - TS$ for the system at constant temperature and pressure.

To visualize the roles of H and TS in determining the equilibrium position, consider the vaporization of a solid in a closed space. The tendency of the system to achieve a low enthalpy would by itself lead to the complete condensation of the vapor phase onto the solid, this being the phase of lower enthalpy. The tendency of the system to achieve a high entropy would by itself lead to the complete vaporization of the solid to the gaseous state, this being the phase of higher entropy. The dependence of H and TS on the fraction of the substance in the vapor phase may be calculated and is illustrated in a general way in Fig. 2.2.* The enthalpy of the

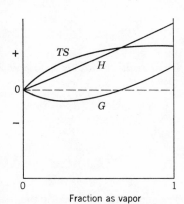

Fig. 2.2 Thermodynamic functions for a crystal-vapor system at constant temperature and pressure.

---

* K. G. Denbigh, *The Principles of Chemical Equilibrium*, Cambridge University Press, Cambridge, 1971, p. 84.

system increases linearly with the fraction in the vapor phase, but the entropy increases more rapidly at first. Thus $H - TS$ will show a minimum value with some fraction of the substance in the vapor phase, and this is the position of the equilibrium. If the concentration in the vapor phase is lower than the equilibrium value, vaporization will occur spontaneously, and if the concentration in the vapor phase is higher, condensation will occur spontaneously. Either of these processes leads to a decrease in the Gibbs free energy.

Equation 2.49 is important because ΔH and ΔS can be determined by direct calorimetric measurements without recourse to an equilibrium measurement, and ΔG can then be calculated. The Gibbs free energy change is useful in calculating the equilibrium constants of chemical reactions and the voltages of electrochemical cells.

2.12 FUNDAMENTAL EQUATIONS FOR CLOSED SYSTEMS

For a closed system of constant composition that may do only pressure-volume work, the first and second laws for a reversible process may be combined by inserting the definition of entropy (equation 2.1) into the differential form of the first law (equation 1.16) to obtain

$$dU = T\,dS - P\,dV \tag{2.50}$$

This equation is often referred to as the *fundamental equation* for a closed system of constant composition.

Since the internal energy U is a state function and may be expressed as a function of S and V, its differential is exact and is given by

$$dU = \left(\frac{\partial U}{\partial S}\right)_V dS + \left(\frac{\partial U}{\partial V}\right)_S dV \tag{2.51}$$

Comparing equations 2.50 and 2.51 we see that

$$\left(\frac{\partial U}{\partial S}\right)_V = T \tag{2.52}$$

and

$$\left(\frac{\partial U}{\partial V}\right)_S = -P \tag{2.53}$$

These equations illustrate the fact that the derivative of an extensive property with respect to another extensive property is an intensive property. These derivatives are useful, and we will now obtain further derivatives of this type by expressing the fundamental equation in terms of H, A, and G.

The differentials for H, A, and G may be obtained from their definitions, equations 1.29, 2.37, and 2.40.

$$dH = dU + P\,dV + V\,dP \tag{2.54}$$

$$dA = dU - T\,dS - S\,dT \tag{2.55}$$

$$dG = dU + P\,dV + V\,dP - T\,dS - S\,dT \tag{2.56}$$

Introducing the expression for dU from equation 2.50, we have

$$dH = T\,dS + V\,dP \tag{2.57}$$

$$dA = -S\,dT - P\,dV \tag{2.58}$$

$$dG = -S\,dT + V\,dP \tag{2.59}$$

These last three equations do not contain new information because H, A, and G are defined in terms of U, P, V, S, and T that are all involved in equation 2.50. Since dH, dA, and dG are exact differentials, these equations are convenient sources of the following derivatives

$$\left(\frac{\partial H}{\partial S}\right)_P = T \tag{2.60}$$

$$\left(\frac{\partial H}{\partial P}\right)_S = V \tag{2.61}$$

$$\left(\frac{\partial A}{\partial T}\right)_V = -S \tag{2.62}$$

$$\left(\frac{\partial A}{\partial V}\right)_T = -P \tag{2.63}$$

$$\left(\frac{\partial G}{\partial T}\right)_P = -S \tag{2.64}$$

$$\left(\frac{\partial G}{\partial P}\right)_T = V \tag{2.65}$$

As an illustration of the usefulness of these derivatives let us consider the last two. Since the entropy of a system is always positive, G decreases with increasing temperature at constant pressure. Since S is greater for a gas than for the corresponding solid, the temperature coefficient of the Gibbs free energy for a gas is much more negative than for the corresponding solid. Since the volume of a system is always positive, G increases with increasing P at constant T. Since V is greater for a gas than for the corresponding solid, the pressure coefficient of G is much larger for a gas than for the corresponding solid.

If the pressure on a substance is changed at constant temperature the change in Gibbs free energy may be obtained by integrating equation 2.65.

$$\int_{G_1}^{G_2} dG = G_2 - G_1 = \int_{P_1}^{P_2} V(P)\,dP \tag{2.66}$$

where we have written the volume as $V(P)$ to emphasize that it is a function of pressure at constant temperature. If this equation is applied to a single substance, the equation of state may be used to express V as a function of P. For a solid or liquid where the molar volume is independent of pressure,

$$\bar{G}_2 - \bar{G}_1 = \bar{V}\int_{P_1}^{P_2} dP = \bar{V}(P_2 - P_1) \tag{2.67}$$

For an ideal gas, $\bar{V} = RT/P$ so that equation 2.66 becomes

$$\int_{\bar{G}_1}^{\bar{G}_2} d\bar{G} = \int_{P_1}^{P_2} \frac{RT\,dP}{P}$$

$$\Delta\bar{G} = \bar{G}_2 - \bar{G}_1 = RT \ln \frac{P_2}{P_1} \tag{2.68}$$

Equations 2.60–2.65 may also be written in terms of the change in the thermodynamic function for a physical change or a chemical reaction. For example, applying equation 2.65 to a transformation from state 1 to state 2 results in:

$$\left(\frac{\partial G_2}{\partial P}\right)_T - \left(\frac{\partial G_1}{\partial P}\right)_T = V_2 - V_1 \tag{2.69}$$

$$\left(\frac{\partial \Delta G}{\partial P}\right)_T = \Delta V \tag{2.70}$$

Example 2.4   Calculate the equilibrium pressure for the conversion of graphite to diamond at 25°. The densities of graphite and diamond may be taken to be 2.25 and 3.51 g cm^{-3}, respectively, independent of pressure, in calculating the change of ΔG with pressure. The necessary enthalpy data and entropy data are to be found in Tables 1.3 and 2.4.

$$C(\text{graphite}) = C(\text{diamond})$$

$$\Delta G° = \Delta H° - T\Delta S°$$

$$= 453.2 - (298)(0.5829 - 1.3609)$$

$$= 685 \text{ cal}$$

$$\left[\frac{\partial \Delta G}{\partial P}\right]_T = \Delta V = 12\left(\frac{1}{3.51} - \frac{1}{2.55}\right) \times 10^{-3} \text{ liters}$$

$$= -1.28 \times 10^{-3} \text{ liters}$$

$$\int_1^2 d\,\Delta G = \int_1^P \Delta V\,dP$$

$$= \Delta G_2 - \Delta G_1 = \Delta V(P - 1)$$

$$0 - 685\,\frac{0.082}{1.987} = -1.28 \times 10^{-3}(P - 1)$$

$$P - 1 = 22,000 \text{ atm}$$

Example 2.5 One mole of an ideal gas at 27.0° expands isothermally and reversibly from 10 atm to 1 atm against a pressure that is gradually reduced. Calculate q and w and each of the thermodynamic quantities $\Delta \bar{U}$, $\Delta \bar{H}$, $\Delta \bar{G}$, $\Delta \bar{A}$, and $\Delta \bar{S}$. Calculations with the ideal gas equation show that the volume expands from 2.462 to 24.62 liters.

Since the process is carried out isothermally and reversibly,

$$w_{\max} = -RT \ln \frac{\bar{V}_2}{\bar{V}_1} = (1.987 \text{ cal K}^{-1} \text{ mol}^{-1})(300.1 \text{ K})(2.303) \log \frac{24.62}{2.462}$$

$$= -1373 \text{ cal mol}^{-1}$$

$$\Delta \bar{A} = w_{\max} = -1373 \text{ cal mol}^{-1}$$

Since the internal energy of an ideal gas is not affected by a change in volume,

$$\Delta \bar{U} = 0$$

$$q = \Delta \bar{U} - w = 0 + 1373 = 1373 \text{ cal}$$

$$\Delta \bar{H} = \Delta \bar{U} + \Delta(P\bar{V}) = 0 + 0 = 0$$

since $P\bar{V}$ is constant for an ideal gas at constant temperature.

$$\Delta \bar{G} = \int_{10}^{1} \bar{V} \, dP = RT \ln \tfrac{1}{10}$$

$$= (1.987 \text{ cal}^{-1} \text{K}^{-1} \text{ mol}^{-1})(300.1 \text{ K})(2.303)(-1)$$

$$= -1373 \text{ cal mol}^{-1}$$

$$\Delta \bar{S} = \frac{q_{\text{rev}}}{T} = \frac{1373 \text{ cal mol}^{-1}}{300.1 \text{ K}} = 4.58 \text{ cal K}^{-1} \text{ mol}^{-1}$$

Also,

$$\Delta \bar{S} = \frac{\Delta \bar{H} - \Delta \bar{G}}{T} = \frac{0 - (-1373 \text{ cal mol}^{-1})}{300.1 \text{ K}} = 4.58 \text{ cal K}^{-1} \text{ mol}^{-1}$$

Example 2.6 One mole of an ideal gas expands isothermally at 27° into an evacuated vessel so that the pressure drops from 10 to 1 atm; that is, it expands from a vessel of 2.462 liters into a connecting vessel such that the total volume is 24.62 liters. Calculate the change in thermodynamic quantities.

This process is isothermal, but it is not reversible.

$w = 0$ because the system as a whole is closed and no external work can be done.

$\Delta \bar{U} = 0$ because the gas is ideal.

$$q = \Delta \bar{U} - w = 0 + 0 = 0.$$

$\Delta \bar{U}$, $\Delta \bar{H}$, $\Delta \bar{G}$, $\Delta \bar{A}$, and $\Delta \bar{S}$ are the same as in example 2.5 because the initial and final states are the same.

2.13 MAXWELL RELATIONS

For an exact differential du

$$du = \left(\frac{\partial u}{\partial x}\right)_y dx + \left(\frac{\partial u}{\partial y}\right)_x dy \tag{2.71}$$

the mixed second partial derivatives are equal.

$$\left[\frac{\partial}{\partial y}\left(\frac{\partial u}{\partial x}\right)_J\right]_x = \left[\frac{\partial}{\partial x}\left(\frac{\partial u}{\partial y}\right)_x\right]_y \tag{2.72}$$

Application of this relation to equations 2.50 and 2.57 to 2.59 yields the Maxwell relations

$$\left(\frac{\partial T}{\partial V}\right)_S = -\left(\frac{\partial P}{\partial S}\right)_V \tag{2.73}$$

$$\left(\frac{\partial T}{\partial P}\right)_S = \left(\frac{\partial V}{\partial S}\right)_P \tag{2.74}$$

$$\left(\frac{\partial S}{\partial V}\right)_T = \left(\frac{\partial P}{\partial T}\right)_V \tag{2.75}$$

$$-\left(\frac{\partial S}{\partial P}\right)_T = \left(\frac{\partial V}{\partial T}\right)_P \tag{2.76}$$

As an example of the usefulness of these equations, 2.75 is applied in the next section.

2.14 THERMODYNAMIC EQUATION OF STATE

The fundamental equation (equation 2.50) for a closed system of constant composition expressed in terms of the internal energy U may be written in the form

$$\left(\frac{\partial U}{\partial V}\right)_T = T\left(\frac{\partial S}{\partial V}\right)_T - P \tag{2.77}$$

Introducing equation 2.75 we have

$$\left(\frac{\partial U}{\partial V}\right)_T = T\left(\frac{\partial P}{\partial T}\right)_V - P \tag{2.78}$$

which is referred to as the *thermodynamic equation of state*. This equation is perfectly general and applies to the thermodynamic properties of any closed system of constant composition. For an ideal gas $P = nRT/V$ so that

$$\left(\frac{\partial P}{\partial T}\right)_V = \frac{nR}{V} \tag{2.79}$$

Substituting this relation in equation 2.78 yields

$$\left(\frac{\partial U}{\partial V}\right)_T = 0 \tag{2.80}$$

Thus the internal energy of an ideal gas is independent of the volume. It is also independent of the pressure at constant temperature, and the enthalpy H is also independent of the volume and pressure at constant temperature. In molecular terms there are negligible interactions between molecules of an ideal gas.

Alternatively we can use equations 2.78 and 2.80 to derive the ideal gas law and thereby show that the temperature introduced in the definition of entropy is the same as that in the ideal gas law.

2.15   INFLUENCE OF TEMPERATURE ON GIBBS FREE ENERGY

The change in Gibbs free energy with temperature is related to the enthalpy change in a simple way that was first derived independently by Gibbs and by Helmholtz. Substitution of equation 2.64 into equation 2.40 yields

$$G = H + T\left(\frac{\partial G}{\partial T}\right)_P \tag{2.81}$$

This equation involves both the Gibbs free energy and the temperature derivative of the Gibbs free energy, and it is more convenient to transform it so that only a temperature derivative appears. This may be accomplished by first differentiating G/T with respect to temperature at constant pressure:

$$\left[\frac{\partial(G/T)}{\partial T}\right]_P = -\frac{G}{T^2} + \frac{1}{T}\left(\frac{\partial G}{\partial T}\right)_P \tag{2.82}$$

Eliminating G from the right-hand side by use of equation 2.81, we have

$$\left[\frac{\partial(G/T)}{\partial T}\right]_P = \frac{-H}{T^2} \tag{2.83}$$

Since $\partial(1/T)/\partial T = -T^{-2}$,

$$\left[\frac{\partial(G/T)}{\partial(1/T)}\right]_P = \left[\frac{\partial(G/T)}{\partial T}\right]_P \frac{\partial T}{\partial(1/T)} = H \tag{2.84}$$

This equation may also be written in terms of ΔG and ΔH to obtain

$$\left[\frac{\partial(\Delta G/T)}{\partial(1/T)}\right]_P = \Delta H \tag{2.85}$$

Thus ΔH for a reaction may be obtained from a plot of $\Delta G/T$ versus $1/T$ as well as by calorimetric measurements. This equation, which is referred to as the Gibbs-Helmholtz equation, is important for the calculation of ΔG at another temperature if it is known at one temperature and ΔH is known (see Section 5.15).

2.16 FUNDAMENTAL EQUATION
FOR OPEN SYSTEMS

If substances are added to a system or taken away from it or if a chemical reaction occurs in a system, the thermodynamic properties of the system change. Such a system is referred to as an open system. In this section we will consider homogeneous open systems because they are simpler than open systems with more than one phase. If a homogeneous system contains k different substances its internal energy U may be considered to be a function of S, V, n_1, n_2, . . . , n_k, where n_i is the number of moles of substance i. Since the differential of the energy U is so conveniently expressed in terms of the differential of the entropy and the volume in equation 2.50, S and V are called the "natural" variables for the energy. The natural variables for the Gibbs free energy are T and P. The total differential of U is

$$dU = \left(\frac{\partial U}{\partial S}\right)_{V,n_i} dS + \left(\frac{\partial U}{\partial V}\right)_{S,n_i} dV + \sum_{i=1}^{k} \left(\frac{\partial U}{\partial n_i}\right)_{S,V,n_j} dn_i \qquad (2.86)$$

where $j \neq i$, which means that in the derivatives in the summation the numbers of moles of all of the components except the one being varied are held constant. The first two derivatives are given by equations 2.52 and 2.53 and so

$$dU = T \, dS - P \, dV + \sum_{i=1}^{k} \mu_i \, dn_i \qquad (2.87)$$

where

$$\mu_i = \left(\frac{\partial U}{\partial n_i}\right)_{S,V,n_j} \qquad (2.88)$$

is referred to as the *chemical potential* of the ith component.

By use of the definitions of H, A, and G it is readily shown that

$$dH = T \, dS + V \, dP + \sum_{i=1}^{k} \mu_i \, dn_i \qquad (2.89)$$

$$dA = -S \, dT - P \, dV + \sum_{i=1}^{k} \mu_i \, dn_i \qquad (2.90)$$

$$dG = -S \, dT + V \, dP + \sum_{i=1}^{k} \mu_i \, dn_i \qquad (2.91)$$

Thus the chemical potential μ_i is given by the four expressions

$$\mu_i = \left(\frac{\partial U}{\partial n_i}\right)_{S,V,n_j} = \left(\frac{\partial H}{\partial n_i}\right)_{S,P,n_j} = \left(\frac{\partial A}{\partial n_i}\right)_{T,V,n_j} = \left(\frac{\partial G}{\partial n_i}\right)_{T,P,n_j} \qquad (2.92)$$

where $j \neq i$.

The fundamental equation for an open system (equation 2.87) may be expressed in integrated form as follows: Let us suppose that the system being considered is increased in size with the temperature, pressure and the relative proportions of the components being held constant. Since the relative proportions of the components are unchanged the chemical potentials are constant, and equation 2.87 may be written

$$\Delta U = T\Delta S - P\Delta V + \sum_{i=1}^{k} \mu_i \Delta n_i \qquad (2.93)$$

If the size of the system is increased by a factor f, $\Delta U = fU - U = (f - 1)U$. Similar equations can be written for S, V, and n_i. Substituting these relations in equation 2.93 yields

$$U = TS - PV + \sum_{i=1}^{k} n_i \mu_i \qquad (2.94)$$

Use of the definitions of H, A, and G yields

$$H = TS + \sum_{i=1}^{k} \mu_i n_i \qquad (2.95)$$

$$A = -PV + \sum_{i=1}^{k} \mu_i n_i \qquad (2.96)$$

$$G = \sum_{i=1}^{k} \mu_i n_i \qquad (2.97)$$

This last equation is of special interest. It shows that the Gibbs free energy of a system is the sum of the contributions of the various components. Since equilibrium is reached at constant temperature and pressure when G attains its *minimum* value we will differentiate this equation (Sections 5.3 and 5.4) to obtain a simple expression for chemical equilibrium in terms of the chemical potentials of the components.

When we differentiate equation 2.94 we obtain

$$dU = T\,dS + S\,dT - P\,dV - V\,dP + \sum_{i=1}^{k} \mu_i\,dn_i + \sum_{i=1}^{k} n_i\,d\mu_i \qquad (2.98)$$

Subtracting equation 2.87 yields the Gibbs-Duhem equation

$$S\,dT - V\,dP + \sum_{i=1}^{k} n_i\,d\mu_i = 0 \qquad (2.99)$$

This shows that the possible variations of the intensive variables T, P, and $\mu_1, \ldots \mu_k$ for a system are restricted. For a two-component system at constant temperature and pressure that contains one mole of material

$$X_1\,d\mu_1 + X_2\,d\mu_2 = 0 \qquad (2.100)$$

$$X_1\,d\mu_1 + (1 - X_1)\,d\mu_2 = 0 \qquad (2.101)$$

where X_1 is the mole fraction of component 1 and $(1 - X_1)$ is the mole fraction of component 2. Thus the change in the chemical potential of component number 2 is not independent of the change in the chemical potential of component number 1.

2.17   THE CHEMICAL POTENTIAL

The thermodynamic quantities U, H, A, and G are extensive properties, but the chemical potential μ is an intensive property. We will see that the chemical potential is like the electrical potential in that it is a driving force. It is the chemical potential that determines whether a substance will undergo a chemical reaction or diffuse from one part of a system to another.

In order to obtain an expression for the chemical potential of a pure substance we start with equation 2.91. Taking mixed partial derivatives as we did in obtaining the Maxwell relations (equations 2.73 to 2.76) we obtain

$$-\left(\frac{\partial S}{\partial n_i}\right)_{P,T,n_j} = \left(\frac{\partial \mu_i}{\partial T}\right)_{P,n_i,n} \tag{2.102}$$

$$\left(\frac{\partial V}{\partial n_i}\right)_{P,T,n_j} = \left(\frac{\partial \mu_i}{\partial P}\right)_{T,n_i,n} \tag{2.103}$$

The quantities on the left are referred to as *partial molar quantities*, and we will represent them by writing a bar on the corresponding extensive thermodynamic property. Partial molar quantities are more fully discussed in the next section.

Using the notation for partial molar quantities

$$-\bar{S}_i = \left(\frac{\partial \mu_i}{\partial T}\right)_{P,n_i,n} \tag{2.104}$$

$$\bar{V}_i = \left(\frac{\partial \mu_i}{\partial P}\right)_{T,n_i,n} \tag{2.105}$$

In words the first equation says that when the temperature is changed at constant pressure and composition, the differential change in μ_i is proportional to the negative of the partial molar entropy of component i. For a pure component $\bar{S}_i$ is positive, and so the chemical potential of a substance decreases when the temperature is raised.

Equation 2.105 says that when the pressure is changed at constant temperature and composition, the differential change in μ_i is proportional to the partial molar volume of the component. For a pure component $\bar{V}_i$ is positive, and so the chemical potential increases with the pressure.

2.18   PARTIAL MOLAR QUANTITIES

We have seen how the partial molar Gibbs free energy (the chemical potential) is useful in the treatment of open systems in which phase transformations or chemical reactions occur or in which material is being added or removed. There are partial molar quantities for any extensive thermodynamic properties, and in this section we will use the partial molar volume to illustrate the computation and use of partial molar quantities.

The volume of an ideal solution is simply the sum of the volumes of the components. This is not true for many real solutions, however; for example, if 100 cm^3 of sulfuric acid is added to 100 cm^3 of water, the final volume is 182 cm^3, and there is an evolution of a considerable quantity of heat. The sulfuric acid and the water interact, and the sulfuric acid ionizes, with the result that the volumes are not additive.

Let V represent the volume of a homogeneous binary solution. Since the volume depends on the numbers of moles n_1 and n_2 of the components and on the pressure and temperature, we can write

$$dV = \left(\frac{\partial V}{\partial n_1}\right)_{T,P,n_2} dn_1 + \left(\frac{\partial V}{\partial n_2}\right)_{T,P,n_1} dn_2 + \left(\frac{\partial V}{\partial P}\right)_{T,n_1,n} dP + \left(\frac{\partial V}{\partial T}\right)_{P,n_1,n_2} dT \quad (2.106)$$

The partial molar volume is defined by

$$\bar{V}_i = \left(\frac{\partial V}{\partial n_i}\right)_{T,P,n_j} \quad (2.107)$$

The subscript n_j means that the number of moles of each component is held constant except for component i. This definition may be stated in words by saying that $\bar{V}_i$ is the change in V per mole of i added, when an infinitesimal amount of this component is added to the solution at constant temperature and pressure. Alternatively it may be said that $\bar{V}_i$ is the change in V when 1 mol of i is added to an infinite amount of the solution at constant temperature and pressure. The partial molar volume of a pure substance is equal to its molar volume, $\bar{V} = V/n$.

Introducing the partial molar volumes defined by equation 2.107 into equation 2.106 and limiting the discussion to constant temperature and pressure, we obtain

$$dV = \bar{V}_1 \, dn_1 + \bar{V}_2 \, dn_2 \quad (2.108)$$

This equation may be integrated at constant composition. At constant composition the mole fractions X_1 and X_2 are constant, and $\bar{V}_1$ and $\bar{V}_2$ are also constant, being independent of the volume of the solution at constant composition. Since $n_i = X_i n$, where n is the total number of moles,

$$dn_i = X_i \, dn \quad (2.109)$$

Substituting these relations into equation 2.108 yields

$$dV = (\bar{V}_1 X_1 + \bar{V}_2 X_2) \, dn \quad (2.110)$$

Since the quantity in parentheses is constant, integration yields

$$V = (\bar{V}_1 X_1 + \bar{V}_2 X_2)n + C \quad (2.111)$$

but the integration constant C is zero since $V = 0$ when $n = 0$. Thus equation 2.111 may be written

$$V = \bar{V}_1 n_1 + \bar{V}_2 n_2 \quad (2.112)$$

The volume V of the solution may be calculated for a given concentration, using this equation, if the partial molar volumes $\bar{V}_1$ and $\bar{V}_2$ at the given concentration are known.

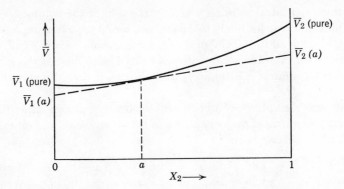

Fig. 2.3 Molar volume of a binary solution versus the mole fraction of one of the components.

To obtain the partial molar volumes of the components of a binary solution it is necessary to measure the densities of solutions over a range of concentrations. The partial molar volumes may be calculated by any one of several methods. In the simplest the volume is plotted versus number of moles of one component with the number of moles of the other held constant, and the slope is measured. The method of intercepts is one of the most graphic for visualizing partial molar quantities. For this method the volume of a mole of solution (i.e., a total of one mole of the two components) is plotted versus the mole fraction of one of the components as shown in Fig. 2.3. The molar volume of the solution is given by

$$\bar{V} = \bar{V}_1 X_1 + \bar{V}_2 X_2 = \bar{V}_1 + (\bar{V}_2 - \bar{V}_1)X_2 \qquad (2.113)$$

where the last form is obtained by introducing $X_1 = 1 - X_2$. If the mole fraction of component 2 is a, the slope of the plot of V versus X_2 is $\bar{V}_2(a) - \bar{V}_1(a)$ as shown in Fig. 2.3. The equation for the dashed tangent line at this mole fraction is

$$\bar{V} = \bar{V}_1(a) + [\bar{V}_2(a) - \bar{V}_1(a)]X_2 \qquad (2.114)$$

By putting $X_2 = 0$ in this equation, $\bar{V} = \bar{V}_1(a)$ so that the intercept at $X_2 = 0$ is the partial molar volume of component 1 when $X_2 = a$. By putting $X_2 = 1$ in this equation, $\bar{V} = \bar{V}_2(a)$ so that the intercept at $X_2 = 1$ is the partial molar volume of component 2 when $X_2 = a$.

Thus we may obtain the partial molar volumes of the two components at $X_2 = a$ by observing the intercepts of the tangent line on the two vertical coordinate lines of the graph. It can be seen that as the composition of the solution approaches pure component 1 or pure component 2 the partial molar volume of that component approaches the molar volume of the pure component. This method can be used to find any partial molar quantity and is simply illustrated here with the volume.

The method for determining partial molar volumes illustrated in Fig. 2.3 is not very accurate because of the inherent difficulty in obtaining the slope of a curve; however, this method of visualizing the partial molar volumes will be found useful later.

2.19 THIRD LAW OF THERMO-DYNAMICS

In 1902 Richards noted that as the temperature was reduced ΔS for some chemical reactions approached zero. The same observation was made by Nernst in 1906. This result suggested, but did not prove, that the entropies of the reactants also approached zero at absolute zero.

The first fully satisfactory statement of the third law was that of Lewis and Randall:*

"If the entropy of each element in some crystalline state be taken as zero at the absolute zero of temperature, every substance has a finite positive entropy; but at the absolute zero of temperature the entropy may become zero, and does so become in the case of perfect crystalline substances."

The third law is in agreement with equation 2.24 since for a perfect crystal at absolute zero the number of equally probable arrangements Ω is unity.

The entropy of imperfect crystals is greater than zero because they can be considered as mixtures and an entropy of mixing (Section 2.8) is involved. This applies whether the crystal is a chemical mixture or has lattice vacancies (Section 19.21) or other defects. Crystals of carbon monoxide have an entropy of 1.1 cal K^{-1} mol^{-1} at absolute zero, as calculated from data on chemical equilibrium by methods we are going to discuss. This entropy at absolute zero is apparently due to disorder in the crystal lattice arising from end-over-end randomness in the arrangement of the adjacent molecules. Thus in solid CO the molecules are arranged CO, OC, CO, CO, OC rather than CO, CO, CO, CO, CO. If the orientation were perfectly random, the crystal might be regarded as a mixed crystal with equal mole fractions of CO and OC. The entropy of the mixed crystal would then be the entropy of mixing. Using equation 2.20 we see that

$$\Delta \bar{S}_{mix} = -R(\tfrac{1}{2} \ln \tfrac{1}{2} + \tfrac{1}{2} \ln \tfrac{1}{2})$$
$$= (1.987 \text{ cal } K^{-1} \text{ mol}^{-1})(2.303)(0.301)$$
$$= 1.38 \text{ cal } K^{-1} \text{ mol}^{-1}$$

The entropy of mixing of isotopically different species is ignored in chemical applications because isotopes are not separated significantly in chemical reactions, and so the entropies of mixing of isotopes in reactants and products cancel.

2.20 DETERMINATION OF THIRD LAW ENTROPIES

We have already seen how to calculate the change in entropy of a system with changing temperature; $\bar{C} \, dT/T$ is simply integrated from one temperature to the

---

* G. N. Lewis and M. Randall, *Thermodynamics and the Free Energy of Chemical Substances*, 1st ed. McGraw-Hill Book Co., New York, 1923, p. 448.

other. If the heat capacity measurements are extended to the neighborhood of absolute zero, the third law entropy of a substance may be obtained at a higher temperature. Since measurements of $\bar{C}$ cannot be carried to 0 K, the Debye function (Section 1.14) is used to represent $\bar{C}$ below the temperature of the lowest measurements.

The third law entropy of a gas at temperature T may be calculated by integrating dq_{rev}/T from 0 K to the desired temperature. If data on the heat capacity and the enthalpy of fusion at the melting point T_m and the enthalpy of vaporization at the boiling point T_b are available, the third law entropy at temperature T may be calculated from

$$\Delta \bar{S}_T^\circ = \int_0^{T_m} \frac{C_P(s)}{T} \, dT + \frac{\Delta H_{\text{fus}}^\circ}{T_m} + \int_{T_m}^{T_b} \frac{C_P(l)}{T} \, dT + \frac{\Delta H_{\text{vap}}^\circ}{T_b} + \int_{T_b}^{T} \frac{C_P(g)}{T} \, dT \quad (2.115)$$

If there are various solid forms with enthalpies of transition between the forms, the corresponding entropies of transition would have to be included in this sum.

As an illustration of the determination of the third law entropy of a substance, the measured heat capacities for SO_2 are shown as a function of T and of log T in Fig. 2.4a* and b. Solid SO_2 melts at 197.64 K, and the heat of fusion is 1769 cal mol[-1]. Liquid SO_2 vaporizes at 263.08 K, and the heat of vaporization is 5960 cal mol[-1]. The calculation of the entropy at 25° is summarized in Table 2.2.

Heat-capacity measurements down to these very low temperatures are made with special calorimeters in which the substance is heated electrically in a carefully insulated system and the input of electrical energy and the temperature are measured accurately.

The attainment of very low temperatures in the laboratory involves successive application of different methods. Vaporization of liquid helium (b.p. 4.2 K) by rapid pumping produces temperatures down to about 0.3 K. Lower temperatures

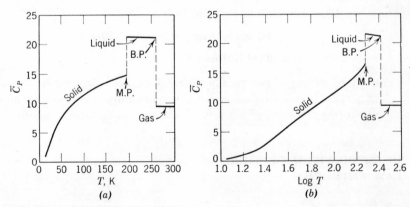

Fig. 2.4 Heat capacity in cal K[-1] mol[-1] of sulfur dioxide at a constant pressure of 1 atm at different temperatures.

* W. F. Giauque and C. C. Stephenson, *J. Am. Chem. Soc.,* **60,** 1389 (1938).

Table 2.2 The Entropy of Sulfur Dioxide

Temperature, K	Method of Calculation	$\Delta \bar{S}$, cal K^{-1} mol^{-1}
0–15	Debye function ($\bar{C}_P$ = constant T^3)	0.30
15–197.64	Graphical, solid	20.12
197.64	Fusion, 1769.1/197.64	8.95
197.64–263.08	Graphical, liquid	5.96
263.08	Vaporization, 5960/263.08	22.06
263.08–298.1	From $\bar{C}_P$ of gas	1.25
		$\bar{S}^\circ_{298.1} = 59.64$

may be reached by use of adiabatic demagnetization. A paramagnetic (Section 16.1) salt like gadolinium sulfate is cooled with liquid helium in the presence of a strong magnetic field. The salt is thermally isolated from its surroundings and the magnetic field is slowly removed. The salt undergoes a reversible adiabatic process in which the atomic spins become disordered. Since the energy must come from the crystal lattice, the salt is cooled. Temperatures of about 0.001 K may be reached in this way. Adiabatic demagnetization of nuclear spins can then be used to obtain temperatures of the order of a millionth of a degree Kelvin.

Table 2.3 gives the third law entropies, $\bar{S}^\circ$, at 25° for a number of elements, compounds, and ions. The standard state for the ions in solution is the hypothetical state of unit activity (Section 7.11) at 1 atm and 25° C. The third law entropy of H$^+$ in dilute aqueous solution is arbitrarily assigned the value of zero.

Example 2.7 Calculate the entropy change for the formation of water vapor at 25° from oxygen and hydrogen.

$$\tfrac{1}{2}O_2(g) + H_2(g) = H_2O(g)$$

$$\Delta S^\circ = 45.106 - 31.211 - \tfrac{1}{2}(49.003)$$

$$= -10.606 \text{ cal K}^{-1} \text{ mol}^{-1}$$

Third law entropies may be calculated for simple molecules from spectroscopic data by use of statistical mechanics (Chapter 17). It is of considerable interest to compare the entropies calculated theoretically with those obtained from heat-capacity measurements on the basis of the third law of thermodynamics. It is found that the agreement is within the experimental error for a large number of compounds, but H$_2$, CO, H$_2$O, N$_2$O, and certain other compounds are exceptions. For these substances the calorimetric values are 1.1–1.5 cal K^{-1} mol^{-1} smaller than the values calculated from spectroscopic data. The origin of the discrepancy for CO has been described above. The discrepancy for H$_2$ has been accounted for on the basis of the existence of ortho and para forms.

Table 2.3[1]   Third Law Entropies at 25° ($\bar{S}^\circ$ in cal K^{-1} mol^{-1})

Elements and Inorganic Compounds

$O_2(g)$	49.003	$NO(g)$	50.339	$AgCl(s)$	22.97
$O_3(g)$	56.8	$NO_2(g)$	57.47	$Fe(s)$	6.49
$H_2(g)$	31.211	$NH_3(g)$	46.01	$Fe_2O_3(s)$	21.5
$H_2O(g)$	45.106	$HNO_3(l)$	37.19	$Fe_3O_4(s)$	35.0
$H_2O(l)$	16.716	$P(g)$	38.98	$Al(s)$	6.769
$He(g)$	30.126	$P(s,\ \text{white})$	10.6	$Al_2O_3(s)$	12.186
$Cl_2(g)$	53.286	$PCl_3(g)$	74.49	$UF_6(g)$	90.76
$HCl(g)$	44.617	$PCl_5(g)$	84.3	$UF_6(s)$	54.45
$Br_2(g)$	58.639	$C(s,\ \text{diamond})$	0.5829	$Ca(s)$	9.95
$Br_2(l)$	36.4	$C(s,\ \text{graphite})$	1.3609	$CaO(s)$	9.5
$HBr(g)$	47.437	$CO(g)$	47.301	$CaCO_3(s)$	22.2
$HI(g)$	49.314	$CO_2(g)$	51.061	$Na(s)$	12.2
$S(\text{rhombic})$	7.62	$Pb(s)$	15.51	$NaF(s)$	14.0
$S(\text{monoclinic})$	7.78	$PbO_2(s)$	18.3	$NaCl(s)$	17.3
$SO_2(g)$	59.40	$PbSO_4(s)$	35.2	$K(s)$	15.2
$SO_3(g)$	61.24	$Hg(g)$	41.80	$KF(s)$	15.91
$H_2S(g)$	49.15	$Hg(l)$	18.5	$KCl(s)$	19.76
$N_2(g)$	45.767	$Ag(s)$	10.206		

Organic Compounds

Methane, $CH_4(g)$	44.50	Propylene, $C_3H_6(g)$	63.80
Ethane, $C_2H_6(g)$	54.85	1-Butene, $C_4H_8(g)$	73.48
Propane, $C_3H_8(g)$	64.51	Acetylene, $C_2H_2(g)$	47.997
n-Butane, $C_4H_{10}(g)$	74.10	Formaldehyde, $CH_2O(g)$	52.26
Isobutane, $C_4H_{10}(g)$	70.42	Acetaldehyde, $C_2H_4O(g)$	63.5
n-Pentane, $C_5H_{12}(g)$	83.27	Methanol, $CH_3OH(l)$	30.3
n-Hexane, $C_6H_{14}(g)$	92.45	Ethanol, $CH_3CH_2OH(l)$	38.4
n-Heptane, $C_7H_{16}(g)$	101.64	Formic acid, $HCO_2H(l)$	30.82
n-Octane, $C_8H_{18}(g)$	110.82	Acetic acid, $CH_3CO_2H(l)$	38.2
Benzene, $C_6H_6(g)$	64.34	Oxalic acid, $(CO_2H)_2(s)$	28.7
Benzene, $C_6H_6(l)$	41.30	Carbon tetrachloride, $CCl_4(l)$	51.25
Ethylene, $C_2H_4(g)$	52.45	Glycine, $C_2H_5O_2(s)$	26.1

Ions in H_2O

H^+	15.606	SO_4^{2-}	4.1	Cu^{2+}	−23.6
OH^-	−18.125	HS^-	14.6	Ag^+	17.67

[1] These data have been obtained from F. D. Rossini, D. D. Wagman, W. H. Evans, S. Levine, and I. Jaffee, "Selected Values of Chemical Thermodynamic Properties," *Natl. Bur. Standards Cir. 500*. U.S. Government Printing Office Washington, D.C., 1952, and F. D. Rossini, K. S. Pitzer, W. J. Taylor, J. P. Ebert, J. E. Kilpatrick, C. W. Beckett, M. G. Williams, and H. G. Werner, "Selected Values of Properties of Hydrocarbons," *Natl. Bur. Standards Circ. C 461*, U.S. Government Printing Office, Washington, D.C., 1947. *Circular 500* is being replaced by a series of revised tables. Data for the first thirty four elements are given in NBS Technical Note 270-3 (1968).

Table 2.3 (*Continued*)

		Ions in H_2O (*Continued*)			
F^-	-17.9	NO_3^-	35.0	Mg^{2+}	-28.2
Cl^-	-2.44	NH_4^+	26.97	Ca^{2+}	-13.2
ClO_4^-	43.5	PO_4^{3-}	-52	Li^+	3.4
Br^-	19.29	CO_3^{2-}	-12.7	Na^+	30.0
I^-	26.14	Zn^{2+}	-25.45	K^+	24.5
S^{2-}	5.3	Cd^{2+}	-14.6		

		Gaseous Atoms			
H	27.3927	Br	41.8052	N	36.6147
F	37.917	I	43.184	C	37.76
Cl	39.4569				

References

F. C. Andrews, *Thermodynamics: Principles and Applications*, Wiley-Interscience, New York, 1971.

H. A. Bent, *The Second Law*, Oxford University Press, Fair Lawn, N.J., 1965.

K. Denbigh, *The Principles of Chemical Equilibrium*, Cambridge University Press, Cambridge, 1971.

J. W. Gibbs, *The Collected Works of J. Willard Gibbs*, Yale University Press, New Haven, 1948.

E. A. Guggenheim, *Modern Thermodynamics by the Methods of Willard Gibbs*, Methuen and Co., London, 1933.

W. Kauzmann, *Thermal Properties of Matter Vol. II, Thermodynamics and Statistics: With Applications to Gases*, W. A. Benjamin, Inc., New York, 1967.

G. Kirkwood and I. Oppenheim, *Chemical Thermodynamics*, McGraw-Hill Book Co., New York, 1961.

I. M. Klotz and R. M. Rosenberg, *Chemical Thermodynamics*, W. A. Benjamin Inc., New York, 1972.

G. N. Lewis and M. Randall revised by K. S. Pitzer and L. Brewer, *Thermodynamics*, McGraw-Hill Book Co., New York, 1961.

P. A. Rock, *Chemical Thermodynamics*, The Macmillan Co., London, 1969.

F. D. Rossini, *Chemical Thermodynamics*, Wiley, New York, 1950.

D. R. Stull, E. F. Westrum, and G. C. Sinke, *The Chemical Thermodynamics of Organic Compounds*, Wiley-Interscience, New York, 1969.

F. T. Wall, *Chemical Thermodynamics*, W. H. Freeman & Co., San Francisco, 1965.

R. E. Wood, *Introduction to Chemical Thermodynamics*, Appleton-Century-Crofts, New York, 1970.

Problems

2.1 A gas is carried through a cyclic process (Carnot cycle) in the following four steps:
1. An isothermal reversible expansion at temperature T_2 in which heat q_2 is absorbed by the gas.

2. An adiabatic ($q = 0$) reversible expansion in which the temperature falls from T_2 to T_1.

3. An isothermal reversible compression at temperature T_1 in which heat q_1 is transferred to the heat reservoir at T_1.

4. An adiabatic ($q = 0$) reversible compression in which the temperature rises from T_1 to T_2, returning the gas to its original conditions.

(a) From the fact that $\Delta S = 0$ for the cyclic process, derive a relation between T_1, T_2, q_1, and q_2.

(b) By use of the first law derive the relation for the efficiency of this heat engine, that is, $-w/q_2$.   *Ans.* (a) $q_1/T_1 = -q_2/T_2$. (b) $(T_2 - T_1)/T_2$.

2.2 Theoretically, how high could a gallon of gasoline lift an automobile weighing 2800 lb against the force of gravity, if it is assumed that the cylinder temperature is 2200 K and the exit temperature 1200 K? (Density of gasoline $= 0.80$ g cm^{-3}; 1 lb $= 453.6$ g; 1 ft $= 30.48$ cm; 1 liter $= 0.2642$ gal. Heat of combustion of gasoline $= 11,200$ cal g^{-1}.)   *Ans.* 17,000 ft.

2.3 A mole of steam is condensed at 100° and the water is cooled to 0° and frozen to ice. What is the entropy change of the water? Consider that the average specific heat of liquid water is 1.0 cal K^{-1} g^{-1}. The heat of vaporization at the boiling point and the heat of fusion at the freezing point are 539.7 and 79.7 cal g^{-1}, respectively.

Ans. $\Delta \bar{S} = -36.9$ cal K^{-1} mol^{-1}.

2.4 Calculate the change in entropy if 350 g of water at 5° is mixed with 500 g of water at 70°, assuming that the specific heat is 1.00 cal K^{-1} g^{-1}.   *Ans.* 4.485 cal K^{-1}.

2.5 Two blocks of the same metal are of the same size but are at different temperatures, T_1 and T_2. These blocks of metal are brought together and allowed to come to the same temperature. Show that the entropy change is given by

$$\Delta S = C_P \ln \left[\frac{(T_1 + T_2)^2}{4T_1 T_2} \right]$$

if C_P is constant. How does this equation show that the change is spontaneous?

Ans. ΔS is positive.

2.6 Calculate the entropy changes for the following processes: (a) melting of 1 mol of aluminum at its melting point, 660° ($\Delta \bar{H}_{fus} = 1.91$ kcal mol^{-1}); (b) evaporation of 2 mol of liquid oxygen at its boiling point, $-182.97°$ ($\Delta \bar{H}_{vap} = 1.630$ kcal mol^{-1}); (c) heating of 10 g of hydrogen sulfide from 50 to 100° at constant pressure ($\bar{C}_P = 7.15 + 0.00332T$).

Ans. (a) 2.05, (b) 36.2, (c) 0.351 cal K^{-1}.

2.7 Calculate the entropy change in joules for a hundredfold expansion of a mole of ideal gas isothermally.   *Ans.* 38.3 J K^{-1} mol^{-1}.

2.8 In the reversible isothermal expansion of an ideal gas at 300 K from 1 to 10 liters, where the gas has an initial pressure of 20 atm, calculate (a) ΔS for the gas and (b) ΔS for all systems involved in the expansion.   *Ans.* (a) 3.72, (b) 0 cal K^{-1}.

2.9 Calculate the change in Gibbs free energy for the process

$$H_2O(l, -10°) = H_2O(s, -10°)$$

The vapor pressure of water at $-10°$ is 2.149 Torr, and the vapor pressure of ice at $-10°$ is 1.950 Torr. The process may be carried out by the following reversible steps:

1. A mole of water is transferred at $-10°$ from liquid to saturated vapor ($P = 2.149$ Torr). $\Delta \bar{G} = 0$, since the two phases are in equilibrium.

2. The water vapor is allowed to expand from 2.149 to 1.950 Torr at $-10°$.

3. A mole of water is transferred at $-10°$ from vapor at $P = 1.950$ Torr to ice at $-10°$.

Ans. $- 50.8$ cal mol^{-1}.

2.10 Calculate $\Delta \bar{S}$ for the formation of a quantity of air containing 1 mole of gas by mixing nitrogen and oxygen. Air may be taken to be 80% nitrogen and 20% oxygen by volume.

Ans. 0.994 cal K^{-1} mol^{-1}.

2.11 Equations 2.75 and 2.76 are important because they express the rate of change of entropy with respect to pressure at constant temperature and the rate of change of entropy with volume at constant temperature in terms of readily measured quantities. Express these derivatives in terms of α and κ, the coefficients of thermal expansion and compressibility (problem 2.16), respectively. (It is necessary to use the cyclic rule

$$\left(\frac{\partial P}{\partial T}\right)_V \left(\frac{\partial T}{\partial V}\right)_P \left(\frac{\partial V}{\partial P}\right)_T = -1$$

to obtain one of the results.)

Ans. $\left(\frac{\partial S}{\partial P}\right)_T = -V\alpha.$ $\left(\frac{\partial S}{\partial V}\right)_T = \frac{\alpha}{\kappa}$

2.12 One mole of an ideal gas is allowed to expand reversibly and isothermally (25°) from a pressure of 1 atm to a pressure of 0.1 atm. (*a*) What is the change in Gibbs free energy? (*b*) What would be the change in Gibbs free energy if the process occurred irreversibly?

Ans. (*a*) −1364, (*b*) −1364 cal.

2.13 (*a*) Calculate the work done against the atmosphere when 1 mol of toluene is vaporized at its boiling point, 111°. The heat of vaporization at this temperature is 86.5 cal g^{-1}. For the vaporization of 1 mole, calculate (*b*) q, (*c*) $\Delta \bar{H}$, (*d*) $\Delta \bar{U}$, (*e*) $\Delta \bar{G}$, (*f*) $\Delta \bar{S}$.

Ans. (*a*) 763, (*b*) 7969, (*c*) 7969, (*d*) 7206, (*e*) 0 cal mol^{-1}, (*f*) 20.7 cal K^{-1} mol^{-1}.

2.14 One liter of an ideal gas at 300 K has an initial pressure of 15 atm and is allowed to expand isothermally to a volume of 10 liters. Calculate (*a*) the maximum work that can be obtained from the expansion, (*b*) ΔU, (*c*) ΔH, (*d*) ΔG, (*e*) ΔA.

Ans. (*a*) 836, (*b*) 0, (*c*) 0, (*d*) −836, (*e*) −836 cal.

2.15 One mole of ammonia (considered to be an ideal gas) initially at 25° and 1 atm pressure is heated at constant pressure until the volume has trebled. Calculate (*a*) q, (*b*) w, (*c*) $\Delta \bar{H}$, (*d*) $\Delta \bar{U}$, (*e*) $\Delta \bar{S}$. (See Table 1.2.)

Ans. (*a*) 6320, (*b*) −1185, (*c*) 6320, (*d*) 5135 cal mol^{-1}, (*e*) 11.23 cal K^{-1} mol^{-1}.

2.16 Show that, by use of the second law, the difference in heat capacities at constant pressure and constant volume

$$\bar{C}_P - \bar{C}_V = \left[P + \left(\frac{\partial \bar{U}}{\partial \bar{V}}\right)_T\right] \left(\frac{\partial \bar{V}}{\partial T}\right)_P$$

may be written in terms of quantities

$$\alpha = \frac{1}{V}\left(\frac{\partial V}{\partial T}\right)_P \quad \text{and} \quad \kappa = -\frac{1}{V}\left(\frac{\partial V}{\partial P}\right)_T$$

that may be determined more easily from experiment, as

$$\bar{C}_P - \bar{C}_V = \frac{T\bar{V}\alpha^2}{\beta}$$

2.17 Calculate the partial molar volume of zinc chloride in 1-molal $ZnCl_2$ solution using the following data:

% by weight of $ZnCl_2$	2	6	10	14	18	20
Density, g cm^{-3}	1.0167	1.0532	1.0891	1.1275	1.1665	1.1866

Ans. 29.3 cm^3 mol^{-1}.

2.18   Calculate the molar entropy of liquid chlorine at its melting point, 172.12 K, from the following data obtained by W. F. Giauque and T. M. Powell:

T, K	15	20	25	30	35	40	50	60
$\bar{C}_P$, cal K^{-1} mol^{-1}	0.89	1.85	2.89	3.99	4.97	5.73	6.99	8.00

T, K	70	90	110	130	150	170	172.12
$\bar{C}_P$, cal K^{-1} mol^{-1}	8.68	9.71	10.47	11.29	12.20	13.17	M.P.

The heat of fusion is 1531 cal mol^{-1}. Below 15 K it may be assumed that $\bar{C}_P$ is proportional to T^3.   *Ans.* 25.8 cal K^{-1} mol^{-1}.

2.19   Using atomic and molecular entropies from Table 2.3, calculate $\Delta S°$ for the following reactions at 25°.

$$(a)\ \ H_2(g) + \tfrac{1}{2}O_2(g) = H_2O(l)$$
$$(b)\ \ H_2(g) + Cl_2(g) = 2HCl(g)$$
$$(c)\ \ \text{Propane}(g) + \text{ethane}(g) = n\text{-pentane}(g) + H_2(g)$$
$$(d)\ \ \text{Methane}(g) + \tfrac{1}{2}O_2(g) = \text{methanol}(l)$$

Ans. (*a*) −38.996, (*b*) 4.737, (*c*) −4.88, (*d*) −38.7 cal K^{-1}.

2.20   Calculate the Gibbs free-energy changes at 25° for the reactions in problem 20.19 when the reactants are in their standard states by use of enthalpy-of-formation data in Table 1.3.   *Ans.* (*a*) −56.690, (*b*) −45.536, (*c*) 11.51, (*d*) −27.59 kcal.

2.21   Calculate the theoretical maximum efficiency with which heat can be converted into work in the following hypothetical turbines: (*a*) steam at 100° with exit at 40°; (*b*) mercury vapor at 300° with exit at 140°; (*c*) steam at 400° and exit at 150°; (*d*) air at 800° and exit at 400°; (*e*) air at 1000° and exit at 400°, special alloys being used; (*f*) helium at 1500° and exit at 400°, heat from atomic energy being used. (See problem 2.1.)

2.22   Calculate the increase in entropy of nitrogen when it is heated from 25° to 1000° (*a*) at constant pressure and (*b*) at constant volume.

2.23   A 2-liter container at 0° contains hydrogen sulfide (assumed to be an ideal gas) at 1 atm pressure. The gas is heated to 100°, the external pressure remaining 1 atm. Calculate (*a*) the heat absorbed, (*b*) the work done, (*c*) ΔU, (*d*) ΔH, and (*e*) ΔS. For hydrogen sulfide, $\bar{C}_P = 7.15 + 0.00332\,T$.

2.24   Compute the entropy difference between 1 mole of liquid water at 25° and 1 mole of water vapor at 100° and 1 atm. The average specific heat of liquid water may be taken as 1 cal K^{-1} g^{-1}, and the heat of vaporization is 540 cal g^{-1}.

2.25   Calculate the entropy change of a mole of helium which undergoes a reversible adiabatic expansion from 25° and 1 atm to a final pressure of $\tfrac{1}{2}$ atm.

2.26   One hundred and fifty grams of ice is added to a kilogram of water at 25° in an isolated system. Calculate the change in entropy if the heat of fusion is 1435 cal mol^{-1} and $\bar{C}_P = $ 18 cal K^{-1} mol^{-1} for liquid water.

2.27   Show that the process

$$H_2O(l, -5°) = H_2O(s, -5°)$$

is a spontaneous process in an isolated system containing in addition to the water a thermostat at −5°. The heat of fusion of water is 79.7 cal g^{-1} at 0°, and the specific heats for water and ice may be taken as 1 cal K^{-1} g^{-1} and 0.5 cal K^{-1} g^{-1}, respectively.

2.28   A 1-liter bulb containing nitrogen at 1 atm pressure and 25° is connected by a tube with a stopcock to a 3-liter bulb containing carbon dioxide at 2 atm pressure. The stopcock is opened, and the gases are allowed to mix until equilibrium is reached. Assuming that the gases are both ideal, what is ΔS for this spontaneous change?

2.29 Two liters of methane under 4 atm pressure and 25° and 4 liters of oxygen under 20 atm and 25° are forced into a 3-liter evacuated reaction vessel, the temperature being maintained at 25°. Calculate the change in entropy of the gases assuming that they are ideal.

2.30 What is the expression for the entropy change and Gibbs free energy change of mixing of three components to form an ideal solution?

2.31 Show that

$$\left(\frac{\partial U}{\partial S}\right)_V = \left(\frac{\partial H}{\partial S}\right)_P$$

$$\left(\frac{\partial H}{\partial P}\right)_S = \left(\frac{\partial G}{\partial P}\right)_T$$

2.32 The vapor pressures of water and ice at $-5°$ are 3.163 and 3.013 Torr, respectively. Calculate $\Delta \bar{G}$ for the transformation of water to ice at $-5°$.

2.33 Assuming the density of water is independent of pressure in the range 1 to 50 atm, what is the change in Gibbs free energy of a mole of water when the pressure is raised this amount?

2.34 Two moles of ideal gas are compressed isothermally from 1 to 5 atm at 100°. (a) What is the Gibbs free-energy change? (b) What would have been the Gibbs free-energy change if the compression had been carried out at 0°?

2.35 At 50° the partial pressure of $H_2O(g)$ over a mixture of $CuSO_4 \cdot 3H_2O(s)$ and $CuSO_4 \cdot H_2O(s)$ is 30 Torr and over a mixture of $CuSO_4 \cdot 3H_2O(s)$ and $CuSO_4 \cdot 5H_2O(s)$ is 47 Torr. Calculate the Gibbs free-energy change for the reaction:

$$CuSO_4 \cdot 5H_2O(s) = CuSO_4 \cdot H_2O(s) + 4H_2O(g)$$

2.36 The heat of vaporization of liquid oxygen at 1 atm is 1630 cal mol^{-1} at its boiling point, $-183°$. For the reversible evaporation of 1 mole of liquid oxygen calculate (a) q, (b) $\Delta \bar{U}$, (c) $\Delta \bar{G}$, and (d) $\Delta \bar{S}$.

2.37 One mole of an ideal gas in 22.4 liters is expanded isothermally and reversibly at 0° to a volume of 224 liters and $\frac{1}{10}$ atm. Calculate (a) w, (b) q, (c) $\Delta \bar{H}$, (d) $\Delta \bar{G}$, (e) $\Delta \bar{S}$ for the gas.

One mole of an ideal gas in 22.4 liters is allowed to expand irreversibly into an evacuated vessel such that the final total volume is 224 liters. Calculate (f) w, (g) q, (h) $\Delta \bar{H}$, (i) $\Delta \bar{G}$ (j) $\Delta \bar{S}$ for the gas.

Calculate (k) ΔS for the system and its surroundings involved in the reversible isothermal expansion and calculate (l) ΔS for the system and its surroundings involved in the irreversible isothermal expansion.

2.38 One-half mole of an ideal monatomic gas initially at 25° and occupying a volume of 2 liters is expanded adiabatically and reversibly to a pressure of 1 atm. The gas is then compressed isothermally and reversibly until its volume is 2 liters at the lower temperature. Calculate (a) q, (b) w, (c) ΔU, (d) ΔH, and (e) ΔS.

2.39 Calculate $(\partial U / \partial V)_T$ for a van der Waals' gas for which

$$P = \frac{RT}{\bar{V} - b} - \frac{a}{\bar{V}^2}$$

2.40 If the partial specific volume v in milliliters per gram is independent of concentration, it is referred to as the apparent specific volume. It is equal to the volume of the solution minus the volume of pure solvent it contains divided by the weight of solute:

$$v = \frac{100 - \left(\frac{100\rho_s - g}{\rho_0}\right)}{g}$$

where g is the number of grams of solute in 100 ml of solution, ρ_s is the density of the solution, and ρ_0 is the density of the solvent. The density of a solution of serum albumin containing 1.54 g of protein per 100 cm³ is 1.0004 g cm⁻³ at 25° ($\rho_0 = 0.99707$ g cm⁻³). Calculate the apparent specific volume.

2.41 A solution of magnesium chloride, $MgCl_2$, in water containing 41.24 g liter⁻¹ has a density of 1.0311 g cm⁻³ at 20°. The density of water at this temperature is 0.99823 g cm⁻³. Calculate (a) the partial specific volume (see the equation in problem 2.40) and (b) the apparent partial molar volume of $MgCl_2$ in this solution.

2.42 Calculations with the Debye formula show the molar entropy of silver iodide to be 1.5 cal K⁻¹ mol⁻¹ at 15 K. From the following data for the molar heat capacity at constant pressure, calculate the molar entropy of silver iodide at 298.1 K.

T, K	$\bar{C}_P$	T, K	$\bar{C}_P$	T, K	$\bar{C}_P$	T, K	$\bar{C}_P$
21.00	3.82	64.44	9.36	145.67	11.92	258.79	13.05
30.53	5.23	88.58	10.00	170.86	12.15	273.23	13.26
42.70	7.09	105.79	11.19	198.89	12.49	287.42	13.48
52.15	8.20	126.53	11.60	228.34	12.76	301.37	13.64

2.43 Calculate the molar entropy of carbon disulfide at 25° from the following heat-capacity data and the heat of fusion, 1049.0 cal mol⁻¹, at the melting point (161.11 K):

T, K	15.05	20.15	29.76	42.22	57.52	75.54	89.37
$\bar{C}_P$, cal K⁻¹ mol⁻¹	1.65	2.87	4.96	6.97	8.50	9.57	10.31

T, K	99.00	108.93	119.91	131.54	156.83	161–298
$\bar{C}_P$, cal K⁻¹ mol⁻¹	10.98	11.59	12.07	12.58	13.53	18.04

2.44 Calculate the entropy changes for the following reactions at 25°:

        (a) S(rhombic) = S(monoclinic)
        (b) Ethanol(l) + $\frac{1}{2}O_2(g)$ = acetaldehyde(g) + $H_2O(l)$
        (c) n-Hexane(g) = benzene(g) + $4H_2(g)$

2.45 When an ideal gas is allowed to expand isothermally in a piston, $\Delta U = q + w = 0$. Thus the work done by the system on the surroundings is equal to the heat transferred from the reservoir to the gas, the efficiency of turning heat into work is 100%. Explain why this is not a violation of the second law.

2.46 (a) What is the maximum work that can be obtained from 1000 cal of heat supplied to a water boiler at 100° if the condenser is at 20°? (b) If the boiler temperature is raised to 150° by the use of superheated steam under pressure, how much more work can be obtained? (See problem 2.1.)

2.47 Calculate the increase in entropy of a mole of silver that is heated at constant pressure from 0 to 30°, if the value of $\bar{C}_P$ in this temperature range is considered to be constant at 6.09 cal K⁻¹ mol⁻¹.

2.48 Calculate the change in entropy of a mole of aluminum which is heated from 600° to 700°. The melting point of aluminum is 660°, the heat of fusion is 94 cal g⁻¹, and the heat capacities of the solid and liquid may be taken as 7.6 and 8.2 cal K^{-1} mol^{-1}, respectively.

2.49 Calculate the differences between the molar entropies of $Hg(l)$ and $Hg(s)$ at $-50°$. The melting point of mercury is $-39°$, and the heat of fusion is 560 cal mol^{-1}. The heat capacity per gram atom of $Hg(l)$ may be taken as $7.1 - 0.0016\,T$, and that of $Hg(s)$ as 6.4 K^{-1} mol^{-1}.

2.50 Derive the expression for the entropy change of a van der Waals' gas that is allowed to expand from volume $\bar{V}_1$ to $\bar{V}_2$ at constant temperature.

2.51 Calculate the entropy change of aluminum when it crystallizes $100°$ below its melting point. The required data are given in problem 2.48. What is the entropy change for the aluminum plus its surroundings at $500°$?

2.52 A mole of helium at $100°$ is mixed with 0.5 mole of neon that is at $0°$. What is ΔS for this change if the initial and final pressures are all 1 atm and the gases are ideal?

2.53 An equation for the pressure P as a function of height h in the atmosphere may be derived easily if it is assumed that the temperature T, acceleration g of gravity, and molecular weight M of air are constant. The molar Gibbs free energy of the gas is a function of pressure and height and must be the same throughout the atmosphere. Thus

$$d\bar{G} = \left(\frac{\partial \bar{G}}{\partial P}\right)_h dP + \left(\frac{\partial \bar{G}}{\partial h}\right)_P dh = 0$$

Show that $(\partial \bar{G}/\partial h)_P = Mg$, and then that

$$P = P_0 e^{-Mgh/RT}$$

2.54 The density of Fe is 7.86 g cm^{-3} at $25°$. Assuming the compressibility is zero, what is the change in Gibbs free energy when the pressure on one mole of Fe (atomic weight 55.85) is isothermally raised from 1 atm to 10 atm?

2.55 Calculate $\Delta G°$ for

$$H_2O(g, 25°) = H_2O(l, 25°)$$

The vapor pressure of water at $25°$ is 23.76 Torr.

2.56 Calculate the Gibbs free-energy change for the expansion of 2 mol of ideal gas from 1 atm to $\frac{1}{10}$ atm at $25°$.

2.57 The change in Gibbs free energy for the conversion of aragonite to calcite at $25°$ is -190 cal mol^{-1}. The density of aragonite is 2.93 g cm^{-3} at $25°$ and the density of calcite is 2.71 g cm^{-3}. At what pressure at $25°$ would these two forms of $CaCO_3$ be in equilibrium?

2.58 Calculate (a) w and (b) $\Delta \bar{G}$ when 1 mol of liquid water is evaporated at its boiling point $100°$ at 1 atm and then expanded in its vapor state to 50 liters at $100°$.

2.59 Ten grams of helium is compressed isothermally and reversibly at $100°$ from a pressure of 2 atm to 10 atm. Calculate (a) q, (b) w, (c) ΔG, (d) ΔA, (e) ΔH, (f) ΔU, (g) ΔS.

2.60 Derive the relation for $\bar{C}_P - \bar{C}_V$ for a gas that follows van der Waals' equation.

2.61 One mole of steam is compressed reversibly to liquid water at the boiling point $100°$. The heat of vaporization of water at $100°$ and 760 Torr is 539.7 cal g^{-1}. Calculate w and q and each of the thermodynamic quantities $\Delta \bar{H}$, $\Delta \bar{U}$, $\Delta \bar{G}$, $\Delta \bar{A}$, and $\Delta \bar{S}$.

2.62 Calculate the entropy changes for the gas plus reservoir in Examples 2.5 and 2.6.

2.63 If z is a function of x and y, it is readily shown that

$$\left(\frac{\partial x}{\partial y}\right)_z \left(\frac{\partial y}{\partial z}\right)_x \left(\frac{\partial z}{\partial x}\right)_y = -1$$

which is referred to as the cyclic rule. Use this rule to show that for a gas

$$\left(\frac{\partial P}{\partial T}\right)_V = \frac{\alpha}{\kappa}$$

where the coefficient of thermal expansivity α and the isothermal compressibility κ are given by

$$\left(\frac{\partial V}{\partial T}\right)_P = V\alpha$$

$$\left(\frac{\partial V}{\partial P}\right)_T = -V\kappa$$

2.64 From tables giving $\Delta \bar{G}_f^\circ$, $\Delta \bar{H}_f^\circ$, and $\bar{C}_P$ for $H_2O(l)$ and $H_2O(v)$ at 298 K calculate (a) the vapor pressure of $H_2O(l)$ at $25°$ and (b) the standard boiling point.

2.65 When 1 mol of water was added to an infinitely large amount of an aqueous methanol solution having a mole fraction of methanol of 0.40, the volume of the solution increased 17.35 ml. When 1 mole of methanol was added to such a solution, the volume increased 39.01 ml. Calculate the volume of solution containing 0.40 mol of methanol and 0.60 mol of water.

2.66 Using the following data, calculate the molar entropy of gaseous isobutene at $25°$ [S. S. Todd and G. S. Parks, *J. Am. Chem. Soc.*, **58**, 134 (1936)]:

$$\bar{S}_{90K} = 10.81 \text{ cal K}^{-1} \text{ mol}^{-1}$$

$$\text{F.P.} = -140.7°$$

$$\Delta \bar{H}_{fus} = 25.22 \text{ cal}^{-1} \text{ g}^{-1}$$

$$\text{B.P.} = -7.1°$$

$$\Delta \bar{H}_{vap} = 96.5 \text{ cal}^{-1} \text{ g}^{-1}$$

$$\bar{C}_P(\text{gas}) \text{ in range } 226\text{–}298 \text{ K} = 20 \text{ cal K}^{-1} \text{ mol}^{-1}$$

Specific Heat (cal g^{-1})

T, K	93.3	105.5	118.9	139.2	166.1	179.8	210.2	253.1
	0.2498	0.2749	0.3056	0.4547	0.4621	0.4681	0.4860	0.5173

2.67 If C_P in the temperature range below 15 K is given by

$$C_P = C_{15}\left(\frac{T}{15}\right)^3$$

show that

$$S_{15} = \frac{C_{15}}{3}$$

2.68 Calculate the entropy changes for the following reactions at $25°$:

(a) $H^+(aq) + OH^-(aq) = H_2O(l)$

(b) $Ag^+(aq) + Cl^-(aq) = AgCl(s)$

(c) $HS^-(aq) = H^+(aq) + S^{2-}(aq)$

2.69 Calculate ΔG° at $25°$ for the following reactions:

(a) $Fe_2O_3(s) + Fe(s) + \frac{1}{2}O_2(g) = Fe_3O_4(s)$

(b) $n\text{-}C_6H_{14}(g) = C_6H_6(g) + 4H_2(g)$

(c) $NaCl(s) + KF(s) = NaF(s) + KCl(s)$

CHAPTER 3

ONE-COMPONENT SYSTEMS

A *phase* is a part of a system, uniform throughout in chemical composition and physical properties, which is separated from other homogeneous parts of the system by boundary surfaces. Since gases are completely miscible with each other, there can be only a single gaseous phase in a system, but there may be many solid phases and several liquid phases.

The phase behavior exhibited by pure substances is quite varied and complicated, but this information may be organized and predictions may be made by use of thermodynamics. For each phase of a pure substance, the relation between P, V, and T is given by the equation of state. After discussing equations of state of gases and critical phenomena we will derive the phase rule and the Clapeyron equation.

Thermodynamics provides relations that apply to phase changes, but it does not tell us the temperature at which a particular phase transition takes place. Someday statistical mechanics (Chapter 17) will hopefully provide a means for calculating transition temperatures, but the difficulties of including intermolecular interactions in the calculation make this possibility somewhat distant. The conditions for liquid-gas phase changes may be obtained by expressing the properties of a liquid and the corresponding gas on either side of a transition point by use of an empirical equation. The van der Waals equation is an example of such an equation.

3.1 EQUATION OF STATE

An equation of state gives the relation between pressure, volume, and temperature for a fixed quantity of a substance. The ideal gas law (Section 1.2) is the simplest example. Much more complicated equations are required to represent pressure-volume-temperature data for real gases. One of the ways of presenting experimental data is to plot the *compressibility factor PV/nRT* versus pressure at constant temperature.

The compressibility factors for H_2 and O_2 at 0° are plotted versus pressure in Fig. 3.1. The behavior of an ideal gas is represented by the horizontal dashed line. In the limit of zero pressure the compressibility factor of every gas is unity because it behaves ideally.

A compressibility factor of less than unity indicates that the gas is more compressible than an ideal gas. All gases show a minimum in the plot of compressibility factor versus pressure if the temperature is low enough. Hydrogen and helium,

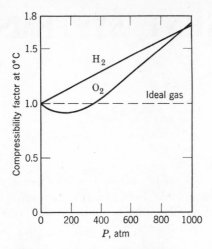

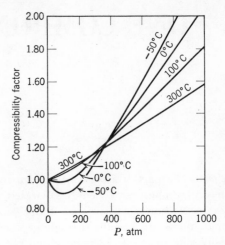

Fig. 3.1 Influence of high pressure on the compressibility factor, $P\overline{V}/RT$, for two gases.

Fig. 3.2 Influence of pressure on the compressibility factor, $P\overline{V}/RT$, for nitrogen at different temperatures.

which have very low boiling points, exhibit this minimum only at temperatures much below 0° C.

The effect of temperature on the change of compressibility factor of N_2 with pressure is shown in Fig. 3.2. At 0° C and low pressures N_2 is more compressible than an ideal gas, but at high pressures it is less compressible.

3.2 PRESSURE-TEMPERATURE-VOLUME DIAGRAM FOR WATER

As an example of the pressure-volume-temperature behavior of a pure substance, the three-dimensional diagram for H_2O is given in Fig. 3.3. Each point on this surface represents an equilibrium state. Projections of this surface on the P–T plane and the P–V plane are shown. There are three two-phase regions on the surface: liquid plus vapor, ice plus vapor, and liquid plus ice. These are ruled surfaces—that is, they may be thought of as being generated by a moving straight line, in this case one perpendicular to the P–T plane. These three surfaces intersect at the triple point A. Thus at the triple point vapor, liquid, and solid are in equilibrium.

The vapor pressure of liquid water is given as a function of temperature by curve AB in the projection on the left (P versus T). At pressures and temperatures represented by points below this line the liquid will vaporize completely, and so this region is labeled vapor. At pressure and temperatures represented by points above AB the vapor is completely condensed to liquid. The upper limit of the vaporization curve at B, which is the critical point, is at 374° and 218 atm for

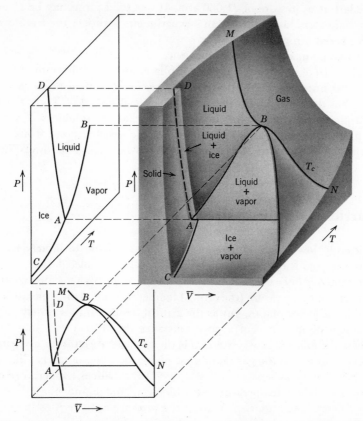

Fig. 3.3 Plot of the pressure-volume-temperature relation for a pure substance like water. The diagram is not drawn to scale.

water. Above this temperature there are no distinguishable vapor and liquid phases.

The line AC is the sublimation curve of ice. Above it lies ice, and below it lies vapor. Only at temperatures and pressures represented by points along this line can ice and water exist in equilibrium. Line AC goes down to absolute zero.

The line AD shows how the melting point of ice depends on the pressure. For most substances, line AD tilts away from the vertical axis of the plot of P versus T rather than toward it, but water is an unusual substance that expands when it freezes. Since liquid water occupies a smaller volume than ice, we can predict, with the aid of Le Chatelier's principle (Section 5.14), that raising the pressure will cause the system at equilibrium to shift toward the liquid form, that is, the freezing point will be lowered.

Investigations conducted by Bridgman to determine the course of the fusion curve AD have revealed the existence of seven different crystalline modifications of ice, all of which, with the exception of ordinary ice, are denser than water. The first of these new forms of ice makes its appearance at a pressure of 2047 atm and the last at a pressure of 21,680 atm. If one solid form may be changed into another solid form, as in Bridgman's experiments with ice, the transformation is called an enantiotropic change.

The triple point of water in the absence of air is at 0.0100° C and 4.58 Torr. In the presence of air at 1 atm the three phases are in equilibrium at 0°, which is one of the defined points on the centigrade scale. The total pressure in this case is 1 atm, but the partial pressure of water vapor is only 4.58 Torr. The lowering of the triple point by air is due to two effects: (a) the solubility of air in liquid water at 1 atm pressure is sufficient to lower the freezing point 0.0024° (Section 4.14), and (b) the increase of pressure from 4.58 Torr to 1 atm lowers the freezing point 0.0075°, as will be shown shortly.

3.3 CRITICAL PHENOMENA*

At sufficiently low temperatures any gas may be made to liquefy by applying pressure, thus reducing the volume and bringing the molecules so close together that the attractive force between them becomes large enough to cause condensation. Below a certain temperature (the critical temperature) there is a meniscus between the liquid and vapor phases, but as the critical temperature is closely approached this meniscus disappears. For a pure substance the critical state may be defined by either of the following two criteria: (1) the critical state is the state of temperature and pressure at which the gas and liquid phases become so nearly alike that they can no longer exist as separate phases; or (2) the critical temperature of a pure liquid is the highest temperature at which gas and liquid phases can exist as separate phases. The critical pressure is the pressure at the critical point, and the critical volume is the molar volume under these conditions.

Only the first of these definitions is applicable to mixtures.† More complicated behavior is encountered with mixtures since the coexisting liquid and vapor phases are, in general, of different composition.

The appearance of the pressure-volume curves in the vicinity of the critical temperature is shown in Fig. 3.4. For isopentane the pressure is plotted against molar volume at the several temperatures indicated on the graph.

Lines on a graph that refer to a specified constant temperature are called *isotherms*. At 553 K the pressure-volume curve is a hyperbolic curve similar to that for an ideal gas. At 473 K the curve has a more complicated shape showing that the molecules attract each other, thus making the volume of the gas smaller than that of an ideal gas, in which no such attraction exists.

At all temperatures below 461 K the isotherms in Fig. 3.4 exhibit horizontal sections. These horizontal lines indicate that an infinitesimal increase in pressure causes a very large decrease in volume, owing to the fact that the gas is liquefying.

---

* J. V. Sengers and A. L. Sengers, "The Critical Region," *Chem. and Engr. News*, June 10, 1968, p. 104.

† L. G. Roof, *J. Chem. Educ.*, **34**, 492 (1957).

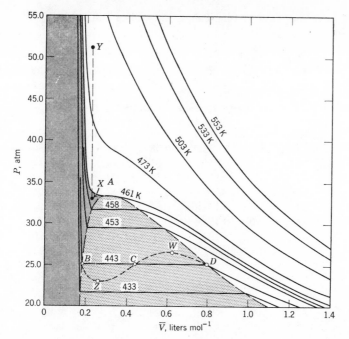

Fig. 3.4 Isotherms for isopentane showing the critical region. The two-phase region is light gray, and the liquid region is darker gray. The gas region is white.

At the right of the diagram outside the two-phase region gas alone is present. At the left of the diagram outside the two-phase region liquid alone is present; and, since the liquid is much less compressible than the gas, the isotherms are much steeper than for the gas. The liquid region is marked with the darker gray. Along the horizontal lines gas and liquid exist together. A gas in equilibrium with the corresponding liquid is generally referred to as a vapor. The region in which vapor and liquid are coexistent is marked with light gray. The vapor pressures of isopentane at various temperatures are given by the horizontal lines in Fig. 3.4. The maximum point A gives the critical temperature, critical pressure, and critical volume. The wavy line $DWCZB$ is discussed in Section 3.4.*

---

* The tie line BD connects the P-V conditions of the liquid and vapor states, which are at equilibrium at 443 K. The difference between the molar Gibbs free energy of the liquid and vapor at constant temperature is given by

$$\overline{G}_g - \overline{G}_l = \int_l^g \overline{V}\, dP$$

which is the integral form of equation 2.65. Integrating by parts

$$\int_l^g \overline{V}\, dP = P_{eq}(\overline{V}_g - \overline{V}_l) - \int_l^g P\, d\overline{V}$$

It is seen that this is the difference between the two areas BCZ and CWD. Since the liquid and vapor are in equilibrium, $\overline{G}_g = \overline{G}_l$, and these two areas have to be equal.

Table 3.1 Critical Constants, Melting Points, and Boiling Points

Gas	T_{MP}, K	T_{BP}, K	T_c, K	$\dfrac{T_{BP}}{T_c}$	P_c, atm	$\bar{V}_c$, 1 mol^{-1}	$\dfrac{P_c \bar{V}_c}{RT_c}$
			Simple Nonpolar Molecules[1]				
He	0.9	4.2	5.3	0.79	2.26	0.0578	0.300
H_2	14.0	20.4	33.3	0.68	12.8	0.0650	0.304
Ne	24.5	27.2	44.5	0.61	25.9	0.0417	0.296
A	83.9	87.4	151	0.58	48	0.0752	0.291
Xe	133	164.1	289.81	0.57	57.89	0.1202	0.293
N_2	63.2	77.3	126.1	0.61	33.5	0.0901	0.292
O_2	54.7	90.1	154.4	0.58	49.7	0.0744	0.292
CH_4	89.1	111.7	190.7	0.59	45.8	0.0990	0.290
CO_2	...	194.6	304.2	0.64	72.8	0.0942	0.274
			Hydrocarbons				
Ethane, C_2H_6	89.98	184.6	305.5	0.60	48.2	0.139	0.267
Propane, C_3H_8	185.5	231.1	370.0	0.62	42.1	0.195	0.270
Isobutane, C_4H_{10}	113.6	261.5	407	0.64	37	0.250	0.276
n-butane C_4H_{10}	134.9	272.7	426	0.64	36	0.250	0.257
n-Hexane, C_6H_{14}	178.8	342.1	507.9	0.67	29.6	0.367	0.260
n-Octane, C_8H_{18}	216.6	397.7	570	0.70	24.7	0.490	0.259
Benzene, C_6H_6	278.6	352.7	561.6	0.63	47.9	0.256	0.265
Cyclohexane, C_6H_{12}	279.7	353.9	554	0.64	40.57	0.312	0.280
Ethylene, C_2H_4	103.7	169.3	282.8	0.60	50.5	0.126	0.274
Acetylene, C_2H_2	191.3	189.5	308.6	0.61	61.6	0.113	0.275
			Polar Molecules[1]				
H_2O	273.1	373.1	647.3	0.58	217.7	0.0566	0.232
NH_3	195.4	239.7	405.5	0.59	112.2	0.0720	0.243
CH_3OH	175.4	337.9	513.2	0.66	78.67	0.118	0.220
CH_3Cl	175.5	249.7	416.3	0.60	65.8	0.148	0.285
C_2H_5Cl	134.2	285.9	460.4	0.62	52	0.196	0.269

[1] Polar and nonpolar molecules are discussed in Section 14.10. Although both types of molecules are electrically neutral, in the polar molecules there is a separation of positive and negative charge in different parts of the molecule and a resulting attraction between the molecules.

If liquid is converted to gas by a series of steps that take it through the two-phase region, a boundary or meniscus will be formed between the gas and liquid phases. However, the liquid may be converted to gas smoothly and continuously without the appearance of a meniscus. If a sample of liquid at point X is heated at constant volume to point Y, there is complete continuity between liquid and gaseous states.

The critical temperature of a pure liquid is usually obtained by observing the temperature at which the meniscus between the gas and liquid phases under pressure disappears when warmed and reappears when cooled. Although the material at the hydrostatic level at which the interface has just disappeared is at the critical temperature and pressure, it is important to realize that the pressure is not the same throughout the sample because of gravity, and the density ρ varies with the height in the sample. Near the critical point $d\rho/dP$ is large (actually it is infinite at the critical point) so that the material in the lower part of the sample is more dense than that in the upper part.

In the neighborhood of the critical point a fluid is so compressible that the acceleration of gravity sets up large differences in density between the top and bottom of a container. Gravity can produce density differences as large as 10% in a column of fluid only a few centimeters high. This makes it difficult to determine PV isotherms near the critical point.

At conditions near the critical point a fluid scatters light strongly. This critical opalescence is the result of large fluctuations in the refractive index that are a consequence of large fluctuations in density.

Table 3.1 gives the melting point T_{MP}, boiling point T_{BP}, and the molar volume $\bar{V}_c$, pressure P_c, and temperature T_c at the critical point for a number of substances. The significance of $P_c\bar{V}_c/RT_c$, the compressibility factor at the critical temperature and volume, is discussed later.

3.4 THE VAN DER WAALS THEORY

In 1879 van der Waals provided an interpretation of the nonideal behavior of gases that also helped understand critical phenomena. He realized that real gases are more compressible than ideal gases because molecules attract each other. The forces that lead to condensation are still referred to as van der Waals forces. He provided for intermolecular attraction by adding to the observed pressure P in the equation of state a term n^2a/V^2, where a is a van der Waals constant and n is the number of moles of gas. At sufficiently high pressure real gases occupy larger volumes than ideal gases, and van der Waals provided for this effect by subtracting an excluded volume nb from the observed volume V to give the actual volume in which the molecules move.

Van der Waals' full equation is

$$\left(P + \frac{n^2a}{V^2}\right)(V - nb) = nRT \tag{3.1}$$

or

$$\left(P + \frac{a}{\bar{V}^2}\right)(\bar{V} - b) = RT \tag{3.2}$$

Table 3.2   Van der Waals' Constants

Gas	a, liter^2 atm mol^{-2}	b, liter mol^{-1}	Gas	a, liter^2 atm mol^{-2}	b, liter mol^{-1}
H_2	0.2444	0.02661	CH_4	2.253	0.04278
He	0.03412	0.02370	C_2H_6	5.489	0.06380
N_2	1.390	0.03913	C_3H_8	8.664	0.08445
O_2	1.360	0.03183	$C_4H_{10}(n)$	14.47	0.1226
Cl_2	6.493	0.05622	C_4H_{10}(iso)	12.87	0.1142
NO	1.340	0.02789	$C_5H_{12}(n)$	19.01	0.1460
NO_2	5.284	0.04424	CO	1.485	0.03985
H_2O	5.464	0.03049	CO_2	3.592	0.04267

When the molar volume $\bar{V}$ is large, both b and $a/\bar{V}^2$ become negligible; and van der Waals' equation reduces to the ideal gas law, $P\bar{V} = RT$.

Van der Waals' constants for a few gases are listed in Table 3.2. They can be calculated from experimental measurements of P, $\bar{V}$, and T or from the critical constants, as shown later in equation 3.8.

We can estimate the excluded volume at sufficiently low pressures that only pairs of molecules are close to each other, triplets and larger clusters being negligible. The volume that is excluded per molecule may be calculated if the molecules are considered to be rigid spheres. As can be seen from Fig. 3.5 the volume that is excluded per pair of molecules is $\frac{4}{3}\pi d^3$, where d is the molecular diameter. Thus the volume excluded *per molecule* is $\frac{2}{3}\pi d^3$. Since the volume of a molecule is $\frac{4}{3}\pi(d/2)^3 = (\pi/6)d^3$, the excluded volume per molecule is just 4 times the volume of a single molecule. As a first approximation then we may expect the constant b to be 4 times the molecular volume times Avogadro's constant.

$$b = 4\left[\frac{4}{3}\pi\left(\frac{d}{2}\right)^3\right]N_A = \frac{2}{3}\pi d^3 N_A \tag{3.3}$$

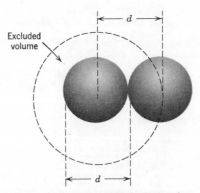

Fig. 3.5   Volume excluded to a pair of rigid spherical molecules.

Multiplying out the terms in van der Waals' equation 3.2 and rearranging in descending powers of $\bar{V}$, we have

$$\bar{V}^3 - \bar{V}^2\left(b + \frac{RT}{P}\right) + \bar{V}\frac{a}{P} - \frac{ab}{P} = 0 \qquad (3.4)$$

At temperatures below the critical temperature this cubic equation has three real solutions, each value of P giving three values of $\bar{V}$. This equation is shown graphically by the dashed line $DWCZB$ on the 440 K isothermal in Fig. 3.4, where the three values of $\bar{V}$ are the intersections B, C, and D on the horizontal line corresponding to a fixed value of the pressure. This dashed calculated line then appears to give a continuous transition from the gaseous phase to the liquid phase, but in reality the transition is abrupt and discontinuous, both liquid and vapor existing along the straight horizontal lines. The theoretical dashed line $DWCZB$ does not correspond to normal physical conditions; for example, the slope of the curve at C is positive, a fact that would lead to the unnatural condition that an increase in pressure produces an increase in volume. It is possible, however, to have pressures of gas in an unstable condition represented by the beginning of the dashed line DW, before the supercooled vapor has a chance to liquefy and bring the pressure down to that of the horizontal line. It is also possible to have liquid under metastable conditions along BZ. At the critical temperature (461 K for isopentane) and pressure there is only one real root, the critical volume $\bar{V}_c$.

The values of van der Waals' constants may be calculated from the critical constants for a gas. As may be seen in Fig. 3.4, there is a horizontal inflection point in the P versus $\bar{V}$ curve at the critical point so that $(\partial P/\partial \bar{V})_{T_c} = 0$ and $(\partial^2 P/\partial \bar{V}^2)_{T_c} = 0$. At the critical temperature van der Waals' equation may be written

$$P = \frac{RT_c}{\bar{V} - b} - \frac{a}{\bar{V}^2} \qquad (3.5)$$

Differentiating with respect to molar volume,

$$\left(\frac{\partial P}{\partial \bar{V}}\right)_{T_c} = \frac{-RT_c}{(\bar{V} - b)^2} + \frac{2a}{\bar{V}^3} \qquad (3.6)$$

$$\left(\frac{\partial^2 P}{\partial \bar{V}^2}\right)_{T_c} = \frac{2RT_c}{(\bar{V} - b)^2} - \frac{6a}{\bar{V}^4} \qquad (3.7)$$

At the critical point these derivatives are both equal to zero, and $\bar{V}$ is replaced by $\bar{V}_c$ and P by P_c.

Equations 3.5, 3.6, and 3.7 may be solved together to obtain

$$a = 3P_c\bar{V}_c^2 \qquad b = \frac{\bar{V}_c}{3} \qquad R = \frac{8P_c\bar{V}_c}{3T_c} \qquad (3.8)$$

The critical constants are expressed in terms of the van der Waals constants by

$$\bar{V}_c = 3b \qquad P_c = \frac{a}{27b^2} \qquad T_c = \frac{8a}{27bR} \tag{3.9}$$

According to equation 3.8 all van der Waals gases should have a compressibility factor $P_c \bar{V}_c / RT$ at the critical point of $\frac{3}{8} = 0.375$. The compressibility factors at the critical points are given for a number of substances in Table 3.1. It may be seen that for simple nonpolar molecules this ratio has a value near 0.29. As shown by the last group of substances, polar molecules give a still greater deviation.

If the relations in 3.8 are substituted into equation 3.2, we obtain

$$\left(\frac{P}{P_c} + \frac{3\bar{V}_c^2}{\bar{V}^2}\right)\left(\frac{\bar{V}}{\bar{V}_c} - \frac{1}{3}\right) = \frac{8}{3}\frac{T}{T_c} \tag{3.10}$$

Defining a set of reduced variables, the reduced pressure P_r, the reduced molar volume $\bar{V}_r$, and the reduced temperature T_r

$$P_r = \frac{P}{P_c} \qquad \bar{V}_r = \frac{\bar{V}}{\bar{V}_c} \qquad T_r = \frac{T}{T_c} \tag{3.11}$$

equation 3.10 becomes

$$\left(P_r + \frac{3}{\bar{V}_r^2}\right)(\bar{V}_r - \tfrac{1}{3}) = \tfrac{8}{3}T_r \tag{3.12}$$

so that all constants connected with the individual nature of the gas have disappeared. This equation does not represent experimental data any better than the original van der Waals equation, but leads to the idea that if different substances are compared at equal fractions of their critical temperatures, pressures, and volumes they will appear more similar. This principle is not exact but is a useful approximation and forms the basis of the Hougen, Watson, and Ragatz[1] generalized compressibility charts.

Van der Waals' equation 3.2 gives a qualitative description of critical behavior, including the large compressibility near the critical point. Van der Waals' theory, in common with other classical theories, predicts that in the neighborhood of the critical point the difference between the density of the liquid or gas and the critical density is proportional to $(T - T_c)^{1/2}$ but the experimental results are closer to $(T - T_c)^{1/3}$. The van der Waals theory also fails to explain the observation that C_V varies with $\log (T - T_c)$ near the critical point.

The experimental results can be accounted for by extension of a theory developed in 1925 by Ising to explain ferromagnetism. He assumed that the magnetic spins are localized on the lattice sites of a regular array and are capable of only two orientations in opposite directions. He assumed that parallel spins attract each other and antiparallel spins repel, but that there are interactions only between nearest neighbors. Thus the cooperative effect of ferromagnetism arises through chains of nearest-neighbor interactions. In the critical fluid case the spins pointing in one direction are replaced by molecules, and spins pointing in the opposite direction are replaced by "holes" or empty sites.

Critical phenomena are also shown by liquid solutions, biopolymers, liquid crystals, alloys, superconductors, and ferromagnetic metals.

---

[1] O. A. Hougen, K. M. Watson, and R. A. Ragatz, *Chemical Process Principles*, Part II, Wiley, New York, 1959, Chapter XII.

3.5   CRITERIA OF EQUILIBRIUM IN TERMS OF INTENSIVE PROPERTIES

Criteria of equilibrium can be stated in terms of the intensive properties T, P, and μ. We are familiar with the fact that for two phases, such as ice and water, to be in equilibrium they must be at the same temperature and pressure. This can be proved by considering that an infinitesimal quantity of heat dq is reversibly transferred from a phase α to a phase β with which it is in equilibrium. The condition for equilibrium in an isolated system is that the entropy of the system not be changed by the transfer. Thus

$$dS = 0 \qquad \text{or} \qquad dS_\alpha + dS_\beta = 0 \tag{3.13}$$

where the phases are designated by subscripts. Since the process is reversible

$$\frac{-dq}{T_\alpha} + \frac{dq}{T_\beta} = 0 \tag{3.14}$$

or

$$T_\alpha = T_\beta \tag{3.15}$$

The fact that the pressure must be the same in two phases at equilibrium may be proved by considering that phase α increases in volume by an infinitesimal volume dV and phase β decreases by the same amount. If the temperature and volume of the whole system are held constant, $dA = 0$ so that

$$dA_\alpha + dA_\beta = 0 \qquad \text{or} \qquad -P_\alpha\, dV + P_\beta\, dV = 0 \tag{3.16}$$

and so

$$P_\alpha = P_\beta \tag{3.17}$$

An additional restriction may be derived by considering the transfer of a small quantity of substance i from phase α to phase β, which are in equilibrium. If the temperature and pressure of the whole system are kept constant, then, since $dG = 0$,

$$dG_\alpha + dG_\beta = 0 \tag{3.18}$$

or using equation 2.91 (Section 2.16)

$$-\mu_{i\alpha}\, dn_i + \mu_{i\beta}\, dn_i = 0 \tag{3.19}$$

or

$$\mu_{i\alpha} = \mu_{i\beta} \tag{3.20}$$

Thus the chemical potential of a component is the same in all phases at equilibrium.

If phases α and β are not in equilibrium and a small quantity dn_i of species i is transferred from phase α to phase β in the direction of approaching equilibrium, we have at constant temperature and pressure

$$dG_\alpha + dG_\beta < 0 \qquad \text{or} \qquad -\mu_{i\alpha}\, dn_i + \mu_{i\beta}\, dn_i < 0 \tag{3.21}$$

If dn_i is positive then

$$\mu_{i\alpha} > \mu_{i\beta} \tag{3.22}$$

Thus, a substance will tend to pass spontaneously from the phase where it has the higher chemical potential to the phase where it has the lower chemical potential. Also a substance will diffuse spontaneously from a region where its concentration and chemical potential are higher into a more dilute solution where its chemical potential is lower. In this respect the chemical potential is like other kinds of potential, electrical, gravitational, etc., in that the spontaneous change is always in the direction from high to low potential. It is really to this property that the chemical potential owes its name.

3.6 PHASE RULE

In 1876, Gibbs* derived a simple relationship between the number of phases in equilibrium, the number of components, and the number of intensive independent variables that must be specified in order to describe the state of the system completely.

The *number of components c* in a system is the smallest number of substances in terms of which the compositions of each of the phases in the system may be described separately. The number of components is smaller than the number of substances s that may be added to form the system because there may be relationships between the concentrations of various substances at equilibrium that make it unnecessary to specify the concentrations of all s substances in describing the system. There are two types of relations: chemical equilibrium expressions and initial conditions. For each independent chemical equilibrium expression, the number of independent concentrations is reduced by one. For example, if solid calcium oxide, solid calcium carbonate, and gaseous carbon dioxide are in equilibrium, the number of independent components is reduced by one by the equilibrium expression for

$$CaCO_3(s) = CaO(s) + CO_2(g) \tag{3.23}\dagger$$

If molecular hydrogen and oxygen are in equilibrium with water there are at most two independent components (H_2O and O_2, H_2O and H_2, or H_2 and O_2) because the concentration of the third is fixed by the equilibrium expression

$$K_p = \frac{p_{H_2}p_{O_2}^{1/2}}{p_{H_2O}} \tag{3.24}$$

---

* J. W. Gibbs, *Trans. Conn. Acad. Arts Sci.*, 1876–1878; *The Collected Works of J. Willard Gibbs*, Vol. 1, Yale University Press, New Haven, reprinted 1948.

† The role of initial conditions in the case of calcium carbonate is more complicated in that two different phases are involved. If pure $CaCO_3(s)$ is added initially and is partially dissociated at equilibrium there are still two components. The equilibrium expression does not result in any new restriction on the concentrations of the dissociation products. L. Katz, *J. Chem. Ed.*, **44**, 282 (1967).

If the initial conditions are specified, the number of components is reduced to one. For example, if the hydrogen and oxygen are formed only from water, there is an additional relationship, $p_{H_2} = 2p_{O_2}$.

If there are s substances, n independent equilibrium conditions, and m relations between concentrations due to initial conditions, the number of components c is given by

$$c = s - n - m \tag{3.25}$$

There are often several possible and equally satisfactory choices of components. The choices are arbitrary, but the *number* of components is an important characteristic of a system.

The *number of degrees of freedom or variance v* of a system is the smallest number of independent variables (pressure, temperature, and concentrations of the various phases) that must be specified to describe completely the state of the system. As we have seen before, to describe the state of a fixed amount of a pure gas it is necessary to specify only two variables, T and P, or P and V, or V and T, because the third variable can be calculated from the equation of state. Thus a pure gas has two degrees of freedom or a variance $v = 2$.

Consider a system in equilibrium that consists of p phases. When we refer to the number of phases, we mean the number of different kinds of phases; for example, a system containing liquid water and many pieces of ice, but no gas phase, has only two phases. If a phase contains c components, its composition may be specified by stating $(c - 1)$ concentrations—one less than the number of components because the concentration of one component can be obtained from $\sum X_i = 1$, where X_i represents the mole fraction of component i. Thus the total number of concentrations to be specified for the whole system is $(c - 1)$ for each of the p phases or $(c - 1)p$ concentrations In general, there are two more variables that have to be considered, temperature and pressure, so that the total number of independent variables is $(c - 1)p + 2$. We do not have to talk about the temperatures and pressures of the different phases separately because they are in equilibrium and so they are all at the same pressure and temperature. If temperature or pressure were held constant, the number of independent variables would be $(c - 1)p + 1$. On the other hand, if the system were affected by both temperature and pressure and another independent variable, such as magnetic field strength, the number of variables would be $(c - 1)p + 3$.

Next we consider the number of relationships that must be satisfied at equilibrium. The chemical potential μ for each component is the same in each phase α, β, γ, etc., and so $\mu_{i,\alpha} = \mu_{i,\beta} = \mu_{i,\gamma} = \cdots$ for component i. There are p phases but only $(p - 1)$ equilibrium relationships for each component. For example, if there are two phases, there is only one equilibrium relationship for each component that gives its distribution between the two phases. Altogether there are c components, each one of which can be involved in an equilibrium between phases. Thus there is a total of $c(p - 1)$ equilibrium relations.

The number of degrees of freedom v is equal to the total number of variables minus the total number of equilibrium relations between these variables; that is,

v is the additional number of variables that must be specified to define the system completely. Thus, for systems in which pressure, temperature, and concentration are the only variables,

$$v = [p(c - 1) + 2] - c(p - 1)$$

or

$$v = c - p + 2 \qquad (3.26)$$

This is the important phase rule of Gibbs.

It can be seen from this equation that, the greater the number of components in a system, the greater is the number of degrees of freedom or variance. On the other hand, the greater the number of phases the smaller is the number of variables such as temperature, pressure, and concentration that must be specified to describe the system completely.

For a given number of components the number of phases is a maximum when the variance is zero. For a one-component system, the maximum number of phases that can be present at equilibrium is $p = c - v + 2 = 3$. For a two-component system, the maximum number of phases is 4, etc.

3.7 PHASE RULE FOR A ONE-COMPONENT SYSTEM

For a one-component system the phase rule becomes $v = 3 - p$. Thus if there is a single phase, $v = 2$ and two independent variables have to be specified to describe the system. For example, if the system consists of water vapor, it is necessary to specify the temperature and pressure to define the system. If there are two phases in equilibrium, $v = 1$ and it is necessary to specify only one variable to describe the system. For example, if the system consists of water vapor in equilibrium with liquid water, it is necessary to specify only the temperature *or* pressure, because for a given temperature there is only one equilibrium pressure and for a given pressure there is only one equilibrium temperature. If there are three phases in equilibrium, $v = 0$ and so for a one-component system three phases can coexist only at a particular temperature and pressure. For example, water vapor, liquid water, and ice can exist together only at 0.0100° C and 4.58 Torr, and so the conditions are completely specified by simply saying that the three phases are present at equilibrium. Such a system is said to be *invariant*. In summary:

$$p = 1 \qquad v = 2 \qquad \text{bivariant}$$
$$p = 2 \qquad v = 1 \qquad \text{univariant}$$
$$p = 3 \qquad v = 0 \qquad \text{invariant}$$

3.8 EXISTENCE OF PHASES IN A ONE-COMPONENT SYSTEM

In order to understand the change from solid to liquid to gas phases when a solid is heated at constant pressure we may consider a plot of chemical potential versus temperature at

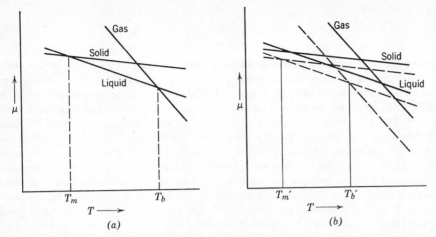

Fig. 3.6 Dependence of chemical potentials of solid, liquid, and gas phases on temperature at constant pressure. The dashed lines in (b) are for a lower pressure. The plots should be slightly concave downward since the entropy increases with increasing temperature, but they have been drawn as straight lines here for simplicity.

constant pressure for the various phases as shown in Fig. 3.6a. As we have already seen (Section 3.5), the stable phase is that with the lowest value of the chemical potential. If two or more phases have the same value of the chemical potential, they will coexist at equilibrium as at the melting point T_m and boiling point T_b. Below the melting point T_m the solid has the lowest chemical potential and is therefore the stable phase. Above the boiling point T_b the gas phase has the lowest chemical potential, and it is the stable phase. Between T_m and T_b, the liquid is the stable phase. It may be seen from this figure that the phase transitions are sudden and there are no indications of a drastic change as the temperature is changed toward the transition point.*

The slopes of the lines giving the chemical potentials of solid, liquid, and gas in Fig. 3.6a are given by (see Section 2.17)

$$\left(\frac{\partial \mu}{\partial T}\right)_P = -\bar{S} \tag{3.27}$$

Since the entropy is positive the slopes are negative, and since $\bar{S}_g > \bar{S}_l > \bar{S}_s$, the slope is more negative for the gas than the liquid and more negative for the liquid than for the solid. The plots of μ versus T are shown as straight lines for simplicity but are curved according to the derivative of equation 3.27.

$$\left(\frac{\partial^2 \mu}{\partial T^2}\right)_P = -\left(\frac{\partial \bar{S}}{\partial T}\right)_P = -\frac{\bar{C}_P}{T} \tag{3.28}$$

At a lower pressure the plots of μ versus T are displaced as shown in Fig. 3.6b. The effect of pressure on the chemical potential of a pure substance at constant temperature is given by

---

* There is another type of transition, called a lambda transition, in which the heat capacity changes rapidly as the transition temperature is approached. This also occurs at the critical point but not elsewhere along the vapor-liquid equilibrium line.

(see Section 2.17)

$$\left(\frac{\partial \mu}{\partial P}\right)_T = \bar{V} \tag{3.29}$$

Since the molar volume is always positive, the chemical potential μ decreases as the pressure is decreased at constant temperature. Since $\bar{V}_g \gg \bar{V}_l, \bar{V}_s$, this effect is much larger for a gas than for a liquid or solid. As shown in Fig. 3.6b, reducing the pressure lowers the boiling point and normally the melting point. The effect on the boiling point is much larger because of the large difference in the molar volumes of gas and liquid. As a result the range of temperature over which the liquid is the stable phase has been reduced. It is evident that at a sufficiently low pressure the curve for the chemical potential of the gas will intercept the solid curve below the temperature where the solid and liquid have the same chemical potential. At this low pressure the solid will sublime rather than melt, that is, it passes directly into the vapor without going through the liquid state as illustrated by dry ice.

At some particular pressure the solid, liquid, and vapor curves will intersect at a point; the temperature and pressure at which the three phases coexist is referred to as the triple point.

3.9 THE CLAPEYRON EQUATION

If two phases of a pure substance are in equilibrium with each other they have the same chemical potential at that temperature and pressure. When the temperature is changed at constant pressure, or the pressure is changed at constant temperature, one of the phases will disappear. However, if the temperature and pressure are both changed in such a way as to keep the two chemical potentials equal to each other the two phases will continue to coexist. The necessary relation for dP/dT was derived by Clapeyron.

If two phases α and β of a pure substance are in equilibrium the chemical potentials, or partial molar Gibbs free energies, are equal.

$$\bar{G}_\alpha = \bar{G}_\beta \tag{3.30}$$

If the pressure and temperature are changed so that equilibrium is maintained, it is necessary that

$$d\bar{G}_\alpha = d\bar{G}_\beta \tag{3.31}$$

Since $\bar{G}$ depends only on P and T, this equation may be written

$$\left(\frac{\partial \bar{G}_\alpha}{\partial P}\right)_T dP + \left(\frac{\partial \bar{G}_\alpha}{\partial T}\right)_P dT = \left(\frac{\partial \bar{G}_\beta}{\partial P}\right)_T dP + \left(\frac{\partial \bar{G}_\beta}{\partial T}\right)_P dT \tag{3.32}$$

Utilizing equation 2.64 and 2.65 this becomes

$$\bar{V}_\alpha \, dP - \bar{S}_\alpha \, dT = \bar{V}_\beta \, dP - \bar{S}_\beta \, dT \tag{3.33}$$

or

$$\frac{dP}{dT} = \frac{\bar{S}_\beta - \bar{S}_\alpha}{\bar{V}_\beta - \bar{V}_\alpha} = \frac{\Delta \bar{S}}{\Delta \bar{V}} = \frac{\Delta \bar{H}}{T \Delta \bar{V}} \tag{3.34}$$

This equation is referred to as the Clapeyron equation, and it may be applied to vaporization, sublimation, fusion, or the transition between two solid phases of a pure substance. The enthalpies of sublimation, fusion, and vaporization at a given temperature are related by

$$\Delta \bar{H}_{sub} = \Delta \bar{H}_{fus} + \Delta \bar{H}_{vap} \tag{3.35}$$

since the heat required to vaporize a given amount of the solid is the same whether this process is carried out directly or by first melting the solid and then vaporizing the liquid.

In using equation 3.34 it is necessary to express the enthalpy change of the process in the same units as the product of pressure and volume change. For this purpose it is useful to calculate the factor for converting calories to liter atmospheres: $(0.08205 \text{ liter atm K}^{-1} \text{ mol}^{-1})/(1.987 \text{ cal K}^{-1} \text{ mol}^{-1}) = 0.04129 \text{ liter atm cal}^{-1}$.

Example 3.1 What is the change in the boiling point of water at $100°$ per Torr change in atmospheric pressure? The heat of vaporization is 539.7 cal g^{-1}, the molar volume of liquid water is 18.78 cm³, and the molar volume of steam is 30.199 liters, all at $100°$ and 1 atm.

$$\frac{dP}{dT} = \frac{\Delta \bar{H}_{vap}}{T(\bar{V}_v - \bar{V}_l)}$$

$$= \frac{(539.7 \text{ cal g}^{-1})(18.02 \text{ g mol}^{-1})(0.04129 \text{ liter atm cal}^{-1})}{(373.1 \text{ K})(30.180 \text{ liter mol}^{-1})}$$

$$= 0.03566 \text{ atm K}^{-1}$$

$$= (0.03566 \text{ atm K}^{-1})(760 \text{ Torr atm}^{-1})$$

$$= 27.10 \text{ Torr K}^{-1}$$

Thus $dT/dP = 0.0369$ K Torr^{-1}

Example 3.2 Calculate the change in pressure required to change the freezing point of water $1°$. At $0°$ the heat of the fusion of ice is 79.7 cal g^{-1}, the density of water is 0.9998 g cm^{-3}, and the density of ice is 0.9168 cm^{-3}. The reciprocals of the densities, 1.0002 and 1.0908, are the volumes in cm³ of 1 g. The volume change upon freezing $(V_l - V_s)$ is therefore -9.06×10^{-5} liter g^{-1}. For small changes ΔH_{fus}, T, and $(V_l - V_s)$ are virtually constant, so that

$$\frac{\Delta P}{\Delta T} = \frac{\Delta H_{fus}}{T(V_l - V_s)}$$

$$= \frac{(79.7 \text{ cal g}^{-1})(0.04129 \text{ liter atm cal}^{-1})}{(273.1 \text{ K})(-9.06 \times 10^{-5} \text{ liter g}^{-1})}$$

$$= -133 \text{ atm K}^{-1}$$

In other words, a pressure of 133 atm is required to lower the freezing point of water $1°$; and the reciprocal,

$$\frac{\Delta T}{\Delta P} = \frac{-1}{133} = -0.0075 \text{ K atm}^{-1}$$

shows that an increase in pressure of 1 atm lowers the freezing point 0.0075 K. The negative sign indicates that an increase in pressure causes a decrease in temperature.

3.10 THE CLAUSIUS-CLAPEYRON EQUATION

For vaporization and sublimation Clausius showed how the Clapeyron equation may be simplified by assuming that the vapor obeys the ideal gas law and by neglecting the volume of a mole of liquid $\bar{V}_l$ in comparison with a mole of vapor $\bar{V}_v$. For example, with water at $100°$, $\bar{V}_v$ is 30.2 liters and $\bar{V}_l$ is 0.0188 liter. Substituting RT/P for $\bar{V}_v$, we have

$$\frac{dP}{dT} = \frac{\Delta \bar{H}_{vap}}{T\bar{V}_v} = \frac{P\Delta \bar{H}_{vap}}{RT^2} \tag{3.36}$$

This differential expression may be employed if the changes in temperature and pressure are small. For example, it is convenient to use the following form in correcting boiling points for barometric fluctuations:

$$\frac{\Delta T}{\Delta P} = \frac{RT^2}{\Delta \bar{H}_{vap} P} \tag{3.37}$$

On rearrangement equation 3.36 becomes

$$\frac{dP}{P} = d \ln P = \frac{\Delta \bar{H}_{vap}}{RT^2} dT \tag{3.38}$$

Integrating on the assumption that $\Delta \bar{H}_{vap}$ is independent of temperature and pressure yields

$$\int d \ln P = \frac{\Delta \bar{H}_{vap}}{R} \int T^{-2} \, dT \tag{3.39}$$

$$\ln P = -\frac{\Delta \bar{H}_{vap}}{RT} + C \tag{3.40}$$

$$\log P = \frac{-\Delta \bar{H}_{vap}}{2.303 RT} + C' \tag{3.41}$$

where C is the integration constant. This suggests that a plot of $\log P$ versus $1/T$ should be linear, and this is born out by data on both vaporization and sublimation, as shown in Fig. 3.7. Over a wide temperature range there are significant deviations from linearity because some of the approximations made in the above derivation do not apply. When logarithms to the base 10 are used, the slope is $-\Delta \bar{H}_{vap}/2.303R$. Thus the heat of vaporization can be calculated, using

$$\Delta \bar{H}_{vap}(\text{in cal mol}^{-1}) = -(\text{slope})(2.303)(1.987 \text{ cal K}^{-1} \text{ mol}^{-1}) \tag{3.42}$$

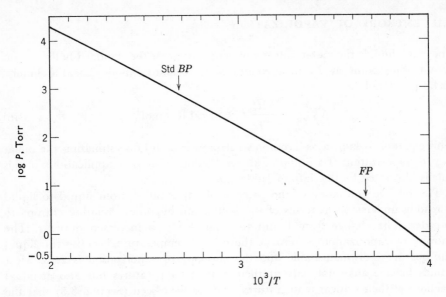

Fig. 3.7 Vapor pressure of water.

Frequently, it is more convenient to use the equation obtained by integrating between limits, P_2 at T_2 and P_1 at T_1, as follows:

$$\int_{P_1}^{P_2} d\ln P = \frac{\Delta \bar{H}_{vap}}{R} \int_{T_1}^{T_2} T^{-2}\, dT \qquad (3.43)$$

$$\ln P_2 - \ln P_1 = \frac{\Delta \bar{H}_{vap}}{R}\left[-\frac{1}{T_2} - \left(-\frac{1}{T_1}\right)\right] \qquad (3.44)$$

$$\log \frac{P_2}{P_1} = \frac{\Delta \bar{H}_{vap}(T_2 - T_1)}{2.303 R T_1 T_2} \qquad (3.45)$$

Using this equation, it is possible to calculate the heat of vaporization or the heat of sublimation from the vapor pressures at two different temperatures. Any units of pressure may be chosen as long as the same units are used for both pressures, and any units of heat may be used as long as $\Delta \bar{H}_{vap}$ and RT have the same units of energy.

Since equation 3.45 for calculating the heat of vaporization was derived on the assumption that the vapor is an ideal gas, the results obtained by its use are no more accurate than the calculations involving the equation $P\bar{V} = RT$.

Another approximation is involved in the assumption that the heat of vaporization is independent of temperature. Over wide temperature ranges, plots of $\log P$ versus $1/T$ are somewhat curved because $\Delta \bar{H}_{vap}$ varies with the temperature. It is possible to calculate the heat of vaporization at any particular temperature from the slope of the curve by drawing a tangent to the curve at that temperature.

3.11 ENTROPY OF VAPORIZATION

For many liquids the molar entropy of vaporization at the standard boiling point (the boiling point at 1 atm pressure) is a constant, about 21 cal K^{-1} mol^{-1} or 88 J K^{-1} mol^{-1}.

$$\Delta \bar{S}_{vap} = \frac{\Delta \bar{H}_{vap}}{T_b} \cong 21 \text{ cal } K^{-1} \text{ mol}^{-1} \qquad (3.46)$$

This equation is known as Trouton's rule and is useful for estimating the molar heat of vaporization of a liquid of known boiling point. The applicability of this rule is illustrated by the data of Table 3.3.

The relative constancy of the entropy of vaporization from liquid to liquid is readily understood in terms of the Boltzmann hypothesis relating entropy to disorder. The change from liquid to vapor leads to increased disorder. The entropy of vaporization is zero at the critical temperature because the liquid and gas are indistinguishable and the enthalpy of vaporization is zero. Most liquids behave alike not only at their critical temperatures but also at equal fractions of their critical temperatures, and we have seen (Section 3.3) that the standard boiling points of many liquids are roughly equal fractions of the critical temperatures. Hence, different liquids should have about the same entropy of

Table 3.3   Entropy of Vaporization at the Standard Boiling Point and Entropy of Fusion at the Freezing Point

Substance	Boiling Temperature °C at 760 Torr	$\Delta \bar{H}_{vap}$, kcal mol^{-1}	$\Delta \bar{S}_{vap}$, cal K^{-1} mol^{-1}	Melting Point, °C	$\Delta \bar{H}_{fus}$, kcal mol^{-1}	$\Delta \bar{S}_{fus}$, cal K^{-1} mol^{-1}
O_2	−182.97	1.630	18.07	−218.76	0.106	1.95
H_2	−252.77	0.216	10.6	−259.20	0.028	2.0
H_2O	100.00	9.7171	26.040	0	1.4363	5.2581
He	−268.944	0.020	4.7	−269.7	0.005	1.5
Cl_2	−34.06	4.878	20.40	−101.00	1.531	8.89
SO_2	−10.02	5.955	22.63	−75.48	1.769	8.95
NH_3	−33.43	5.581	23.28	−77.76	1.351	6.914
CH_3OH	64.7	8.43	24.95	−97.90	0.757	4.32
CCl_4	76.7	7.17	20.5	−22.9	0.60	2.4
C_2H_5OH	78.5	9.22	26.22	−114.6	1.200	7.57
PbI_2	872	24.8	21.7	412	5.2	7.6
Methane, CH_4	−161.49	1.955	17.51	−182.48	0.225	2.48
Ethane, C_2H_6	−88.63	3.517	19.06	−183.27	0.6834	7.603
n-Butane, C_4H_{10}	−0.50	5.352	19.63	−138.350	1.114	8.263
Benzene, C_6H_6	80.10	7.353	20.81	5.533	2.351	8.436
Acetic acid, CH_3CO_2H	118.3	5.82	14.8	16.61	2.80	9.66

vaporization at their boiling point, provided that there is no association or dissociation upon vaporization. For substances like water and alcohols which form hydrogen bonds (Section 14.8) the entropy of vaporization is greater than 21 cal K^{-1} mol^{-1}. Hydrogen and helium, which boil at only a little above absolute zero, might well be expected to show large departures from this rule. Acetic acid and carboxylic acids, in general, have abnormally low heats of vaporization, since the vapor consists of double molecules and still more energy would be required to break them up into single molecules comparable with those of other gases.

For nonpolar liquids Trouton's rule provides a means of estimating the vapor pressure at a given temperature, provided that the standard boiling point is known.

Example 3.3   The normal boiling point of *n*-hexane is 69.0°. Estimate (*a*) its molar heat of vaporization and (*b*) its vapor pressure at 60°.

$$(a)\ \Delta \bar{H}_{vap} \cong (21\ \text{cal K}^{-1}\ \text{mol}^{-1})(342\ \text{K}) = 7190\ \text{cal mol}^{-1}$$

The experimental value is 6896 cal mol^{-1}
  Using equation 3.45

$$(b)\ \log \frac{760}{P_1} = \frac{(7190\ \text{cal mol}^{-1})(9\ \text{K})}{(2.303)(1.987\ \text{cal K}^{-1}\ \text{mol}^{-1})(333\ \text{K})(342\ \text{K})}$$

$$P_1 = 571\ \text{Torr}$$

The experimental value is 555.9 Torr.

The entropy of fusion is not so constant from substance to substance as the entropy of vaporization. The enthalpies and entropies of fusion of a number of substances at their melting points are given in Table 3.3. The entropies of fusion of elongated molecules are especially large. This is connected with the fact that there is a large increase in the number of configurations and motions when such a molecule passes from the crystal into solution, provided that the elongated molecules do not rotate freely in the crystalline form.

References

W. E. Addison, *The Allotropy of the Elements*, American Elsevier, New York, 1968.

K. Denbigh, *The Principles of Chemical Equilibrium*, Cambridge University Press, Cambridge 1971.

R. R. Dreisbach, *Physical Properties of Chemical Compounds*, American Chemical Society, Washington D.C., 1955.

T. E. Jordan, *Vapor Pressure of Organic Compounds*, Interscience Publishers, New York, 1954.

W. N. Lacey and B. H. Sage, *Thermodynamics of One-Component Systems*, Academic Press, New York, 1957.

J. Timmermans, *Physico-Chemical Constants of Pure Organic Compounds*, Elsevier Publishing Co., New York, 1956.

Problems

3.1 The density of ammonia was determined at various pressures by weighing the gas in large glass bulbs. The values of the density in grams per liter at $0°$ were as follows: 0.77169 at 1 atm, 0.51185 at $\frac{2}{3}$ atm, 0.38293 at $\frac{1}{2}$ atm, 0.25461 at $\frac{1}{3}$ atm. (a) What is the molecular weight of ammonia? (b) If the atomic weight of hydrogen is taken as 1.008, what is the atomic weight of nitrogen?   *Ans.* (a) 17.030, (b) 14.006.

3.2 What is the pressure in newtons per square meter exerted by a column of mercury 76 cm high at a location where the acceleration of gravity is 9.80665 m s^{-2}? The density of mercury is 13.5951 g cm^{-3} at 0°C.   *Ans.* 101,325 N m^{-2}.

3.3 The critical temperature of carbon tetrachloride is 283.1°. The densities in grams per cm^3 of the liquid ρ_l and vapor ρ_v at different temperatures are as follows:

t	100°	150°	200°	250°	270°	280°
ρ_l	1.4343	1.3215	1.1888	0.9980	0.8666	0.7634
ρ_v	0.0103	0.0304	0.0742	0.1754	0.2710	0.3597

What is the critical molar volume of CCl_4? It is found that the mean of the densities of the liquid and vapor does not vary rapidly with temperature and can be represented by

$$\frac{\rho_l + \rho_v}{2} = AT + B$$

where A and B are constants. The extrapolated value of the average density at the critical temperature is the critical density. The molar volume $\bar{V}_c$ at the critical point is equal to the molecular weight divided by the critical density.   *Ans.* 276 cm^3 mol^{-1}.

3.4 Using van der Waals' equation, calculate the pressure exerted by 1 mol of carbon dioxide at $0°$ in a volume of (a) 1.00 liter, (b) 0.05 liter. (c) Repeat the calculations at $100°$ and 0.05 liter.   *Ans.* (a) 19.82, (b) 1621, (c) 2739 atm.

3.5 Calculate the molar volume of methane at $0°$ and 50 atm using (a) the ideal gas law and (b) van der Waals' equation. In the calculation in part (b), the solution of a cubic equation may be avoided by use of a successive approximation method. The van der Waals equation is written as follows:

$$\bar{V} = \frac{RT}{P + (a/\bar{V}^2)} + b$$

The value of $\bar{V}$ obtained from the ideal gas law is substituted in the right-hand side of this equation, and an approximate value of $\bar{V}$ is calculated. This value is then substituted into the right-hand side of the equation to obtain a still more accurate value of $\bar{V}$. This process is continued until the calculated value of $\bar{V}$ is essentially the same as that substituted in the right-hand side of the equation.   *Ans.* (a) 0.448, (b) 0.391 liter mol^{-1}.

3.6 What is the maximum number of phases that can be in equilibrium at constant temperature and pressure in one-, two-, and three-component systems.   *Ans.* $p = 3, 4, 5$.

3.7 In the *gas-saturation method* for measuring the vapor pressure a measured current of dry air or other gas is bubbled slowly through a weighed amount of the liquid or solid whose vapor pressure is to be determined, so that the gas stream becomes saturated. The liquid or solid is maintained at constant temperature, and its loss in weight is measured, or the vapor may be removed from the gas stream in an absorption tube or cold trap and weighed. Show that if a volume v' of inert gas (measured before saturation and at barometric pressure P) picks up g grams of vapor of molecular weight M, the vapor pressure p of the liquid is given by

$$p = \frac{gRT/Mv'}{1 + (gRT/Mv'P)}$$

3.8 Ten liters of air was bubbled through carbon tetrachloride at 20°. The loss in weight of the liquid was 8.698 g. Calculate the vapor pressure of CCl_4. (See problem 3.7.)

Ans. 0.120 atm.

3.9 Liquid mercury has a density of 13.690 g cm^{-3}, and solid mercury has a density of 14.193 g cm^{-3}, both being measured at the melting point, $-38.87°$, under 1 atm pressure. The heat of fusion is 2.33 cal g^{-1}. Calculate the melting points of mercury under a pressure of (a) 10 atm and (b) 3540 atm. The observed melting point under 3540 atm is $-19.9°$.

Ans. (a) $-38.81°$, (b) $-16°$.

3.10 The heats of vaporization and of fusion of water are 595 cal g^{-1} and 79.7 cal g^{-1} at 0°. The vapor pressure of water at 0° is 4.58 Torr. Calculate the sublimation pressure of ice at $-15°$, assuming that the enthalpy changes are independent of temperature.

Ans. 1.245 Torr.

3.11 Propene has the following vapor pressures:

T, K	150	200	250	300
P, Torr	3.82	198.0	2074	10,040

From these data calculate (a) the heat of vaporization and (b) the vapor pressure at 225 K by a graphical method. *Ans.* (a) 4670 cal mol^{-1}, (b) 741 Torr.

3.12 n-Propyl alcohol has the following vapor pressures:

t, °C	40	60	80	100
P, Torr	50.2	147.0	376	842.5

Plot these data so as to obtain a nearly straight line, and calculate (a) the heat of vaporization and (b) the boiling point at 760 Torr. *Ans.* (a) 10.7 kcal mol^{-1}, (b) 98°.

3.13 The heat of vaporization of ether is 88.39 cal g^{-1} at its boiling point, 34.5°. (a) Calculate the rate of change of vapor pressure with temperature dP/dT, at the boiling point. (b) What is the boiling point at 750 Torr? (c) Estimate the vapor pressure at 36.0°.

Ans. (a) 26.5 Torr K^{-1}, (b) 34.1°, (c) 800 Torr.

3.14 For uranium hexafluoride the vapor pressures (in Torr) for the solid and liquid are given by

$$\log P_s = 10.648 - 2559.5/T$$

$$\log P_l = 7.540 - 1511.3/T$$

Calculate the temperature and pressure of the triple point. *Ans.* 64°, 1122 Torr.

3.15 Trouton's rule is useful in estimating the vapor pressure of a substance for which only the standard boiling point is known. For example, at what temperature would you expect aniline to boil in a vacuum still at 20 Torr pressure? The standard boiling point is 185°.

Ans. 68°.

3.16 What mass may be supported by a balloon containing 1000 liters of helium at 25° and 1 atm pressure? The average molecular weight of air may be taken to be 28.8.

3.17 The coefficient of thermal expansivity α is defined by

$$\alpha = \frac{1}{V}\left(\frac{\partial V}{\partial T}\right)_P$$

and the isothermal compressibility κ is defined by

$$\kappa = -\frac{1}{V}\left(\frac{\partial V}{\partial P}\right)_T$$

Calculate these quantities for an ideal gas.

3.18 Show that to a first approximation the equation of state of a gas which dimerizes to a small extent is given by

$$\frac{P\bar{V}}{RT} = 1 - \frac{K_c}{\bar{V}}$$

where K_c is the equilibrium constant for the formation of dimer.

$$2A = A_2 \qquad K_c = \frac{(A_2)}{(A)^2}$$

3.19 The densities of liquid and vapor methyl ether in grams per cubic centimeter at various temperatures are as follows:

°C	30	50	70	100	120
ρ_l	0.6455	0.6116	0.5735	0.4950	0.4040
ρ_v	0.0142	0.0241	0.0385	0.0810	0.1465

Calculate the critical density and temperature. (See problem 3.3.)

3.20 Calculate the number of grams of hydrogen in a vessel of 500 cm³ capacity when hydrogen is forced in at 100 atm at 200°, using (a) simple gas laws and (b) van der Waals' equation.

3.21 Ice has the unusual property of a melting point that is lowered by increasing pressure. Thus, one can skate on ice provided that the pressure exerted by one's skates is great enough to liquefy the ice under them.

  Would a 75-kg man whose skates contact the ground with an area of 0.1 cm² be able to skate at −3° C?

3.22 From the critical temperature and critical pressure of helium given in Table 3.1, calculate van der Waals' constants.

3.23 A room 5 by 10 by 4 m is filled with air containing some water vapor. The temperature is 20°, and the relative humidity is 60%. The vapor pressure or water at 20° is 17.36 Torr. (The relative humidity is the partial pressure of water vapor divided by the vapor pressure of liquid water at that temperature.) How many grams of water are contained in the air of this room?

3.24 Calculate the loss in weight of a sample of water held at 25° when 10 liters of dry air is bubbled through slowly. The vapor pressure of water at this temperature is 23.76 Torr. (See problem 3.7.)

3.25 The vapor pressure of toluene is 60 Torr at 40.3° and 20 Torr at 18.4°. Calculate (a) the heat of vaporization and (b) the vapor pressure at 25°.

3.26 If $\Delta\bar{C}_P = \bar{C}_{P,\text{vap}} - \bar{C}_{P,\text{liq}}$ is independent of temperature, then

$$\Delta\bar{H}_{\text{vap}} = \Delta\bar{H}_{0,\text{vap}} + T\Delta\bar{C}_P$$

where $\Delta\bar{H}_{0,\text{vap}}$ is the hypothetical enthalpy of vaporization at absolute zero. Since $\Delta\bar{C}_P$ is negative, $\Delta\bar{H}_{\text{vap}}$ decreases as the temperature increases. Show that if the vapor is an ideal gas the vapor pressure is given as a function of temperature by

$$\log P = \frac{-\Delta\bar{H}_{0,\text{vap}}}{2.303RT} + \frac{\Delta\bar{C}_P}{R}\log T + \text{constant}$$

3.27 The vapor pressure of 2,2-dimethyl-1-butanol is given by the expression

$$\log P = \frac{-4849.3}{T} - 14.701 \log T + 53.1187$$

where P is expressed in Torr. Calculate the heat of vaporization (a) at $25°$, and (b) at the boiling point, $136.7°$. (See problem 3.26.)

3.28 The vapor pressure of bromobenzene is 1 Torr at $2.9°$ and 20 Torr at $53.8°$. Calculate the boiling point. What are the possible reasons for the difference from the experimental value of $156.2°$?

3.29 Estimate the vapor pressure of ice at the temperature of solid carbon dioxide ($-78°$ at 1 atm pressure of CO_2), assuming that the heat of sublimation is constant. The heat of sublimation of ice is 676 cal g^{-1}, and the vapor pressure of ice is 4.58 Torr at $0°$.

3.30 The sublimation pressure of solid CO_2 is 1 Torr at $-134.3°$ and 20 Torr at $-114.4°$. Calculate the heat of sublimation.

3.31 The vapor pressure of solid benzene, C_6H_6, is 2.24 Torr at $-30°$ and 24.5 Torr at $0°$, and the vapor pressure of liquid C_6H_6 is 46.3 Torr at $10°$ and 118.5 Torr at $30°$. From these data, calculate (a) the triple point of C_6H_6 and (b) the heat of fusion of C_6H_6.

3.32 The boiling point of n-butyl chloride is $77.96°$. (a) Using this as the only experimental datum available, estimate the vapor pressure at $50°$ and compare this estimate with that calculated from the empirical equation (in which P is expressed in Torr):

$$\log P = -\frac{1763}{T} + 7.912$$

(b) Calculate the heat of vaporization per gram from this equation.

3.33 Oil diffusion pumps may be used to attain pressures of 10^{-6} Torr rather readily. If the gas present is nitrogen and the temperature $25°$ calculate the density in grams per liter.

3.34 Ordinary carbon contains 98.9% of an isotope with an atomic weight of 12.00 and 1.1% of an isotope with an atomic weight of 13.00. By a certain physical-chemical operation the concentration of this isotope is increased from 1.1 to approximately 2.0%. It is planned to determine the exact percentage of this heavier isotope by measuring the density of carbon dioxide with a gas-density balance. If this concentration of 2% must be known to 1 part in 500 for the experiment planned, how accurately in parts per million must the gas density of the CO_2 be measured? To what fraction of a degree must the temperature be known at about $25°$ and to what fraction of a Torr must the pressure be known at about 740 Torr to achieve this accuracy?

3.35 Calculate the critical temperature and volume of hydrogen from the following densities (in grams per cubic centimeter) of liquid ρ_l and vapor ρ_v.

°C	-246	-244	-242	-241
ρ_l	0.061	0.057	0.051	0.047
ρ_v	0.0065	0.0095	0.014	0.017

(See problem 3.3.)

3.36 Calculate the pressure exerted by 1 mol of carbon dioxide in $\frac{1}{2}$ liter at $25°$, using (a) the ideal gas law and (b) van der Waals' equation.

3.37 Show that the entropy of a van der Waals gas can be expressed as

$$S(V, T) = nR \ln (V - nb) + f(T)$$

where $f(T)$ is a function of T only (in other words f is independent of V). Why—physically—is the volume contribution to S less than that for an ideal gas?

3.38 (a) How many tons of water can be evaporated from a square mile of moist land on a clear summer day, if it is assumed that the limiting factor is the supply of solar heat, which amounts to about 1.0 cal min^{-1} cm^{-2} for an 8-hr day? The temperature is 25°, the vapor pressure of water is 23.7 Torr, and the heat of vaporization is 582 cal g^{-1}. One mile = 1.609 km. (b) How many liters of air are required to hold this much water if the temperature is 25°?

3.39 Twenty liters of dry air are bubbled slowly through liquid CCl$_4$, which is held at 30.0°. The barometric pressure is 760 Torr, and 28.6 g of CCl$_4$ ($M = 153.8$) are vaporized. Calculate the vapor pressure of CCl$_4$. (See problem 3.7.)

3.40 A block of ice is placed in a lake of pure water, forced 100 ft below the surface, and maintained in a quiet, steady position. What will be the temperature of the surface of the ice? Assume that the densities of the ice and water are not changed by the pressure and are 0.9106 and 1.000 g cm^{-3}, respectively.

3.41 The heat of vaporization of water at 0° is 10,720 cal mol^{-1}, and the heat of sublimation of ice is 12,120 cal mol^{-1}. Given the fact that the triple point is at 0.0099° and a pressure of 4.58 Torr, calculate the vapor pressure of water at 15° and of ice at $-15°$ using the Clausius-Clapeyron equation. These three points are used to construct a plot of log P versus $1/T$. The fusion curve may be drawn in as a vertical line above the triple point, since the effect of pressures in this range is negligible. The regions in the diagram are then labeled.

3.42 The standard boiling point of cyclohexane is 80.7°, and the heat of vaporization is 7.19 kcal mol^{-1}. Calculate (a) the boiling point at 650 Torr pressure and (b) the vapor pressure at 25°.

3.43 What is the boiling point of water on a mountain where the barometer reading is 660 Torr? The heat of vaporization of water may be taken to be 9.72 kcal mol^{-1}.

3.44 At 0° ice absorbs 79.7 cal g^{-1} in melting; water absorbs 595 cal g^{-1} in vaporizing. (a) What is the heat of sublimation of ice at this temperature? (b) At 0° the vapor pressure of both ice and water is 4.58 Torr. What is the rate of change of vapor pressure with temperature dP/dT for ice and liquid water at this temperature? (c) Estimate the vapor pressures of ice and of liquid water at $-5°$.

3.45 (a) Calculate the vapor pressure of water at 200° using the Clausius-Clapeyron equation and assuming that $\Delta \bar{H}_{vap}$ is 9.7171 kcal mol^{-1} independent of temperature. (b) Calculate the vapor pressure, allowing for the fact that $\Delta \bar{C}_P = -10$ cal K^{-1} mol^{-1}. The directly measured value is 11661.2 Torr. (See problem 3.26.)

3.46 Using data from Table 3.1, calculate the sublimation pressure of lead iodide at 350°.

3.47 The sublimation pressures of solid Cl$_2$ are 2.64 Torr at $-112°$ and 0.26 Torr at $-126.5°$. The vapor pressures of liquid Cl$_2$ are 11.9 Torr at $-100°$ and 58.7 Torr at $-80°$. Calculate (a) $\Delta \bar{H}_{sub}$, (b) $\Delta \bar{H}_{vap}$, (c) $\Delta \bar{H}_{fus}$, and (d) the triple point.

3.48 Show that the observed boiling point T_{obs}, K, at a barometric pressure of P Torr may be corrected approximately to the standard boiling point T_{std}, K, at 760 Torr by the following equation: $T_{std} = T_{obs} + (T_{obs}/8000)(760 - P)$

3.49 Show that the effect of external hydrostatic pressure P on the vapor pressure p of a liquid of partial molar volume $\bar{V}_l$ is given by

$$RT \frac{d \ln p}{dP} = \bar{V}_l$$

provided the vapor behaves as an ideal gas. *Hint:* When the pressure is changed at constant temperature, $dG_l = dG_{vap}$ so that $\bar{V}_{gas} \, dp = \bar{V}_l \, dP$.

CHAPTER 4

PHASE EQUILIBRIA

Now that we have had an introduction to phase equilibria with one-component systems we discuss the phase behavior of systems containing two or more components. We will start with ideal binary liquid solutions and will find that the thermodynamic quantities for mixing are the same as we found earlier for ideal gas solutions. The equation for the lowering of the vapor pressure by a nonvolatile solute in an ideal solution is derived, and osmotic pressure is discussed. Nonideal solutions are then considered, and the calculation of activity coefficients is presented in detail.

In the second part of the chapter, phase phenomena involving the separation of solid phases is discussed. This includes the formation of congruently and incongruently melting compounds, the formation of solid solutions, and gas-solid equilibria.

4.1 BINARY SYSTEMS

The phase behavior of a binary system may be represented by a three-dimensional diagram of temperature, pressure, and mole fraction. Such a diagram is shown in Fig. 4.1 for a binary system in the region where only vapor and a single liquid phase exist. The complete diagram for a binary system would include the region where there is equilibrium between solid and liquid phases as well and perhaps a region of partial miscibility of the two liquids. The right-hand side of Fig. 4.1 is simply a plot of the vapor pressure of toluene versus temperature, and the left-hand side is a plot of vapor pressure of benzene versus temperature. The upper curved surface inside the box gives the total vapor pressure as a function of temperature and mole fraction of toluene in the liquid; it is labeled l. The lower curved surface, which is mostly hidden and has dashed lines drawn on it at 0.2, 0.4, 0.6, 0.8 mole fraction toluene, gives the compositions of the vapor phases in equilibrium with the liquid phases; it is therefore labeled v. Under conditions represented by points above the l surface, only the liquid phase exists. Under conditions represented by points below the v surface, only the vapor phase exists. Under conditions represented by points in the bladelike region between the surface, vapor and liquid phases coexist.

The phase rule for a binary system is $v = 2 - p + 2 = 4 - p$ if T and P are variables, $v = 2 - p + 1 = 3 - p$ if only P or T is variable, and $v = 2 - p$

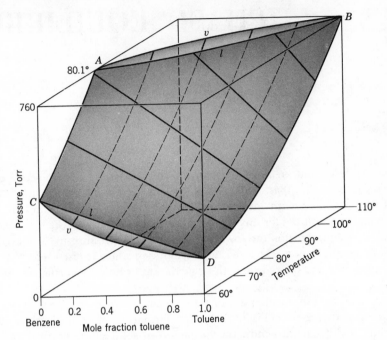

Fig. 4.1 Three-dimensional phase diagram for the benzene-toluene system.

if neither P nor T is variable. The variances v may therefore be summarized as follows:

Variance v of a Binary System

No. of Phases	T and P Variable	T or P Variable	T and P Constant
1	3	2	1
2	2	1	0
3	1	0	—
4	0	—	—

Thus, considering the system represented in Fig. 4.1 under conditions such that there is a single phase, the system is trivariant; and the temperature, pressure, and mole fraction, X, of one of the components must be specified to describe the system completely. If two phases are present, as will be the case between surfaces in a P–T–X plot, only two variables need be specified to define the system. For example, if the pressure and temperature are specified, the compositions of the liquid and vapor phases are given by the phase diagram. The relative amounts of the two phases are not fixed by only specifying the temperature and pressure, but the phase rule is not concerned with the relative amounts of phases. Since two variables have to be specified we say the system is bivariant. If three phases are present, the system is univariant. If four phases are present, the system is invariant, that is, there is only one temperature, pressure, and composition at which the four phases can exist together in equilibrium in a two-component system.

To simplify the description of binary systems we will make extensive use of sections through the complete three-dimensional diagram. We will first consider sections at constant temperature, for example, the front face of Fig. 4.1.

4.2 VAPOR PRESSURE OF AN IDEAL SOLUTION

The vapor pressures of benzene and toluene over binary solutions of these two components at constant temperature are shown in Fig. 4.2a. The total vapor pressure is the sum of these two partial pressures.

It is evident that the vapor pressure of benzene over solutions of benzene and toluene is directly proportional to the mole fraction of benzene in the solution, and the proportionality constant is the vapor pressure of pure benzene. A similar statement may be made about the vapor pressure of toluene. This generalization, discovered by Raoult in 1884, is referred to as Raoult's law.

The components of a binary solution may be referred to as 1 and 2, so that p_1 represents the partial pressure of component 1 above the solution and p_2 the partial pressure of component 2. Raoult's law may be written

$$p_1 = X_1 p_1^\circ \tag{4.1}$$

$$p_2 = X_2 p_2^\circ \tag{4.2}$$

where p_1° and p_2° are the vapor pressures of pure component 1 and pure component 2, respectively, at the equilibrium temperature. Actually air is present, and the total pressure is 1 atm. The presence of the air does not make a significant difference, but the vapor pressure of a liquid does depend slightly on the applied pressure (Problem 3.49).

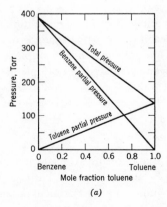
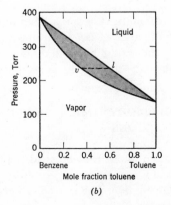

Fig. 4.2 Benzene-toluene at 60°. (a) Partial and total pressures. (b) Liquid and vapor compositions.

Solutions that obey Raoult's law are referred to as *ideal* solutions. It will be evident from later data that even for nonideal solutions Raoult's law always applies to a component in the limit as its mole fraction approaches 1.

The mole fraction of a component in the vapor is equal to its pressure fraction in the vapor. Since for ideal solutions the partial pressures of the components in the vapor may be readily calculated using equations 4.1 and 4.2, the mole fraction of a component in the vapor may be calculated using

$$X_{1,\text{vap}} = \frac{p_1}{p_1 + p_2} = \frac{X_1 p_1^\circ}{X_1 p_1^\circ + X_2 p_2^\circ} = \frac{X_1 p_1^\circ}{X_1(p_1^\circ - p_2^\circ) + p_2^\circ} \tag{4.3}$$

Example 4.1 The vapor pressures of pure benzene and toluene at 60° are 385 and 139 Torr, respectively. Calculate the partial pressures of benzene and toluene, the total vapor pressure of the solution, and the mole fraction of toluene in the vapor above a solution with 0.60 mole fraction toluene.

$$p_{\text{benzene}} = (0.40)(385 \text{ Torr}) = 154.0 \text{ Torr}$$
$$p_{\text{toluene}} = (0.60)(139 \text{ Torr}) = \underline{83.4 \text{ Torr}}$$
$$p_{\text{total}} = 237.4 \text{ Torr}$$

$$X_{\text{toluene,vap}} = \frac{83.4}{237.4} = 0.351$$

In Fig. 4.2b the total pressure and the composition of the vapor are plotted versus the mole fraction of toluene. This figure is identical with the front face of Fig. 4.1. It is noted that the top or "liquid" line in Fig. 4.2b is the same as in Fig. 4.2a. The lower curve is obtained by plotting the total pressure of the vapor along the vertical axis and the mole fraction of the vapor along the horizontal axis. The vapor curve gives the pressures at which vapors of a given composition first form a liquid phase when the pressure is increased.

At pressures and compositions above the shaded region the system is a liquid; at pressures and compositions below the shaded region the system is a vapor. In the shaded region the system is a mixture of liquid and vapor. The compositions of the liquid and vapor phases in equilibrium with each other in the shaded region are given by the ends of horizontal lines, called *tie lines*, of which one example is given. The point labeled *v* on the vapor curve has been calculated in example 4.1.

4.3 THERMODYNAMICS OF IDEAL SOLUTIONS

Information about the chemical potential of a component in a solution may be obtained by measuring the equilibrium pressure of the vapor of that component above the solution. This is possible because at equilibrium the chemical potential of a component *i* is the same in the vapor phase and the liquid phase.

$$\mu_{i,\text{soln}} = \mu_{i,\text{vap}} \tag{4.4}$$

For an ideal gaseous mixture the chemical potential of component i in the gas phase is given by equation 5.6,

$$\mu_{i,\text{vap}} = \mu_i^\circ + RT \ln p_i \tag{4.5}$$

where p_i is the partial pressure of i, and μ_i° is a function of temperature only and is the chemical potential of i when $p_i = 1$ atm.

If the solution is ideal, Raoult's law is obeyed and the chemical potential of the ith component in the vapor may be expressed in terms of its mole fraction X_i in the liquid by substituting Raoult's law into equation 4.5.

$$\mu_{i,\text{vap}} = \mu_i^\circ + RT \ln X_i p_i^\circ = \mu_i^\circ + RT \ln p_i^\circ + RT \ln X_i \tag{4.6}$$

$$= \mu_i^* + RT \ln X_i \tag{4.7}$$

where $\mu_i^* = \mu_i^\circ + RT \ln p_i^\circ$ is a constant at a given temperature and total pressure. Substituting equation 4.7 in equation 4.4

$$\mu_{i,\text{soln}} = \mu_i^* + RT \ln X_i \tag{4.8}$$

From this equation it may be seen that μ_i^* is the chemical potential of component i when $X_i = 1$, that is for the pure component. This equation may be taken as the definition of an ideal solution.

The thermodynamic treatment of ideal liquid solutions is very much like that for ideal gaseous solutions given in Section 5.2, and the same equations apply to ideal solid solutions. As in the case of gases we can use equation 2.97 to calculate the Gibbs free energy of two liquids before and after they are mixed to form an ideal solution. The Gibbs free energy of n_1 moles of liquid 1 and n_2 moles of liquid 2 before mixing is

$$G_{\text{initial}} = n_1 \mu_1^* + n_2 \mu_2^* \tag{4.9}$$

After the liquids have been mixed to form an ideal solution, the chemical potentials of the components are given by equation 4.8, and the Gibbs free energy of the solution is given by

$$G_{\text{final}} = n_1 \mu_1^* + n_2 \mu_2^* + RT(n_1 \ln X_1 + n_2 \ln X_2) \tag{4.10}$$

Thus the Gibbs free energy of mixing is

$$\Delta G_{\text{mix}} = G_{\text{final}} - G_{\text{initial}} = RT(n_1 \ln X_1 + n_2 \ln X_2)$$

$$= nRT(X_1 \ln X_1 + X_2 \ln X_2) \tag{4.11}$$

where $n = n_1 + n_2$. This equation is the same as equation 5.14 for the mixing of ideal gases, and so the plot of ΔG_{mix} for the mixing of two liquids to give an ideal solution is the same as that given in Fig. 5.3 for the mixing of two ideal gases.

The entropy of mixing of two liquids to form an ideal solution may be obtained, as we did with gases, by use of equation 2.64.

$$\Delta S_{\text{mix}} = -nR(X_1 \ln X_1 + X_2 \ln X_2) \tag{4.12}$$

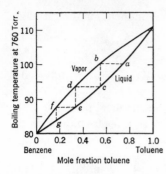

Fig. 4.3 Benzene-toluene boiling points; liquid and vapor compositions. The liquid boils at the temperature given by the lower curve.

Thus the plot given in Fig. 5.3 for ΔS_{mix} applies to the mixing of two liquids to form an ideal solution as well as to the mixing of two ideal gases.

It may similarly be shown that there is no volume change on mixing to form an ideal solution ($\Delta V_{mix} = 0$), and no heat is evolved or absorbed ($\Delta H_{mix} = 0$).

4.4 BOILING-POINT DIAGRAMS
OF BINARY SOLUTIONS

The preceding discussions have been concerned with vapor-pressure isotherms. We now consider plots of boiling temperature at a given pressure versus mole fraction.

A solution boils when the sum of the partial pressures of the components becomes equal to the applied pressure. The boiling points of benzene-toluene solutions are given by the lower line in Fig. 4.3.

The relation between the vapor-pressure diagram for benzene-toluene (Fig. 4.2b) and the boiling-point diagram (Fig. 4.3) is shown in Fig. 4.1 as a three-dimensional diagram. The front face of this diagram is identical with Fig. 4.2b and the top face with Fig. 4.3. The vapor curve always lies below the liquid curve in plots of vapor pressure versus composition, and above the liquid curve in plots of boiling point versus composition.

The boiling-point diagram can be calculated for two liquids that form ideal solutions, provided the vapor pressures are known for the two pure liquids at temperatures between their boiling points. This is illustrated in example 4.2.

Example 4.2   Calculate the composition of the benzene-toluene solution that will boil at 1 atm pressure at 90°, assuming that the solution is ideal. Also calculate the vapor composition. At 90° benzene has a vapor pressure of 1022 Torr, and toluene has a vapor pressure of 406 Torr. The mole fraction of benzene in the liquid that will boil at 90° is obtained from

$$760 = 1022 X_B + 406(1 - X_B)$$

$$X_B = 0.574$$

The mole fraction of benzene in the vapor is equal to its pressure fraction in the vapor, which is given by

$$X_{B,\text{vap}} = \frac{1022 X_B}{760} = 0.772$$

Other points on the liquid and vapor curves in Fig. 4.1 may be calculated in the same way. For nonideal solutions the points have to be obtained experimentally.

4.5 FRACTIONAL DISTILLATION

When a binary solution is partially vaporized, the component that has the higher vapor pressure is concentrated in the vapor phase, thus producing a difference in composition between the liquid and the equilibrium vapor. This vapor may be condensed, and the vapor obtained by partially vaporizing this condensate is still further enriched in the more volatile component. In *fractional distillation* this process of successive vaporization and condensation is carried out in a fractionating column. Figure 4.3 shows that a solution of 0.75 mole fraction toluene and 0.25 mole fraction benzene boils at 100° under 1 atm pressure as indicated by point *a*. The equilibrium vapor is richer in the more volatile compound, benzene, and has the composition *b*. This vapor may be condensed by lowering the temperature along the line *bc*. If a small fraction of this condensed liquid is vaporized, the first vapor formed will have the composition corresponding to *d*. This process of vaporization and condensation may be repeated many times, with the result that a vapor fraction rich in benzene is obtained.

Each vaporization and condensation represented by the line *abcde* corresponds to an idealized process in that only a small fraction of the vapor is condensed and only a small fraction of the condensate is revaporized. It is more practical to effect the separation by means of a distillation column, such as the bubble-cap column illustrated in Fig. 4.4.

Each layer of liquid on the plates of the column is equivalent to the boiling liquid in a distilling flask, and the liquid on the plate above it is equivalent to the condenser. The vapor passes upward through the bubble caps, where it is partially condensed in the liquid and mixed with it. Part of the resulting solution is vaporized in this process and is condensed in the next higher layer, while part of the liquid overflows and runs down the tube to the next lower plate. In this way there is a continuous flow of redistilled vapor coming out the top and a continuous flow of recondensed liquid returning to the boiler at the bottom. To make up for this loss of material from the distilling column, fresh solution is fed into the column, usually at the middle. The column is either well insulated or surrounded by a controlled heating jacket so that there will not be too much condensation on the walls. The whole system reaches a steady state in which the composition of the solution on each plate remains unchanged as long as the composition of the liquid in the distilling pot remains unchanged.

A distillation column may alternatively be packed with material that provides efficient contact between liquid and vapor and occupies only a small volume so that there is free space to permit a large throughput of vapor. Helices of glass,

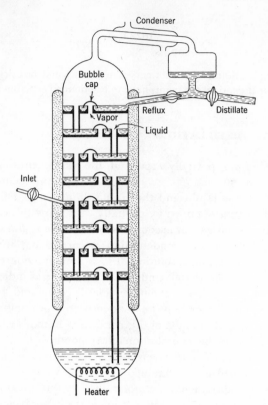

Fig. 4.4 Bubble-cap fractionating column.

spirals of screen, and different types of packing* are used with varying degrees of efficiency.

The efficiency of a column is expressed in terms of the equivalent number of theoretical plates. The number of *theoretical plates* in a column is equal to the number of successive infinitesimal vaporizations at equilibrium required to give the separation which is actually achieved. The number of theoretical plates depends somewhat on the reflux ratio, the ratio of the rate of return of liquid to the top of the column to the rate of distilling liquid off. The number of theoretical plates in a distillation column under actual operating conditions may be obtained by counting the number of equilibrium vaporizations required to achieve the separation actually obtained with the column.

Suppose that in distilling a solution of benzene and toluene with a certain distillation column it is found that distillate of composition g is obtained when the composition of the liquid in the boiler is given by a, in Fig. 4.3. Such a distillation is

---

* F. Daniels, J. W. Williams, P. Bender, R. A. Alberty, C. D. Cornwell, and J. E. Harriman, *Experimental Physical Chemistry*, McGraw-Hill Book Co., New York, 1970, T. P. Carney, *Laboratory Fractional Distillation*, The Macmillan Co., New York, 1949.

equivalent to three simple vaporizations and condensations, as indicated by steps *abc*, *cde*, and *efg*. Since the distilling pot itself corresponds to one theoretical plate, the column has two theoretical plates.

4.6 LOWERING OF THE VAPOR PRESSURE BY A NONVOLATILE SOLUTE

If the solute has a negligibly small vapor pressure, the vapor pressure of the solution is simply that of the solvent. For ideal solutions the vapor pressure of the solvent is given by Raoult's law

$$p_1 = X_1 p_1^\circ = (1 - X_2)p_1^\circ \tag{4.13}$$

where subscript 1 refers to the solvent and the subscript 2 to the nonvolatile solute. For nonideal solutions this equation applies when X_2, the mole fraction of the solute, is sufficiently small so that the effect of the solute on the character of the solvent is negligible. Equation 4.13 may be rearranged to give

$$\frac{p_1^\circ - p_1}{p_1^\circ} = X_2 \tag{4.14}$$

As a result of the lowering of the vapor pressure of a solvent by a nonvolatile solute, the boiling point is elevated. The molecular weight of the solute may be calculated from the elevation of the boiling point.* The elevation of the boiling point of water by an ideal one molar solute is 0.513 K.

4.7 OSMOTIC PRESSURE

When a solution is separated from the solvent by a semipermeable membrane which is permeable to solvent but not to solute, the solvent flows through the membrane into the solution, where the chemical potential of the solvent is lower. This process is known as osmosis. This flow of solvent through the membrane can be prevented by applying a sufficiently high pressure to the solution. The osmotic pressure Π is the pressure difference across the membrane required to prevent spontaneous flow in either direction across the membrane.

The phenomenon of osmotic pressure was described by Abbé Nollet in 1748, and Pfeffer, a botanist, made the first direct measurements in 1877. Van't Hoff analyzed Pfeffer's data on the osmotic pressure of sugar solutions and found empirically that an equation quite analogous to the ideal gas law gave approximately the behavior of dilute solutions, namely, $\Pi \bar{V} = RT$, where $\bar{V}$ is the volume of solution containing a mole of solute. The origin of the pressure is quite different

---

* F. Daniels, J. W. Williams, P. Bender, R. A. Alberty, C. D. Cornwell, and J. E. Harriman, *Experimental Physical Chemistry*, McGraw-Hill Book Co., New York, 1970.

from that for a gas, however, and the equation of the form of the ideal gas equation is applicable only in the limit of low concentrations.

At equilibrium the chemical potential of pure solvent at 1 atm pressure (μ_1°) is equal to the chemical potential of solvent in the compressed solution (μ_1').

$$\mu_1^{\circ} = \mu_1' \qquad (4.15)$$

The chemical potential of the solvent in the solution is reduced by the solute but it is increased by the applied pressure. When the applied pressure is equal to the osmotic pressure, these two effects exactly counterbalance one another, so that equation 4.15 is obeyed.

The effect on the chemical potential of the solvent of adding solute and changing the pressure is given by

$$d\mu_1 = \left(\frac{\partial \mu_1}{\partial P}\right)_{T,X_2} dP + \left(\frac{\partial \mu_1}{\partial X_2}\right)_{T,P} dX_2 \qquad (4.16)$$

where $d\mu_1$ is the total change in the chemical potential of the solvent when the pressure is changed by dP and the mole fraction of the solute is changed by dX_2. From equation 2.65, we have

$$\left(\frac{\partial \bar{G}_1}{\partial P}\right)_{T,X_2} = \left(\frac{\partial \mu_1}{\partial P}\right)_{T,X_2} = \bar{V}_1 \qquad (4.17)$$

where $\bar{V}_1$ is the partial molar volume of the solvent.

For an ideal solution the chemical potential of the solvent in the solution is given by

$$\mu_1 = \mu_1^* + RT \ln (1 - X_2) \qquad (4.18)$$

Differentiating with respect to X_2 at constant T and P yields

$$\left(\frac{\partial \mu_1}{\partial X_2}\right)_{T,P} = \frac{-RT}{1 - X_2} \qquad (4.19)$$

Substituting equations 4.17 and 4.19 into 4.16 yields

$$d\mu_1 = \bar{V}_1 \, dP - \frac{RT \, dX_2}{1 - X_2} \qquad (4.20)$$

Since the pressure and concentration are changed in such a way that the chemical potential is constant, $d\mu_1 = 0$ and

$$\bar{V}_1 \, dP = \frac{RT \, dX_2}{1 - X_2} = \frac{-RT \, dX_1}{X_1} \qquad (4.21)$$

Integrating from 1 atm to $(1 + \Pi)$ atm, and from $X_1 = 1$ or $\ln X_1 = 0$ to $\ln X_1$

assuming that $\bar{V}_1$ is independent of pressure and concentration,

$$\int_1^{1+\Pi} \bar{V}_1 \, dP = -RT \int_{\ln X_1=0}^{\ln X_1} d \ln X_1$$

$$\bar{V}_1 \Pi = -RT \ln X_1 = -RT \ln (1 - X_2) \tag{4.22}$$

At sufficiently high dilution the logarithmic term of equation 4.22 may be expanded according to

$$\ln (1 + x) = x - \tfrac{1}{2}x^2 + \tfrac{1}{3}x^3 - \cdots \qquad (-1 < x < 1) \tag{4.23}$$

When only the first term in the series is retained, equation 4.22 becomes

$$\bar{V}_1 \Pi = RTX_2 \tag{4.24}$$

Since the solution is dilute, $X_2 = n_2/n_1$ and $\bar{V}_1 = V/n_1$, where V is the volume of the solution. Thus equation 4.24 may be written

$$\Pi V = n_2 RT \tag{4.25}$$

or

$$\Pi = \frac{cRT}{M} \tag{4.26}$$

where c is the concentration of solute in grams per unit volume and M is the molecular weight of the solute. This is the approximate equation that van't Hoff found empirically. It is evident from the approximations introduced why this equation cannot hold for concentrated solutions.

Since the osmotic pressure depends on the number of molecules, the molecular weight determined in this way is the number average molecular weight defined in equation 20.8 if the solute has a distribution of molecular weights.

4.8 VAPOR PRESSURE OF NONIDEAL SOLUTIONS

For nonideal solutions, which are much more common than ideal solutions, the vapor pressure of a component is not directly proportional to its concentration on the mole fraction scale over the whole range of concentrations. The vapor pressure of a component is proportional to its mole fraction at low concentrations, but there the proportionality constant is not equal to the vapor pressure of the pure liquid. Negative deviations from Raoult's law are found, as illustrated in Fig. 4.5 for solutions of acetone and chloroform, and positive deviations are also found, as illustrated in Fig. 4.7 for solutions of acetone and carbon disulfide. The values of the partial pressures and the total pressure calculated using Raoult's law are plotted in Figs. 4.5 and 4.7 as dashed lines. It is noted that for both these solutions the vapor pressure of the component present at higher concentration approaches the values given by Raoult's law as its mole fraction approaches unity. Other types of deviations from Raoult's law are also found. A component may show positive deviations in dilute solutions and negative deviations in concentrated solutions or vice versa.*

---

* M. L. McGlashan, *J. Chem. Ed.*, **40**, 516 (1963).

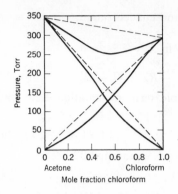

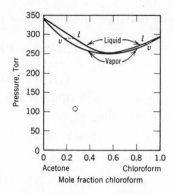

Fig. 4.5 Acetone-chloroform; partial and total pressures at 35.2°.

Fig. 4.6 Acetone-chloroform; liquid and vapor compositions at 35.2°.

The interaction between acetone and chloroform that leads to negative deviations from Raoult's law is due to the formation of a weak hydrogen bond between the oxygen of the acetone and the hydrogen of the chloroform. A hydrogen bond is a bond between two molecules, or two parts of one molecule, that results from the sharing of a proton between two atoms, one of which is usually fluorine, oxygen, or nitrogen (Section 14.8). Thus:

$$Cl—\overset{\displaystyle Cl}{\underset{\displaystyle Cl}{C}}—H\cdots\cdots O=\overset{\displaystyle CH_3}{\underset{\displaystyle CH_3}{C}}$$

If the positive deviations from Raoult's law are large enough, the molecules of the two types squeeze each other out and immiscibility of the two liquids results. If the vapor pressure of a component above a rather dilute solution approaches that of

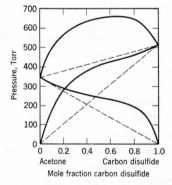

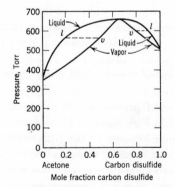

Fig. 4.7 Acetone-carbon disulfide; partial and total pressures at 35.2°.

Fig. 4.8 Acetone-carbon disulfide; liquid and vapor compositions at 35.2°.

the pure component, the conditions are favorable for phase separation. Phase separation occurs when the Gibbs free energy of the two-phase system is lower than that of the homogeneous system.

The vapor curve in Fig. 4.6 has a minimum and in Fig. 4.8 has a maximum. At the compositions corresponding with these extremes, the vapor and liquid phases have the same composition. Figure 4.8 shows that the vapor in equilibrium with a solution of 0.65 mole fraction carbon disulfide has the same composition as the liquid. For solutions containing less than 0.65 mole fraction carbon disulfide the vapor is richer in carbon disulfide, but for solutions containing more than 0.65 mole fraction carbon disulfide the vapor phase is richer in acetone.

Solutions that exhibit a maximum or a minimum in the vapor pressure curves exhibit a minimum or a maximum in the boiling point curves. When a boiling-point curve has a maximum or a minimum, the solutions having the maximum or minimum boiling points are called azeotropes. These solutions distill without change in composition because the liquid and the vapor have the same composition. Many examples of azeotropic solutions are known.* Ethanol, which boils at 78.3°, and water form a minimum-boiling azeotrope that boils at 78.174° and contains 4.0 % water by weight. Hydrochloric acid, which boils at −80°, and water form a maximum-boiling azeotrope at 108.584° that contains 20.222 % HCl by weight.

4.9 HENRY'S LAW

Although the acetone-chloroform and acetone-carbon disulfide solutions do not obey Raoult's law, it is evident from Figs. 4.5 and 4.7 that the partial pressure of the component present at lower concentration is directly proportional to its mole fraction X_2 for dilute solutions.

$$p_2 = X_2 K_2 \tag{4.27}$$

The subscript 2 indicates that the solute (in this case simply the component at lower concentration) is being considered. This equation is referred to as Henry's law and the constant K_2 is referred to as the Henry's law constant. In dilute solutions the environment of the minor component is constant, and its escaping tendency is proportional to its mole fraction. For nonideal solutions Henry's law holds for the solute in the same range where Raoult's law holds for the solvent. For ideal solutions $K_2 = p_2^\circ$, and Henry's law becomes identical with Raoult's law.

The value of the Henry's-law constant K_2 is obtained by plotting the ratio p_2/X_2 versus X_2 and extrapolating to $X_2 = 0$. Such a plot is shown later in Fig. 4.10.

It is convenient to express the solubilities of gases in liquids by use of Henry's law constants. A few gas solubilities at 25° are summarized in this way in Table 4.1. Up to a pressure of 1 atm Henry's law holds within 1–3 % for many slightly soluble gases.

---

* L. H. Horsley and co-workers, *Azeotropic Data, Advances in Chemistry Series*, American Chemical Society, Washington, D.C., 1963.

Table 4.1   Henry's Law Constants[1] for Gases at 25°

	Solvent	
Gas	Water	Benzene
H_2	5.34×10^7	2.75×10^6
N_2	6.51×10^7	1.79×10^6
O_2	3.30×10^7	
CO	4.34×10^7	1.22×10^6
CO_2	1.25×10^6	8.57×10^4
CH_4	31.4×10^6	4.27×10^5
C_2H_2	1.01×10^6	
C_2H_4	8.67×10^6	
C_2H_6	23.0×10^6	

[1] $K_2 = p_2/X_2$. The partial pressure of the gas is given in Torr, and the concentration units are mole fractions.

Example 4.3   Using the Henry's law constant, calculate the solubility of carbon dioxide in water at 25° at a partial pressure of CO_2 over the solution of 760 Torr. Assume that a liter of solution contains practically 1000 grams of water.

$$K = \frac{p_2}{X_2} = 1.25 \times 10^6 = \frac{760}{(CO_2)}\left((CO_2) + \frac{1000}{18.02}\right)$$

Since (CO_2) may be considered negligible in comparison with the number of moles of water, 1000/18.02,

$$(CO_2) = \frac{(760)(55.49)}{1.25 \times 10^6} = 3.38 \times 10^{-2} \text{ M}$$

The solubility of a gas in liquids usually decreases with increasing temperature, since heat is generally evolved in the solution process. There are numerous exceptions, however, especially with the solvents liquid ammonia, molten silver, and many organic liquids. It is a common observation that a glass of cold water, when warmed to room temperature, shows the presence of many small air bubbles.

The solubility of an unreactive gas is due to intermolecular attractive forces between gas molecules and solvent molecules. There is a good correlation between the solubilities of gases in solvents at room temperature and their boiling points. Substances with low boiling points (He, H_2, N_2, Ne, etc.) have weak intermolecular attractions and are therefore not very soluble in liquids.

The solubility of gases in water is usually decreased by the addition of other solutes, particularly electrolytes. The extent of this "salting out" varies considerably with different salts, but with a given salt the relative decrease in solubility is nearly the same for different gases. The solubility of liquids and solids in water also shows this salting-out phenomenon.

4.10 ACTIVITY AND ACTIVITY COEFFICIENT†

The concept of an ideal solution forms such a useful basis for comparison that it is advantageous in dealing with nonideal solutions to use equations of the same form as for ideal solutions. This is accomplished by introducing the activity a_i and the activity coefficient γ_i, which are defined by

$$\mu_i = \mu_i^* + RT \ln a_i \tag{4.28}$$

$$= \mu_i^* + RT \ln f_i X_i \tag{4.29}$$

where μ_i is the chemical potential of component i. Note that activities are pure numbers; concentration units, if any, come in through the definition of the standard state.

The chemical potential μ_i^*, in the standard state where $a_i = 1$, is a function of temperature and pressure only, whereas f_i is a function of concentration as well as temperature and pressure. The activity a_i for a nonelectrolyte is simply equal to the product of an activity coefficient and a concentration.

$$a_i = f_i X_i \tag{4.30}$$

A more complicated relation is needed for electrolytes, as will be seen in Chapter 6. To complete the definition of the activity coefficient f_i it is necessary to specify the conditions under which f_i becomes equal to unity. There are two ways of doing this because Raoult's law is approached as $X_i \to 1$ and Henry's law is approached as $X_i \to 0$.

Convention I. If the components of the solution are liquids, the activity coefficient of each component may be taken to approach unity as its mole fraction approaches unity.

$$f_i \to 1 \quad \text{as} \quad X_i \to 1 \tag{4.31}$$

Since the logarithmic term in equation 4.29 vanishes under these limiting conditions, μ_i^* is equal to the Gibbs free energy of a mole of pure i at the temperature and pressure under consideration. If both components follow Raoult's law over the whole range of concentration (as they do in an ideal solution), their activity coefficients will be equal to unity over the whole range of concentration if Convention I is used.

---

† Different symbols for activity coefficients are used, depending on the concentration scale used as a basis for the activity a.

$$a = fX$$
$$a = \gamma m$$
$$a = y(B)$$

where X is mole fraction, m is molality (moles per kg of solvent), and (B) is concentration in moles per liter.

Convention II. It is convenient to use this convention if it is not possible to vary the mole fractions of both components up to unity. For example, one component may be a gas or a solid. For such solutions a different convention is applied for solvent and solute. The activity coefficient of the solvent is given by Convention I.

$$f_{solvent} \to 1 \qquad as \qquad X_{solvent} \to 1 \qquad\qquad (4.32)$$

Usually the component that is present in the higher concentration is taken as the solvent. The activity coefficient for the solute is taken to approach unity as its mole fraction approaches zero,

$$f_{solute} \to 1 \qquad as \qquad X_{solute} \to 0 \qquad\qquad (4.33)$$

If the activity coefficient of the solute is to approach unity at infinite dilution, μ_i^* for the solute in equation 4.29 must be the chemical potential of pure solute in a hypothetical standard state in which the solute at unit concentration has the properties that it would have at infinite dilution.

4.11 CALCULATION OF ACTIVITY COEFFICIENTS FOR BINARY LIQUID SYSTEMS

At equilibrium between solution and vapor phases the following equation can be written for each component:

$$\mu_{i,soln} = \mu_{i,vap} \qquad\qquad (4.34)$$

Assuming the vapor phase is ideal so that equation 4.5 can be used, and using equation 4.29,

$$\mu_i^* + RT \ln f_i X_i = \mu_i^\circ + RT \ln p_i \qquad\qquad (4.35)$$

By rearranging, we obtain

$$p_i = f_i(X_i K_i) \qquad\qquad (4.36)$$

where the Henry law constant K_i is given by

$$K_i = e^{(\mu_i^* - \mu_i^\circ)/RT} \qquad\qquad (4.37)$$

If $f_i \to 1$ as $X_i \to 1$, K_i must be equal to the vapor pressure of the pure component p_i° so that equation 4.36 becomes

$$p_i = f_i(X_i p_i^\circ) \qquad\qquad (4.38)$$

As $X_i \to 1$ this becomes Raoult's law

$$p_i = X_i p_i^\circ \qquad\qquad (4.39)$$

It is evident from equation 4.38 that for solutions that show positive deviations from Raoult's law, f_i is greater than unity, and for solutions that show negative deviations from Raoult's law, f_i is less than unity.

The activity coefficients of ether and acetone in ether-acetone solutions may be calculated from the data of Fig. 4.9.* If these substances formed ideal solutions, the

---

* J. Sameshima, *J. Am. Chem. Soc.*, **40**, 1498 (1918).

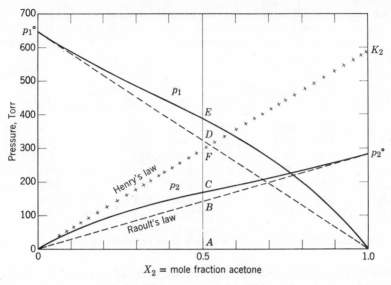

Fig. 4.9 Partial pressures of ether-acetone solutions at 30°.

partial pressure of acetone above a solution containing 0.5 mole fraction acetone would be given by the length AB, which is $0.5p_2^\circ$, acetone being referred to as component 2. The actual partial pressure is represented by the length AC. Solving equation 4.38 for the activity coefficient of acetone, we have

$$f_2 = \frac{p_2}{X_2\, p_2^\circ} = \frac{AC}{AB} = 1.19 \qquad (4.40)$$

Similarly, according to Convention I the activity coefficient of ether (component 1) at 0.5 mole fraction is given by

$$f_1 = \frac{p_1}{X_1\, p_1^\circ} = \frac{AE}{AD} =\,' 1.21 \qquad (4.41)$$

The activity coefficients of both components, calculated in this way at other concentrations by use of Convention I, are summarized in Table 4.2. It will be noted that, as the mole fraction of either component approaches unity, its activity coefficient approaches unity, since the vapor pressure asymptotically approaches that given by Raoult's law.

The calculation of the activity coefficients of acetone on the basis of convention II is accomplished by use of the line passing through F in Fig. 4.9. This line is tangent to the vapor-pressure curve for acetone in the limit as the mole fraction of acetone approaches zero, and its slope is equal to the value of the Henry's-law constant for acetone in ether, obtained by extrapolating the apparent Henry's law constant defined by

$$K_2' = \frac{p_2}{X_2} \qquad (4.42)$$

$p_1 \qquad p_2$

Table 4.2   Activity Coefficients for Acetone-Ether Solutions at 30°

| Mole Fraction Acetone X_2 | Convention I | | | | | | Convention II[1] | |
| | Ether | | | Acetone | | | Acetone | |
	p_1	$X_1 p_1^\circ$	f_1	p_2	$X_2 p_2^\circ$	f_2	$K_2 X_2$	f_2
0	646	646	1.00	0	0	...	0	(1.000)
0.2	535	517	1.03	90	56.6	1.59	117.6	0.766
0.4	440	388	1.13	148	113.2	1.31	235.2	0.630
0.5	391	323	1.21	168	141.5	1.19	294	0.575
0.6	332	258.4	1.28	190	170.0	1.12	353	0.538
0.8	202	129.2	1.56	235	226	10.4	471	0.499
1.0	0	0	...	283	283	1.00	588	(0.481)

[1] The activity coefficients for ether are the same as those calculated by Convention I.

to infinite dilution of the acetone. The extrapolation of this ratio for the data of Fig. 4.9 is illustrated in Fig. 4.10, where values of p_2/X_2 are plotted versus X_2. It is found that the Henry's-law constant at infinite dilution (K_2) has a value of 588 Torr at this temperature of 30°.

If acetone obeyed Henry's law with this value of the constant over the entire concentration range, its vapor pressure at 0.5 mole fraction would be represented by the length AF in Fig. 4.9. The actual partial pressure is represented by the length AC, so that according to equation 4.38 the activity coefficient of acetone at mole fraction 0.5 is given by

$$f_2 = \frac{p_2}{K_2 X_2} = \frac{AC}{AF} = 0.575 \tag{4.43}$$

Thus, when Convention II is used, the extent to which the activity coefficient differs from unity is a measure of the deviation from Henry's law. The activity coefficients

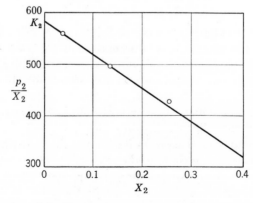

Fig. 4.10   Evaluation of the Henry's-law constant, K_2, for acetone in ether-acetone solutions at 30°.

of acetone calculated in this way are also summarized in Table 4.2. The activity coefficients for the "solvent" ether remain the same as calculated before with Convention I.

Although the numerical values of the activity coefficients of acetone depend on which convention is employed, the same result is obtained in any thermodynamic calculation using these activity coefficients, independent of whether Convention I or II is chosen. These thermodynamic calculations involve two different concentrations, and the standard reference state cancels out. The magnitude of the activity coefficient also depends on the concentration scale used; for example, the molal scale would require a different standard state from that needed with the mole-fraction scale.

4.12 TWO-COMPONENT SYSTEMS CONSISTING OF SOLID AND LIQUID PHASES

The simplest type of binary system consisting of only solid and liquid phases is encountered where the components are completely miscible in the liquid state and completely immiscible in the solid state, so that only the pure solid phases separate out on cooling solutions. Such a phase diagram is illustrated in Fig. 4.11. The diagram is for a constant pressure sufficiently high that no vapor phase is present in this temperature range. Such a diagram may be determined by studying the rate of cooling of solutions of various compositions.

When a liquid consisting of one component is cooled, the plot of temperature versus time has a nearly constant slope. At the temperature at which the solid crystallizes out, however, the cooling curve becomes horizontal if the cooling is slow enough. The halt in the cooling curve results from the heat evolved when the liquid

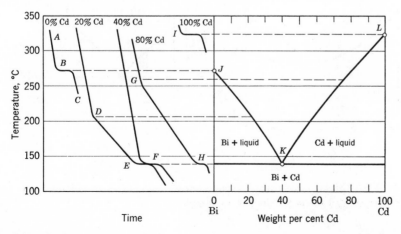

Fig. 4.11 Cooling curves and the temperature-concentration phase diagram for the system bismuth-cadmium.

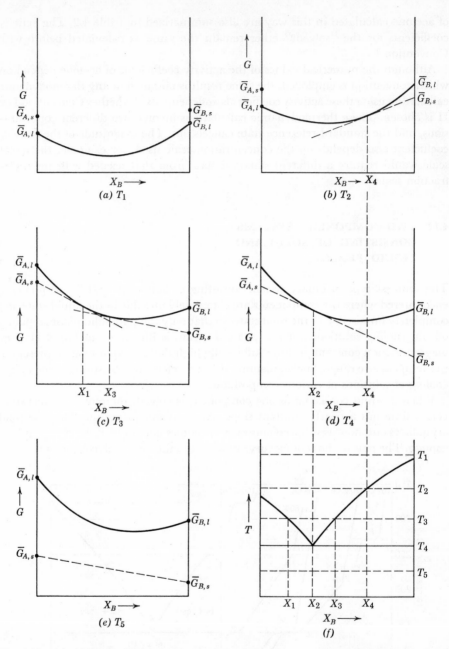

Fig. 4.12 (a)–(e) Molar Gibbs energy versus mole fraction at a series of temperatures and constant pressure. (f) Corresponding phase diagram.

solidifies. This is shown by the cooling curves for bismuth (labeled 0% Cd) and cadmium in Fig. 4.11 at 273° and 323°, respectively.

When a *solution* is cooled, there is a change in slope of the cooling curve at the temperature at which one of the components begins to crystallize out. The change in slope is due to the evolution of heat by the progressive crystallization of the solid as the solution is cooled and to the change in heat capacity. Such changes in slope are evident in the cooling curves for 20% cadmium and 80% cadmium. These curves also show horizontal sections, both at 140°. At this temperature both solid cadmium and solid bismuth come out together. The temperature at which this occurs is called a *eutectic temperature*. A solution of cadmium and bismuth containing 40% cadmium shows a single plateau F at 140°, and so this is the eutectic composition.

The temperatures at which new phases appear, as indicated by the cooling curves, are then transferred to the temperature-composition diagram as shown at the right in Fig. 4.11. In the area above JKL there is one liquid phase and $v = c - p + 1 = 2 - 1 + 1 = 2$. Along JK, bismuth freezes out, and along LK, cadmium freezes out. Thus along line JK and in the area under it and down to the eutectic temperature K there are two phases, solid bismuth and a solution having a composition that is determined by the temperature. Since $v = 2 - 2 + 1 = 1$, the system is univariant. When either the temperature or the composition of the liquid phase is specified, the other may be found from the diagram on the line JK. Also, along the line KL and in the area under it there are two phases, solid cadmium and solution, and accordingly $v = 1$.

At the eutectic point K there are three phases—solid bismuth, solid cadmium, and liquid solution containing 40% cadmium. Then $v = 2 - 3 + 1 = 0$, and so this is an invariant point. There is only one temperature and one composition of solution at which these three phases can exist together at equilibrium at a given constant pressure.

The area below the eutectic temperature K is a two-phase area in which solid bismuth and solid cadmium are present, and $v = 2 - 2 + 1 = 1$. Only the temperature need be specified to describe the system completely at a given constant pressure. The ratio of bismuth to cadmium may change, but there is only a mixture of pure solid bismuth and pure solid cadmium, and there is no need to specify any concentration. The eutectic has a fine grain structure, but it is not a phase; it is a mixture of two solid phases.

4.13 PLOTS OF GIBBS FREE ENERGY AND PHASE DIAGRAMS

A consideration of plots of Gibbs free energy versus mole fraction can illuminate the nature of phase diagrams. Figure 4.12 illustrates the effect of varying the temperature. The first five diagrams are isothermal sections through a three-dimensional plot of Gibbs free energy versus temperature and mole fraction at a constant pressure high enough to eliminate the vapor phase. The plot of Gibbs free energy versus mole fraction for one mole of binary solution has the general shape shown in Fig. 4.12a. The molar free energies of pure liquid A and B are represented by $\bar{G}_{A,l}$ and $\bar{G}_{B,l}$. The molar free energies of pure solid A and B at the same temperature and pressure are represented by $\bar{G}_{A,s}$ and $\bar{G}_{B,s}$.

The free energies of two-phase mixtures of the two liquids would be given by a straight line between $\bar{G}_{A,l}$ and $\bar{G}_{B,l}$ in any of these diagrams. Since the free energy curve for homogeneous solutions lies below this line at all concentrations, the solution has the lower free energy and the two liquids are completely miscible. The points in Fig. 4.12 may be calculated for ideal solutions.

Since the free energies of the solids are higher than the liquids at temperatures above the freezing points for both pure substances; the liquids, having lower molar free energies, are stable with respect to the solids.

By the method of intercepts described in Section 2.18 the partial molar free energies of the two components in any solution may be obtained as intercepts on the two ordinate lines. The converse of this is that a straight line from a point on the ordinate that is tangent to the curve gives the composition of the solution in which this component has the partial molar free energy given by the point on the ordinate. For example, at the temperature T_2 of Fig. 4.12b the partial molar free energy of B in a solution having a mole fraction of X_4 in B is equal to the molar free energy of solid B, which is represented by $\bar{G}_{B,s}$.

The partial molar Gibbs free energies, or chemical potentials, of a component in two phases must be equal if the phases are in equilibrium. Therefore, solution X_4 is in equilibrium with pure solid B. In the concentration range $X_B = X_4$ to 1, the free energy of a two-phase mixture, given by the straight dashed line in Fig. 4.12b is lower than that of the solution. Therefore, in this region there exist two phases rather than a homogeneous solution.

The Gibbs free energies of the pure solids and liquids decrease as the temperature is raised because $(\partial G/\partial T)_P = -S$ and the entropy is positive. The molar free energies of the liquids decrease more rapidly than those of the solids because the entropies of the liquids are greater than the entropies of the solids. At temperature T_3 only solutions in the range X_1 to X_3 are stable with respect to two-phase mixtures. At temperature T_4 there is one point of contact between the free energy curve for the solution and a straight line between the molar free energies of the two pure solids. Thus the partial molar free energy of B in the solution having a mole fraction of X_2 and the molar free energy of solid B are equal; the same is true for A. Thus solid A, solid B, and solution X_2 are in equilibrium, and this mixture is referred to as a eutectic. At T_5 the partial molar free energies of both components of the solution are always greater than the molar free energies of the pure solids, and so a mixture of the two pure solids is thermodynamically stable.

4.14 FREEZING POINT LOWERING

It is evident from Fig. 4.11 that the addition of cadmium lowers the freezing point of bismuth along line JK, and that the addition of bismuth lowers the freezing point of cadmium along line LK. Alternatively we may consider that JK is the solubility curve for bismuth in cadmium, and LK is the solubility curve for cadmium in bismuth. If the solutions are ideal and if the phases that separate are pure solids, the equations for these lines may be readily derived. When there is equilibrium between solid and solution phases the chemical potentials of component 1 (the one forming the solid) must be the same in both phases.

$$\mu_1(s) = \mu_1(\text{soln}) \tag{4.44}$$

$$= \mu_1^* + RT \ln X_1 \tag{4.45}$$

where X_1 is the mole fraction of 1 in solution. Rearranging we obtain

$$\frac{\mu_1(s)}{T} - \frac{\mu_1^*}{T} = R \ln X_1 \tag{4.46}$$

Then by differentiating equation 4.46 with respect to absolute temperature at constant pressure, we have

$$\left[\frac{\partial(\mu_1(s)/T)}{\partial T}\right]_P - \left[\frac{\partial(\mu_1^*/T)}{\partial T}\right]_P = R \frac{\partial \ln X_1}{\partial T} \tag{4.47}$$

Using equation 2.83, we see that

$$\frac{-\bar{H}_s}{T^2} + \frac{\bar{H}_1^*}{T^2} = R \frac{\partial \ln X_1}{\partial T} = \frac{\Delta \bar{H}_{\text{fus},1}}{T^2} \tag{4.48}$$

since $\bar{H}_1^* - \bar{H}_s = \Delta \bar{H}_{\text{fus},1}$. Integrating from mole fraction X_1 at temperature T to $X_1 = 1$ at the freezing point of the pure component $T_{0,1}$, assuming that $\Delta \bar{H}_{\text{fus},1}$ is independent of T, we then have

$$\int_{X_1}^{X_1=1} d \ln X_1 = \int_T^{T_{0,1}} \frac{\Delta \bar{H}_{\text{fus},1}}{RT^2} dT \tag{4.49}$$

$$-\ln X_1 = \frac{\Delta \bar{H}_{\text{fus},1}(T_{0,1} - T)}{RTT_{0,1}} \tag{4.50}$$

or

$$T = \frac{T_{0,1}}{1 - \dfrac{RT_{0,1}}{\Delta \bar{H}_{\text{fus},1}} \ln X_1} \tag{4.51}$$

This equation gives the temperature T at which pure solid 1 is in equilibrium with liquid solution of mole fraction X_1. The freezing point of the component that freezes out is $T_{0,1}$, and its heat of fusion is $\Delta \bar{H}_{\text{fus},1}$. Since equation 4.51 contains no parameters for the solvent we can see that the solubility in mole fraction units is the same in all solvents which form ideal solutions.

Exercise I   Assuming that bismuth and cadmium form ideal solutions, plot lines JK and LK, using equation 4.51. The heat of fusion of cadium is 1450 cal mol^{-1}, and the heat of fusion of bismuth is 2500 cal mol^{-1}.

Equation 4.50 may be written as follows for small freezing point depressions

$$-\ln X_1 = \frac{\Delta \bar{H}_{\text{fus},1}\Delta T_f}{RT_{0,1}^2} \tag{4.52}$$

This equation has been derived for ideal solutions but is applicable to nonideal solutions, provided that the mole fraction of the solvent is very close to unity. For dilute solutions, $-\ln X_1$ can be represented by the first several terms of a power series in X_2, the mole fraction of the solute, as shown in equation 4.23. For sufficiently low concentrations of solute, the second and higher terms of this series are negligible, and so equation 4.52 may be written

$$\Delta T_f = \frac{RT_{0,1}^2}{\Delta \bar{H}_{\text{fus},1}} X_2 \tag{4.53}$$

In a discussion of the elevation of the boiling point, the concentration of the solute is generally given in terms of molal concentration m (that is, moles of solute per 1000 g of solvent) rather than of mole fraction. The relation between these concentrations is

$$X_2 = \frac{n_2}{n_1 + n_2} = \frac{m}{1000/M_1 + m} \xrightarrow{m \to 0} \frac{m}{1000/M_1} \tag{4.54}$$

where M_1 is the molecular weight of the solvent. The last form is applicable to dilute solutions for which the number of moles of solute is negligible in comparison with the number of moles of solvent.

Substituting the last form of equation 4.54 into equation 4.53 and rearranging we obtain

$$\Delta T_f = \frac{RT_{0,1}^2 M_1 m}{1000\,\Delta \bar{H}_{\text{fus},1}} = K_f m \tag{4.55}$$

and

$$K_f = \frac{RT_{0,1}^2 M_1}{1000\,\Delta \bar{H}_{\text{fus},1}} \tag{4.56}$$

$$\Delta T_f = K_f \frac{w_2}{M_2}\frac{1000}{w_1} \tag{4.57}$$

Example 4.4 Calculate the freezing point constant K_f for water. The enthalpy of fusion is 79.7 cal g^{-1} at 273.1 K.

$$K_f = \frac{RT_{0,1}^2 M_1}{1000\,\Delta \bar{H}_{\text{fus},1}} = \frac{(1.987)(273.1)^2(18.02)}{(1000)(18.02)(79.7)} = 1.86 \text{ K molal}^{-1}$$

According to this value of K_f, solute added to 1000 g of water will lower the freezing point 1.86 K molal^{-1}, but the relation holds only for dilute solutions. Even a 1-molal solution is too concentrated, and the depression will be something less than 1.86 K.

Studies of freezing point depression are useful for determining the molecular weight of solutes, but care must be taken not to supercool the solution. The foregoing relations apply only to ideal solutions. Information about activity coefficients may be obtained by studying the freezing points of more concentrated solutions.

The solubility of one substance in another may be considered as an apparent equilibrium constant. For the solubility of a solid in a solvent,

$$A(s) + \text{solvent} = A(\text{sat. soln.}) \tag{4.58}$$

the apparent equilibrium constant may be taken as the mole fraction of A in the saturated solution; $K = X_2$. The true equilibrium constant would be the activity of A in the saturated solution. The heat of solution to form a saturated solution may be calculated from the temperature coefficient of the solubility by using an equation from Section 5.14

$$\frac{\partial \ln X_2}{\partial T} = \frac{\Delta \bar{H}_{\text{sat}}}{RT^2} \tag{4.59}$$

where $\Delta \bar{H}_{\text{sat}}$ is the differential heat of solution of A in the saturated solution calculated assuming the activity coefficient of the solute in the saturated solution is unity at each temperature.

4.15 CONGRUENTLY MELTING COMPOUND

The components of a binary system may react to form a solid compound that exists in equilibrium with liquid over a range of composition. If the formation of a compound leads to a maximum in the temperature-composition diagram as illustrated by Fig. 4.13 for the zinc-magnesium system, we say there is a congruently melting compound. The composition that corresponds to the maximum temperature is the composition of the compound. On the mole percent scale such maxima are achieved at 50%, 33%, 25%, etc., corresponding to integer ratios of the components of 1:1, 1:2, 1:3, etc. Figure 4.13 looks very much like two phase diagrams of the type we have discussed placed side by side, but there is a difference. The liquidus curve has a horizontal tangent (zero slope) at the melting point of the congruently melting compound $MgZn_2$, while the slope is not zero at the melting points of the pure components.* This means that if a congruently melting compound AB exists in an A-B system, additions of very small amounts of A and B to the compound will not lower the melting or freezing point.

Example 4.5 Six-tenths mole of Mg and 0.40 mol of Zn are heated to 650°, represented by point J in Fig. 4.13. Describe what happens when this solution is cooled down to 200°, as indicated by the vertical line. (The experiment would have to be done in an inert atmosphere to prevent oxidation by air.) At 470° point K is reached and solid

---

* A. F. Berndt and D. J. Diestler, *J. Phys. Chem.*, **72**, 2263 (1968).

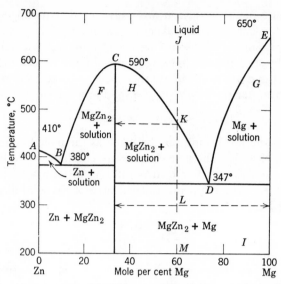

Fig. 4.13 Temperature-composition diagram, showing a maximum for the system zinc-magnesium.

MgZn$_2$ is thrown out of solution. The freezing point is gradually lowered as the solution becomes richer in Mg. Finally, at 347°, when the liquid is 74 mole % in Mg and 26 mole % in Zn, the whole solution freezes, and solid MgZn$_2$ and solid Mg come out together.

From this temperature down to 200° there is no further change in the phases. At all temperatures below 347° there are pure solids Mg and MgZn$_2$.

4.16 INCONGRUENTLY MELTING COMPOUND

Instead of melting, a compound may decompose into another compound and a solution at a definite temperature. This melting point, called an incongruent melting point, is illustrated in Fig. 4.14, which shows part of the phase diagram for the sodium sulfate water system.

When pure Na$_2$SO$_4$·10H$_2$O is heated, it undergoes a transition at 32.38° to give anhydrous Na$_2$SO$_4$ and solution of composition C. The line BC gives the solubility of Na$_2$SO$_4$·10H$_2$O in water, and the line CD gives the solubility of Na$_2$SO$_4'$ in water. Figure 4.14 explains the discontinuity in the solubility curve BCD for sodium sulfate in water.

When the three phases Na$_2$SO$_4$, Na$_2$SO$_4$·10H$_2$O, and saturated solution are in equilibrium with each other at constant pressure, the system is invariant.

4.17 SOLID SOLUTIONS

Often pure solid freezes out of a solution, but at other times a solid solution freezes out. A continuous series of solid solutions may be formed, as illustrated

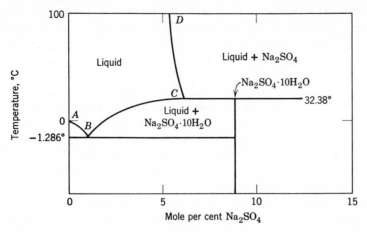

Fig. 4.14 Part of the phase diagram for Na_2SO_4—H_2O showing incongruent melting of $Na_2SO_4 \cdot 10H_2O$ to rhombic anhydrous Na_2SO_4.

in Fig. 4.15 for platinum and gold. The two lines in this diagram give the compositions of the liquid solutions (upper line) and solid solutions (lower line) which are in equilibrium with each other. When these diagrams are studied, it is convenient to remember that the liquid phase is richer in that component of mixture which has the lower melting point.

Above the upper line of Fig. 4.15 the two metals exist in liquid solutions; below the lower line the two metals exist in solid solutions. The upper curve is the freezing point curve for the liquid, and the lower one is the melting-point curve for the solid. The space between the two curves represents mixtures of the two— one liquid solution and one solid solution in equilibrium. For example, a mixture containing 50 mole % gold and 50 mole % platinum, when brought to equilibrium at 1400°, will consist of two phases, a solid solution containing 70 mole % platinum and a liquid solution containing 28 mole % platinum. If the original mixture

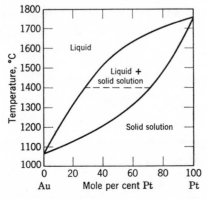

Fig. 4.15 Phase diagram for gold-platinum showing solid solutions.

contained 60 mole % platinum, there would still be the same two liquid and solid solutions at 1400° of the same compositions, 70 and 28 mole %, but there would be a relatively greater amount of the solid solution that contains 70 mole % platinum.

The fractional crystallization of solid solutions is seriously complicated by the fact that the attainment of equilibrium is much slower in solid solutions than in liquid solutions. It takes a considerable length of time, particularly at low temperatures, for a change in concentration at the surface to affect the concentration at a point in the interior of the solid solution. Slow diffusion does take place, however.

In view of the use of the freezing point as a criterion of purity it is important to note that when solid solutions are formed the freezing point may be *raised* by the presence of the other component.

Figure 4.15 is analogous to the phase diagram for two miscible liquids and vapor, as shown in Fig. 4.3. Systems exhibiting solid solution behavior may show maxima or minima in their melting curves that have nothing to do with the formation of compounds.

Many properties of alloys, ceramics, and structural materials depend on the presence of solid solutions. The hardening and tempering of steel involve the existence of solid solutions of carbon in different iron-carbon compounds. The solid solution stable at the high temperatures is hard; to retain this hardness, the proper compositions and temperatures are obtained, as indicated by the phase diagrams, and the steel is quenched quickly in oil or water, so that it does not have time to form the solid solution which is stable at lower temperatures. Reheating the steel to a somewhat lower temperature gives an opportunity for partial conversion to the softer solid solution which is stable at the lower temperature. In this way the steel may be given different degrees of hardening.

Sometimes partial miscibility is encountered in the solid state just as it is in the liquid state. The silver-copper system is an example of partial miscibility of solids. As shown by Fig. 4.16, at 800° copper dissolves in solid silver to the extent of 6% by weight, and silver dissolves in copper to the extent of 2% by weight. At the eutectic point, pure copper and silver do not crystallize out, but saturated solid solutions do. The regions α and β represent continuously variable solid solutions, and the variance v is 2 in these regions because there is a single phase.

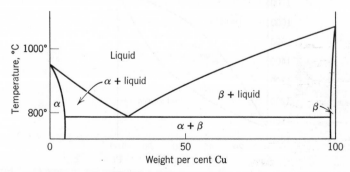

Fig. 4.16 Phase diagram for silver-copper showing partial miscibility of solid solutions.

4.18   TERNARY SYSTEMS

For a ternary system the phase rule yields $v = 5 - p$. If there is a single phase, then $v = 4$, and so a complete geometrical representation would require the use of four-dimensional space. If the pressure is constant a three-dimensional representation may be used. If both the temperature and pressure are constant, then $v = 3 - p$, and so the system may be represented in two dimensions and

$$p = 1 \qquad v = 2 \qquad \text{bivariant}$$
$$p = 2 \qquad v = 1 \qquad \text{univariant}$$
$$p = 3 \qquad v = 0 \qquad \text{invariant}$$

The equilibrium diagram for a ternary system at constant T and P may be plotted in rectangular coordinates by plotting the weight fraction of one component horizontally and the weight fraction of another vertically. The weight fraction of the third may be calculated from the fact that the sum of the three weight fractions is unity. To obtain a symmetrical representation for three components in a plane, however, an equilateral triangle is used.

In an equilateral triangle, the sum of the distances from any given point to the three sides, along the perpendiculars to the sides, is equal to the height of the triangle. The distance from each apex to the center of the opposite side of the equilateral triangle is divided into 100 parts, corresponding to percentage composition, and the composition corresponding to a given point is readily obtained by measuring the perpendicular distance to the three sides. For example, in Fig. 4.17 point O represents a mixture with a gross composition of 50% by weight acetic acid, 10% by weight vinyl acetate, and 40% by weight water.

Of the many possible kinds of ternary systems we will consider only certain types formed by three liquids and by a liquid and two solids.

4.19   SYSTEMS FORMED BY THREE LIQUIDS

If two pairs of the liquids are completely miscible and one pair is partially miscible, a diagram of the type illustrated in Fig. 4.17 is obtained. This figure represents the system water–acetic acid–vinyl acetate at 25° and atmospheric pressure.*

When water is added to vinyl acetate along the line BC the water dissolves at first, forming a homogeneous solution. However, as more water is added, saturation is reached at composition x, and there are two liquid phases, vinyl acetate saturated with water and a little water saturated with vinyl acetate, having the composition z. As more water is added, the amount of the z phase increases and that of the x phase decreases, but the composition of each phase remains always the same. Finally, when the percent of water exceeds that given

---

* J. C. Smith, *J. Phys. Chem.*, **45**, 1301 (1941).

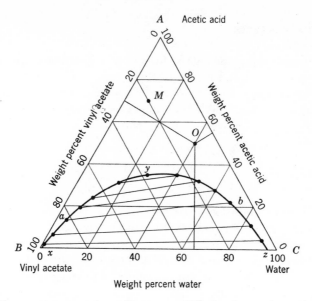

Fig. 4.17 Three-component phase diagram at 25° showing regions of miscibility and immiscibility.

by z, there is only one liquid phase, an unsaturated solution of vinyl acetate in water. At all compositions between x and z there are two liquid phases with compositions x and z.

If acetic acid, which is miscible with vinyl acetate and water in all proportions, is added, it is distributed between the two layers, forming two ternary solutions of vinyl acetate, water, and acetic acid that are in equilibrium with each other, provided that the gross composition of the mixture falls in the region below the xyz curve. For example, if the gross composition lies on the line ab, the two phases that are in equilibrium are represented by points a and b.

Other tie lines are shown for other gross compositions; usually tie lines are not parallel to each other or to a side of the triangle. The compositions of the two phases that are in equilibrium with each other, corresponding to the inter-section of the tie line with the curves xy and zy, have to be determined experiment-ally. As more acid is added, the two phases become more alike, and the tie lines become shorter. Ultimately, when the compositions of the two solutions become identical, the tie line shrinks to the single point y. Point y is a *critical point*, since further addition of acetic acid will result in the formation of a single homogeneous phase. Any point under the curve represents a ternary mixture that will separate into two liquid phases; any point above the curve represents a single homogeneous liquid phase.

If there are two liquid phases, as in the area below the line xyz, the variance is 1 so that it is necessary to specify the percentage of one component in only one phase to describe the system completely. The percentage of the other com-ponents in this phase can be obtained from the intersection of this percentage with the line *xyz*, and the composition of the other phase can be obtained from the intersection of the other end of the tie line with the line *xyz*. For example, if one phase in the two-phase system in Fig. 4.17 contains 5 % water, the composition of this phase is given by point *a* and the composition of the other phase by point *b*.

4.20 DISTRIBUTION OF A SOLUTE BETWEEN TWO PHASES

When a substance is added to a two-phase liquid mixture, it is in general distributed with different equilibrium concentrations in the two phases. The distribution of acetic acid between the water-rich and vinyl acetate-rich phases may be calculated from the data of Fig. 4.17. It is apparent from this figure that the ratio of the concentrations of acetic acid in the two phases, as given by the ends of the tie lines, changes with the amount of acetic acid added. However, if the amount of solute added is sufficiently small, it is often found that the distribution coefficient, which is defined as the ratio of the concentrations of the solute in the two phases, is relatively independent of concentration. In some cases the distribution coefficient depends markedly on concentration because the solute exists in dissociated or associated forms in one of the phases. For example, hydrochloric acid dissolves in water to give H^+ and Cl^- ions, but in benzene it is not dissociated into ions. Other solutes, as, for example, benzoic acid, associate in a nonpolar solvent like benzene to give double molecules, as determined by boiling-point or freezing-point measurements, but they do not associate in a polar solvent like water or ether. The association is due to the formation of hydrogen bonds.

Extraction with immiscible solvents finds many practical applications. Organic compounds are frequently more soluble in hydrocarbons than in water and can be extracted from water into a hydrocarbon phase. The presence of other solutes may profoundly affect the distribution ratio either by forming some complex compound with the solute or by changing the character of the solvent.

Substances having slightly different distribution coefficients between two immiscible solvents may be separated by means of successive extractions. Craig* and others have designed apparatus for carrying out multiple extractions systematically and automatically. Many substances, particularly biological products, which are difficult to separate by other procedures have been separated by these methods.

Substances having slightly different distribution coefficients may also be separated by a column operation in which one liquid phase is held by a finely divided solid with a large surface area and the other liquid phase flows through the column. This process is referred to as *partition chromatography* and is closely related to chromatography experiments depending on adsorption (Section 8.10). In column operation the equivalent of many theoretical plates, each equivalent to a batch extraction, is obtained. The components of a mixture emerge at the

---

* L. C. Craig and N. L. Craig in, *Technique of Organic Chemistry*, Vol. III, ed. by A. Weissberger, Interscience Publishers, New York, 1956.

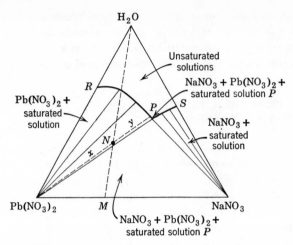

Fig. 4.18   Phase diagram for the system lead nitrate-sodium nitrate-water at 25°.

bottom of the column one by one at different times if their distribution coefficients are sufficiently different.

4.21   SYSTEMS INVOLVING TWO SOLIDS AND A LIQUID

Figure 4.18 shows an example of such a system at 25° and atmospheric pressure in which no compounds are formed.† Solutions along RP are saturated with $Pb(NO_3)_2$, and solutions along PS are saturated with $NaNO_3$. At the intersection of these two solubility curves (point P) the solution is saturated with respect to both $Pb(NO_3)_2$ and $NaNO_3$; there are three phases in equilibrium, and so this point is invariant if the temperature and pressure are constant. A few tie lines are shown in the two-phase regions.

A diagram such as Fig. 4.18 is useful in deciding how to obtain the maximum amount of pure substance from a mixture. For example, if water is added to mixture M, the gross composition moves along the dashed line toward the H_2O apex. If only a small amount of water is added, the phases $Pb(NO_3)_2$, $NaNO_3$, and solution of composition P will be present. If sufficient water is added to reach point N, the only solid phase at equilibrium will be $Pb(NO_3)_2$. If the solution is heated to get all the $NaNO_3$ into solution and then cooled to 25°, the solid phase will be pure $Pb(NO_3)_2$. The composition of the mother liquor would be only slightly different from P. The relative amounts of $Pb(NO_3)_2$ and mother liquor are given by y/x. Thus it is seen that if more water is added the recovery of $Pb(NO_3)_2$ will be reduced.

---

† The determination of this particular phase diagram has been described in detail by E. L. Heric, *J. Chem. Educ.*, **35,** 510 (1958).

In a phase diagram like Fig. 4.18, two phase regions are indicated by the tie lines that show the compositions of phases that are in equilibrium. The blank regions may represent either one phase or three phases, and so it is important to understand how to distinguish between these two types of areas. One-phase regions, like the solution region, have curved boundaries, whereas boundaries of the three-phase regions are straight so that these regions have a triangular shape.

References

A. Alper, *Phase Diagrams*, Volumes I, II, and III, Academic Press, New York, 1970.

R. S. Bradley and D. C. Munro, *High Pressure Chemistry*, Pergamon Press, New York, 1965.

K. G. Denbigh, *The Principles of Chemical Equilibrium*, Cambridge University Press, Cambridge, 1971.

A. Findlay, A. N. Campbell, and N. O. Smith, *The Phase Rule and Its Applications*, Dover Publications, New York, 1951.

A. W. Francis, *Liquid-Liquid Equilibriums*, Interscience Publishers, New York, 1963.

R. M. Garrels and C. L. Christ, *Solutions, Minerals, and Equilibria*, Harper and Row, New York, 1965.

E. Hala, J. Pick, V. Fried, and O. Vilim, *Vapor-Liquid Equilibrium* (translated by G. Standart), Pergamon Press, New York, 1958.

J. H. Hildebrand and R. L. Scott, *Solubility of Non-Electrolytes*, Reinhold Publishing Corp., New York, 1950.

L. H. Horsley, *Azeotropic Data*, Advances in Chemistry Series, American Chemical Society, Washington, D.C., 1963.

A. Reisman, *Phase Equilibria*, Academic Press, New York, 1970.

A. G. Williamson, *An Introduction to Non-Electrolyte Solutions*, Wiley, New York, 1967.

J. Zernike, *Chemical Phase Theory*, N. V. Uitgevers-Maatschappij Æ Kluwer, Deventer, 1955.

Problems

4.1 Ethanol and methanol form very nearly ideal solutions. The vapor pressure of ethanol is 44.5 Torr, and that of methanol is 88.7 Torr, at 20°. (*a*) Calculate the mole fraction of methanol and ethanol in a solution obtained by mixing 100 grams of each. (*b*) Calculate the partial pressures and the total vapor pressure of the solution. (*c*) Calculate the mole fraction of methanol in the vapor.

$$Ans. \ (a) \ X_{C_2H_5OH} = 0.410; \ X_{CH_3OH} = 0.590.$$
$$(b) \ p_{C_2H_5OH} = 18.2, p_{CH_3OH} = 52.3, P_{total} = 70.5 \ Torr.$$
$$(c) \ 0.741.$$

4.2 Calculate the (*a*) enthalpy, (*b*) entropy, and (*c*) Gibbs free energy of mixing of 1 mol of benzene and 2 mol of toluene at 25°.   *Ans.* (*a*) 0 cal. (*b*) 3.79 cal K^{-1}. (*c*) −1131 cal.

4.3 At 100° benzene has a vapor pressure of 1357 Torr, and toluene has a vapor pressure of 558 Torr. Assuming that these substances form ideal binary solutions with each other, calculate the composition of the solution that will boil at 1 atm at 100° and the vapor composition.   *Ans.* $X_{benzene,liq} = 0.253$, $X_{benzene,vap} = 0.451$.

4.4 The following table gives mole % acetic acid in aqueous solutions and in the equilibrium vapor at the boiling point of the solution at 1 atm:

B.P., °C		118.1	113.8	107.5	104.4	102.1	100.0
Mole %	Liquid	100	90.0	70.0	50.0	30.0	0
acetic acid	Vapor	100	83.3	57.5	37.4	18.5	0

Calculate the minimum number of theoretical plates for the column required to produce an initial distillate of 28 mole % acetic acid from a solution of 80 mole % acetic acid.   *Ans.* 3.

4.5 If two liquids (1 and 2) are completely immiscible, the mixture will boil when the sum of the two partial pressures exceeds the applied pressure: $P = p_1^\circ + p_2^\circ$. In the vapor phase the ratio of the mole fractions of the two components is equal to the ratio of their vapor pressures.

$$\frac{p_1^\circ}{p_2^\circ} = \frac{X_1}{X_2} = \frac{g_1 M_2}{g_2 M_1}$$

where g_1 and g_2 are the masses of components 1 and 2 in the vapor phase, and M_1 and M_2 are their molecular weights. The boiling point of the immiscible liquid system naphthalene-water is 98° under a pressure of 733 Torr. The vapor pressure of water at 98° is 707 Torr. Calculate the weight percent of naphthalene in the distillate.   *Ans.* 20.7%.

4.6 The vapor pressure of a solution containing 13 g of a nonvolatile solute in 100 g of water at 28° is 27.371 Torr. Calculate the molecular weight of the solute, assuming that the solution is ideal. The vapor pressure of water at this temperature is 28.065 Torr.

Ans. 92.3 g mol^{-1}.

4.7 Ten grams of benzene, 10 g of toluene, and 10 g of naphthalene are mixed together to give a homogeneous solution. If it is assumed that the solution is ideal, how many grams of toluene will be vaporized by passing through 10 liters of air at 30° if the vapor pressure of toluene at this temperature is 36.7 Torr, that of benzene is 118.5 Torr, and that of naphthalene is negligible?   *Ans.* 0.617 g.

4.8 Calculate the lowering of the vapor pressure of water at 20° by the addition of 54.1 g of mannitol ($M = 182.2$) per 1000 g of water. The vapor pressure of water at this temperature is 17.54 Torr. The experimental value is 0.0922 Torr [J. C. W. Frazer, B. F. Lovelace, T. H. Rogers, *J. Am. Chem. Soc.*, **42**, 1793 (1920)].   *Ans.* 0.0935 Torr.

4.9 The osmotic pressure of an aqueous solution of 1 g of sucrose ($M = 342$) per 100 cm^3 of solution is 0.649 atm at 0°. Calculate the expected osmotic pressure, using the simple form of the osmotic-pressure equation.   *Ans.* 0.655 atm.

4.10 Calculate the osmotic pressure of a 1 M sucrose solution in water from the fact that at 30° the vapor pressure of the solution is 31.207 Torr. The vapor pressure of water at 30° is 31.824 Torr. The density of pure water at this temperature (0.99564 g cm^{-3}) may be used to estimate $\bar{V}_1$ for a dilute solution. To do this problem, Raoult's law is introduced into equation 4.22.*   *Ans.* 26.9 atm.

---

* This problem may be solved to a sufficiently high degree of accuracy with a slide rule rather than logarithm tables by writing the logarithmic term as a series. For small values of x

$$\ln (1 + x) \cong x$$

Thus,

$$\ln \frac{31.824}{31.207} = \ln \left(1 + \frac{31.824 - 31.207}{31.207}\right)$$

$$= \ln (1 + 0.0198) \cong 0.0198$$

4.11 For a solution of n-propanol and water, the following partial pressures in Torr are measured at 25°. Draw a complete pressure-composition diagram, including the total pressure. What is the composition of the vapor in equilibrium with a solution containing 0.5 mole fraction of n-propanol?

$X_{n\text{-propanol}}$	p_{H_2O}	$p_{n\text{-propanol}}$	$X_{n\text{-propanol}}$	p_{H_2O}	$p_{n\text{-propanol}}$
0	23.76	0	0.600	19.9	15.5
0.020	23.5	5.05	0.800	13.4	17.8
0.050	23.2	10.8	0.900	8.13	19.4
0.100	22.7	13.2	0.950	4.20	20.8
0.200	21.8	13.6	1.000	0.00	21.76
0.400	21.7	14.2			

Ans. $X_{n\text{-propanol,vap}} = 0.406.$

4.12 Using the Henry law constants in Table 4.1, calculate the percentage (by volume) of oxygen and nitrogen in air dissolved in water at 25°. The air in equilibrium with the water at 1 atm pressure may be considered to be 20% oxygen and 80% nitrogen by volume.
Ans. 33% oxygen, 67% nitrogen.

4.13 The following data on ethanol-chloroform solutions at 35° were obtained by G. Scatchard and C. L. Raymond [*J. Am. Chem. Soc.*, **60**, 1278 (1938)]:

$X_{EtOH,liq}$	0	0.2	0.4	0.6	0.8	1.0
$X_{EtOH,vap}$	0.0000	0.1382	0.1864	0.2554	0.4246	1.0000
Total pressure, Torr	295.11	304.22	290.20	257.17	190.19	102.78

Calculate the activity coefficients of ethanol in these solutions according to convention I.
Ans. 2.04, 1.315, 1.067, 0.983, 1.000.

4.14 Using the data in problem 4.11, calculate the activity coefficients of water and n-propanol at 0.20, 0.40, 0.60, and 0.80 mole fraction n-propanol, using convention II and considering n-propanol to be the solvent.
Ans. $X_1 = 0.20, 0.40, 0.60, 0.80;$
$f_1 = 3.12, 1.63, 1.19, 1.02;$
$f_2 = 0.314, 0.417, 0.574, 0.773.$

4.15 Calculate the solubility of naphthalene at 25° in any solvent in which it forms an ideal solution. The melting point of naphthalene is 80°, and the heat of fusion is 4610 cal mol^{-1}. The actual measured solubility of naphthalene in benzene is $X_1 = 0.296$.
Ans. $X_1 = 0.297.$

4.16 The data for the solubility of urea in water are given in the following table. Calculate the differential heat of solution of urea in its saturated solution in water at 100°.

X_{urea}	1.000	0.9004	0.8190	0.7217	0.5680	0.4741
t, °C	132.6	123.2	115.3	104.4	84.4	68.5

Ans. 3.2 kcal mol^{-1}.

4.17 If 68.4 g of sucrose ($M = 342$) is dissolved in 1000 g of water: (*a*) What is the vapor pressure at 20°? (*b*) What is the freezing point? The vapor pressure of water at 20° is 17.363 Torr.
Ans. (*a*) 17.30 Torr. (*b*) $-0.372°.$

4.18 A certain number of grams of a given substance in 100 g of benzene lowers the freezing point by 1.28°. The same weight of solute in 100 g of water lowers the freezing point by 1.395°. If the substance has its normal molecular weight in benzene and is completely dissociated in water, into how many ions does a molecule of this substance dissociate when placed in water? (The freezing point constant K_f for benzene is 5.12 K molal^{-1}.

Ans. 3.

4.19 The phase diagram for magnesium-copper at constant pressure shows that two compounds are formed: $MgCu_2$ which melts at 800°, and Mg_2Cu, which melts at 580°. Copper melts at 1085°, and Mg at 648°. The three eutectics are at 9.4% by weight Mg (680°), 34% by weight Mg (560°), and 65% by weight Mg (380°). Construct the phase diagram. State the variance for each area and eutectic point.

Ans. In the liquid region $v = 2$, in the two-phase regions $v = 1$, and at the eutectic points $v = 0$.

4.20 Sketch the phase diagram for thallium and mercury with freezing points plotted against percent by weight. Use the following facts: Hg melts at $-39°$; the compound Tl_2Hg_5 melts at 15°; Tl melts at 303°; Tl lowers the freezing point of Hg down to a minimum of $-60°$ at a composition of 8% by weight Tl; the eutectic point for Tl and Tl_2Hg_5 is 0.4° at a composition corresponding to 41% by weight Tl. Label the phases present and state the variance at each area and eutectic point.

Ans. In the liquid region $v = 2$, in the two-phase regions $v = 1$, and at the eutectic points $v = 0$.

4.21 For the ternary system benzene-isobutanol-water at 25° and 1 atm the following compositions have been obtained for the two phases in equilibrium:

Water-Rich Phase		Benzene-Rich Phase	
Isobutanol, wt. %	Water, wt. %	Isobutanol, wt. %	Benzene, wt. %
2.33	97.39	3.61	96.20
4.30	95.44	19.87	79.07
5.23	94.59	39.57	57.09
6.04	93.83	59.48	33.98
7.32	92.64	76.51	11.39

Plot these data on a triangular graph, indicating the tie lines. (*a*) Estimate the compositions of the phases that will be produced from a mixture of 20% isobutanol, 55% water, and 25% benzene. (*b*) What will be the composition of the principal phase when the first drop of the second phase separates when water is added to a solution of 80% isobutyl alcohol in benzene?   *Ans.* (*a*) H_2O layer: 5.23% isobutanol, 94.5% H_2O. Benzene layer: 39.57% isobutanol, 57.09% benzene.

(*b*) 10% H_2O; 72% isobutanol; 18% benzene.

4.22 Picric acid is distributed between benzene and water as indicated by the following equilibrium concentrations in moles per liter:

Aqueous phase, $c \times 10^3$	2.08	3.27	7.01	10.1
Benzene phase, $c \times 10^3$	0.932	2.25	10.1	19.9

Picric acid exists as nondissociated and nonassociated $C_6H_2(NO_2)_3OH$ in benzene.

What conclusions can you draw regarding dissociation or association in the water phase?

> *Ans.* Picric acid dissolved in water dissociates into two ions.

4.23 The following data are available for the system nickel sulfate-sulfuric acid-water at 25°. Sketch the phase diagram on triangular coordinate paper, and draw appropriate tie lines.

Liquid Phase

$NiSO_4$, wt. %	H_2SO_4, wt. %	Solid Phase
28.13	0	$NiSO_4 \cdot 7H_2O$
27.34	1.79	$NiSO_4 \cdot 7H_2O$
27.16	3.86	$NiSO_4 \cdot 7H_2O$
26.15	4.92	$NiSO_4 \cdot 6H_2O$
15.64	19.34	$NiSO_4 \cdot 6H_2O$
10.56	44.68	$NiSO_4 \cdot 6H_2O$
9.65	48.46	$NiSO_4 \cdot H_2O$
2.67	63.73	$NiSO_4 \cdot H_2O$
0.12	91.38	$NiSO_4 \cdot H_2O$
0.11	93.74	$NiSO_4$
0.08	96.80	$NiSO_4$

4.24 At 25° the vapor pressures of chloroform and carbon tetrachloride are 199.1 and 114.5 Torr, respectively. If the liquids form an ideal solution, (*a*) what is the composition of the vapor in equilibrium with a solution containing 1 mol of each; (*b*) what is the total vapor pressure of the mixture?

4.25 Ethylene dibromide and propylene dibromide form very nearly ideal solutions. Plot the partial vapor pressure of ethylene dibromide ($p° = 172$ Torr), the partial vapor pressure of propylene dibromide ($p° = 127$ Torr) and the total vapor pressure of the solution versus the mole fraction of ethylene dibromide at 80°. (*a*) What will be the composition of the vapor in equilibrium with a solution containing 0.75 mole fraction of ethylene dibromide? (*b*) What will be the composition of the liquid phase in equilibrium with ethylene dibromide-propylene dibromide vapor containing 0.50 mole fraction of each?

4.26 Ten grams of benzene are mixed with 10 g of toluene at 25°. Calculate the (*a*) entropy of mixing and (*b*) Gibbs free energy of mixing.

4.27 At 1 atm pressure propane boils at $-42.1°$ and *n*-butane boils at $-0.5°$; the following vapor-pressure data are available:

Temperature, °C	-31.2	-16.3
Vapor pressure, propane	1200 Torr	2240 Torr
Vapor pressure, *n*-butane	200 Torr	400 Torr

Assuming that these substances form ideal binary solutions with each other, (*a*) calculate the mole fractions of propane at which the solution will boil at 1 atm pressure at $-31.2°$ and $-16.3°$. (*b*) Calculate the mole fractions of propane in the equilibrium vapor at these temperatures. (*c*) Plot the temperature-mole fraction diagram at 1 atm, using these data.

4.28 The following table gives the mole per cent of *n*-propanol ($M = 60.1$) in aqueous solutions and in the vapor at the boiling point of the solution at 760 Torr pressure.

Mole % n-Propanol

Liquid	Vapor	B.P., °C
0	0	100.0
2.0	21.6	92.0
6.0	35.1	89.3
20.0	39.2	88.1
43.2	43.2	87.8
60.0	49.2	88.3
80.0	64.1	90.5
100.0	100.0	97.3

With the aid of a graph of these data calculate the mole fraction of n-propanol in the first drop of distillate when the following solutions are distilled with a simple distilling flask that gives one theoretical plate: (a) 87 g of n-propanol and 211 g of water; (b) 50 g of n-propanol and 5.02 g of water.

4.29   The vapor pressure of the immiscible liquid system diethylaniline-water is 760 Torr at 99.4°. The vapor pressure of water at that temperature is 744 Torr. How many grams of steam are necessary to distill 100 g of diethylaniline? (See problem 4.5.)

4.30   The vapor pressure of water at 25° (23.756 Torr) is lowered 0.071 Torr by the addition of 1.53 g of a nonvolatile substance to 100 g of water. Calculate the molecular weight of the solute, using Raoult's law. (See equation 4.14 for the most convenient form of Raoult's law for this calculation.)

4.31   The vapor pressure of water at 25° is 23.756 Torr. Calculate the vapor pressure of solutions containing (a) 6.01 g of urea, NH_2CONH_2, (b) 9.4 g of phenol, C_6H_5OH, and (c) 6.01 g of urea + 9.4 g of phenol per 1000 g of water, assuming no chemical action between the two substances. (d) Calculate (c), assuming that a stable compound is formed containing 1 mol of the urea to 1 mol of phenol.

4.32   Derive an equation for the lowering of the vapor pressure of a solvent by the addition of two nonvolatile solutes of mole fraction X_2 and X_3.

4.33   An aqueous solution of maltose at 25° has a vapor pressure of 23.476 Torr, whereas pure water has a vapor pressure of 23.756 Torr. What is the osmotic pressure of the solution? (See problem 4.10.)

4.34   Plot the following boiling-point data for benzene-ethanol solutions and estimate the azeotropic composition:

B.P., °C	78	75	70	70	75	80
Mole fraction of benzene						
In liquid	0	0.04	0.21	0.86	0.96	1.00
In vapor	0	0.18	0.42	0.66	0.83	1.00

State the range of mole fractions of benzene for which pure benzene could be obtained by fractional distillation at 1 atm.

4.35   A 10-liter tank of methane at 740 Torr total pressure and 25° contains 1 liter of water. How many grams of methane are dissolved in the water?

4.36   The Henry law constants for oxygen and nitrogen in water at 0° are 1.91×10^7 Torr and 4.09×10^7 Torr, respectively. Calculate the lowering of the freezing point of water by dissolved air with 80% N_2 and 20% O_2 by volume at 1 atm pressure.

4.37 Using the data in problem 4.11, calculate the activity coefficients of water and n-propanol at 0.20, 0.40, 0.60, and 0.80 mole fraction of n-propanol, using Convention II and considering water to be the solvent.

4.38 Using the data of the following table, which gives vapor pressures in Torr at 35.2°, calculate the activity coefficients of acetone and chloroform at 35.2°, following Convention I, for mole fractions of chloroform of 0.20, 0.40, 0.60, and 0.80.

X_{CHCl_3}	p_{CHCl_3}	$p_{acetone}$
0	0	344
0.2	34	270
0.4	82	183
0.6	148	102
0.8	225	42
1.00	293	0

4.39 Calculate the solubility of anthracene ($M = 178.2$) in toluene ($M = 92.1$) at 100°. The heat of fusion of anthracene is 6900 cal mol^{-1}, and the melting point of anthracene is 217°. The actual solubility is 0.0592 on the mole-fraction scale. How do you explain the discrepancy?

4.40 Calculate the solubility of cadmium in bismuth at 250°C. The melting point of cadmium is 323° and the heat of fusion at the melting point is 1450 cal mol^{-1}.

4.41 Calculate the solubility of monoclinic sulfur in carbon tetrachloride at 25°. That of rhombic sulfur is 0.84 g per 100 g of CCl$_4$. Sulfur exists in both solutions as S$_8$. The Gibbs free energy of formation of monoclinic sulfur is 23 cal mol^{-1} greater than that of rhombic sulfur at 25°.

4.42 The following cooling curves have been found for the system antimony–cadmium:

Cd, % by weight	0	20	37.5	47.5	50	58	70	93	100
First break in curve, °C	—	550	461	—	419	—	400	—	—
Continuing constant temp,°C	630	410	410	410	410	439	295	295	321

Construct a phase diagram, assuming that no breaks other than these actually occur in any cooling curve. Label the diagram completely, and give the formula of any compound formed. State the variance in each area and at each eutectic point.

4.43 The melting points of magnesium and nickel are 651° and 1450°, respectively. As Ni is added to Mg, the freezing point is lowered until a eutectic point is reached at 510° and 28% by weight Ni. The other phase that separates contains 54.7% by weight Ni. As the weight percent Ni is increased past 28%, the temperature of first-phase separation rises to a maximum of 1180°. Above 770° and 38% Ni, the solid phase that separates out contains 83% Ni. The phase containing 83% Ni melts sharply at 1180°. There is a eutectic point at 88% Ni and 1080°. Draw the phase diagram and indicate the phases present in each region.

4.44 The following data are obtained by cooling solutions of magnesium and nickel:

Ni, wt. %	0	10	28	38	60	83	88	100
Inflection in cooling curve, °C	—	608	—	770	1050	—	—	—
Plateau in cooling curve, °C	651	510	510	510	770	1180	1080	1450

It is found that in addition cooling solutions containing between 28 and 38% Ni deposit

Mg_2Ni, whereas solutions containing between 38 and 82% Ni deposit $MgNi_2$. Plot the phase diagram.

4.45 The following are the compositions of the phases in equilibrium with each other in the system methylcyclohexane–aniline–n-heptane at 1 atm and 25°. Draw a triangular diagram for the system, including tie lines, and compute the exact composition of the first infinitesimal increment of the new liquid phase that forms when a sufficient quantity of pure aniline is added to a 40% solution of methylcyclohexane in n-heptane to give separation into two phases.

Hydrocarbon Layer		Aniline Layer	
Methylcyclohexane, wt. %	n-Heptane, wt. %	Methylcyclohexane, wt. %	n-Heptane, wt. %
0.0	92.0	0.0	6.2
9.2	83.0	0.8	6.0
18.6	73.4	2.7	5.3
33.8	57.6	4.6	4.5
46.0	45.0	7.4	3.6
59.7	30.7	9.2	2.8
73.6	16.0	13.1	1.4
83.3	4.4	15.6	0.6
88.1	0.0	16.9	0.0

4.46 The distribution coefficient at 25° of lactic acid between water and chloroform, c_{CHCl_3}/c_{H_2O}, is 0.0203 when concentrations are expressed in moles per liter. (a) How much lactic acid will be extracted from 100 cm³ of a 0.8 molar solution of lactic acid in $CHCl_3$ by shaking with 100 ml of H_2O? (b) How much will be extracted if the 100 cm³ of $CHCl_3$ is shaken first with 50 cm³ of H_2O and later with another 50 cm³ of water?

4.47 The following data are available for the system Na_2SO_4-$Al_2(SO_4)_3$-H_2O at 42°. Draw the phase diagram on triangular coordinate paper, and draw appropriate tie lines.

Liquid Phase		
Na_2SO_4, wt. %	$Al_2(SO_4)_3$, wt. %	Solid Phase
33.20	0	Na_2SO_4
32.00	1.52	Na_2SO_4
31.79	1.87	Na_2SO_4
28.75	1.71	$Na_2SO_4 \cdot Al_2(SO_4)_3 \cdot 14H_2O$
24.47	2.84	$Na_2SO_4 \cdot Al_2(SO_4)_3 \cdot 14H_2O$
16.81	5.63	$Na_2SO_4 \cdot Al_2(SO_4)_3 \cdot 14H_2O$
10.93	10.49	$Na_2SO_4 \cdot Al_2(SO_2)_3 \cdot 14H_2O$
4.72	17.11	$Na_2SO_4 \cdot Al_2(SO_4)_3 \cdot 14H_2O$
1.75	18.59	$Al_2(SO_4)_3$
0	16.45	$Al_2(SO_4)_3$

4.48 From the phase diagram for $Pb(NO_3)_2$-$NaNO_3$-H_2O in Fig. 5.18, what solid phase will crystallize out when water is evaporated (a) from a solution containing 10% $NaNO_3$ and 30% $Pb(NO_3)_2$, (b) from a solution containing 20% $NaNO_3$ and 5% $Pb(NO_3)_2$?

4.49 Benzene and toluene form very nearly ideal solutions. At 80° the vapor pressures of benzene and toluene are as follows: benzene, $p° = 753$ Torr; toluene, $p° = 290$ Torr. (a) For a solution containing 0.5 mole fraction of benzene and 0.5 mole fraction of toluene, what is the composition of the vapor and the total vapor pressure at 80°? (b) What is the composition of the liquid phase in equilibrium at 80° with benzene-toluene vapor having 0.75 mole fraction benzene?

4.50 At 25° the vapor pressure of carbon tetrachloride, CCl_4 is 143 Torr and that of chloroform, $CHCl_3$, is 199 Torr. These liquids form very nearly ideal solutions with each other. If 1 mol of CCl_4 and 3 mol of $CHCl_3$ are mixed, what will be the mole fraction of CCl_4 in the vapor phase and the total vapor pressure of the solution?

4.51 Derive the equation for the Gibbs free-energy change when a mole of substance is transferred from a large amount of an ideal solution where its mole fraction is X_1 to another where its mole fraction is X_2. The Gibbs free-energy change is calculated for each of the steps: (a) transfer to the vapor in equilibrium with solution 1, (b) change in pressure of the vapor to the equilibrium vapor pressure of solution 2, and (c) transfer from the equilibrium vapor to the liquid phase of mole fraction X_2.

4.52 The boiling points of o-xylene and p-xylene at 760 Torr pressure are 144.4° and 138.4°, respectively. Assuming that these substances form ideal binary solutions and that for each the change in boiling point per Torr change in pressure is 0.05° in this temperature range, calculate several points on the boiling-point line and several points on the corresponding vapor line.

4.53 Calculate the composition of an ethanol-methanol solution which will boil at a pressure of 60 Torr at 20°, assuming the solution is ideal. The vapor pressure of ethanol at this temperature is 44.5 Torr and that of methanol is 88.4 Torr.

4.54 From the data given in the following table construct a complete temperature-composition diagram for the system ethanol-ethyl acetate for 760 Torr pressure. A solution containing 0.8 mol fraction of ethanol, EtOH, is distilled completely at 760 Torr. (a) What is the composition of the first vapor to come off? (b) That of the last drop of liquid to evaporate? (c) What would be the values of these quantities if the distillation were carried out in a cylinder provided with a piston so that none of the vapor could escape?

$X_{EtOH,liq}$	$X_{EtOH,vap}$	B.P., °C	$X_{EtOH,liq}$	$X_{EtOH,vap}$	B.P., °C
0	0	77.15	0.563	0.507	72.0
0.025	0.070	76.7	0.710	0.600	72.8
0.100	0.164	75.0	0.833	0.735	74.2
0.240	0.295	72.6	0.942	0.880	76.4
0.360	0.398	71.8	0.982	0.965	77.7
0.462	0.462	71.6	1.000	1.000	78.3

4.55 Calculate the molecular weight of nitrobenzene from the fact that when a mixture of nitrobenzene and water is distilled at a pressure of 731.9 Torr, the distillation temperature is 98.2° and the weight ratio of nitrobenzene to water in the distillate is 0.188. The vapor pressure of water at this temperature is 712.4 Torr. (See problem 4.5.)

4.56 Calculate the vapor pressure at 25° of an aqueous solution of urea ($M = 60.06$) containing 5 g of urea per 1000 g of water. The vapor pressure of water at this temperature is 23.756 Torr, and ideal-solution behavior may be assumed.

4.57 Calculate the osmotic pressure of 0.16 M sodium chloride at 37°, assuming complete dissociation.

4.58 (*a*) Assuming ideal solutions, what molar concentration of solutes at 20° is required to raise by osmosis a column of solution having a density of approximately 1.0 g cm^{-3} to a height of 100 ft? (*b*) What is the vapor pressure of the solution at 20°? The vapor pressure of pure water at 20° is 17.363 Torr. (See problem 4.10.)

4.59 Two 2-liter vessels are connected with a tube and stopcock of negligible volume. Initially the first bulb contains 10 g of water and is at 10°. The other bulb contains ammonia at a pressure of 5 atm and is at 0°. Calculate the total pressure in the system when the stopcock is opened and the whole apparatus is brought to equilibrium at 25°. The solubility of ammonia in water at 25° is 27.011 mol per 1000 g of water at 1 atm. Second-order effects may be neglected.

4.60 By use of the data of the following table, which gives pressures in Torr at 35.2° for carbon disulfide-acetone solutions, calculate the activity coefficients (following convention I) for acetone and carbon disulfide at 35.2° for a solution containing 0.6 mol fraction of carbon disulfide.

X_{CS_2}	p_{CS_2}	$p_{acetone}$
0	0	344
0.2	280	290
0.4	378	255
0.6	425	230
0.8	460	190
1.00	512	0

4.61 Using the data in problem 4.11, calculate the activity coefficients of water and *n*-propanol at 0.20, 0.40, 0.60, and 0.80 mol fraction of *n*-propanol using convention I.

4.62 What mole percent impurity is required to lower the freezing point of benzene 0.05°, provided that pure benzene crystallizes out?

4.63 Calculate the temperature at which pure cadmium is in equilibrium with a Bi-Cd solution containing 0.846 mole fraction Cd.

4.64 The solubility of succinamide ($M = 116.1$) in water is 0.5 g per 100 g of H$_2$O at 15° and 11 g per 100 g of H$_2$O at 100°. Estimate the differential heat of solution of succinamide in its saturated solution in water.

4.65 State the phases present and the variance at *B, F, G, H, J*, and *L* in Fig. 4.13.

4.66 Show that in a solvent extraction a given volume of the insoluble solvent will extract more solute from a given volume of aqueous solution if the solvent is divided into several small parts and used in a series of successive extractions. For a given concentration and distribution ratio, calculate the amount of solute extracted from 100 cm^3 of aqueous solution by 100 cm^3 of solvent. Repeat the calculation, using two extractions with 50 ml each and then *n* extractions with 100/*n* cm^3 of solvent in each extraction.

4.67 At 25° the solubility of KNO$_3$ in pure H$_2$O is 46.2% by weight, the solubility of NaNO$_3$ in pure H$_2$O is 52.2% by weight, and NaNO$_3$, KNO$_3$, and saturated solution are in equilibrium when the composition of the solution is H$_2$O, 31.3%, KNO$_3$, 28.9%, and NaNO$_3$, 39.8%. No crystalline hydrates or double salts are formed. Sketch this system on a triangular diagram, labeling the areas in which you would expect to find (*a*) only solution; (*b*) a mixture of solution and solid KNO$_3$; (*c*) a mixture of solution and solid NaNO$_3$; (*d*) a mixture of solid KNO$_3$, solid NaNO$_3$, and solution.

CHAPTER 5

CHEMICAL EQUILIBRIUM

In the preceding chapters we applied thermodynamics mainly to physical processes. We now apply these ideas to chemical reactions, starting with reactions of ideal gases. The thermodynamic treatment of chemical equilibrium illustrates the usefulness of the concept of chemical potential.

The idea of the reversibility of chemical reactions was first stated clearly in 1799 by C. Berthollet, while he was acting as scientific adviser to Napoleon in Egypt. He noted the deposits of sodium carbonate in certain salt lakes and concluded that they were produced by the high concentration of sodium chloride and dissolved calcium carbonate, the reverse of the laboratory experiment in which sodium carbonate reacts with calcium chloride to precipitate calcium carbonate. In 1863 the influence that the concentrations of ethyl alcohol and acetic acid have on the concentration of ethyl acetate was reported by M. Berthelot and Saint-Gilles.

In 1864 Guldberg and Waage showed experimentally that in chemical reactions a definite equilibrium is reached that can be approached from either direction. They were apparently the first to realize that there is a mathematical relation between the concentrations of reactants and products at equilibrium. In 1877 van't Hoff suggested that in the equilibrium expression for the hydrolysis of ethyl acetate the concentrations of each reactant should appear to the first power, each reactant having a coefficient of unity in the balanced chemical equation. Since Guldberg and Waage used the term "active masses" to mean concentration, the term "mass action expression" is still used, although we now realize that chemical equilibria can be expressed accurately only in terms of activities.

5.1 THERMODYNAMICS OF IDEAL GASES

Since the chemical potential μ of a pure substance is equal to its molar Gibbs free energy $\bar{G}$, equations 2.64 and 2.65 may be written

$$\left(\frac{\partial \mu}{\partial T}\right)_P = -\bar{S} \tag{5.1}$$

$$\left(\frac{\partial \mu}{\partial P}\right)_T = \bar{V} \tag{5.2}$$

Since for an ideal gas $\bar{V} = RT/P$,

$$\left(\frac{\partial \mu}{\partial P}\right)_T = \frac{RT}{P} \tag{5.3}$$

This equation may be integrated from a defined standard pressure P° to any pressure P.

$$\int_{\mu^\circ}^{\mu} d\mu = RT \int_{P^\circ}^{P} \frac{dP}{P} \tag{5.4}$$

$$-\mu^\circ + RT \ln \frac{P}{P^\circ} \tag{5.5}$$

The chemical potential for a mole of gas at the standard pressure is represented by μ°. The standard chemical potential μ° will have a different value at different temperatures. The logarithmic dependence of the chemical potential of an ideal gas on the pressure of the gas is illustrated in Fig. 5.1. The pressures of gases are commonly measured in atmospheres and the standard pressure P° is taken as one atmosphere. Thus equation 5.5 is usually written

$$\mu = \mu^\circ + RT \ln P \tag{5.6}$$

but we must remember that P really stands for P/P°, which is dimensionless. We cannot take the logarithm of 10 atm, only the logarithm of a pure number.

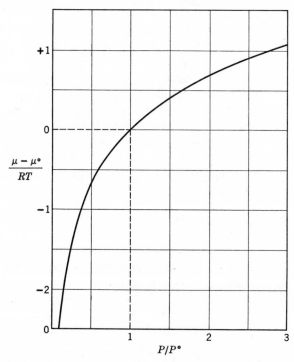

Fig. 5.1   Dependence of the chemical potential of an ideal gas on the pressure of the gas.

5.2 THERMODYNAMICS OF IDEAL GAS MIXTURES

A gas mixture is ideal if the chemical potential μ_i of each component of the mixture follows the equation

$$\mu_i = \mu_i^\circ + RT \ln X_i P = \mu_i^\circ + RT \ln p_i \tag{5.7}$$

where X_i is the mole fraction of component i and P is the total pressure of the mixture. We define $X_i P$ as the partial pressure p_i of the ith component. The derivatives of the chemical potential of the ith component of a mixture have been given earlier in equation 2.104 and 2.105. Applying the following equation

$$\left(\frac{\partial \mu_i}{\partial P} \right)_{T, n_i, n_j} = \bar{V}_i \tag{5.8}$$

to equation 5.7 yields

$$\bar{V}_i = \frac{RT}{P} \tag{5.9}$$

Thus the partial molar volume $\bar{V}_i$ of an ideal gas in an ideal gas mixture is equal to the volume it would occupy at a pressure equal to that on the mixture. The total volume of the mixture is given by (see Section 2.18)

$$V = \sum n_i \bar{V}_i = \frac{RT}{P} \sum n_i \tag{5.10}$$

Thus the total volume of an ideal gas mixture follows the ideal gas law with the number of moles equal to the total number of moles of gas present.

In equation 2.97 we showed that the Gibbs free energy of a system is the sum of the contributions of the various components.

$$G = \sum n_i \mu_i \tag{5.11}$$

Let us use this equation to calculate the Gibbs free energy of mixing two gases each initially at one atmosphere, as shown in Fig. 5.2. The Gibbs free energy of the two gases in the initial state is

$$G_{\text{initial}} = n_1 \mu_1^\circ + n_2 \mu_2^\circ \tag{5.12}$$

If the gases mix to form an ideal mixture the Gibbs free energy is obtained by using equation 5.7 in equation 5.11.

$$G_{\text{final}} = n_1 \mu_1^\circ + n_2 \mu_2^\circ + RT(n_1 \ln X_1 + n_2 \ln X_2) \tag{5.13}$$

Thus the Gibbs free energy of mixing is

$$\Delta G_{\text{mix}} = G_{\text{final}} - G_{\text{initial}} = RT(n_1 \ln X_1 + n_2 \ln X_2)$$
$$= nRT(X_1 \ln X_1 + X_2 \ln X_2) \tag{5.14}$$

where $n = n_1 + n_2$. Since mole fractions are less than unity, the logarithmic

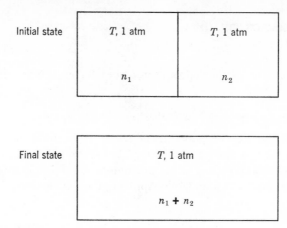

Fig. 5.2 Mixing of ideal gases. The partition between n_1 moles of gas 1 at T and 1 atm and n_2 moles of gas 2 at T and 1 atm is withdrawn so that the gases can mix.

terms are negative and $\Delta G_{\mathrm{mix}} < 0$. This corresponds with the fact that the mixing of gases is a spontaneous process at constant temperature and pressure.

The Gibbs free energy of mixing two ideal gases is plotted versus the mole fraction of one of the gases in Fig. 5.3a. The greatest Gibbs free energy of mixing is obtained for $X_1 = X_2 = \tfrac{1}{2}$.

Since according to equation 2.64

$$\left(\frac{\partial \Delta G_{\mathrm{mix}}}{\partial T}\right)_P = -\Delta S_{\mathrm{mix}} \tag{5.15}$$

$$\Delta S_{\mathrm{mix}} = -nR(X_1 \ln X_1 + X_2 \ln X_2) \tag{5.16}$$

This is the same as equation 2.20 obtained earlier by a different method. The dependence of the entropy of mixing on the mole fraction of one of the components is shown in Fig. 5.3b.

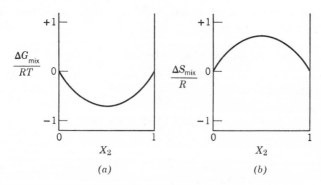

$$(a) \qquad\qquad (b)$$

Fig. 5.3 Thermodynamic quantities for the mixing of two ideal gases to form one mole of mixture.

Since according to equation 2.49

$$\Delta G_{\text{mix}} = \Delta H_{\text{mix}} - T\Delta S_{\text{mix}} \tag{5.17}$$

we find that

$$\Delta H_{\text{mix}} = 0 \tag{5.18}$$

Thus when ideal gases are mixed at constant temperature and pressure there is no heat effect. This corresponds with the fact that molecules of ideal gases do not attract or repel each other. Thus from an energy standpoint it makes no difference whether the gases are separated or mixed. The driving force for mixing arises exclusively from the change in entropy. From the view point of statistical mechanics (Chapter 17) the mixed state is found at equilibrium because it is more probable, as discussed in Section 2.9 for diffusion in idealized crystals.

Since by equation 2.65

$$\left(\frac{\partial \Delta G_{\text{mix}}}{\partial P}\right)_T = \Delta V_{\text{mix}} \tag{5.19}$$

we find that for the mixing of ideal gases

$$\Delta V_{\text{mix}} = 0 \tag{5.20}$$

Example 5.1 Calculate the changes in the thermodynamic quantities G, S, H, and V for the mixing of half a mole of oxygen with half a mole of nitrogen at $25°$ and 1 atm, assuming that they are ideal gases.

$$\Delta G_{\text{mix}} = nRT(X_1 \ln X_1 + X_2 \ln X_2)$$
$$= (1 \text{ mol})(2.303)(1.987 \text{ cal K}^{-1}\text{ mol}^{-1})(298.15 \text{ K})(0.5 \log 0.5 + 0.5 \log 0.5)$$
$$= -410 \text{ cal}$$
$$\Delta S_{\text{mix}} = -\Delta G_{\text{mix}}/T$$
$$= 1.376 \text{ cal K}^{-1}$$
$$\Delta H_{\text{mix}} = 0$$
$$\Delta V_{\text{mix}} = 0$$

5.3 THERMODYNAMICS OF A SIMPLE GAS REACTION

As a simple example of a chemical equilibrium let us consider the isomerization of an ideal gas A to ideal gas B.

$$\text{A}(g) = \text{B}(g) \tag{5.21}$$

According to equation 5.11 the Gibbs free energy of the reaction mixture at any extent of reaction is

$$G = n_{\text{A}}\mu_{\text{A}} + n_{\text{B}}\mu_{\text{B}} \tag{5.22}$$

where n_{A} is the number of moles of A and n_{B} is the number of moles of B. If the reaction is started with one mole of A, the numbers of moles of A and B at a later time are given by

$$n_A = 1 - \xi \tag{5.23}$$

$$n_B = \xi \tag{5.24}$$

where ξ (xi) is called the extent of reaction. Thus

$$G = (1 - \xi)\mu_A + \xi\mu_B \tag{5.25}$$

The chemical potentials of A and B in the ideal gaseous mixture are given by

$$\mu_A = \mu_A^\circ + RT \ln X_A + RT \ln P$$
$$= \mu_A^\circ + RT \ln (1 - \xi) + RT \ln P \tag{5.26}$$

$$\mu_B = \mu_B^\circ + RT \ln X_B + RT \ln P$$
$$= \mu_B^\circ + RT \ln \xi + RT \ln P \tag{5.27}$$

Substituting these equations into equation 5.25.

$$G = (1 - \xi)\mu_A^\circ + \xi\mu_B^\circ + RT \ln P + RT[(1 - \xi) \ln (1 - \xi) + \xi \ln \xi] \tag{5.28}$$

where the last group of terms is simply the Gibbs free energy of mixing $(1 - \xi)$ moles of A with ξ moles of B. Figure 5.4 gives a plot of G versus ξ. The first three terms in equation 5.28 give the linear function represented by the dashed line. It is the mixing term that causes the minimum in the plot of G versus extent of reaction ξ. At constant temperature and pressure the criterion of equilibrium is that the Gibbs free energy is a minimum (see Section 2.11). Thus starting with A the Gibbs free energy can decrease along the curve until ξ_{eq} moles of B have been formed. Starting with B the Gibbs free energy can decrease until $(1 - \xi_{eq})$ moles of A have formed. Even though B has the lower value of the molar Gibbs

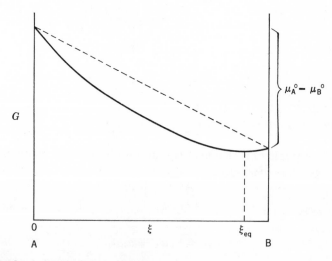

Fig. 5.4 Gibbs free energy of the reaction system $A(g) = B(g)$ versus extent of reaction ξ at constant temperature and pressure.

free energy $(\mu_A^\circ > \mu_B^\circ)$, the system can achieve a lower Gibbs free energy by having some A present at equilibrium with the resulting Gibbs free energy of mixing. Generalizing from this example we can say that no chemical reaction of gases really goes to completion; nevertheless it may be very difficult to detect reactants at equilibrium if the products have a very much lower Gibbs free energy.

The thermodynamic relation for equilibrium may be obtained by differentiation.

$$\left(\frac{\partial G}{\partial \xi}\right)_{T,P} = 0 \tag{5.29}$$

Setting the derivative of equation 5.28 equal to zero yields

$$\mu_A^\circ + RT \ln (X_A)_{eq} = \mu_B^\circ + RT \ln (X_B)_{eq} \tag{5.30}$$

$$\mu_{A,eq} = \mu_{B,eq} \tag{5.31}$$

Thus at equilibrium the chemical potential of the product is equal to the chemical potential of the reactant.

Rearranging equation 5.30 yields

$$\left(\frac{X_B}{X_A}\right)_{eq} = e^{(\mu_A^\circ - \mu_B^\circ)/RT} = K \tag{5.32}$$

Since the chemical potentials of A and B at the standard pressure are constants at a constant temperature, the ratio X_B/X_A at equilibrium is a constant at a constant temperature. Thus the equilibrium constant K embodies the idea that the Gibbs free energy is a minimum at equilibrium at constant temperature and pressure.

5.4 DERIVATION OF THE GENERAL EQUILIBRIUM EXPRESSION

We showed earlier (in Section 1.18) that a generalized chemical reaction may be represented by

$$0 = \sum \nu_i A_i \tag{5.33}$$

At any time during the reaction the Gibbs free energy of the reaction mixture is given by equation 2.97, which is

$$G = n_1 \mu_1 + n_2 \mu_2 + \cdots$$
$$= \sum_i n_i \mu_i \tag{5.34}$$

where n_i represents the number of moles of substance i at a particular extent of reaction. When a chemical reaction occurs the changes in the number of moles of the various reactants and products are related through the stoichiometric coefficients of the balanced chemical equation 5.33. If a reaction system initially contains n_{io} moles of i, the number n_i of moles of i at a later time may be written

$$n_i = n_{io} + \nu_i \xi \tag{5.35}$$

Where ξ is the extent of reaction. If a product is not present initially, the extent of reaction is simply the number of moles that have been produced divided by the stoichiometric coefficient. The extent of reaction is defined so that it is initially zero and is always positive.

Following the procedure used in the preceeding section we may express the Gibbs free energy of a reaction system in terms of the extent of reaction ξ by substituting equation 5.35 into equation 5.34.

$$G = n_{1o}\mu_1 + n_{2o}\mu_2 + \cdots + (\nu_1\mu_1 + \nu_2\mu_2 + \cdots)\xi$$
$$= \sum_i n_{io}\mu_i + \xi \sum_i \nu_i\mu_i \tag{5.36}$$

The derivative of the Gibbs free energy with respect to the extent of reaction at constant temperature and pressure is

$$\left(\frac{dG}{d\xi}\right)_{T,P} = \sum \nu_i\mu_i = \Delta G \tag{5.37}$$

We will refer to this derivative simply as ΔG, and so it is important to emphasize what we mean by this symbol. The change in Gibbs free energy ΔG for a chemical reaction is the change in G per mole of a product with stoichiometric coefficient 1, or per $|\nu_1|$ moles for a reactant with stoichiometric coefficient ν_1, *for an infinitesimal transfer of reactant to product.* The transfer is accomplished without a change in concentration and so ΔG is the Gibbs free energy change for reactants at *any* specified chemical potentials to products at *any* specified chemical potentials. If you do not like to think about an infinitesimal transfer, think about the transfer of a mole of a reactant with stoichiometric coefficient 1 in a very large reaction mixture so that there is no significant change in any of the concentrations.

If a chemical reaction is at equilibrium the change in Gibbs free energy for a transfer from reactants to products is zero; $\Delta G = 0$. Thus equation 5.37 leads to the following general condition for chemical equilibrium.

$$\sum \nu_i\mu_i = 0 \tag{5.38}$$

where the chemical potentials in this equation are the values at equilibrium. This general condition applies to all chemical reactions whether they involve gases, liquids, solids, or solutions.

5.5 CHEMICAL EQUILIBRIUM IN
IDEAL GASES

Since the chemical potential of an ideal gas is independent of the presence of other gases, we may substitute equation 5.7 into equation 5.37 to obtain an expression for the change in Gibbs free energy for a reaction of ideal gases.

$$\Delta G = \nu_1\mu_1^\circ + \nu_2\mu_2^\circ + \cdots + RT \ln p_1^{\nu_1} p_2^{\nu_2} \cdots$$
$$= \sum_i \nu_i\mu_i^\circ + RT \ln \prod_i p_i^{\nu_i}$$
$$= \Delta G^\circ + RT \ln \prod_i p_i^{\nu_i} \tag{5.39}$$

where $\prod\limits_{i}$ represents the product of the partial pressures p_i of the various gases, each raised to the power that corresponds with its stoichiometric coefficient in the balanced chemical equation. (Remember that the stoichiometric coefficients of reactants are negative numbers and the stoichiometric coefficients of products are positive numbers.) The quantity $\Delta G°$ is referred to as the *standard change in Gibbs free energy*. It is the difference between the Gibbs free energy of the numbers of moles of products indicated by the balanced chemical equation with each of the gases at its standard pressure (usually 1 atm) and the Gibbs free energy of the numbers of moles of reactants indicated by the balanced chemical equation with each of the reactant gases at its standard pressure. Thus $\Delta G°$ refers to the complete conversion of reactants to products.

If the reactants and products are present at their equilibrium pressures, $\Delta G = 0$ and equation 5.39 becomes

$$\Delta G° = -RT \ln \left(\prod_{i} p_i^{v_i} \right)_{\text{eq}}$$
$$= -RT \ln K_p \qquad (5.40)$$

where

$$K_p = \left(\prod_{i} p_i^{v_i} \right)_{\text{eq}} \qquad (5.41)$$

Since $\Delta G°$ is a definite quantity at a given temperature, the product of equilibrium pressures each raised to the appropriate power is a constant. For ideal gases K_p is a function of temperature only

Equation 5.41 makes it appear that K_p will have pressure units if there are different numbers of moles of gaseous reactants and products. However, as pointed out in connection with equation 5.6, the pressure in the expression for the chemical potential is really to be understood as the *ratio* of the actual pressure to the defined pressure of the standard state. Thus the equilibrium constant is, strictly speaking, a dimensionless quantity. In some cases it is convenient to use pressure units to remind us of the standard state since different defined standard states lead to different numerical values of K_p if the number of moles of gaseous reactants and products are different.

Example 5.2 Given that the equilibrium constant and standard Gibbs free energy change for the ammonia synthesis reaction at $400°$ are given by

(1) $N_2(g) + 3H_2(g) = 2NH_3(g)$

$$K_1 = \frac{p_{NH_3}^2}{p_{N_2} p_{H}^3} = 1.64 \times 10^{-4}$$

$$\Delta G_1° = -RT \ln K_i = -(1.987 \text{ cal K}^{-1} \text{ mol}^{-1})(673 \text{ K})(2.303) \log (1.64 \times 10^{-4})$$
$$= 11{,}657 \text{ cal mol}^{-1}$$

when pressures are expressed in atmospheres, calculate the equilibrium constants and standard free energy changes for

(2) $\frac{1}{2}N_2(g) + \frac{3}{2}H_2(g) = NH_3(g)$

and

(3) $2NH_3(g) = N_2(g) + 3H_2(g)$

$$K_2 = \frac{p_{NH_3}}{p_{N_2}^{1/2} p_{H_2}^{3/2}} = (1.64 \times 10^{-4})^{1/2} = 1.28 \times 10^{-2}$$

$$\Delta G_2^\circ = -(1.987 \text{ cal K}^{-1} \text{ mol}^{-1})(673 \text{ K})(2.303) \log (1.28 \times 10^{-2})$$
$$= 5829 \text{ cal mol}^{-1}$$

$$K_3 = \frac{p_{N_2} p_{H_2}^3}{p_{NH_3}^2} = \frac{1}{1.64 \times 10^{-4}} = 0.6098 \times 10^4$$

$$\Delta G_3^\circ = -(1.987 \text{ cal K}^{-1} \text{ mol}^{-1})(673 \text{ K})(2.303) \log (0.6098 \times 10^4)$$
$$= -11,657 \text{ cal mol}^{-1}$$

This example illustrates the necessity of knowing the balanced chemical equation in order to interpret the numerical value of an equilibrium constant. It is clear that if a reaction is reversed the equilibrium constant becomes the reciprocal of that for the first reaction; the standard Gibbs free energy change has the same magnitude but the opposite sign.

5.6 CHEMICAL EQUILIBRIUM IN NONIDEAL GASES

Nonideal gases do not follow the equations we have been using; that is, their chemical potential is not a simple logarithmic function of their pressure. In order to treat the thermodynamic properties of nonideal gases it is convenient to define the fugacity f, which is like the pressure of an ideal gas in that the chemical potential is a linear function of the logarithm of the fugacity.

$$\mu = RT \ln f + B(T) \tag{5.42}$$

where $B(T)$ is a characteristic of the substance and is a function only of the temperature. As the pressure of a nonideal gas approaches zero it behaves more and more like an ideal gas so that

$$\frac{f}{P} \to 1 \quad \text{as} \quad P \to 0 \tag{5.43}$$

For most gases f/P does not differ very much from unity at one atmosphere pressure.

The equilibrium expression for nonideal gases is obtained from the general equation 5.38 by substituting equation 5.42 for the chemical potential in terms of the fugacity f. Following the procedure of the preceeding section we obtain

$$\Delta G^\circ = -RT \ln \left(\prod_i f_i^{\nu_i} \right)_{eq} \tag{5.44}$$
$$= -RT \ln K_f$$

with

$$K_f = \left(\prod_i f_i^{v_i} \right)_{eq}$$ (5.45)

For nonideal gases $\Delta G°$ refers to the change in Gibbs free energy for the reactants at unit fugacity going to products at unit fugacity.

Since all gases approach ideality as their pressures approach zero, the value of K_f may be determined in experiments at very low pressure. The calculation of equilibrium extents of reaction that satisfy equation 5.45 is complicated by the fact that the fugacity of a gas is affected by the presence of other gases. However, by use of approximations it is possible to make useful estimates of the fugacity*.

5.7 K_p, K_c, AND K_X

The equilibrium expression for a reaction of ideal gases may also be written in terms of the concentrations of the reactants and products at equilibrium. Substituting $p = cRT$, where c represents the number of moles of a gas per liter, in equation 5.41, we have

$$K_p = K_c (RT)^{\Sigma v_i}$$ (5.46)

where

$$K_c = \prod_i (c_i)_{eq}^{v_i}$$ (5.47)

and $\sum_i v_i$ is the difference between the number of moles of gaseous products and the number of moles of gaseous reactants. If the number of moles of gaseous products is equal to the number of moles of gaseous reactants, $\sum_i v_i = 0$ and $K_p = K_c$.

Sometimes it is advantageous to express the equilibrium constants in terms of mole fractions X_i rather than pressures or concentrations.

$$K_X = \prod_i (X_i)_{eq}^{v_i}$$ (5.48)

This equilibrium constant is related to K_p in the following way:

$$K_p = \prod_i (X_i P)_{eq}^{v_i} = P^{\Sigma v_i} \prod_i (X_i)_{eq}^{v_i} = P^{\Sigma v_i} K_X$$ (5.49)

where P is the total pressure of the reacting gases. Alternatively, P may be taken as the total pressure, including inert gases, but then the numbers of moles in inert gases must be included in the calculations of mole fractions. The value of K_p (or K_c) for a reaction of ideal gases does not depend on the pressure. But if $\sum v_i \neq 0$, the value of the equilibrium constant K_X expressed in terms of mole fractions does *depend on the pressure*, according to equation 5.49.

---

* G. N. Lewis and M. Randall, rev. by K. S. Pitzer and L. Brewer, *Thermodynamics*, McGraw-Hill Book Co., New York, 1961, p. 153.

Example 5.3   Calculate $\Delta G°$ for the reaction $N_2(g) + 3H_2(g) = 2NH_3(g)$ at $400°$ using K_c. Interpret the different values of $\Delta G°$ calculated from K_p and K_c.

$$K_c = K_p(RT)^{-\Sigma \nu_i}$$
$$= (1.64 \times 10^{-4})[(0.08205 \text{ liter atm K}^{-1} \text{ mol}^{-1})(673 \text{ K})]^2$$
$$= 0.500$$

$$\Delta G° = -RT \ln K_c$$
$$= -(1.987 \text{ cal K}^{-1} \text{ mol}^{-1})(673 \text{ K})(2.303) \log 0.500$$
$$= 927 \text{ cal mol}^{-1}$$

This is the change in Gibbs free energy when 1 mol of N_2 in the ideal gas state at the concentration of 1 mol per liter and 3 mol of H_2 in the ideal gas state at 1 mol per liter react to form 2 mol of NH_3 in the ideal gas state at 1 mol per liter.

$\Delta G_1°$ is the change in Gibbs free energy when 1 mol of N_2 in the ideal gas state at 1 atm, and 3 mol of H_2 in the ideal gas state at 1 atm react to form 2 mol of NH_3 in the ideal gas state at 1 atm.

5.8   DETERMINATION OF EQUILIBRIUM CONSTANTS

If the initial concentrations of the reactants are known, it is necessary to determine the concentration of only one reactant or product at equilibrium to be able to calculate the concentrations or pressures of the others by means of the balanced chemical equation. Chemical methods can be used for such analyses only when the reaction may be stopped at equilibrium, as by a very sudden chilling to a temperature where the rate of further chemical change is negligible, or by destruction of a catalyst. Otherwise, the concentrations will shift during the chemical analysis.

Measurements of physical quantities, such as density, pressure, light absorption refractive index, and electrical conductivity, are especially useful for the determination of the concentration of reactants at equilibrium, since it is unnecessary to "stop" the reaction.

It is essential to know that equilibrium has been reached before the analysis of the mixture can be used for calculating the equilibrium constant. The following criteria for the attainment of equilibrium at constant temperature are useful.

1. The same equilibrium constant should be obtained when the equilibrium is approached from either side.
2. The same equilibrium constant should be obtained when the concentrations of reacting materials are varied over a wide range.

The determination of the density of a partially dissociated gas provides one of the simplest methods for measuring the extent to which the gas is dissociated. When a gas dissociates, more molecules are produced, and at constant temperature and pressure the volume increases. The density at constant pressure then decreases, and the difference between the density of the undissociated gas and that of the partially dissociated gas is directly related to the degree of dissociation.

Let α represent the fraction dissociated, so that $1 - \alpha$ denotes the fraction remaining undissociated. If we start with 1 mol of gas the number of moles of gaseous products in the balanced chemical equation is $1 + \sum \nu_i$. Therefore the number of moles of gas present at equilibrium is

$$(1 - \alpha) + (1 + \sum \nu_i)\alpha = 1 + \alpha \sum \nu_i \qquad (5.50)$$

Since the density of an ideal gas at constant pressure and temperature is *inversely* proportional to the number of moles for a given weight, the ratio of the density ρ_1 of the undissociated gas to the density ρ_2 of the partially dissociated gas is given by the expression

$$\frac{\rho_1}{\rho_2} = 1 + \alpha \sum \nu_i \qquad (5.51)$$

$$\alpha = \frac{\rho_1 - \rho_2}{\rho_2 \sum \nu_i} \qquad (5.52)$$

It is always advantageous to check an equation with some simple calculations. With reference to equation 5.52, if there is no dissociation, then $\alpha = 0$ and $\rho_1 = \rho_2$; if dissociation is complete, then $\alpha = 1$, $\rho_2 \sum \nu_i = \rho_1 - \rho_2$ and $\rho_1 = (1 + \sum \nu_i)\rho_2$.

Molecular weights may be substituted in equation 5.52 for the densities of gases to which they are proportional at constant temperature and pressure, giving

$$\alpha = \frac{M_1 - M_2}{M_2 \sum \nu_i} \qquad (5.53)$$

where M_1 is the molecular weight of the undissociated gas, and M_2 is the average molecular weight of the gases when the gas is partially dissociated.

The dissociation of nitrogen tetroxide is represented by the equation

$$N_2O_4 = 2NO_2$$

and

$$K_p = \frac{p^2_{NO_2}}{p_{N_2O_4}} \qquad (5.54)$$

If α represents the degree of dissociation $(1 - \alpha)$ is proportional to the number of moles of undissociated N_2O_4; 2α is proportional to the number of moles of NO_2; and $(1 - \alpha) + 2\alpha$ or $1 + \alpha$ is proportional to the total number of moles.

If the total pressure of N_2O_4 plus NO_2 is P, the partial pressures are:

$$p_{N_2O_4} = \frac{1 - \alpha}{1 + \alpha} P \quad \text{and} \quad p_{NO_2} = \frac{2\alpha}{1 + \alpha} P$$

Then,

$$K_p = \frac{\left(\dfrac{2\alpha}{1+\alpha}P\right)^2}{\dfrac{1-\alpha}{1+\alpha}P} = \frac{4\alpha^2 P}{1-\alpha^2} \tag{5.55}$$

In this reaction there is an increase in volume at constant pressure; each mole of gas that dissociates produces 2 mol of gas. According to the principle of Le Châtelier (Sections 5.14 and 5.16) an increase of pressure will cause the reaction to shift toward N_2O_4 since the equilibrium shifts in the direction that tends to minimize the effect of the applied change.

Example 5.4 If 1.588 g of nitrogen tetroxide gives a total pressure of 760 Torr when partially dissociated in a 500-cm^3 glass vessel at 25°, what is the degree of dissociation α? What is the value of K_p? What is the degree of dissociation at a total pressure of 0.5 atm?

$$M_2 = \frac{RT}{P}\frac{g}{V} = \frac{(0.08205 \text{ liter atm K}^{-1} \text{ mol}^{-1})(298.1 \text{ K})(1.588 \text{ g})}{(1 \text{ atm})(0.500 \text{ liter})}$$

$$= 77.68 \text{ g mol}^{-1}$$

$$\alpha = \frac{92.02 - 77.68}{77.68} = 0.1846$$

$$K_p = \frac{4\alpha^2 P}{1-\alpha^2} = \frac{(4)(0.1846)^2(1)}{1-(0.1846)^2} = 0.141$$

The degree of dissociation at 0.5 atm is obtained as follows:

$$K_p = 0.141 = \frac{4\alpha^2(0.5)}{1-\alpha^2}$$

$$0.141(1-\alpha^2) = 2\alpha^2$$

$$\alpha = 0.257$$

If the product molecules are different, the calculation of K_p is altered; this is illustrated for the reaction

$$PCl_5 = PCl_3 + Cl_2$$

When 1 mole of PCl_5 dissociates, there will be at equilibrium $1-\alpha$ mole of PCl_5, α mole of PCl_3, and α mole of Cl_2, where α is the degree of dissociation. The total number of moles is

$$[1-\alpha] + \alpha + \alpha = 1 + \alpha$$

If P is the total pressure due to PCl_5, PCl_3, and Cl_2, the partial pressures are

$$p_{PCl_5} = \frac{1-\alpha}{1+\alpha}P \qquad p_{PCl_3} = \frac{\alpha}{1+\alpha}P \qquad p_{Cl_2} = \frac{\alpha}{1+\alpha}P$$

and

$$K_p = \frac{p_{PCl_3}\, p_{Cl_2}}{p_{PCl_5}} = \frac{\left(\dfrac{\alpha}{1+\alpha}\,P\right)\left(\dfrac{\alpha}{1+\alpha}\,P\right)}{[(1-\alpha)/(1+\alpha)]P} = \frac{\alpha^2 P}{1-\alpha^2} \tag{5.56}$$

It should be noted that this equation differs from equation 5.55 by the lack of the factor 4.

This reaction is an interesting one to discuss in more detail. Qualitatively, it can be seen that increasing the total pressure of the three reactants will decrease the degree of dissociation, because the undissociated PCl_5 occupies the smaller volume. If chlorine is added, p_{Cl_2} increases, and since K_p remains constant, p_{PCl_3} must diminish and p_{PCl_5} must increase. The degree of dissociation is decreased also by the addition of PCl_3. In general, the dissociation of any substance is repressed by the addition of its dissociation products. The addition of an inert gas at constant volume has no effect on the dissociation because the partial pressures of the gases involved in the reaction are not affected by the presence of another gas provided that the gases behave ideally.

Another classical example of an equilibrium in gases is the dissociation of hydrogen iodide gas,

$$2HI = H_2 + I_2$$

Very careful measurements were made* in which quartz vessels of known volume were filled with hydrogen iodide at a measured pressure and heated in an electric thermostat at 425.1° for several hours until equilibrium was established. The vessels were then chilled quickly and analyzed for iodine by titration with sodium thiosulfate. The concentration of hydrogen at equilibrium is equal to that of the iodine. The concentration of hydrogen iodide at equilibrium was obtained by subtracting twice the iodine concentration from the initial hydrogen iodide concentration. The equilibrium concentrations in moles per liter are shown in Table 5.1. The first three sets of data were obtained by starting with hydrogen iodide. The last five were obtained by starting from the other side of the equilibrium, weighing the initial quantity of iodine, measuring the pressure of hydrogen, and titrating the iodine after equilibrium was reached. The close check between the two sets of data shows that equilibrium was reached in every case.

It is interesting that a change in pressure does not alter the equilibrium in this gaseous reaction. In terms of the partial pressures of the components of this gaseous system, instead of the concentrations, we have

$$\frac{p_{H_2}\, p_{I_2}}{p_{HI}^2} = K_p \tag{5.57}$$

Now, if the total pressure on the system is increased to n times its original value, all the partial pressures are increased in the same proportion, and

$$\frac{(np_{H_2})(np_{I_2})}{n^2 p_{HI}^2} = K_p \tag{5.58}$$

* A. H. Taylor, Jr., and R. H. Crist, *J. Am. Chem. Soc.*, **63**, 1381 (1941).

Table 5.1 Equilibrium Between Hydrogen, Iodine, and Hydrogen Iodide at 698.2 K.
$K_c = c_{H_2}c_{I_2}/c_{HI}^2$

$c_{I_2} \times 10^3$, mol/l.	$c_{H_2} \times 10^3$, mol/l.	$c_{HI} \times 10^3$, mol/l.	$K_c \times 10^2$
0.4789	0.4789	3.531	1.839
1.1409	1.1409	8.410	1.840
0.4953	0.4953	3.655	1.836
1.7069	2.9070	16.482	1.827
1.2500	3.5600	15.588	1.831
0.7378	4.5647	13.544	1.836
2.3360	2.2523	16.850	1.853
3.1292	1.8313	17.671	1.835

which is equivalent to the original expression, since n cancels out. The equilibrium is thus seen to be independent of the pressure. This independence applies only to those reactions for which the number of moles of gaseous reactants is equal to the number of moles of gaseous products. In this case $K_p = K_c = K_X$.

5.9 REACTION EQUILIBRIUM IN SOLUTION

The fugacities f of components of condensed phases are frequently difficult to determine; this is particularly true for solids that are practically nonvolatile. However, since we are generally only interested in *changes* in thermodynamic quantities we only need *relative* fugacities. The relative fugacity or activity a_i of a substance is defined as the ratio of the fugacity f_i in some state to the fugacity f_i° in some standard state at the same temperature.

$$a_i = \frac{f_i}{f_i^\circ} \tag{5.59}$$

The activity is a measure of the difference in the chemical potential of the chosen and the standard states because, according to equation 5.42,

$$\mu_i - \mu_i^\circ = RT\ln f_i + B_i(T) - [RT\ln f_i^\circ + B_i(T)]$$
$$= RT\ln\frac{f_i}{f_i^\circ}$$
$$= RT\ln a_i \tag{5.60}$$

Using this equation we can show that in general an equilibrium constant is given by

$$K = \prod_i (a_i)_{eq}^{\nu_i} \tag{5.61}$$

For a reaction in ideal solution, the activity coefficients (Section 4.10) are equal to unity, and on the mole fraction scale

$$K = \prod_i (X_i)_{eq}^{\nu_i} \qquad (5.62)$$

The equilibrium constants for reactions in solution may be markedly affected by the solvent, even if the solvent is not a direct participant in the reaction, because the solvent may affect the activity coefficients of the reactants and products quite differently. A number of equilibria in solution will be discussed in Chapter 7.

Example 5.5 One mole of acetic acid is mixed with 1 mol of ethanol at 25°, and after equilibrium is reached a titration with standard alkali solution shows that 0.667 mol of acetic acid has reacted to form ethyl acetate according to

$$CH_3CO_2H + C_2H_5OH = CH_3CO_2C_2H_5 + H_2O$$

Calculate the apparent equilibrium constant in terms of mole fraction.

$$K = \frac{X_{CH_3CO_2C_2H_5}X_{H_2O}}{X_{CH_3CO_2H}X_{C_2H_5OH}} = \frac{(0.667/2)(0.667/2)}{[(1.000 - 0.667)/2][(1.000 - 0.667)/2]}$$
$$= 4.00$$

The total number of moles appears in the denominator of each term and therefore cancels. When 0.500 mol of ethanol is added to 1.000 mol of acetic acid at 25°, how much ester x will be formed at equilibrium?

$$K = 4.00 = \frac{x^2}{(1.000 - x)(0.500 - x)} \qquad x = 0.422 \qquad or \qquad 1.577 \, mol$$

Two solutions of a quadratic equation are possible, but in such problems one solution is incompatible with the physical-chemical facts. In this example it is impossible to produce more moles of ester than the original number of moles of ethanol, and so the value 1.577 is impossible. Actually 0.422 mol of ester and 0.422 mol of water are formed, and (0.500 − 0.422) or 0.078 mol of ethanol and (1.000 − 0.422) or 0.578 mol of acetic acid remain unreacted. An experimental value of 0.414 mol of ester was obtained in the laboratory. Additional determinations are as follows:

Moles ethanol added to 1 mol of acetic acid	0.080	0.280	2.240	8.000
Moles of ethyl acetate calculated from K	0.078	0.232	0.864	0.945
Moles of ethyl acetate found experimentally	0.078	0.226	0.836	0.966

In example 5.5 the number of moles of reactants and of products is the same, and the calculation is simplified. In reactions where $\sum \nu_i$ is not zero the total number of moles will not cancel out.

In some cases it is convenient to write the equilibrium expression in terms of molal or molar concentrations, but even for *ideal* solutions such equilibrium constants will be independent of concentration only in dilute solutions. The reasons for this is that molar or molal concentrations are proportional to mole fractions only in dilute solutions.

5.10   EQUILIBRIA INVOLVING SOLIDS

The expression for chemical equilibrium in 5.38 is perfectly general. If pure solids (or pure immiscible liquids) are involved, their chemical potentials are constants independent of the extent of reaction so long as they are not used up before equilibrium is reached. For example, for the reaction

$$CaCO_3(s) = CaO(s) + CO_2(g) \tag{5.63}$$

at equilibrium

$$\mu^{\circ}_{CaCO_3(s)} = \mu^{\circ}_{CaO(s)} + \mu^{\circ}_{CO_2(g)} + RT \ln p_{CO_2} \tag{5.64}$$

$$\Delta G^{\circ} = \mu^{\circ}_{CaO(s)} + \mu^{\circ}_{CO_2(g)} - \mu^{\circ}_{CaCO_3(s)} = -RT \ln p_{CO_2}$$

$$= -RT \ln K \tag{5.65}$$

so that

$$K_p = p_{CO_2} \tag{5.66}$$

The equilibrium constant of such a reaction is independent of the amount of pure solid (or liquid) phase, provided only that it is present at equilibrium. Table 5.2 gives the equilibrium pressures of CO_2 above $CaCO_3 + CaO$ at various temperatures.

If the partial pressure of CO_2 over $CaCO_3$ is maintained lower than K_p at a constant temperature, all the $CaCO_3$ is converted into CaO and CO_2. On the other hand, if the partial pressure of CO_2 is maintained higher than K_p, all the CaO is converted into $CaCO_3$. In this respect equilibria involving pure solids (or liquids) are different from other chemical equilibria which would simply go to a new equilibrium position, and not to completion, if the partial pressure of one of the reactants or products is maintained constant.

If solid or liquid solutions are formed (for example, if CaO and $CaCO_3$ were somewhat mutually soluble), the position of the equilibrium would depend on the concentration or, more specifically, the activities of the components in the equilibrium solid solution.

5.11   CALCULATION OF ΔG°

There are three ways that ΔG° for a reaction may be obtained: (1) ΔG° may be calculated from measured equilibrium constants using equation 5.40, (2) ΔG°

Table 5.2   Dissociation Pressures of Calcium Carbonate

Temp, °C	500	600	700	800
P, atm	9.3×10^{-5}	2.42×10^{-3}	2.92×10^{-2}	0.220
Temp, °C	897	1000	1100	1200
P, atm	1.000	3.871	11.50	28.68

may be calculated from $\Delta H°$ obtained calorimetrically and $\Delta S°$ obtained from third law entropies using

$$\Delta G° = \Delta H° - T \, \Delta S° \tag{5.67}$$

and (3) $\Delta G°$ may be calculated using statistical mechanics (Chapter 17) and certain information about molecules obtained from spectroscopic data (see Free-Energy Function, Section 5.18). The latter two methods are especially important for reactions that proceed so slowly that direct measurements of equilibrium cannot be made or for conditions that are difficult to achieve experimentally. For example, methane appears to be a stable substance at room temperature, and carbon and hydrogen appear to be unreactive toward each other. It is not possible then to measure the equilibrium among the three at room temperature, but the equilibrium constant may be calculated from the third law entropies and enthalpies of formation.

Example 5.6 Calculate the equilibrium constant for the following reaction at 25°:

$$C(graphite) + 2H_2(g) = CH_4(g)$$

From Table 1.3, $\Delta H° = -17,889$ cal. From Table 2.4, the entropy change for this reaction is

$$\Delta S° = 44.50 - 2(31.211) - 1.3609$$
$$= -19.28 \text{ cal K}^{-1}$$
$$\Delta G° = \Delta H° - T \, \Delta S° = -17,889 - (298.1)(-19.28)$$
$$= -12,140 \text{ cal}$$

The Gibbs free-energy change obtained in this way is identical with that shown in Table 5.3.

The equilibrium constant may be calculated from

$$-12,140 = -(1.987)(298.1)(2.303) \log K_p$$
$$\log K_p = 8.91$$
$$K_p = 8.1 \times 10^8$$

Although the equilibrium constant for this reaction is very large at room temperature, it is not possible to carry the reaction out since no catalyst is known. At high temperatures where a catalyst would probably not be required the equilibrium constant for the reaction is unfavorable for the synthesis of methane from its elements.

5.12 STANDARD GIBBS FREE ENERGY OF FORMATION

By adding and subtracting the reactions for which $\Delta G°$ is known, the Gibbs free-energy changes may be obtained for many other reactions, but the most convenient way to tabulate the extensive data on chemical equilibrium is by means of the Gibbs *free energy of formation* $\Delta \bar{G}_f°$, which is defined similarly to the enthalpy of formation (Section 1.18). The Gibbs free energy of formation of a

Table 5.3 Standard Gibbs Free Energies of Formation at 25° ($\Delta \bar{G}_f^\circ$ in kcal mol^{-1})

Elements and Inorganic Compounds

$O_3(g)$	39.06	$C(s, \text{diamond})$	0.6850
$H_2O(g)$	-54.6357	$CO(g)$	-32.8079
$H_2O(l)$	-56.6902	$CO_2(g)$	-94.2598
$HCl(g)$	-22.769	$PbO_2(s)$	-52.34
$Br_2(g)$	0.751	$PbSO_4(s)$	-193.89
$HBr(g)$	-12.72	$Hg(g)$	7.59
$HI(g)$	0.31	$AgCl(s)$	-26.224
$S(\text{monoclinic})$	0.023	$Fe_2O_3(s)$	-177.1
$SO_2(g)$	-71.79	$Fe_3O_4(s)$	-242.4
$SO_3(g)$	-88.52	$Al_2O_3(s)$	-376.77
$H_2S(g)$	-7.892	$UF_6(g)$	-485
$NO(g)$	20.719	$UF_6(s)$	-486
$NO_2(g)$	12.390	$CaO(s)$	-144.4
$NH_3(g)$	-3.976	$CaCO_3(s)$	-269.78
$HNO_3(l)$	-19.100	$NaF(s)$	-129.3
$P(g)$	66.77	$NaCl(s)$	-91.785
$PCl_3(g)$	-68.42	$KF(s)$	-127.42
$PCl_5(g)$	-77.59	$KCl(s)$	-97.592

Organic Compounds

Methane, $CH_4(g)$	-12.140	Propylene, $C_3H_6(g)$	14.990
Ethane, $C_2H_6(g)$	-7.860	1-Butene, $C_4H_8(g)$	17.217
Propane, $C_3H_8(g)$	-5.614	Acetylene, $C_2H_2(g)$	50.000
n-Butane, $C_4H_{10}(g)$	-3.754	Formaldehyde, $CH_2O(g)$	-26.3
Isobutane, $C_4H_{10}(g)$	-4.296	Acetaldehyde, $CH_3CHO(g)$	-31.96
n-Pentane, $C_5H_{12}(g)$	-1.96	Methanol, $CH_3OH(l)$	-39.73
n-Hexane, $C_6H_{14}(g)$	0.05	Ethanol, $CH_3CH_2OH(l)$	-41.77
n-Heptane, $C_7H_{16}(g)$	2.09	Formic acid, $HCO_2H(l)$	-82.7
n-Octane, $C_8H_{18}(g)$	4.14	Acetic acid, $CH_3CO_2H(l)$	-93.8
Benzene, $C_6H_6(g)$	30.989	Oxalic acid, $(CO_2H)_2(s)$	-166.8
Benzene, $C_6H_6(l)$	29.756	Carbon tetrachloride, $CCl_4(l)$	-16.4
Ethylene, $C_2H_4(g)$	16.282	Glycine, $H_2NCH_2CO_2H(s)$	-88.61

Ions in Water

H^+	0.000	SO_4^{2-}	-177.34	Cu^{2+}	15.53
OH^-	-37.595	HS^-	3.01	Ag^+	18.430
F^-	-66.08	NO_3^-	-26.41	Mg^{2+}	-108.99
Cl^-	-31.350	NH_4^+	-19.00	Ca^{2+}	-132.18
ClO_4^-	-2.57	PO_4^{3-}	-245.1	Li^+	-70.22
Br^-	-24.574	CO_3^{2-}	-126.22	Na^+	-62.589
I^-	-12.35	Zn^{2+}	-35.184	K^+	-67.466
S^{2-}	20.0	Cd^{2+}	-18.58		

Gaseous Atoms

H	48.575	Br	19.69	N	81.471
F	14.2	I	16.766	C	160.845
Cl	25.192				

These data have been obtained from F. D. Rossini, D. D. Wagman, W. H. Evans, S. Levine, and I. Jaffe, "Selected Values of Chemical Thermodynamic Properties," *Natl. Bur. of Standards Circ. 500*, U.S. Government Printing Office, Washington, D.C., 1952. *Circular 500* is being replaced by a series of revised tables. Data for the first 34 elements are given in NBS Technical Note 270-3 (1968).

substance is the Gibbs free-energy change for the reaction in which the substance in its standard state is formed from its elements in their standard states.

Table 5.3 gives some Gibbs free energies of formation taken from the larger tables of the National Bureau of Standards. The Gibbs free-energy change for a reaction (represented by equation 5.33) may be calculated from the Gibbs free energies of formation by use of an equation of the form

$$\Delta G° = \sum v_i \Delta \bar{G}_{f,i} \tag{5.68}$$

The equilibrium constant for the reaction may then be calculated by use of an equation of the form of equation 5.40. The way in which Gibbs free-energy changes for reactions at other temperatures are calculated from the values at 25° is discussed in a later section. The use of Table 5.3 in calculating Gibbs free-energy changes and equilibrium constants is illustrated by example 5.7.

Example 5.7 Calculate $\Delta G°$ and K_p at 25° for

$$CO(g) + H_2O(g) = CO_2(g) + H_2(g)$$

$$\Delta G° = (-94.2598 + 0) - (-32.8079 - 54.6357)$$

$$= -6.8162 \text{ kcal mol}^{-1}$$

$$\Delta G° = -(1.987)(298.1)(2.303) \log K_p$$

$$K_p = \frac{p_{H_2} p_{CO_2}}{p_{CO} p_{H_2O}} = 1.02 \times 10^5$$

5.13 SPONTANEITY OF CHEMICAL REACTIONS

The equilibrium constant may be expressed in terms of $\Delta H°$ and $\Delta S°$ using equations 5.40 and 5.67

$$K = e^{-\Delta G°/RT} = e^{\Delta S°/R} e^{-\Delta H°/RT} \tag{5.69}$$

From this equation we can see what is required of $\Delta S°$ and $\Delta H°$ in order for $K > 1$, that is, for the reactants in their standard states to be converted spontaneously to the products in their standard states. If $\Delta H°$ is negative and $\Delta S°$ is positive, the reaction will certainly be spontaneous. If $\Delta H°$ is positive and $\Delta S°$ is negative, the reaction cannot be spontaneous. If $\Delta H°$ and $\Delta S°$ have the same sign, the reaction may or may not be spontaneous, depending on the magnitudes of $\Delta S°$ and $\Delta H°$ and T. For example, in a reaction in which bonds are broken, $\Delta H°$ is positive, and, since the dissociated products have a greater degree of randomness, $\Delta S°$ is positive. At low temperatures $K < 1$ because the $\Delta H°$ term predominates. As the temperature is raised and $T \Delta S°$ in equation 5.67 becomes larger, however, the entropy factor becomes more important and at a sufficiently high temperature $K > 1$. Conversely, for the reverse reaction (that is, one in which bonds are formed), for which $\Delta H°$ is negative and $\Delta S°$ is negative, $K > 1$ at sufficiently low temperatures and $K < 1$ at sufficiently high temperatures.

We have been discussing $\Delta G°$, but of course the thermodynamic criteria of whether a given process will occur is ΔG, not $\Delta G°$. For our hypothetical reaction 5.33, equation 5.39 gives the change is Gibbs free energy for reactants at specified partial pressures and products at specified partial pressures.

If $\Delta G°$ has a positive value, the reactants in their standard states will not react spontaneously to give products in their standard states. By increasing the pressures or concentrations of the reactants or decreasing the pressures or concentrations of the products, however, it may be possible to make ΔG negative so that the reaction will proceed spontaneously.

Example 5.8   Calculate the change in Gibbs free energy for the production of $2NO_2(g)$ at 1 atm from $N_2O_4(g)$ at 10 atm at $25°$. As may be calculated from the data in example 5.4, the change in Gibbs free energy for the reaction

$$N_2O_4(g) = 2NO_2(g)$$

is $+1161$ cal, so that this reaction is not spontaneous at $25°$. Equation 5.39 becomes

$$\Delta G = \Delta G° + RT \ln \frac{p_{NO_2}^2}{p_{N_2O_4}}$$

$$= +1161 + (1.987)(298)(2.303) \log \frac{1^2}{10}$$

$$= -212 \text{ cal}$$

Thus the production of $NO_2(g)$ at 1 atm from $N_2O_4(g)$ at 10 atm is a spontaneous process. Thus a continuous process would be thermodynamically feasible if N_2O_4 was maintained at 10 atm and NO_2 was withdrawn by some method so that its partial pressure was maintained at 1 atm.

5.14   EFFECT OF TEMPERATURE ON CHEMICAL EQUILIBRIUM

The effect of temperature on chemical equilibrium is determined by $\Delta H°$, as shown by the Gibbs-Helmholtz equation (Section 2.15). Over moderate ranges of temperature $\Delta H°$ is frequently relatively independent of temperature. This happens when $\Delta C_P°$ is relatively small, and, when this is the case, $\Delta S°$ is also

Table 5.4 Equilibrium Constants for the Reaction $N_2(g) + O_2(g) = 2NO(g)$

Temp, K	1900	2000	2100	2200	2300	2400	2500	2600
$K_p \times 10^4$	2.31	4.08	6.86	11.0	16.9	25.1	36.0	50.3

relatively independent of temperature. Under these conditions the effect of temperature is most easily discussed in terms of equation 5.69, which may be written

$$\ln K = -\frac{\Delta H^\circ}{RT} + \frac{\Delta S^\circ}{R} \qquad (5.70)^*$$

According to this equation a plot of $\ln K$ versus $1/T$ has a slope of $-\Delta H^\circ/R$ and an intercept at $1/T = 0$ of $\Delta S^\circ/R$. If a straight line is not obtained ΔH° and ΔS° depend on the temperature. The value of ΔH° at any temperature may be calculated from the slope of the plot at that temperature.

Data are given in Table 5.4 for the equilibrium constants at different temperatures for the reaction $N_2(g) + O_2(g) = 2NO(g)$.

In Fig. 5.5 the values of $\log K_p$ are plotted against $1/T$, and it is evident that a straight line is produced. The enthalpy of reaction ΔH° in the range 1900–2600 K can be calculated from the slope of the line, as follows:

$$\Delta H^\circ = -\text{slope} \times 2.303\ R$$
$$= -(-9510)(2.303)(1.987)$$
$$= 43{,}500 \text{ cal mol}^{-1}$$

The intercept of Fig. 5.5 at $1/T = 0$ can be calculated from the experimental value of K_p at some temperature and the slope. The intercept may be used to calculate the standard entropy change ΔS° according to equation 5.70.

$$\frac{\Delta S^\circ}{2.303\ R} = 5.365$$
$$\Delta S^\circ = (5.365)(2.303)(1.987)$$
$$= 24.5 \text{ cal K}^{-1} \text{ mol}^{-1}$$

---

* The dependence of K and T may be derived from the Gibbs-Helmholtz equation (Section 2.15).

$$\frac{\partial(\Delta G^\circ/T)}{\partial(1/T)} = \Delta H^\circ$$

Substituting $\Delta G^\circ = -RT \ln K$, we obtain

$$\frac{\partial \ln K}{\partial(1/T)} = \frac{-\Delta H^\circ}{R}$$

Assuming that ΔH° is independent of temperature, the indefinite integral of this equation is

$$\ln K = \frac{-\Delta H^\circ}{RT} + C$$

where C is an integration constant.

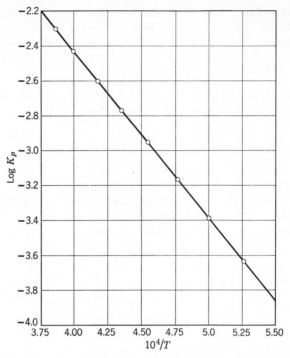

Fig. 5.5 Log K_p plotted against reciprocal absolute temperature for the reaction $N_2(g)$ + $O_2(g) = 2NO(g)$. The standard enthalpy change for the reaction is calculated from the slope of the straight line.

If $\Delta H°$ is independent of temperature, equation 5.70 may be written for two different temperatures. Subtracting one from the other yields

$$\log \frac{K_2}{K_1} = \frac{\Delta H°(T_2 - T_1)}{2.303 \, RT_1T_2} \tag{5.71}$$

Example 5.9 Calculate the enthalpy change for the reaction $N_2(g) + O_2(g) = 2NO(g)$ from the equilibrium constants given in Table 5.4 for 2000 K and 2500 K.

$$\log \frac{K_{2500 \, K}}{K_{2000 \, K}} = \log \frac{3.60 \times 10^{-3}}{4.08 \times 10^{-4}} = \frac{\Delta H°(2500 - 2000)}{(2.303)(1.987)(2500)(2000)}$$

and

$$\Delta H° = 43,300 \text{ cal mol}^{-1}$$

According to Le Châtelier's principle, when an equilibrium system is perturbed, the equilibrium will always be displaced in such a way as to oppose the applied change. When the temperature of an equilibrium system is raised, this change cannot be prevented by the system, but what happens is that the equilibrium shifts in such a way that more heat is required to heat the reaction mixture to the higher temperature than would have been required if the mixture were inert. In other words, when the temperature is raised the equilibrium shifts in the direction which causes an absorption of heat. This conclusion should be verified by thinking about the signs of the quantities in equation 5.70.

To state Le Châtelier's principle in a completely unambiguous way it is necessary to introduce more advanced thermodynamic concepts. The correct formulations and the difficulties with simpler formulations have been discussed by de Heer.*

The present section has dealt with reactions over ranges of temperature where ΔH can be considered independent of temperature. We must now discuss the equations that result when the change of ΔH with temperature must be considered.

5.15 INFLUENCE OF TEMPERATURE ON GIBBS FREE-ENERGY CHANGES

Often the Gibbs free energies of formation $\Delta \bar{G}_f^\circ$ and enthalpies of formation $\Delta \bar{H}_f^\circ$ are available for $25°$, and the values $\bar{C}_P$ for the reactants and products are known over a range of temperature. With this information it is possible to calculate $\Delta G°$ at any temperature in the range where the $\bar{C}_P$ data are valid, and of course from $\Delta G°$ the equilibrium constant may be calculated using $\Delta G° = -RT \ln K$.

The enthalpy of reaction at temperature T is given by equation 1.60, which is

$$\Delta H_T = \Delta H_0 + \int_0^T \Delta \bar{C}_P \, dT \qquad (5.72)$$

where ΔH_0 is the hypothetical enthalpy of reaction at absolute zero. Since the equations used to represent $\bar{C}_P$ as a function of temperature are not valid down to absolute zero, ΔH_0 would not be the actual enthalpy change at absolute zero. However, if this equation is used only to calculate ΔH in the temperature range where the empirical equations for $\bar{C}_P$ are valid, no error is introduced. Since $\Delta \bar{C}_P = \Delta a + (\Delta b) T + (\Delta c) T^2 + \cdots$, it is readily shown that

$$\Delta H = \Delta H_0 + (\Delta a) T + \tfrac{1}{2}(\Delta b) T^2 + \tfrac{1}{3}(\Delta c) T^3 + \cdots \qquad (5.73)$$

Substituting this value of ΔH into equation 2.83, which is the convenient form of the Gibbs-Helmholtz equation, and integrating, it is readily shown that

$$\Delta G = \Delta H_0 - (\Delta a) T \ln T - \tfrac{1}{2}(\Delta b) T^2 - \tfrac{1}{6}(\Delta c) T^3 + \cdots + IT \qquad (5.74)$$

where I is an integration constant.

The Gibbs free-energy change for a reaction may be calculated from equation 5.74 if (1) the heat capacity of each reactant and product is known as a function of temperature from $25°$ to the desired temperature (that is, the values of the

---

* J. de Heer, *J. Chem. Educ.*, **34**, 375 (1957); **35**, 133 (1958).

constants a, b, and c have been determined for each reactant and product); (2) the heat of reaction ΔH is known at one temperature so that ΔH_0 may be evaluated; and (3) the value of ΔG is known at one temperature so that the integration constant I may be calculated.

Example 5.10   Calculate $\Delta G°$ and the equilibrium constant at 1000 K for the water-gas reaction

$$C(\text{graphite}) + H_2O(g) = CO(g) + H_2(g)$$

From Table 5.3,

$$\Delta G°_{298 \text{ K}} = -32.8079 - (-54.6357)$$

$$= 21.8278 \text{ kcal mol}^{-1}$$

From Table 1.3,

$$\Delta H°_{298 \text{ K}} = -26.4157 - (-57.7979)$$

$$= 31.3822 \text{ kcal mol}^{-1}$$

According to the empirical equations for the heat capacities of gases, Table 1.2, and $\bar{C}_{P,\text{graphite}} = 3.81 + 1.56 \times 10^{-3}T$,

$$\Delta \bar{C}_P = \bar{C}_{P,H_2} + \bar{C}_{P,CO} - \bar{C}_{P,H_2O} - \bar{C}_{P,\text{graphite}}$$

$$= 2.29 - 2.30 \times 10^{-3}T - 0.077 \times 10^{-7}T^2$$

$$\Delta H_0 = \Delta H - 2.29T - \tfrac{1}{2}(-2.30 \times 10^{-3})T^2 - \tfrac{1}{3}(-0.077 \times 10^{-7})T^3$$

$$= 31,382.2 - (2.29)(298.1) + (1.5 \times 10^{-3})(298.1)^2$$

$$+ (0.026 \times 10^{-7})(298.1)^3$$

$$= 30,801 \text{ cal mol}^{-1}$$

Substituting in equation 5.74, we have

$$21,827.8 = 30,801 - (2.29)(298.1)(2.303)(2.474) - \tfrac{1}{2}(-2.30 \times 10^{-3})(298.1)^2$$

$$-\tfrac{1}{6}(-0.077 \times 10^{-7})(298.1)^3 + 298.1I$$

$$I = -17.4$$

and

$$\Delta G° = 30,801 - 5.26T \log T + 1.15 \times 10^{-3}T^2 + 0.013 \times 10^{-7}T^3 - 17.4T$$

This general equation may now be used to calculate the Gibbs free-energy change for the reaction at any temperature in the range for which the heat-capacity equations are valid (300–1500 K). At 1000 K this general equation yields

$$\Delta G°_{1000K} = -1330 \text{ cal mol}^{-1}$$

The equilibrium constant for the reaction at 1000 K may be calculated from

$$\Delta G°_{1000 \text{ K}} = -RT \ln K_p = -1330 = -(1.987)(1000)(2.303) \log K_p$$

$$K_p = 1.96$$

It should be noted that the equilibrium constant is greater than unity at 1000 K, although it is far smaller than unity at room temperature.

5.16   THE EFFECT OF PRESSURE ON EQUILIBRIUM

The value of K_p for a reaction is not affected by pressure if the reactants and products are ideal gases. Therefore for ideal gases no shift in equilibrium is produced by the addition of an inert gas provided that the volume is held constant. When an inert gas is added at constant total pressure, however, the volume increases and there is a decrease in the partial pressures of reactants and products and a consequent shift in equilibrium, provided that $\sum \nu_i \neq 0$, just as if the equilibrium mixture of gases had been allowed to expand into a larger volume. If $\sum \nu_i = 0$, alterations in volume have no effect if the gases are ideal.

For ideal gases the effect of pressure on the equilibrium constant expressed in terms of mole fractions is given by $K_X = K_p P^{-\sum \nu_i}$ (Section 5.7). Taking the natural logarithms of both sides of this equation and then differentiating with respect to total pressure at constant temperature, we see that

$$\ln K_X = \ln K_p - \ln P \sum \nu_i \tag{5.75}$$

$$\left(\frac{\partial \ln K_X}{\partial P} \right)_T = \frac{-\sum \nu_i}{P} = \frac{-\Delta V}{RT} \tag{5.76}$$

Thus for a reaction with $\sum \nu_i > 0$ (i.e., the number of moles of gaseous products is greater than the number of moles of gaseous reactants), an increase in pressure at constant temperature leads to a decrease in K_X. Thus when the pressure is raised the equilibrium is displaced in the direction of volume contraction. This is an illustration of Le Châtelier's principle.

Example 5.11   At $250°$ PCl_5 is 80% dissociated at a pressure of 1 atm, and so $K_p = 1.78$. What is the percentage dissociation at equilibrium after sufficient nitrogen has been added at constant pressure to produce a nitrogen partial pressure of 0.9 atm? The total pressure is maintained at 1 atm. The sum of the partial pressures of the reacting gases is 0.1 atm.

$$K_p = \frac{\alpha^2 P}{1 - \alpha^2} = 1.78 = \frac{\alpha^2 0.10}{1 - \alpha^2}$$

$$\alpha = 0.973 \quad \text{or} \quad 97.3\%$$

5.17   THEORETICAL CALCULATION OF EQUILIBRIUM CONSTANTS

To suggest at this stage that the equilibrium constant for a gas reaction may be calculated from spectroscopic data may be a bit mysterious. The following derivation is intended to make this a little less mysterious. As we will see later in connection with quantum mechanics, a molecule may exist in any of a series of energy levels. It is the distribution of energy levels that determines the equilibrium state of the system. This may be most simply illustrated for an isomerization reaction of ideal gases: $A = B$. The energy level patterns of A and B are represented schematically in Fig. 5.6. For real molecules the energy level

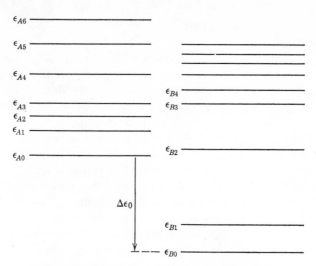

Fig. 5.6   Molecular energy levels of A and B.

pattern would be much more complicated. For A and B the energy in the lowest state is taken as ϵ_{A0} and ϵ_{B0}. Molecules A and B in their lowest energy states have different energies, and this difference is represented by $\Delta\epsilon_0$. The energy change at absolute zero is $\Delta\epsilon_0$, and the $A = B$ reaction is exothermic at absolute zero.

At equilibrium, molecules are distributed between the various energy levels available to the system (i.e. between all levels of molecule A and all levels of molecule B) according to the Boltzmann distribution (Section 17.3). The equilibrium constant is equal to the ratio of the total number of molecules in the various B levels to the total number of molecules in the various A levels. If N_{A0} represents the number of A molecules in the lowest level, N_{A1} the number in the first level, etc, the total number N_A of molecules in A levels is given by

$$
\begin{aligned}
N_A &= N_{A0} + N_{A1} + N_{A2} + \cdots \\
&= N_{A0} + N_{A0}e^{-\epsilon_{A1}/kT} + N_{A0}e^{-\epsilon_{A2}/kT} + \cdots \\
&= N_{A0}[1 + e^{-\epsilon_{A1}/kT} + e^{-\epsilon_{A2}/kT} + \cdots] \\
&= N_{A0}\sum_{i=0}^{\infty} e^{-\epsilon_{Ai}/kT}
\end{aligned}
\tag{5.77}
$$

Similarly for B,

$$
N_B = N_{B0}\sum_{i=0}^{\infty} e^{-\epsilon_{Bi}/kT}
\tag{5.78}
$$

Summations of the type given in equations 5.77 and 5.78 are referred to as partition functions. The Boltzmann constant is represented by k.

Since there is no change in volume in this gaseous isomerization reaction, the equilibrium constant K_p is simply equal to the ratio of the number of B molecules to A molecules at equilibrium: $K_p = N_B/N_A$. Substituting equations 5.77 and 5.78, we have

$$
K_p = \frac{N_{B0}\sum_{i=0}^{\infty} e^{-\epsilon_{Bi}/kT}}{N_{A0}\sum_{i=0}^{\infty} e^{-\epsilon_{Ai}/kT}}
\tag{5.79}
$$

The ratio of the number of B molecules in their lowest energy state to the number of A molecules in their lowest energy state is determined by $\Delta\epsilon_0$.

$$\frac{N_{B0}}{N_{A0}} = e^{-\Delta\epsilon_0/kT} \tag{5.80}$$

Since $\Delta\epsilon_0$ is negative in Fig. 5.6, the ground level for B will be more highly populated than for A. Substituting equation 5.80 in equation 5.79, we have

$$K = e^{-\Delta\epsilon_0/kT} \frac{\sum\limits_{i=0}^{\infty} e^{-\epsilon_{B_i}/kT}}{\sum\limits_{i=0}^{\infty} e^{-\epsilon_{A_i}/kT}} \tag{5.81}$$

The equilibrium constant for a reaction may be calculated if the energy levels of the reactants and products are known and $\Delta\epsilon_0$ is known. Exact calculations for simple molecules will be shown later.

We have already discussed how the enthalpy change and entropy change determine the equilibrium constant for a reaction. The first factor $e^{-\Delta\epsilon_0/kT}$ is an enthalpy factor which favors the product B if $\Delta\epsilon_0$ is negative. This factor becomes less important as the temperature is raised. The second factor, the ratio of partition functions, favors the product B if it has *more* low-lying energy levels than A. This ratio may be considered to be an entropy or probability factor.

5.18 FREE-ENERGY FUNCTION*

The quantity most convenient to calculate equilibrium constants, using spectroscopic data or using third law entropies, is the *free-energy function*, $(\bar{G}_T^\circ - \bar{H}_0^\circ)/T$. The values of the free-energy function do not change much with temperature, and tables giving values every 100 or 500 K are sufficient for interpolation and extrapolation, as may be seen in Table 5.5. The difference between the free-energy function of the products and reactants may be written

$$\Delta\left(\frac{\bar{G}_T^\circ - \bar{H}_0^\circ}{T}\right) = \frac{\Delta G_T^\circ}{T} - \frac{\Delta H_0^\circ}{T} \tag{5.82}$$

In accordance with previous conventions, ΔG° and ΔH_0° are used to represent changes of the thermodynamic quantities for the overall reaction, whereas the bars are used for the molar quantities. Then

$$\frac{\Delta G_T^\circ}{T} = \frac{\Delta H_0^\circ}{T} + \Delta\left(\frac{\bar{G}_T^\circ - \bar{H}_0^\circ}{T}\right) \tag{5.83}$$

Since $\Delta G_T^\circ = -RT \ln K$,

$$\ln K = 2.303 \log K = -\frac{1}{R}\left[\frac{\Delta H_0^\circ}{T} + \Delta\left(\frac{\bar{G}_T^\circ - \bar{H}_0^\circ}{T}\right)\right] \tag{5.84}$$

To calculate the value of an equilibrium constant at a given temperature, it is necessary to know the value of ΔH_0° as well as the values of the free-energy function. The enthalpy of

---

* W. F. Giauque, *J. Am. Chem. Soc.*, **52**, 4808 (1930); J. L. Margrave, *J. Chem. Educ.*, **32**, 520 (1955).

Table 5.5   Free-Energy Function (based on $\bar{H}_0^\circ$), $\bar{H}_{298}^\circ - \bar{H}_0^\circ$, and $\Delta\bar{H}_0^\circ$

Elements

	$-(\bar{G}_T^\circ - \bar{H}_0^\circ)/T$, cal K^{-1} mol^{-1}					$\bar{H}_{298}^\circ - \bar{H}_0^\circ$ kcal mol^{-1}	$\Delta\bar{H}_0^\circ$, kcal mol^{-1}
	298.15 K	500 K	1000 K	1500 K	2000 K		
Br(g)	36.84	39.41	42.85	44.89	46.36	1.481	26.90
Br$_2$(g)	50.85	54.99	60.80	64.31	66.83	2.325	8.37
Br$_2$(l)	25.5	—	—	—	—	3.240	0
C(g)	32.52	35.20	38.73	40.77	42.21	1.56	169.2
C$_2$(g)	39.14	43.68	49.80	53.33	55.86	2.530	196 ± 6
C(graphite)	0.53	1.16	2.78	4.19	5.38	0.251	0
Cl(g)	34.43	37.06	40.69	42.83	44.34	1.499	28.54
Cl$_2$(g)	45.93	49.85	55.43	58.85	61.34	2.194	0
H(g)	22.42	24.99	28.44	30.45	31.88	1.481	51.62
H$_2$(g)	24.42	27.95	32.74	35.59	37.67	2.024	0
I$_2$(g)	54.18	58.46	64.40	67.96	70.52	2.148	15.66
I$_2$(s)	17.18	—	—	—	—	3.154	0
N(g)	31.65	43.22	37.66	39.67	41.10	1.481	112.54
N$_2$(g)	38.82	42.42	47.31	50.28	52.48	2.072	0
O(g)	33.08	35.84	39.46	41.54	43.00	1.607	58.98
O$_2$(g)	42.06	45.68	50.70	53.81	56.10	2.075	0

Gaseous Inorganic Compounds

	$-(\bar{G}_T^\circ - \bar{H}_0^\circ)/T$, cal K^{-1} mol^{-1}					$\bar{H}_{298}^\circ - \bar{H}_0^\circ$, kcal mol^{-1}	$\Delta\bar{H}_0^\circ$, kcal mol^{-1}
	298.15 K	500 K	1000 K	1500 K	2000 K		
HCl	37.72	41.31	46.16	49.08	51.23	2.065	22.019
HBr	40.53	44.12	48.99	51.95	54.13	2.067	−8.1
HI	42.40	45.99	50.90	53.90	56.11	2.069	6.7
NaCl	47.18	51.25	57.05	60.56	63.09	2.295	−42.4
NaBr	49.75	53.91	59.77	63.30	65.86	2.342	−33.2
NaI	51.51	55.72	61.64	65.20	67.75	2.379	−20.6
PCl$_3$	61.66	68.87	80.07	—	—	3.84	−65.9
PCl$_5$	66.76	76.30	91.56	—	—	5.22	−87.2
HO	36.82	40.48	45.39	48.30	50.41	2.106	10 ± 0.3
H$_2$O	37.17	41.29	47.01	50.60	53.32	2.368	−57.107
NH$_3$	37.99	42.28	48.63	53.03	56.56	2.37	−9.37
NO$_2$	49.19	53.60	60.23	64.58	67.88	2.465	8.68
SO$_2$	50.82	55.38	62.28	66.82	70.2	2.519	−70.36

Table 5.5 (*Continued*)

Gaseous Carbon Compounds

	$-(\bar{G}_T - \bar{H}_0^\circ)/T$, cal K^{-1} mol^{-1}					$\bar{H}_{298}^\circ - \bar{H}_0^\circ$, kcal mol^{-1}	$\Delta \bar{H}_0^\circ$, kcal mol^{-1}
	298.15 K	500 K	1000 K	1500 K	2000 K		
CO	40.25	43.86	48.77	51.78	53.99	2.073	-27.202
CO$_2$	43.56	47.67	54.11	58.48	61.85	2.238	-93.969
CH$_4$	36.46	40.75	47.65	52.84	57.1	2.397	-15.99
CH$_3$Cl	47.45	52.06	59.78	65.54	—	2.489	-17.7
CH$_3$OH	48.13	53.14	61.58	—	—	2.731	-45.47
HCOOH	50.72	55.60	63.99	70.17	75.14	2.601	-88.65
C$_2$H$_2$	39.98	44.51	52.01	57.23	61.33	2.392	54.33
C$_2$H$_4$	43.98	48.74	57.29	63.94	69.46	2.525	14.52
C$_2$H$_6$	45.27	50.77	61.11	69.46	—	2.856	-16.52
C$_2$H$_5$OH	56.20	62.82	75.28	85.15	—	3.39	-52.41
C$_3$H$_6$	52.95	59.32	71.57	81.43	—	3.237	8.47
C$_3$H$_8$	52.73	59.81	74.10	85.86	—	3.512	-19.48
n-C$_4$H$_{10}$	58.54	67.91	86.60	101.95	—	4.645	-23.67
C$_6$H$_6$	52.93	60.24	76.57	90.45	—	3.401	24.00

reaction at absolute zero ΔH_0° is calculated from measurements of the heat of reaction at constant pressure at one temperature and the heat capacities down to nearly absolute zero.

Example 5.12 Calculate the equilibrium constant for the formation of methanol from synthesis gas (CO + H$_2$) at 1000 K.

$$CO(g) + 2H_2(g) = CH_3OH(g)$$

Using equation 5.84, we have

$$\log K = \frac{-1}{(2.303)(1.987)}\left[\frac{(-45,470 + 27,202)}{1000} - 61.58 + 48.77 + 2(32.74)\right]$$

$$= -7.518$$

$$K = 3.03 \times 10^{-8}$$

Calculations of equilibrium constants at high temperatures using the free-energy function are of great value in predicting the temperature at which a reaction must be carried out in order for the equilibrium to be favorable.

The amount of "isoöctane" available for antiknock fuel has been greatly increased by use of the reaction

$$C_4H_{10}(g, \text{isobutane}) + C_4H_8(g, \text{isobutene}) = C_8H_{18}(g, \text{isoöctane}) \qquad (5.85)$$

The variation of log K for this reaction with temperature, as calculated from various thermodynamic data by Rossini,* is shown by line A in Fig. 5.7. This reaction cannot be

---

* F. D. Rossini, *J. Wash. Acad. Sci.*, **39**, 249 (1949).

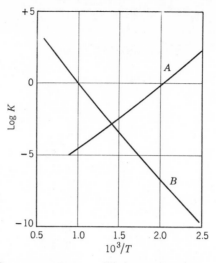

Fig. 5.7   Influence of temperature on equilibrium constants.

$$(A) \quad C_4H_{10}(g) + C_4H_8(g) = C_8H_{18}(g)$$
$$(B) \quad C_4H_8(g) = C_4H_6(g) + H_2(g)$$

carried out at high temperatures because of the unfavorable equilibrium constant. Its use at temperatures as low as room temperature, where the equilibrium is quite favorable, was made possible by the discovery of commercially usable catalysts (sulfuric acid and hydro-fluoric acid).

An example of a reaction for which the equilibrium is favorable only at rather high temperatures is the dehydrogenation of 1-butene to yield 1,3-butadiene.

$$
\begin{array}{c}
\text{H H H H} \\
| \ | \ | \ | \\
\text{H—C=C—C—C—H}(g) = \text{H—C=C—C=C—H}(g) + H_2(g) \\
| \ | \\
\text{H H}
\end{array}
\qquad (5.86)
$$

The variation of $\log K$ for this reaction with temperature, as calculated from various thermo-dynamic data by Rossini, is shown by line B in Fig. 5.7. For such reactions there is generally no need to search for a catalyst because the reaction is thermodynamically feasible at high temperatures where reaction rates are generally fast. However, at high temperatures other reactions may become sufficiently important to reduce the equilibrium yield of a desired product.

References

K. G. Denbigh, *The Principles of Chemical Equilibrium*, Cambridge University Press, Cambridge, 1971.

R. E. Dickerson, *Molecular Thermodynamics*, W. A. Benjamin Co., New York, 1969.

I. M. Klotz and R. M. Rosenberg, *Chemical Thermodynamics*, W. A. Benjamin Co., New York, 1972.

G. N. Lewis, M. Randall, revised by K. S. Pitzer, and L. Brewer, *Thermodynamics*, McGraw-Hill Book Co., New York, 1961.

A. B. Pippard, *Elements of Classical Thermodynamics*, Cambridge University Press, Cambridge, 1960.

P. A. Rock, *Chemical Thermodynamics*, The Macmillan Co., London, 1969.

F. D. Rossini, K. S. Pitzer, W. J. Taylor, J. P. Ebert, J. E. Kilpatrick C. W. Beckett, M. G. Williams, and H. G. Werner, "Selected Values of Properties of Hydrocarbons," *Natl. Bur. Standards Circ. C461*, U.S. Government Printing Office, Washington D.C., 1947.

F. D. Rossini, D. D. Wagman, W. H. Evans, S. Levine, and I. Jaffe, "Selected Values of Chemical Thermodynamics Properties," *Natl. Bur. Standards Circ. 500*, U.S. Government Printing Office, Washington D.C., 1952.

F. T. Wall, *Chemical Thermodynamics*, W. H. Freeman and Co., San Francisco, 1965.

Problems

5.1 What are the Gibbs free energy of mixing and entropy of mixing of $\frac{1}{2}$ mol of ideal gas A with $\frac{1}{2}$ mol of ideal gas B at 25° C? $\qquad$ *Ans.* -410.6 cal. 1.377 cal K^{-1}.

5.2 At 25° $K_p = 1.7 \times 10^{12}$ for the reaction

$$SO_2(g) + \tfrac{1}{2}O_2(g) = SO_3(g)$$

Calculate (a) K_p and (b) K_c at this temperature for the reaction

$$2SO_3(g) = 2SO_2(g) + O_2(g)$$

$\qquad$ *Ans.* (a) 0.35×10^{-24}, (b) 1.4×10^{-26}.

5.3 At 55° and 1 atm the average molecular weight of partially dissociated N_2O_4 is 61.2 g mol^{-1}. Calculate (a) α and (b) K_p for the reaction $N_2O_4(g) = 2NO_2(g)$. (c) Calculate α at 55° if the total pressure is reduced to 0.1 atm. $\qquad$ *Ans.* (a) 0.503, (b) 1.36, (c) 0.876.

5.4 For the reaction $N_2O_4(g) = 2NO_2(g)$, K_p at 25° is 0.141. What pressure would be expected if 1 g of liquid N_2O_4 were allowed to evaporate into a liter vessel at this temperature? Assume that N_2O_4 and NO_2 are ideal gases. $\qquad$ *Ans.* 0.351 atm.

5.5 Under what total pressure at equilibrium must PCl_5 be placed at 250° to obtain a 30% conversion into PCl_3 and Cl_2? For the reaction, $PCl_5(g) = PCl_3(g) + Cl_2(g)$, $K_p = 1.78$ at 250°. $\qquad$ *Ans.* 18.0 atm.

5.6 For the reaction $A(g) + B(g) = AB(g)$, $\Delta G° = -2000$ cal at 27°. Under what total pressure must an equimolecular gaseous mixture of A and B be placed to produce a 40% conversion into AB? $\qquad$ *Ans.* 0.063 atm.

5.7 The equilibrium constant for the reaction $SO_2(g) + \tfrac{1}{2}O_2(g) = SO_3(g)$ at 727° is given by

$$K_p = \frac{p_{SO_3}}{p_{SO_2}p_{O_2}^{1/2}} = 1.85$$

What is the ratio p_{SO_3}/p_{SO_2} (a) when the partial pressure of oxygen at equilibrium is 0.3 atm, (b) when the partial pressure of oxygen at equilibrium is 0.6 atm? (c) What is the effect on the equilibrium if the total pressure of the mixture of gases is increased by forcing in nitrogen at constant volume?

$\qquad$ *Ans.* (a) 1.01, (b) 1.44, (c) No effect if the gases behave ideally.

5.8 A $1:3$ mixture of nitrogen and hydrogen was passed over a catalyst at $450°$. It was found that 2.04% by volume of ammonia was formed when the total pressure was maintained at 10 atm [A. T. Larson and R. L. Dodge, *J. Am. Chem. Soc.*, **45**, 2918 (1923)]. Calculate the value of K_p for $\frac{3}{2}H_2 + \frac{1}{2}N_2 = NH_3$ at this temperature.

Ans. $K_p = 6.44 \times 10^{-3}$.

5.9 The equilibrium constant for the reaction $N_2O_4 = 2NO_2$ in chloroform solution at $0°$ is 4.5×10^{-7} when the equilibrium constant is expressed in mole fractions and the concentration of N_2O_4 is below 0.1 mole fraction. If 0.02 mol of N_2O_4 is dissolved in 1 mol of chloroform, how many moles of NO_2 will be produced at equilibrium?

Ans. 9.6×10^{-5} mol.

5.10 The equilibrium constant for the association of benzoic acid to a dimer in dilute benzene solutions is as follows at $43.9°$:

$$2C_6H_5COOH = (C_6H_5COOH)_2 \qquad K_c = 2.7 \times 10^2 \; M^{-1}$$

Calculate $\Delta G°$, and state its meaning.

Ans. -3530 cal. This is the decrease in Gibbs free energy when 2 mol of monomer at unit activity on the molar scale is converted to 1 mol of dimer at unit activity on the molar scale.

5.11 For the reaction forming the triphenyl methyl free radical $(\phi_3C\cdot)$

$$\phi_3CC\phi_3 = 2\phi_3C\cdot$$

in $CHCl_3$ at $25°$, $\Delta H° = 11.60$ kcal mol^{-1}, and $\Delta S° = 23.4$ cal K^{-1} mol^{-1}. The dot represents the unpaired electron in the free radical. Calculate (a) K_c and (b) the molar concentration of $\phi_3C\cdot$ if 1 mole of $\phi_3CC\phi_3$ is added per liter. *Ans.* (a) 4.1×10^{-4}, (b) 2.0×10^{-2} M.

5.12 At 1273 K and at a total pressure of 30 atm the equilibrium in the reaction $CO_2(g) + C(s) = 2CO(g)$ is such that 17 mole $\%$ of the gas is CO_2. (a) What percentage would be CO_2 if the total pressure were 20 atm? (b) What would be the effect on the equilibrium of adding N_2 to the reaction mixture in a closed vessel until the partial pressure of N_2 is 10 atm? (c) At what pressure of the reactants will 25% of the gas be CO_2?

Ans. (a) 12.5%, (b) No effect, (c) 54 atm.

5.13 For the reaction

$$\tfrac{1}{2}SnO_2(s) + H_2(g) = \tfrac{1}{2}Sn(s) + H_2O(g)$$

the total pressure of the system is 32.0 Torr at $750°$, and the partial pressure of water is 23.7 Torr. (a) Calculate K_p for this reaction. For the reaction

$$H_2(g) + CO_2(g) = CO(g) + H_2O(g)$$

K_p has a value of 0.771 at $750°$. (b) Calculate K_p for the reaction

$$\tfrac{1}{2}SnO_2(s) + CO(g) = \tfrac{1}{2}Sn(s) + CO_2(g)$$

Ans. (a) 2.85, (b) 3.71.

5.14 When platinum is heated in the presence of chlorine gas, the following reaction takes place:

$$Pt(s) + Cl_2(g) = PtCl_2(g)$$

At 1000 K, $\Delta G°_{1000} = 14$ kcal. If the pressure of Cl_2 is 1 atm, what will be the partial pressure of $PtCl_2$? *Ans.* 8.7×10^{-4} atm.

5.15 Calculate (a) K_p and (b) $\Delta G°$ for the following reaction at $20°$:

$$CuSO_4\cdot4NH_3(s) = CuSO_4\cdot2NH_3(s) + 2NH_3(g)$$

The equilibrium pressure of NH_3 is 62 Torr.

Ans. (*a*) 6.66×10^{-3} atm^2. (*b*) 2920 cal.

5.16 Calculate the equilibrium constant for the isomerization of *n*-butane to isobutane at 25°:

$$n\text{-Butane}(g) = \text{isobutane}(g)$$

using data on the Gibbs free energies of formation. *Ans.* $K_p = 2.54$.

5.17 (*a*) Calculate the Gibbs free energy of formation of urea, $CO(NH_2)_2(s)$, from the following data:

$$CO_2(g) + 2NH_3(g) = H_2O(g) + CO(NH_2)_2(s) \quad \Delta G^\circ_{298} = 456 \text{ cal mol}^{-1}$$
$$H_2O(g) = H_2(g) + \tfrac{1}{2}O_2(g) \quad \Delta G^\circ_{298} = 54,636 \text{ cal mol}^{-1}$$
$$C(\text{graphite}) + O_2(g) = CO_2(g) \quad \Delta G^\circ_{298} = -94,260 \text{ cal mol}^{-1}$$
$$N_2(g) + 3H_2(g) = 2NH_3(g) \quad \Delta G^\circ_{298} = -7752 \text{ cal mol}^{-1}$$

(*b*) Calculate $\Delta \bar{G}^\circ_f$ for urea at 298.1 K from $\Delta \bar{S}^\circ_f$, which is -109.05 cal K^{-1} mol^{-1}, and $\Delta \bar{H}^\circ_f$, which is $-79,634$ cal mol^{-1}. *Ans.* (*a*) $-46,920$, (*b*) $-47,126$ cal mol^{-1}.

5.18 From the $\Delta \bar{G}^\circ_f$ of $Br_2(g)$ at 25°, calculate the vapor pressure of $Br_2(l)$. The pure liquid at 1 atm and 25° is taken as the standard state. *Ans.* 214 Torr.

5.19 The equilibrium constant for the gas reaction A = B is 0.10 at 27°. Calculate (*a*) ΔG° and (*b*) ΔG for the production of 1 mol of B at a pressure of 1 atm from A at a pressure of 20 atm. (*c*) Is the reaction spontaneous under the latter conditions? *Ans.* (*a*) $+1373$ cal mol^{-1}, (*b*) -413 cal mol^{-1}, (*c*) Yes.

5.20 The following reaction takes place in the presence of aluminum chloride:

$$\text{Cyclohexane}(l) = \text{Methylcyclopentane}(l)$$

At 25° $K_c = 0.143$, and at 45° $K_c = 0.193$. From these data calculate: (*a*) ΔG° at 25°, (*b*) ΔH°, (*c*) ΔS°. *Ans.* (*a*) 1152, (*b*) 2820 cal mol^{-1}, (*c*) 5.59 cal K^{-1} mol^{-1}.

5.21 The following data apply to the reaction $Br_2(g) = 2Br(g)$:

T, K	1123	1173	1223	1273
$K_p \times 10^3$	0.403	1.40	3.28	7.1

Determine by graphical means the enthalpy change when 1 mol of Br_2 dissociates completely at 1200 K. *Ans.* 47.6 kcal mol^{-1}.

5.22 Mercuric oxide dissociates according to the reaction $2HgO(s) = 2Hg(g) + O_2(g)$. At 420° the dissociation pressure is 387 Torr, and at 450° it is 810 Torr. Calculate (*a*) the equilibrium constants, and (*b*) the enthalpy of dissociation per mole of HgO. *Ans.* (*a*) 0.0196, 0.1794 atm^3, (*b*) 36,750 cal mol^{-1}.

5.23 The vapor pressure of water above mixtures of $CuCl_2 \cdot H_2O(s)$ and $CuCl_2 \cdot 2H_2O(s)$ is given as a function of temperature in the following table:

t, °C	17.9	39.8	60.0	80.0
P, atm	0.0049	0.0247	0.120	0.322

(*a*) Calculate ΔH° for the reaction

$$CuCl_2 \cdot 2H_2O(s) = CuCl_2 \cdot H_2O(s) + H_2O(g)$$

(*b*) Calculate ΔG° for the reaction at 60.0°. (*c*) Calculate ΔS° for the reaction at 60.0°. *Ans.* (*a*) 13,700 cal mol^{-1}, (*b*) 1410 cal mol^{-1}, (*c*) 36.9 cal K^{-1} mol^{-1}.

5.24   Show that for gaseous equilibria where $K_p = K_c(RT)^{\Sigma \nu_i}$

$$\frac{\partial \ln K_c}{\partial T} = \frac{\Delta U^{\circ}}{RT^2}$$

5.25   At 2000° water is 2% dissociated into oxygen and hydrogen at a total pressure of 1 atm. (a) Calculate K_p. (b) Will the degree of dissociation increase or decrease if the pressure is reduced? (c) Will the degree of dissociation increase or decrease if argon gas is added, holding the total pressure equal to 1 atm? (d) Will the degree of dissociation change if the pressure is raised by addition of argon at constant volume to the closed system containing partially dissociated water vapor? (e) Will the degree of dissociation increase or decrease if oxygen gas is added while holding the total pressure constant at 1 atm?

    Ans. (a) 2.03×10^{-3} atm$^{1/2}$, (b) Increase, (c) Increase, (d) No change, (e) Decrease.

5.26   Chlorine gas is partially dissociated to chlorine atoms at elevated temperatures. Calculate the degree of dissociation of Cl_2 gas at (a) 1250 K and (b) 1750 K when the total pressure is 0.1 atm.   *Ans.* (a) 1.33×10^{-2}, (b) 0.379.

5.27   Calculate the degree of dissociation of $H_2O(g)$ at 2000 K and 1 atm pressure, using the free-energy function.   *Ans.* 1.41×10^{-6}.

5.28   What are ΔG and ΔS for the mixing of nitrogen and oxygen at 25° to produce air with 20% by volume oxygen?

5.29   In determining equilibrium constants for reactions with large or small constants, the analytical method usually puts a limit on the magnitude of the constant which can be experimentally determined. For a reaction of the type A = B it is found that there is less than 1 part per 1000 of B at equilibrium at 25°. Calculate the *minimum* value for ΔG°_{298} for this reaction.

5.30   For the reaction

$$CH_4(g) + 2H_2S(g) = CS_2(g) + 4H_2(g)$$

$K_p = 2.05 \times 10^9$ at 25°. Calculate (a) K_p and (b) K_c at this temperature for

$$2H_2(g) + \tfrac{1}{2}CS_2(g) = H_2S(g) + \tfrac{1}{2}CH_4(g)$$

5.31   The dissociation of N_2O_4 is represented by $N_2O_4(g) = 2NO_2(g)$. If the density of the equilibrium gas mixture is 3.174 g liter^{-1} at a total pressure of 1 atm at 24°, what minimum pressure would be required to keep the degree of dissociation of N_2O_4 below 0.1 at this temperature?

5.32   At 250°, 1 liter of partially dissociated phosphorus pentachloride gas, at 1 atm, weighs 2.690 g. Calculate the degree of dissociation α and the equilibrium constant K_p.

5.33   The reaction

$$2NOCl(g) = 2NO(g) + Cl_2(g)$$

comes to equilibrium at 1 atm total pressure and 227° when the partial pressure of the nitrosyl chloride, NOCl, is 0.64 atm. Only NOCl was present initially. (a) Calculate ΔG° for this reaction. (b) At what total pressure will the partial pressure of Cl_2 be 0.1 atm?

5.34   For the reaction

$$2HI(g) = H_2(g) + I_2(g)$$

at 698.6 K, $K_p = 1.83 \times 10^{-2}$. (a) How many grams of hydrogen iodide will be formed when 10 g of iodine and 0.2 g of hydrogen are heated to this temperature in a 3 liter vessel? (b) What will be the partial pressures of H_2, I_2, and HI?

5.35   At $400°$ $K_p = 78.1$ for the reaction

$$NH_3(g) = \tfrac{1}{2}N_2(g) + \tfrac{3}{2}H_2(g)$$

Show that the fraction α of NH_3 dissociated at a total pressure P is given by

$$\alpha = \frac{1}{\sqrt{1 + kP}}$$

and calculate the value of k in this equation.

5.36   Amylene, C_5H_{10}, and acetic acid react to give the ester according to the reaction

$$C_5H_{10} + CH_3COOH = CH_3COOC_5H_{11}$$

What is the value of K_c if 0.00645 mol of amylene and 0.001 mol of acetic acid dissolved in 845 cm³ of a certain inert solvent react to give 0.000784 mol of ester?

5.37   The following thermodynamic quantities for α- and β-pinene have been obtained at $25°$ from heat-of-combustion measurements and low-temperature heat-capacity determinations (H. A. McGee, Jr.):

	$\Delta \bar{H}_f^\circ$, kcal mol^{-1}	$\bar{S}^\circ$, cal K^{-1} mol^{-1}
DL-α-pinene(l)	-3.6 ± 0.3	70.95 ± 0.11
DL-β-pinene(l)	-1.5 ± 0.3	69.48 ± 0.15

Calculate the Gibbs free-energy change and equilibrium constant on the mole-fraction scale at $25°$ for the reaction.

$$\text{DL-}\alpha\text{-pinene}(l) = \text{DL-}\beta\text{-pinene}(l)$$

Indicate the magnitude of the uncertainty in these values resulting from the indicated uncertainties in the thermodynamic values

5.38   For the reaction

$$C(s) + 2H_2(g) = CH_4(g)$$

at $1000°$, $K_p = 0.263$. Calculate the total pressure at equilibrium when 0.100 mole of CH_4 is placed in a volume of 2 liters at $1000°$.

5.39   Ten grams of calcium carbonate is placed in a container of 1-liter capacity and heated to $800°$. (*a*) How many grams of $CaCO_3$ remain undecomposed? (*b*) If the amount of $CaCO_3$ were 20 g, how much would remain undecomposed?

5.40   A gas mixture containing 97 mole % water and 3 mole % hydrogen is heated to 1000 K. Will the equilibrium mixture react with nickel at 1000 K to produce nickel oxide?

$$Ni(s) + \tfrac{1}{2}O_2(g) = NiO(s) \qquad \Delta G_{1000}^\circ = 35,400 \text{ cal}$$
$$H_2(g) + \tfrac{1}{2}O_2(g) = H_2O(g) \qquad \Delta G_{1000}^\circ = -45,600 \text{ cal}$$

5.41   In the formation of 1 mol of $AgI(s)$ from solid silver and solid iodine, 14,815 cal is evolved at $25°$. From specific heat measurements at low temperature the molar entropy of $AgI(s)$ at $25°$ has been found to be 27.6 cal K^{-1} mol^{-1}, whereas the atomic entropy of silver at $25°$ is 10.2 and that of iodine is 13.3 cal K^{-1} mol^{-1}. Calculate the Gibbs free energy of formation of $AgI(s)$ at $25°$.

5.42   Using the fact that $K_p = 0.141$ atm at $25°$ for $N_2O_4(g) = 2NO_2(g)$ and the Gibbs free energy of formation of NO_2 in Table 5.3, calculate the Gibbs free energy of formation of N_2O_4.

5.43   Calculate the equilibrium constants at 25° for the following reactions:

$$CO(g) + 2H_2(g) = CH_3OH(g)$$

$$C_2H_4(g) + H_2O(g) = C_2H_5OH(g)$$

The vapor pressures of methanol and ethanol at 25° are 0.16 and 0.078 atm, respectively. These vapor pressures may be used to calculate the differences in Gibbs free energies of formation of the liquids and corresponding gases.

5.44   For the reaction

$$N_2(g) + 3H_2(g) = 2NH_3(g)$$

$K_p = 1.64 \times 10^{-4}$ at 400°. Calculate (a) $\Delta G°$ and (b) ΔG when the pressures of N_2 and H_2 are maintained at 10 atm and 30 atm, respectively, and NH_3 is removed at a partial pressure of 3 atm. (c) Is the reaction spontaneous under the latter conditions?

5.45   The equilibrium constant K_p for $H_2(g) = 2H(g)$ is 1.52×10^{-7} at 1800 K and 3.10×10^{-6} at 2000 K. Calculate $\Delta H°$ for this temperature range.

5.46   The average molecular weights M_{avg} of equilibrium mixtures of NO_2 and N_2O_4 at 1 atm total pressure are given in the following table at three temperatures:

t, °C	25	45	65
M_{avg}	77.64	66.80	56.51

(a) Calculate the degree of dissociation of N_2O_4 and the equilibrium constant at each of these temperatures. (b) Plot $\log K_p$ against $1/T$ and calculate $\Delta H°$ for the dissociation of N_2O_4. (c) Calculate the equilibrium constant at 35°. (d) Calculate the degree of dissociation α for N_2O_4 at 35° when the total pressure is 0.5 atm.

5.47   The partial pressure of oxygen in equilibrium with a mixture of silver oxide and silver is given by

$$\log p = 6.2853 - 2859/T$$

where p is expressed in atmospheres. Calculate the temperature at which silver oxide should begin to decompose when it is heated in air at 760 Torr pressure.

5.48   For the formation of nitric oxide

$$N_2(g) + O_2(g) = 2NO(g)$$

K_p at 2126.9° is 2.5×10^{-3}. (a) In an equilibrium mixture containing 0.1 atm partial pressure of N_2 and 0.1 atm partial pressure of O_2, what is the partial pressure of NO? (b) In an equilibrium mixture of N_2, O_2, NO, CO_2, and other inert gases at 2126.9° and 1 atm total pressure, 80% by volume of the gas is N_2 and 16% is O_2. What is the percent by volume of NO? (c) What is the total partial pressure of the inert gases?

5.49   When N_2O_4 is allowed to dissociate to form NO_2 at 25° at a total pressure of 1 atm, it is 18.5% dissociated at equilibrium, and so $K_p = 0.141$. (a) If N_2 is added to the system at constant volume, will the equilibrium shift? (b) If the system is allowed to expand as N_2 is added at a constant total pressure of 1 atm, what will be the equilibrium degree of dissociation when the N_2 partial pressure is 0.6 atm.

5.50   Calculate the equilibrium constant for the dissociation of gaseous oxygen at 2000 K.

$$O_2(g) = 2O(g)$$

5.51   The equilibrium vapor pressure of tantalum has been determined as 6.216×10^{-9}

atm at 2624 K, as 9.692×10^{-8} atm at 2839 K, and as 3.655×10^{-7} atm at 2948 K. The free-energy functions for solid and gaseous Ta are:

T, K	$-\left(\dfrac{\bar{G}_T^\circ - \bar{H}_0^\circ}{T}\right)_{\text{solid}},$ cal K^{-1} mol^{-1}	$-\left(\dfrac{\bar{G}_T^\circ - \bar{H}_0^\circ}{T}\right)_{\text{gas}},$ cal K^{-1} mol^{-1}
2624	18.12	51.38
2839	18.67	51.92
2948	18.94	52.18

Calculate the heat of sublimation of Ta at 0 K. Compare this with the heat obtained from the slope of a log p versus $1/T$ plot. Why do these heats differ?

5.52 Calculate ΔG and ΔS in joules for mixing 2 mol of H_2 with 1 mol of O_2 at 25° under conditions where no chemical reaction occurs.

5.53 At 670 K

$$H_2(g) + D_2(g) = 2HD(g) \qquad K_p = 3.78$$

where D is deuterium. Calculate K_p and K_c at this temperature for the reaction

$$HD(g) = \tfrac{1}{2}H_2(g) + \tfrac{1}{2}D_2(g)$$

5.54 At 55° $K_p = 1.36$ for $N_2O_4(g) = 2NO_2(g)$. (a) How many moles of N_2O_4 must be added to a 10-liter vessel at 55° for the concentration of NO_2 to be 0.1 M? (b) How many moles of N_2O_4 must be added to a previously evacuated 10-liter vessel at 55° for the total pressure to be 2 atm?

5.55 An evacuated tube containing 5.96×10^{-3} M of solid iodine is heated to 973 K. The experimentally determined pressure is 0.490 atm [M. L. Perlman and G. K. Rollefson, J. Chem. Phys., **9**, 362 (1941)]. Assuming that the gases are ideal, calculate K_p for $I_2(g) = 2I(g)$.

5.56 Show that for a reaction of the type

$$AB(g) = A(g) + B(g)$$

the total pressure at which AB is 50% dissociated will be numerically equal to three times the equilibrium constant K_p.

5.57 For the reaction

$$CO(g) + H_2O(g) = CO_2(g) + H_2(g)$$

$K_p = 10.0$ at 690 K, and the enthalpy change ΔH° is $-10,200$ cal mol^{-1}. Calculate the partial pressures of each of the gases in an equilibrium mixture obtained from 0.400 mol of CO and 0.200 mol of H_2O in a volume of 5 liters at 800 K.

5.58 What will be the degree of dissociation of phosphorus pentachloride at 250° when 0.1 mol is placed in a 3-liter vessel containing chlorine at 0.5 atm pressure? Let $x =$ the number of moles of PCl_3 formed or additional moles of Cl_2 formed by the dissociation of the PCl_5.

5.59 How many moles of phosphorus pentachloride must be added to a liter vessel at 250° to obtain a concentration of 0.1 mol of chlorine per liter?

Given:

$$K_p = \frac{p_{PCl_3} p_{Cl_2}}{p_{PCl_5}} = 1.78$$

5.60   Under what pressure must an equimolar mixture of chlorine and phosphorus trichloride be placed at 250° in order to obtain an 80% conversion of the phosphorus trichloride into phosphorus pentachloride?

5.61   A liter reaction vessel containing 0.233 mol of N_2 and 0.341 mol of PCl_5 is heated to 250°. The total pressure at equilibrium is 28.95 atm. Assuming that all the gases are ideal, calculate K_p for the only reaction that occurs:

$$PCl_5(g) = PCl_3(g) + Cl_2(g)$$

5.62   Calculate the total pressure that must be applied to a mixture of 3 parts of hydrogen and 1 part nitrogen to give a mixture containing 10% ammonia at 400°. At 400°, $K_p = 1.64 \times 10^{-4}$ for the reaction $N_2(g) + 3H_2(g) = 2NH_3(g)$.

5.63   Derive an equation for K_p in terms of α and the total pressure P for the reaction $2A(g) = 2B(g) + C(g)$. Show that α varies inversely as the cube root of P, when P/K_p is large.

5.64   At 500° $K_p = 5.5$ for the reaction

$$CO(g) + H_2O(g) = CO_2(g) + H_2(g)$$

If a mixture of 1 mol of CO and 5 mol of H_2O is passed over a catalyst at this temperature what will be the equilibrium mole fraction of H_2O?

5.65   Show that the equilibrium constant K_p for a reaction $A + B = 2C$ determined in the gas phase is related to the equilibrium constant K_X for the reaction in the solution phase by

$$\frac{K_p}{K_X} = \frac{K_C^2}{K_A K_B}$$

where K_A, K_B, and K_C are the Henry's-law constants for the solubility of A, B, and C in the solvent.

5.66   Air (20% O_2) will not oxidize silver at 200° and 1 atm pressure. Make a statement about the equilibrium constant for the reaction

$$2Ag(s) + \tfrac{1}{2}O_2(g) = Ag_2O(s)$$

5.67   The solubility of hydrogen in a molten iron alloy is found to be proportional to the square root of the partial pressure of hydrogen. How can this be explained?

5.68   For the reaction

$$2NaHCO_3(s) = Na_2CO_3(s) + CO_2(g) + H_2O(g)$$

$K_p = 0.23$ when the pressures are expressed in atmospheres at 100°. On a day when the barometric pressure was 740 Torr, the room temperature 27°, and the relative humidity 0.70, 20 g of solid $NaHCO_3$ was placed in a 5-liter flask and sealed with the air from the room. It was then brought up to 100°. (a) What was the partial pressure of CO_2 at equilibrium in the flask (if the CO_2 of the air is neglected)? The vapor pressure of water at 27° is 26.8 Torr. (b) What was the pressure in the flask?

5.69   A certain optically active organic compound slowly racemized in solution, and eventually the solution showed no optical rotation, the D and L forms being at equal concentrations. When the temperature of the solution was varied over wide limits, there was no return of optical activity. What facts can be deduced about $\Delta G°$ and $\Delta H°$ for the racemization reaction?

5.70   The dissociation of ammonium carbamate takes place according to the reaction

$$(NH_2)CO(ONH_4)(s) = 2NH_3(g) + CO_2(g)$$

When an excess of ammonium carbamate is placed in a previously evacuated vessel, the partial pressure generated by NH_3 is twice the partial pressure of the CO_2, and the partial pressure of $(NH_2)CO(ONH_4)$ is negligible in comparison. Show that

$$K_p = (p_{NH_3})^2 p_{CO_2} = \tfrac{4}{27} P^3$$

where P is the total pressure.

5.71 From the entropy values and the enthalpies of formation given in Chapters 1 and 2, together with the value 29.09 cal K^{-1} mol^{-1} for the entropy of Ag_2O at 25°, calculate $\Delta G°$ and K for the reaction

$$2Ag_2O(s) = 4Ag(s) + O_2(g)$$

at 25°. What is the dissociation pressure of Ag_2O at 25°?

5.72 Calculate the equilibrium constants K_p for the following reactions at 25°:

$$(a) \ \ P(g) + 2\tfrac{1}{2}Cl_2(g) = PCl_5(g)$$

$$(b) \ \ UF_6(g) = U(s) + 3F_2(g)$$

$$(c) \ \ H_2S(g) + 2O_2(g) = H_2O(g) + SO_3(g)$$

5.73 If 1 mol of CO_2 is mixed with 2 mol of NH_3, what total pressure must be applied to obtain 0.5 mol of urea at equilibrium at 298 K if H_2O remains in the vapor phase? The Gibbs free energy of formation of urea(s) is $-47,120$ cal mol^{-1}, and the Gibbs free energies of formation of the other substances may be obtained from Table 5.3.

5.74 The measured density of an equilibrium mixture of N_2O_4 and NO_2 at 15° and 1 atm is 3.62 g liter^{-1}, and the density at 75° and 1 atm is 1.84 g liter^{-1}. What is the enthalpy change of the reaction $N_2O_4(g) = 2NO_2(g)$?

5.75 For the reaction

$$M(s) = M(g) \qquad \Delta H_1$$

$$M(s) = \tfrac{1}{2}M_2(g) \qquad \Delta H_2$$

$$\tfrac{1}{2}M_2(g) = M(g) \qquad \Delta H_D$$

find the relation between ΔH_1, ΔH_2, and ΔH_D. If $\Delta H_2 > \Delta H_1$, will the pressure of monomer (M) or dimer (M_2) gas molecules above $M(s)$ increase more rapidly as the temperature is increased?

5.76 Calculate ΔH for the reaction

$$CaCO_3(s) = CaO(s) + CO_2(g)$$

in the range 1000–1200° from the data of Table 5.2.

5.77 For the reaction

$$Fe_2O_3(s) + 3CO(g) = 2Fe(s) + 3CO_2(g)$$

the following values of K_p are known:

t, °C	100	250	1000
K_p	1100	100	0.0721

At 1120° for the reaction $2CO_2(g) = 2CO(g) + O_2(g)$, $K_p = 1.4 \times 10^{-12}$ atm. What equilibrium partial pressure of O_2 would have to be supplied to a vessel at 1120° containing solid Fe_2O_3 just to prevent the formation of Fe?

5.78 A mixture of N_2 and H_2 is passed over a catalyst at 400° to obtain the equilibrium conversion to NH_3. The equilibrium constant at this temperature for $N_2(g) + 3H_2(g) = 2NH_3(g)$ is $K_p = 1.64 \times 10^{-4}$. (a) Calculate the total pressure required to produce 5.26 mole % NH_3 if an equimolar mixture of N_2 and H_2 is passed over the catalyst. (b) If a mixture of 2 mol of N_2 per mol of H_2 is passed over the catalyst, what total pressure will be required to produce 5.26 mole % NH_3 at equilibrium? The fact that a higher pressure is required when a larger proportion of N_2 is used may appear to be in violation of Le Châtelier's principle, but it is not in conflict with the correct statement of this principle [J. de Heer, *J. Chem. Educ.*, **34**, 375 (1957)].

5.79 Using the free-energy function, calculate the equilibrium constant at 1000 K for the reaction

$$CO(g) + H_2O(g) = CO_2(g) + H_2(g)$$

5.80 Which process will be most important in the vaporization of $NiCl_2$: (a) vaporization of gaseous $NiCl_2$ molecules or (b) decomposition to $Ni(s)$ and $Cl_2(g)$? The equilibrium gas pressure over $NiCl_2(s)$ is 10^{-3} atm at 934 K. The following data are available:

	$\dfrac{-(\bar{G}_T^\circ - \bar{H}_{298}^\circ)}{T}$, cal K^{-1} mol^{-1}			$\Delta \bar{H}_{298}^\circ$,
	298.1 K	500 K	1000 K	kcal mol^{-1}
$NiCl_2(s)$	25.6	27.1	34.5	−73
$Ni(s)$	7.12	7.82	10.68	0
$Cl_2(g)$	53.31	54.25	57.65	0
$Ni(g)$	43.53	44.17	46.46	101.75

5.81 Methane may be converted into acetylene at 1500 K according to the reaction

$$2CH_4(g) = C_2H_2(g) + 3H_2(g)$$

(a) Calculate K_p at this temperature. (b) What is the percent conversion at 1 atm?

CHAPTER 6

ELECTROMOTIVE FORCE

Measurements of electromotive forces of electrochemical cells provide thermodynamic quantities for the chemical reactions occuring in the cells. The measurement of electromotive force at zero current is a thermodynamic measurement because the cell reaction can be reversed by an infinitesimal change in the applied electric potential.

The current of an electrochemical cell results from a chemical reaction in which electrons are taken up at one electrode and released at the other. The reaction is made up of two parts, an oxidation reaction occurring at the anode where electrons (ne) are given up by a reduced form Red to produce an oxidized form Ox,

$$\text{Red} = \text{Ox} + ne$$

as, for example, $\text{Zn} = \text{Zn}^{++} + 2e$, and a reduction reaction at the cathode involving another set of oxidized and reduced forms,

$$ne + \text{Ox}' = \text{Red}'$$

as, for example, $2e + \text{Cu}^{++} = \text{Cu}$. The net reduction obtained by adding together the two half-reactions is

$$\text{Red} + \text{Ox}' = \text{Red}' + \text{Ox}$$

or for our specific half reactions, $\text{Zn} + \text{Cu}^{++} = \text{Zn}^{++} + \text{Cu}$. An electrochemical cell is a device for keeping the oxidation and reduction parts of the reaction separated, so that the current may be used to perform useful work. The equilibrium electromotive force of the cell, or difference in electrical potential between the two electrodes when no current flows, depends on the equilibrium constant for the chemical reaction that occurs in the cell, and the activities of reactants and products. Thus measurements of electromotive force may be used to obtain thermodynamic data, and thermodynamic data from other sources may be used to calculate electromotive forces.

An understanding of the conversion of chemical energy into electrical energy is important for work with batteries, fuel cells, electroplating, corrosion, electrorefining (e.g., the production of aluminum) and electroanalytical techniques. This chapter discusses the thermodynamics of such processes.

191

6.1 COULOMB'S LAW, ELECTRIC POTENTIAL, AND UNITS

The practical unit of electric current is the ampere A, which is a base unit in the SI system and is defined in the Appendix. The practical unit of electric charge is the ampere-second or coulomb C. When calculations with Coulomb's law are carried out in SI units the law is written

$$F = \frac{1}{4\pi\epsilon_0} \frac{Q_1 Q_2}{\epsilon r^2} \tag{6.1}$$

where F is the force acting on each of the charges Q_1 and Q_2, r is the distance between the charges, ϵ is the dielectric constant, and ϵ_0 is the permittivity of free space. The value of ϵ_0 is 8.85419×10^{-12} C^2 N^{-1} m^{-2}, and $1/4\pi\epsilon_0 = 0.898755 \times 10^{10}$ N m^2 C^{-2}. When Q_1 and Q_2 are expressed in coulombs C, and the distance is expressed in the meters, the force F is in newtons N.

The electric potential φ is the work required to bring a unit charge from a region where its potential is defined as zero to a given position. If a unit test charge Q_1 is brought from an infinite distance to a distance r from Q_2

$$\varphi = -\int_\infty^r \frac{Q_2 \, dr}{4\pi\epsilon_0\epsilon r^2} = \frac{Q_2}{4\pi\epsilon_0\epsilon r} \tag{6.2}$$

In the SI system electric potential is measured in joules per coulomb. This unit is referred to as a volt and will be abbreviated as V. Table 6.1 summarizes the symbols and definitions of the units that we will be using.

Table 6.1   Electric Units in the SI System

Quantity	Unit	Symbol	Definition
Electric current I	ampere	A	See Appendix
Electric charge Q	coulomb	C	A s
Electric potential difference φ	volt	V	J A^{-1} s^{-1} = J C^{-1}
Electric resistance R	ohm	Ω	V A^{-1}

The difference in electric potential between two points in the same phase or between two pieces of the same chemical substance is measured by the work required to move a unit test charge from one point to the other. It is not possible to measure the difference in electric potential between two different chemical substances because there are local interactions of the test charge with the chemically different surrounding media. Thus it is impossible to measure the difference in potential between an electrode and the solution with which it is in contact. The difference in potential between two electrodes is measured, and one of the electrodes is frequently referred to as a reference electrode (e.g., the hydrogen electrode or calomel electrode).

6.2 ELECTROCHEMICAL CELLS

Figure 6.1A shows a simple cell in which gaseous hydrogen is bubbled over a piece of platinized platinum immersed in the hydrochloric acid solution to form a hydrogen electrode. The other electrode consists of a silver wire coated with a deposit of silver chloride. When the two electrodes are connected through a resistor, as illustrated in the figure, a current flows. The hydrogen molecules give up electrons to the platinum to form hydrogen ions, and silver ions from the AgCl react with electrons (coming through the wire) to produce metallic silver. The difference in electrical potential between the electrodes is due to the fact that H_2 has a greater tendency to give up electrons in the presence of H^+ than Ag does in the presence of Cl^-.

The electrode at which the *oxidation* half-reaction occurs is the *anode*, and the electrode at which the *reduction* half-reaction occurs is the *cathode*. Thus, when an electrochemical cell discharges spontaneously, electrons flow through the external circuit from anode to cathode as illustrated in Fig. 6.1.*

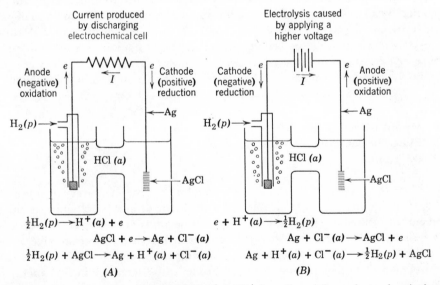

Fig. 6.1 Operation of the $H_2(p) \mid HCl(a) \mid AgCl \mid Ag$ cell as (A) an electrochemical cell or (B) an electrolysis cell. Activity is represented by a.

---

* Conventionally the "current" I is designated as a flow of positive current from the positive electrode through the wire to the negative electrode. In discussing electrochemical reactions the flow of negative electrons, which is in the opposite direction, is emphasized.

The cell reaction may be reversed by applying a voltage higher than the reversible electromotive force of the cell. The chemical processes at the electrodes are then reversed: the hydrogen electrode becomes the cathode, and the Ag-AgCl electrode the anode, as shown in Fig. 6.1*B*. Whether a cell is operating as a discharging battery or as an electrolysis cell the "anode" is the electrode at which oxidation occurs and electrons are given up to that electrode; the "cathode" is the electrode at which reduction occurs and electrons are taken up by a reactant in solution.

If in the experiment described in Fig. 6.1*A* the electrochemical cell were operated long enough to deliver 1 faraday of electricity, designated by F, the following processes would take place:

1. One faraday, 96,485 coulombs or ampere-seconds, which is Avogadro's number of electrons, would travel through the external circuit to the Ag-AgCl electrode.
2. One-half mole of hydrogen gas would be consumed with the production of 1 mol of hydrogen ion.
3. One mole of AgCl would be converted into 1 mol of silver plus 1 mol of chloride ion.

As long as the circuit is closed, these reactions go on producing hydrochloric acid and metallic silver at the expense of H_2 and AgCl.

The electromotive force of a reversible cell may be measured under conditions such that by an infinitesimal change of the applied voltage the cell can be changed from a discharging battery to an electrolysis cell. To obtain this reversible electromotive force all steps in the cell reaction must be at equilibrium. Not all cells are reversible. In some, irreversible processes occur so that it is not possible to reverse the chemical reaction by an infinitesimal change in the applied voltage.

6.3 MEASUREMENT OF THE ELECTROMOTIVE FORCE OF CELLS

To obtain the reversible electromotive force of a cell, the measurement must be made without drawing appreciable current. If an ordinary voltmeter is used and current I flows through the cell, there will be a potential drop IR, where R is the resistance of the cell, which will be subtracted from the electromotive force of the cell. The passage of a current also causes concentration changes at the electrodes, and this polarization of the electrodes causes a change in electromotive force.

The reversible electromotive force of a cell may be measured by use of a potentiometer, for which the basic circuit diagram is illustrated in Fig. 6.2. A battery B is connected in series with a resistor R and an adjustable resistance R'. A cell of known electromotive force E is also connected across R through a galvanometer G and a tapping key K. The cell is connected so that its electromotive force opposes

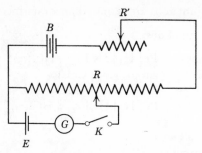

Fig. 6.2 Circuit diagram for a potentiometer that may be used to measure the voltage E of an unknown cell by adjusting the sliding contact on the resistor R so that no current flows, as indicated by a zero reading on the galvanometer G.

that of battery B. It is convenient to consider that resistance R is a slide wire graduated evenly with numbers so that the sliding contact may be set at the known voltage of the cell. A Weston cadmium cell with an electromotive force of 1.0184 V at 25° is often used. Resistance R' is then adjusted so that when key K is tapped no current is indicated by the galvanometer. The key is tapped rather than locked down so that no appreciable chemical change occurs in the cell. After the potentiometer has been calibrated, a cell of unknown voltage is substituted at E, and the sliding contact moved until the galvanometer indicates zero current when the key is tapped.

6.4 NOTATION FOR ELECTROCHEMICAL CELLS

Many different types of reversible electrodes have been studied. A metal may be immersed in a solution containing its ions, as in the zinc electrode $Zn \mid Zn^{2+}$. Or an inert electrode such as platinum or gold may be immersed in a solution of oxidizing and reducing ions, as in the ferrous-ferric electrode $Pt \mid Fe^{2+}, Fe^{3+}$. The vertical line indicates a contact between two phases such as exists between a metal and a solution, or an interface between two unmixed liquid solutions, such as a solution of zinc sulfate touching a solution of copper sulfate. A comma is used to separate different ions or molecules which exist in the same phase.

At the contact between two different electrolytic solutions there is a small junction potential in addition to the potential difference due to the two electrodes. If the contact between the two electrolytic solutions is made by means of an inert electrolyte, for example, potassium chloride, the inert electrolyte section is called a salt bridge and denoted by $\|$. Cells may be classified as follows:

I. Cells without liquid junctions:

Different electrodes	$Pt \mid H_2 \mid HCl \mid Cl_2 \mid Pt$
Concentration cell with amalgams	$(Hg + 10\% \ Cd) \mid CdCl_2 \mid$
	$(Hg + 1\% \ Cd)$

Cells without liquid junctions are required for exact thermodynamic treatment.

II. Cells with liquid junctions:

Concentration cells	$Pt \mid H_2 \mid HCl(c_1) \mid HCl(c_2) \mid H_2 \mid Pt$
Different electrodes	$Zn \mid ZnCl_2 \mid CuCl_2 \mid Cu$
Cells with salt bridges	$Zn \mid ZnCl_2 \parallel CuCl_2 \mid Cu$
	$Pt \mid H_2 \mid HCl(c_1) \parallel HCl(c_2) \mid H_2 \mid Pt$

6.5 JUNCTION POTENTIALS

A junction between two electrolyte solutions contributes a potential to a cell. If, for example, a concentrated solution of hydrochloric acid forms a junction with a dilute solution, both hydrogen ions and chloride ions diffuse from the concentrated solution into the dilute solution. The hydrogen ion moves faster, and thus the dilute solution becomes positively charged because of an excess of hydrogen ions. The more concentrated solution is left with an excess of chloride ions and thus acquires a negative charge. The actual separation of charge is very small, but the potential difference produced is appreciable.

In general, it may be stated that the difference of potential resulting from the junction of two solutions is caused by the difference in the rates of diffusion of the two ions, *the more dilute solution acquiring a charge corresponding to that of the faster-moving ion.*

The difference in potential resulting from the junction of two liquids must be eliminated or corrected to obtain electrode potentials. The most convenient way of minimizing the liquid junction potential is by means of a salt bridge of potassium chloride connecting the two solutions of different electrolytes. Potassium chloride is often used because the mobilities (Section 11.5) of the two ions are about the same. When potassium chloride is not feasible, as for example, in a cell containing silver nitrate, a salt bridge of ammonium nitrate is substituted.

6.6 THERMODYNAMICS OF ELECTROCHEMICAL CELLS

The Gibbs free-energy changes for reactions occurring in electrochemical cells may be readily calculated from the *reversible* electromotive force. If a cell is exactly balanced against an external electromotive force so that no charging or discharging of the cell is taking place, and we imagine that an infinitesimal quantity of electricity is allowed to pass through the cell, the reversible electrical work at constant temperature and pressure, or Gibbs free-energy change, is equal to the product of the voltage and the quantity of electricity. The quantity of electrical charge corresponding to the molar quantities indicated in the balanced chemical equation is zF, where z is the charge number for the cell reaction and F is the Faraday constant ($96{,}485 \ C \ mol^{-1}$). When the chemical reaction occurs as written, the quantity of electric charge that flows is zF. If this quantity of electrical charge is transported through a potential difference of E volts, the amount of work required is given by zFE. Since this electrical change does not involve pressure-volume work and is carried out isothermally, the change in Gibbs free energy is given by

$$\Delta G = -zFE \tag{6.3}$$

where E is the potential difference, which by convention is taken as positive. Since ΔG is negative for a spontaneous cell reaction and E for a spontaneously discharging cell is taken as positive, the negative sign must be used in equation 6.3. The electromotive force of a cell does not depend on the stoichiometric coefficients in the balanced chemical reaction, but the change in Gibbs free energy ΔG does depend on z, which in turn depends on how the stoichiometric equation is written.

If the Faraday constant is expressed in C mol^{-1}, the electrical work calculated from equation 6.3 is expressed in joules mol^{-1}. The value of ΔG in calories is obtained by dividing by 4.184 J cal^{-1}. Alternatively the faraday may be expressed as 96,485 C mol^{-1}/4.184 J cal^{-1} = 23,060 cal V^{-1} mol^{-1}, since a joule is a volt-coulomb.

The entropy change for the cell reaction may be calculated from the temperature coefficient of the electromotive force since (see Section 2.12)

$$\left(\frac{\partial \Delta G}{\partial T}\right)_P = -\Delta S \tag{6.4}$$

Introducing equation 6.3 we have

$$zF\left(\frac{\partial E}{\partial T}\right)_P = \Delta S \tag{6.5}$$

The enthalpy change for the cell reaction may be calculated by substituting equations 6.3 and 6.5 into

$$\Delta H = \Delta G + T\Delta S$$
$$= -zFE + zFT\left(\frac{\partial E}{\partial T}\right)_P \tag{6.6}$$

Thus from measurements of the reversible electromotive force of a cell at a series of temperatures it is possible to calculate ΔG, ΔS, and ΔH for the cell reaction. The great accuracy of electrical measurements often makes the determination of thermodynamic quantities by this method more exact than the direct determination of equilibrium constants or the calorimetric determination of enthalpies of reaction.

Example 6.1 The electromotive force of the cell

$$Cd \mid CdCl_2 \cdot 2\tfrac{1}{2}H_2O, \text{ sat. solution} \mid AgCl \mid Ag$$

at 25° is 0.67533 V, and the temperature coefficient is -6.5×10^{-4} V K^{-1}. Calculate the values of ΔG, ΔS, and ΔH at 25° for the reaction

$$Cd(s) + 2AgCl(s) = 2Ag(s) + CdCl_2 \cdot 2\tfrac{1}{2}H_2O(s)$$

$$\Delta G = -(2)(23,060 \text{ cal V}^{-1} \text{ mol}^{-1})(0.67533 \text{ V})$$

$$= -31,150 \text{ cal mol}^{-1}$$

$$\Delta S = (2)(23,060 \text{ cal V}^{-1} \text{ mol}^{-1})(-6.5 \times 10^{-4} \text{ V K}^{-1})$$

$$= -30.0 \text{ cal K}^{-1} \text{ mol}^{-1}$$

$$\Delta H = -(2)(23,060)(0.67533) - (2)(23,060)(298.1)(6.5 \times 10^{-4})$$

$$= -40,090 \text{ cal mol}^{-1}$$

Alternatively the thermodynamic quantities may be calculated in SI units

$$\Delta G = -(2)(96,485 \text{ C mol}^{-1})(0.67533 \text{ V})$$

$$= -130.32 \text{ kJ mol}^{-1}$$

$$\Delta S = (2)(96,485 \text{ C mol}^{-1})(-6.5 \times 10^{-4} \text{ V K}^{-1})$$

$$= -125.43 \text{ J K}^{-1} \text{ mol}^{-1}$$

$$\Delta H = -(2)(96,485)(0.67533) - (2)(96,485)(298.1)(6.5 \times 10^{-4})$$

$$= -167.72 \text{ kJ mol}^{-1}$$

Direct calorimetric measurements give $\Delta H = -39.530 \text{ kcal mol}^{-1}$ or $-165.39 \text{ kJ mol}^{-1}$ for this reaction.

6.7 FUNDAMENTAL EQUATION FOR AN ELECTROCHEMICAL CELL

As shown in section 1.18 a chemical reaction may be represented by

$$0 = \sum v_i A_i \tag{6.7}$$

In analogy with equation 5.39 the Gibbs free energy change for the reaction is given by

$$\Delta G = \Delta G^\circ + RT \ln \prod_i a_i^{v_i} \tag{6.8}$$

where the a's represent the activities of the reactants and products under a given set of conditions. Activities may be regarded as effective concentrations, and they are defined in such a way that equation 6.8 is obeyed. Substituting $\Delta G = -zFE$, and $\Delta G^\circ = -zFE^\circ$ for the reactants and products in their standard states, we have

$$E = E^\circ - \frac{RT}{zF} \ln \prod_i a_i^{v_i} \tag{6.9}$$

The standard electromotive force for the cell, E°, is the electromotive force for the cell in which the activities of the reactants and products of the cell reaction are each equal to unity. At 25°

$$E = E° - \frac{(8.314)(298.1)(2.303)}{z(96,485)} \log \prod_i a_i^{\nu_i}$$

$$E = E° - \frac{0.0591}{z} \log \prod_i a_i^{\nu_i} \qquad (6.10)$$

This equation is often referred to as the Nernst equation. When the activities are all unity the logarithmic term vanishes and $E = E°$.

If the activities of the reactants and products correspond with those of an equilibrium mixture, equation 6.9 becomes

$$E° = \frac{RT}{zF} \ln K \qquad (6.11)$$

where K is the equilibrium constant for reaction 6.7.

Since special equations are involved in expressing the activities of electrolytes, we must now consider these expressions.

6.8   ACTIVITY OF ELECTROLYTES

When a solute does not dissociate, the activity is equal to the product of the concentration and an activity coefficient. Using Convention II (Section 4.11), the activity coefficient approaches unity at infinite dilution. When the solute is an electrolyte which is considered to be completely dissociated in solution, the expression for the activity becomes more complicated.

The chemical potential of a strong, completely dissociated, electrolyte MX is equal to the sum of the chemical potentials of the ions M^+ and X^-.

$$\mu_{MX} = \mu_{M^+} + \mu_{M^-} \qquad (6.12)$$

$$\mu_{MX}° + RT \ln a_{MX} = \mu_{M^+}° + RT \ln a_{M^+} + \mu_{X^-}° + RT \ln a_{X^-} \qquad (6.13)$$

where $\mu_{MX}°$ is the chemical potential of MX at unit activity, $\mu_{M^+}°$ is the chemical potential of the cation at unit activity, and $\mu_{X^-}°$ is the chemical potential of the anion at unit activity. Since $\mu_{MX}° = \mu_{M^+}° + \mu_{X^-}°$, equation 6.13 shows that

$$a_{MX} = (a_{M^+})(a_{X^-}) \qquad (6.14)$$

The activities of the cation and anion may be expressed as products of the molal concentrations m and the activity coefficients γ_+ and γ_- of the cation and anion.

$$a_{M^+} = m\gamma_+ \quad \text{and} \quad a_{X^-} = m\gamma_- \qquad (6.15)$$

Then

$$a_{MX} = (m\gamma_+)(m\gamma_-) = m^2\gamma_\pm^2 \qquad (6.16)$$

where $\gamma_\pm$ represents the mean ionic activity coefficient for the 1–1 electrolyte (that is, one having a univalent cation and a univalent anion). It follows from equation 6.16 that

$$\gamma_\pm = (\gamma_+\gamma_-)^{\frac{1}{2}} \tag{6.17}$$

The mean ionic activity coefficient is important since it can be determined experimentally, whereas the individual ion activity coefficients γ_+ and γ_- cannot. The mean ionic activity coefficient is taken to approach unity as the concentration of MX approaches zero.

The expression for the mean ionic activity coefficient becomes more complicated for molecules with polyvalent ions. If the strong electrolyte is $M_{\nu_+}X_{\nu_-}$, where ν_+ is the number of cations and ν_- is the number of anions, then

$$\mu_{M_{\nu_+}X_{\nu_-}} = \nu_+\mu_M + \nu_-\mu_X \tag{6.18}$$

which leads to

$$a_{M_{\nu_+}X_{\nu_-}} = (a_M)^{\nu_+}(a_X)^{\nu_-} \tag{6.19}$$

The activity coefficients of the cation and anion are equal to the ratios of their activities to their concentrations.

$$\gamma_+ = a_M/m_M = a_M/\nu_+ m \qquad \gamma_- = a_X/m_X = a_X/\nu_- m \tag{6.20}$$

Then, using these relations to eliminate a_M and a_X from equation 6.19,

$$a_{M_{\nu_+}X_{\nu_-}} = (\nu_+ m\gamma_+)^{\nu_+}(\nu_- m\gamma_-)^{\nu_-} = (m_\pm\gamma_\pm)^{\nu_++\nu_-} \tag{6.21}$$

By rearranging terms in this equation, it is seen that the mean ionic molality $m_\pm$ and mean ionic activity coefficient $\gamma_\pm$ are given by

$$m_\pm = m(\nu_+^{\nu_+}\nu_-^{\nu_-})^{1/(\nu_++\nu_-)} \tag{6.22}$$

$$\gamma_\pm = (\gamma_+^{\nu_+}\gamma_-^{\nu_-})^{1/(\nu_++\nu_-)} \tag{6.23}$$

The mean ionic molality of a 1–1 electrolyte like NaCl is equal to m, of a 2–1 electrolyte like $CaCl_2$ is equal to $4^{1/3}m$, of a 2–2 electrolyte like $CuSO_4$ is equal to m, and of a 3–1 electrolyte like $LaCl_3$ is equal to $27^{1/4}m$, as may be deduced from equation 6.23. The numbers 1, 2, and 3 refer to the number of charges on the cation and anion.

In equations where the activities of electrolytes occur, these activities may be replaced by expressions involving the molality m and the mean ionic activity coefficient $\gamma_\pm$.

Example 6.2   Write the expressions for the activities of NaCl, $CaCl_2$, $CuSO_4$, and $LaCl_3$ in terms of their molalities and mean ionic activity coefficients.

$$a_{NaCl} = m^2\gamma_\pm^2$$
$$a_{CaCl_2} = 4m^3\gamma_\pm^3$$
$$a_{CuSO_4} = m^2\gamma_\pm^2$$
$$a_{LaCl_3} = 27m^4\gamma_\pm^4$$

6.9 IONIC STRENGTH

Electrolytes containing ions with multiple changes have larger effects on the activity coefficients of ions than electrolytes containing only singly charged ions. In order to express electrolyte concentrations in a way that takes this into account, G. N. Lewis* introduced the ionic strength I defined by

$$I = \tfrac{1}{2} \sum_i m_i z_i^2 = \tfrac{1}{2}(m_1 z_1^2 + m_2 z_2^2 + \cdots) \qquad (6.24)$$

where the summation is continued over all the different ionic species in the solution and m is the molal concentration. The greater effectiveness of ions of higher charge in reducing the activity coefficient is provided for by multiplying their concentrations by the square of their valence. According to equation 6.24, the ionic strength of a 1–1 electrolyte is equal to its molality. The ionic strength for a 1–2 electrolyte is 3m and for a 2–2 electrolyte is 4m.

6.10 DEBYE-HÜCKEL THEORY†

The activity coefficient of an electrolyte depends markedly on the concentration. In dilute solutions the interaction between ions is a simple Coulombic attraction or repulsion, and this interaction extends considerably further through the solution than other intermolecular forces. At infinite dilution the distribution of ions in an electrolytic solution can be considered to be completely random because the ions are too far apart to exert any attraction on each other, and the activity coefficient of the electrolyte is unity. At higher concentrations where the ions are closer together, however, the Coulomb attractive and repulsive forces become important. Because of this interaction of ions the concentration of positive ions is slightly higher in the neighborhood of a negative ion, and the concentration of negative ions is slightly higher in the neighborhood of a positive ion, than in the bulk solution. Because of the attractive forces between an ion and its surrounding ionic atmosphere, the activity coefficient of the electrolyte is reduced. This effect is greater for ions of high charge and is greater in solvents of lower dielectric constant where the electrostatic interactions are stronger.

Debye and Hückel were able to show that the activity coefficient γ_i of an ion species i with z_i electronic charges is given by

$$\log \gamma_i = -A z_i^2 I^{1/2} \qquad (6.25)$$

where I is the ionic strength and

$$A = \frac{1}{2.303} \left(\frac{2\pi N_A m_{solv}}{V} \right)^{1/2} \left(\frac{e^2}{4\pi \epsilon_0 \epsilon k T} \right)^{3/2} \qquad (6.26)$$

where m_{solv} is the mass of solvent in volume V.

---

* G. N. Lewis and M. Randall, revised by K. S. Pitzer and L. Brewer, *Thermodynamics*, McGraw-Hill Book Co., New York, 1961, p. 335.

† P. Debye and E. Hückel, *Physik, Z.*, **24,** 185, 305 (1923).

Example 6.3 Calculate the value of the coefficient A in the Debye-Hückel equation for aqueous solutions at 298.15 K. The dielectric constant ϵ of water at this temperature is 78.54.

$$A = \frac{1}{2.303}\left(\frac{2\pi(6.022 \times 10^{23}\ \text{mol}^{-1})(997\ \text{kg})}{1.000\ \text{m}^3}\right)^{\frac{1}{2}}$$

$$\times \left[\frac{(1.602 \times 10^{-19}\ \text{C})^2(0.8988 \times 10^{10}\ \text{N m}^2\ \text{C}^{-2})}{(78.54)(1.3807 \times 10^{-23}\ \text{J K}^{-1})(298.15\ \text{K})}\right]^{\frac{3}{2}}$$

$$= 0.509\ \text{kg}^{\frac{1}{2}}\ \text{mol}^{-\frac{1}{2}}$$

Since the ionic strength I has the units mol kg^{-1}, the units of $I^{\frac{1}{2}}$ and A cancel, as they must, to give a logarithm.

Equation 6.26 gives the activity coefficient of a single ion, but the quantity that is accessible to experimental determination is the mean ionic activity coefficient, which for the electrolyte $M_{\nu_+}X_{\nu_-}$ is by equation 6.23

$$\gamma_\pm = (\gamma_+^{\nu_+}\gamma_-^{\nu_-})^{1/(\nu_+ + \nu_-)} \tag{6.27}$$

Taking the logarithm of equation 6.27 we have

$$\log \gamma_\pm = \frac{1}{(\nu_+ + \nu_-)}(\nu_+ \log \gamma_+ + \nu_- \log \gamma_-) \tag{6.28}$$

Substituting equation 6.25 for each activity coefficient, we have

$$\log \gamma_\pm = -0.509\sqrt{I}\left(\frac{\nu_+ z_+^2 + \nu_- z_-^2}{\nu_+ + \nu_-}\right) \tag{6.29}$$

Introducing $\nu_+ z_+ = \nu_- z_-$, we see that

$$\log \gamma_\pm = -0.509 z_+ z_- \sqrt{I} \tag{6.30}$$

The Debye-Hückel theory has been of great value in interpreting the properties of electrolyte solutions. It is a limiting law at low concentrations in the same sense that the ideal gas law is a limiting law at low pressures. At high values of the ionic strength the activity coefficient of an electrolyte usually increases with increasing ionic strength. Equation 6.30 is in excellent agreement with experiment up to an ionic strength of about 0.01, but large deviations are encountered even at this ionic strength if the product of the valence of the highest charged ion of the salt and the valence of the oppositely charged ion of the electrolyte medium is greater than about 4.

Applications of the Debye-Hückel theory are illustrated in the next section and in Section 7.10.

6.11 CELLS WITHOUT LIQUID JUNCTIONS

Such cells can be given an exact thermodynamic treatment and are useful for determining activity coefficients.

Since a cell without liquid junction contains a single electrolyte solution, the two electrodes must be chosen so that one is reversible with respect to a cation of the electrolyte and the other with respect to an anion of the electrolyte. For example, if the electrolyte is hydrochloric acid, one electrode would be the hydrogen electrode and the other a chlorine or silver chloride electrode. In the latter case the cell may be represented by

$$\text{Pt} \mid H_2(g) \mid HCl(m) \mid AgCl \mid Ag$$

The cell reaction is

$$\tfrac{1}{2}H_2(g) + AgCl(s) = HCl(m) + Ag(s) \tag{6.31}$$

According to equation 6.9, the electromotive force for this cell is given by

$$E = E^\circ - \frac{RT}{zF} \ln \frac{a_{HCl}}{p_{H_2}^{\frac{1}{2}}} \tag{6.32}$$

assuming that H_2 may be treated as an ideal gas. If the pressure of hydrogen is 1 atm and equation 6.16 is introduced,

$$E = E^\circ - \frac{2.303\, RT}{zF} \log\, m^2 \gamma_\pm^2 \tag{6.33}$$

The mean ionic activity coefficient of hydrochloric acid is represented by $\gamma_\pm$, and m is the molality. The electromotive force of the cell when the activity of hydrochloric acid is unity and the pressure of hydrogen is 1 atm is represented by E° called the standard electromotive force of the cell.

As it stands, equation 6.33 contains two unknown quantities, E° and $\gamma_\pm$. These may be obtained by determining the electromotive force of this cell over a range of hydrochloric acid concentrations, including dilute solutions. Rearranging equation 6.33 and substituting numerical values for 25° gives

$$E + 0.1183 \log m = E^\circ - 0.1183 \log \gamma_\pm \tag{6.34}$$

The exponents in equation 6.33 have been placed in front of the logarithmic term, giving $(2)(0.05916) = 0.1183$. Since at infinite dilution $m = 0$, $\gamma_\pm = 1$, and $\log \gamma_\pm = 0$, it can be seen that, when $E + 0.1183 \log m$ is plotted against m, the extrapolation of $E + 0.1183 \log m$ to $m = 0$ will give E°.

To make a satisfactory extrapolation, use is made of the Debye–Hückel theory to furnish a function that will give nearly a straight line. The following expression is used for the mean ionic activity coefficient of a 1–1 electrolyte in dilute aqueous solutions at 25°:

$$\log \gamma_\pm = -0.509\sqrt{m} + bm$$

where b is an empirical constant.

Substituting in equation 6.34 and rearranging terms, we have

$$E + 0.1183 \log m - 0.0602m^{\frac{1}{2}} = E' = E^\circ - (0.1183b)m \tag{6.35}$$

According to this equation, the left-hand side, which we will designate as E', will give a straight line when it is plotted against m, and the intercept at $m = 0$ is E°.

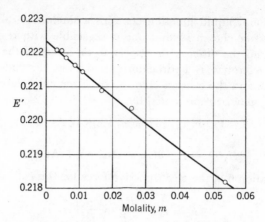

Fig. 6.3   Determination of the silver-silver chloride electrode potential by extrapolation of a function of the potential of the cell Pt | H_2(1 atm) | HCl(m) | AgCl | Ag to infinite dilution.

In Fig. 6.3, E' is plotted against m. The extrapolated value is 0.2224 V. This is the electromotive force the cell would deliver with the hydrochloric acid at unit activity, and it is also the standard electrode potential of the silver-silver chloride electrode, since the other electrode is the standard hydrogen electrode for which the standard electrode potential is zero by definition. Similar cells have been used for determining the standard electrode potentials for other electrodes.

The value of $E°$ having been determined, the activity coefficient of hydrochloric acid at any other concentration may be calculated from the electromotive force of the cell containing hydrochloric acid at that concentration.

Example 6.4   Calculate the mean ionic activity coefficient of 0.1 molal hydrochloric acid at 25° from the fact that the electromotive force of the cell described in this section is 0.3524 V at 25°. Substituting in equation 6.34, we have

$$0.3524 = 0.2224 - 0.1183 \log \gamma_\pm - 0.1183 \log 0.1$$

$$\log \gamma_\pm = (-0.3524 + 0.2224 + 0.1183)/0.1183 = -0.0989$$

$$\gamma_\pm = 0.798$$

In this general manner the activity coefficients of the electrolytes plotted in Fig. 6.4 have been determined. It should be noted that at high concentrations of electrolytes activity coefficients may be considerably greater than unity.

Activity coefficients may be calculated from a number of different types of measurements, including vapor pressure, freezing-point lowering, boiling-point elevation, osmotic pressure, distribution coefficients, equilibrium constants, solubility, and electromotive force. All different methods for determining activity coefficients must lead to the same value for a given solution.

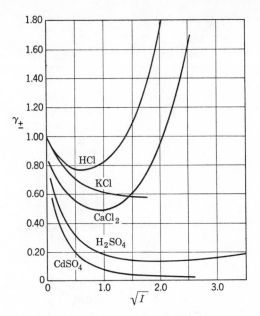

Fig. 6.4 Dependence of the mean ionic activity coefficient $\gamma_{\pm}$ on $I^{1/2}$ for electrolytes at 25°.

6.12 ELECTRODE POTENTIALS

The relative electrode potential of any electrode may be determined by combining the electrode with a standard hydrogen electrode, preferably without a liquid junction, and measuring the voltage of the cell as a function of concentration. If one electrode consists of a metal in contact with a solution containing its ions with charge $n+$, the cell may be represented by

$$\text{Pt} \mid \text{H}_2(g,\ 1\ \text{atm}) \mid \text{H}^+ \parallel M^{n+} \mid M \tag{6.36}$$

The corresponding cell reaction is written according to conventions that are given shortly.

$$\tfrac{1}{2}\text{H}_2 + \frac{1}{n} M^{+n} = \text{H}^+ + \frac{1}{n} M \tag{6.37}$$

The term *standard electrode potential* is used to designate the potential that would be obtained with the constituents present at unit activity. *The standard electrode potential of an electrode is given a positive value if this electrode is more positive than the standard hydrogen electrode and a negative sign if it is more negative than the standard hydrogen electrode.* The electrode with the more positive electrode potential is attached to the positive terminal of the potentiometer. The positive electrode is the one at which reduction occurs and the negative electrode is the one at which oxidation occurs if the cell is allowed to react spontaneously.

Table 6.2[1] Standard Electrode Potentials at 25°

Electrode	$E°$	Half Cell Reaction
Li^+ | Li	−3.045	$Li^+ + e = Li$
K^+ | K	−2.925	$K^+ + e = K$
Rb^+ | Rb	−2.925	$Rb^+ + e = Rb$
Na^+ | Na	−2.714	$Na^+ + e = Na$
Mg^{2+} | Mg	−2.37	$\frac{1}{2}Mg^{2+} + e = \frac{1}{2}Mg$
Pu^{3+} | Pu	−2.07	$\frac{1}{3}Pu^{3+} + e = \frac{1}{3}Pu$
Th^{4+} | Th	−1.90	$\frac{1}{4}Th^{4+} + e = \frac{1}{4}Th$
Np^{3+} | Np	−1.86	$\frac{1}{3}Np^{3+} + e = \frac{1}{3}Np$
Al^{3+} | Al	−1.66	$\frac{1}{3}Al^{3+} + e = \frac{1}{3}Al$
Zn^{2+} | Zn	−0.763	$\frac{1}{2}Zn^{2+} + e = \frac{1}{2}Zn$
Fe^{2+} | Fe	−0.440	$\frac{1}{2}Fe^{2+} + e = \frac{1}{2}Fe$
Cr^{3+}, Cr^{2+} | $Pt^{2,3}$	−0.41	$Cr^{3+} + e = Cr^{2+}$
Cd^{2+} | Cd	−0.403	$\frac{1}{2}Cd^{2+} + e = \frac{1}{2}Cd$
Tl^+ | Tl	−0.3363	$Tl^+ + e = Tl$
Br^- | $PbBr_2(s)$, Pb	−0.280	$\frac{1}{2}PbBr_2 + e = \frac{1}{2}Pb + Br^-$
Co^{2+} | Co	−0.277	$\frac{1}{2}Co^{2+} + e = \frac{1}{2}Co$
Ni^{2+} | Ni	−0.250	$\frac{1}{2}Ni^{2+} + e = \frac{1}{2}Ni$
I^- | $AgI(s)$, Ag	−0.151	$AgI + e = Ag + I^-$
Sn^{2+} | Sn	−0.140	$\frac{1}{2}Sn^{2+} + e = \frac{1}{2}Sn$
Pb^{2+} | Pb	−0.126	$\frac{1}{2}Pb^{2+} + e = \frac{1}{2}Pb$
D^+ | D_2, Pt	−0.0034	$D^+ + e = \frac{1}{2}D_2$
H^+ | H_2, Pt	0.0000	$H^+ + e = \frac{1}{2}H_2$
Ti^{4+}, Ti^{3+} | Pt	0.04	$Ti^{4+} + e = Ti^{3+}$
Br^- | $AgBr(s)$, Ag	0.095	$AgBr + e = Ag + Br^-$
Sn^{4+}, Sn^{2+} | Pt	0.15	$\frac{1}{2}Sn^{4+} + e = \frac{1}{2}Sn^{2+}$
Cu^{2+}, Cu^+ | Pt	0.153	$Cu^{2+} + e = Cu^+$
Cl^- | $AgCl(s)$, Ag	0.2224	$AgCl + e = Ag + Cl^-$
Cl^- | $Hg_2Cl_2(s)$, Hg^4	0.268	$\frac{1}{2}Hg_2Cl_2 + e = Hg + Cl^-$
Cu^{2+} | Cu	0.337	$\frac{1}{2}Cu^{2+} + e = \frac{1}{2}Cu$
H^+ | $C_2H_4(g)$, $C_2H_6(g)$, Pt	0.52	$H^+ + \frac{1}{2}C_2H_4(g) + e = \frac{1}{2}C_2H_6(g)$
Cu^+ | Cu	0.521	$Cu^+ + e = Cu$
I^- | $I_2(s)$, Pt	0.5355	$\frac{1}{2}I_2 + e = I^-$
H^+, quinhydrone(s) | Pt	0.6996	$\frac{1}{2}C_6H_4O_2 + H^+ + e = \frac{1}{2}C_6H_6O_2$
Fe^{3+}, Fe^{2+} | Pt	0.771	$Fe^{3+} + e = Fe^{2+}$
Hg_2^{2+} Hg	0.789	$\frac{1}{2}Hg_2^{2+} + e = Hg$
Ag^+ | Ag |	0.7991	$Ag^+ + e = Ag$
Hg^{2+}, Hg_2^{2+} | Pt	0.920	$Hg^{2+} + e = \frac{1}{2}Hg_2^{2+}$
Pu^{4+}, Pu^{3+} | Pt	0.97	$Pu^{4+} + e = Pu^{3+}$
Br^- | $Br_2(l)$ | Pt	1.0652	$\frac{1}{2}Br_2(l) + e = Br^-$

[1] All ions are at unit activity in water, and all gases are at 1 atm.

[2] The symbol Pt represents an inert electrode like platinum.

[3] The order of writing the ions in the electrolyte solution is immaterial.

[4] The electromotive force of the normal calomel electrode is 0.2802 V and of the calomel electrode containing saturated KCl is 0.2415 V.

Table 6.2   (continued)

$Tl^{3+}, Tl^+ \mid Pt$	1.250	$\frac{1}{2}Tl^{3+} + e = \frac{1}{2}Tl^+$
$Cl^- \mid Cl_2(g) \mid Pt$	1.3595	$\frac{1}{2}Cl_2(g) + e = Cl^-$
$Pb^{2+} \mid PbO_2 \mid Pb$	1.455	$\frac{1}{2}PbO_2 + 2H^+ + e = \frac{1}{2}Pb^{2+} + H_2O$
$Au^{3+} \mid Au$	1.50	$\frac{1}{3}Au^{3+} + e = \frac{1}{3}Au$
$Ce^{4+}, Ce^{3+} \mid Pt$	1.61	$Ce^{4+} + e = Ce^{3+}$
$Co^{3+}, Co^{2+} \mid Pt$	1.82	$Co^{3+} + e = Co^{2+}$
$F^- \mid F_2(g) \mid Pt$	2.87	$\frac{1}{2}F_2(g) + e = F^-$
$HF(aq) \mid F_2(g) \mid Pt$	3.06	$H^+ + \frac{1}{2}F_2(g) + e = HF(aq)$

All electrodes may be arranged in a table according to their standard electrode potentials, which are determined as described in the preceding section by means of an extrapolation to find the electromotive force at unit activity. Table 6.2 gives standard electrode potential $E°$ of a number of electrodes at 25°. It is to be noted that these electrodes are written in the order ion-electrode and that *standard electrode potentials are standard reduction potentials* as recommended by IUPAC (International Union of Pure and Applied Chemistry). The electrode reaction is written as a reduction, that is, the addition of electrons.

The magnitude of the standard electrode potential is a measure of the tendency of the half-reaction to occur in the direction of reduction. The lower a half reaction is in the last column of Table 6.2, the greater is the tendency of the oxidized form to accept electrons and be reduced. The higher a half reaction is in the table, the greater is the tendency of the reduced form to donate electrons and be oxidized. For example, the active metals sodium and potassium have very large negative standard electrode potentials and strong tendencies to donate electrons.

The reduced form of any element or ion at unit activity will reduce the oxidized form of any element or ion at unit activity which has a less negative, that is, more positive, standard electrode potential. Table 6.2 of standard electrode potentials may be used for calculations of the electromotive forces of electrochemical cells and the equilibrium constants of the corresponding chemical reactions, as will be shown presently.

6.13 CONVENTIONS FOR ELECTRO-
CHEMICAL CELLS AND
ELECTRODE REACTIONS

The following two rules are convenient for calculating the standard electromotive force of a cell, determining the cell reaction, and deciding whether the cell reaction is spontaneous as written. The standard electromotive force $E°$ of a cell is the electromotive force when all constituents are at unit activity.

1. *The standard electromotive force of a cell is equal to the standard electrode potential of the right-hand electrode minus the standard electrode potential of the left-hand electrode.*

$$E° = E°_{\text{right}} - E°_{\text{left}}$$

It is to be noted that the electrode at the right is automatically written in the order ion-electrode as it is in Table 6.2, and that, according to the geometry of the cell, the electrode at the left must be written in the order electrode-ion.

For the cell

$$Zn \mid Zn^{2+} \parallel Cu^{2+} \mid Cu$$

$$E^\circ = 0.337 - (-0.763) = 1.100 \text{ V} \qquad \text{at } 25^\circ \tag{6.38}$$

For the cell

$$Cu \mid Cu^{2+} \parallel Zn^{2+} \mid Zn$$

$$E^\circ = -0.763 - (0.337) = -1.100 \text{ V} \qquad \text{at } 25^\circ \tag{6.39}$$

2. *The reaction taking place at the left electrode is written as an oxidation reaction, and the reaction taking place at the right electrode is written as a reduction reaction. The cell reaction is the sum of these two reactions.* For cell 6.38

$$\text{Left electrode:} \qquad Zn = Zn^{2+} + 2e \quad \text{(oxidation)} \tag{6.40}$$

$$\text{Right electrode:} \qquad Cu^{2+} + 2e = Cu \quad \text{(reduction)} \tag{6.41}$$

$$\text{Cell reaction:} \qquad Zn + Cu^{2+} = Zn^{2+} + Cu \tag{6.42}$$

These equations could equally well be multiplied through by $\frac{1}{2}$ so that the cell reaction would correspond to a one-electron change and would be written

$$\tfrac{1}{2}Zn + \tfrac{1}{2}Cu^{2+} = \tfrac{1}{2}Zn^{2+} + \tfrac{1}{2}Cu \tag{6.43}$$

For cell 6.39 the cell reaction would be written

$$Cu + Zn^{2+} = Zn + Cu^{2+} \tag{6.44}$$

or

$$\tfrac{1}{2}Cu + \tfrac{1}{2}Zn^{2+} = \tfrac{1}{2}Zn + \tfrac{1}{2}Cu^{2+} \tag{6.45}$$

Since $\Delta G^\circ = -zFE^\circ$, a cell reaction is spontaneous when E° is positive. Thus reaction 6.42 is spontaneous when Cu^{2+} and Zn^{2+} are at unit activity because $E^\circ = 1.100 \text{ V}$ and $\Delta G^\circ = -(2)(96,485 \text{ C mol}^{-1})(1.100 \text{ V}) = -212.3 \text{ kJ mol}^{-1}$. Reaction 6.44 is not spontaneous at unit activities of reactants and products because $E^\circ = -1.100 \text{ V}$ and $\Delta G^\circ = 212.3 \text{ kJ mol}^{-1}$. Multiples of a cell reaction may be taken without changing the electromotive force E°, but ΔG° depends on how the chemical equation is written as indicated by the factor z in equation 6.3. For example, the Gibbs free-energy change for reaction 6.43 is

$$\Delta G^\circ = -(1)(96,485)(1.100) = -106.15 \text{ kJ mol}^{-1}.$$

For a cell in which the reactants are not at unit activity, the criterion of spontaneous reaction is that $\Delta G = -zFE$ is negative.

Example 6.5 Calculate E° and the equilibrium constant K at 25° for the cell

$$Cd \mid Cd^{2+} \parallel Cu^{2+} \mid Cu$$

and determine the cell reaction and its equilibrium constant.

Reduction at right	$Cu^{2+} + 2e = Cu(s)$	$E^\circ_{Cu^{2+};Cu} = 0.337$
Oxidation at left	$Cd(s) = Cd^{2+} + 2e$	$E^\circ_{Cd^{2+};Cd} = -0.403$

$$Cd(s) + Cu^{2+} = Cd^{2+} + Cu(s) \qquad E^\circ = 0.740 \text{ V}$$

Since E° is positive, the reaction is spontaneous as written when reactants and products are at unit acitivity, and $Cd(s)$ will precipitate $Cu(s)$ from a solution in which $a_{Cu^{2+}} = 1$ to produce Cd^{2+} at unit activity. The Gibbs free-energy change for this reaction may be calculated as follows:

$$\Delta G^\circ = -zFE^\circ = -(2)(96{,}485)(0.740)/4.184 = -34{,}200 \text{ cal mol}^{-1}$$

The equilibrium constant for the cell reaction is readily calculated from the Gibbs free-energy change using equation 6.11.

$$\log K = \frac{zE^\circ}{2.303RT} = \frac{(2)(0.740)}{0.0591} = 25.0$$

$$K = \frac{a_{Cd^{2+}}}{a_{Cu^{2+}}} = 10^{25}$$

Example 6.6 Calculate E°, ΔG°, and K for the following reaction at 25°:

$$\tfrac{1}{2}Cu(s) + \tfrac{1}{2}Cl_2(g) = \tfrac{1}{2}Cu^{2+} + Cl^-$$

$$\text{Oxidation} \quad \tfrac{1}{2}Cu(s) = \tfrac{1}{2}Cu^{2+} + e$$

$$\text{Reduction} \quad \tfrac{1}{2}Cl_2(g) + e = Cl^-$$

Since the oxidation reaction is to occur at the left electrode and the reduction at the right electrode, the cell is written

$$Cu \mid Cu^{2+} \parallel Cl^- \mid Cl_2(g) \mid Pt$$

$$E^\circ = E^\circ_{Cl^-|Cl_2|Pt} - E^\circ_{Cu^{2+}|Cu} = 1.360 - 0.337 = 1.023 \text{ V}$$

$$\Delta G^\circ = -zFE^\circ = -\frac{(96{,}485)(1.023)}{(4.1840)} = -23{,}590 \text{ cal}$$

$$\log K = \frac{zE^\circ}{0.0591} = \frac{1.023}{0.0591} = 17.3$$

$$K = \frac{a_{Cl^-}a_{Cu^{2+}}{}^{1/2}}{P_{Cl_2}{}^{1/2}} = 2 \times 10^{17}$$

Equation 6.10 can be used to calculate the electromotive force of a cell when the ions are not at unit activity. It is not possible to determine the activities of single ions, but, for some purposes it is possible to replace activities by concentrations. In other cases the activity coefficients for ions of the same charge type tend to cancel in equation 6.10. This is a good approximation when these ions are in the same solution, or if they are in different solutions of the same ionic strength.

6.14   DETERMINATION OF pH

The concentrations of hydrogen ions in aqueous solutions range from about 1 molar in 1 M HCl to about 10^{-14} molar in 1 M NaOH. Because of this wide range of concentrations Sorenson adopted an exponential notation in 1909. He defined pH as the negative exponent of 10 that gives the hydrogen-ion concentration. Thus

$$(H^+) = 10^{-pH}$$

or

$$pH = -\log(H^+) \tag{6.46}$$

Electromotive-force methods are useful for studying hydrogen-ion concentrations; however, since the activity of a single ion cannot be determined, the modern definition of pH is expressed operationally in terms of the method of measuring it rather than by equation 6.46. The pH of a solution may be determined by use of a cell such as the following:

$$Pt \mid H_2(p) \mid H^+(a_{H^+}) \parallel Cl^- \mid Hg_2Cl_2 \mid Hg$$

The electromotive force of this cell may be considered to be made up of three contributions:

$$E = 0.2802 - 0.0591 \log (a_{H^+}/p_{H_2}^{1/2}) + E_{\text{liquid junction}} \tag{6.47}$$

where the contribution by the normal calomel electrode is 0.2802 V at 25°. Although the activities of single ions cannot be determined, equation 6.50 is often used with the assumption that $E_{\text{liquid junction}} = 0$. If $p_{H_2} = 1$ atm,

$$E - 0.2802 = -0.0591 \log a_{H^+} \tag{6.48}$$

Hydrogen-ion activities obtained in this way are of great practical use even though they are based upon an approximation.

Taking $pH = -\log a_{H^+}$, equation 6.48 becomes

$$E - 0.2802 = 0.0591 \text{ pH}$$

or

$$pH = \frac{E - 0.2802}{0.0591} \tag{6.49}$$

A glass electrode consists of a reversible electrode, such as a calomel or Ag-AgCl electrode, in a solution of constant pH inside a thin membrane of a special glass. The thin glass bulb of this electrode is immersed in the solution to be studied along with a reference calomel electrode to form the cell indicated by

$$Ag \mid AgCl \mid Cl^-, H^+ \mid \text{glass membrane} \mid \text{solution} \parallel \text{calomel electrode}$$

It is found experimentally that the potential of such a glass electrode varies with the activity of hydrogen ions in the same way as the hydrogen electrode, that is, 0.0591 V per pH unit at 25°. An ordinary potentiometer cannot be used to measure the voltage of such a cell because of the high resistance of the glass membrane, and so an electronic voltmeter must be employed. Electronic devices have been developed that make it possible to measure pH values to ± 0.01 pH unit with easily portable apparatus. The pH meter, as it is often called, is calibrated by means of a buffer of known pH before it is used to measure the pH of an unknown solution. The theory and use of the glass electrode have been fully treated by Bates.*

The glass electrode has become the most useful electrode for determining the pH of a solution. It is not affected by oxidizing or reducing agents and is not easily poisoned. It is especially useful in biochemical investigations.

The pCO_2 electrode is a useful clinical tool. This electrode consists of a glass electrode in contact with a solution of fixed bicarbonate concentration that is separated from the sample solution by a polymer membrane permeable to CO_2. When CO_2 diffuses into the bicarbonate solution it is hydrated to H_2CO_3 (Section 7.8) and then rapidly ionizes to $HCO_3^- + H^+$ with a consequent change in pH.

6.15 BATTERIES

It is useful to distinguish between primary and secondary batteries and fuel cells. The objective of a battery is to store electrical energy until it is needed. The objective of a fuel cell (sometimes called an electrochemical energy converter) is to convert the Gibbs free energy of a chemical reaction directly into electrical energy. An example of a primary battery is the familiar dry cell containing zinc, ammonium chloride, and manganese dioxide. A primary battery is essentially a fuel cell in that the reactants are used up and are not regenerated by charging. An example of a secondary battery is the lead storage battery that can be recharged many times by applying a higher voltage in the opposite direction. In a fuel cell the reactants (usually gases) are fed into their respective chambers continuously when electrical energy is desired.

The Leclanché dry cell with a voltage of about 1.6 V consists of a zinc can containing a carbon electrode surrounded by manganese dioxide and graphite immersed in a starch paste containing zinc chloride and an excess of solid ammonium chloride. The electrode reactions may be written

$$Zn + 2NH_3 = Zn(NH_3)_2^{2+} + 2e$$

$$2MnO_2 + 2NH_4^+ + 2e = Mn_2O_3 + H_2O + 2NH_3$$

where $Zn(NH_3)_2^{2+}$ represents the average of a mixture of complex ions having different numbers of ammonia molecules.

The lead storage battery consists of an electrode of lead and an electrode of lead dioxide immersed in sulfuric acid. Each plate has a rough surface exposing a

---

* R. G. Bates, *Determination of pH: Theory and Practice*, Wiley, New York, 1973.

large area, and the two are held close together in rigid frames. The cell reaction is

$$Pb + HSO_4^- = PbSO_4(s) + H^+ + 2e$$
$$PbO_2 + 3H^+ + HSO_4^- + 2e = PbSO_4(s) + 2H_2O$$

$$Pb + PbO_2 + 2H^+ + 2HSO_4^- \underset{\text{Charge}}{\overset{\text{Discharge}}{\rightleftharpoons}} 2PbSO_4(s) + 2H_2O$$

The important thing about this cell reaction is the fact that it can be reversed. In the charging process additional spongy lead is redeposited on one electrode, lead dioxide is deposited on the other and the sulfuric acid is regenerated.

6.16 FUEL CELLS

Fuel cells may be classified according to the temperature range in which they operate: low temperature (25–100° C), medium temperature (100–500° C), high temperature (500–1000° C), and very high temperature (above 1000° C). The advantage of using high temperatures is that catalysts for the various steps in the process are not so necessary. Polarization of a fuel cell reduces the current. Polarization is the result of slow reactions or processes such as diffusion in the cell.

Figure 6.5 indicates the construction of a hydrogen-oxygen fuel cell with a solid electrolyte, which is an ion exchange membrane. The membrane is impermeable to the reactant gases, but is permeable to hydrogen ions, which carry the current between the electrodes. In order to keep the resistance of the cell as low as possible, the membrane is kept as thin as possible. In order to facilitate the operation of the cell at 40–60° C the electrodes are covered with finely divided platinum that functions as a catalyst. Water is drained out of the cell during operation. Fuel cells of this general type have been used successfully in the space program and are quite efficient. Their disadvantages for large-scale commercial application are that hydrogen presents storage problems, and platinum is an expensive catalyst. Cheaper catalysts have been found for higher temperature operation of hydrogen-oxygen fuel cells.

Fuel cells that use hydrocarbons and air have been developed, but their power per unit weight is too low to make them practical in ordinary automobiles. Better catalysts are needed. Fuel cells offer the possibility of achieving high thermodynamic efficiency in the conversion of Gibbs free energy into mechanical work. Internal combustion engines at best convert only the fraction $(T_2 - T_1)/T_2$ of the heat of combustion into mechanical work. In this relation, which comes from the second law of thermodynamics, T_2 is the temperature of the gas during expansion and T_1 is the temperature of the exhaust.

A hydrogen-oxygen fuel cell may have an acidic or alkaline electrolyte. The half-cell reactions are

$$H_2(g) = 2H^+ + 2e$$
$$\tfrac{1}{2}O_2(g) + 2H^+ + 2e = H_2O(l)$$

$$H_2(g) + \tfrac{1}{2}O_2(g) = H_2O(l)$$

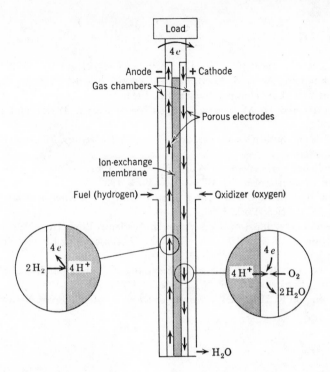

Fig. 6.5 Hydrogen-oxygen fuel cell with an ion exchange membrane (from J. O'M. Bockris and S. Srinivasan, *Fuel Cells, Their Electrochemistry*, McGraw-Hill Book Co., New York, 1969).

or

$$H_2(g) + 2OH^- = 2H_2O(l) + 2e$$

$$\frac{1}{2}O_2(g) + H_2O(l) + 2e = 2OH^-$$

$$\overline{H_2(g) + \frac{1}{2}O_2(g) = H_2O(l)}$$

Example 6.7   Calculate the standard electromotive force of the hydrogen-oxygen fuel cell from the standard Gibbs free energy of formation of $H_2O(l)$.

$$\Delta G° = -zFE°$$

$$-56{,}690 \text{ cal mol}^{-1} = -2(23{,}060 \text{ cal V}^{-1} \text{ mol}^{-1})E°$$

$$E° = 1.229 \text{ V}$$

This chapter was concerned with electrochemical reactions at equilibrium. Practical applications involve the flow of significant electrical current. To treat these phenomena the ideas of chemical kinetics have to be introduced.

References

R. G. Bates, *Determination of pH: Theory and Practice*, Wiley, New York, 1973.

A. J. de Bethune and N. A. S. Loud, *Standard Aqueous Electrode Potentials and Temperature Coefficients at 25°*, C. A. Hampel, Skokie, Ill., 1964.

J. O'M. Bockris and D. M. Drazic, *Electro-Chemical Science*, Taylor and Francis Ltd., London, 1972.

J. O'M. Bockris and S. Srinivasan, *Fuel Cells: Their Electrochemistry*, McGraw-Hill Book Co., New York, 1969.

E. A. Guggenheim, *Thermodynamics*, Wiley, New York, 1967.

W. J. Hamer, *Electrolytic Solutions*, Wiley, New York, 1959.

H. S. Harned and B. B. Owen, *The Physical Chemistry of Electrolytic Solutions*, Reinhold Publishing Corp., New York, 1958.

D. J. G. Ives and G. J. Janz, *Reference Electrodes*, Academic Press, Inc.. New York, 1961.

G. F. A. Kortüm and J. O'M. Bockris, *Textbook of Electrochemistry*, Elsevier Publishing Co., New York, 1951.

W. M. Latimer, *The Oxidation States of the Elements and Their Potentials in Aqueous Solutions*, Prentice-Hall, Englewood Cliffs, N.J., 1952.

R. A. Robinson and R. H. Stokes, *Electrolyte Solutions*, Academic Press, New York, 1959.

Problems

6.1 How much work is required to bring two protons from an infinite distance of separation to 1 Å (10^{-8} cm)? Calculate the answer in joules using the protonic charge 1.602×10^{-19} C. What is the work in kcal mol^{-1} for a mole of proton pairs?

Ans. 23.07×10^{-19} J. 331.9 kcal mol^{-1}.

6.2 A small dry battery of zinc and ammonium chloride weighing 85 g will operate continuously through a 4-Ω resistance for 450 min before its voltage falls below 0.75 V. The initial voltage is 1.60, and the effective voltage over the whole life of the battery is taken to be 1.00. Theoretically, how many miles above the earth could this battery be raised by the energy delivered under these conditions? *Ans.* 5.04 miles.

6.3 The voltage of the cell

$$\text{Pb} \mid \text{PbSO}_4 \mid \text{Na}_2\text{SO}_4 \cdot 10\text{H}_2\text{O(sat)} \mid \text{Hg}_2\text{SO}_4 \mid \text{Hg}$$

is 0.9647 at 25°. The temperature coefficient is 1.74×10^{-4} V K^{-1}. Calculate (*a*) the enthalpy change for the reaction and (*b*) the change in Gibbs free energy for the reaction

$$\text{Pb}(s) + \text{Hg}_2\text{SO}_4(s) = \text{PbSO}_4(s) + 2\text{Hg}(l)$$

Ans. (*a*) $-42,100$, (*b*) $-44,500$ cal mol^{-1}.

6.4 Derive an expression for the activity of a 1–2 electrolyte (like Na$_2$SO$_4$) in terms of the mean ionic activity coefficient and the molality. *Ans.* $a_{\text{Na}_2\text{SO}_4} = 4m^3\gamma_\pm^3$.

6.5 What is the ionic strength of each of the following solutions: (*a*) 0.1 M NaCl, (*b*) 0.1 M Na$_2$C$_2$O$_4$, (*c*) 0.1 M CuSO$_4$, (*d*) a solution containing 0.1 M Na$_2$HPO$_4$ and 0.1 M NaH$_2$PO$_4$? *Ans.* (*a*) 0.1 (*b*) 0.3, (*c*) 0.4, (*d*) 0.4.

6.6 For 0.002 M CaCl$_2$ at 25° use the Debye-Hückel limiting law to calculate the activity coefficients of Ca^{++} and Cl$^-$. What is the mean ionic activity coefficient for the electrolyte?

Ans. $\gamma_{\text{Ca}^{++}} = 0.695$. $\gamma_{\text{Cl}^-} = 0.913$. $\gamma_\pm = 0.834$.

6.7 Given the mean ionic activity coefficients of HCl(aq) at 0.01 and 0.05 m as 0.906 and 0.833 at 25°; calculate the emf of the cell

$$\text{Pt} \mid \text{H}_2(1 \text{ atm}) \mid \text{HCl}(m) \mid \text{Cl}_2(1 \text{ atm}) \mid \text{Pt}$$

where m is (a) 0.01 and (b) 0.05.   *Ans.* (a) 1.602, (b) 1.5225 V.

6.8 Given the cell at 25°,

$$\text{Pb} \mid \text{Pb}^{2+}(a = 1) \parallel \text{Ag}^+(a = 1) \mid \text{Ag}$$

(a) calculate the voltage; (b) write the cell reaction; and (c) calculate the Gibbs free-energy change. (d) Which electrode is positive?

Ans. (a) 0.925 V, (b) $\frac{1}{2}\text{Pb}(s) + \text{Ag}^+(a = 1) = \frac{1}{2}\text{Pb}^{2+}(a = 1) + \text{Ag}(s)$,
(c) $-21,330$ cal mol^{-1}, (d) Silver electrode.

6.9 Given the reaction

$$\text{Fe}^{2+}(a = 1) + \text{Ce}^{4+}(a = 1) = \text{Fe}^{3+}(a = 1) + \text{Ce}^{3+}(a = 1)$$

(a) Write the cell that corresponds to this reaction. (b) Calculate $E°$. (c) Calculate $\Delta G°$. (d) Which electrode is positive?

Ans. (a) Pt $\mid$ Fe$^{2+}(a = 1)$, Fe$^{3+}(a = 1) \parallel$ Ce$^{3+}(a = 1)$, Ce$^{4+}(a = 1) \mid$ Pt,
(b) 0.839 V, (c) $-19,350$ cal mol^{-1}, (d) Right electrode.

6.10 (a) Write the cell reaction for the cell

$$\text{Zn} \mid \text{ZnCl}_2(a = 0.5) \mid \text{AgCl} \mid \text{Ag}$$

(b) Calculate $E°$. (c) Calculate E for the cell as written in (a). (d) Calculate ΔG in J mol^{-1}. (e) Calculate $\Delta G°$ in J mol^{-1}.

Ans. (a) Zn + 2AgCl = 2Ag + ZnCl$_2(a = 0.5)$, (b) 0.985 V,
(c) 0.994 V, (d) -191.9 kJ mol^{-1}, (e) -190.2 kJ mol^{-1}.

6.11 (a) Calculate the voltage of the following cell at 25°:

$$\text{Zn} \mid \text{Zn}^{2+}(a = 0.0004) \parallel \text{Cd}^{2+}(a = 0.2) \mid \text{Cd}$$

(b) Write the cell reaction. (c) Calculate the value of the Gibbs free-energy change involved in the reaction.

Ans. (a) 0.4398 V, (b) Zn(s) + Cd$^{2+}(a = 0.2)$ = Zn$^{2+}(a = 0.0004)$ + Cd(s),
(c) $-20,287$ cal mol^{-1}.

6.12 (a) Diagram the cell for the reaction

$$\text{H}_2(g, 1 \text{ atm}) + \text{I}_2(s) = 2\text{HI}(aq, a = 1)$$

(b) Calculate $E°$. (c) Calculate $\Delta G°$. (d) Calculate K. (e) What differences would there be if the reaction had been written

$$\tfrac{1}{2}\text{H}_2(g, 1 \text{ atm}) + \tfrac{1}{2}\text{I}_2(s) = \text{HI}(aq, a = 1)$$

Ans. (a) Pt $\mid$ H$_2(g, 1 \text{ atm}) \mid$ HI($aq, a = 1$) $\mid$ I$_2(s) \mid$ Pt, (b) 0.5355 V,
(c) $-24,697$ cal mol^{-1}, (d) 1.27×10^{18},
(e) a and b are the same; $\Delta G° = -12,325$ cal mol^{-1},
$K = 1.13 \times 10^9$.

6.13 (a) Calculate the equilibrium constant at 25° for the reaction

$$\text{Fe}^{2+} + \text{Ag}^+ = \text{Ag} + \text{Fe}^{3+}$$

(b) Calculate the concentration of silver ion at equilibrium (assuming that concentrations may be substituted for activities) for an experiment in which an excess of finely divided metallic silver is added to a 0.05 M solution of ferric nitrate. *Ans.* (a) 3.0, (b) 0.0443 M.

6.14 Using standard electrode potentials at 25°, calculate the equilibrium constant for the reaction

$$Sn + Pb^{2+} = Pb + Sn^{2+}$$

Ans. 2.97.

6.15 Given the cell at 25°

$$Pt \mid X_2 \mid X^-(a = 0.1) \parallel X^-(a = 0.001) \mid X_2 \mid Pt$$

where X is an unknown halogen. (a) Write the cell reaction. (b) Which electrode is negative? (c) What is the voltage of the cell? (d) Is the reaction spontaneous?

Ans. (a) $X^-(a = 0.1) = X^-(a = 0.001)$, (b) Left, (c) 0.11832 V, (d) Yes.

6.16 Devise an electromotive force cell for which the cell reaction is

$$AgBr(s) = Ag^+ + Br^-$$

Calculate the equilibrium constant (usually called the solubility product) for this reaction at 25°. *Ans.* $10^{-11.85}$.

6.17 When hydrogen electrode and normal calomel electrode are immersed in a solution at 25° a potential of 0.664 V is obtained. Calculate (a) the pH and (b) the hydrogen-ion activity. *Ans.* (a) 6.50, (b) 3.16×10^{-7}.

6.18 Ammonia may be used as the anodic reactant in a fuel cell. The reactions occurring at the electrodes are

$$NH_3 + 3OH^- = \tfrac{1}{2}N_2 + 3H_2O + 3e$$

$$O_2 + 2H_2O + 4e = 4OH^-$$

What is the electromotive force of this fuel cell at 25°? *Ans.* 1.17 V.

6.19 How much work in kcal mol^{-1} can in principle be obtained when an electron is brought to 5 Å from a proton?

6.20 Calculate the emf at 25° of the cell

$$Ag \mid AgCl(s) \mid NaCl(a = 1) \mid Hg_2Cl_2(s) \mid Hg$$

from the following Gibbs free energies of formation:

Ag^+	$+18,488$ cal mol^{-1}
$AgCl(s)$	$-26,187$
Na^+	$-62,588$
Cl^-	$-31,367$
Hg_2Cl_2	$-50,274$

6.21 The voltage of the standard Weston cadmium cell is 1.0186 V at 20° and decreases 4.06×10^{-5} V for each degree rise in temperature above 20°. (a) State these facts with a single mathematical equation, and calculate (b) ΔH for the cell reaction per mole of cadmium consumed in J, and (c) the change in Gibbs free energy per mole of cadmium consumed in J.

6.22 Derive an expression for the activity of a 3–1 electrolyte (like $FeCl_3$) in terms of the mean ionic activity coefficient and the molality.

6.23 Calculate the electromotive force of the following cell at 25°

$$Li \mid LiCl(0.1 \text{ m}) \mid Cl_2(1 \text{ atm}) \mid Pt$$

assuming that the mean ionic activity coefficient is 0.77.

6.24   The solubility of Ag_2CrO_4 in water is 8.00×10^{-5} M at $25°$, and its solubility in 0.04 molar $NaNO_3$ is 8.84×10^{-5} M. What is the mean ionic activity coefficient of Ag_2CrO_4 in 0.04 M $NaNO_3$?

6.25   The cell $Pt \mid H_2(1 \text{ atm}) \mid HBr(m) \mid AgBr \mid Ag$ has been studied by H. S. Harned, A. S. Keston, and J. G. Donelson [*J. Am. Chem. Soc.*, **58**, 989 (1936)]. The following table gives the emf's obtained at $25°$:

m	0.01	0.02	0.05	0.10
E	0.3127	0.2786	0.2340	0.2005

Calculate (a) $E°$ and (b) the activity coefficient for a 0.10 molal solution of hydrogen bromide.

6.26   The most electropositive metal is lithium, and the most electronegative element is fluorine. Calculate the reversible electromotive force of the Li-F_2 cell.

6.27   Given the cell at $25°$

$$Cd \mid Cd^{2+}(a = 1) \parallel I^-(a = 1) \mid I_2(s) \mid Pt$$

(a) Write the cell reaction. (b) Calculate E. (c) Calculate $\Delta G°$. (d) Which electrode is positive?

6.28   Given the cell at $25°$,

$$Pt \mid Cl_2(g) \mid Cl^-(a = 1) \parallel Tl^{3+}(a = 1), \ Tl^+(a = 1) \mid Pt$$

(a) Write the cell reaction. (b) Calculate E. (c) Calculate $\Delta G°$. (d) Which electrode is positive?

6.29   (a) Design a cell in which the reaction is

$$\tfrac{1}{2}Cl_2(g, 1 \text{ atm}) + Br^-(a = 1) = Cl^-(a = 1) + \tfrac{1}{2}Br_2(l)$$

Calculate (b) the voltage at $25°$ and (c) the Gibbs free-energy change of the reaction at $25°$.

6.30   (a) Write the representation for the cell for which the reaction is

$$Cu + 2Ag^+(a = 1) = Cu^{2+}(a = 1) + 2Ag$$

(b) Calculate the emf of this cell. (c) What is $\Delta G°$ for the cell reaction? (d) Which electrode is positive?

6.31   The value of $E°$ for the electrode Pt, $O_2(g) \mid OH^-$ cannot be measured directly because the electrode is not reversible. Calculate $E°$ and $\Delta G°$ at $25°$ for the electrode reaction

$$\tfrac{1}{4}O_2(g) + \tfrac{1}{2}H_2O + e = OH^-(a = 1)$$

from the known Gibbs free-energy changes for the reactions

$$H_2(g) + \tfrac{1}{2}O_2(g) = H_2O(l)$$
$$H_2O(l) = H^+(a = 1) + OH^-(a = 1)$$
$$\tfrac{1}{2}H_2(g) = H^+(a = 1) + e$$

6.32   (a) Calculate the voltage at $25°$ of the cell

$$Pt \mid Ti^{3+}(a = 0.3), \ Ti^{4+}(a = 0.5) \parallel Ce^{3+}(a = 0.7), \ Ce^{4+}(a = 0.002) \mid Pt$$

(b) Write the cell reaction. (c) Calculate ΔG for the cell reaction as written. (d) Calculate $E°$. (e) Calculate $\Delta G°$. (f) Calculate K.

6.33 Using Table 6.1 of electrode potentials, calculate the equilibrium constants at 25° for the following reactions in aqueous solution and give the corresponding equilibrium expressions:

$$(a) \ \ \text{Fe} + 2\text{CrCl}_3 = \text{FeCl}_2 + 2\text{CrCl}_2$$
$$(b) \ \ \text{Fe(NO}_3)_2 + \tfrac{1}{2}\text{Hg}_2(\text{NO}_3)_2 = \text{Fe(NO}_3)_3 + \text{Hg}$$

6.34 Calculate the equilibrium constant at 25° for the reaction

$$2\text{H}^+ + \text{D}_2(g) = \text{H}_2(g) + 2\text{D}^+$$

from the electrode potential for $\text{D}^+ \mid \text{D}_2$, Pt, which is -3.4 mV at 25°.

6.35 A thallium amalgam of 4.93% Tl in mercury and another amalgam of 10.02% Tl are placed in separate legs of a glass cell and covered with a solution of thallous sulfate to form a concentration cell. The voltage of the cell is 0.029480 V at 20° and 0.029971 V at 30°. (a) Which is the negative electrode? (b) What is the heat of dilution per mole of Tl when Hg is added at 30° to change the concentration from 10.02% to 4.93%? (c) What is the voltage of the cell at 40°?

6.36 A hydrogen electrode and a normal calomel electrode give a voltage of 0.435 when placed in a certain solution at 25°. (a) What is the pH of the solution? (b) What is the value of a_{H^+}?

6.37 The oxidation of methanol may be utilized to produce an electric current in a fuel cell. Calculate the reversible electromotive force at 25°.

6.38 Calculate $E^\circ_{\text{Zn}^{2+}|\text{Zn}}$ at 25° from the following data:

$\text{Zn}^{2+}(a = 1)$	$\Delta \bar{H}^\circ_f = -36.43$ kcal mol^{-1}	$\bar{S}^\circ = -25.45$ cal K^{-1} mol^{-1}
$\text{Zn}(s)$		$\bar{S}^\circ = 9.95$ cal K^{-1} mol^{-1}
$\text{H}^+(a = 1)$	$\Delta \bar{H}^\circ_f = 0.000$	$\bar{S}^\circ = 0.000$
$\text{H}_2(p = 1)$		$\bar{S}^\circ = 31.211$ cal K^{-1} mol^{-1}

6.39 (a) Write the reaction that occurs when the cell

$$\text{Zn} \mid \text{ZnCl}_2(0.555 \text{ m}) \mid \text{AgCl} \mid \text{Ag}$$

delivers current and calculate (b) ΔG, (c) ΔS, and (d) ΔH at 25° for this reaction. At 25° $E = 1.015$ V and $(\partial E / \partial T)_P = -4.02 \times 10^{-4}$ V K^{-1}.

6.40 Calculate the energy in kcal mol^{-1} required to separate a positive and negative charge from 3 Å to infinity in (a) a vacuum, (b) a solvent of dielectric constant 10, and (c) water at 25°, which has a dielectric constant of approximately 80.

6.41 Give the expressions for the mean ionic activity coefficients of LiCl, AlCl$_3$, and MgSO$_4$ in terms of the single ion activities and the molality.

6.42 A solution of NaCl has an ionic strength of 0.24. (a) What is its concentration? (b) What concentration of Na$_2$SO$_4$ would have the same ionic strength? (c) What concentration of MgSO$_4$?

6.43 The voltage of the cell

$$\text{Ag} \mid \text{AgBr} \mid \text{MBr(aq soln)} \mid \text{Hg}_2\text{Br}_2 \mid \text{Hg}$$

has been determined accurately by T. W. Dakin and D. T. Ewing [*J. Am. Chem. Soc.*, **62**, 2280 (1940)], and the values of $E^\circ_{\text{Br}^-|\text{AgBr}|\text{Ag}}$ have been determined by B. B. Owen and

L. Foering [*J. Am. Chem. Soc.*, **58**, 1575 (1936)]. The data are as follows:

Temp., °C	E_{cell}	$E^\circ_{Br^-|AgBr|Ag}$
15	0.06492	0.07586
25	0.06804	0.07121
35	0.07116	0.06591

(a) Calculate $E^\circ_{Br^-|Hg_2Br_2|Hg}$ at each temperature, and find equations that give E_{cell} and $E^\circ_{Br^-|Hg_2Br_2|Hg}$ as functions of temperature. (b) Calculate ΔG° at 25° for the reaction $2Ag + Hg_2Br_2 = 2AgBr + 2Hg$. (c) Calculate ΔH for this reaction. (d) Calculate ΔS for this reaction. (e) Using the Gibbs free energy of formation of silver bromide, $\Delta \bar{G}^\circ_f = -22{,}935$ cal mol^{-1}, show that the free energy of formation of mercurous bromide is $-42{,}732$ cal mol^{-1}.

6.44   Design cells without liquid junction that could be used to determine the activity coefficients of aqueous solutions of (a) NaOH and (b) H_2SO_4. Give the equations relating voltage to mean ionic activity coefficient.

6.45   Derive the expression for the electromotive force of the cell

$$Pt \mid H_2(g, 1 \text{ atm}) \mid KH_2PO_4(m_1), Na_2HPO_4(m_2) \parallel NaX(m_3) \mid AgX(s) \mid Ag$$

Substitute the equilibrium expression for the second dissociation of phosphoric acid and describe how the thermodynamic dissociation constant K for that dissociation could be obtained from electromotive force measurements at constant temperature.

6.46   Calculate the voltage at 25° for the fuel cell based on the reaction

$$H_2(1 \text{ atm}) + \tfrac{1}{2}O_2(1 \text{ atm}) = H_2O(l)$$

using (a) the Gibbs free energy of formation of water, and (b) the oxygen electrode potential calculated in problem 6.31.

6.47   Calculate the voltage for the half-cell reactions:

(a) $H_2(1 \text{ atm}) = 2H^+(a = 10^{-7}) + 2e$

(b) $H_2(1 \text{ atm}) + 2OH^-(a = 1) = 2H_2O(l) + 2e$

6.48   A small, efficient battery for hearing aids contains zinc and mercury electrodes. The mercury is mixed with mercuric oxide, and the electrolyte is potassium hydroxide. (a) Write a net reaction for the cell in which hydroxyl ions and zinc are consumed, mercury is deposited, and potassium zincate (K_2ZnO_2) is formed. (b) Write the half-cell reactions.

6.49   Given the cell at 25°

$$Zn \mid Zn^{2+}(a = 0.001) \parallel I^-(a = 0.1) \mid I_2(s) \mid Pt$$

(a) Write the cell reaction. (b) Calculate the voltage of the cell. (c) Calculate the equilibrium constant for the cell reaction.

6.50   (a) Calculate the equilibrium constant at 25° for the reaction

$$Sn^{4+} + 2Ti^{3+} = 2Ti^{4+} + Sn^{2+}$$

(b) When 0.01 mol of Sn^{2+} ion is added to 1.0 mol of Ti^{4+} ion in 1000 g of water, what will be the concentration of Ti^{3+} ions (if is is assumed for the calculation that the activities are equal to the concentrations)?

6.51   (a) Diagram the cell that corresponds to the reaction $Ag^+ + Cl^- = AgCl$. Calculate at 25° (b) E°, (c) $\Delta \bar{G}^\circ$, (d) K.

6.52   For the reaction

$$2Fe^{3+} + 2Hg = 2Fe^{2+} + Hg_2^{2+}$$

the equilibrium constant is 0.018 at 25° and 0.054 at 35°. Calculate the value of $E°$ at 45° for the cell that corresponds to this reaction.

6.53   Develop an equation for calculating the voltage of a cell consisting of two chlorine electrodes at different partial pressures immersed in a solution of a chloride. Calculate the voltage at 25° if the chlorine of one electrode is at 1 atm, and the other is at 0.70 atm with nitrogen added to bring the total pressure to 1 atm.

6.54   Show that for the concentration cell

$$X \mid X^-(a_1) \parallel X^-(a_2) \mid X$$

where X^- is a negative ion, the equation for the electromotive force of the cell is

$$E = -\frac{RT}{F} \ln \frac{a_2}{a_1}$$

6.55   At 25° the potential of the cell

$$Ag \mid AgI \mid KI(1\ M) \parallel AgNO_3(0.001\ M) \mid Ag$$

is 0.72 V. The mean ionic activity coefficient of 1 M KI may be taken as 0.65, and of 0.001 M $AgNO_3$ as 0.98. (*a*) What is the solubility of AgI? (*b*) What is the solubility of AgI in pure water?

6.56   When the pH of a solution is measured by means of a hydrogen electrode against a calomel electrode, the partial pressure of the hydrogen is usually not exactly 1 atm. At 25° how low can the partial pressure of the hydrogen be before a correction is necessary, if an error of 0.01 pH unit is allowable?

6.57   A hydrogen electrode and calomel cell are used to determine the pH of a solution on a mountain where the barometric pressure is 500 Torr. The hydrogen is allowed to bubble out of the electrode at the atmospheric pressure prevailing there. If the pH is calculated to be 4.00, what is the correct pH of the solution?

6.58   Calculate the standard electromotive force of an ethane-oxygen fuel cell at 25°.

CHAPTER 7

IONIC EQUILIBRIA AND BIOCHEMICAL REACTIONS

Electrochemical cells offer one of the best methods for studying equilibria of acids, bases, and complex ions. Such reactions are often studied in aqueous solution, and water is a remarkable solvent that plays a very important role in the dissociation of acids and bases. Thermodynamics helps us understand why these reactions occur to the extent that they do.

Biochemical reactions involve weak acids and complex ions. The thermodynamics of biochemical reactions are handled in a rather different way from the thermodynamics of reactions in the gas phase or in solutions of nonelectrolytes.

The binding of oxygen by hemoglobin illustrates the very special properties a protein may have in a binding reaction. In contrast with most experience with small molecules, the affinity of a protein for a ligand may increase as more ligands are bound.

7.1 ACID AND BASE EQUILIBRIA

According to the proton theory of acids, which was developed by Brönsted and Bjerrum in Denmark and Lowry in England in 1923, an acid is a substance that yields protons and a base is a substance that accepts protons. Thus the dissociation of an acid HA always produces a proton and a base A^-.

$$HA = H^+ + A^- \tag{7.1}$$

Here HA and A^- are referred to as a conjugate acid-base pair. The proton is just the nucleus of a hydrogen atom, and this small particle has a very great tendency to attach itself to molecules of the solvent. Thus the proton does not exist free in solution, but is in combination with the solvent. In water the proton may form a hydronium ion, H_3O^+, or other complex species. Since the state of the proton in aqueous solution is not exactly known, the symbol H^+ will be used to represent the hydrated hydrogen ion.

The extent to which an acid dissociates in different solvents depends on the affinity of the solvent for protons as well as the "intrinsic" acid strength of the acid. Thus an acid may dissociate very completely in liquid ammonia and very incompletely in acetic acid, since ammonia has a much greater attraction for protons than has acetic acid.

The solvation of protons leads to a leveling effect on the strength of all strong acids in a given solvent. All acids that are completely dissociated in water therefore have nearly the same acid strength in water because the actual acid is the hydrated hydrogen ion. On the other hand, various acids that are strong in water have quite different strengths in acetic acid, which is not as basic a solvent and does not combine so strongly with protons.

The dissociation constant of a weak acid HA may be defined exactly or by use of approximations. For exact studies it is defined by

$$K = \frac{a_{H^+}a_{A^-}}{a_{HA}} = \frac{(H^+)(A^-)y_{\pm}^2}{(HA)y_{HA}} \tag{7.2}$$

where a is activity and $y_{\pm}$ is the mean ionic activity coefficient for the dissociated acid $H^+ + A^-$ on the moles per liter concentration scale. For many practical purposes, such as preparing buffers and interpreting titration curves and in calculating biochemical equilibria, it is more convenient to use the apparent acid dissociation constant K', which is expressed in terms of concentrations. By rearranging equation 7.2 we obtain

$$K' = K\frac{y_{HA}}{y_{\pm}^2} = \frac{(H^+)(A^-)}{(HA)} \tag{7.3}$$

The apparent acid dissociation constant K' depends on the salt concentration. As the electrolyte and acid concentrations approach zero, K' approaches K. In this chapter we will use a K without the prime to represent both the limiting value of the dissociation constant and the value at a particular ionic strength because both are equilibrium constants from the standpoint of thermodynamics. In considering the dissociation constants of weak acids it is often convenient to use the pK, which is the negative base 10 logarithm of the dissociation constant: $pK = -\log K$.

For low ionic strength values the dependence of the apparent acid dissociation constant on ionic strength may be calculated using the Debye-Hückel theory as shown in Section 7.10.

7.2 IONIC EQUILIBRIA OF MONOPROTIC ACIDS

We will consider the calculation of (H^+) of any mixture of a monoprotic acid HA, like acetic acid, and its completely dissociated salt. The following equations allow us to calculate the titration curve up to the equivalence point.

The first step is to write down all of the equations that have to be satisfied simultaneously. These are of the following types:

1. The equilibrium expressions.
2. The statement of electrical neutrality of the solution.
3. The conservation of substances added to the solution.

If to make a liter of solution we use c_a moles of monoprotic weak acid HA and c_s moles of the salt NaA, the equations that have to be satisfied simultaneously are as follows:

1. The equilibrium expressions are

$$K = (H^+)(A^-)/(HA)$$
$$K_w = (H^+)(OH^-)$$

2. Assuming that the salt is completely dissociated, electrical neutrality of the solution requires that

$$(H^+) + c_s = (A^-) + (OH^-)$$

3. Conservation of added weak acid and salt requires that

$$c_a + c_s = (HA) + (A^-)$$

The elimination of (HA), (A^-), and (OH^-) between these equations yields

$$(H^+) + c_s = \frac{c_a + c_s}{1 + (H^+)/K} + \frac{K_w}{(H^+)} \tag{7.4}$$

This is a cubic equation in (H^+) that may be solved for (H^+) for any mixture of the weak acid HA and the salt NaA.

If $c_s \gg (H^+)$ and $K_w/(H^+)$, equation 7.4 becomes

$$K = \frac{(H^+)c_s}{c_a} \tag{7.5}$$

This is simply the equilibrium constant expression, because the addition of the salt represses the ionization of the acid, so that to a very good approximation the concentration of undissociated acid is equal to the molar concentration of added weak acid (c_a); and the addition of the acid represses the hydrolysis of the salt, so that to a very good approximation the concentration of anions of the weak acid is equal to the molar concentration of the salt c_s. When we introduce the definitions of pH and pK, this equation becomes

$$pH = pK + \log \frac{c_s}{c_a} \tag{7.6}$$

To the approximation used in deriving equations 7.5 and 7.6, the pK for a weak acid is the pH of a solution containing equimolar quantities of salt and acid.

Example 7.1 Calculate the pH at 25° of a solution containing 0.10 M sodium acetate and 0.03 M acetic acid. The apparent pK for acetic acid at this ionic strength is 4.57.

$$pH = 4.57 + \log (0.10/0.030) = 5.09$$

Equation 7.4 may also be used to calculate the pH of a solution of the pure salt; this is the pH of the equivalence point in a titration. A salt formed from a strong base and a strong acid gives a pH of 7 when dissolved in water, but a salt of a weak acid and a strong base gives a pH greater than 7 because the anions of the weak acid react with water molecules to produce molecules of weak acid and hydroxyl ions. A salt of a weak base and a strong acid gives a pH less than 7 because the cation of the weak base reacts with water to produce a molecule of weak base and a proton. This reaction is referred to as hydrolysis.

If c_a is set equal to zero in equation 7.4, it may be solved for c_s to obtain

$$c_s = \frac{KK_w}{(H^+)^2} - K + \frac{K_w}{(H^+)} - (H^+) \tag{7.7}$$

If the concentration c_s of the salt is much greater than K, (OH^-), and (H^+) the last three terms may be neglected, and the hydrogen ion concentration may be calculated from

$$(H^+) = \left(\frac{KK_w}{c_s}\right)^{1/2} \tag{7.8}$$

Example 7.2 Calculate the hydrogen-ion concentration and pH of 0.1 M sodium acetate at 25°. The value of K_a at this ionic strength is 2.69×10^{-5}.

$$(H^+) = \left(\frac{K_w K}{c_s}\right)^{1/2} = \left[\frac{(10^{-14})(2.69 \times 10^{-5})}{0.1}\right]^{1/2}$$

$$= 1.64 \times 10^{-9}$$

$$pH = -\log (1.64 \times 10^{-9}) = 8.79$$

This is the pH at the equivalence point in a titration of 0.1 M acetic acid with a concentrated solution of a strong base; this is the point at which the number of moles of base added is equal to the number of moles of acetic acid.

A final special case of equation 7.4 is obtained when $c_s = 0$, that is, when only weak acid has been added to the solution.

$$c_a = \left[1 + \frac{(H^+)}{K}\right]\left[(H^+) - \frac{K_w}{(H^+)}\right] \tag{7.9}$$

This cubic equation may be solved for the hydrogen-ion concentration corresponding to the stoichiometric concentration c_a of added weak acid.

Under certain circumstances, approximations may be introduced to simplify the numerical calculations of the hydrogen-ion concentration. For example, if $(H^+) > 10^{-6}$, then $K_w/(H^+) < 10^{-8}$ and may be neglected in the second term of equation 7.9. When this is done, equation 7.9 becomes

$$K = \frac{(H^+)^2}{c_a - (H^+)} \tag{7.10}$$

This quadratic equation is easier to solve for (H^+) than equation 7.9.

The solution of a quadratic equation may be avoided if the weak acid is only slightly dissociated, so that $c_a \gg (H^+)$, and equation 7.10 becomes

$$(H^+) = (Kc_a)^{1/2} \tag{7.11}$$

Example 7.3   Calculate the hydrogen-ion concentration and pH of 0.1 M acetic acid at 25°. Given $K = 1.75 \times 10^{-5}$.

$$(H^+) = (Kc_a)^{1/2} = [(1.75 \times 10^{-5})(0.1)]^{1/2}$$

$$= 1.32 \times 10^{-3} \text{ M}$$

$$pH = -\log (H^+) = 2.88$$

Since $(H^+) > 10^{-6}$ and is about 1 % of c_a, we are justified in using equation 7.11.

7.3   TITRATION CURVES OF WEAK ACIDS

The determination of titration curves offers one of the most convenient means for obtaining the ionization constants of weak acids and bases. The titration curve for 0.004 M acetic acid titrated with a concentrated solution of sodium hydroxide at 25° is given in Fig. 7.1a. The pH is plotted horizontally, and the number of moles of sodium hydroxide added to 1 mol of acid is plotted vertically. The pH at the midpoint of the titration is equal to the pK_a of acetic acid at this salt concentration, 4.75. The pH at any other point from about 5 to 95 % neutralization may be calculated with equation 7.6. As the end point is approached the pH changes rapidly, and the exact end point is at the pH of a solution of sodium acetate, which may be calculated as shown in example 7.2.

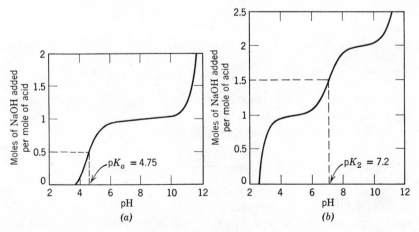

Fig. 7.1   (a) Titration of 0.004 M acetic acid with sodium hydroxide at 25°. (b) Titration of 0.004 M phosphoric acid with sodium hydroxide at 25°.

The titration curve for 0.004 M phosphoric acid titrated with a concentrated solution of sodium hydroxide is given in Fig. 7.1b. The third ionization is so weak that it cannot be studied in dilute aqueous solutions. The first ionization of phosphoric acid is sufficiently strong so that the shape of the titration curve in the first step cannot be calculated by means of equation 7.6 since the assumptions made in its derivation are not valid. However, the second ionization is sufficiently weak that the shape of the second step is in agreement with values calculated from equation 7.6. The midpoint of the second step of the titration is at pH 7.2, which is the value of pK_2.

In the neighborhood of the pKs, titration curves have their greatest slope as illustrated by the plots in Fig. 7.1. These are called *buffering* regions because the pH changes most slowly when acid or base is added. At 1 pH unit away from the pK a buffer is about 33 % as effective (see problem 7.44).

Buffer solutions are important in analytical chemistry, in general laboratory work, and particularly in biochemistry, where equilibria and rates are very dependent upon the pH. Blood, milk, and various other animal fluids are highly buffered with bicarbonate ions and carbonic acid and with proteins. The pH of human blood in a normal person is approximately 7.4. Ordinarily variations are less than 0.1 of a pH unit, and an increase or decrease of as much as 0.4 is fatal. Rates of most enzymatic reactions depend markedly on the pH.

7.4 IONIZATION OF BASES

Problems involving the ionization of weak bases may be solved in the same way as for weak acids by using the K_a expression for the *conjugate acid* of the base. The conjugate acid is the acid that is formed when a proton reacts with the base. For a base B like NH_3 that does not contain a hydroxyl group

$$BH^+ = H^+ + B$$

$$K_a = \frac{(H^+)(B)}{(BH^+)} \tag{7.12}$$

On the other hand, for bases it is more common to use K_b, which is defined by

$$B + H_2O = OH^- + BH^+$$

$$K_b = \frac{(OH^-)(BH^+)}{(B)} = \frac{K_w}{K_a} \tag{7.13}$$

where $K_w = (H^+)(OH^-)$. The concentration of water is left out of the expression for the equilibrium constant in equation 7.13 and in the expression for the dissociation of water because the water concentration is constant in dilute aqueous solutions.

For a base BOH that involves a hydroxyl group the acidic ionization may be written as follows:

$$B^+ + H_2O = H^+ + BOH$$

$$K_a = \frac{(H^+)(BOH)}{(B^+)} \tag{7.14}$$

If this expression is divided into $K_w = (H^+)(OH^-)$, then

$$K_b = \frac{K_w}{K_a} = \frac{(OH^-)(B^+)}{(BOH)} \tag{7.15}$$

which is the equilibrium constant for

$$BOH = B^+ + OH^- \tag{7.16}$$

The product of K_a and K_b is equal to K_w, as shown by equations 7.13 and 7.15.

7.5 AMINO ACIDS

Glycine, $H_2NCH_2CO_2H$, is an example of a substance containing both a basic and an acidic group. In view of what has been said in the preceding section it is convenient to consider an ampholyte like glycine to be a dibasic acid. The acid dissociations of glycine are represented as follows, with the most acidic form at the left and the most basic form at the right:

$$
\begin{array}{c}
k_1 \qquad\qquad {}^+H_3NCH_2CO_2^- \qquad\qquad k_3 \\
{}^+H_3NCH_2CO_2H \qquad\qquad\qquad H_2NCH_2CO_2^- \\
k_2 \qquad\qquad H_2NCH_2CO_2H \qquad\qquad k_4
\end{array}
\tag{7.17}
$$

The dissociation constants k_1, k_2, k_3, and k_4 are often called microscopic constants, and the constants K_1 and K_2 obtainable from titrations are called macroscopic constants. The equilibrium constant k_z represents the ratio of the dipolar ion to the uncharged molecule. The titration data do not distinguish between the dipolar ion and the uncharged molecule. The first acid dissociation constant K_1 is given by

$$K_1 = \frac{(H^+)[({}^+H_3NCH_2CO_2^-) + (H_2NCH_2CO_2H)]}{({}^+H_3NCH_2CO_2H)}$$

$$= k_1 + k_2 = 10^{-2.35} \tag{7.18}$$

where k_1 and k_2 are the equilibrium constants for the two possible dissociations of a first proton. The second acid-dissociation constant is given by

$$K_2 = \frac{(H^+)(H_2NCH_2CO_2^-)}{[({}^+H_3NCH_2CO_2^-) + (H_2NCH_2CO_2H)]}$$

$$= \frac{1}{(1/k_3) + (1/k_4)} = 10^{-9.78} \tag{7.19}$$

where k_3 and k_4 are the equilibrium constants for the two possible dissociations of a second proton. This same formulation would apply to any dibasic acid. In addition the principle of detailed balancing (Section 10.13) yields a further relationship between the values of the four microscopic constants.

$$k_1 k_3 = k_2 k_4 \qquad (7.20)$$

Since K_1 and K_2 are known (values are given for 25° in Table 7.1) we need a value of only one of the four microscopic constants in order to calculate values

Table 7.1 Thermodynamic functions for acid dissociations at 25°[1]

	pK	$\Delta G°$	$\Delta H°$	$\Delta S°$	$\Delta C_P°$
Water (K_w)	13.997	19,089	13,519	−18.7	−47
Acetic acid	4.756	6,486	−92	−22.1	−37
Chloroacetic acid	2.861	3,901	−1,158	−17.0	−40
Butyric acid	4.82	6,574	−693	−24.4	
Succinic acid, pK_1	4.207	5,740	762	−16.7	−32
Succinic acid, pK_2	5.636	7,693	−108	−26.1	−52
Carbonic acid, pK_1	6.352	8,666	2,240	−21.6	−90
Carbonic acid, pK_2	10.329	14,092	3,603	−35.2	−65
Phosphoric acid, pK_1	2.148	2,930	−1,828	−16.0	−37
Phosphoric acid, pK_2	7.198	9,823	987	−29.6	−54
Glycerol-2-phosphoric acid, pK_1	1.335	1,820	2,893	−15.8	−78
Glycerol-2-phosphoric acid, pK_2	6.650	9,069	−412	−31.8	−54
Ammonium ion	9.245	12,614	12,480	−0.4	0
Methylammonium ion	10.615	14,484	13,088	−4.7	8
Dimethylammonium ion	10.765	14,687	11,859	−9.5	23
Trimethylammonium ion	9.791	13,358	8,815	−15.2	44
Tris(hydroxymethyl)aminomethane	8.076	11,018	10,900	−0.3	
Glycine, pK_1	2.350	3,205	1,156	−6.9	−32
Glycine, pK_2	9.780	13,340	10,550	−9.4	−12
Glycylglycine, pK_1	3.148	4,140	862	−12.9	−40
Glycylglycine, pK_2	8.252	11,260	10,600	−2.0	−10

[1] $\Delta G°$ and $\Delta H°$ are in cal mol^{-1}. $\Delta S°$ and $\Delta C_p°$ are in cal K^{-1} mol^{-1}. These values apply at zero ionic strength and are obtained by extrapolation of experimental data at higher ionic strengths. From J. Edsall and J. Wyman, *Biophysical Chemistry*, Academic Press, New York, 1958.

of the other three. Assuming that k_2 has the same value as the dissociation constant of the methyl ester of glycine

$$^+H_3NCH_2CO_2CH_3 = H^+ + H_2NCH_2CO_2CH_3$$
$$k_2 = 10^{-7.7} \qquad (7.21)$$

Thus $k_1 = 10^{-2.35}$ from equation 7.18 and

$$\frac{k_4}{k_3} = \frac{10^{-2.35}}{10^{-7.7}} \qquad (7.22)$$

from equation 7.20. Substituting this ratio in equation 7.19 yields $k_4 = 10^{-4.43}$ so that $k_3 = 10^{-9.78}$.

Thus the ratio k_z of dipolar ion to neutral molecule is

$$k_z = \frac{k_1}{k_2} = \frac{10^{-2.35}}{10^{-7.7}} = 10^{5.35} \tag{7.23}$$

This dissociation accordingly goes almost exclusively by the top path; this is not true for all amino acids.

An amino acid is said to be *isoelectric* at the pH at which there are equal concentrations of the positively and negatively charged forms. Setting

$$(^{+}\text{H}_3\text{NCH}_2\text{CO}_2\text{H}) = (\text{H}_2\text{NCH}_2\text{CO}_2^{-})$$

and substituting from equations 7.18 and 7.19 yields

$$(\text{H}^+)^2_{\text{isoelectric}} = K_1 K_2$$

which may be written

$$\text{pH}_I = \tfrac{1}{2}(\text{p}K_1 + \text{p}K_2) \tag{7.24}$$

where pH_I is the pH of the isoelectric point. For glycine the isoelectric point is

$$\tfrac{1}{2}(2.35 + 9.78) = 6.06$$

Equation 7.20 may be written in the form

$$\text{p}k_4 - \text{p}k_1 = \text{p}k_3 - \text{p}k_2 \tag{7.25}$$
$$4.43 - 2.35 = 9.78 - 7.7$$

This equation says that the strengthening of the carboxyl dissociation by the positive charge on the amino group is equal to the weakening of the dissociation of the NH_3^+ group by the negative charge on the carboxyl group.

7.6 COMPLEX IONS

A number of metal ions combine with anions or neutral molecules to form ions which are referred to as complex ions. For example, solutions of CdI_2 contain the species CdI^+, CdI_2, CdI_3^-, and CdI_4^{2-} in equilibrium with Cd^{2+} and I^-. Other examples of complex ions are Ag(CN)_2^-, $\text{Cu(NH}_3)_4^{2+}$, and FeOH^{2+}. In general terms a substance L, called a ligand, is bound by a metal ion M to give a complex ion.

The successive association constants for the formation of ML, ML_2, $\ldots$; ML_i are

$$M + L = ML \qquad K_1 = (ML)/(M)(L)$$
$$ML + L = ML_2 \qquad K_2 = (ML_2)/(ML)(L)$$

$$\cdot \qquad\qquad\qquad \cdot$$
$$\cdot \qquad\qquad\qquad \cdot$$
$$\cdot \qquad\qquad\qquad \cdot$$

$$\underline{ML_{i-1} + L = ML_i \qquad K_i = (ML_i)/(ML_{i-1})(L)}$$
$$M + iL = ML_i \qquad \beta_i = K_1 K_2 \cdots K_i = (ML_i)/(M)(L)^i$$

where β_i is the equilibrium constant for the overall reaction. The average number $\bar{n}$ of ligands L bound per metal ion is given by

$$\bar{n} = \frac{(ML) + 2(ML_2) + \cdots + i(ML_i)}{(M) + (ML) + (ML_2) + \cdots + (ML_i)} \tag{7.26}$$

Introducing the equilibrium expressions, we see that

$$\bar{n} = \frac{K_1(L) + 2K_1K_2(L)^2 + \cdots + iK_1K_2 \cdots K_i(L)^i}{1 + K_1(L) + K_1K_2(L)^2 + \cdots + K_1K_2 \cdots K_i(L)^i} \tag{7.27}$$

The values of the successive equilibrium constants may be obtained by determining $\bar{n}$ as a function of (L). If the ligand has acidic or basic properties, its concentration may be determined by means of a hydrogen electrode in solutions containing known total concentrations of the acidic or basic substance. For example, the concentration of free ammonia in solutions containing cupric nitrate and ammonium nitrate buffer has been determined by glass-electrode measurements.[*] If the concentration of ammonium salt is known and the hydrogen-ion concentration is measured, the concentration of ammonia may be calculated from

$$(NH_3) = \frac{K(NH_4^+)}{(H^+)} \tag{7.28}$$

Since the concentration of added ammonia, c_{NH_3}, is known, the average number $\bar{n}$ of NH_3 molecules bound per copper ion is

$$\bar{n} = \frac{c_{NH_3} - (NH_3)}{c_{Cu^{2+}}} \tag{7.29}$$

where $c_{Cu^{2+}}$ is the total concentration of copper ions. Figure 7.2a gives a plot of the number of NH_3 molecules bound per copper ion for various values of $-\log (NH_3)$. Since the equilibrium constants for the binding of successive NH_3 molecules are not greatly different, there are no steps in this plot. However, the

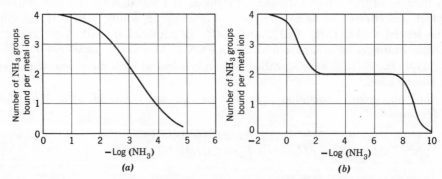

Fig. 7.2 Formation curves for ammine complexes: (*a*) with Cu^{2+} at 30° in 2 M NH_4NO_3; (*b*) with Hg^{2+} at 22° in 2 M NH_4NO_3.

---

[*] J. Bjerrum, *Metal Ammine Formation in Aqueous Solution*, P. Haase & Son, Copenhagen, 1941.

values of the successive association constants may be calculated from these data by use of equation 7.27; they are $K_1 = 10^{4.15}$, $K_2 = 10^{3.50}$, $K_3 = 10^{2.89}$, and $K_4 = 10^{2.13}$.

In the formation of the tetrammine mercuric ion complex, the first two ammonia molecules are bound much more strongly than the last two. In other words, the complex $Hg(NH_3)_2^{2+}$ is formed at very low ammonia concentrations, as shown in Fig. 7.2b.

The concentrations of colored complexes may be determined spectrophotometrically. Measurements of solubility may be used to study complex-ion equilibria in certain cases.

7.7 THERMODYNAMIC FUNCTIONS FOR ACID DISSOCIATIONS

The value of K may be obtained by extrapolation to zero ionic strength by use of the Debye-Hückel theory (Section 7.10). By use of electromotive force cells and such an extrapolation the acid dissociation constants of a number of weak acids have been determined over a range of temperature. From the temperature dependence it is possible to calculate $\Delta H°$, $\Delta S°$, and $\Delta C_P°$ for the dissociation. The values of these thermodynamic quantities are given for a number of weak acids in aqueous solutions at 25° in Table 7.1.

The fact that $\Delta S°$ for dissociation of most weak acids is negative is surprising at first. Since the dissociation products have more freedom to move around than the molecule or ion, it might be expected that the entropy of the products would be greater than the entropy of the undissociated acid and that $\Delta S°$ would have a positive sign. However, the experimental results show that there is a decrease in entropy and, correspondingly, an increase in "order" in the dissociation. This results from a participation of water molecules in the reaction which is not indicated by the balanced equation in 7.1. Water molecules, being dipoles, tend to be oriented in the neighborhood of ions. Thus the dissociation of an acid results in the orientation of a number of water molecules, and the consequent decrease in $S°$ overshadows the increase in $S°$ resulting from the formation of two particles from one. The effects that lead to a negative value of $\Delta S°$ tend to oppose dissociation. For a number of weak acids in Table 7.1 the value of $\Delta H°$ is very small, and so the standard entropy change largely determines the value of the pK according to $2.3RTpK = \Delta H° - T \Delta S°$.

There is almost no entropy change in the dissociation of ammonium ion

$$NH_4^+ = H^+ + NH_3 \qquad (7.30)$$

because there is no change in the number of ions in the reaction, so that the dissociation causes little change in water structure. For the methylated ammonium ions the water molecules are not so organized around the methylated ammonium ion as around NH_4^+, and, therefore, there is a decrease in entropy upon dissociation; that is, $\Delta S°$ has a negative value.

7.8 CARBON DIOXIDE, CARBONIC ACID, AND BICARBONATE ION THERMODYNAMICS

Only a small fraction of the CO_2 dissolved in water is hydrated to carbonic acid.

$$CO_2 + H_2O = H_2CO_3 \qquad (7.31)$$

The hydration constant K_h is written

$$K_h = \frac{(H_2CO_3)}{(CO_2)} \qquad (7.32)$$

At 25° C $K_h = 0.00258$. The standard enthalpy change and the standard entropy change are both unfavorable for the hydration reaction, as shown by the values in Table 7.2; nevertheless, CO_2 has to be hydrated in order for the dissolved

Table 7.2 Thermodynamic Quantities at 25° and Zero Ionic Strength[1,2]

Reaction	pK	$\Delta G°$	$\Delta H°$	$\Delta S°$	$\Delta C_P°$
$K_h = (H_2CO_3)/(CO_2)$	2.59	3,530	1,130	-8	-63
$K_{H_2CO_3} = \dfrac{(H^+)(HCO_3^-)}{(H_2CO_3)}$	3.77	5,170	1,010	-14	-28
$K_1 = \dfrac{(H^+)(HCO_3^-)}{(CO_2)+(H_2CO_3)}$	6.352	8,666	2,240	-21.6	-90
$K_2 = \dfrac{(H^+)(CO_3^{2+})}{(HCO_3^-)}$	10.329	14,092	3,603	-35.2	-65

[1] From J. T. Edsall, CO_2: *Chemical, Biochemical, and Physiological Aspects*, NASA SP–188, 1969.
[2] ΔG and ΔH in cal mol^{-1}, ΔS and ΔC_P in cal K^{-1} mol^{-1}.

CO_2 produced by metabolism in our tissues to be converted to HCO_3^- and be transported to our lungs. The entropy change is negative because joining two molecules to form one reduces the translational and rotational degrees of freedom of the system. ΔC_P is negative because $H_2O + CO_2$ can absorb more heat per degree increase in temperature than H_2CO_3.

Carbonic acid is a somewhat stronger acid than acetic acid.

$$H_2CO_3 = H^+ + HCO_3^- \qquad (7.33)$$

$$K_{H_2CO_3} = \frac{(H^+)(HCO_3^-)}{(H_2CO_3)} \qquad (7.34)$$

At 25° the acid dissociation constant $K_{H_2CO_3} = 1.72 \times 10^{-4}$ (pK = 3.77). However, this is *not* the acid dissociation constant you will find if you look in tables, and it is *not* the one you would use in the laboratory to do calculations on bicarbonate buffers. The reason is that in the laboratory you are not ordinarily able to distinguish between dissolved CO_2 and H_2CO_3 and simply lump these two species together by counting all dissolved CO_2 as H_2CO_3.

The first acid dissociation constant K_1 for carbonic acid is defined as

$$K_1 = \frac{(H^+)(HCO_3^-)}{(H_2CO_3) + (CO_2)} = \frac{(H^+)(HCO_3^-)}{(H_2CO_3)[1 + (CO_2)/(H_2CO_3)]} \quad (7.35)$$

Thus

$$K_1 = \frac{K_{H_2CO_3}}{1 + 1/K_h} \quad (7.36)$$

Calculating K_1 from the values of K_h and $K_{H_2CO_3}$ at $25°$ that are given above we obtain $K_1 = 4.45 \times 10^{-7}$ ($pK = 6.352$). This is the pK you would find in titrating dissolved CO_2.

Bicarbonate ion undergoes an acid dissociation with a pK of 10.329 at $25°$

$$HCO_3^- = H^+ + CO_3^{2-}$$

$$K_2 = \frac{(H^+)(CO_3^{2-})}{(HCO_3^-)} \quad (7.37)$$

Thus there are negligible concentrations of carbonate ion CO_3^{2-} under normal physiological conditions. For reaction 7.37 the value of $\Delta S°$ is more unfavorable for dissociation than for reaction 7.35 because the higher charged CO_3^{2-} ion is more effective in organizing water molecules around it than the HCO_3^- ion.

7.9 STATISTICAL EFFECTS

In comparing the strengths of weak acids with more than one dissociable group there is a statistical effect to be taken into account. As a simple example consider the two acid dissociation constants for a dibasic acid that has two acidic groups of the same kind that are so far apart that they are really independent and identical.

$$H_2A = H^+ + HA^- \quad K_1 = \frac{(H^+)(HA^-)}{(H_2A)} \quad (7.38)$$

$$HA^- = H^+ + A^{2-} \quad K_2 = \frac{(H^+)(A^{2-})}{(HA^-)} \quad (7.39)$$

We can look at this dissociation from a microscopic point of view and represent the intermediate form by HA if the "right" proton dissociates and by AH if the "left" proton dissociates. Since the acid groups are identical

$$K = \frac{(H^+)(HA^-)}{(H_2A)} = \frac{(H^+)(AH^-)}{(H_2A)} = \frac{(H^+)(A^{2-})}{(HA^-)} = \frac{(H^+)(A^{2-})}{(AH^-)} \quad (7.40)$$

Thus

$$K_1 = \frac{(H^+)[(HA^-) + (AH^-)]}{(H_2A)} = 2K \tag{7.41}$$

$$K_2 = \frac{(H^+)(A^{2-})}{(HA^-) + (AH^-)} = \frac{1}{1/K + 1/K} = \frac{K}{2} \tag{7.42}$$

so that $K_1 = 4K_2$. The first acid dissociation constant is four times larger than the second because of this statistical effect.*

7.10   EFFECT OF IONIC STRENGTH ON DISSOCIATION CONSTANTS

At low ionic strength values (several hundredths for monovalent weak acids and several thousandths for highly charged weak acids) the dependence of pK on ionic strength may be calculated using the Debye-Hückel theory (Section 6.10) and simple extensions of that theory. According to the Güntelberg equation

$$\log y_i = \frac{-z_i^2 A I^{1/2}}{1 + I^{1/2}} \tag{7.43}$$

where y_i is the activity coefficient of ion i with charge z_i (in units of electron charge), I is ionic strength on the molar scale, and A is the Debye-Hückel constant (0.5091 at 25°). This equation differs from the Debye-Hückel theory by the $1 + I^{1/2}$ in the denominator. Although this is a useful improvement, coefficients of $I^{1/2}$ other than unity give better results for particular weak acid-electrolyte combinations.

For the dissociation of an acid with n negative charges

$$HA^{-n} = H^+ + A^{-(n+1)} \tag{7.44}$$

the thermodynamic dissociation constant K is given by

$$K = \frac{a_{H^+}(A^{-(n+1)})y_{-(n+1)}}{(HA^{-n})y_{-n}} = K' \frac{y_{-(n+1)}}{y_{-n}} \tag{7.45}$$

Substituting equation 7.43 and rearranging yields

$$pK_I = pK'_{I=0} - (2n + 1) A I^{1/2}/(1 + I^{1/2}) \tag{7.46}$$

7.11   THERMODYNAMICS OF BIO-CHEMICAL REACTIONS

The thermodynamic quantities for biochemical reactions are handled in a rather different way from other reactions. The fact that the reactants and products exist

---

* For the treatment of a larger number of identical and independent sites see C. Tanford, *Physical Chemistry of Macromolecules*, Wiley, New York, 1961, p. 532.

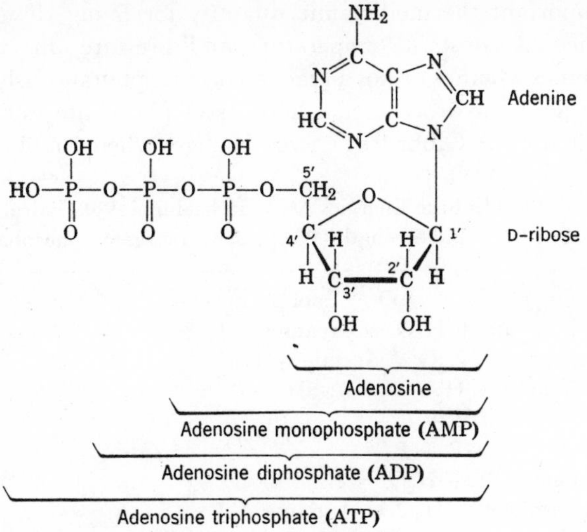

Fig. 7.3 Structure of adenosine triphosphate, ATP.

in various degrees of ionization and complexation is usually ignored, and a single symbol is used to represent the sum of the concentrations of these various species. For example, the hydrolysis of adenosine triphosphate (Fig. 7.3) to adenosine diphosphate and inorganic phosphate is usually represented by

$$ATP + H_2O = ADP + P \tag{7.47}$$

and the equilibrium constant and standard Gibbs free energy of hydrolysis are represented by

$$K_{obs} = \frac{(ADP)(P)}{(ATP)} \tag{7.48}$$

$$\Delta G^\circ_{obs} = -RT \ln K_{obs} \tag{7.49}$$

The equilibrium constant is referred to as the observed constant because it is the value obtained in the laboratory without knowing the relative concentrations of the various ionized and complexed forms of the reactants and products. The concentration of water is omitted because it is a constant in the dilute aqueous solutions which are used.

In reaction 7.47, and in the equilibrium expression, the symbol ATP refers to the sum of all of the various ionized and complexed forms that exist in the equilibrium solution, and the same comment applies to ADP and P. As a consequence the value of K_{obs} at a given temperature and electrolyte concentration is a function of pH and concentration of metal ions that are complexed by ATP, ADP, or P. Since K_{obs} is a function of pH and metal ion concentration, the values of the various thermodynamic quantities are also functions of pH and metal ion concentration. This type of equilibrium constant will be related to the more familiar types of equilibrium constants after we have considered the acid dissociations and metal ion complexation reactions of ATP and related compounds.

The most important thermodynamic quantity for living things is the Gibbs free energy since at constant temperature and pressure this thermodynamic quantity determines whether or not a reaction occurs spontaneously. The standard Gibbs free energies of hydrolysis of a number of phosphate esters are given in Table 7.3. These are the Gibbs free energy changes when 1 mol of the ester at a

Table 7.3   Standard Gibbs Free Energies ΔG_{obs}° in kcal mol^{-1} of Hydrolysis at 25°, pH 7, pMg 4 at 0.2 Ionic Strength (P represents inorganic phosphate)

Phosphoenolpyruvate + H_2O = Enol pyruvate + P	−14.8
Creatine phosphate + H_2O = Creatine + P	−10.4
Acetyl phosphate + H_2O = Acetate + P	−10.3
CoA-S-phosphate + H_2O = CoA-SH + P	−9.0
ATP + H_2O = ADP + P	−9.5
ADP + H_2O = AMP + P	−8.8
Pyrophosphate (PP) + H_2O = 2P	−8.2
Arginine phosphate + H_2O = Arginine + P	−7.0
Glucose-6-phosphate + H_2O = Glucose + P	−3.0
Fructose-1-phosphate + H_2O = Fructose + P	−3.0
AMP + H_2O = Adenosine + P	−3.0
Glycerol-1-phosphate + H_2O = Glycerol + P	−2.2

concentration of 1 mol per liter is hydrolyzed to give the indicated products, each at one molar concentration, at 25°, pH 7 and pMg 4, where pMg = $-\log (Mg^{2+})$. The Gibbs free energy changes for phosphate transfer reactions may be obtained by adding and subtracting these reactions. Since the equilibrium constant may be calculated for any pair of reactions, such a table summarizes equilibrium data for a very large number of phosphate transfer reactions. Similar tables may be constructed for the transfer of pyrophosphate (pyrophosphoric acid is $H_4P_2O_7$) or any other group.

Example 7.4   What is the equilibrium constant K_{obs} at pH 7 and pMg 4 for

Creatine phosphate + ADP = Creatine + ATP
since

$$\text{Creatine phosphate} + H_2O = \text{Creatine} + P \qquad \Delta G_{obs}^{\circ} = -10.4 \text{ kcal mol}^{-1}$$
$$\text{ADP} + P = \text{ATP} + H_2O \qquad \Delta G_{obs}^{\circ} = 9.5 \text{ kcal mol}^{-1}$$

For the given reaction given $\Delta G_{obs}^{\circ} = -10.4 + 9.5 = -0.9$ kcal mol^{-1}. Using $\Delta G_{obs}^{\circ} = -RT \ln K_{obs}$,

$$K_{obs} = 4.6 = \frac{(\text{Cr})(\text{ATP})}{(\text{CrP})(\text{ADP})}$$

As shown in this example a reaction with a more negative Gibbs free energy change may be used to drive a phosphorylation reaction with a less negative

Gibbs free energy change. Such reactions are said to be coupled and the coupling is provided by the enzyme (Section 10.26) that catalyzes the phosphate transfer reaction. The reaction given in Example 7.4 probably occurs by creatine phosphate transferring a phosphate group to the enzyme that subsequently transfers the phosphate group to ADP. In this way the large standard Gibbs free energy change for the first reaction is conserved rather than being dissipated as it would be if creatine phosphate were simply hydrolyzed. This is an absolutely essential process for living things because this is the way a thermodynamically spontaneous reaction can drive a nonspontaneous reaction that synthesizes a needed compound.

There is an analogy between the transfer of phosphate in these reactions and the transfer of electrons between reactions studied in electrochemistry. The intensity of phosphate transfer depends on the pH and pMg at constant temperature, and so we will now turn to a discussion of how K_{obs} depends on these variables.

7.12 THERMODYNAMICS OF THE HYDROLYSIS OF ATP

At pH values well above the neutral range ATP exists in aqueous solution as a -4 ion. As the pH is reduced to pH 7 ATP picks up a proton and becomes a -3 ion. As the pH is reduced to pH 4, ATP picks up another proton and becomes a -2 ion. Since ATP^{4-} is so highly charged, it has a strong tendency to bind cations from the medium. Significant concentrations of complexes with Na^+ and K^+ are formed at 0.1 M concentrations of the ions. Therefore, the acid dissociation constants of ATP are most conveniently determined in media having bulky cations such as $(n\text{-propyl})_4N^+$ that are bound much less strongly. The pKs for ATP, ADP, adenosine-5-monophosphate AMP, and inorganic phosphate determined in 0.2 ionic strength $(n\text{-propyl})_4NCl$ at $25°$ are summarized in Table 7.4. The first proton to go on the most basic ion is referred to as number 1.

Table 7.4 Thermodynamic Quantities for Acid Dissociation at $25°$ and 0.2 Ionic Strength

		pK	$\Delta G°$, kcal mol^{-1}	$\Delta H°$, kcal mol^{-1}	$\Delta S°$, cal K^{-1} mol^{-1}
$H_2ATP^{2-} = H^+ + HATP^{3-}$	K_{2ATP}	4.06	5.55	0	-18.6
$HATP^{3-} = H^+ + ATP^{4-}$	K_{iATP}	6.95	9.48	-1.68	-37.4
$H_2ADP^{1-} = H^+ + HADP^{2-}$	K_{2ADP}	3.93	5.36	1.0	-14.6
$HADP^{2-} = H^+ + ADP^{3-}$	K_{1ADP}	6.88	9.39	-1.37	-36.1
$H_2AMP^{1-} = H^+ + HAMP^{2-}$	K_{2AMP}	3.75	5.10	1.0	-13.8
$HAMP^{1-} = H^+ + AMP^{2-}$	K_{1AMP}	6.45	8.80	-0.85	-32.4
$H_2PO_4^{1-} = H^+ + HPO_4^{2-}$	K_{2P}	6.78	9.25	0.80	-28.4
$HP_2O_7^{3-} = H^+ + P_2O_7^{4-}$	K_{1PP}	8.95	12.21	0.40	-39.6
$H_2P_2O_7^{2-} = H^+ + HP_2O_7^{3-}$	K_{2PP}	6.12	8.35	0.11	-27.6

The values of $\Delta H°$ were obtained by determining pKs at two or more temperatures. Since the enthalpy changes are small in all cases, it is the entropy changes that determine the differences in strength of these acids. The magnitude of the standard entropy change correlates with the charge and charge distribution of the acid. The larger the charge of the base formed in acid dissociation the more negative is $\Delta S°$ as a result of the greater hydration of the more highly charged ions.

The various phosphate ions bind Mg^{2+} and Ca^{2+}, and the dissociation constants of the complexes with Mg^{2+} expressed as pK values are given in Table 7.5 along

Table 7.5 Thermodynamic Quantities for the Dissociation of Complexes with Mg^{2+} at 25° and 0.2 Ionic Strength

		pK	$\Delta G°$, kcal mol^{-1}	$\Delta H°$, kcal mol^{-1}	$\Delta S°$, cal K^{-1} mol^{-1}
$MgATP^{2-} = Mg^{2+} + ATP^{4-}$	K_{MgATP}	4.00	5.46	−3.3	−29.4
$MgHATP^{1-} =$					
$\quad Mg^{2+} + HATP^{1-}$	K_{MgHATP}	1.49	2.04	−1.9	−13.2
$MgADP^{1-} = Mg^{2+} + ADP^{3-}$	K_{MgADP}	3.01	4.11	−3.6	−25.9
$MgHADP^{0} =$					
$\quad Mg^{2+} + HADP^{2-}$	K_{MgHADP}	1.45	1.98	−2.0	−13.3
$MgAMP^{0} = Mg^{2+} + AMP^{2-}$	K_{MgAMP}	1.69	2.31	−2.9	−17.5
$MgHPO_4^{0} = Mg^{2-} + HPO_4^{2-}$	K_{MgP}	1.88	2.56	−2.9	−18.3
$MgP_2O_7^{2-} = Mg^{2+} + P_2O_7^{4-}$	K_{MgPP}	5.41	7.39	−3.5	−36.5
$MgHP_2O_7^{1-} =$					
$\quad Mg^{2+} + HP_2O_7^{3-}$	K_{MgHPP}	3.06	4.17	−3.5	−25.7

with the other thermodynamic quantities. A simple physical interpretation of the dissociation constants is that they are equal to the molar concentration of free Mg^{2+} ion at which half of the ligand (that is the anion in this case) is complexed. The dissociation constants of the magnesium complexes may be determined by acid titrations in the presence of various concentrations of magnesium salts.

Heat is evolved in the dissociation of $MgATP^{2-}$ ($\Delta H° = -3.3$ kcal mol^{-1}), so that the energy term *favors* dissociation, but the entropy term is negative and *does not favor* dissociation ($\Delta S° = -29.4$ cal K^{-1} mol^{-1}). The entropy change strongly favors *association*. Why? When association occurs H_2O of hydration is liberated, and there is an increase in randomness.

Equation 7.48 for K_{obs} may be written in terms of the concentrations of the various ionic species that are important in the range of pH or pMg of interest. In order to simplify the equations that will be derived we will not consider all of the species in Tables 7.4 and 7.5 but only those that have to be considered when pH > 5.5 and pMg > 1.5 so that only one acid dissociation and one magnesium ion complex need be considered for each reactant and product. In this case equation 7.48 can be written

$$K_{obs} = \frac{[(ADP^{3-}) + (MgADP^{1-}) + (HADP^{2-})][(HPO_4^{2-}) + (MgHPO_4^{0}) + (H_2PO_4^{1-})]}{[(ATP^{4-}) + (MgATP^{2-}) + (HATP^{3-})]}$$

$$(7.50)$$

If (ADP^{3-}) is taken out of the first term, (HPO_4^{2-}) is taken out of the second, and (ATP^{4-}) is taken out of the denominator term, the equilibrium constants for the various acid and magnesium complex dissociations may conveniently be introduced to obtain

$$K_{obs} = \frac{(ADP^{3-})(HPO_4^{2-})}{(ATP^{4-})} \frac{\left[1 + \dfrac{(Mg^{2+})}{K_{MgADP}} + \dfrac{(H^+)}{K_{1ADP}}\right]\left[1 + \dfrac{(Mg^{2+})}{K_{MgP}} + \dfrac{(H^+)}{K_{2P}}\right]}{[1 + (Mg^{2+})/K_{MgATP} + (H^+)/K_{1ATP}]}$$

$$(7.51)$$

The hydrolysis of ATP may be expressed in terms of particular ionized species by

$$ATP^{4-} + H_2O = ADP^{3-} + HPO_4^{2-} + H^+ \tag{7.52}$$

for which the equilibrium constant expression is

$$K = \frac{(ADP^{3-})(HPO_4^{2-})(H^+)}{(ATP^{4-})} \tag{7.53}$$

For reaction 7.52, the thermodynamic quantities are $\Delta G° = 0.28\,\text{kcal mol}^{-1}$, $\Delta H° = -4.70\,\text{kcal mol}^{-1}$, and $\Delta S° = -16.7\,\text{cal K}^{-1}\,\text{mol}^{-1}$. Substituting equation 7.53 in equation 7.51 yields

$$K_{obs} = \frac{K[1 + (Mg^{2+})/K_{MgADP} + (H^+)/K_{1ADP}][1 + (Mg^{2+})/K_{MgP} + (H^+)/K_{2P}]}{(H^+)[1 + (Mg^{2+})/K_{MgATP} + (H^+)/K_{1ATP}]}$$

$$(7.54)$$

Since K is independent of (H^+) and (Mg^{2+}), K_{obs} is given as a function of these variables by equation 7.54. A function of two variables may be represented by a surface, and in this case it is convenient to represent $\log K_{obs}$ as a function of pH and pMg. The contour lines on this surface are shown in Fig. 7.4. In calculating this surface* all of the ionic equilibria in Tables 7.4 and 7.5 have been taken into account. In this range of pH and pMg the value of K_{obs} varies over 3 powers of 10.

At high pH in the absence of Mg^{2+} the value of K_{obs} increases by a factor of 10 for each increase of pH of one unit. This results from the fact that under these conditions a mole of H^+ is liberated for each mole af ATP^{4-} hydrolyzed, as shown in equation 7.52.

The expression of K_{obs} in terms of equilibrium constants and concentrations of H^+ and Mg^+ is the key to the calculation of the standard Gibbs free energy change $\Delta G°_{obs}$, the standard enthalpy change $\Delta H°_{obs}$, and the standard entropy change $\Delta S°_{obs}$.

$$\Delta G°_{obs} = -RT \ln K_{obs} \tag{7.55}$$

$$\Delta H°_{obs} = \left[\frac{\partial(\Delta G°_{obs}/T)}{\partial(1/T)}\right]_{pH,pMg} = RT^2\left[\frac{\partial \ln K_{obs}}{\partial T}\right]_{pH,pMg} \tag{7.56}$$

$$\Delta S°_{obs} = -\left[\frac{\partial \Delta G°_{obs}}{\partial T}\right]_{pH,pMg} = R\left[\frac{\partial T \ln K_{obs}}{\partial T}\right]_{pH,pMg} \tag{7.57}$$

---

* R. A. Alberty, *J. Biol. Chem.*, **244**, 3290 (1969).

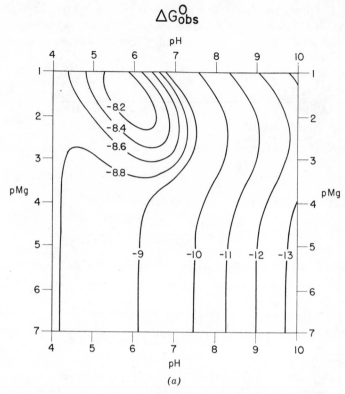

Fig. 7.4 Contour diagrams for thermodynamic quantities for ATP + H_2O = ADP + P at 25° as a function of pH and pMg. The values of (a) ΔG°_{obs}, (b) ΔH°_{obs}, and (c) $T\Delta S^\circ_{obs}$ are in kcal mol^{-1}. The principle electrolyte present is 0.2 ionic strength tetra-n-propyl ammonium chloride.

The change in standard Gibbs free energy ΔG°_{obs} is equal to the molar Gibbs free energy of ADP in hypothetical one molar aqueous solution having a particular pH and pMg plus that of hypothetical one molar orthophosphate minus that of hypothetical one molar ATP all at 25° C in 0.2 ionic strength tetra-n-propyl ammonium chloride. We say hypothetical because we would not apply these calculations to such high concentrations of the reactants. The dependence of ΔG°_{obs} on pH and pMg is given in Fig. 7.4.

Carrying out the required differentiations for ΔH°_{obs} yields

$$\Delta H^\circ_{obs} = \Delta H^\circ - \frac{[(Mg^{2+})/K_{MgADP}]\,\Delta H^\circ_{MgADP} + [(H^+)/K_{1ADP}]\,\Delta H^\circ_{1ADP}}{1 + (Mg^{2+})/K_{MgADP} + (H^+)/K_{1ADP}}$$

$$- \frac{[(Mg^{2+})/K_{MgP}]\,\Delta H^\circ_{MgP} + [(H^+)/K_{2P}]\,\Delta H^\circ_{2P}}{1 + (Mg^{2+})/K_{MgP} + (H^+)/K_{2P}}$$

$$+ \frac{[(Mg^{2+})/K_{MgATP}]\,\Delta H^\circ_{MgADP} + [(H^+)/K_{1ATP}]\,\Delta H^\circ_{1ATP}}{1 + (Mg^{2+})/K_{MgATP} + (H^+)/K_{1ATP}} \qquad (7.58)$$

$$\Delta H^O_{obs}$$

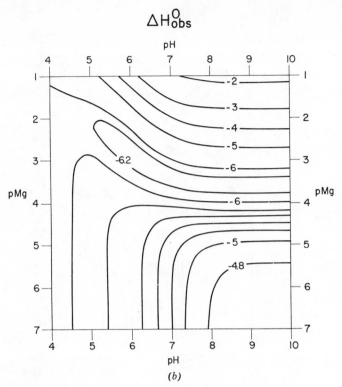

Fig. 7.4   (*contd.*)

where $\Delta H^\circ = -4.70 \text{ kcal mol}^{-1}$ is the standard enthalpy change for reaction 7.52, and the other ΔH° values are those of the acid and complex dissociations in Tables 7.4 and 7.5. The heat of hydrolysis of ATP at constant pressure varies with pH and pMg because the heats of the various dissociation reactions are involved to different extents at different pH and pMg values. The dependence of the heat of hydrolysis of ATP at constant pressure on pH and pMg is shown in Fig. 7.4. When this reaction is carried out in a buffer solution in a calorimeter an additional heat effect has to be taken into account; the H^+ produced reacts with the buffer to give an amount of heat $-n_{\rm H} \Delta H^\circ_{\rm B}$, where $\Delta H^\circ_{\rm B}$ is the enthalpy of dissociation of the buffer acid and $n_{\rm H}$ is the number of moles of acid produced. The Mg^{2+} produced may also react with a component of the solution to produce a heat effect.

The expression for ΔS°_{obs} may be obtained by substituting equations 7.55 and 7.58 into

$$\Delta G^\circ_{obs} = \Delta H^\circ_{obs} - T\Delta S^\circ_{obs} \tag{7.59}$$

The roles of these various thermodynamic quantities in determining K_{obs} changes with pH and pMg.*

---

* R. A. Alberty, *J. Chem. Ed.*, **46,** 713 (1969).

$$T\Delta S^{0}_{obs}$$

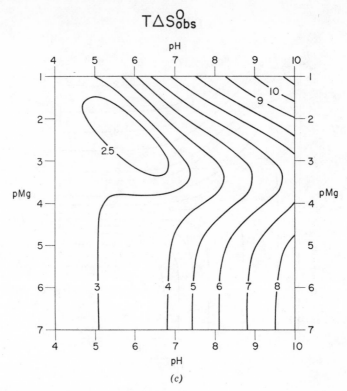

Fig. 7.4   (contd.)

The biochemical equilibria we have been discussing involve small molecules, but many biochemical equilibria involve macromolecules such as proteins and nucleic acids. As an example we will consider the binding of oxygen by hemoglobin.

7.13 BINDING OF OXYGEN BY MYOGLOBIN AND HEMOGLOBIN

Hemoglobin is the oxygen transport protein in many species. It has a molecular weight of 64,000, and each molecule contains four heme groups, four atoms of iron, and binds four molecules of oxygen when it is saturated. Myoglobin is an oxygen storage protein that is found in muscle. Its molecular weight is 16,000, and each molecule contains one heme group, one atom of iron, and binds one molecule of oxygen when it is saturated. Myoglobin was the first protein for which the detailed molecular structure was obtained by X-ray diffraction (Kendrew 1959). The detailed molecular structure of hemoglobin has also been obtained by X-ray diffraction. Hemoglobin is really a tetramer, and the four chains in it

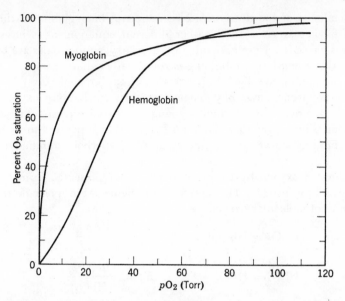

Fig. 7.5 Oxygen dissociation curves for myoglobin and hemoglobin at pH 7.4 and 38°.

have molecular weights of 16,000 each and are closely related to myoglobin both in terms of the amino acid composition and three-dimensional conformation.

The extent of oxygenation of myoglobin may be determined spectrophotometrically. The degree of saturation is shown as a function of the partial pressure of oxygen in Fig. 7.5. The shape of this plot is exactly what is expected for the simple equilibrium

$$MbO_2 = Mb + O_2 \qquad (7.60)$$

where Mb represents a molecule of myoglobin. The dissociation constant is defined by

$$K = \frac{(Mb)p_{O_2}}{(MbO_2)} \qquad (7.61)$$

where p_{O_2} is partial pressure oxygen in the gas phase. Conservation of myoglobin requires that

$$(Mb)_o = (Mb) + (MbO_2) \qquad (7.62)$$

where $(Mb)_o$ is the total molar concentration of myoglobin. Combining equations 7.61 and 7.62 yields

$$\frac{(MbO_2)}{(Mb)_o} = \frac{1}{1 + K/p_{O_2}} \qquad (7.63)$$

This is the equation of the myoglobin line in Fig. 7.5. At low p_{O_2} the fractional saturation is proportional to p_{O_2}, and at high p_{O_2}, the percent saturation approaches 100 asymptotically.

The dependence of the degree of oxygenation of hemoglobin on the partial pressure of oxygen is given by a rather different equation, as is indicated by Fig. 7.5. This curve is very remarkable in that as oxygen molecules are bound by hemoglobin the strength of the binding *increases*, rather than decreases as in most equilibria when successive ligands (in this case, oxygen molecules) are bound. This effect is of tremendous physiological importance because oxyhemoglobin dissociates oxygen over a much smaller change in partial pressure of oxygen than oxymyoglobin. The oxygenation curve for hemoglobin is referred to as a sigmoid (S-shaped) binding curve to contrast with the hyperbolic binding curve for myoglobin.

The binding of oxygen by hemoglobin may be represented formally by the following equilibria in which Hb represents the hemoglobin molecule that binds four molecules of molecular oxygen.

$$HbO_2 = Hb + O_2 \qquad K_1 = \frac{(Hb)p_{O_2}}{(HbO_2)} \tag{7.64}$$

$$HbO_4 = HbO_2 + O_2 \qquad K_2 = \frac{(HbO_2)p_{O_2}}{(HbO_4)} \tag{7.65}$$

$$HbO_6 = HbO_4 + O_2 \qquad K_3 = \frac{(HBO_4)p_{O_2}}{(HbO_6)} \tag{7.66}$$

$$HbO_8 = HbO_6 + O_2 \qquad K_4 = \frac{(HbO_6)p_{O_2}}{(HbO_8)} \tag{7.67}$$

The fractional saturation Y is given by

$$Y = \frac{(HbO_2) + 2(HbO_4) + 3(HbO_6) + 4(HbO_8)}{4[(Hb) + (HbO_2) + (HbO_4) + (HbO_6) + (HbO_8)]} \tag{7.68}$$

Substituting equations 7.64 to 7.67 yields

$$Y = \frac{p_{O_2}/K_1 + 2p_{O_2}^2/K_1K_2 + 3p_{O_2}^3/K_1K_2K_3 + 4p_{O_2}^4/K_1K_2K_3K_4}{4[1 + p_{O_2}/K_1 + p_{O_2}^2/K_1K_2 + p_{O_2}^3/K_1K_2K_3 + p_{O_2}^4/K_1K_2K_3K_4]} \tag{7.69}$$

The values of K_1 and K_4 can be calculated from experimental data, but the values of K_2 and K_3 cannot be determined with any accuracy because as soon as the species HbO_4 becomes important the species HbO_6 also becomes important.

The increase in oxygen binding affinity is a cooperative effect (Section 20.4) that results from changes in the three-dimensional arrangement of the peptide chains in hemoglobin.* This is referred to as an allosteric effect, and such effects are also encountered with enzymes.

Rather than using equation 7.69, binding data for hemoglobin are often expressed in terms of the constants n and K in the Hill equation.

$$Y = \frac{Kp^n}{1 + Kp^n} \tag{7.70}$$

---

* See R. E. Dickerson and I. Geis, *The Structure and Action of Proteins*, Harper and Row, New York, 1969.

This equation may be rearranged to

$$\frac{Y}{1-Y} = Kp^n \qquad (7.71)$$

For human hemoglobin at pH 7, a value of n of about 2.8 is obtained by plotting $\log [Y/(1-Y)]$ versus $\log p$ and using the middle straight part. The values of the constants in the Hill equation cannot be given a simple interpretation unless n is an integer.

Example 7.5 If n molecules of a ligand L combine with a molecule of protein to form PL_n without intermediate steps, derive the relation between the fractional saturation Y and the concentration of L.

$$P + nL = PL_n$$

$$K = \frac{(P)(L)^n}{(PL_n)} \qquad (P)_o = (P) + (PL_n)$$

$$\frac{[(P)_o - (PL_n)](L)^n}{(PL_n)} = K$$

$$(P)_o(L)^n = (PL_n)[K + (L)^n]$$

$$Y = \frac{(PL_n)}{(P)_o} = \frac{1}{1 + K/(L)^n} = \frac{(L)^n/K}{1 + (L)^n/K}$$

This equilibrium represents a cooperative effect in that as soon as one ligand molecule is bound, the other $(n-1)$ ligand molecules are also bound.

Hemoglobin has another remarkable property that makes it even more effective in the oxygen transport system. When hemoglobin is oxygenated at pH 7.4 it dissociates 0.6 mol of H^+ for each oxygen molecule bound. In the lungs H^+ liberated by hemoglobin reacts with bicarbonate ion to make H_2CO_3 that dissociates to produce CO_2 that diffuses into the air space in the lungs. The dissociation of H_2CO_3 to $H_2O + CO_2$ is slow compared with the rate of flow of blood in the lung, and in the red cell this reaction is catalyzed by the enzyme carbonic anhydrase. In the capillaries this process is reversed; the CO_2 produced metabolically produces H_2CO_3, which dissociates to HCO_3^- and H^+; the latter is absorbed by hemoglobin as a part of the reaction of dissociating oxygen. This so-called Bohr effect results from the fact that the acid dissociation constants of oxyhemoglobin are different from deoxyhemoglobin.

References

R. G. Bates, *Determination of pH*, Wiley, New York, 1973.
R. P. Bell, *Acids and Bases*, Methuen and Co., London, 1969.
R. E. Dickerson and I. Geis, *The Structures and Action of Proteins*, Harper and Row, New York, 1969.

J. Edsall and J. Wyman, *Biophysical Chemistry*, Academic Press, New York, 1958.

H. S. Harned and B. B. Owen, *The Physical Chemistry of Electrolytic Solutions*, Reinhold Publ. Corp., New York, 1958.

I. Klotz, *Energy Changes in Biochemical Reactions*, Academic Press, New York, 1967.

A. L. Lehninger, *Bioenergetics*, W. A. Benjamin, Inc., New York, 1965.

Lars G. Sillén, *Stability Constants of Metal-Ion Complexes*, 2nd ed., Chemical Society, London, 1964.

Problems

7.1 Calculate the pH of (*a*) a 0.1 M solution of *n*-butyric acid, (*b*) a solution containing 0.05 M butyric acid and 0.05 M sodium butyrate, and (*c*) a 0.1 M solution of pure sodium butyrate. Using these data, sketch the titration curve for 0.1 M butyric acid that is titrated with a strong base so concentrated that the volume of the solution may be considered to remain constant ($K = 1.48 \times 10^{-5}$ at 25°).   *Ans.* (*a*) 2.91, (*b*) 4.83, (*c*) 8.91.

7.2 A buffer contains 0.01 mol of lactic acid (p$K = 3.60$) and 0.05 mol of sodium lactate per liter. (*a*) Calculate the pH of this buffer. (*b*) Five milliliters of 0.5 N hydrochloric acid is added to a liter of the buffer. Calculate the change in pH. (*c*) Calculate the pH change to be expected if this quantity of acid is added to 1 liter of a solution of a strong acid of the same initial pH.   *Ans.* (*a*) 4.30, (*b*) 0.12, (*c*) 1.7.

7.3 Calculate the number of moles per liter of Na_2HPO_4 and NaH_2PO_4 which should be used to prepare a 0.10 ionic strength (equation 6.24) buffer of pH 7.30. At this ionic strength the second pK of phosphoric acid may be taken as 6.84.

$Ans.$ (Na_2HPO_4) = 0.0299 M, (NaH_2PO_4) = 0.0104 M.

7.4 In order to determine the ionization constant of the weak monobasic acid dimethyl arsinic acid, a solution was titrated with a solution of sodium hydroxide, using a pH meter. After 17.3 cm³ of NaOH had been added, the pH was 6.23. It was found that 27.6 cm³ was required to neutralize the acid solution completely. Calculate the pK value.   *Ans.* 6.00.

7.5 At 40° the ionization constant of NH_4OH is 2.0×10^{-5}. (*a*) What is the OH^--ion concentration in 0.1 M NH_4OH? (*b*) What is the OH^--ion concentration of a solution 0.1 M with respect to NH_4OH which is also 0.1 M with respect to NH_4Cl?

Ans. (*a*) 1.4×10^{-3}, (*b*) 2.0×10^{-5}.

7.6 (*a*) What is the hydrogen-ion concentration of a 0.5 M solution of NH_4Br at 25°? The K_b of NH_4OH is 1.8×10^{-5}. (*b*) What is the pH?   *Ans.* (*a*) 1.67×10^{-5}, (*b*) 4.78.

7.7 Five grams of lactic acid, $CH_3CHOHCO_2H$, is diluted with water to 1 liter. What is the concentration of hydrogen ions at 25°? The dissociation constant of $CH_3CHOHCO_2H$ is 1.36×10^{-4} at this temperature.   *Ans.* 2.68×10^{-3} M.

7.8 At what pH is the average net charge on a lysine molecule zero? That is, what is its isoelectric point?

$$pK_1 = 2.16(-CO_2H)$$

$$pK_2 = 9.18(\alpha - NH_3^+)$$

$$pK_3 = 10.79(\epsilon - NH_3^+)$$

(Simply set up the equation for calculating (H^+), but do not attempt to calculate (H^+) because the equation is cubic.)   *Ans.* $2(H^+)^3 + K_1(H^+)^2 - K_1K_2K_3 = 0$.

7.9 Given the following thermodynamic quantities calculate the pH of the midpoint of titration of the first acid group of carbonic acid at 25° C.

	$\Delta H°$	$\Delta S°$
	kcal mol^{-1}	cal K^{-1} mol^{-1}
$CO_2 + H_2O = H_2CO_3$	1.13	-8
$H_2CO_3 = H^+ + HCO_3^-$	1.01	-14

Ans. 6.39.

7.10 Estimate pK_1 and pK_2 for H_3PO_4 at 25° and 0.1 ionic strength. The values at zero ionic strength are

$$pK_1 = 2.148$$
$$pK_2 = 7.198$$

Ans. 2.026, 6.831.

7.11 Will 0.01 M creatine phosphate react with 0.01 M adenosine diphosphate to produce 0.04 M creatine and 0.02 M adenosine triphosphate at 25°, pH 7, pMg 4? What concentration of ATP can be formed if the other reactants are maintained at the indicated concentrations?   *Ans.* No. 1.13×10^{-2} M.

7.12 In a series of biochemical reactions the product in one reaction is a reactant in the next. This has the effect that spontaneous reactions drive nonspontaneous reactions. For example, reaction 2 follows reaction 1.

1. L-malate = fumarate + H_2O $\Delta G°_{obs} = +700$ cal

2. fumarate + ammonia = asparate $\Delta G°_{obs} = -3720$ cal

The $\Delta G°$ values are for pH 7 and 37°, and the state of ionization of the reactants is ignored. In reaction 1, (H_2O) is to be taken as 1. If the ammonia concentration is 10^{-2} M, calculate (asparate)/(L-malate) at equilibrium.   *Ans.* 1.3.

7.13 The experimental values of pK_1 for very dilute ATP at 25° and 0.2 ionic strength are 6.95 in tetra-n-propyl ammonium chloride and 6.41 in NaCl. What is the standard Gibbs free energy change for the reaction

$$NaATP^{3-} = Na^+ + ATP^{4-}$$

assuming $HATP^{3-}$ does not bind Na^+ significantly?   *Ans.* 1.5 kcal mol^{-1}.

7.14 0.01 M NaH_2PO_4 dissolved in 0.2 M $(CH_3CH_2)_4NCl$ is titrated with $(CH_3CH_2)_4NOH$ at 25°. The midpoint of the titration curve is at pH 6.80. An identical solution is made 0.05 M in $MgCl_2$ and the titration is repeated. This time the midpoint is pH 6.37. What is the dissociation constant of $MgHPO_4$?   *Ans.* 2.96×10^{-2}.

7.15 Biochemistry textbooks give $\Delta G°_{obs} = -4.80$ kcal mol^{-1} for the hydrolysis of ethyl acetate at pH 7 and 25°. Experiments in acid solution show that

$$\frac{(CH_3CH_2OH)(CH_3CO_2H)}{(CH_3CO_2CH_2CH_3)} = 14$$

where concentrations are in moles per liter. What is the value of $\Delta G°_{obs}$ obtained from this equilibrium quotient? The pK of acetic acid = 4.60 at 25° C.   *Ans.* -4.85 kcal mol^{-1}.

7.16 Given $\Delta G° = +11.8$ kcal for

$$ATP^{4-} + H_2O = AMP^{2-} + P_2O_7^{4-} + 2H^+$$

calculate $\Delta G°_{obs}$ at pH 7 and 25° and 0.2 ionic strength. See Table 7.4.

Ans. -9.8 kcal mol^{-1}.

7.17 The cleavage of fructose 1,6-diphosphate (FDP) to dihydroxyacetone phosphate (DHP) and glyceraldehyde 3-phosphate (GAP) is one of a series of reactions most organisms use to obtain energy. At 37° and pH 7, ΔG°_{obs} for the reaction FDP = DHP + GAP is 5.73 kcal mol^{-1}. What is ΔG_{obs} in an erythrocyte in which (FDP) = 3 μM, (DHP) = 138 μM, and (GAP) = 18.5 μM?   *Ans.* 1.54 kcal mol^{-1}.

7.18 The hydrolysis of adenosine triphosphate ATP to adenosine diphosphate ADP and inorganic phosphate at pH 8 and 25°

$$ATP^{-4} + H_2O = ADP^{-3} + HPO_4^{-2} + H^+$$

has a standard enthalpy change of -3 kcal. The standard enthalpy changes of acid dissociation of $HATP^{-3}$, $HADP^{-2}$, and $H_2PO_4^{-1}$ are -2, 0, and $+2$ kcal mol^{-1}, respectively. Calculate the standard enthalpy change for the reaction

$$HATP^{-3} + H_2O = HADP^{-2} + H_2PO_4^-$$

Ans. -7 kcal mol^{-1}.

7.19 The percent saturation of a sample of human hemoglobin was measured at a series of oxygen partial pressures at 20° C, pH 7.1, 0.3 M phosphate buffer and 3×10^{-4} M heme.

p_{O_2}, Torr	Percent Saturation
2.95	4.8
5.9	20
8.9	45
18.8	78
22.4	90

Calculate the values of n and K in the Hill equation.   *Ans.* 2.4, 3.6×10^{-3}.

7.20 What concentrations of sodium acetate and acetic acid should be used to prepare an acetate buffer of pH 5.10 and 0.1 ionic strength ($K_{HAc} = 2.7 \times 10^{-5}$)?

7.21 The ionization constant of nitrous acid, HNO_2, is 4×10^{-4} at 25°. What is the pH of a solution made up with 0.1 mol HNO_2 and 0.03 mol of sodium nitrite in 1 liter?

7.22 Calculate the pH of a 0.1 M solution of pyridinium perchlorate in water at 25°. For pyridine $K_b = 1.6 \times 10^{-9}$.

7.23 For benzoic acid at 25° the ionization constant is 7.3×10^{-5}. Calculate the pH of 0.001 M solution of benzoic acid at 25°.

7.24 The ionization constants at 25° for acetic acid, lactic acid, and bromoacetic acid are 1.8×10^{-5}, 1.4×10^{-4}, and 1.4×10^{-3}, respectively. Calculate the degree of dissociation α for a 0.01 M solution of each of these acids (*a*) by the approximate method (assuming $1 - \alpha = 1$) and (*b*) by the exact method (see Section 7.2).

7.25 Calculate the pH of 0.1 M sodium isobutyrate at 25° ($K = 0.98 \times 10^{-5}$).

7.26 Sketch the titration curve of aspartic acid indicating the approximate pH values obtained when monosodium aspartate and disodium aspartate are dissolved in water.

7.27 Calculate the pH of 10^{-5} M acetic acid in water at 25°, using the exact equation rather than the approximate form. What result would have been obtained if the approximate form had been used?

7.28 Calculate the pH of a 0.1 M solution of potassium cyanide at 25° ($K = 7.2 \times 10^{-10}$).

7.29 Given the following thermodynamic quantities calculate the dissociation constant for $NH_4OH = NH_4^+ + OH^-$ at $25°$ C.

	$\Delta H°$ kcal mol^{-1}	$\Delta S°$ cal K^{-1} mol^{-1}
$NH_4^+ = NH_3 + H^+$	12,480	-0.4
$H_2O = H^+ + OH^-$	13,519	-18.7

7.30 The ionization constant of ammonium hydroxide is 1.4×10^{-5} at $0°$ and 2×10^{-5} at $40°$. What is the average heat of ionization of NH_4OH in this range of temperature?

7.31 Calculate the normal pH of human blood from the fact that the normal concentration of HCO_3^- is 0.026 M and the normal partial pressure of CO_2 in the alveoli of the lungs is 40 Torr. The pK_1 value for H_2CO_3 at $38°$ C and 0.15 ionic strength is 6.1. The Henry law constant K for the solubility of CO_2 in water is 3.30×10^{-5} M Torr^{-1} where $(CO_2) = Kp_{CO_2}$.

7.32 Using the Debye-Hückel theory, estimate the apparent pK for acetic acid at 0.01 ionic strength. At $25°$ the thermodynamic pK value is 4.76. It is assumed that the activity coefficient for undissociated acetic acid is unity at this value of the ionic strength.

7.33 In the living cell two reactions may be coupled together by having a common intermediate. This is true for the following two reactions that are enzyme catalyzed,

creatine + inorg. phosphate = creatine phosphate $\Delta G° = +11$ kcal

ATP = ADP + inorg. phosphate $\Delta G° = -8$ kcal

where ATP and ADP are adenosinetriphosphate and adenosinediphosphate, respectively. The phosphate ionizations are ignored and the $\Delta G°$ values are for pH 7.5 and $25°$. If in a steady state in a living cell $(ATP) = 10^{-3}$ M and $(ADP) = 10^{-4}$ M, calculate the maximum value of the ratio (creatine phosphate)/(creatine).

7.34 Will cyclic AMP at 0.01 M react with ADP and P at 0.01 M at pH 7 and $25°$ to form ATP and AMP, each at 0.01 M?

CAMP + H_2O = AMP $\Delta G°_{obs} = -10.0$ kcal mol^{-1}

ADP + P = ATP + H_2O $\Delta G°_{obs} = +9.5$ kcal mol^{-1}

7.35 How many grams of ATP have to be hydrolyzed to ADP to lift 100 lb 100 ft if the available Gibbs free energy can be converted into mechanical work with 100% efficiency? It is assumed that $(ATP) = (ADP) = (P) = 0.01$ M and that $\Delta G°_{obs}$ is -9.5 kcal mol^{-1} at $25°$.

7.36 The pK for the dissociation of $CaATP^{2-}$ at $25°$ in 0.2 M (n-propyl)$_4$NCl is 3.60. The pK for

$$HATP^{3-} = H^+ + ATP^{4-}$$

is 6.95. Calculate the apparent pK of this ATP ionization when ATP is titrated in a solution containing 0.1 M $CaCl_2$.

7.37 If $\Delta G°_{obs}$ for the hydrolysis of acetyl phosphate ($CH_3CO_2PO_3H_2$) is -10.3 kcal mol^{-1} at $25°$ and pH 7, what is the value at pH 4? It may be assumed that the pK of acetyl phosphate in the neighborhood of pH 7 is identical with pK_2 of orthophosphate.

7.38 At sufficiently high pH (pH > 8) and sufficiently high pMg (pMg > 5), the value of $\Delta G°_{obs}$ for

$$pyrophosphate + H_2O = 2\ phosphate$$

can be calculated with only the following information. At 25° and 0.2 ionic strength

$$P_2O_7^{4-} + H_2O = 2HPO_4^{2-} \qquad \Delta G° = -10.65 \text{ kcal mol}^{-1}$$
$$HP_2O_7^{3-} = H^+ + P_2O_7^{4-} \qquad \Delta G° = 12.21 \text{ kcal mol}^{-1}$$
$$MgP_2O_7^{2-} = Mg^{2+} + P_2O_7^{4-} \qquad \Delta G° = 7.39 \text{ kcal mol}^{-1}$$

What is the value of the equilibrium constant

$$\frac{(\text{phosphate})^2}{(\text{pyrophosphate})}$$

at pH 9 and pMg 6?

7.39 At sufficiently high pH (pH > 7.5) and sufficiently high pMg (pMg > 3.5) the effect of (Mg^{2+}) on the heat of hydrolysis of ATP and on the maximum non-PV work at 25° may be calculated with only the following information:

	$\Delta H°$ kcal mol^{-1}	$\Delta S°$ cal K^{-1} mol^{-1}
$ATP^{4-} + H_2O = ADP^{3-} + HPO_4^{2-} + H^+$	-4.7	-16.7
$MgATP^{2-} = Mg^{2+} + ATP^{4-}$	-3.3	-29.4

Calculate $\Delta H°_{obs}$ and $\Delta G°_{obs}$ at pH 8, pMg 4, and pH 8, pMg 8.

7.40 The partial pressure of oxygen required to half saturate hemoglobin at pH 7.4 is 28 Torr. If the partial pressure of oxygen in the alveolar spaces of the lungs is 100 Torr, and the partial pressure in the capillaries is 40 Torr, what percentage of the total oxygen carrying capacity of hemoglobin is being used if n in the Hill equation is 2.7?

7.41 Calculate the pH at 25° of a buffer solution containing 50 cm³ of 0.2 M potassium hydrogen phthalate and 45.45 cm³ of 0.2 M sodium hydroxide solution, all diluted to 200 cm³. The dissociation constant for the second hydrogen of phthalic acid at 25° is 3.1×10^{-6}.

7.42 A buffer contains 0.04 mol of Na_2HPO_4 per liter and 0.02 mol of NaH_2PO_4 per liter. (a) Calculate the pH, using pK = 6.84, which is the pK corresponding to the ionic strength of the solution. (b) One cm³ of 1 M HCl is added to a liter of the buffer. Calculate the change in pH. (c) Calculate the pH change to be expected if this quantity of HCl is added to 1 liter of pure water that has a pH of 7.

7.43 The titration curve of a weak monobasic acid is plotted as equivalents of strong base added per mole of weak acid versus pH. Show by use of differentiation that the slope of this plot at the inflection point (pH = pK) is equal to 2.303/4.

7.44 Show that the slope of the titration curve of a monobasic weak acid is given by

$$\frac{d\alpha}{d\text{pH}} = \frac{2.303K(H^+)}{[K + (H^+)]^2}$$

where α is the degree of neutralization.

7.45 For the acid dissociation of acetic acid $\Delta H°$ is approximately zero at room temperature in H_2O. For the acidic form of aniline, which is approximately as strong an acid as acetic acid, $\Delta H°$ is approximately $+5$ kcal mol^{-1}. Calculate $\Delta S°_{298}$ for each of the following reactions:

$$CH_3CO_2H = H^+ + CH_3CO_2^- \qquad \text{p}K = 4.75$$
$$C_6H_5NH_3^+ = H^+ + C_6H_5NH_2 \qquad \text{p}K = 4.63$$

How do you interpret these entropy changes? What compensates for the increase in entropy expected from the increase in number of molecules in the reaction?

7.46 For the reaction

$$Fe^{+3} + OH^- = FeOH^{+2}$$

at $25°$, $\Delta H° = -3.0$ kcal mol^{-1}, and $\Delta S° = 44$ cal K^{-1} mol^{-1}. Calculate K for

$$Fe(H_2O)^{+3} = H^+ + FeOH^{+2}$$

Also calculate $\Delta H°$ and $\Delta S°$ for this acid dissociation.

7.47 What is the maximum concentration of ATP that can be formed from acetyl phosphate and ADP each at 0.01 M and pH 7 and pMg 4 at $25°$, assuming that the ambient concentration of acetate is also 0.01 M? Given

$$acetyl\ P + H_2O = acetate + P \qquad \Delta G°_{obs} = -10.3\ kcal\ mol^{-1}$$
$$ADP + P = ATP + H_2O \qquad \Delta G°_{obs} = +9.5\ kcal\ mol^{-1}$$

7.48 The equilibrium ratio of sodium fumarate to sodium L-malate at pH 7 is 0.22 at $25°$. If in a living tissue the concentration of fumarate is maintained at 0.001 M by other enzymatic reactions and the concentration of L-malate is maintained at 0.01 M, can L-malate be spontaneously dehydrated to fumarate?

7.49 At pH 7 and pMg 4 what value of pCa is required to put half the ATP in the form CaATP^{-2}? At 0.2 ionic strength and $25°$?

$$HATP^{3-} = H^+ + ATP^{4-} \qquad pK = 6.95$$
$$MgATP^{2-} = Mg^{2+} + ATP^{4-} \qquad pK = 4.00$$
$$CaATP^{2-} = Ca^{2+} + ATP^{4-} \qquad pK = 3.60$$

7.50 Derive the equations for the number r_A of molecules of A bound per molecule of P and the number r_B of molecules of B bound per molecule of P for the reactions

$$P + A \overset{K_A}{=} PA$$
$$P + B \overset{K_B}{=} PB$$

where the Ks represent dissociation constants. Show that

$$\frac{\partial r_A}{\partial \ln (B)} = \frac{\partial r_B}{\partial \ln (A)}$$

This relation expresses the linkage between the binding of A and B.

7.51 Given the following system of reactions

$$
\begin{array}{ccc}
AH + B & \overset{K_1}{\rightleftharpoons} & CH \\
K_{AH} \Updownarrow & & \Updownarrow K_{CH} \\
A + B & \underset{K_2}{\rightleftharpoons} & C
\end{array}
$$

where the Ks represent dissociation constants. (a) Calculate the dependence on hydrogen ion concentration of the apparent equilibrium constant

$$K_{obs} = \frac{[(C) + (CH)]}{[(A) + (AH)](B)}$$

(b) What is the relationship between the four dissociation constants?

CHAPTER 8

SURFACE THERMODYNAMICS

In our earlier consideration of thermodynamic properties we ignored surface effects. But molecules or atoms in a surface are in a different environment from molecules or atoms in the bulk phase, and if material is finely divided, surface effects may be quite significant. A cubic meter of material divided into 10^{-6} m cubes has a surface area of 2.3 sq miles.

The unsymmetrical force field at a surface gives rise to a surface tension parallel with the surface, a tendency of molecules to be oriented in a surface, and a capacity to bind other molecules at the surface either physically or chemically.

There are many practical applications of surface thermodynamics in understanding the lowering of surface tension by solutes, adsorption by solids, chromatography, colloids, and surface catalysis.

8.1 SURFACE TENSION

A liquid surface tends to contract to the minimum area as a result of unbalanced forces of molecular attraction at the surface. The molecules at the surface are attracted into the body of the liquid because the attraction of the underlying molecules is greater than the attraction by the vapor molecules on the other side of the surface. This inward attraction causes the surface to contract if it can and gives rise to a force in the plane of the surface. Surface tension is responsible for the formation of spherical droplets, the rise of water in a capillary, and the movement of a liquid through a porous solid. Solids also have surface tensions, but it is harder to measure them. Crystals tends to form with faces with the lowest surface tensions.

The surface tension of a liquid, γ, is the force per unit length on the surface that opposes the expansion of the surface area. This definition is illustrated by the idealized experiment in Fig. 8.1, where the movable bar is pulled with force F to expand a liquid film that is stretched like a soap-bubble film on a wire frame. The surface tension can be calculated from

$$\gamma = \frac{F}{2l} \tag{8.1}$$

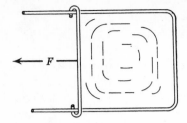

Fig. 8.1 Illustration of the surface tension of a liquid.

where l is the length of the bar, and the factor 2 is introduced because there are two liquids surfaces, one at the front and one at the back.*

The surface tension of a liquid may be measured by a variety of methods.† Since the surface tension affects the equilibrium shape of liquid surfaces, analysis of drop or bubble shape may be used to determine surface tension. The rise of liquid in a capillary or the pull on a thin vertical plate partially immersed in the liquid may be determined and used to calculate the surface tension quite accurately. Less accurate values of the surface tension may be obtained from measurements on moving liquid surfaces. These methods include studies of liquid jets, ripples, drop weight, and the force required to rupture a surface.

At a liquid-solid-gas interface there is a characteristic contact angle θ. This angle is measured in the liquid. The contact angle depends on the nature of the three phases, and its measurement is sometimes complicated by a difference between advancing and receding contact angles.

Some liquids, like water, wet the walls of a glass capillary tube and so the contact angle θ is zero; others, like mercury, do not and so the contact angle θ is 180°. When a liquid wets the tube, the liquid adhering to the walls pulls the body of the liquid up; but when the liquid does not wet the tube, surface tension pulls the body of the liquid down. The use of the capillary method is complicated by the fact that θ is not usually 0° or 180°.

A capillary tube of radius r, as shown in Fig. 8.2, is immersed in a vessel of liquid whose density is ρ. The liquid wets the tube, and the liquid rises. It continues to rise until the force due to surface tension pulling the liquid upward is counterbalanced by the force of gravity pulling it downward. The height of the equivalent cylinder of liquid which is supported is represented by h. The actual meniscus is curved, but to a high degree of approximation h is equal to the height of the lowest point of the meniscus plus one-third of the radius of the capillary if $\theta = 0°$. Then the downward force is $\pi r^2 h \rho g$, where g is the acceleration of gravity.

The surface makes an angle θ with the walls, and only the vertical component of this force is effective in pulling the liquid upward in the capillary. The surface

---

* In the literature surface tensions are generally expressed in dyne cm^{-1} and surface energies in erg cm^{-2}. A dyne is the force required to give a mass of 1 g an acceleration of 1 cm s^{-2}. Thus a dyne is 10^{-5} N. The surface tension of water at 25° is $(71.97 \text{ dyne } cm^{-1})(10^{-5} \text{ N dyne}^{-1})(10^2 \text{ cm } m^{-1}) = 71.97 \times 10^{-3} \text{ N } m^{-1}$ or $71.97 \times 10^{-3} \text{ J } m^{-2}$, since $J = N \text{ m}$.

† A. W. Adamson, *Physical Chemistry of Surfaces*, 2nd ed., Interscience-Wiley, New York, 1967.

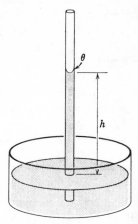

Fig. 8.2 Rise of a liquid in a capillary tube.

tension acts along the whole circumference of the surface. Thus, the upward force is $2\pi r\gamma \cos\theta$. At equilibrium the upward and downward forces are equal, and

$$2\pi r\gamma \cos\theta = \pi r^2 h\rho g \tag{8.2}$$

or

$$\gamma = \frac{h\rho g r}{2\cos\theta} \tag{8.3}$$

For many situations, including water against glass, the contact angle θ is 0, and $\cos\theta$ is 1. Then,

$$\gamma = \tfrac{1}{2}h\rho g r \tag{8.4}$$

The surface tension of a liquid decreases as the temperature rises, and becomes very small a few degrees below the critical temperature. It is zero at the critical temperature. The greater thermal agitation at the higher temperatures reduces the attractive force pulling the molecule inward.

The surface tensions of a few common liquids, measured in air, at different temperatures, are shown in Table 8.1. The surface tensions of liquid metals and molten salts are large in comparison with those of organic liquids. For example, the surface tension of mercury at $0°$ is 480.3×10^{-3} N m^{-1}, and that of silver at $800°$ is 800×10^{-3} N m^{-1}.

Table 8.1 Surface Tension of Liquids (in N m^{-1} or J m^{-2})

Temperature, °C	H_2O	CCl_4	C_6H_6	$C_6H_5NO_2$	C_2H_5OH	CH_3COOH
0	0.07564	0.0290	0.0316	0.0464	0.0240	0.0295
25	0.07197	0.0261	0.0282	0.0432	0.0218	0.0271
50	0.06791	0.0231	0.0250	0.0402	0.0198	0.0246
75	0.0635	0.0202	0.0219	0.0373	—	0.0220

The interface between two mutually saturated immiscible liquids contracts because of the interfacial tension. In principle, this interfacial tension can be measured by all the methods used to measure surface tension, but since interfacial tensions are even more sensitive to impurities than surface tensions, such measurements are more difficult to make.

8.2 THERMODYNAMICS OF A ONE-COMPONENT SYSTEM WITH A SURFACE

The atoms or molecules in a surface are in a different environment from the atoms or molecules in the bulk of a phase. In order to define the thermodynamic properties associated with the surface separately we consider a large homogeneous phase surrounded by a surface of area $\mathscr{A}$. This surface is so slightly curved that it may be considered as a plane. The enthalpy and entropy of the bulk phase are represented by H_b and S_b, and the enthalpy and entropy of the surface per unit area are represented by h^σ and s^σ. Thus the total thermodynamic quantities are given by

$$H = H_b + \mathscr{A}h^\sigma \tag{8.5}$$

$$S = S_b + \mathscr{A}s^\sigma \tag{8.6}$$

$$G = G_b + \mathscr{A}g^\sigma \tag{8.7}$$

where

$$G_b = H_b - TS_b \tag{8.8}$$

and

$$g^\sigma = h^\sigma - Ts^\sigma \tag{8.9}$$

The thermodynamic properties of the surface are the excesses of the measured thermodynamic properties that are due to the presence of the surface.

For a one-component system the surface tension γ is equal to the reversible work required to increase the area of the surface at constant temperature and pressure, per unit area. Thus

$$dw_s = \gamma \, d\mathscr{A} \tag{8.10}$$

Since the reversible work under these conditions is equal to dG

$$\left(\frac{\partial G}{\partial \mathscr{A}}\right)_{T,P} = \gamma = g^\sigma \tag{8.11}$$

where the second relation is obtained from equation 8.7. Since

$$\left(\frac{\partial g^\sigma}{\partial T}\right)_P = -s^\sigma \tag{8.12}$$

$$\left(\frac{\partial \gamma}{\partial T}\right)_P = -s^\sigma \tag{8.13}$$

Substituting equations 8.11 and 8.13 in equation 8.9

$$h^\sigma = \gamma - T\left(\frac{\partial \gamma}{\partial T}\right)_P \qquad (8.14)$$

Since the surface tension for a pure substance decreases as the temperature is raised, h^σ, the surface enthalpy, is larger than the surface tension. Since force per unit length and work per unit area have the same dimensions, the surface tension has the same numerical value whether it is expressed in N m^{-1} or J m^{-2}.

For water at 20° C, $\gamma = 0.07275$ J m^{-2}, $(\partial \gamma / \partial T)_P = -1.48 \times 10^{-4}$ J m^{-2} K^{-1}, and so the surface enthalpy is 0.1162 J m^{-2}. This is the decrease in enthalpy associated with the destruction of 1 m^2 of liquid surface. The determination of the heat evolved when a liquid surface is destroyed forms the basis for a method[*] to determine the area of a finely divided crystalline solid. The solid is suspended in the saturated vapor of a liquid until it becomes coated with an adsorbed film. At equilibrium, when the vapor pressures of the adsorbed liquid and the bulk liquid are the same, the surface energy of the liquid on the particles becomes equal to that of the liquid in bulk. The crystals are then dropped into the liquid, and the large surface area of the liquid on the particles is destroyed. Heat is evolved, and the temperature change is measured.

8.3 NUCLEATION OF CONDENSATION

In order for a vapor to condense in the absence of foreign surfaces it is necessary for small clusters of molecules to form and to grow and finally coalesce to form the bulk phase. This does not happen if the vapor pressure is only slightly higher than the equilibrium vapor pressure because the very small droplets that are formed first have a higher vapor pressure. However, when the vapor pressure has been increased sufficiently over the equilibrium value general condensation of droplets occurs.

The thermodynamics of condensation may be discussed in terms of the following two-step process:

$$n\text{A(gas, } P) = n\text{A(gas, } P^\circ) = n\text{A(small liquid droplet of radius } r) \qquad (8.15)$$

where n is number of molecules and P° is the vapor pressure of the liquid in small droplets. The change in Gibbs free energy for the first step is

$$\Delta G = -nkT \ln \frac{P}{P^\circ} \qquad (8.16)$$

(See Section 2.12.) The change in Gibbs free energy for the second step would be zero if surface effects were negligible, but since $4\pi r^2$ of surface has to be formed,

---

[*] W. D. Harkins and G. Jura, *J. Am. Chem. Soc.*, **66**, 1362 (1944).

there is an increase in Gibbs free energy of $4\pi r^2\gamma$. Thus the total Gibbs free energy change for the condensation is

$$\Delta G = -nkT\ln\frac{P}{P^\circ} + 4\pi r^2\gamma \qquad (8.17)$$

The number of molecules n may also be expressed

$$n = \tfrac{4}{3}\pi r^3\left(\frac{N_A\rho}{M}\right) \qquad (8.18)$$

where ρ is the density of the liquid, M its molecular weight and N_A is Avogadro's constant. Thus equation 8.17 becomes

$$\Delta G = -\left(\frac{4\pi r^3\rho}{3M}\right)RT\ln\frac{P}{P^\circ} + 4\pi r^2\gamma \qquad (8.19)$$

For a given degree of supersaturation the plot of ΔG versus r has the shape indicated in Fig. 8.3. At the critical radius r_c the Gibbs free energy change becomes smaller as the particle increases in size. Thus a particle of this size grows spontaneously. The critical radius may be obtained by differentiation

$$\frac{\partial\Delta G}{\partial r} = -\left(\frac{4\pi r^2\rho}{M}\right)RT\ln\frac{P}{P^\circ} + 8\pi r\gamma = 0 \qquad (8.20)$$

thus

$$\ln\frac{P}{P^\circ} = \frac{2M\gamma}{RT\rho r_c} \qquad (8.21)$$

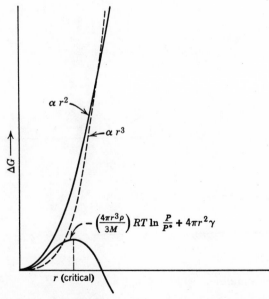

Fig. 8.3 Gibbs free energy change for the formation of a droplet of radius r from vapor with a constant degree of supersaturation.

where r_c is the critical radius. This equation gives the vapor pressure P of a droplet of radius r_c. We can think of small droplets as having higher vapor pressures than the bulk liquid because molecules are not drawn into the interior by so many near neighbors. If the sign of one side of this equation is changed it yields the vapor pressure of the concave surface of a liquid in a capillary. We can think of the surface of a liquid in a capillary that it wets as having a lower vapor pressure than the bulk liquid because molecules in the surface are drawn into the interior by more near neighbors than in a flat surface.

Both experiment and theory show that the degree of supersaturation $(P/P°)$ required to produce the formation of droplets is about 4 for many liquids. Equation 8.21 shows that the critical radius for water is about 8 Å.*

8.4   CONTACT ANGLE
AND   ADHESION

Figure 8.4 shows a liquid in equilibrium with a solid surface and gas. For some liquid-solid pairs the contact angle θ is zero, and the liquid spreads until it covers the whole solid surface. For other liquid-solid pairs the contact angle θ is greater than 90°, and the liquid stands up in little droplets and does not wet the solid surface.

At equilibrium the three interfacial tensions must balance along the line of contact. Thus

$$\gamma_{lg} \cos \theta + \gamma_{sl} = \gamma_{sg} \tag{8.22}$$

or

$$\cos \theta = \frac{\gamma_{sg} - \gamma_{sl}}{\gamma_{lg}} \tag{8.23}$$

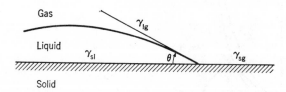

Fig. 8.4   A liquid in equilibrium with solid and vapor.

---

* Similar phenomena are involved in the freezing of liquids. Water without impurities or foreign surfaces may be cooled to −40° C before nucleation begins spontaneously.

This equation does not hold if $\gamma_{sl} > \gamma_{sl} + \gamma_{lg}$ because in this case the system has the lowest free energy if the solid is completely wet. It also does not hold if $\gamma_{sl} > \gamma_{sl} + \gamma_{lg}$ so that the solid is not wet at all.

8.5 SURFACE TENSION OF SOLUTIONS

The addition of a solute to a solvent may lower the surface tension considerably; but if the solute causes an increase in surface tension, the effect is small because the solute is forced out of the surface layer, as will be explained presently. Solutes

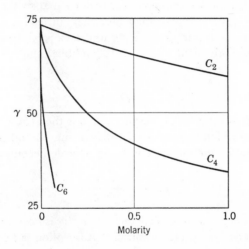

Fig. 8.5 Influence of concentration on the surface tension of aqueous solutions of fatty acids, including acetic (C_2), butyric (C_4), and hexanoic (C_6).

are classified as "capillary active" or "capillary inactive" on the basis of their effect on the surface tension. For the aqueous solution-air interface, inorganic electrolytes, salts of organic acids, bases of low molecular weight, and certain nonvolatile nonelectrolytes such as sugar and glycerin are *capillary inactive. Capillary-active* solutes are organic acids, alcohols, esters, ethers, amines, ketones, etc. The effect of capillary-active substances on the surface tension of water may be very great, as illustrated in Fig. 8.5. Soaps and detergents are especially effective in lowering the surface tension or interfacial tension. They form surface films on dirt particles in washing. Since the addition of fatty acids, for example, lowers the surface tension (surface free energy), fatty acids tend to be concentrated spontaneously in the surface layer. Gibbs derived an equation that relates the adsorption on the surface to the change in surface tension.

8.6 GIBBS' EQUATION

To be able to define the amount of adsorption at a surface it is necessary to have a clear definition of the location of the surface. In the surface region average properties change continuously with distance perpendicular to the surface, but a hypothetical boundary surface can be located in the following way. In a one-component system the surface is located so that the concentration of one bulk phase times its volume plus the similar product for the other phase gives the total amount of material present in the system. In a two-component system the interface can be located in the same way for one of the components. We will refer to this component as number 1, and it is usually chosen as the component present in larger amount. This locates the surface, and it is found that the sum of products of concentration and volume for component 2 is not equal to the total amount of component 2 in the system. There may be either an excess or a deficiency of component 2. The excess of component 2 per unit area is referred to as the surface concentration and is represented by Γ_2. In 1876 Gibbs* showed that

$$\Gamma_2 = -\frac{1}{RT}\frac{\partial \gamma}{\partial \ln a_2} \tag{8.24}$$

where γ is the surface or interfacial tension, and a_2 is the activity of component 2. In dilute solutions in which the activity coefficient may be taken equal to unity, equation 8.24 written for the molar concentration scale becomes

$$\Gamma_2 = -\frac{c}{RT}\frac{\partial \gamma}{\partial c} \tag{8.25}$$

If a solute causes a decrease in surface tension ($\partial \gamma / \partial c$ is negative), equation 8.25 shows that there is adsorption on the surface. Adsorption is here a spontaneous process because the Gibbs free energy of the system is thereby lowered.

If a solute causes an increase in surface tension ($\partial \gamma / \partial c$ is positive), its surface concentration is negative. The solute here avoids the surface region, and the effect on the surface tension is small. For example, soaps reduce the surface tension of water greatly, but no substance has been discovered to raise the surface tension of water more than a few per cent.

8.7 SURFACE PRESSURE

If the second component in a system is very insoluble and concentrates in the surface, the lowering of the surface tension can be measured directly by a method devised by Langmuir. In this method the force on a floating barrier between film-covered surface and clean-water surface is measured. A horizontal tray

---

* For the derivation of this equation see A. W. Adamson, *Physical Chemistry of Surfaces*, Wiley-Interscience, New York, 1967.

coated with a lacquer, which is not wet by water, is filled with clean water to a level slightly higher than the edges. Movable strips laid across the tray are used to sweep impurities off the surface. A small amount of film-forming substance is added to the surface, as, for example, a solution of stearic acid in a volatile solvent that evaporates after spreading rapidly over the surface. The film is then compressed between a light floating barrier and one of the moveable strips, and the pressure of the film against the floating barrier is measured by use of a torsion balance.

The force on the movable barrier may be looked at in another way. On the clean-water side the force per unit length pulling the barrier in the direction to reduce the area of the clear water surface is γ_0. On the side with the film the force per unit length pulling the barrier in the direction to reduce the area of the film covered surface is γ. The measured surface pressure π on the floating barrier is simply the difference between those surface tensions

$$\pi = \gamma_0 - \gamma \tag{8.26}$$

For dilute solutions it is frequently found that the surface tension decreases linearly with the concentration

$$\gamma = \gamma_0 - bc \tag{8.27}$$

where γ_0 is the surface tension of the solvent. Thus according to equation 8.26 the film pressure is $\pi = bc$. The Gibbs equation 8.25 may be used to determine the surface excess concentration Γ. Substituting $-d\gamma/dc = b$ and $\pi = bc$ in equation 8.25 yields

$$\pi = \Gamma RT = \frac{1}{\sigma} kT \tag{8.28}$$

where Γ is the excess number of moles of solute per unit area of the surface and σ is the area per excess molecule of solute in the surface. This behavior is generally shown by low molecular weight fatty acids. For other substances the film pressure π varies with area per mole in the surface in more complicated ways reminiscent of nonideal gases.

Some surface films behave more like solids. No significant pressure is detected in a monolayer of stearic acid until the film has been confined to a certain area; then the surface pressure rises rapidly and further decreasing the area causes the film to crumple. Thus stearic acid films behave like two-dimensional solids. The movement of the film when it is pushed or blown may be made visible by dusting the surface with some inert powder.

The reason that stearic acid forms a surface film may be explained as follows. The carboxyl group, being a polar group, has a strong affinity for water, whereas the hydrocarbon chain does not, and as a result stearic acid is very insoluble in water. At the surface the carboxyl "heads" can be dissolved in the water phase, whereas the hydrocarbon "tails" stick up out of the surface. The correctness of this interpretation is confirmed by calculating the cross-sectional area of the stearic acid molecule and its length, assuming that the film is a monolayer.

Example 8.1   It is found that 0.106 mg of stearic acid covers 500 cm² of water surface at the point where the surface pressure just begins to rise sharply. Given the molecular weight (284) and density (0.85 g cm⁻³) of stearic acid, estimate the cross-sectional area a per stearic acid molecule, and the thickness t of the film.

$$500 \text{ cm}^2 = \frac{(0.106 \times 10^{-3} \text{ g})}{(284 \text{ g mol}^{-1})} (6.02 \times 10^{23} \text{ mol}^{-1})a$$

$$a = 22 \times 10^{-16} \text{ cm}^2 \text{ or } 22 \text{ Å}^2$$

$$(500 \text{ cm}^2)t = \frac{0.106 \times 10^{-3} \text{ g}}{0.85 \text{ g cm}^{-3}}$$

$$t = 25 \times 10^{-8} \text{ cm or } 25 \text{ Å}$$

8.8   ADSORPTION BY SOLIDS

Adsorption occurs on the surface of a solid because of the attractive forces of the atoms or molecules in the surface of the solid. The potential energy of a surface plus a molecule decreases as the molecule approaches the surface, and this may be represented by a curve like that shown in Fig. 9.7, which represents the potential energy of two atoms as a function of distance. The adsorbed molecules may be considered to form a two-dimensional phase. In the two-dimensional phase, the molecule may still retain two degrees of translational freedom, and in that case the adsorbed film is said to be mobile. If the molecule loses all its translational freedom, the film is immobile.

It is convenient to distinguish between *physical adsorption* and *chemisorption*. The forces causing physical adsorption are the same type as those which cause the condensation of a gas to form a liquid and are generally referred to as van der Waals' forces. The heat evolved in the physical-adsorption process is of the order of magnitude of the heat evolved in the process of condensing the gas, and the amount adsorbed may correspond to several monolayers. Physical adsorption is readily reversed by lowering the pressure of the gas or the concentration of the solute, and the extent of physical adsorption is smaller at higher temperatures.

Chemisorption involves the formation of chemical bonds. It is therefore more specific than physical adsorption, but it is not possible to make a sharp distinction between these two kinds of adsorption in borderline cases. In chemisorption the bonding may be so tight that the original species may not be recovered. For example, when oxygen is adsorbed on graphite, heating the surface in a vacuum yields carbon monoxide. The rate of chemisorption may be fast or slow depending upon the activation energy (Section 10.15). For example, physical adsorption of a gas may occur at a low temperature and the gas may become chemisorbed when the temperature is raised. It is difficult to obtain a perfectly clean solid surface. Chemisorbed gases cannot be removed by simply exposing a solid surface to a vacuum.

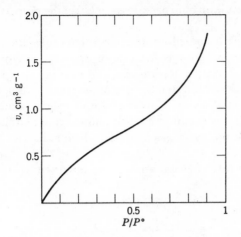

Fig. 8.6 Adsorption isotherm of nitrogen on finely divided potassium chloride at 89.9 K. The ordinate gives the volume (at $0°$ C and 1 atm) adsorbed on 1 g of solid absorbent, and the abscissa gives the pressure of nitrogen as a fraction of the saturated vapor pressure. (A. D. Keenan and J. M. Holmes, *J. Phys. Colloid Chem.*, **53**, 1309 (1949).)

The extent of adsorption depends greatly on the specific nature of the solid and of the molecules being adsorbed and is a function of pressure (or concentration) and temperature. If the gas being physically adsorbed is below its critical point, it is customary to plot the amount adsorbed per gram of adsorbent versus $P/P°$, where $P°$ is the vapor pressure of the bulk liquid adsorbate at the temperature of the experiment. Such an adsorption isotherm is shown in Fig. 8.6. The amount of gas adsorbed was determined by measuring the volume v of gas taken up by the adsorbent at various pressures and at a constant temperature of 89.9 K. As $P/P°$ approaches unity, the amount adsorbed increases rapidly because at $P/P° = 1$ bulk condensation can occur. If the temperature of the experiment is above the critical temperature of the gas, $P/P°$ cannot be calculated. When the adsorption of various gases is compared above their critical temperature, it is generally found that the amount adsorbed is smaller the lower the critical temperature.

To discuss the theory of adsorption we first consider an oversimplified model treated by Langmuir.

8.9 LANGMUIR THEORY OF ADSORPTION

Langmuir* considered the surface of a solid to be made up of elementary spaces, each of which could adsorb one gas molecule. He assumed that all the elementary spaces are identical in their affinity for a gas molecule and that the

---

* I. Langmuir, *J. Am. Chem. Soc.*, **38**, 2267 (1916); **40**, 1361 (1918).

presence of a gas molecule on one space does not affect the properties of neighboring spaces.

If θ is the fraction of the surface occupied by gas molecules, the rate of evaporation from the surface is $r\theta$, where r is the rate of evaporation from the completely covered surface at a certain temperature. The rate of adsorption of molecules on the surface is proportional to the fraction of the area that is not covered $(1 - \theta)$ and to the pressure of the gas. Thus the rate of condensation is expressed by $k(1 - \theta)P$, where k is a constant at a given temperature and includes a factor to allow for the fact that not every gas molecule which strikes an unoccupied space will stick.

At equilibrium the rate of evaporation of the adsorbed gas is equal to the rate of condensation.

$$r\theta = k(1 - \theta)P \tag{8.29}$$

$$\theta = \frac{kP}{r + kP} = \frac{1}{1 + r/kP} \tag{8.30}$$

Since the volume v of gas adsorbed is proportional to θ, equation 8.30 may be written as

$$v = \frac{v_m}{1 + k'/P} \tag{8.31}$$

where v_m is the volume of gas adsorbed when the entire surface is covered and $k' = r/k$. Thus v is directly proportional to P at very low pressures where $k'/P \gg 1$. As the pressure is increased, the volume adsorbed increases and approaches the value v_m asymptotically.

It is convenient to determine the constants k' and v_m by plotting $1/v$ versus $1/P$, since

$$\frac{1}{v} = \frac{1}{v_m} + \frac{k'}{v_m P} \tag{8.32}$$

Data such as that of Fig. 8.6 do not show asymptotic saturation and do not give a linear plot except at low pressures. Thus adsorption on solids is more complicated than the Langmuir theory indicates.

The derivation of the Langmuir adsorption isotherm involves five implicit assumptions: (1) the adsorbed gas behaves ideally in the vapor phase, (2) the adsorbed gas is confined to a monomolecular layer, (3) the surface is homogeneous, that is, the affinity of each binding site for gas molecules is the same, (4) there is no lateral interaction between adsorbate molecules, (5) the adsorbed gas molecules are localized, that is, they do not move around on the surface. Although the first assumption is good at low pressure, the second one nearly always breaks down as the pressure of the adsorbed gas is increased. As the gas pressure approaches the saturation vapor pressure, the vapor condenses without limit on all surfaces if the contact angle θ is zero, as shown by Fig. 8.6. The third assumption is poor because real surfaces are heterogeneous; the affinity for gas molecules is different on different crystalline faces and different kinds of binding sites are introduced by edges, cracks, and crystal imperfections. Heterogeneity leads to a decrease in binding energy as the surface coverage increases. The incorrectness of the fourth assumption was first shown experimentally when it was found that in certain cases the heat of adsorption may increase with the surface concentration of adsorbed molecules. This effect, which is the opposite of that expected to result from surface heterogeneity, is caused by lateral attractions of adsorbed molecules. The fifth assumption is incorrect because there are several kinds of evidence that surface films may be mobile.

In physical adsorption molecules of vapor may be adsorbed to the depth of many monolayers. An adsorption isotherm equation has been derived by Brunauer, Emmett, and Teller (BET)* to provide for multilayer adsorption. It is assumed that the surface possesses uniform, localized sites and that adsorption at one site does not affect adsorption at neighboring sites, just as in the Langmuir theory. It is further assumed that molecules can be adsorbed in second, third, . . . , and nth layers with the surface area available for the nth layer equal to the coverage of the $(n - 1)$th layer. The energy of adsorption in the first layer, E_1, is assumed to be constant, and the energy of adsorption in succeeding layers is assumed to be E_L, the energy of liquefaction of the gas.

By use of these assumptions it is possible to derive an equation that yields the volume of gas that is adsorbed when the entire adsorbent surface is covered with a complete unimolecular layer.

The surface area occupied by a single molecule of adsorbate on the surface may be estimated from the density of the liquefied adsorbate. For example, the area occupied by a nitrogen molecule at $-195°$ is estimated to be 16.2 Å^2 on the assumption that the molecules are spherical and that they are close packed in the liquid. Thus, from the measured value of v_m, the surface area of the adsorbent may be calculated. This method is widely used in the determination of the surface area of solid catalysts and adsorbents. The area values determined in this way seem, in general, to be perfectly satisfactory in spite of the rough approximations of the theory.

Example 8.2 The volume of nitrogen gas v_m (measured at 760 Torr and $0°$) required to cover a sample of silica gel with a unimolecular layer is 129 cm^3 g^{-1} of gel. Calculate the surface area per gram of the gel if each nitrogen molecule occupies 16.2 Å^2.

$$\frac{(0.129 \text{ liter g}^{-1})(6.02 \times 10^{23} \text{ mol}^{-1})(16.2 \text{ Å}^2)(10^{-10} \text{ m Å}^{-1})^2}{(22.4 \text{ liter mol}^{-1})} = 560 \text{ m}^2 \text{ g}^{-1}$$

8.10 CHROMATOGRAPHY†

In 1906 Tswett discovered that, if a solution of chlorophyll from leaves is poured on the top of a column of a suitable adsorbent and a solvent is used to wash the

---

* S. Brunauer, P. H. Emmett, and E. Teller, *J. Am. Chem. Soc.*, **60,** 309 (1938).

† H. G. Cassidy, *Fundamentals of Chromatography*, Wiley-Interscience, New York, 1957; J. C. Giddings, *Dynamics of Chromatography, Part I, Principles and Theory*, Marcel Dekker, Inc., New York, 1965; L. Fischer, *An Introduction to Gel Chromatography*, North-Holland, London, 1969.

pigments down the column, the colored band is resolved into a series of bands moving at different rates. Each component of the original mixture is represented by a band and may be obtained in a pure form by collecting the various solutions as they come from the column or by cutting the column into segments and eluting the pigment from the adsorbent in each segment. This method is widely used for separation of substances difficult to separate by other methods. It has proved of great value in the detection and separation of biological materials. When used for separating radioactive materials, it is possible to detect and separate extremely small quantities of such materials.

Chromatography is a versatile and powerful tool for achieving separations. The separation process may be based on adsorption, distribution between gas and liquid or liquid and liquid phases, or differences in the rate of diffusion into a gel depending upon size; this group of processes is referred to as sorption. A mixture is passed through a column that contains a finely divided stationary phase. Molecules in the mixture go through a series of random steps alternating between sorption and desorption. The more time a molecules spends in the sorbed state the more slowly it moves through the column. Thus the position of an emergent peak in a chromatogram is primarily controlled by its equilibrium sorption. The width of a peak depends on kinetic and mass transfer (diffusion, flow, turbulence) processes. The slower the flow in a chromatographic column the nearer each step of the separation is to equilibrium but the more each peak is broadened by diffusion. The flow rate may be increased if equilibration can be speeded. Equilibration can be speeded by using a more finely divided sorbent, but then higher pressures are required to get suitable flow rates.*

8.11   COLLOIDS

In 1861 in the course of his investigations on diffusion in solutions, Thomas Graham drew a distinction between "colloids" such as proteins, gums and polysaccharides and "crystalloids," which are substances of lower molecular weight. He noted that colloids did not diffuse through membranes whereas crystalloids did, and he gave the name *dialysis* to this process of separation by diffusion. The essential difference between colloids and crystalloids is their relative particle sizes. It is generally considered that colloids range in diameter from about 10 Ångstroms to 10,000 Å (from 0.001 to 1 μm). Simple molecules and crystalloids are less than 10 Å in diameter. Particles larger than 10,000 Å can be seen with a microscope and they are regarded as beyond the colloidal size-range.

It is convenient to distinguish between colloids consisting of (1) small particles having the same structure as the bulk solid or liquid, (2) aggregates formed from smaller molecules and (3) molecules so large in size that their dimensions are in the colloidal size range. A dispersion of a finely divided solid, such as gold, or liquid, such as benzene, is an example of the first type. Soaps and detergents are examples of the second type. They consist of organic molecules,

---

* J. C. Giddings, *J. Chem. Ed.*, **44**, 704 (1967).

with both hydrophobic ("water-hating") and hydrophilic ("water-loving") parts, which aggregate to form micelles. In these micelles, which may contain as many as 100 molecules, the hydrophobic parts are together on the inside and the hydrophilic parts are on the outside. Proteins and high polymers are examples of the third type. These substances consist of molecules held together by covalent bonds; the important characteristic they have in common with colloidal particles is their size. Proteins and other biological macromolecules like deoxynucleic acid (DNA) are of importance for understanding the chemical processes involved in living organisms. Synthetic high polymers are of increasing importance in industry.

Lyophobic colloids are those having little attraction for the solvent medium; examples are colloidal gold and inorganic precipitates. Lyophilic colloids are those having a strong attraction for the solvent medium; examples are proteins in water and polystyrene in benzene.

The most important factors in stabilizing colloids are electric charge, adsorption of large molecules, and a strong attraction for the solvent.* Either the existence of a net electric charge that causes the particles to repel each other or a coating of large molecules, which prevents the particles from adhering to each other, may be sufficient to keep the dispersed particles in the colloidal state. In lyophobic colloids the stability of the colloid depends chiefly on the fact that the charged particles repel each other.

References

N. K. Adam, *The Physics and Chemistry of Surfaces*, Oxford University Press, Oxford, 1949.
A. W. Adamson, *Physical Chemistry of Surfaces*, 2nd ed., Wiley-Interscience, New York, 1967.
J. J. Bikerman, *Surface Chemistry*, Academic Press, New York, 1958.
W. E. Garner, ed., *Chemisorption*, Butterworths Scientific Publications, London, 1957.
S. Gregg and K. Sing, *Adsorption, Surface Area and Porosity*, Academic Press, New York, 1967.
N. B. Hannay, *Solid-State Chemistry*, Prentice-Hall, Inc., Englewood Cliffs, N.J., 1967, Chapter 10.
S. Ross and J. P. Oliver, *On Physical Adsorption*, Interscience Publishers, New York, 1964.
D. J. Shaw, *Introduction to Colloid and Surface Chemistry*, Butterworth, London, 1966.
G. A. Somorjai, *Principles of Surface Chemistry*, Prentice-Hall, Inc., Englewood Cliffs, N.J., 1972.
E. J. W. Verwey and J. Th. Overbeek, *Theory of the Stability of Lyophobic Colloids*, Elsevier Publ. Co., New York, 1948.

Problems

8.1 Estimate the surface tension of a metal given that the heat of sublimation is 10^5 cal mol^{-1} and that there are 10^{15} atoms cm^{-2} in the surface. It may be assumed that the structure is close packed so that each atom has 12 nearest neighbors. With this information only an approximate answer can be calculated.   *Ans.* 3.5 J m^{-2}.

---

* E. J. W. Verwey and J. Th. G. Overbeek, *Theory of the Stability of Lyophobic Colloids*, Elsevier Publishing Co., New York, 1948.

8.2 The surface tension of toluene at $20°$ is 0.0284 N m^{-1}, and its density at this temperature is 0.866 g cm^{-3}. What is the radius of the largest capillary that will permit the liquid to rise 2 cm?   *Ans.* 0.0335 cm.

8.3 (*a*) The surface tension of water against air at 1 atm is given in the following table for various temperatures:

t, °C	20	22	25	28	30
γ, N m^{-1}	0.07275	0.07244	0.07197	0.07150	0.07118

Calculate the surface enthalpy at $25°$ in calories per square centimeter. (*b*) If a finely divided solid whose surface is covered with a very thin layer of water is dropped into a container of water at the same temperature, heat will be evolved. Calculate the heat evolution for 10 grams of a powder having a surface area of 200 m^2 g^{-1} [W. D. Harkins and G. Jura, *J. Am. Chem. Soc.*, **66**, 1362 (1944)].   *Ans.* (*a*) 2.78×10^{-6} cal cm^{-2}, (*b*) 55.6 cal.

8.4 The surface tension of liquid copper at 1535 °C is 1.30 N m^{-1}, and the temperature coefficient is -3.1×10^{-4} N m^{-1} K^{-1}. What is the surface enthalpy?   *Ans.* 1.86 J m^{-2}.

8.5 Water vapor is rapidly cooled to $25°$ to find the degree of supersaturation required to spontaneously nucleate water droplets. It is found that the vapor pressure of water must be four times its equilibrium vapor pressure. (*a*) Calculate the radius of a water droplet formed at this degree of supersaturation. (The surface tension of water at $25°$ is 0.07197 N m^{-1}.) (*b*) How many water molecules are there in the droplet?   *Ans.* (*a*) 7.6 Å, (*b*) 62.

8.6 A solution of palmitic acid ($M = 256$) in benzene contains 4.24 g of acid per liter. When this solution is dropped on a water surface, the benzene evaporates and the palmitic acid forms a monomolecular film of the solid type. If we wish to cover an area of 500 cm^2 with a monolayer, what volume of solution should be used? The area occupied by one palmitic acid molecule may be taken to be 21 Å^2.   *Ans.* 0.0239 cm^3.

8.7 A protein with a molecular weight of 60,000 forms an ideal gaseous film on water. What area of film per milligram of protein will produce a pressure of 0.005 N m^{-1} at $25°$?   *Ans.* 82.7 cm^2.

8.8 The acid $CH_3(CH_2)_{13}CO_2H$ forms a nearly ideal gaseous monolayer on water at $25°$. Calculate the weight of acid per 100 cm^2 required to produce a film pressure of 1 N m^{-1}.   *Ans.* 9.86×10^{-7} g.

8.9 The following table gives the number of milliliters (v) of nitrogen (reduced to $0°$ and 1 atm) adsorbed per gram of active carbon at $0°$ at a series of pressures:

P, Torr	3.93	12.98	22.94	34.01	56.23
v, cm^3 g^{-1}	0.987	3.04	5.08	7.04	10.31

Plot the data according to the Langmuir isotherm, and determine the constants.   *Ans.* $k' = 156$ Torr, $v_m = 40$ cm^3 g^{-1}.

8.10 Calculate the surface area of a catalyst that adsorbs 103 cm^3 of nitrogen (calculated at 760 Torr and $0°$) per gram in order to form a monolayer. The adsorption is measured at $-195°$, and the effective area occupied by a nitrogen molecule on the surface is 16.2 Å^2 at this temperature.   *Ans.* 449 m^2.

8.11 The pressures of nitrogen required to cause the adsorption of 1.0 cm^3 g^{-1} ($25°$, 1 atm) of gas on P-33 graphitized carbon black are 0.18 Torr at 77.5 K and 2.2 Torr at 90.1 K. Calculate the heat of adsorption at this fraction of surface coverage using the Clausius-Clapeyron equation.   *Ans.* 2.8 kcal mol^{-1}.

8.12 Acetone has a density of 0.790 g cm^{-3} at $20°$ and rises to a height of 2.56 cm in a capillary tube having a radius of 0.0235 cm. What is the surface tension of the acetone at this temperature?

8.13 (a) If a liquid surface is expanded, will there be a heating effect or a cooling effect? (b) Given the surface-tension data for water in problem 8.3, calculate the surface enthalpy.

8.14 The surface tension of liquid nitrogen at 75 K is 0.00971 N m^{-1}, and the temperature coefficient of the surface tension is -23×10^{-4} N m^{-1} K^{-1}. What is the surface enthalpy?

8.15 When n-butanol vapor is cooled to 0° C it is found that the degree of supersaturation has to be about 4 in order for droplets to nucleate spontaneously. The surface tension of n-butanol at 0° is 0.0261 N m^{-1}. (a) What are the radii of droplets formed at this degree of supersaturation. (b) How many molecules are there in a droplet?

8.16 Suppose that you want to measure the reduction of the vapor pressure of water caused by the curvature of the surface produced in a small capillary. In order to obtain a 10% reduction of the vapor pressure at 25° C, how small a capillary should be used? The radius of curvature of the surface is equal to the radius of the capillary, and for this calculation a negative sign is inserted on one side of the equation. (The surface tension of water at 25° is 0.07197 N m^{-1}.)

8.17 One hundred grams of oleic acid, $C_{17}H_{33}COOH$, is poured on the surface of a clean lake, where the spreading film can be seen if the water is rippled by a gentle wind or marked with rain drops. The cross-sectional area of the molecule is about 22 Å^2. What will be the maximum diameter in meters of a circular film produced in this way?

8.18 A certain substance forms a surface film that obeys the ideal two-dimensional gas law. Calculate the excess surface concentration required to cause a surface tension lowering of 0.01 N m^{-1} at 25°.

8.19 The adsorption of nitrogen on mica is as follows:

P	0.28	0.61	1.73
x/m	12.0	19.0	28.2

(P is given in N cm^{-2}, and x/m in cubic millimeters of gas at 20° and 760 Torr adsorbed on 24.3 g of mica, having a surface of 5750 cm^2.) (a) Determine the constants of the Langmuir equation. (b) Calculate the value of x/m at $P = 2.38$. (The experimental value is 3.08.)

8.20 Magnesium oxide adsorbs silica from water, and this adsorption follows the Freundlich equation.

$$\frac{x}{m} = kc^n$$

where x is the amount adsorbed on mass m of adsorbent, c is the concentration of the substance being adsorbed, and k and n are constants. Magnesium oxide may be used to remove silica from boiler scale. Make a log-log plot of the following data, and calculate the constants of the Freundlich equation. Calculate the parts per million (ppm) of magnesium oxide needed to reduce the residual silica to 2.9 ppm.

MgO, ppm	0	75	100	200
Residual SiO$_2$, ppm	26.2	9.2	6.2	1.0
SiO$_2$ removed, ppm	0	17.0	20.0	25.2

8.21 The diameter of the hydrogen molecule is about 2.7 Å. If an adsorbent has a surface of 850 m^2 cm^{-3} and 95% of the surface is active, how much H$_2$ (measured at standard conditions) could be adsorbed by 100 ml of the adsorbent? It may be assumed that the adsorbed molecules just touch in a plane and are arranged so that four adjacent spheres have their centers at the corners of a square.

8.22 Is it possible for a substance to have a surface tension that goes through a minimum as the temperature is raised?

8.23 The following table gives data on the adsorption of benzene by graphitized carbon black (P-33) at three surface coverages and three temperatures.

v(cm^3 g^{-1} at 25°, 1 atm)	0.2	0.4	0.6
Temperature		Pressure (Torr)	
0°	0.1	0.2	0.3
35.0°	0.6	1.3	2.2
49.45°	1.3	2.9	5.3

Calculate the heat of adsorption q at each degree of coverage of the surface using the Clausius-Clapeyron equation and extrapolate this heat to zero coverage. (S. Ross and J. P. Oliver, *On Physical Adsorption*, Wiley-Interscience, New York, 1964, p. 238.)

8.24 Mercury does not wet a glass surface. Draw a figure analogous to Fig. 8.2 for mercury and glass and calculate the relative positions of the mercury surfaces, if the diameter of the capillary is (*a*) 0.1 mm and (*b*) 2 mm. The density of mercury is 13.5 g cm^{-3}. The surface tension of mercury at 25° is 0.520 N m^{-1}.

8.25 What pressure in atmospheres is required to prevent water from rising in a 10^{-4} cm diameter capillary at 25° and 1 atm?

8.26 Calculate the vapor pressure of a water droplet at 25° that has a radius of 20 Å. The vapor pressure of a flat surface of water is 23.76 Torr at 25°.

8.27 One gram of a certain activated charcoal has a surface area of 1000 m^2. If complete surface coverage is assumed, as a limiting case, how much ammonia, at 25° and 1 atm, could be adsorbed on the surface of 45 g of activated charcoal? The diameter of the NH_3 molecule is 3×10^{-10} m, and it is assumed that the molecules just touch each other in a plane so that four adjacent spheres have their centers at the corners of a square.

8.28 Two gases, *A* and *B*, compete for the binding sites on the surface of an adsorbent. Show that the fraction of the surface covered by *A* molecules is

$$\theta_A = \frac{b_A P_A}{1 + b_A P_A + b_B P_B}$$

8.29 The adsorption of ammonia on charcoal is studied at 30° and 80°. It is found that the pressure required to cause the adsorption of a certain amount of NH_3 per gram of charcoal is 106 Torr at 30° and 560 Torr at 80°. Calculate the heat of adsorption using the Clausius-Clapeyron equation.

PART
TWO

DYNAMICS

This part of the book is concerned with the rates with which various processes occur. We first consider the velocities of molecules in ideal gases because this helps us to understand the rates of chemical reactions and transport processes (diffusion, thermal conductivity, and viscosity) in the gas phase. Simple kinetic theory of gases also helps us to understand thermodynamic properties in molecular terms.

Chemical kinetics, which is the study of the rate of approach of chemical systems to equilibrium, is another area of dynamics. From the study of rates, information about the mechanism of a chemical reaction may be obtained. From the study of the effect of temperature on rates, information may be obtained about activation energy requirements and the energies of intermediate states in the reaction.

Part Two ends with a consideration of irreversible processes in solution: viscosity, electrical conductivity, and diffusion. These time-dependent processes are important for learning about the properties of solutions.

CHAPTER 9

KINETIC THEORY OF GASES

The laws of mechanics may be used to calculate the properties of bulk matter at two levels. At the first level (kinetic theory to be discussed in this chapter) rather simple mathematical averaging techniques are used. At the second level (statistical mechanics to be discussed in Chapter 17) a more abstract statistical approach is used. With a very simple model of a gas it is possible to derive the ideal gas law and the distribution of molecular velocities by use of kinetic theory. The magnitudes of the heat capacities of gases may be calculated up to the point where quantum effects are encountered. Thus kinetic theory helps us to understand thermodynamic properties in molecular terms. But kinetic theory goes beyond this and helps us to understand the rates of various processes. By introduction of the concept of collision cross section it is possible to calculate the frequency of molecular collisions and of the rates of transfer of mass, energy, and momentum in a gas for simple models.

9.1 MODEL OF A PERFECT GAS

Elementary kinetic theory is based on the assumption that molecules do not attract or repel each other and therefore travel in straight lines between collisions. It is also based on the assumption that collisions between molecules and with the wall are perfectly elastic. An elastic collision is one in which there is no change in kinetic energy; thus the molecules do not absorb energy into internal motions, and the wall does not absorb energy from the gas. Boyle's law (PV is a constant for a fixed mass of gas at a constant temperature), the Maxwell distribution of molecular velocities, and the classical theory of heat capacities can be derived from this simple model without any assumptions about molecular size except that the gas contains many molecules and the molecules fill a negligibily small fraction of the volume. The molecules of real gases attract and repel each other in a way that we will describe later, and chemical reactions are examples of inelastic collisions.

According to kinetic theory the pressure exerted by a gas on the walls of the container is equal to the average force per unit area exerted by the impacts of many molecules over a period of time.

Since the velocity of a molecule has both direction and magnitude it may be represented by a vector v. The magnitude of this vector is represented by the scalar v, and we will refer to v as the *speed* of the molecule. The velocity v may be

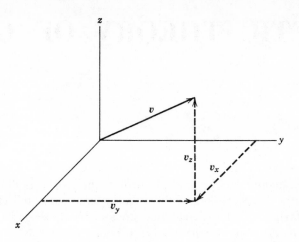

Fig. 9.1 Vector components v_x, v_y, and v_z, of the vector velocity v. The magnitude of the speed v and the magnitudes of the three components v_x, v_y, v_z are scalars.

considered to be made up of components in the x, y, and z directions, as shown in Fig. 9.1. The magnitudes v_x, v_y, and v_z of these components are related to the speed v by

$$v^2 = v_x^2 + v_y^2 + v_z^2 \tag{9.1}$$

In order to calculate the pressure exerted by a gas let us consider N molecules of mass m in a rectangular container with sides of length a, b, and c, as illustrated in Fig. 9.2. Molecule number 1 will move back and forth colliding with face A

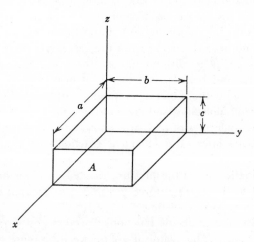

Fig. 9.2 Rectangular container with sides of length a, b, and c.

with a time interval Δt between collisions of

$$\Delta t = \frac{2a}{v_{x1}} \tag{9.2}$$

where v_{x1} is the component of the velocity in the x direction.

The time-averaged pressure exerted by molecule 1 can be calculated using Newton's second law: force is equal to mass times acceleration. Thus the contribution of molecule number 1 to the force on side A in direction x is $m(dv_{x1}/dt) = d(mv_{x1})/dt$, the rate of transfer of momentum. Since molecule number 1 approaches the surface with momentum mv_{x1} and bounces off with momentum $-mv_{x1}$ in the x direction, the transfer of momentum to side A in the x direction is $2mv_{1x}$, and the average rate of change of momentum is

$$\text{force} = \frac{2\,mv_{x1}}{\Delta t} = \frac{mv_{x1}^2}{a} \tag{9.3}$$

using equation 9.2. Since the area of side A is bc, the force per unit area or pressure produced by molecule 1 is

$$P_1 = \frac{mv_{x1}^2}{abc} = \frac{mv_{x1}^2}{V} \tag{9.4}$$

where V is the volume of the container.

The total pressure is the sum of the pressures exerted by the N molecules.

$$P = P_1 + P_2 + \cdots = \sum_{i=1}^{N} P_i = \sum_{i=1}^{N} \frac{mv_{xi}^2}{V} = \frac{Nm\langle v_x^2 \rangle}{V} \tag{9.5}$$

where the mean square velocity $\langle v_x^2 \rangle$ in the x direction is defined by

$$\langle v_x^2 \rangle = \frac{1}{N} \sum_{i=1}^{N} v_{xi}^2 \tag{9.6}$$

Since the x components of velocities are not changed by elastic collisions with the walls, the mean square velocity $\langle v_x^2 \rangle$ is a constant, and equation 9.5 expresses Boyle's law that PV is a constant for a fixed mass of gas at a fixed temperature.

In order to write equation 9.5 in terms of the mean square speed $\langle v^2 \rangle$, let us see how $\langle v^2 \rangle$ is related to the component velocities. The squares of the speeds of the individual molecules are given by equation 9.1 so that

$$
\begin{aligned}
v_1^2 &= v_{x1}^2 + v_{y1}^2 + v_{z1}^2 \\
v_2^2 &= v_{x2}^2 + v_{y2}^2 + v_{z2}^2 \\
v_3^2 &= v_{x3}^2 + v_{y3}^2 + v_{z3}^2
\end{aligned} \tag{9.7}
$$

$$
\begin{array}{c} \cdot \\ \cdot \\ \cdot \end{array}
$$

Thus

$$\sum_{i=1}^{N} v_i^2 = \sum_{i=1}^{N} v_{xi}^2 + \sum_{i=1}^{N} v_{yi}^2 + \sum_{i=1}^{N} v_{zi}^2 \tag{9.8}$$

Dividing each term in this equation by N yields the following relation between the mean square velocities

$$\langle v^2 \rangle = \langle v_x^2 \rangle + \langle v_y^2 \rangle + \langle v_z^2 \rangle \tag{9.9}$$

Since molecular motion is random, and the orientation of the coordinate system is chosen arbitrarily,

$$\langle v_x^2 \rangle = \langle v_y^2 \rangle = \langle v_z^2 \rangle \tag{9.10}$$

so that

$$\langle v^2 \rangle = 3\langle v_x^2 \rangle \tag{9.11}$$

Substituting this relation in equation 9.5 yields

$$\begin{aligned} PV &= \tfrac{1}{3} Nm\langle v^2 \rangle \\ &= \tfrac{2}{3}(\tfrac{1}{2} Nm\langle v^2 \rangle) \\ &= \tfrac{2}{3} U_{tr} \end{aligned} \tag{9.12}$$

where the translational energy U_{tr} of our idealized gas is the total kinetic energy $\tfrac{1}{2} Nm\langle v^2 \rangle$.

If we apply equation 9.12 to a mole of ideal gas

$$P\bar{V} = \tfrac{2}{3}\bar{U}_{tr} \tag{9.13}$$

Since $P\bar{V} = RT$,

$$\bar{U}_{tr} = \tfrac{3}{2} RT \tag{9.14}$$

Thus we arrive at the conclusion that the kinetic energy of an ideal gas is directly proportional to the thermodynamic temperature. The thermodynamic temperature is a measure of the average kinetic energy of gas molecules. At 300 K the kinetic energy amounts to $\tfrac{3}{2}(1.987 \text{ cal K}^{-1} \text{ mol}^{-1})(300 \text{ K}) = 894 \text{ cal mol}^{-1}$. At 1273 K it amounts to 3794 cal mol^{-1}. It is important to notice that for an ideal gas the translational kinetic energy is independent of the volume or pressure and the molecular weight or type of molecule and depends only on the temperature. Thus a helium atom has the same translational kinetic energy, on the average, as a heavy hydrocarbon molecule at the same temperature.

9.2 DISTRIBUTION OF MOLECULAR VELOCITIES

The molecules of a gas are continually undergoing collisions and changing velocity and direction, but in a sample containing many molecules the velocity distribution is constant. To illustrate what we mean by velocity distribution let us consider the distribution of velocities in the x direction for a sample of gas containing a very large number N of molecules. The distribution of velocity components in the x direction, v_x, at any instant may be described by dividing the span of velocities into equal intervals Δv_x and determining the numbers of molecules whose v_x values fall in the different intervals. It is convenient to plot as the ordinate, not the fraction of molecules in a given velocity interval, but the fraction divided by the width of the interval, as shown in Fig. 9.3a. When this is done the *area* of each rectangle in the bar graph gives the *fraction* of the molecules in the specified interval.

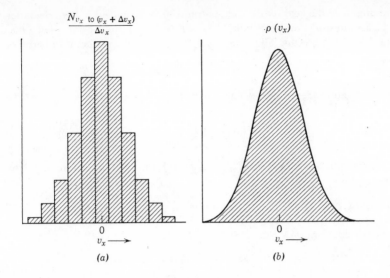

Fig. 9.3 (a) Fraction of the molecules in a certain velocity range divided by the width of the velocity range plotted against average velocity. (b) Continuous distribution curve for the distribution of velocities in the x direction.

The accuracy of this representation can be increased by reducing the interval Δv_x. When this is done the bar graph approaches a smooth curve. In the limit of smaller and smaller intervals the fraction of molecules divided by the velocity interval is represented by $\rho(v_x)$, the *probability density* function for molecular velocities. The *fraction* of molecules having velocities within the range v_x to $v_x + dv_x$ is then $\rho(v_x)\,dv_x$, which is the *probability* that the velocity is in this range. The dv_x appears in the probability $\rho(v_x)\,dv_x$ because the fraction of molecules in the velocity range of width dv_x is proportional to the width of the range dv_x. A plot of the probability density $\rho(v_x)$ versus v_x such as that shown in Fig. 9.3b completely describes the distribution of molecular velocities in a single direction. The fraction of molecules with x component of velocity between any two given values is equal to the area under the curve between these two values of v_x. The mathematical forms of the velocity distributions in the three directions are the same because these are arbitrarily chosen directions, and the gas is isotropic.

In deriving the equation for $\rho(v_x)$ we will use a method first used by Maxwell in 1859. In order to do that we need to consider the probability density $F(v_x, v_y, v_z)$ for velocity components $v_x, v_y,$ and v_z. The quantity $F(v_x, v_y, v_z)$ is defined so that $F(v_x, v_y, v_z)\,dv_x\,dv_y\,dv_z$ is the fraction of the molecules that simultaneously have v_x between v_x and $v_x + dv_x$, v_y between v_y and $v_y + dv_y$, and v_z between v_z and $v_z + dv_z$. Maxwell arrived at the functional form of $F(v_x, v_y, v_z)$ by realizing that the velocity components are *independent* and that the choice of directions of the coordinate axes is *arbitrary*. Since the velocity components in the three directions are independent, the probability density $F(v_x, v_y, v_z)$ is simply the product of the probability densities in the three directions.

$$F(v_x, v_y, v_z) = \rho(v_x)\rho(v_y)\rho(v_z) \tag{9.15}$$

Furthermore, since the velocity distribution must be independent of the orientation of the x, y, and z axes in space, the probability density $F(v_x, v_y, v_z)$ can depend only on the magnitude v of the velocity vector. Thus $F(v_x, v_y, v_z) = g(v)$, and equation 9.15 becomes

$$g(v) = \rho(v_x)\rho(v_y)\rho(v_z) \tag{9.16}$$

Differentiating both sides of this equation with respect to v_x yields

$$\frac{\partial}{\partial v_x}[g(v)]_{v_y, v_z} = \left(\frac{dg(v)}{dv}\right)\left(\frac{\partial v}{\partial v_x}\right)_{v_y, v_z} = \rho(v_y)\rho(v_z)\frac{d\rho(v_x)}{dv_x} \tag{9.17}$$

Differentiating equation 9.1 yields $\partial v/\partial v_x = v_x/v$. Substituting this relation in the preceding equation

$$\frac{dg(v)}{v\, dv} = \rho(v_y)\rho(v_z)\frac{d\rho(v_x)}{v_x\, dv_x} \tag{9.18}$$

Dividing both sides by $g(v)$ yields

$$\frac{dg(v)}{g(v)v\, dv} = \frac{d\rho(v_x)}{\rho(v_x)v_x\, dv_x} \tag{9.19}$$

This process may also be carried out with v_y and v_z so that we obtain

$$\frac{d\rho(v_x)}{\rho(v_x)v_x\, dv_x} = \frac{d\rho(v_y)}{\rho(v_y)v_y\, dv_y} = \frac{d\rho(v_z)}{\rho(v_z)v_z\, dv_z} \tag{9.20}$$

Since a function of v_x alone is equal to a function of v_y alone, which is equal to a function of v_z alone, each must separately equal a constant, which we will represent by $-\lambda$. Thus

$$\frac{d\rho(v_x)}{\rho(v_x)} = -\lambda v_x\, dv_x \tag{9.21}$$

$$d[\ln \rho(v_x)] = -\frac{\lambda}{2}\, d[v_x^2] \tag{9.22}$$

Integration yields

$$\rho(v_x) = A e^{-\lambda v_x^2/2} \tag{9.23}$$

The constant λ is a positive quantity; otherwise $\rho(v_x)$ would increase without limit. The value of the constant A may be determined from the fact that the probability that v_x is between $-\infty$ and $+\infty$ is unity

$$1 = \int_{-\infty}^{+\infty} \rho(v_x)\, dv_x = A \int_{-\infty}^{+\infty} e^{-\lambda v_x^2/2}\, dv_x \tag{9.24}$$

Since

$$\int_{-\infty}^{\infty} e^{-\alpha x^2}\, dx = \left(\frac{\pi}{\alpha}\right)^{\frac{1}{2}} \tag{9.25}$$

$A = (\lambda/2\pi)^{\frac{1}{2}}$, and so the normalized velocity distribution in the x direction is given by

$$\rho(v_x) = \left(\frac{\lambda}{2\pi}\right)^{\frac{1}{2}} e^{-\lambda v_x^2/2} \tag{9.26}$$

In order to determine the value of λ we can make use of the fact that we already have an expression for the translational kinetic energy of an ideal gas; according to equation 9.14, $\bar{U}_{tr} = 3RT/2$. Since this is the translational kinetic energy for motion in three dimensions, the average translational kinetic energy per molecule for motion in the x direction is

$$\epsilon_x = \tfrac{1}{2}m\langle v_x^2 \rangle = \tfrac{1}{2}kT \tag{9.27}$$

where the Boltzmann constant k is equal to the ideal gas constant divided by the Avogadro constant N_A.

$$k = \frac{R}{N_A} = \frac{1.987 \text{ cal K}^{-1} \text{ mol}^{-1}}{6.022 \times 10^{23} \text{ mol}^{-1}}$$

$$= 3.300 \times 10^{-24} \text{ cal K}^{-1}$$

$$= 1.380 \times 10^{-23} \text{ J K}^{-1} \tag{9.28}$$

Thus we may think of the Boltzmann constant as the gas constant per molecule or as the conversion factor between the temperature scale and the molecular energy scale.

The average square of the velocity in the x direction may be calculated using equation 9.26 in

$$\langle v_x^2 \rangle = \int_{-\infty}^{\infty} v_x^2 \rho(v_x) \, dv_x$$

$$= \frac{1}{\lambda} \tag{9.29}$$

since

$$\int_{-\infty}^{\infty} x^2 e^{-\alpha x^2} \, dx = \frac{\pi^{1/2}}{2\alpha^{3/2}} \tag{9.30}$$

Thus $\lambda = m/kT$, and equation 9.26 becomes

$$\rho(v_x) = \left(\frac{m}{2\pi kT}\right)^{1/2} e^{-mv_x^2/2kT} \tag{9.31}*$$

A plot of $\rho(v_x)$ versus v_x has the form shown by Fig. 9.3b. Since the probability that a molecule will have a velocity between v_x and $v_x + dv_x$ is $\rho(v_x) \, dv_x$, it can be seen that the most probable velocity in the x direction is zero. The probabilities of velocities in the positive and

---

* This expression for the velocity distribution in the x direction may also be derived by substituting the expression for the kinetic energy of a molecule in the x direction ($\epsilon_x = \tfrac{1}{2}mv_x^2$) into the Boltzmann equation (Section 17.3)

$$N_i = Ae^{-\epsilon_i/kT}$$

and normalizing to determine the value of A. Another simple example of the application of the Boltzmann equation is the so-called barometric formula for the pressure P in an isothermal atmosphere.

$$P = P_0 e^{-mgh/kT}$$

where P_0 is the pressure at height $h = 0$, and mgh is the potential energy of a single molecule at height h. Here m is the mass of a molecule and g is the acceleration of gravity.

negative directions are, of course equal, the gas as a whole being stationary. The probabilities of very high velocities in either the positive or negative x direction are very small.

The distribution functions for $\rho(v_y)$ and $\rho(v_z)$ have the same functional form as $\rho(v_x)$.

9.3 PROBABILITY OF MOLECULAR SPEEDS

Now that we have calculated the probability densities in the x, y, and z directions we can calculate the fraction of molecules that simultaneously have component velocities in the ranges v_x to $v_x + dv_x$, v_y to $v_y + dv_y$, and v_z to $v_z + dv_z$. Using equations 9.15 and 9.31 this fraction is

$$F(v_x, v_y, v_z)\, dv_x\, dv_y\, dv_z = \left(\frac{m}{2\pi kT}\right)^{3/2} e^{-m(v_x{}^2+v_y{}^2+v_z{}^2)/2kT}\, dv_x\, dv_y\, dv_z \qquad (9.32)$$

since $e^a e^b e^c = e^{a+b+c}$. However, for most purposes we are interested in the fraction of the molecules having a speed in the range v to $v + dv$, independent of the direction. If the origins of the velocity vectors of all the molecules in a sample of gas are brought to the origin of a coordinate system, we are interested in the molecules having speed vectors ending in a spherical shell of radius v around the origin with thickness dv. The volume of this shell is $4\pi v^2\, dv$, and so the fraction of the molecules with velocities v in the range v to $v + dv$ is given by

$$f(v)\, dv = \left(\frac{m}{2\pi kT}\right)^{3/2} e^{-mv^2/2kT} 4\pi v^2\, dv \qquad (9.33)$$

This equation, which was first derived by Maxwell in 1860, is of fundamental importance to the kinetic theory of gases.

A plot of $f(v)$ versus the magnitude of the molecular speed v is shown in Fig. 9.4 for oxygen gas at 100, 300, 500, and 1000 K. The probability that a molecule has its speed between any two values is given by the area under the curve between these two values of the speed. In contrast with Fig. 9.3b for the velocity components in a given direction, it can be seen that the probability of a zero speed is zero. Figure 9.3b is based on equation 9.31, which does not have the term v^2 contained in equation 9.33. Small speeds are favored by the exponential factor in equation 9.33; but high speeds are favored by the v^2 factor. The plot of $f(v)$ versus v is approximately parabolic near the origin. At higher speeds the probability decreases toward zero because the exponential term decreases much more rapidly than v^2 increases. Thus very few molecules have very high or very low speeds. The fraction of the molecules having speeds greater than 10 times the most probable speed is 9×10^{-42} at any temperature. The Avogadro constant times this fraction is a very small number. As the temperature is increased, the most probable speed (which is given by the maximum in the plot) moves to higher speeds and the distribution of speeds also becomes broader, but the relative shape is independent of temperature.

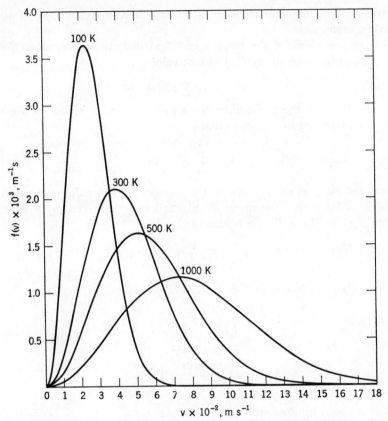

Fig. 9.4 Probability density of various speeds v for oxygen at 100, 300, 500, and 1000 K, calculated using equation 9.33.

9.4 TYPES OF AVERAGE SPEEDS

Since there is a distribution of speeds, various types of averages may be considered. The most probable speed v_p is obtained by setting $df(v)/dv$ equal to zero.

$$\frac{df(v)}{dv} = \left(\frac{m}{2\pi kT}\right)^{3/2} e^{-mv^2/2kT}\left[8\pi v + 4\pi v^2\left(\frac{-mv}{kT}\right)\right] \tag{9.34}$$

This derivative vanishes at $v = v_p$, where

$$v_p = \left(\frac{2kT}{m}\right)^{1/2} = \left(\frac{2RT}{M}\right)^{1/2} \tag{9.35}$$

The arithmetic mean speed $\langle v \rangle$ is obtained by summing all the speeds and dividing by the total number of molecules N.

$$\langle v \rangle = \frac{1}{N}\sum_{i=1}^{N} v_i \tag{9.36}$$

The symbol $\sum v_i$ indicates the sum of the speeds $v_1, v_2, v_3, \ldots, v_N$ of all the N individual molecules.

We may also think of the mean speed $\bar{v}$ as the summation of each possible velocity times the probability $P(v)$ of that velocity.

$$\langle v \rangle = \sum v P(v) \tag{9.37}$$

As discussed earlier, the probability that a molecule has a velocity between v and $v + dv$ is represented by $f(v)\, dv$ so that

$$\langle v \rangle = \int_0^\infty v f(v)\, dv \tag{9.38}$$

We may use an integration because the number of molecules in a macroscopic sample is so large that the speed distribution may be taken to be continuous. Substituting the Maxwell distribution

$$\langle v \rangle = 4\pi \left(\frac{m}{2\pi k T} \right)^{3/2} \int_0^\infty e^{-mv^2/2kT} v^3\, dv \tag{9.39}$$

Substituting $mv^2/2kT = x^2$ and $dv = 2kT x\, dx / mv$ and using

$$\int_0^\infty x^3 e^{-x^2}\, dx = \tfrac{1}{2} \tag{9.40}$$

we obtain

$$\langle v \rangle = \left(\frac{8kT}{\pi m} \right)^{1/2} = \left(\frac{8RT}{\pi M} \right)^{1/2} \tag{9.41}$$

The root-mean-square velocity $\langle v^2 \rangle^{1/2}$ is defined by

$$\langle v^2 \rangle^{1/2} = \left(\frac{1}{N} \sum_{i=1}^N v_i^2 \right)^{1/2} \tag{9.42}$$

The symbol $\sum v_i^2$ indicates the sum of the squares of the velocities, $v_1, v_2, v_3, \ldots, v_N$ of all the N individual molecules. Since the velocity distribution is continuous, the root-mean-square velocity is obtained by multiplying each velocity squared by the probability of that velocity, integrating over all velocities, and taking the square root:

$$\langle v^2 \rangle^{1/2} = \left[\int_0^\infty v^2 f(v)\, dv \right]^{1/2} \tag{9.43}$$

Substituting the Maxwell distribution and using the relation

$$\int_0^\infty x^4 e^{-x^2}\, dx = \tfrac{3}{8}\sqrt{\pi} \tag{9.44}$$

we obtain

$$\langle v^2 \rangle^{1/2} = \left(\frac{3kT}{m} \right)^{1/2} = \left(\frac{3RT}{M} \right)^{1/2} \tag{9.45}$$

It can be seen that at a given temperature each of these average velocities is inversely proportional to the square root of the molecular weight. Lighter molecules move more rapidly so that their average kinetic energies are exactly equal to those for the heavier molecules. The velocity of sound in a gas is of the same order of magnitude as these molecular velocities.*

Example 9.1 Calculate the most probable speed v_p, the arithmetic mean speed $\langle v \rangle$, and the root-mean-square speed $\langle v^2 \rangle^{1/2}$ for hydrogen molecules at $0°$ C.

$$v_p = (2RT/M)^{1/2}$$
$$= [(2)(8.314 \text{ J K}^{-1} \text{ mol}^{-1})(273 \text{ K})/(2.016 \times 10^{-3} \text{ kg mol}^{-1})]^{1/2}$$
$$= 1.50 \times 10^3 \text{ m s}^{-1}$$

$$\langle v \rangle = (8RT/\pi M)^{1/2}$$
$$= [(8)(8.314 \text{ J K}^{-1} \text{ mol}^{-1})(273 \text{ K})/(3.1416)(2.016 \times 10^{-3} \text{ kg mol}^{-1})]^{1/2}$$
$$= 1.69 \times 10^3 \text{ m s}^{-1}$$

$$\langle v^2 \rangle^{1/2} = (3RT/M)^{1/2}$$
$$= [(3)(8.314 \text{ J K}^{-1} \text{ mol}^{-1})(273 \text{ K})/(2.016 \times 10^{-3} \text{ kg mol}^{-1})]^{1/2}$$
$$= 1.84 \times 10^3 \text{ m s}^{-1}$$

The root-mean-square speed of a hydrogen molecule at $0°$ is 4120 mph, but at ordinary pressures it travels only an exceedingly short distance before colliding with another molecule and changing direction.

The distribution of molecular velocities may be determined experimentally using the molecular beam method. A beam of molecules moving in a vacuum is collimated by slits and passed through a velocity selector that allows only those

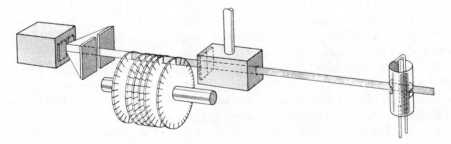

Fig. 9.5 Molecular beam apparatus. Starting at the left the parts are: effusion source, collimating slit, slotted-disk-type velocity selector, scattering chamber and detector.

---

* The speed of the pressure wave depends on the speed of the gas molecules themselves. Since the sound wave causes adiabatic heating it is not surprising that the speed of sound in a gas depends on the heat capacity. A detailed analysis (R. D. Present, *Kinetic Theory of Gases*, McGraw-Hill Book Co., New York, 1958) shows that the speed of sound in a gas is $(C_P RT/C_V M)^{1/2}$.

molecules in a narrow range of velocity to reach the detector. A velocity selector may be constructed from a system of toothed disks on a shaft that is rotated at an accurately controlled velocity with slots appropriately arranged. A typical molecular beam apparatus for studying velocity distributions as well as molecular collision cross sections (Section 9.10) is shown in Fig. 9.5. Using apparatus such as this the Maxwell-Boltzmann velocity distribution has been checked experimentally.

9.5 HEAT CAPACITY AND EQUIPARTITION OF ENERGY

Equation 9.45 for the root-mean-square velocity confirms equation 9.14 that the translational energy of an ideal gas is given by

$$\bar{U}_{tr} = \tfrac{3}{2}RT \tag{9.46}$$

The heat capacity at constant volume for the idealized gas is

$$\bar{C}_V = \left(\frac{\partial \bar{U}}{\partial T}\right)_V = \tfrac{3}{2}R$$

$$= 2.98 \text{ cal K}^{-1} \text{ mol}^{-1} \tag{9.47}$$

and $\bar{C}_P$ is given by equation 1.46 so that

$$\bar{C}_P = \bar{C}_V + R = 4.97 \text{ cal K}^{-1} \text{ mol}^{-1} \tag{9.48}$$

These values are in excellent agreement with the values of $\bar{C}_V$ and $\bar{C}_P$ for the noble gases given in Table 1.2. Polyatomic gases can absorb energy in rotational and vibrational motion, which we need to discuss now.

Equation 9.46 may be given the following interpretation: since three independent coordinates are needed to describe translational motion (e.g., the three Cartesian components of v), we can say that a molecule has three degrees of translational freedom and each contributes $\tfrac{1}{2}RT$ to the translational energy. This is an example of the principle of equipartition. According to this principle, energy is absorbed into each of the degrees of freedom of a molecule.

A molecule containing N nuclei has $3N$ degrees of freedom because $3N$ coordinates are required to locate the nuclei in space. Three coordinates are required to locate the center of mass of the molecule, and so there are $3N-3$ internal degrees of freedom. A diatomic molecule or a linear polyatomic molecule has two rotational degrees of freedom because two angles are required to orient the axis of the molecule with respect to a coordinate system. Thus for a diatomic or linear molecule there are $3N-5$ vibrational degrees of freedom. For a diatomic molecule there is a single vibration, and for a linear triatomic molecule there are four (see Section 15.9). For a nonlinear polyatomic molecule there are three rotational degrees of freedom because three angles are required to orient the molecule with respect to a coordinate system. For such molecules there are $3N-6$ vibrational degrees of freedom.

Each degree of translational or rotational freedom contributes $\frac{1}{2}R$ to $\bar{C}_V$. Thus for a rigid diatomic molecule we would expect

$$\bar{C}_V = \tfrac{5}{2}R = 4.97 \text{ cal K}^{-1} \text{ mol}^{-1}$$
$$\bar{C}_P = \tfrac{7}{2}R = 6.95 \text{ cal K}^{-1} \text{ mol}^{-1}$$

These predictions of elementary kinetic theory are borne out quite well by the experimental values for N_2, H_2, and O_2 in Table 1.2 at 25° C.

According to classical kinetic theory each vibrational degree of freedom contributes R to $\bar{C}_V$. The reason for this larger contribution by vibrational degrees of freedom is that a vibration has both kinetic and potential energy associated with it, and each contributes $\frac{1}{2}R$. Thus according to classical kinetic theory the molar heat capacities of a vibrating diatomic molecule would be expected to be

$$\bar{C}_V = \tfrac{7}{2}R = 6.95 \text{ cal K}^{-1} \text{ mol}^{-1}$$
$$\bar{C}_P = \tfrac{9}{2}R = 8.94 \text{ cal K}^{-1} \text{ mol}^{-1}$$

As shown in Table 1.2, the heat capacity of Cl_2 is not this large, even though its heat capacities are larger than expected for a rigid diatomic molecule. Thus we conclude that at room temperature the principle of equipartition applies to translation and rotation, but not to vibration. This is an example of the failure of classical mechanics when applied to molecules. We will defer the discussion of the resolution of this problem by quantum mechanics to Chapters 12 and 17.

9.6 COLLISIONS WITH A WALL OR OPENING

The number of collisions per unit time per unit area of wall may be derived by considering Fig. 9.6 that depicts the volume containing all the molecules with certain components v_x, v_y, and v_z that will strike area A in time δt. The volume of the cylinder is equal to its height, $v_z \, \delta t$, times the area of the end A. Thus the number of molecules in the cylinder is $nAv_z \, \delta t$, where n is the concentration of molecules. The probability that the molecules will have the required velocity components is given by equation 9.31. The number of collisions is therefore

$$n(Av_z \, \delta t)F(v_x, v_y, v_z) \, dv_x \, dv_y \, dv_z = n(Av_z \, \delta t)\rho(v_x)\rho(v_y)\rho(v_z) \, dv_x \, dv_y \, dv_z \quad (9.49)$$

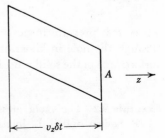

Fig. 9.6 Oblique cylinder containing all the molecules with certain components v_x, v_y, v_z, which will strike area A in time δt.

where the probability densities are given by equations like equation 9.31. We must now integrate over all values of the components of the velocities to obtain the total number of collisions. The number ν of collisions per unit area of wall per unit time is

$$\nu = n \int_0^\infty \int_{-\infty}^{+\infty} \int_{-\infty}^{+\infty} v_z \rho(v_x) \rho(v_y) \rho(v_z) \, dv_x \, dv_y \, dv_z \tag{9.50}$$

The integration for the z component is only for positive velocities since we are only interested in molecules that are moving toward the surface. Since the probability densities in equation 9.31 have been normalized, the integrations over v_x and v_y simply yield unity so that

$$\nu = n \left(\frac{m}{2\pi kT}\right)^{1/2} \int_0^\infty v_z e^{-mv_z^2/kT} \, dv_z$$

$$= n \sqrt{\frac{kT}{2\pi m}}$$

$$= n \sqrt{\frac{RT}{2\pi M}} \tag{9.51}$$

This is also the number of molecules per square meter per second that would escape through a small hole in the vessel containing the gas. We refer to a hole that is small compared with the mean free path (Section 9.8) so that the spatial and velocity distributions of the molecules inside the vessel are not appreciably affected by the effusion loss. If the pressure on the other side of the hole is so low that no molecules return, equation 9.51 gives the flux (molecules $m^{-2} s^{-1}$) out of the hole. The rates of flow of two gases through a small aperture are inversely proportional to the square root of the molecular weight. The proportionality is exact if the cross section of the aperture is small compared with the average distance through which a molecule travels before colliding with another molecule.

Equation 9.51 is the basis for the Knudsen method for measuring the vapor pressures of solids. The mass w of molecules striking an opening per unit area per unit time is $w = \nu M/N_A$. Introducing the ideal gas law in the form $P = nRT/N_A$, where n is the concentration of molecules, and equation 9.51, we obtain

$$P = w \left(\frac{2\pi RT}{M}\right)^{1/2} \tag{9.52}$$

where P is pressure in newtons per square meter, and w is the rate of effusion through the hole in kilograms per square meter per second. A sufficiently large surface area of the solid must be exposed to maintain the saturation vapor pressure.

Example 9.2   The vapor pressure of solid beryllium was measured by R. B. Holden, R. Speiser, and H. L. Johnston [*J. Am. Chem. Soc.*, **70**, 3897 (1948)] using a Knudsen cell. The effusion hole was 0.318 cm in diameter, and they found a weight loss of 9.54 mg in 60.1 min at a temperature of 1457 K. Calculate the vapor pressure.

$$P = w \sqrt{\frac{2\pi RT}{M}}$$

$$= \frac{(9.54 \times 10^{-6} \text{ kg})}{\pi (0.159 \times 10^{-2} \text{ m})^2 (60 \times 60.1 \text{ s})} \sqrt{\frac{2\pi (8.314 \text{ J K}^{-1} \text{ mol}^{-1})(1457 \text{ K})}{(9.013 \times 10^{-3} \text{ kg mol}^{-1})}}$$

$$= 0.968 \text{ N m}^{-2}$$

$$= \frac{0.968 \text{ N m}^{-2}}{101{,}325 \text{ N m}^{-2} \text{ atm}^{-1}}$$

$$= 9.55 \times 10^{-6} \text{ atm}$$

9.7 MOLECULAR COLLISIONS IN A GAS

The preceding results have been obtained without any consideration of collisions between molecules. However, since the mean free path and the transport properties (diffusion, heat conductivity, and viscosity) are determined by molecular collisions, it is necessary to introduce the concept of molecular size. Molecules of a real gas repel each other at very short distances and attract each other at greater distances, so that when they approach they interact in a very complex fashion. For certain purposes, however, a useful first approximation may be obtained by assuming that the molecules are rigid, noninteracting spheres with diameter σ. In deriving an approximate equation for the number of collisions per second it is convenient to assume that the molecules all travel with the same speed, the arithmetic mean velocity $\langle v \rangle$.

If the molecular diameter is σ, two identical molecules will just touch when the distance separating their centers is σ. Thus a moving molecule will collide with other molecules whose centers come within a distance of σ. The quantity $\pi\sigma^2$ is called the *collision cross section* for the rigid spherical molecule because it is the cross-sectional area of an imaginary sphere surrounding the molecule into which the center of another molecule cannot penetrate. A molecule moving with a constant velocity of v will sweep out $\pi\sigma^2\langle v \rangle$ m^3 s^{-1} and strike $\pi\sigma^2\langle v \rangle n$ molecules per second, where n is the concentration of molecules.

In this simple calculation we have assumed that one molecule moves with a velocity $\langle v \rangle$ and that the other molecules are stationary. To look at the calculation this way we need to use the average relative velocity—the average velocity the observer would measure if he were sitting on the molecule being struck. The average relative velocity exceeds the actual average velocity $\langle v \rangle$ by the factor $\sqrt{2}$ for collisions between like molecules, as may be shown by a detailed calculation. Thus the number of collisions per molecule per second is given by

$$z = \sqrt{2}\pi\sigma^2\langle v \rangle n \tag{9.53}$$

Since there are n molecules per unit volume, there will be $\sqrt{2}\pi\sigma^2\langle v\rangle n^2$ molecules per unit volume undergoing collision per unit time. The number Z of collisions per unit time per unit volume is given by

$$Z = \frac{\sqrt{2}}{2}\pi\sigma^2\langle v\rangle n^2 \tag{9.54}$$

since the number of collisions is one-half the number of identical molecules colliding.

Example 9.3   For oxygen at 25° calculate the number z of collisions per second per molecule, the number Z of collisions per cubic meter per second, and the number of moles of collisions per liter per second at a pressure of 1 atm. The molecular diameter of oxygen in 3.61 Å, or 3.61×10^{-10} m, as determined in a manner to be described shortly.

$$n = \frac{N}{V} = \frac{PN_A}{RT} = \frac{(1\text{ atm})(6.022 \times 10^{23}\text{ mol}^{-1})(10^3\text{ liter m}^{-3})}{(0.082051\text{ liter atm K}^{-1}\text{ mol}^{-1})(298\text{ K})}$$

$$= 2.46 \times 10^{25}\text{ molecules m}^{-3}$$

$$\langle v\rangle = (8RT/\pi M)^{1/2}$$

$$= [(8)(8.314\text{ J K}^{-1}\text{ mol}^{-1})(298\text{ K})/\pi(32 \times 10^{-3}\text{ kg mol}^{-1})]^{1/2}$$

$$= 444\text{ m s}^{-1}$$

$$z = \sqrt{2}\pi\sigma^2\langle v\rangle n$$

$$= (1.414)(3.14)(3.61 \times 10^{-10}\text{ m})^2(444\text{ m s}^{-1})(2.46 \times 10^{25}\text{ mol m}^{-3})$$

$$= 6.32 \times 10^9\text{ s}^{-1}$$

which is the number of collisions per molecule per second.

$$Z = \frac{\sqrt{2}}{2}\pi\sigma^2\langle v\rangle n^2$$

$$= (0.707)(3.14)(3.61 \times 10^{-10}\text{ m})^2(444\text{ m s}^{-1})(2.46 \times 10^{25}\text{ m}^{-3})^2$$

$$= 7.77 \times 10^{34}\text{ collisions m}^{-3}\text{ s}^{-1}$$

$$= \frac{(7.77 \times 10^{34}\text{ collisions m}^{-3}\text{ s}^{-1})(10^{-3}\text{ m liter}^{-1})}{6.022 \times 10^{23}\text{ mol}^{-1}}$$

$$= 1.29 \times 10^8\text{ mol liter}^{-1}\text{ s}^{-1}$$

which is the number of moles of collisions per liter per second.

Actual chemical reaction rates are usually far smaller than the collision rates, indicating that the collisions must involve certain energies and orientations to lead to reaction.

9.8   THE MEAN FREE PATH

The mean free path is the average distance traversed by a molecule between collisions. In a second a molecule will on the average traverse $\langle v\rangle$ m and suffer

z collisions. Thus from equation 9.53 the mean free path is

$$l = \frac{\langle v \rangle}{z} = \frac{1}{\sqrt{2}\pi\sigma^2 n} = \frac{kT}{\sqrt{2}\pi\sigma^2 P} \qquad (9.55)$$

where the last form of this equation is obtained by substituting the ideal gas law in the form $n = P/kT$ where n is the concentration of molecules. It is seen that at constant density the mean free path is independent of temperature (assuming σ to be independent of temperature) and at constant pressure the mean free path is directly proportional to the absolute temperature.

At pressures so low that the mean free path becomes comparable with the dimensions of the containing vessel, the flow properties of the gas become markedly different.

Example 9.4 For oxygen at $25°$, calculate the mean free path at (*a*) 1 atm pressure and (*b*) 10^{-3} Torr.

(*a*) From example 9.3, $n = 2.46 \times 10^{25}$ molecules m^{-3}

$$l = [(1.414)(3.14)(3.61 \times 10^{-10} \text{ m})^2(2.46 \times 10^{25} \text{ m}^{-3})]^{-1}$$
$$= 7.02 \times 10^{-8} \text{ m}$$

(*b*)
$$n = (2.46 \times 10^{25} \text{ m}^{-3})(10^{-3}/760)$$
$$= 3.24 \times 10^{19} \text{ molecules m}^{-3}$$

$$l = [(1.414)(3.14)(3.61 \times 10^{-10} \text{ m})^2(3.24 \times 10^{19} \text{ m}^{-3})]^{-1}$$
$$= 0.053 \text{ m} = 5.3 \text{ cm}$$

9.9 MOLECULAR INTERACTIONS

It is a useful first approximation to treat molecules as rigid and perfectly elastic spheres, but more realistic calculations have to take into account the fact that molecules attract each other at longer distances as well as repel each other at shorter distances. The attractive force usually varies inversely with the seventh power of the intermolecular distance (Section 14.15) for large separations. The repulsive force varies inversely with a higher power of the intermolecular distance.

The interaction of two molecules or atoms may be represented by a plot of potential energy versus intermolecular distance as shown in Fig. 9.7. The potential energy V is the negative of the work required to bring one of the molecules from infinity (where the potential energy is defined as zero) to distance r. The force F, which one molecule exerts on the other, is the negative derivative of the potential energy.

$$F = -dV/dr \qquad (9.56)$$

To the right of the minimum dV/dr is positive so that the force is negative; this means that the molecules are attracted toward each other. To the left of the minimum dV/dr is negative so that the force is positive; this means that the molecules repel each other.

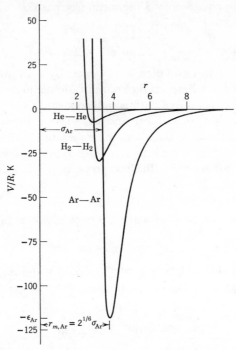

Fig. 9.7   Potential energy expressed as V/R for the interaction between two molecules. The intermolecular distance r is expressed in Angstroms Å $= 10^{-10}$ m.

The following functional form that is often used to represent the intermolecular potential was first introduced by Lennard-Jones.

$$V = 4\epsilon \left[\left(\frac{\sigma}{r} \right)^{12} - \left(\frac{\sigma}{r} \right)^6 \right] \tag{9.57}$$

In this so-called "6–12 equation" for the potential the inverse sixth power dependence comes from the theory for the van der Waals' attraction (Section 14.15), but the twelfth power dependence for the repulsion is chosen empirically. The significance of the parameters ϵ and σ is shown in Fig. 9.7. The depth of the potential well is ϵ. The distance between atoms or molecules at which $V = 0$ is σ.

The values of parameters ϵ and σ may be obtained from scattering experiments and from studies of transport properties (Section 9.11).

Example 9.5   By differentiation of the expression for the Lennard-Jones potential, show that the distance r_m at the minimum where $dV/dr = 0$ is $r_m = 2^{1/6}\sigma$. Substituting this relationship in equation 9.57 show that the Lennard-Jones potential may also conveniently be given by

$$V = \epsilon \left[\left(\frac{r_m}{r} \right)^{12} - 2 \left(\frac{r_m}{r} \right)^6 \right]$$

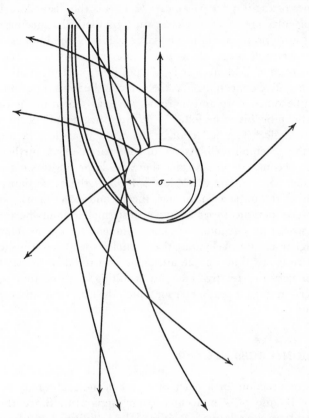

Fig. 9.8 Trajectories for two spherically symmetrical particles with one relative kinetic energy and several values of the impact parameter b. The particles are assumed to interact according to the Lennard-Jones potential, given by equation 9.57. The collisions are represented in relative coordinates. These trajectories are for low velocity collisions. The two molecules are assumed to approach with a relative kinetic energy one tenth as large as the energy parameter ϵ of the Lennard-Jones potential. (From Walter Kauzmann, *Kinetic Theory of Gases*, copyright (c) 1966 by W. A. Benjamin, Menlo Park, California.)

For molecules, in contrast with atoms, the intermolecular forces depend on the relative orientation of the molecules, in addition to the distance from each other.

The effects of intermolecular attractions and repulsions on elastic molecular collisions are illustrated in Fig. 9.8 for two molecules that are spherically symmetrical. In an actual intermolecular collision both molecules would recoil, but it is more convenient to consider molecular collisions in a coordinate system fixed at the center of mass of the colliding particles that allows us to consider the collision as if it were one molecule scattered by a fixed center. This system is used in Fig. 9.8. As shown by the figure molecules may approach each other with different values of the impact parameter b. This is the distance between centers when the molecules are at their closest on hypothetical trajectories that would occur if there were no interaction. For large values of the impact parameter the molecules attract each other along the whole trajectory, and the deflection is negative by definition (although it is not possible experimentally to distinguish positive from negative deflections). As the value of the impact parameter decreases the deflection becomes more and more negative, as shown in the diagram, until the repulsive force begins to be felt. The greatest negative deflection θ_r is known as the rainbow angle.* The deflection at the rainbow angle is greater the greater the depth of the potential well. As the impact parameter is further reduced the repulsive force becomes dominant and there are large positive deflections. For a head-on collision ($b = 0$) the deflection is 180°. If there are many collisions at random values of the impact parameter, theory and experiment show that there is a concentration of trajectories at angles just smaller than the rainbow angle. The rainbow scattering phenomena is less prominent for higher relative velocities because at higher relative velocities the particles spend less time close to each other, and there is less time for the attractive and repulsive forces to produce a deflection. For very low relative energies "orbiting" occurs, that is, θ_r becomes infinite at a certain impact parameter for which the attractive and repulsive forces exactly balance.

9.10 COLLISION CROSS SECTION

The collision cross section for a particular pair of molecules may be determined experimentally by use of a molecular beam apparatus. Since the total cross section Q depends on the relative velocities of the colliding molecules it is necessary to use a velocity selected beam as illustrated in Fig. 9.5.

To define the collision cross section Q consider a beam of A particles directed through a region of uniform concentration of B particles. As a simplification we will assume that the A particles are all moving with the same velocity, and the B particles are at rest. The intensity I_A of the beam of A particles is the number of A particles crossing a unit area perpendicular to the beam per unit time. The decrease dI_A in intensity of the beam of A particles because of collisions with B particles per unit time per unit area perpendicular to the beam is proportional to the intensity I_A of the beam of A particles, the concentration n_B of B particles in the scattering volume, and the length dx along the path of the beam. Neglecting collisions involving three or more particles,

$$dI_A = -Q_{AB}I_A n_B \, dx \tag{9.58}$$

where Q_{AB} is the collision cross section. The bigger the collision cross section the greater the attenuation of the beam of A particles due to collisions scattering

---

* The rainbow angle receives its name from the similarity with the refraction of light by water drops that produce rainbows.

A particles out of the beam. Equation 9.58 may be integrated over the length of the region of uniform density of B particles.

$$\int_{I_A^0}^{I_A} \frac{dI_A}{I_A} = -Q_{AB} n_B \int_0^L dx \qquad (9.59)$$

$$I_A = I_A^0 e^{-Q_{AB} n_B L} \qquad (9.60)$$

This equation is analogous to Beer's law for the absorption of light (Section 15.12).

The classical cross section for molecules interacting according to the Lennard-Jones potential goes to infinity as the relative velocity goes to zero. This happens because the potential does not become zero except for infinite separation of the molecules. Of course, it does not make sense to have an infinite cross section, and the explanation of this unphysical prediction was provided by quantum mechanics. When collision cross sections are calculated quantum mechanically, infinite values are not obtained.

In this section we have assumed that collisions are elastic, but they may also be inelastic (energy is absorbed into internal motions) or chemical reaction may occur.

9.11 TRANSPORT PHENOMENA IN GASES

If a gas is not uniform with respect to composition, temperature, and velocity, transport processes occur until the gas does become uniform. The transport of matter in the absence of bulk flow is referred to as *diffusion*. The transport of heat from regions of high temperature to regions of lower temperature without convection is referred to as *thermal conductivity*, and the transfer of momentum from a region of higher velocity to a region of lower velocity gives rise to the phenomenon of *viscosity*. In each case the rate of flow is proportional to the rate of change of some property with distance, a so-called gradient.

The flux of matter due to diffusion is proportional to the concentration gradient. Thus the flux J_{iz} of component i in the z direction in terms of quantity per unit area per unit time due to its concentration gradient in that direction, dn_i/dz (where n_i is the concentration of molecules i) is given by

$$J_{iz} = -D \frac{dn_i}{dz} \qquad (9.61)$$

The proportionality constant is the diffusion coefficient D. If dn_i/dz has the units of mol m^{-4}, D has the units of m^2 s^{-1}. The negative sign comes from the fact that if n_i increases in the positive z direction, dn_i/dz is positive, but the flux is in the negative z direction because the flow is in the direction of lower concentrations.

The diffusion coefficient for the diffusion of one gas into another may be determined by use of a cell such as that shown schematically in Fig. 9.9a. The heavier gas is placed in chamber A and the lighter in chamber B. The sliding

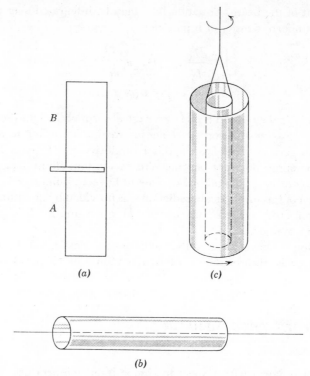

Fig. 9.9   Schematic diagrams of apparatus for the measurements of (a) the diffusion coefficient D, (b) the coefficient of thermal conductivity λ, and (c) the coefficient of viscosity η of gases.

partition is withdrawn for a definite interval of time. From the average composition of one chamber or the other, after a time interval, D may be calculated.

The transport of heat is referred to as thermal conductivity and is due to a gradient in temperature. Thus the flux of energy q_z in the z direction due to the temperature gradient in that direction is given by

$$q_z = -\lambda \frac{dT}{dz} \qquad (9.62)$$

where the proportionality constant λ is the coefficient of thermal conductivity. When q_z has the units of cal m^{-2} s^{-1} and dT/dz has the units of K m^{-1}, λ has the units of cal m^{-1} s^{-1} K^{-1}. The negative sign in equation 9.62 indicates that if dT/dz is positive, the flow of heat is in the negative z direction, which is the direction of lower temperature.

The determination of the coefficient of thermal conductivity by the hot-wire method is illustrated schematically in Fig. 9.9b. The outer cylinder is kept at a constant temperature by a thermostatically controlled bath. The tube is filled with the gas under investigation, and the fine wire at the axis of the tube is heated electrically. When a steady state is achieved, the temperature of the wire is measured by determining its electrical resistance. The coefficient of thermal conductivity is calculated from the temperature of wire and wall, the heat dissipation, and the dimensions of the apparatus.

Thermal diffusion is the flux of material due to a temperature gradient of dT/dz. The fact that the thermal diffusion coefficients depend on mass makes it possible to separate isotopes by use of this effect.

9.12 COEFFICIENT OF VISCOSITY

Viscosity is a measure of the resistance that a fluid offers to an applied shearing force. Consider what happens to the fluid between parallel planes, illustrated in Fig. 9.10, when one of these planes is moved in the y direction at a constant speed relative to the other while maintaining a constant distance between the planes (coordinate z). The planes are considered to be very large so that edge effects may be ignored. The layer of fluid immediately adjacent to the moving plane moves with the velocity of this plane. The layer next to the stationary plane is stationary; in between the velocity usually changes linearly with distance, as shown. The velocity gradient, that is, the rate of change of velocity with respect to distance measured *perpendicular* to the direction of flow, is represented by dv_y/dz. The coefficient of viscosity η is defined by the equation

$$F = -\eta \frac{dv_y}{dz} \qquad (9.63)$$

Here F is the force per unit area required to move one plane relative to the other. The negative sign comes from the fact that, if F is in the $+y$ direction, the velocity v_y decreases in successive layers away from the moving plane and dv_y/dz is negative. If F has the units of kg m s^{-2}/m^2 and dv_y/dz has the units of m s^{-1}/m, the coefficient of viscosity η has the units of kg m^{-1} s^{-1}. The SI unit of viscosity is the Pascal second. The Pascal (abbreviated Pa) is the name of the SI unit of pressure that is one newton per square meter. Since 1 N = 1 kg m s^{-2}, 1 Pa s = 1 kg m^{-1} s^{-1}. A

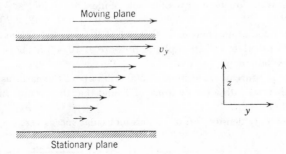

Fig. 9.10 Velocity gradient in a fluid due to a shearing action.

fluid has a viscosity of 1 Pa s if a force of 1 N is required to move a plane of 1 m²
at a velocity of 1 m s⁻¹ with respect to a plane surface a meter away and parallel
with it.*

Although the coefficient of viscosity is conveniently defined in terms of this
hypothetical experiment, it is easier to measure it by determining the rate of flow
through a tube, the torque on a disk that is rotated in the fluid, or other experimental arrangement. An experimental arrangement is illustrated in Fig. 9.9c.
The outer cylinder is rotated at a constant velocity by an electric motor. The
inner coaxial cylinder is suspended on a torsion wire. A torque is transmitted to
the inner cylinder by the fluid, and this torque is calculated from the angular
twist of the torsion wire.

References

R. B. Bird, W. E. Stewart, and E. N. Lightfoot, *Transport Phenomena*, Wiley, New York,
1960.
S. Chapman and T. G. Cowling, *The Mathematical Theory of Non-Uniform Gases*, 3rd ed.,
Cambridge University Press, 1970.
H. J. M. Hanley, *Transport Phenomena in Fluids*, Marcel Dekker, New York, 1969.
J. O. Hirschfelder, C. F. Curtiss, and R. B. Bird, *The Molecular Theory of Gases and Liquids*,
Wiley, New York, 1954.
J. H. Jeans, *Introduction to the Kinetic Theory of Gases*, Cambridge University Press, 1940.
W. Jost, *Diffusion in Solids, Liquids, Gases*, Academic Press, New York, 1960.
W. Kauzmann, *Kinetic Theory of Gases*, W. A. Benjamin Inc., New York, 1966.
E. H. Kennard, *Kinetic Theory of Gases*, McGraw-Hill Book Co., New York, 1938.
R. D. Present, *Kinetic Theory of Gases*, McGraw-Hill Book Co., New York, 1958.

Problems

9.1 If the diameter of a gas molecule is 4 Å, and each is imagined to be in a separate cube,
what is the length of the side of the cube in molecular diameters at 0° and pressures of (a)
1 atm and (b) 10⁻³ Torr.

> *Ans.* (a) 8.3 molecular diameters, (b) 760 molecular diameters.

9.2 At 0° C what is the kinetic energy of a mole of ideal gas in joules? What are C_V and
C_P in J K⁻¹ mol⁻¹?   *Ans.* 3405 J mol⁻¹. 12.5, 20.8 J K⁻¹ mol⁻¹.

9.3 Assuming that the atmosphere is isothermal at 0° and that the average molecular
weight of air is 29, calculate the atmospheric pressure at 20,000 ft above sea level, using
the Boltzmann distribution.   *Ans.* 0.466 atm.

9.4 Calculate the pressure at an altitude of 500 miles above the earth assuming that the
atmosphere is isothermal with a temperature of 0° and that the average molecular weight
is 29, independent of height.   *Ans.* 1.8×10^{-44} atm.

9.5 Plot the probability density $f(v)$ of various molecular speeds versus speed for oxygen
at 25°.

---

* The cgs unit of viscosity is the poise that is 1 g s⁻¹ cm⁻¹. 0.1 Pa s = 1 poise.

9.6 Calculate the average speed and the root-mean-square speed for the following set of molecules: 10 molecules moving at 5×10^2 m s^{-1}, 20 molecules moving 10×10^2 m s^{-1}, and 5 molecules moving 15×10^2 m s^{-1}.   *Ans.* 9.28×10^2, 9.82×10^2 m s^{-1}.

9.7 Calculate the ratio of the root-mean-square speed to the arithmetic mean velocity. to the most probable speed.   *Ans.* $1.225:1.128:1.000$.

9.8 Calculate the number of collisions per square centimeter per second of oxygen molecules with a wall at a pressure of 1 atm and $25°$.   *Ans.* 2.73×10^{23}.

9.9 A Knudsen cell containing crystalline benzoic acid ($M = 122$) is carefully weighed and placed in an evacuated chamber thermostated at $70°$ for 1 hr. The circular hole through which effusion occurs is 0.60 mm in diameter. Calculate the sublimation pressure of benzoic acid at $70°$ in Torr from the fact that the weight loss is 56.7 mg.

Ans. 0.16 Torr.

9.10 A 6:1 mixture by volume of neon and argon is allowed to effuse through a small orifice into an evacuated space. What is the composition of the mixture that first passes through?   *Ans.* 89.4% Ne, 10.6% Ar.

9.11 Large vacuum chambers have been built for testing space vehicles at 10^{-8} Torr. Calculate (*a*) the mean-free path of nitrogen at this pressure and (*b*) the number of molecular impacts per m^2 of wall per second at $25°$. $\sigma_{N_2} = 3.75$ Å.

Ans. (*a*) 4.94×10^3 m, (*b*) 3.84×10^{16} m^{-2} s^{-1}.

9.12 Oxygen is contained in a vessel at 2 Torr pressure and $25°$. Calculate (*a*) the number of collisions between molecules per second per cubic meter, (*b*) the mean free path. $\sigma_{O_2} = 3.61$ Å.   *Ans.* (*a*) 5.43×10^{29} s^{-1} m^{-3}, (*b*) 2.66×10^{-5} m.

9.13 The Lennard-Jones parameters for nitrogen are $\epsilon/k = 95.1$ K and $\sigma = 3.70$ Å. Plot the potential energy (expressed as U/R in K) for the interaction of two molecules of nitrogen.

9.14 For Ne the parameters of the Lennard-Jones 6-12 potential are $\epsilon/k = 35.6$ K and $\sigma = 2.75$ Å. Plot V in kcal mol^{-1} versus r and calculate the distance r_m where $dV/dr = 0$.

9.15 Calculate the mean free path of nitrogen in angstrom units at 1 atm and $25°$.

Ans. 650 Å.

9.16 Consider an atomic beam of potassium passing through a scattering gas of Ar contained in a cell of 1 cm length at $0°$. Assuming an effective total cross section Q of 600 Å^2 for the system potassium-argon, calculate the pressure of argon in Torr required to produce an attenuation of the beam of 25%.   *Ans.* 1.4×10^{-4} Torr.

9.17 Calculate the kinetic energy of a mole of an ideal gas at $500°$ in (*a*) J and (*b*) calories.

9.18 At what temperature is the translational kinetic energy of ideal gas molecules 5 kcal mol^{-1}?

9.19 At sea level and $25°$ what is the change in barometric pressure in (*a*) centimeters of mercury and (*b*) inches of mercury per 1000-ft change in altitude?

9.20 Consider two types of particles that may be assumed to be rigid spheres: type A with radius r_A and type B with radius r_B. Picturing the B particles motionless and bombarded with A particles, show that the collision cross-section Q_{AB} as defined in equation 9.58 is $\pi(r_A + r_B)^2$.

9.21 Calculate the root-mean-square, arithmetic mean, and most probable speeds for oxygen molecules at $25°$.

9.22 Suppose that a gas contains 10 molecules having an instantaneous speed of $2 \times 10_2$ m s^{-1}, 30 molecules with a speed of 4×10^2 m s^{-1}, and 15 molecules with a speed of 6×10^2 m s^{-1}. Calculate $\bar{v}$, v_p, and $\langle v^2 \rangle^{1/2}$.

9.23 What is the ratio of the number of molecules having twice the most probable speed to the number having the most probable speed?

9.24　Calculate the root-mean-square speed of smoke particles of mass 10^{-13} gram in their Brownian motion in air at $25°$.

9.25　The vapor pressure of naphthalene ($M = 128.16$) is 0.133 Torr at $30°$. Calculate the weight loss in a period of 2 hr of a Knudsen cell filled with naphthalene and having a round hole 0.50 mm in diameter.

9.26　Calculate the root-mean-square speed of (a) carbon tetrachloride molecules at $100°$; (b) ammonia molecules at $100°$. (c) How many times longer will it take for a millimole of CCl_4 vapor to effuse out of a small opening than a millimole of NH_3? (d) How many times longer will it take for a milligram of NH_3 to effuse out of a small opening than a milligram of CCl_4 vapor?

9.27　(a) Calculate the number of collisions per second undergone by a single nitrogen molecule in nitrogen at 1 atm pressure and $25°$. (b) What is the number of collisions per cubic centimeter per second? What is the effect on the number of collisions (c) of doubling the absolute temperature at constant pressure and (d) of doubling the pressure at constant temperature?

9.28　The surface of a metal has 10^{15} atoms cm^{-2} that can react with oxygen. Assuming that one oxygen atom reacts with each metal atom and that all oxygen molecules stick, calculate the time for sufficient oxygen to be transferred to the surface by collisions of oxygen molecules if the partial pressure of oxygen is 10^{-9} Torr and the temperature is $25°$ C.

9.29　(a) At a pressure of 10^{-10} Torr, how many molecules are there per cubic centimeter at $0°$? (b) What is the mean free path for oxygen molecules at this pressure? $\sigma_{O_2} = 3.61$ Å.

9.30　The Lennard-Jones parameters for argon are $\epsilon/k = 122$ K and $\sigma = 3.40$ Å. Plot the potential energy (expressed as U/R in K) for the interaction of two molecules of argon.

9.31　A molecular beam of CsCl is passed through a 1-cm path of *cis*-dichloroethylene gas at $0°$ and a pressure of 2.9×10^{-5} Torr, and the attenuation of the CsCl beam is found to be 25%. In another experiment with *trans*-dichloroethylene as the scattering gas at the same temperature and pressure an attenuation of 13.4% is observed. Calculate the effective total cross-sections for the collision of CsCl with the (a) *cis* and (b) *trans* compounds. (c) Why would you expect the interaction with the *cis* compound to be greater? [H. Schumacher, R. B. Bernstein, and E. W. Rothe, *J. Chem. Phys.*, **33**, 584 (1960).]

9.32　Calculate the root-mean-square speed of oxygen molecules having a kinetic energy of 2 kcal mol^{-1}. At what temperature would this be the root-mean-square velocity?

9.33　What is the mean atmospheric pressure in Denver, Colorado, which is a mile high, assuming an isothermal atmosphere at $25°$ C. Air may be taken as 20% O_2 and 80% N_2.

9.34　Calculate the root-mean-square, arithmetic mean, and most probable speeds for butane molecules at $100°$.

9.35　Calculate the velocity of sound in air at $25°$ C. (See Section 9.4.)

9.36　Calculate the velocity of sound in (a) He and (b) N_2 at $25°$ C.

9.37　Calculate the probability densities of argon molecules with kinetic energies of $kT/2$, kT, and $2kT$, at 1000 K.

9.38　The vapor pressure of water at $25°$ is 23.7 Torr. (a) If every water molecule which strikes the surface of liquid water sticks, what is the rate of evaporation of molecules from a square centimeter of surface. (b) Using this result find the rate of evaporation in g cm^{-2} min^{-1} of water into perfectly dry air.

9.39　R. B. Holden, R. Speiser, and H. L. Johnston [*J. Am. Chem. Soc.*, **70**, 3897 (1948)] found the rate of loss of weight of a Knudsen effusion cell containing finely divided beryllium to be 19.8×10^{-7} g cm^{-2} s^{-1} at 1320 K and 1210×10^{-7} g cm^{-2} s^{-1} at 1537 K. Calculate ΔH_{sub} for this temperature range.

9.40   It takes 30 min for the pressure of a certain evacuated vessel of 1-liter volume to rise from 0.001 to 0.003 Torr by leakage of air through a pinhole in the glass. How long will it take for chlorine at the same temperature and external pressure to leak in and raise the pressure from 0.001 to 0.003 Torr? The average molecular weight of air is 28.8.

9.41   Calculate the number of acetaldehyde (CH_3CHO) molecules colliding per milliliter per second at 800 K and 760 Torr. The molecular diameter may be taken as 5 Å.

9.42   The pressure in interplanetary space is estimated to be of the order of 10^{-16} Torr. Calculate (*a*) the average number of molecules per cubic centimeter, (*b*) the number of collisions per second per molecule, and (*c*) the mean free path in miles. Assume that only hydrogen atoms are present and that the temperature is 1000 K.

9.43   (*a*) Calculate the mean free path for hydrogen gas ($\sigma = 2.47$ Å) at 1 atm and 10^{-3} Torr at 25°. (*b*) Repeat the calculation for chlorine gas ($\sigma = 4.96$ Å).

9.44   For methane the parameters for the Lennard-Jones 6-12 potential are $\epsilon/k = 14.82$ K and $\sigma = 3.82$ Å. Plot V in kcal mol^{-1} versus r and calculate r_m.

9.45   The collision diameter for helium is 2.07 Å and for methane is 4.14 Å. How does the mean free path for He compare with that for CH_4 under the same conditions?

9.46   The effective total cross section for collisions of Cs atoms with CH_4 molecules at 0° is $Q = 700$ Å^2. Calculate the pressure (in Torr) of methane which would attenuate a Cs atomic beam 10% over a path of 2 cm.

9.47   (*a*) Show that the fraction of ideal gas molecules of mass m at a temperature T having more than l times the most probable speed is independent of m and T. (*b*) What fraction of ideal gas molecules travel at less than one-half the most probable speed? (*Hint:* the integral is not that hard to do if you remember the Taylor expansion of e^x.)

CHAPTER 10

CHEMICAL KINETICS

Chemical kinetics is concerned with the rates of reactions and the mechanisms by which reactions occur. A mechanism is a detailed description of the individual steps of the reaction. To learn about the mechanism of a reaction, the changes in rate due to variations of concentrations of reactants, products, catalysts, and inhibitors are studied. Important information may also be obtained from studies of the effect on rate of changing the temperature, solvent, electrolyte concentration or isotopic composition. Since a mechanism is a hypothesis to explain experimental facts it is not necessarily unique, even if it accounts for all of the facts; however, some mechanisms may be definitely excluded using kinetic data.

If two substances are mixed together, there may be many different products which are all possible according to thermodynamics, and the relative rates of competing reactions are often more important considerations than equilibria in determining the final products. By altering the concentrations and the temperature and by using specific catalysts, it is possible to change the relative amounts of the various products. Only if one of the reactions goes much faster than all the others can a high yield of a single product be obtained.

10.1 REACTION RATE

In terms of the symbols introduced in Section 1.18 a general chemical reaction

$$\nu_1 A_1 + \nu_2 A_2 = \nu_3 A_3 + \nu_4 A_4 \tag{10.1}$$

may be represented by

$$0 = \sum \nu_i A_i \tag{10.2}$$

by giving the stoichiometric coefficients of the products positive signs and the stoichiometric coefficients of the reactants negative signs. The extent of reaction $\xi(xi)$ is defined by $n_i = n_{i0} + \nu_i \xi$, where n_{i0} is the initial number of moles of a reactant or product and n_i is the number of moles at some later time; thus $dn_i = \nu_i \, d\xi$. The reaction rate is defined as the rate of change of the extent of reaction.

$$\frac{d\xi}{dt} = \frac{1}{\nu_i} \frac{dn_i}{dt} \tag{10.3}$$

If there are intermediates these various rates may not be equal; however, if the

300

intermediates are present only at very low concentrations the various rates will be equal. The reaction rate defined in this way is independent of the choice of reactant and may be used even when there are volume changes during the reaction or two or more phases are present.

If the volume is constant during the reaction the reaction rate may be expressed in terms of the rate of change of the concentration of one of the reactants or products

$$\frac{d\xi}{dt} = \frac{V}{\nu_i}\frac{d(A_i)}{dt} \tag{10.4}$$

where (A_i) is the concentration of A_i and V is the volume in which the reaction is occurring.

In practice reaction rates are frequently expressed in terms of rate of change of concentration, or pressure, of one of the reactants or products. In this chapter we will refer to $d(A_i)/dt$ as a reaction rate with the realization that this is not satisfactory if there are intermediates at appreciable concentrations or a volume change in the reaction.

If the stoichiometric coefficients are not all unity it is obviously very important to specify the concentration of a particular reactant or product. For example for the reaction

$$A + 2B = AB_2$$

$$-\frac{d(A)}{dt} = \frac{1}{2}\left\{-\frac{d(B)}{dt}\right\} = \frac{d(AB_2)}{dt} \tag{10.5}$$

10.2   EXPERIMENTAL METHODS FOR REACTION RATES

Different experimental methods are used depending on the reaction rate. If the reaction is sufficiently slow the reactants may simply be mixed, and the concentration of one of the reactants or products is followed using a spectroscopic or other physical method. The advantage of physical methods of analysis over chemical methods for this purpose is that the reaction does not have to be stopped. In some cases it is possible to stop a reaction by sudden cooling or by adding a reagent that rapidly consumes a catalyst or a reactant.

If a reaction occurs so rapidly that there are significant concentration changes during the process of mixing the reactants, a flow method may be used. An early example was the study of the reaction of hemoglobin and oxygen.[*] A hemoglobin solution was forced into one arm of a Y-mixer and a solution of oxygen in water into the other. In this way it is possible to mix liquids or gases in about 10^{-3} s. In the stopped-flow method, reagents are forced into the mixer, and then the flow is brought to a sudden stop, and observations are made of the extent of reaction. In the continuous flow method the solutions are mixed and flow down

---

[*] H. Hartridge and F. J. W. Roughton, *Proc. Roy. Soc.*, **A, 104**, 395 (1923).

the tube at a steady rate; the extent of reaction is constant at any given distance down the tube but increases with distance from the mixing chamber.

Some reactions occur in much less than 10^{-3} s, and so their kinetics may not be studied by mixing methods. The time range has been extended down to about 10^{-9} s by the use of relaxation methods developed by M. Eigen and coworkers[*] in Göttingen, Germany. A solution in equilibrium is perturbed by rapidly changing one of the independent variables (usually temperature or pressure) on which the equilibrium depends. The change of the system to the new equilibrium is then followed by use of a rapidly responding physical method; for example, light absorption or electrical conductivity.

Equilibria may be shifted by changing the temperature (if $\Delta H \neq 0$), by changing the pressure (if $\Delta V \neq 0$), or by dilution (if $\Delta v \neq 0$). A solution may be heated in a microsecond (10^{-6} s) by discharging a large electrical capacitor through a special conductivity cell containing the sample, or the pressure may be reduced suddenly by allowing high-pressure gas to escape through a rupture disk. Figure 10.1 is a schematic diagram of a temperature-jump apparatus in which an increase in temperature in a small volume of solution is produced by passing a large current for about 1 μ s. If there is a single reaction the return to equilibrium at the new higher temperature is represented by

$$\Delta C = \Delta C_0 e^{-t/\tau} \qquad (10.6)$$

where τ is the relaxation time and ΔC_0 is the displacement of the concentration of one of the reactants from its equilibrium value at $t = 0$. If several reactions are involved in the return to equilibrium, ΔC is expressed by a sum of exponential terms with different relaxation times. The relaxation time τ is the time required for ΔC to drop to $1/e$ of its initial value ΔC_0.

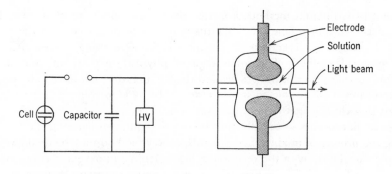

Fig. 10.1 Schematic diagram of a temperature-jump apparatus.

[*] M. Eigen and L. DeMaeyer, in A. Weissberger, *Technique of Organic Chemistry*, Vol. VIII, Part II, Chapter XVIII, Wiley-Interscience, New York, 1963.

I. Amdur and G. G. Hammes, *Chemical Kinetics*, McGraw-Hill Book Co., New York, 1966.

G. H. Czerlinski, *Chemical Relaxation*, M. Dekker, Inc., New York, 1966.

D. N. Hague, *Fast Reactions*, Wiley-Interscience, New York, 1971.

In another form of relaxation method an independent parameter is varied in a sinusoidal manner, and the phase lag in the response of the chemical system is measured. In a sound wave both the pressure and temperature vary, and if the frequency is in the neighborhood of $1/\tau$ there will be enhanced sound absorption.

The line width in various types of spectroscopic experiments gives information about the rates of molecular processes. For example nuclear magnetic resonance (NMR) spectroscopy may be used to determine the rate of a chemical reaction. Electron spin resonance (ESR) spectroscopy may also be used if the chemical environment of an unpaired electron is affected by a reaction.

Photochemical reactions may be initiated rapidly by a flash lamp or by a light pulse from a laser.

To study certain gas reactions at high temperatures it is necessary to heat the gas to the higher temperature very quickly because the reaction occurs rapidly. This may be accomplished by means of a shock tube* in which a shock wave is used to heat the gas suddenly. A tube is divided into two sections separated by a diaphragm that can be ruptured. The gas to be studied is placed on one side of the diaphragm, and a driver gas at a higher pressure on the other. When the diaphragm is ruptured, a shock wave passes through the reacting gas, heating it suddenly to a higher temperature. In some reactions the extent of reaction may be determined as a function of time after passage of the shock wave by measuring the absorption of a beam of light passing perpendicularly across the tube.

10.3   RATE LAW

A rate law gives the dependence of the rate upon concentrations of reactants, products, catalysts and inhibitors. For many reactions the rate is proportional to integer powers of the concentrations of reactants if the volume of the reaction mixture is constant and the concentrations of intermediates are negligible.

$$\frac{d(A_i)}{dt} = k(A_1)^{n_1}(A_2)^{n_2} \cdots \tag{10.7}$$

If $n_1 = 1$, the reaction is said to be first order in A_1, and if $n_1 = 2$ the reaction is said to be second order in A_1. For a rate law of this simple form the overall reaction order is the sum of the exponents $\sum n_i$. The proportionality factor k is referred to as the rate constant and, according to equation 10.7, it has the units $C^{1-\sum n_i}$ time^{-1}. If the reaction is first order, k is usually given in s^{-1} or min^{-1}. If the reaction is second order overall, k is usually given in M^{-1} s^{-1}.

It is important to note that the exponents n_i in the rate law are not the stoichiometric coefficients ν_i of the balanced chemical equation, but must be determined in rate experiments.

Not all rate laws have the simple form of equation 10.7; they may be a more complicated function of the concentrations of reactants, products, catalysts, and inhibitors. If a reaction can go by two paths, for example a catalyzed path and an

---

* D. Britton, N. Davidson and G. Schott, *Discussions Faraday Soc.*, **17,** 58 (1954).

uncatalyzed path, the rate law will consist of two terms, one for each path (see Section 10.25). A complete rate law contains the equilibrium expression for the reaction; it contains positive and negative terms and when the rate is set equal to zero, the equilibrium expression is obtained. However, many reactions go so far toward completion that only the rate law of the forward reaction is obtained. Before discussing the relation between rate law and equilibrium further we need to discuss the commonly occurring rate laws and the determination of the rate constant k.

10.4   FIRST-ORDER REACTIONS

The rate equation for a first-order reaction

$$-\frac{d(A)}{dt} = k(A) \tag{10.8}$$

may be integrated after it is written in the form

$$-\frac{d(A)}{(A)} = k\,dt \tag{10.9}$$

If the concentration of A is $(A)_1$ at t_1 and $(A)_2$ at t_2,

$$-\int_{(A)_1}^{(A)_2} \frac{d(A)}{(A)} = k\int_{t_1}^{t_2} dt \tag{10.10}$$

$$\ln \frac{(A)_1}{(A)_2} = k(t_2 - t_1) \tag{10.11}$$

An especially useful form of this equation is obtained if t_1 is taken to be zero, and the initial concentration is represented by $(A)_0$.

$$\ln \frac{(A)_0}{(A)} = kt \tag{10.12}$$

or

$$(A) = (A)_0 e^{-kt} \tag{10.13}$$

or

$$\log (A) = \frac{-kt}{2.303} + \log (A)_0 \tag{10.14}$$

The last form indicates that the rate constant k may be calculated from a plot of $\log (A)$ versus t; the slope of such a plot is $-k/2.303$.

The calculation of first-order rate constants may be illustrated for the decomposition of nitrogen pentoxide.* Nitrogen pentoxide decomposes completely

---

* F. Daniels and E. H. Johnston, *J. Am. Chem. Soc.*, **43**, 53 (1921); H. Eyring and F. Daniels, *J. Am. Chem. Soc.*, **52**, 1472 (1930).

in the gas phase, or when dissolved in inert solvents, at a rate which is conveniently measured at room temperature. The reaction is strictly first order (except at very low pressures), and the end products are oxygen and a mixture of nitrogen tetroxide and nitrogen dioxide. The following equation represents the overall reaction:

$$N_2O_5 \rightarrow N_2O_4 + \tfrac{1}{2}O_2$$
$$\Updownarrow$$
$$2NO_2 \qquad\qquad (10.15)$$

For every molecule of oxygen produced, two molecules of nitrogen pentoxide have decomposed. It will be shown later (Section 10.12), however, that this reaction is much more complicated, having several intermediate steps. It may be suggested that in chemical kinetics the only simple reactions are the ones that have not been thoroughly studied.

When a solution of nitrogen pentoxide in carbon tetrachloride decomposes, the nitrogen tetroxide and nitrogen dioxide remain in solution while the oxygen escapes and may be measured in a gas buret. The reaction vessel is carefully thermostated, and it is agitated to prevent supersaturation of the oxygen.

Experimental data for the decomposition of nitrogen pentoxide dissolved in carbon tetrachloride at 45° are plotted in Fig. 10.2.

In Fig. 10.2a the concentration of nitrogen pentoxide calculated from the volume of oxygen evolved is plotted against time, and it is seen that the concentration decreases with time, rapidly at first, then more slowly, and finally approaches zero asymptotically.

In Fig. 10.2b the straight line produced by plotting the logarithm of the concentration against time shows that the reaction is first order and follows strictly the relation given by equation 10.14.

The value of the first-order rate constant, $6.22 \times 10^{-4}\,s^{-1}$, is obtained by multiplying the slope of the line in Fig. 10.2b by -2.303 as required by equation 10.14.

It is evident from equations 10.11 and 10.12 that to determine the rate constant for a first-order reaction it is only necessary to determine the *ratio* of the concentrations at two times. Quantities proportional to concentration may be substituted for concentrations in these equations, since the proportionality constants cancel.

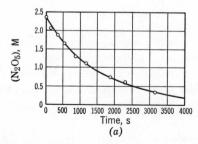 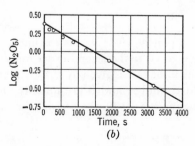

Fig. 10.2 First-order decomposition of N_2O_5.

For example, suppose that an ester is hydrolyzed by an acid and the reaction is followed by titrating aliquots of equal size with sodium hydroxide solution. The difference $V_\infty - V$, where V is the volume of sodium hydroxide required at any time and V_∞ is the volume required when the reaction has gone to completion, is proportional to the concentration of ester *remaining unhydrolyzed* at any time. Thus the first-order rate constant could be calculated from

$$k = \frac{2.303}{t_2 - t_1} \log \frac{V_\infty - V_1}{V_\infty - V_2} \tag{10.16}$$

The half-life $t_{1/2}$ of a reaction is the time required for half of the reactant to disappear. For a first-order reaction the half life is independent of the initial concentration. Thus 50% of the substance remains after one half-life, 25% remains after two half-lives, 12.5% after three, etc. The relation between the half-life and the rate constant is obtained from equation 10.12.

$$k = \frac{2.303}{t_{1/2}} \log \frac{1}{\frac{1}{2}} = \frac{0.693}{t_{1/2}} \tag{10.17}$$

The rate of a first-order reaction may also be expressed by giving the relaxation time τ defined in equation 10.6. The half time $t_{1/2}$ and relaxation time τ are related by

$$\tau = \frac{1}{k} = \frac{t_{1/2}}{0.693} \tag{10.18}$$

10.5   SECOND-ORDER REACTIONS

The simplest example of a second-order reaction is one in which the rate is proportional to the square of the concentration of a reactant. The rate law may be integrated after arranging it in the form

$$-\frac{d(A)}{(A)^2} = k \, dt \tag{10.19}$$

If the concentration is $(A)_0$ at $t = 0$ and (A) at time t, integration yields

$$kt = \frac{1}{(A)} - \frac{1}{(A)_0} \tag{10.20}$$

Thus a plot of $1/(A)$ versus t is linear for such a second-order reaction, and the slope is equal to the second-order rate constant. This integrated rate equation also applies if the rate is given by $k(A)(B)$, the stoichiometry is represented by $A + B = $ products, and A and B are initially at the same concentration. As may be seen from equation 10.20, the half-life for such a second-order reaction is given by

$$t_{1/2} = \frac{1}{k(A)_0} \tag{10.21}$$

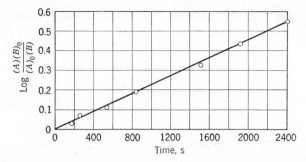

Fig. 10.3 A linear plot for a second-order reaction, the hydrolysis of ethyl acetate at 25°.

If the rate is given by $k(A)(B)$ but the reactants are not present initially at equal concentrations and the stoichiometry is given by

$$aA + bB \rightarrow \text{products}$$

then the integrated rate equation is

$$kt = \frac{a}{[b(A)_0 - a(B)_0]} \ln \frac{(A)(B)_0}{(A)_0(B)} \tag{10.22}$$

If the stoichiometric coefficients of A and B are unity, equation 10.22 may be written

$$\log \frac{(A)(B)_0}{(A)_0(B)} = \frac{[(A)_0 - (B)_0]kt}{2.303} \tag{10.23}$$

The saponification of an ester in alkaline solution is an illustration of a second-order reaction.

$$CH_3COOC_2H_5 + OH^- = CH_3COO^- + C_2H_5OH$$

Solutions of ester and alkali are placed in separate flasks in a thermostat and then mixed. At frequent intervals, a portion of the reaction mixture is removed, discharged into a measured volume of standard acid, and back-titrated with standard alkali. The decrease in concentration of OH^- ions is a measure of the extent of the reaction. A final titration, after the reaction is completed, is necessary to determine the number of equivalents of hydroxide ion left after the reaction is complete and thus by difference to obtain the number of moles per liter of ethyl acetate that were present at the beginning of the experiment. The reaction may be followed also by measuring the change in the electrical conductivity of the system or by measuring the slight increase in volume of the solution as the reaction proceeds.

Experimental data for the saponification of 0.00486 M ethyl acetate by 0.00980 M sodium hydroxide at 25° are plotted in Fig. 10.3 by the method suggested by equation 10.23.

Example 10.1 Calculate the second-order rate constant for the saponification of ethyl acetate at 25° from the data of Fig. 10.3. The slope of the line is 2.29×10^{-4} s^{-1}. According to equation 10.23 the second-order rate constant may be calculated from

$$k = \frac{(2.303)(\text{slope})}{[(A)_0 - (B)_0]} = \frac{(2.303)(2.29 \times 10^{-4} \text{ s}^{-1})}{(0.00494 \text{ M})}$$

$$= 0.107 \text{ M}^{-1} \text{s}^{-1}$$

10.6 THIRD-ORDER REACTIONS

The integration of a third-order rate equation is readily accomplished for the case that the reaction is third order in the concentration of one reactant, or there are three substances initially at the same concentration, reacting with the stoichiometry $A + B + C$.

Integrating

$$\int_{(A)_0}^{(A)} \frac{d(A)}{(A)^3} = -k \int_0^t dt \qquad (10.24)$$

yields

$$kt = \frac{1}{2}\left[\frac{1}{(A)^2} - \frac{1}{(A)_0^2}\right] \qquad (10.25)$$

10.7 ZERO-ORDER REACTIONS

These are reactions in which the rate is unaffected by changes in the concentrations of one or more reactants because it is determined by some limiting factor other than concentration, such as the amount of light absorbed in a photochemical reaction or the amount of catalyst in a catalytic reaction. Then

$$-\frac{d(A)}{dt} = k \qquad (10.26)$$

A catalytic reaction might be first order in catalyst and zero order in reactant.

Integration of equation 10.26 yields

$$kt = (A)_0 - (A) \qquad (10.27)$$

The value of k calculated in this way may include constants corresponding to the intensity of light or the amount of catalyst.

10.8 RESTRICTIONS OF THERMODYNAMICS ON RATE LAWS

When the kinetics of both the forward and reverse reactions may be studied it is found that the rate law may be expressed as a difference between two terms.

For example for a reaction

$$A + B = C$$

it may be found experimentally that

$$\frac{d(C)}{dt} = k_f(A)(B) - k_r(C) \tag{10.28}$$

At equilibrium $d(C)/dt = 0$ so that

$$\frac{(C)}{(A)(B)} = \frac{k_f}{k_r} = K \tag{10.29}$$

Thus the equilibrium constant for this reaction is equal to the ratio of the rate constant for the forward reaction to the rate constant for the reverse reaction.

However, it is important to understand that thermodynamics does not require that the complete rate law (equation 10.28) have the form shown. Thermodynamics does require that the rate be positive in the direction of decreasing free energy and that the rate be zero at equilibrium. This condition would be satisfied by

$$\frac{d(C)}{dt} = k_f(A)^2(B)^2 - k_r(C)^2 \tag{10.30}$$

so that

$$\frac{(C)}{(A)(B)} = \left(\frac{k_f}{k_r}\right)^{\frac{1}{2}} = K \tag{10.31}$$

A number of other rate laws might be written, or even found experimentally, that would also yield the correct equilibrium expression. The restrictions of thermodynamics on the form of the rate equation are described more completely by Denbigh.*

10.9 MECHANISMS OF CHEMICAL REACTIONS

Most chemical reactions proceed through a series of steps. Even when the rate law is simple, a series of steps may be involved. One of the objectives of kinetic studies is to identify the intermediate steps because only in this way can we understand how the reaction occurs. The individual steps are referred to as *elementary reactions*. A description of these steps is called the *mechanism* of the reaction. In discussing a mechanism we refer to the *molecularity* of the steps. The molecularity is the number of reactant molecules in an elementary step. The individual steps of a mechanism are referred to as *unimolecular*, *bimolecular*, and *termolecular*, depending upon whether one, two, or three molecules are involved as reactants. For elementary reactions the molecularity (uni-, bi-, and ter-) and the order (first,

---

* K. Denbigh, *The Principles of Chemical Equilibrium*, 3rd ed., Cambridge University Press, Cambridge, 1971, p. 442.

second, and third) are the same, but these names are not synonyms at the level of the overall rate law. For example, a unimolecular step in a mechanism is first order, but a first-order reaction is not necessarily unimolecular, as we will soon see in Section 10.12.

The discovery of the mechanism of a reaction cannot be described briefly because many types of empirical and theoretical ideas are involved. Kinetic data, however, are very useful for determining a mechanism. Information about mechanism may also be obtained by use of isotopes to determine the paths of various atoms through the reaction and by use of spectroscopy to identify intermediates. Relaxation methods may be used to identify and study the very fast steps in a mechanism that occur while the slow steps are essentially stationary. The mechanism must yield the observed rate law and account satisfactorily for the net chemical change.

Before considering more complicated mechanisms we will consider the effects of reversible and consecutive steps in a mechanism.

10.10 REVERSIBLE ELEMENTARY REACTIONS

As the simplest example of a reversible step let us consider

$$A \underset{k_2}{\overset{k_1}{\rightleftharpoons}} B \tag{10.32}$$

The rate law for this reversible elementary reaction is

$$\frac{d(A)}{dt} = -k_1(A) + k_2(B) \tag{10.33}$$

If initially only A is present,

$$
\begin{aligned}
\frac{d(A)}{dt} &= -k_1(A) + k_2[(A)_0 - (A)] \\
&= k_2(A)_0 - (k_1 + k_2)(A) \\
&= -(k_1 + k_2)\left[(A) - \frac{k_2}{k_1 + k_2}(A)_0\right] \\
&= -(k_1 + k_2)[(A) - (A)_{eq}] \tag{10.34}
\end{aligned}
$$

where the expression for $(A)_{eq}$ is obtained as follows:

$$\frac{(B)_{eq}}{(A)_{eq}} = \frac{(A)_0 - (A)_{eq}}{(A)_{eq}} = \frac{k_1}{k_2} \tag{10.35}$$

and

$$(A)_{eq} = \frac{k_2}{k_1 + k_2}(A)_0 \tag{10.36}$$

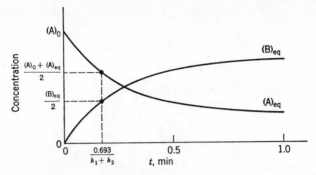

Fig. 10.4   Reversible first-order reaction starting with A at concentration $(A)_0$. The values of the rate constants are $k_1 = 3$ min^{-1} and $k_2 = 1$ min^{-1}.

Integrating equation 10.34 yields

$$-\int_{(A)_0}^{(A)} \frac{d(A)}{(A) - (A)_{eq}} = (k_1 + k_2) \int_0^t dt \qquad (10.37)$$

$$\ln \frac{(A)_0 - (A)_{eq}}{(A) - (A)_{eq}} = (k_1 + k_2)t \qquad (10.38)$$

For such a reaction the concentrations of A and B as functions of time are illustrated in Fig. 10.4. The concentration of A will be halfway to its equilibrium value in a time of $0.693/(k_1 + k_2)$.

Thus a plot of $-\log [(A) - (A)_{eq}]$ versus time is linear, and $(k_1 + k_2)$ may be calculated from the slope. It should be especially noted that the rate of approach to equilibrium in this reaction is determined by the sum of the rate constants of the forward and reverse reactions, not by the rate constant for the forward reaction. Since the ratio k_1/k_2 may be calculated from the equilibrium concentrations by use of equation 10.35, the values of k_1 and k_2 may be obtained.

10.11   CONSECUTIVE FIRST-ORDER
REACTIONS

Consecutive reactions occur when the product of a reaction undergoes further reaction. Two consecutive first-order reactions may be represented by

$$A \xrightarrow{k_1} B \xrightarrow{k_2} C \qquad (10.39)$$

To determine the way in which the concentrations of the compounds in such a mechanism depend on time, the rate equations are first written down for each substance. It is then necessary to obtain the solution of these simultaneous differential equations. For the foregoing reactions the rate equations are as follows:

$$\frac{d(A)}{dt} = -k_1(A) \qquad (10.40)$$

$$\frac{d(B)}{dt} = k_1(A) - k_2(B) \qquad (10.41)$$

$$\frac{d(C)}{dt} = k_2(B) \qquad (10.42)$$

It will be assumed that, at $t = 0$, $(A) = (A)_0$, $(B) = 0$, and $(C) = 0$. The rate equation for A is readily integrated to obtain

$$(A) = (A)_0 e^{-k_1 t} \qquad (10.43)$$

Substitution of this expression into equation 10.41 yields

$$\frac{d(B)}{dt} = k_1(A)_0 e^{-k_1 t} - k_2(B) \qquad (10.44)$$

which may be integrated to obtain

$$(B) = \frac{k_1(A)_0}{(k_2 - k_1)} [e^{-k_1 t} - e^{-k_2 t}] \qquad (10.45)$$

Because of conservation of the number of moles, $(A)_0 = (A) + (B) + (C)$ at any time, and so the concentration of C is given by

$$(C) = (A)_0 - (A) - (B) = (A)_0 \left[1 + \frac{1}{k_1 - k_2} (k_2 e^{-k_1 t} - k_1 e^{-k_2 t}) \right] \qquad (10.46)$$

The concentrations of A, B, and C are shown in Fig. 10.5 for $(A)_0 = 1$ M, $k_1 = 0.1$ hr^{-1}, and $k_2 = 0.05$ hr^{-1}.

If the course of the reaction were followed by analyzing for A, curve A would be obtained; if it were followed by measuring the concentration of the end product C, curve C would result; and, finally, if only the intermediate product B were determined, it would be found that its concentration would rise to a maximum and then fall off, as shown by curve B. The actual rate of production of C is seen to be quite complicated, and the existence of an induction period or time lag at the beginning of the reaction is evident.

Many examples of consecutive first-order reactions are found among the disintegration reactions of the radioactive nuclei. These radioactive reactions can be expressed with exactness by simple first-order equations.

If either one or both of the steps in the reaction A → B → C are reversible, more complicated expressions are obtained.*

---

* For these equations and also discussion of other special cases of the mechanism A ⇌ B, see R. A. Alberty and W. G. Miller, *J. Chem. Phys.*, **26**, 1231 (1957).

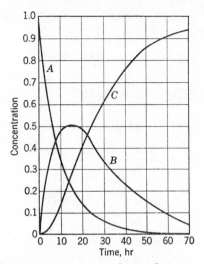

Fig. 10.5   Consecutive first-order reactions: $A \xrightarrow{k_1} B \xrightarrow{k_2} C$; $k_1 = 0.10$ hr^{-1}, $k_2 = 0.05$ hr^{-1}.

10.12   MORE COMPLICATED MECHANISMS

The rate equations may be integrated for a few mechanisms involving reversible, consecutive, and parallel steps of first and second orders. The rate equations for each reactant and intermediate are written down and solved simultaneously. For more complicated mechanisms, however, the differential equations cannot be solved explicitly, and it is necessary to use computers to obtain numerical solutions or to introduce experimentally suitable approximations to simplify the mathematics. Valuable information can be obtained by writing down the rate equations for a complex reaction even if the equations cannot be solved explicitly. Through them it is often possible to understand why there are sometimes induction periods, why complex reactions may approach first order, second order, or a fractional order, and how these apparent orders can change during the course of the reaction.

By the use of approximations it may be possible to obtain the overall rate laws for quite complicated mechanisms. As an illustration, consider the decomposition of N_2O_5 that follows the mechanism*

$$N_2O_5 \underset{k_2}{\overset{k_1}{\rightleftharpoons}} NO_2 + NO_3 \qquad (10.47)$$

$$NO_2 + NO_3 \xrightarrow{k_3} NO + O_2 + NO_2 \qquad (10.48)$$

$$NO + NO_3 \xrightarrow{k_4} 2NO_2 \qquad (10.49)$$

---

* H. S. Johnston, *Gas Phase Reaction Rate Theory*, The Ronald Press Co., New York, 1966, Chapt. 1.

The rate equations for the intermediates NO_3 and NO are

$$\frac{d(NO_3)}{dt} = k_1(N_2O_5) - (k_2 + k_3)(NO_2)(NO_3) - k_4(NO)(NO_3) \quad (10.50)$$

$$\frac{d(NO)}{dt} = k_3(NO_2)(NO_3) - k_4(NO)(NO_3) \quad (10.51)$$

Since the concentrations of NO and NO_3 are never very great during the reaction,

$$\frac{d(NO)}{dt} \ll - \frac{d(N_2O_5)}{dt}$$

and

$$\frac{d(NO_3)}{dt} \ll - \frac{d(N_2O_5)}{dt}$$

It is therefore useful to use the steady-state approximation and set the derivatives in equations 10.50 and 10.51 equal to zero. This leads to the following expressions for the steady-state concentrations of NO and NO_3:

$$(NO) = \left(\frac{k_3}{k_4}\right)(NO_2) \quad (10.52)$$

$$(NO_3) = \frac{k_1(N_2O_5)}{(k_2 + 2k_3)(NO_2)} \quad (10.53)$$

The rate equation for the decomposition of N_2O_5 is

$$\frac{d(N_2O_5)}{dt} = -k_1(N_2O_5) + k_2(NO_2)(NO_3) \quad (10.54)$$

Substituting equation 10.53 yields

$$\frac{d(N_2O_5)}{dt} = - \frac{2k_1k_3}{k_2 + 2k_3}(N_2O_5) \quad (10.55)$$

so that the overall reaction behaves in a first-order manner, as shown experimentally in Fig. 10.2b.

A more drastic assumption that is sometimes useful is to assume that a reaction has a rate determining step. If one of the steps is rate determining, all the steps in the mechanism up to that point remain at equilibrium. Thus the concentrations of all of the substances involved in these prior steps may be calculated from the equilibrium expressions.

10.13   PRINCIPLE OF DETAILED BALANCING

In addition to the ways in which thermodynamics restricts the values of rate constants, there is another restriction that does not come from thermodynamics.

This is the principle of detailed balancing that says that at equilibrium the forward rate of *each* reaction in a mechanism is equal to the reverse rate of that same reaction in the mechanism.* The application of this principle becomes important when a reaction has more than one path.

Suppose the transformation of A to B occurs by a reversible first-order reaction and a reversible second-order reaction involving hydrogen ion:

$$A \underset{k_2}{\overset{k_1}{\rightleftharpoons}} B \tag{10.56}$$

$$A + H^+ \underset{k_4}{\overset{k_3}{\rightleftharpoons}} B + H^+ \tag{10.57}$$

The rate equation for the formation B is

$$\frac{d(B)}{dt} = k_1(A) - k_2(B) + k_3(A)(H^+) - k_4(B)(H^+) \tag{10.58}$$

Since at equilibrium $d(B)/dt = 0$, equation 10.58 leads to

$$K = \frac{(B)_{eq}}{(A)_{eq}} = \frac{k_1 + k_3(H^+)}{k_2 + k_4(H^+)} \tag{10.59}$$

But this is a paradoxical result, since the equilibrium constant for A = B does not depend on the hydrogen-ion concentration. This paradox is eliminated, and the dependence of K on the concentration of hydrogen ion is removed, by the principle of detailed balancing. According to this principle, *the forward and reverse rates for each path must be equal at equilibrium.* Thus at equilibrium

$$k_1(A) = k_2(B)_{eq} \tag{10.60}$$

and

$$k_3(A)_{eq}(H^+) = k_4(B)_{eq}(H^+) \tag{10.61}$$

Substituting equations 10.60 and 10.61 into equation 10.59, it is found that the terms in hydrogen-ion concentration disappear as they should. From equations 10.60 and 10.61 it can be seen that

$$K = \frac{(B)_{eq}}{(A)_{eq}} = \frac{k_1}{k_2} = \frac{k_3}{k_4} \tag{10.62}$$

Thus the ratio k_3/k_4 is necessarily equal to the ratio k_1/k_2.

For the mechanism

$$
\begin{array}{c}
A \underset{k_{-1}}{\overset{k_1}{\rightleftharpoons}} B \\
k_3 \underset{k_{-3}}{} \; k_{-2} \underset{}{} k_2 \\
C
\end{array}
\tag{10.63}
$$

---

* L. Onsager, *Phys. Rev.*, **37,** 405 (1931).

thermodynamics simply requires that at equilibrium the concentrations of A, B, and C be independent of time. This could be accomplished by the cyclic mechanism

$$A \longrightarrow B$$

(10.64)

$$C$$

However, this mechanism of maintaining equilibrium is prohibited by the principle of detailed balancing, which leads directly to the relation $k_1 k_2 k_3 = k_{-1} k_{-2} k_{-3}$; thus the 6 rate constants in mechanism 10.63 are not independent.

If a rate equation has a sum of terms for the forward reaction, indicating multiple paths, the principle of detailed balancing requires that *each* term for the forward reaction be balanced by a thermodynamically appropriate term for the reverse reaction at equilibrium.

10.14   EFFECT OF TEMPERATURE

If the temperature range is not too great, the dependence of rate constants on temperature can usually be represented by an empirical equation proposed by Arrhenius in 1889.

$$k = A e^{-E_a/RT} \tag{10.65}$$

where A is the *pre-exponential factor* and E_a is the *activation energy*. The pre-exponential factor A has the same units as the rate constant, and for a first-order reaction the usual unit is s^{-1}. Since this is the unit of a frequency, A is sometimes referred to as a frequency factor. Equation 10.65 may be written in logarithmic form.

$$\log k = \frac{-E_a}{2.303R} \frac{1}{T} + \log A \tag{10.66}$$

According to this equation, a straight line should be obtained when the logarithm of the rate constant is plotted against the reciprocal of the absolute temperature. Differentiating equation 10.66 with respect to temperature,

$$\frac{d \ln k}{dT} = \frac{E_a}{RT^2} \tag{10.67}$$

and integrating between limits,

$$\log \frac{k_2}{k_1} = \frac{E_a}{2.303R} \left(\frac{T_2 - T_1}{T_1 T_2} \right) \tag{10.68}$$

The rate constants for the decomposition of gaseous nitrogen pentoxide* at different temperatures are given in the second column of Table 10.1 in s^{-1}. The values of $\log k$ are plotted against $1/T$ in Fig. 10.6.

* F. Daniels and E. H. Johnston, *J. Am. Chem. Soc.* **43,** 53 (1921).

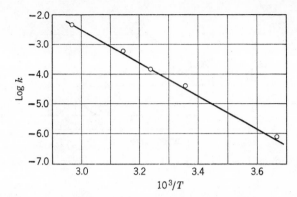

Fig. 10.6 Plot of log k versus $1/T$ for the decomposition of N_2O_5, from which the Arrhenius activation energy E_a may be calculated.

The slope of the line in Fig. 10.6 is -5400 K, and E_a has the value of -2.303 $R(\text{slope}) = 24{,}700$ cal mol^{-1}. Equation 10.66 becomes

$$\log k = -\frac{24{,}700}{(2.303)(1.987)}\frac{1}{T} + 13.638 \tag{10.69}$$

and equation 10.65 becomes

$$k = 4.35 \times 10^{13} e^{-24{,}700/1.987T}\,\text{s}^{-1} \tag{10.70}$$

The values of the rate constants k in reciprocal seconds, as calculated using this equation, are given in the third column of Table 10.1, and the half-lives are given in the last column.

Table 10.1 Rate Constants for the Decomposition of Gaseous Nitrogen Pentoxide at Different Temperatures

t, °C	$k_{obs} \times 10^5$	$k_{calc} \times 10^5$	Half-Life[1]
65	487 s^{-1}	463 s^{-1}	2.49 m
55	150	151	7.64 m
45	49.8	45.9	25.2 m
35	13.5	12.9	89.5 m
25	3.46	3.33	5.78 h
0	0.0787	0.0733	10.9 d

[1] h, hour; d, day; m, minute.

It should be realized that for any given frequency factor there is a fairly narrow range of activation energies that will give reaction rates in the range measurable by conventional techniques; that is, with half-lives from one minute to 10 days. For example, if $A = 10^{13}$ s^{-1} and the temperature is 298 K, reactions with

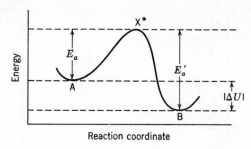

Fig. 10.7 Diagrammatic representation of the activation process in an elementary reaction
A $\rightleftharpoons$ B.

activation energies less than about 20 kcal mol^{-1} will be too fast to study with
ordinary methods, and reactions with activation energies greater than about
25 kcal mol^{-1} will be too slow.

The Arrhenius activation energies for the forward and reverse reactions are
related through the change in internal energy for the overall reaction. If the rate
law for a reaction of ideal gases is written in terms of concentrations, the rate
constants k for the forward reaction and k' for the reverse reaction are related to
K_c by

$$K_c = \frac{k}{k'} = K_p(RT)^{-\Sigma \nu_i} \tag{10.71}$$

with the proviso discussed in Section 10.8. The relation to K_p comes from equation
5.46 Taking the logarithm of equation 10.71 and differentiating with respect to
absolute temperature we obtain

$$\frac{d \ln k}{dT} - \frac{d \ln k'}{dT} = \frac{d \ln K_p}{dT} - \frac{1}{T} \sum \nu_i \tag{10.72}$$

Using equations 10.67 and 1.29 we obtain

$$E_a - E_a' = \Delta H^\circ - RT \sum \nu_i = \Delta U^\circ \tag{10.73}$$

Thus the difference between the activation energies for the forward and reverse
reactions is equal to the change in internal energy U for the overall reaction.

The reverse reaction B $\rightarrow$ A is endothermic (that is, ΔU is positive) for the
reaction illustrated in Fig. 10.7. We can see from the diagram that the activation
energy for an endothermic reaction must be at least as large as the change in
internal energy ΔU for the overall reaction.

10.15 THE ACTIVATED COMPLEX

The ideas discussed in the preceding section are summarized graphically in
Fig. 10.7 for an elementary reaction A $\rightleftharpoons$ B. In order for an A molecule to react

it must receive an additional amount of energy E_a to convert it to an activated complex X* that can then react to form product B. In order for a B molecule to react it must receive a larger amount of energy E_a' to convert it to the activated complex X* which can then react to form A. The reaction coordinate used as abscissa in Fig. 10.7 might be the length of a chemical bond which changes in going from A to B.

Thus we can think of the activation process in terms of going over a mountain pass from one valley to the next. The difference in elevation of the floors of the two valleys is analogous to the difference in internal energy U between reactant A and product B. The height of the mountain pass measured from one of the valleys is analogous to the activation energy. The greater the activation energy, the fewer are the collisions involving sufficient energy to cause reaction at a given temperature, and the slower is the reaction.

The activated complex is not simply an intermediate compound. It is a molecule in the process of breaking or forming bonds. An ordinary molecule can undergo a variety of vibrational motions and can be thought of as existing in a potential energy minimum such as A and B in Fig. 10.7. However, the activated complex X* is not stable because it exists at a potential energy maximum. Both the forward and reverse reactions go through the same activated complex; if a bond is in the process of being broken in X* for the reaction $A \rightarrow B$, the bond is in the process of being formed in X* for the reaction $B \rightarrow A$.

Eyring* has given a quantitative treatment of the activated complex theory that is very useful for interpreting and predicting reaction rates. This theory provides a means for calculating the concentrations of activated complexes and the rate at which they are converted to products. Since the activated complex is in the process of flying apart, one of its vibrations is in the process of becoming a translation. This vibration is assumed to have its classical energy RT/N_A (Section 9.5). In quantum theory (Section 12.4) we will see that the energy of a molecular vibration is $h\nu$ so that in this case $h\nu = RT/N_A$. The vibration frequency ν is taken as the rate with which activated complexes move across the potential energy barrier. For example, consider the reaction of A with B to form the activated complex $AB^{\ddagger}$. The reaction rate is equal to the rate of passage over the barrier $RT/N_A h$ times the concentration of the activated complex $(AB^{\ddagger})$.

$$\frac{-d(A)}{dt} = \kappa \frac{RT}{N_A h} (AB^{\ddagger}) \qquad (10.74)$$

where κ, the transmission coefficient, is the probability that a molecule once across the barrier will go on and not return. In calculations the transmission coefficient κ is generally taken equal to unity.

In the Eyring theory it is assumed that activated complexes are in equilibrium

* H. Eyring, *J. Chem. Phys.*, **3**, 107 (1935); M. G. Evans and M. Polanyi, *Trans. Far. Soc.*, **31**, 875 (1935); H. Pelzer and E. Wigner, *Z. physik. Chem.*, **B15**, 445 (1932); S. Glasstone, K. J. Laidler, and H. Eyring, *The Theory of Rate Processes*, McGraw-Hill Book Co., New York, 1941.

with reactants.

$$\frac{(AB^{\ddagger})}{(A)(B)} = K^{\ddagger} \tag{10.75}$$

Therefore ideas of thermodynamics and statistical mechanics may be used to calculate $K^{\ddagger}$. In using statistical mechanics, allowance is made for the fact that $AB^{\ddagger}$ has one less vibrational degree of freedom than an ordinary molecule since one vibrational motion is being transformed into a translation. Thus for a non-linear activated complex with N nuclei there are $3N-7$ normal modes rather than $3N-6$, and for a linear complex there are $3N-6$ rather than $3N-5$ (see Section 9.5). Substituting equation 10.75 into equation 10.74, we have

$$\frac{-d(A)}{dt} = \kappa \, \frac{RT}{N_A h} \, K^{\ddagger}(A)(B) \tag{10.76}$$

Thus the rate constant for this second-order reaction is given by

$$k = \kappa \, \frac{RT}{N_A h} \, K^{\ddagger} \tag{10.77}$$

This is a general result although we have derived it for a second-order reaction.

The statistical mechanical calculation of rate constants using equation 10.77 has been very useful, but the main problem is that the structure, vibrational frequencies, etc., of the activated complex are not known.

The Eyring theory actually includes the collision theory discussed in Section 10.17 as a special case. This may be shown by calculating the equilibrium constant for the formation of the activated complex, using the partition functions for rigid spherical molecules.

The equilibrium constant $K^{\ddagger}$ for the formation of the activated complex may be expressed in terms of thermodynamic quantities for the assumed equilibrium between reactants and activated complex. The equilibrium constant $K^{\ddagger}$ may be expressed in terms of the standard Gibbs free-energy change $\Delta G^{\ddagger\circ}$ for the activation process. Since $\Delta G^{\ddagger\circ} = -RT \ln K^{\ddagger}$, and $\Delta G^{\ddagger\circ} = \Delta H^{\ddagger\circ} - T \Delta S^{\ddagger\circ}$

$$K^{\ddagger} = e^{-\Delta G^{\ddagger\circ}/RT} = e^{\Delta S^{\ddagger\circ}/R} e^{-\Delta H^{\ddagger\circ}/RT} \tag{10.78}$$

Thus

$$k = \frac{\kappa RT}{N_A h} e^{\Delta S^{\ddagger\circ}/R} e^{-\Delta H^{\ddagger\circ}/RT} \tag{10.79}$$

where k is the rate constant of the reaction, κ is the transmission coefficient, $\Delta S^{\ddagger\circ}$ is the entropy of activation, and $\Delta H^{\ddagger\circ}$ is the enthalpy of activation.

10.16 CALCULATION OF ENTHALPIES OF ACTIVATION AND ENTROPIES OF ACTIVATION

Because of the T in the frequency factor of the Eyring equation, the activation enthalpy $\Delta H^{\ddagger\circ}$ is not to be identified with the Arrhenius activation energy E_a.

The relationship between $\Delta H^{\ddagger\circ}$ and E_a may be obtained by calculating the slope of a plot of $\ln k$ versus $1/T$ according to equation 10.79.

$$\frac{d \ln k}{dT} = \frac{1}{T} + \frac{\Delta H^{\ddagger\circ}}{RT^2} \tag{10.80}$$

$$\frac{d(1/T)}{dT} = -\frac{1}{T^2} \tag{10.81}$$

$$\frac{d \ln k}{d(1/T)} = \frac{d \ln k}{dT}\frac{dT}{d(1/T)} = \frac{-(\Delta H^{\ddagger\circ} + RT)}{R} \tag{10.82}$$

As shown earlier, the slope of the plot of $\ln k$ versus $1/T$ gives the value of $-E_a/R$ according to the Arrhenius equation. Thus

$$E_a = \Delta H^{\ddagger\circ} + RT \tag{10.83}$$

Substitution of $\Delta H^{\ddagger\circ} = E_a - RT$ into equation 10.79 and comparison with the empirical Arrhenius equation

$$k = Ae^{-E_a/RT} \tag{10.84}$$

shows that the frequency factor A in the Arrhenius equation has the following significance when the transmission coefficient κ is taken as unity:

$$A = e\frac{RT}{N_A h} e^{\Delta S^{\ddagger\circ}/R} \tag{10.85}$$

Thus the entropy of activation $\Delta S^{\ddagger\circ}$ may be calculated from the preexponential factor in the Arrhenius equation.

Example 10.2 For the rearrangement of 1-ethyl propenyl allyl malonitrile to 1-ethyl-2-methyl-4-pentenylidene malonitrile the Arrhenius activation energy in the neighborhood of $130°$ is 25,900 cal mol^{-1}, so that $\Delta H^{\ddagger\circ} = 25,900 - (1.987)(403) = 25,100$ cal mol^{-1}. The first-order rate constant at $130°$ is 9.12×10^{-4} s^{-1}. The entropy of activation is calculated as follows:

$$9.12 \times 10^{-4} = \frac{RT}{N_A h} e^{\Delta S^{\ddagger\circ}/R} e^{-25,100/(1.987)(403)}$$

$$\frac{RT}{N_A h} e^{\Delta S^{\circ\ddagger}/R} = 3.74 \times 10^{10}$$

$$\Delta S^{\ddagger\circ} = 2.303\ R \log \frac{N_A h(3.74 \times 10^{10})}{RT}$$

$$= -10.8 \text{ cal } K^{-1} mol^{-1}$$

For many unimolecular, bond-breaking gas reactions $\Delta S^{\ddagger\circ} = 0$ because the activated complex is so much like the original reactants and there is very little change in configuration in going from the reactants to the activated complex.

In this case $e^{\Delta S\ddagger°/R} = 1$, and at room temperatures

$$A = \frac{eRT}{N_A h} = \frac{(2.718)(8.31 \text{ J K}^{-1}\text{ mol}^{-1})(300 \text{ K})}{(6.02 \times 10^{23}\text{ mol}^{-1})(6.63 \times 10^{-34}\text{ J s})}$$
$$= 1.7 \times 10^{13} \text{ s}^{-1} \tag{10.86}$$

This is the order of magnitude of vibration frequencies in molecules.

If the activation of the molecule involves a rearrangement of atoms or a change in configuration, there will be a change in entropy and $e^{\Delta S\ddagger°/R}$ is not unity. The values of $\Delta S\ddagger°$ are rarely large enough to give a value of more than 10^2 or less than 10^{-2} to the term $e^{\Delta S\ddagger°/R}$, and so frequency factors $e(RT/N_A h)e^{\Delta S\ddagger°/R}$ may range from about 10^{11} to 10^{15}. If there is an increase in rotational and vibrational freedom in the activated complex, $\Delta S\ddagger°$ is positive, and the preexponential factor is larger than 1.7×10^{13} s^{-1}. If there is a decrease in rotational and vibrational freedom in the activated complex, $\Delta S\ddagger°$ is negative, and the preexponential factor is smaller than 1.7×10^{13} s^{-1}.

10.17   BIMOLECULAR GAS REACTIONS

The internal dynamics of gas reactions are conceptually simpler than solution reactions and surface reactions, but they are still quite complicated. We will begin to consider gas reactions with bimolecular reactions and then proceed to unimolecular reactions, termolecular reactions, and chain reactions.

One of the experimental problems in studying gas reactions is that they may be catalyzed by the surface of the reaction vessel. If we want to study a homogeneous gas reaction, therefore, it is important to find conditions under which the surface-catalyzed reaction (heterogeneous reaction) is negligible. In general a catalyst lowers the activation energy, and so the rate of a surface-catalyzed reaction does not increase as rapidly with increasing temperature as the homogeneous reaction. Consequently, at sufficiently high temperatures the homogeneous reaction is bound to be faster.

The Arrhenius parameters and steric factors p (defined in equation 10.90) for a number of bimolecular reactions of the type

$$A + BC \rightarrow AB + C \tag{10.87}$$

are summarized in Table 10.2. These parameters vary a good deal from reaction to reaction, and we would like to be able to calculate them from information about the individual molecules. Elementary kinetic theory provides a first approximation. We have already calculated (Section 9.7) the number of collisions Z of identical spherical molecules per unit volume per unit time:

$$Z = \frac{\sqrt{2}}{2}\pi\sigma^2\langle v\rangle n^2 = 2\left(\frac{\pi kT}{m}\right)^{1/2}\sigma^2 n^2 \tag{10.88}$$

Table 10.2 Arrhenius Parameters and Steric Factors for Bimolecular Reactions[1]

Reaction	T range	$\log_{10} A$	E_a	p
$H + D_2 \rightarrow HD + D$	300–750	10.69	9.39	0.088
$D + H_2 \rightarrow HD + H$	250–750	10.64	7.61	0.094
$H + HCl \rightarrow H_2 + Cl$	200–500	10.36	3.50	0.039
$H + Cl_2 \rightarrow HCl + Cl$	273–5200	11.0	5.3	0.074
$O + O_3 \rightarrow O_2 + O_2$	273–900	10.08	4.79	0.037
$Cl + H_2 \rightarrow HCl + H$	250–450	10.08	4.3	0.020

[1] W. C. Gardiner, *Rates and Mechanisms of Chemical Reactions*, W. A. Benjamin, Inc., New York, 1969.

where σ is the collision diameter for the spherical molecules, $\langle v \rangle$ is the arithmetic mean velocity $(8kT/\pi m)^{1/2}$, Section 9.6, and n is the number of molecules per unit volume. Thus the number of collisions is proportional to the square of the concentration of the molecules.

The second-order rate constant for a gas reaction involving no activation energy may be calculated from equation 10.88. Second-order rate constants are ordinarily expressed in $M^{-1} s^{-1}$ since concentrations are expressed in moles liter^{-1}. The concentration c in moles liter^{-1} is related to the concentration n in molecules m^{-3} by $c = 10^{-3} n/N_A$. The rate of reaction dc/dt in moles liter^{-1} s^{-1} is therefore related to the rate of collisions Z in molecules m^{-3} s^{-1} by $dc/dt = 10^{-3} Z/N_A$. The second-order rate constant is defined by

$$k = \frac{dc/dt}{c^2} = \frac{10^{-3} Z/N_A}{(10^{-3}/N_A)^2 n^2} = \frac{10^3 N_A Z}{n^2} \tag{10.89}$$

The second-order rate constants calculated by substituting equation 10.88 in this equation are of the order of 10^{10}–10^{12} $M^{-1} s^{-1}$ for gas molecules of average size and cross section. The fact that the second-order rate constants for most gas reactions are much smaller than this indicates that reaction does not occur in each collision. A collision may be ineffective in producing chemical change for two reasons: (1) the collision is not sufficiently energetic to supply the activation energy, and (2) the colliding molecules are not oriented in such a way that they can react with each other.

The fact that a molecule has a high translational energy does not make it unstable, but if this energy is converted into internal energies of the parts of a molecule in a collision, the molecule may be activated in the collision. The energy which is effective in promoting reaction is not the total kinetic energy of the two colliding molecules, but rather the kinetic energy corresponding to the component of the relative velocity of the two molecules along the line of their centers at the moment of collision. This is the energy with which the two molecules are pressed together. Assuming that the kinetic energy of this component of the relative velocity must be greater than some minimum value E_0, the threshold energy per mole, it can be shown that the fraction of all collisions with a line-of-centers component of the kinetic energy greater than E_0 is given by the Boltzmann factor $e^{-E_0/RT}$.*

Thus according to gas-collision theory the second-order rate constant should be given by

$$k = \frac{10^3 N_A Z p}{n^2} e^{-E_0/RT}$$

$$= 2 \times 10^3 N_A p \left(\frac{\pi RT}{M}\right)^{1/2} \sigma^2 e^{-E_0/RT} \qquad (10.90)$$

if the reacting molecules are the same. The expression for Z is given by equation 10.88. The steric factor p is introduced to allow for the fact that a certain orientation may be required for the reaction to occur.

Because of the $T^{1/2}$ factor preceding the exponential, equation 10.90 predicts a different temperature dependence than the Arrhenius equation. Since the Arrhenius activation energy may be calculated from $E_a = -R[d \ln k/d(1/T)]$, applying this operation to equation 10.90 yields

$$E_a = E_0 + \frac{RT}{2} \qquad (10.91)$$

assuming that the steric factor p and the reaction cross section σ are independent of temperature. Thus if $E_a \gg RT$, the difference between the Arrhenius activation energy E_a and the threshold energy of this simple kinetic theory is not significant.

The values of steric factors p given in Table 10.2 have been calculated from $p = \pi\sigma_R^2/\pi\sigma^2$, where σ is the collision diameter calculated from transport properties and σ_R is the reaction diameter required in equation 10.79 to give the experimentally determined rate constant k. The table shows that p decreases as the complexity of the reactants increases.

The simple collision theory of bimolecular reactions is not a really satisfactory theory in that it does not provide the means for calculating the threshold energy for reaction E_0 or the steric factor p. However, this equation has played an important role in the development of the theory of kinetics and provides a simple visualization of a bimolecular gas reaction.

Example 10.3 The decomposition of gaseous hydrogen iodide is second order, and the second-rate constant at $393.7°$ C is 2.6×10^{-4} $M^{-1}s^{-1}$. The Arrhenius activation energy is found to be 45.6 kcal mol^{-1}. Calculate the second-order rate constant for this reaction to be expected from the collision theory. The collision diameter may be taken as 3.5 Å, and the orientation factor p as unity. The molecular weight of HI is 127.9 g mol^{-1}.

Using equations 10.90 and 10.91

$$k = 2(10^3 \text{ liters m}^{-3})(6.02 \times 10^{23} \text{mol}^{-1}) \left[\frac{\pi(8.31 \text{ J K}^{-1} \text{ mol}^{-1})(666.8 \text{ K})}{127.9 \times 10^{-3} \text{ kg mol}^{-1}}\right]^{1/2} (3.5 \times 10^{-10} \text{m})^2$$

$$\times e^{-(44,938 \text{ cal mol}^{-1})/(1.987 \text{ cal K}^{-1} \text{mol}^{-1})(666.8 \text{K})}$$

$$= 1.0 \times 10^{-4} \text{ M}^{-1} \text{s}^{-1}$$

The agreement with the experimental value of 2.6×10^{-4} M^{-1} s^{-1} is all that can be expected from such an oversimplified theory.

---

* For a much more complete discussion of bimolecular reactions see W. C. Gardiner, Jr., *Rates and Mechanisms of Chemical Reactions*, W. A. Benjamin, Inc., New York, 1969.

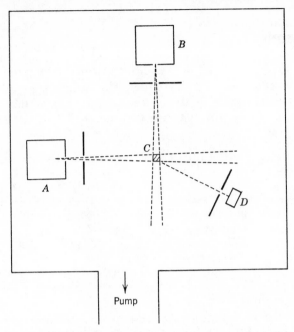

Fig. 10.8 Schematic diagram for a molecular beam apparatus for studying the reaction of molecules from source A with molecules from source B. Products are detected at D.

The molecular beam method provides the means for obtaining a great deal more information about bimolecular reactions. In the simplest type of apparatus illustrated schematically in Fig. 10.8, A and B are sources of beams of the two reactants that collide in region C. These collisions occur in a chamber evacuated with a high-speed pump so that the only collisions are between the molecules from A and B. Product molecules and elastically scattered reactant molecules are detected at D. The effect of changing the angle of approach may be studied by moving A or B, and the effect of the relative velocity of the reactants may be studied by use of velocity selectors (see Fig. 9.5) on the beams as they leave A and B. The orientation of molecules as they collide is also important; the effect of orientation on rate may be studied for molecules with dipole moments (Section 14.13) because molecules may be oriented by use of an electric field. The rate constants for gas reactions average over all directions of approach of the two molecules and for collisions of various energies. Also the colliding molecules may have various amounts of vibrational and rotational energy, and the probability of reaction will also depend on the internal state of the molecules. The effects of these various factors on the probability of reaction may be studied separately in molecular beam experiments.

10.18 UNIMOLECULAR GAS REACTIONS

When certain gases are heated they undergo thermal decomposition or isomerization. These reactions are often first order. The Arrhenius parameters for a few first-order gas isomerization or decomposition reactions at their high pressure limits are given in Table 10.3. The fact that such reactions are first order was a

Table 10.3   Arrhenius Parameters for k_∞ for Unimolecular Gas Reactions

Reactant	Product	$\log A (\text{s}^{-1})$	$E_a (\text{kcal mol}^{-1})$
cis-2-butene	*trans*-2-butene	13.8	63
cyclopropane	propylene	15.2	65
N_2O_5	$NO_2 + NO_3$	13.6	24.7
C_2H_6	$2CH_3$	16.5	88
CO_2	$CO + O$	11.3	110

puzzling observation because it is expected that the activation energy for reaction would come from a bimolecular collision. In 1922 Lindemann proposed the following explanation of this effect. He pointed out that when a molecule is energized by a bimolecular collision there may be a time lag before decomposition or isomerization, and that during this time lag the energized molecule may lose its extra energy in a second bimolecular collision. Since such reactions are usually studied by diluting the reacting gas A with an excess of inert gas M, the energizing and deenergizing collisions are mostly collisions of A with M as indicated in the following mechanism

$$A + M \xrightarrow{k_e} A^* + M \tag{10.92}$$

$$A^* + M \xrightarrow{k_{de}} A + M \tag{10.93}$$

$$A^* \xrightarrow{k_{uni}} \text{products} \tag{10.94}$$

Here A^* represents an energized A molecule. Any collision of A^* with M is assumed to be deenergizing, and the bimolecular rate constant for this process is k_{de}. The decomposition of A^* is unimolecular with a rate constant of k_{uni}.

When a molecule absorbs a large amount of energy in a collision the translational, rotational, and vibrational energy is increased. Electronic excitation is involved only in those rare cases where there are low lying electronic levels. It is primarily the vibrational energy that plays a role in tearing the molecule apart. This energy can be transferred from one type of vibrational motion to another. A molecule with a large amount of vibrational energy may vibrate for a short time and then suddenly dissociate because the energy in one particular vibrational mode at that instant is so great that the molecule flies apart.

Since A^* is never present at a very high concentration the rate law for this mechanism may be derived assuming A^* is in a steady state (Section 10.12).

$$\frac{d(A^*)}{dt} = k_e(A)(M) - [k_{de}(M) + k_{uni}](A^*) = 0 \qquad (10.95)$$

Thus

$$-\frac{d(A)}{dt} = k_{uni}(A^*) = \frac{k_{uni}k_e(A)(M)}{k_{de}(M) + k_{uni}} = k(A) \qquad (10.96)$$

At sufficiently low pressures $k_{de}(M) \ll k_{uni}$ and

$$-\frac{d(A)}{dt} = k_e(A)(M) \qquad (10.97)$$

Under these conditions the reaction is second order, and all A^* molecules decompose to products before they can be deenergized in a collision with a second M molecule. At sufficiently high pressures $k_{de}(M) \gg k_{uni}$ and

$$-\frac{d(A)}{dt} = \frac{k_{uni}k_e}{k_{de}}(A) = k_\infty(A) \qquad (10.98)$$

Under these conditions the reaction is first order in A. The concentration of A^* is at its equilibrium value, $(A^*) = (A)k_e/k_{de}$, at the high-pressure limit, and so the rate of reaction is determined by the equilibrium constant for the production of A^* and by k_{uni}.

Experimental data on unimolecular reactions deviate significantly from equation 10.96, but the basic idea of the theory is correct and modern extensions of the theory are able to account for the experimental results quite satisfactorily.

10.19 TERMOLECULAR REACTIONS

The recombination of atoms to form a diatomic molecule is a third-order reaction. For example, the recombination of halogen atoms follows the rate law

$$\frac{d(X_2)}{dt} = k(X)^2(M) \qquad (10.99)$$

where X is a halogen atom and M is a third body that may be a halogen molecule or a molecule of an added gas. The recombination of halogen atoms may be studied after flash photolysis (Section 18.6) of gaseous I_2 or Br_2.*

The recombination of atoms is third order since the reaction is exothermic and this energy must be carried off by the products. When two atoms combine to form a diatomic molecule the heat of reaction remains in the product molecule which

---

* R. L. Strong, J. C. Chien, P. E. Graf, and J. E. Willard, *J. Chem. Phys.*, **26**, 1287 (1957).

therefore has enough energy to dissociate and may do so before the excess energy is lost in the next collision. If the recombination occurs in a three-body collision, however, the heat of reaction may be carried off as kinetic energy of the third body M.

In contrast to other reactions the rate of recombination of atoms may be slower at a higher temperature than at a lower temperature, for given concentrations of atoms and third body M.

The apparent activation energy is -1.4 kcal mol^{-1} for the recombination of iodine atoms in argon. The negative temperature coefficient may result from a mechanism involving the formation of a complex

$$I + M = IM \qquad (10.100)$$

which precedes the reaction step

$$IM + I \underset{k_{-1}}{\overset{k_1}{\rightleftharpoons}} I_2 + M \qquad (10.101)$$

The explanation of the inverse temperature dependence of the iodine atom-recombination reaction is that the formation of IM is exothermic, so that the concentration of IM decreases as the temperature is raised.

If the rate of combination of two atoms or radicals is proportional to the concentration of some third molecule M, the rate of decomposition of the product into radicals must also be proportional to (M) according to the principle of detailed balancing. In physical terms a collision with a third body M is required to provide the energy for dissociation of I_2. This is expressed by the rate equations for the preceding two-step mechanism, assuming that the first step is in rapid equilibrium.

$$\frac{d(I_2)}{dt} = k_1(I)(IM) - k_{-1}(I_2)(M) \qquad (10.102)$$

Since $(IM) = K(I)(M)$,

$$\frac{d(I_2)}{dt} = k_1 K(I)^2(M) - k_{-1}(I_2)(M) \qquad (10.103)$$

where the concentration of the third body M appears in both terms. At equilibrium the rate is zero and $(I_2)/(I)^2 = k_1 K/k_{-1}$.

10.20   FREE-RADICAL REACTIONS

Free radicals are reactive molecules (or atoms) with unpaired electrons. This term is not applied to stable species like Fe^{3+} and O_2 in spite of the fact that their paramagnetic behavior (Section 16.1) demonstrates that they have unpaired electrons. At very high temperatures organic molecules may be partially dissociated into free radicals, and hexaphenyl ethane is partially dissociated into two triphenyl methyl radicals at room temperature, as proved by Gomberg in 1900. His measurements of the freezing-point depression of solvents showed dissociation of certain solutes into smaller units even though the solutions were not electrically conducting. Alkyl free radicals in the gas phase may be prepared by the thermal decomposition of metal alkyls. For example, methyl radicals $CH_3\cdot$ may be obtained from

$$Pb(CH_3)_4 \rightarrow Pb + 4CH_3\cdot$$

When this reaction is carried out by flowing gas down a heated glass tube, a lead mirror is produced on the inside of the tube. If a gas containing free radicals is passed over such a mirror, the mirror is removed. The removal of the lead mirror by methyl radicals is just the reverse of this reaction. During flow down a tube the radicals disappear by colliding with the wall or by combining with each other. Free radicals may also be produced for research purposes by the absorption of light of sufficiently short wavelength or by passing a gas through an electrical discharge.

Since a radical has an unpaired electron, its reaction with a molecule having paired electrons gives rise to another radical. In this way the reactive center is maintained and can give rise to a chain of reactions. We may ask why such a reaction ever stops. Sometimes, as a matter of fact, the chain reaction does not stop until all the material is consumed. At other times, however, the chain is broken when one of the activated molecules in it collides with the wall of the containing vessel or with another radical to form a neutral molecule. The length of the chain, that is, the number of molecules reacting per molecule activated, is determined by the relative rates of the chain-propagating and the chain-breaking reactions.

Many thermal decomposition reactions of hydrocarbons, ethers, aldehydes, and ketones appear to go by free-radical chain reactions. In 1935 F. O. Rice and K. F. Herzfeld showed how free-radical chain mechanisms could be devised for these reactions that would lead to simple over-all kinetics. The free radicals involved in these reactions include CH_3, C_2H_5, and H. The presence of free radicals in a number of such reactions has been shown by the removal of metallic mirrors, by catalyzing a reaction, such as the polymerization of an olefin which is known to go by a chain mechanism, or by inhibition by a compound such as nitric oxide or propylene. If each molecule of the inhibitor stops a chain and each chain produces a great many molecules of product, it is obvious that mere traces of inhibitors may have a pronounced effect. For example, the oxidation of sulfite ion in solution by oxygen is inhibited markedly by the addition of traces of alcohols.

The kinetics of the reaction

$$H_2 + X_2 = 2HX \qquad (10.104)$$

where X_2 is I_2, Br_2, or Cl_2, are of interest because the mechanisms of the reactions are quite different for the three halogens. The reaction of H_2 with I_2 had been considered to be a classical bimolecular reaction, but is now known to be more complicated. The reaction of H_2 with Br_2 has a very complicated rate expression. The reaction of H_2 with Cl_2 is an explosive chain reaction.

In the temperature range 200–300° C the rate of reaction of H_2 and Br_2 is given by

$$\frac{d(\text{HBr})}{dt} = \frac{k(\text{H}_2)(\text{Br}_2)^{1/2}}{1 + k'(\text{HBr})/(\text{Br}_2)} \tag{10.105}$$

If no HBr is initially present, the initial velocity $d(\text{HBr})/dt$ is directly proportional to $(\text{H}_2)(\text{Br}_2)^{1/2}$. If sufficient HBr is added initially so that $k'(\text{HBr})/(\text{Br}_2) \gg 1$, the intial velocity is directly proportional to $(\text{H}_2)(\text{Br}_2)^{3/2}/(\text{HBr})$, and it is seen that doubling the initial HBr concentration cuts the initial velocity in half.

The reaction of hydrogen with halogens X_2 (the overall reaction represented by the first reaction below) may be explained in terms of the following chain mechanism:

$$\text{H}_2 + \text{X}_2 \underset{k_{-1}}{\overset{k_1}{\rightleftharpoons}} 2\text{HX} \tag{10.106}$$

Initiation (forward reaction) and Termination (reverse reaction)

$$\text{X}_2 \underset{k_{-2}}{\overset{k_2}{\rightleftharpoons}} 2\text{X} \tag{10.107}$$

Propagation (forward reaction) and Inhibition (reverse reaction)

$$\text{X} + \text{H}_2 \underset{k_{-3}}{\overset{k_3}{\rightleftharpoons}} \text{HX} + \text{H} \tag{10.108}$$

Propagation

$$\text{H} + \text{X}_2 \underset{k_{-4}}{\overset{k_4}{\rightleftharpoons}} \text{HX} + \text{X} \tag{10.109}$$

Assuming the atoms are in a steady state, the rate law for this mechanism is

$$\frac{d(\text{HX})}{dt} = 2k_1 \left[(\text{H}_2)(\text{X}_2) - \frac{(\text{HX})^2}{K} \right] \left[1 + \frac{\dfrac{k_3}{k_1}\sqrt{\dfrac{k_2}{k_{-2}(\text{X}_2)}}}{1 + \dfrac{k_{-2}(\text{HX})}{k_4(\text{X}_2)}} \right] \tag{10.110}$$

where $K = (\text{HX})_{\text{eq}}^2/(\text{H}_2)_{\text{eq}}(\text{X}_2)_{\text{eq}}$. There are two terms in the rate law for the forward reaction and two terms in the rate law for the reverse reaction because there are two paths for the reaction—path 10.106 and the path provided by steps 10.107, 10.108, and 10.109.

The various rate constants are quite different for the different halogens, and consequently the steady-state rate law is different for the different overall reactions. The form of equation 10.110 for the forward reaction, for which $(\text{HX})^2/K \ll (\text{H}_2)(\text{X}_2)$, is in agreement with the experimental findings.

In 1967 Sullivan* suggested, on the basis of photochemical experiments, that

---

* J. H. Sullivan, *J. Chem. Phys.*, **46**, 73 (1967).

the reaction of H_2 with I_2 occurs either by a concerted attack of two iodine atoms on a hydrogen molecule

$$2I + H_2 \underset{k_{-2}}{\overset{k_2}{\rightleftharpoons}} 2HI \tag{10.111}$$

or a two-step mechanism

$$M + I + H_2 = IH_2 + M \tag{10.112}$$

$$IH_2 + I \xrightarrow{k_4} 2HI \tag{10.113}$$

in which the first step remains at equilibrium and the second step is rate determining. If the reaction of H_2 and I_2 is studied under conditions where the dissociation of I_2 remains in equilibrium the rate laws for the forward reactions in 10.106, 10.111, and 10.112 plus 10.113 are all the same. At equilibrium $(I) = K(I_2)^{1/2}$ and so the rate law for the forward step of reaction 10.111 is

$$\frac{d(HI)}{dt} = 2k_2(I)^2(H_2) = 2k_2K^2(I_2)(H_2) \tag{10.114}$$

and the rate law for reaction 10.113 is

$$\frac{d(HI)}{dt} = 2k_4(I)(IH_2) = 2k_4K'(I)^2(H_2) = 2k_4K'K^2(I_2)(H_2) \tag{10.115}$$

where $K' = (IH_2)/(I)(H_2)$. This is a good illustration of different mechanisms yielding the same rate law. The fact that the reaction goes by mechanism 10.111 or 10.112 and 10.113 rather than 10.104 was shown by measuring the reaction rate with photochemically produced iodine atoms at temperatures low enough that the rate of reaction was negligible. The activation energy found in this way is in agreement with that found at higher temperatures with the purely thermal reaction.

The reaction of hydrogen and chlorine may be initiated by light, and this photochemical reaction may involve chains of a million molecules, and proceed explosively (Section 18.5).

10.21 BRANCHING-CHAIN REACTIONS

Certain reactions which are evidently propagated by a chain reaction show characteristics which cannot be explained in terms of a simple chain reaction. For example, at about 550° stoichiometric hydrogen-oxygen mixtures react very slowly at pressures below 1 Torr. As the pressure is increased the reaction rate increases slowly, but at a pressure of about 1 Torr, depending on the volume of the vessel, there is a sudden explosion. On the other hand, if the gases are at a considerably higher pressure the rate is again quite low. Hinshelwood found that, if hydrogen at 200 Torr and oxygen at 100 Torr are placed in a 300-cm³ quartz vessel at 550°, the rate of reaction is quite slow and becomes slower if the pressure

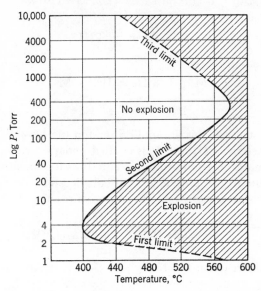

Fig. 10.9 Explosion limits of a stoichiometric oxygen-hydrogen mixture.

is further reduced to 100 Torr. If the pressure is reduced to 98 Torr however, an explosion occurs. Finally, as the total pressure is increased above the explosion zone, the reaction rate increases until it becomes so fast that the reaction mixture may be said to explode. These data are represented by Fig. 10.9, which shows the pressure-temperature conditions for the explosion limits for a stoichiometric hydrogen-oxygen mixture.

To explain such complicated behavior it is necessary to introduce the idea of branching chains, that is, chains in which one radical gives rise to more than one radical so that the number of radicals increases in a geometric progression. The reaction is apparently initiated by the process

$$H_2 + O_2 \rightarrow H_2O + O \qquad \Delta H° \sim 1.5 \text{ kcal mol}^{-1} \qquad (10.116)$$

The production of oxygen atoms leads to the following three reactions.

$$O + H_2 \rightarrow OH + H \qquad \Delta H° \sim 2 \text{ kcal mol}^{-1} \qquad (10.117)$$

$$H + O_2 \rightarrow OH + O \qquad \Delta H° \sim 16 \text{ kcal mol}^{-1} \qquad (10.118)$$

$$OH + H_2 \rightarrow H_2O + H \qquad \Delta H° \sim -15 \text{ kcal mol}^{-1} \qquad (10.119)$$

There is a branching chain because for every O atom produced in reaction 10.116 an extra H atom is produced by reactions 10.117, 10.118, and 10.119, which also yield a replacement oxygen atom. The H atoms and OH radicals, which permit the propagation of a chain, may be destroyed by collision with the wall. If more than one of the two radicals formed per H from H_2 are destroyed at the wall, there will not be a geometric increase in the number of radicals and an explosion, but simply a steady state in which the concentrations of the various radicals remain constant. This is the situation below the first explosion limit. As the pressure is increased, the fraction of radicals removed by diffusion to the wall decreases so that more than one radical is produced per H atom formed, and the number of radicals builds up in a geometric progression, producing an explosion. This explanation is in accord with the fact that the explosion occurs at a lower pressure in a larger reaction vessel, and that the addition of glass beads to the reaction vessel raises the pressure required for explosion. Adding inert gases lowers the explosion limit by slowing down diffusion of radicals to the wall. The transition from slow reaction to explosion is abrupt because, if one radical leads to the formation of slightly more than one radical, the number of radicals increases rapidly.

At pressures just above the second explosion limit the rate of reaction again becomes slow. The pressure and temperature at this limit are independent of the size of the reaction vessel, indicating that the quenching of the reaction chains occurs in collisions of radicals with gas molecules rather than with the wall. The second limit is displaced to lower partial pressures of oxygen and hydrogen by the addition of inert gases, the lighter inert gases being more effective in preventing explosions.

The kinetics of the gaseous oxidation of hydrocarbons is related to the oxidation of hydrogen in that branching-chain reactions are involved but is much more complicated. To reduce the too rapid propagation of chains, which leads to "knock" in internal-combustion engines, tetraethyl lead, which reacts readily with atoms and radicals and breaks the chains, is sometimes added to gasoline.

Chain reactions often occur together with another path for the reaction. The activation energy is generally higher for a chain reaction, and so as the temperature is raised, the rate of the chain reaction increases more rapidly than that of the other path, and eventually the chain reaction predominates.

10.22 DIFFUSION CONTROLLED REACTIONS IN LIQUIDS

There are significant differences between the kinetics of chemical reactions occurring in the gas and liquid phases. In the gas phase, molecules undergo isolated collisions with each other. Since the mean free paths are long compared with the molecular diameter, it is very unlikely that two gas molecules that have collided will collide with each other again. In the liquid phase two molecules that collide are likely to undergo successive collisions because they are surrounded by other molecules that tend to form a *cage* around them. This succession of collisions is called an encounter. Because of the cage effect it is likely that when two reactive molecules undergo an encounter they will collide with each other enough times that reaction is highly probable. For aqueous solutions it has been estimated that the cage lifetime for a pair of noninteracting molecules is of the order 10^{-12} to 10^{-11} s, during which time they may undergo 10 to 10^5 collisions with each other. If this is a sufficient number of collisions for reaction, the reaction will be diffusion controlled; that is, the rate of the reaction will depend on the rate with which the reactants can diffuse together.

Another difference between solution reactions and gas reactions is that the former often involve ions but the latter almost never do. In liquids, ions are solvated and tend to be surrounded by ions of the opposite charge. Accordingly, changes in the dielectric constant of the medium and changes in the concentration and nature of other ions present have significant effects on the rates of reactions in the liquid phase.

The theory of unimolecular reactions in the gas phase (Section 10.18) has emphasized the creation of nonequilibrium vibrational states of reactive molecules. In the liquid phase there are so many collisions that the molecules tend to remain near equilibrium in terms of their translational, vibrational, and rotational energies.

The maximum rate with which reactants can diffuse together in liquids may be calculated using the macroscopic theory of diffusion and experimentally determined diffusion coefficients of the reactants. The elementary theory of diffusion-controlled reactions was developed in 1917 by Smoluchowski in connection with his theoretical study of the coagulation of gold sols.

The diffusion coefficient D is defined in terms of Fick's first law, which is given in Section 9.11. The diffusion coefficients of low molecular weight solutes in aqueous solution at $25°$ are of the order of 10^{-9} m^2 s^{-1}. Experimental methods for determining diffusion coefficients in dilute aqueous solutions are discussed in Section 11.10. The diffusion coefficients for ions may be calculated from their mobilities (Section 11.11).

Smoluchowski considered spherical particles with radii R_1 and R_2 that could be considered to react when they diffused within a distance $R_{12} = R_1 + R_2$ of each other.

We may imagine one reactant molecule stationary and serving as a sink. Since $C = 0$ at distance R_{12}, a spherically symmetrical concentration gradient is set up. The flux through this concentration gradient is calculated and is expressed as a second-order rate constant k_a for association by

$$k_a = 4 \times 10^3 \pi N_A (D_1 + D_2) R_{12} f \qquad (10.120)$$

where the diffusion coefficients of the reactants are expressed in m^2 s^{-1}, R_{12} in m and k in M^{-1} s^{-1}. The electrostatic factor f is the factor by which the reaction is speeded up if the reactants have opposite charges and attract each other or is slowed down if the reactants have the same charge and repel each other. If the ionic strength is so low that ion atmospheres may be neglected, f is given by

$$f = \frac{z_1 z_2 e^2}{4\pi\epsilon_0 \epsilon k T R_{12}} \left[\exp\left(\frac{z_1 z_2 e^2}{4\pi\epsilon_0 \epsilon k T R_{12}} \right) - 1 \right]^{-1} \qquad (10.121)$$

where the charges on the ions are $z_1 e$ and $z_2 e$, ϵ is the dielectric constant, and ϵ_0 is the permittivity of free space. More details on the derivation of the equations for diffusion controlled reactions are given by Amdur and Hammes.*

---

* I. Amdur and G. G. Hammes, *Chemical Kinetics*, McGraw-Hill Book Co., New York, 1966.

If $D_1 + D_2 = 10^{-9} \, \text{m}^2 \, \text{s}^{-1}$ and $R_{12} = 5 \times 10^{-10} \, \text{m}$, $k_a \cong 4 \times 10^9 \, \text{M}^{-1} \, \text{s}^{-1}$. This value of the second-order rate constant cannot be exceeded in water at room temperature unless the reactants have higher diffusion coefficients or the reactants attract each other electrostatically.

Example 10.4 Show that if A and B can be represented by spheres of the same radius that react when they touch the second-order rate constant is given by

$$k_a = \frac{8 \times 10^3 \, RT}{3\eta} \, \text{M}^{-1} \, \text{s}^{-1}$$

For water at 25° C, $\eta = 8.95 \times 10^{-4} \, \text{kg m}^{-1} \, \text{s}^{-1}$. Calculate k at 25° C.

$$k_a = 4 \times 10^3 \, \pi N_A (D_1 + D_2) R_{12} f$$

For spheres $D = RT/N_A 6\pi\eta r$ (equation 20.28)

$$k_a = 4 \times 10^3 \pi N_A \left(\frac{2RT}{N_A 6\pi\eta r} \right) (2r)$$

$$= \frac{8 \times 10^3 \, RT}{3\eta}$$

At 25°

$$k_a = \frac{8(10^3 \, \text{mol m}^{-3} \, \text{M}^{-1})(8.314 \, \text{J K}^{-1} \, \text{mol}^{-1})(298 \, \text{K})}{3(8.95 \times 10^{-4} \, \text{kg m}^{-1} \, \text{s}^{-1})}$$

$$= 7.4 \times 10^9 \, \text{M}^{-1} \, \text{s}^{-1}$$

The temperature coefficients for diffusion-controlled reactions in water are small because they correspond with the temperature coefficient of the viscosity of liquid water ($E_a = 2 \, \text{kcal mol}^{-1}$).

In Göttingen Eigen and his coworkers have developed relaxation methods (Section 10.2) to measure the fastest reactions in aqueous solutions. In 1955 Eigen and DeMaeyer measured the second-order rate constant for the combination of hydrogen ions with hydroxyl ions in aqueous solution; at 25°, $k = 1.4 \times 10^{11} \, \text{M}^{-1} \, \text{s}^{-1}$. Substitution of this value in equation 10.120, along with the proper electrostatic factor yields $R_{12} = 7.5$ Å. Since this reaction radius is equal to about three O—H bond distances, the proton apparently "tunnels" to the hydroxyl ion once it is this close. The reaction of H^+ with OH^- is the fastest reaction that occurs in aqueous solution because of this "tunneling" and the fact that H^+ and OH^- have anomalously high mobilities in aqueous solution (Section 11.6).

The rate constants k_a for some diffusion controlled association reactions of H^+ and OH^- are given in Table 10.4, as measured by Eigen and coworkers. The relation between these rate constants and the experimentally determined relaxation time is derived in the next section. Since the association reactions are diffusion controlled the difference in acid dissociation constants for various weak acids can be attributed primarily to the rate of dissociation of protons. The reaction of

Table 10.4   Rate Constants at 25° for Elementary Reactions in Dilute Aqueous Solutions (k_a applies to the association reaction and k_d to the dissociation reaction)

	k_a $M^{-1}\,s^{-1}$	k_d s^{-1}
$H^+ + OH^- \rightleftharpoons H_2O$	1.4×10^{11}	2.5×10^{-5}
$D^+ + OD^- \rightleftharpoons D_2O$	8.4×10^{10}	2.5×10^{-6}
$H^+ + F^- \rightleftharpoons HF$	1.0×10^{11}	7×10^{7}
$H^+ + CH_3CO^- \rightleftharpoons CH_3CO_2H$	4.5×10^{10}	7.8×10^{5}
$H^+ + C_6H_5CO_2^- \rightleftharpoons C_6H_5CO_2H$	3.5×10^{10}	2.2×10^{6}
$H^+ + NH_3 \rightleftharpoons NH_4^+$	4.3×10^{10}	24.6
$H^+ + C_3N_2H_4 \rightleftharpoons C_3N_2H_5^+$ (imidazole)	1.8×10^{10}	1.1×10^{3}
$OH^- + NH_4^+ \rightleftharpoons NH_3 + H_2O$	3.4×10^{10}	6×10^{5}
$OH^- + C_3N_2H_5^+ \rightleftharpoons C_3N_2H_4 + H_2O$ (imidazole)	2.5×10^{10}	2.5×10^{3}
$OH^- + {}^+H_3NCH_2CO_2^- \rightleftharpoons H_2NCH_2CO_2^- + H_2O$ (glycine)	1.4×10^{10}	8.4×10^{5}

hydroxyl ions with various weak acids is also approximately diffusion controlled because the rate constants are in accord with equation 10.120.

10.23   RELAXATION TIME FOR A SIMPLE REACTION

When a reaction is displaced slightly from equilibrium the return to equilibrium is first order in the displacement from equilibrium. For example, consider the reaction

$$A + B \underset{k_{-1}}{\overset{k_1}{\rightleftharpoons}} C \qquad (10.122)$$

for which the rate equation is

$$\frac{d(C)}{dt} = k_1(A)(B) - k_{-1}(C) \qquad (10.123)$$

At equilibrium

$$0 = k_1(A)_{eq}(B)_{eq} - k_{-1}(C)_{eq} \qquad (10.124)$$

Equations 10.123 may be written in terms of displacements $\Delta(C)$ from the final equilibrium concentrations by introducing

$$(A) = (A)_{eq} - \Delta(C) \qquad (10.125)$$

$$(B) = (B)_{eq} - \Delta(C) \qquad (10.126)$$

$$(C) = (C)_{eq} + \Delta(C) \qquad (10.127)$$

Thus

$$\frac{d\Delta(C)}{dt} = k_1[(A)_{eq} - \Delta(C)][(B)_{eq} - \Delta(C)] - k_{-1}[(C)_{eq} + \Delta(C)] \qquad (10.128)$$

If the displacement from equilibrium $\Delta(C)$ is small

$$\frac{d\Delta(C)}{dt} = -\{k_1[(A)_{eq} + (B)_{eq}] + k_{-1}\}\Delta(C) = -\frac{\Delta(C)}{\tau} \qquad (10.129)$$

where equation 10.124 has been used and the term in $[\Delta(C)]^2$ has been neglected because $\Delta(C)$ is small. Thus the rate of approach to equilibrium is proportional to the displacement from equilibrium $\Delta(C)$. It is customary to use the relaxation time τ (equation 10.6) to characterize the rate of return to equilibrium. From equation 10.129 we see that

$$\tau = \{k_{-1} + k_1[(A)_{eq} + (B)_{eq}]\}^{-1} \qquad (10.130)$$

Thus k_{-1} and k_1 may be obtained as slope and intercept of a plot τ^{-1} versus $(A)_{eq} + (B)_{eq}$.

Example 10.5 When a sample of pure water in a small conductivity cell is heated suddenly with a pulse of microwave radiation, equilibrium in the water dissociation reaction does not exist at the new higher temperature until additional dissociation occurs. It is found that the relaxation time for the return to equilibrium at $25°$ is $36~\mu s$. Calculate k_1 and k_{-1}.

$$H_2O \underset{k_{-1}}{\overset{k_1}{\rightleftharpoons}} H^+ + OH^-$$

$$\tau = \frac{1}{k_1 + k_{-1}[(H^+) + (OH^-)]}$$

$$K = \frac{(H^+)(OH^-)}{(H_2O)} = \frac{k_1}{k_{-1}} = \frac{10^{-14}}{55.5} = 1.8 \times 10^{-16}$$

Eliminating k_1, we have

$$\tau = \frac{1}{k_{-1}[K + (H^+) + (OH^-)]} = \frac{1}{k_{-1}[(1.8 \times 10^{-16}) + (2 \times 10^{-7})]}$$

$$= 36 \times 10^{-6}~s$$

$$k_{-1} = 1.4 \times 10^{11}~M^{-1}\,s^{-1}$$

$$k_1 = Kk_{-1} = (1.8 \times 10^{-16})(1.4 \times 10^{11}) = 2.5 \times 10^{-5}~s^{-1}$$

If the reaction being studied involves two steps, there will be two independent rate equations. If the reactions are both near to equilibrium, these equations may be linearized, and the two linear differential equations will yield two relaxation times. The return to equilibrium will then be given by the sum of two exponential terms. In general the number of exponential terms is equal to the number of independent reactions.*

---

* I. Amdur and G. G. Hammes, *Chemical Kinetics*, McGraw-Hill Book Co., New York, 1966, p. 138.

10.24   CATALYSIS

A substance that increases the rate of a chemical reaction without being used up
in the overall reaction is called a *catalyst*. There are many different kinds of
catalysts and many different mechanisms by which catalysts operate. The
catalyst goes through a cycle in which it is used and regenerated so that it is
used over and over again. In homogeneous catalysis the catalyst is a molecule or
ion in homogeneous solution; in heterogeneous catalysis the reaction occurs on
a surface. The rate of a catalytic reaction is usually proportional to the concentration of the catalyst or to the area of the surface.

The catalyst operates by providing another path for the reaction that has a
higher rate than that available in the absence of the catalyst. The rate may be
greater because of lower activation energies or higher frequency factors or both.

Catalysts speed up both the forward and reverse reactions in such a way that
the equilibrium constant for the overall reaction is not affected. If this were
not true, a perpetual motion machine could be constructed using a catalyst to
regenerate a reactant for an electrochemical cell.

Catalysis may occur in the gas phase as well as in solution or at a surface. An
example was discovered by Johnston* in his studies of the chemical reactions in the
upper atmosphere that are involved in the decomposition of ozone. These studies
are important because nitrogen oxides from the exhaust of supersonic planes
provide a catalytic pathway for the decomposition of ozone.

$$NO + O_3 = O_2 + NO_2 \tag{10.131}$$

$$NO_2 + O = NO + O_2 \tag{10.132}$$

$$\overline{O_3 + O = 2O_2} \tag{10.133}$$

10.25   ACID AND BASE CATALYSIS

Acids and bases catalyze many reactions in which they are not consumed.
Suppose the rate of disappearance of a substance S (often called the substrate of
the catalytic reaction) is first order in S; $-d(S)/dt = k(S)$. The first-order rate
constant k for the reaction in a buffer solution may be a linear function of (H^+)
(OH^-), (HA), and (A^-), where HA is the weak acid in the buffer and A^- is
the corresponding anion.

$$k = k_0 + k_{H^+}(H^+) + k_{OH^-}(OH^-) + k_{HA}(HA) + k_{A^-}(A^-) \tag{10.134}$$

In this expression k_0 is the first-order rate constant at sufficiently low concentrations of all of the catalytic species H^+, OH^-, HA, and A^-. The so-called
catalytic coefficients k_{H^+}, k_{OH^-}, k_{HA}, and k_{A^-} may be calculated from experiments

* H. S. Johnston, *Science*, **173**, 517 (1971).

with different concentrations of these species. If only the term $k_{H^+}(H^+)$ is important the reaction is said to be subject to specific hydrogen-ion catalysis. If the term $k_{HA}(HA)$ is important the reaction is said to be subject to *general acid catalysis*, and if the term $k_{A^-}(A^-)$ is important the reaction is said to be subject to *general base catalysis*.

By considering two types of catalytic mechanisms, we can see how different types of terms arise in equation 10.134. In the first mechanism a proton is transferred from an acid AH to the substrate S, and then the acid form of the substrate reacts with a water molecule to form the product P.

$$S + AH^+ \underset{k_{-1}}{\overset{k_1}{\rightleftharpoons}} SH^+ + A$$

$$SH^+ + H_2O \overset{k_2}{\longrightarrow} P + H_3O^+ \tag{10.135}$$

Assuming that SH^+ is in a steady state, then

$$\frac{d(SH^+)}{dt} = 0 = k_1(S)(AH^+) - [k_{-1}(A) + k_2](SH^+) \tag{10.136}^*$$

The rate of appearance of product is given by

$$\frac{d(P)}{dt} = k_2(SH^+) = \frac{k_1 k_2 (S)(AH^+)}{k_{-1}(A) + k_2} \tag{10.137}$$

where the second form is obtained by solving equation 10.136 for (SH^+). If $k_2 \gg k_{-1}(A)$,

$$\frac{d(P)}{dt} = k_1(S)(AH^+) \tag{10.138}$$

and the reaction is said to be general acid catalyzed. However, if $k_2 \ll k_{-1}(A)$,

$$\frac{d(P)}{dt} = \frac{k_1 k_2 (S)(AH^+)}{k_{-1}(A)} = \frac{k_1 k_2}{k_{-1}K} (S)(H^+) \tag{10.139}$$

where the second form is obtained by inserting $K = (A)(H^+)/(AH^+)$. Here the reaction appears to be specifically hydrogen-ion catalyzed although the proton is initially transferred from AH^+.

In the second mechanism the acid form of the substrate reacts with a base A rather than a water molecule.

$$S + AH^+ \underset{k_{-1}}{\overset{k_1}{\rightleftharpoons}} SH^+ + A$$

$$SH^+ + A \overset{k_2}{\longrightarrow} P + AH^+ \tag{10.140}$$

---

* Note (H_2O) is not written after k_2 because in dilute aqueous solutions (H_2O) cannot be appreciably changed, and so k_2 represents a first order rate constant.

The steady-state treatment of this mechanism leads to

$$\frac{d(P)}{dt} = k_2(SH^+)(A) = \frac{k_1 k_2(S)(AH^+)}{k_{-1} + k_2} \qquad (10.141)$$

This mechanism leads to general acid catalysis whether $k_{-1} \gg k_2$ or $k_{-1} \ll k_2$.

10.26 ENZYME CATALYSIS

The most amazing catalysts are the enzymes, which catalyze the multitudinous reactions in living organisms. Enzymes are proteins; this means that they are large polypeptides with definite three-dimensional structures, as described in Section 20.1. Some enzymes catalyze a single reaction. An example is fumarase, which catalyzes the hydration of fumarate to L-malate.

$$ \qquad (10.142)$$

The reactants in an enzyme reaction are generally referred to as substrates. This reaction may be represented by $S = P$ since the concentration of water is constant. Other enzymes catalyze a class of reactions of a given type like ester hydrolysis. To operate some enzymes require particular metal ions or coenzymes. An example of an enzymatic reaction involving a coenzyme is the alcohol dehydrogenase reaction

$$\text{alcohol} + NAD^+ = \text{aldehyde} + NADH + H^+ \qquad (10.143)$$

where NAD^+ is nicotinamide adenine dinucleotide in the oxidized form and NADH is the reduced form. In the cell there is a means of oxidizing NADH to NAD^+ so that it can participate over and over again in the oxidation of alcohol.

The rate of a reaction catalyzed by an enzyme is usually found to be directly proportional to the concentration of the enzyme. When the concentration of a substrate is varied it is usually found that the rate is first order in substrate at low substrate concentration and approaches zero order in substrate as the substrate concentration is increased; this is illustrated in Fig. 10.10a. Actually the observed rates of the overall reaction are steady-state rates. When solutions of enzyme and substrate are mixed there is initially a very fast reaction of the enzyme with the substrate that can only be observed using special methods (Cf., relaxation methods in Section 10.2). The nature of the steady state reaction can be understood by considering the simplest type of mechanism for the overall reaction $S = P$.

$$E + S \underset{k_2}{\overset{k_1}{\rightleftharpoons}} X \underset{k_4}{\overset{k_3}{\rightleftharpoons}} E + P \qquad (10.144)$$

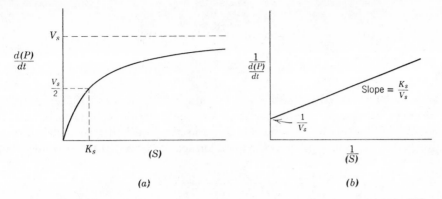

Fig. 10.10 (a) Plot of initial steady-state velocity of appearance of product P at different initial concentrations of substrate S. (b) Plot of reciprocal initial velocity versus reciprocal substrate concentration.

where E is the enzymatic site and X is an intermediate, often referred to as the enzyme-substrate complex. The rate equations for this mechanism are

$$\frac{d(X)}{dt} = k_1(E)(S) - (k_2 + k_3)(X) + k_4(E)(P) \tag{10.145}$$

$$\frac{d(P)}{dt} = k_3(X) - k_4(E)(P) \tag{10.146}$$

These two rate equations cannot be solved to obtain analytic expressions for (E), (S), (X), and (P) as functions of time, but these concentrations may be calculated by use of a computer for specific values of the four rate constants.

Since enzymatic reactions are generally studied with enzyme concentrations (strictly speaking molar concentrations of enzymatic sites) much lower than the concentrations of substrates, it is a good approximation to assume that the enzymatic reaction is in a steady state in which $d(X)/dt = 0$. Since $(E) + (X) = (E)_0$, where $(E)_0$ is the initial concentration of enzymatic sites, equation 10.145 may be written

$$\frac{d(X)}{dt} = k_1[(E)_0 - (X)](S) - (k_2 + k_3)(X) + k_4[(E)_0 - (X)](P) = 0 \tag{10.147}$$

Eliminating (X) between equations 10.146 and 10.147 we obtain

$$\frac{d(P)}{dt} = \frac{\dfrac{V_S}{K_S}(S) - \dfrac{V_P}{K_P}(P)}{1 + \dfrac{(S)}{K_S} + \dfrac{(P)}{K_P}} \tag{10.148}$$

where

$$V_S = k_3(E)_0 \qquad V_P = k_2(E)_0$$

$$K_S = \frac{k_2 + k_3}{k_1} \qquad K_P = \frac{k_2 + k_3}{k_4}$$

Equation 10.148 is written in this way because V_S, V_P, K_S, and K_P represent the collections of kinetic parameters that may be determined experimentally. If only S is added to the enzyme solution and the rate is determined before an appreciable amount of product has been formed, equation 10.148 becomes

$$\frac{d(P)}{dt} = \frac{V_S}{1 + \dfrac{K_S}{(S)}} \qquad (10.149)$$

If $(S) \ll K_S$, the steady-state velocity is directly proportional to (S), so that the reaction is first order in substrate as shown in Fig. 10.10a. If $(S) \gg K_S$, the reaction is zero order in S. In both cases the reaction is first order with respect to the total concentration of enzymatic sites.

The quantity $V_S = k_3(E)_0$ is called the maximum velocity for the enzymatic reaction, and k_3 is called the turnover number for the forward reaction. This latter quantity is greater than 10^8 min^{-1} for catalase, which catalyzes the decomposition of H_2O_2 to $H_2O + \frac{1}{2}O_2$, and is about 10 min^{-1} for chymotrypsin, which catalyzes the hydrolysis of a number of esters and amides.

The quantity $K_S = (k_2 + k_3)/k_1$ is called the Michaelis constant for the substrate, and it is evident from equation 10.149 that it is equal to the concentration of substrate required to give half the maximum velocity. The values of V_S and K_S may be obtained from a plot of $[d(P)/dt]^{-1}$ versus $(S)^{-1}$, which should be linear for this rate law, as shown in Fig. 10.10b.

The way in which V_S and K_S depend on pH, salt concentration, coenzyme concentration, etc., gives further information about the enzymatic mechanism.

At equilibrium the two terms in the numerator of equation 10.148 are equal and so

$$K = \frac{(P)_{eq}}{(S)_{eq}} = \frac{V_S K_P}{V_P K_S} \qquad (10.150)$$

Thus, the kinetic parameters for the forward and reverse reactions are not independent, but are related through the equilibrium constant K.

Compounds that are structurally related to the substrate or product often combine with the catalytic site of the enzyme and cause inhibition; that is, the enzyme catalyzed reaction is slowed down by the inhibitor. Since the substrate and the inhibitor compete for the same site the effect of the inhibitor may be reduced by raising the substrate concentration. This type of competitive inhibition may be represented by the mechanism

$$\text{E} + \text{S} \rightleftharpoons \text{ES} \rightarrow \text{E} + \text{P}$$

$$\text{E} + \text{I} \rightleftharpoons \text{EI} \qquad (10.151)$$

where I is the inhibitor. The steady-state rate law is

$$\frac{d(P)}{dt} = \frac{V_S}{1 + \dfrac{K_S}{(S)}\left[1 + \dfrac{(I)}{K_I}\right]} \qquad (10.152)$$

where K_I is the dissociation constant of EI into E and I.

In general an enzymatic reaction has an optimum pH; the maximum velocity V decreases as the pH is raised or lowered from the optimum pH. In the neutral pH range the effects are generally reversible, but proteins are irreversibly denatured at extreme pH values. The reversible effects of pH on kinetics may arise from changes in the degree of ionization of the substrate, but if the ionization of the substrate does not change in the region of interest the effect is attributable to the ionization of the enzyme-substrate complex. If the enzyme-substrate complex exists in three states with different numbers of protons, and if only the intermediate form breaks down to give product, the expression for the effect of pH on the maximum velocity may be derived from

$$\begin{array}{c} \text{ES} \\ K_{bES}\Big\updownarrow \\ \text{HES} \xrightarrow{\;k\;} \text{enzyme} + \text{product} \\ K_{aES}\Big\updownarrow \\ \text{H}_2\text{ES} \end{array} \qquad (10.153)$$

where K_{aES} and K_{bES} are acid dissociation constants, and k is the rate constant for the rate-determining step. Since

$$\begin{aligned} (E)_0 &= (ES) + (HES) + (H_2ES) \\ &= (HES)[1 + (H^+)/K_{aES} + K_{bES}/(H^+)] \end{aligned} \qquad (10.154)$$

$$V = k(HES) = \frac{k(E)_0}{1 + (H^+)/K_{aES} + K_{bES}/(H^+)} \qquad (10.155)$$

If this is the mechanism a plot of V versus pH is a symmetrical bell-shaped curve. Since a protein has many dissociable acid groups it is perhaps surprising that the experimental results are sometimes as simple as they are.

There is a simple explanation of why two acid groups in the catalytic site might have the total effect on the kinetics that is represented by equation 10.153. If the catalytic function inolves both an acidic function and a basic function then we would expect H_2ES to be inactive because the basic site is occupied by a proton and ES to be inactive because it cannot donate a proton to the substrate. Only HES can yield product because it has one group that can donate a proton and another that can accept a proton.

Mechanisms like the ones we have been discussing can give a variety of complicated effects; that is, the rate law may have a very complicated form. But in general (ES) can increase no more rapidly than (S). However, there are a number of experimental situations where substrates or inhibitors produce larger changes than that. In other words curves analogous to the sigmoid oxygen binding curve for hemoglobin (Section 7.13) are obtained. This has been found to be especially true of enzymes of importance in the regulation of metabolic pathways. These cooperative effects are encountered with multisite enzymes, not single site enzymes, because the cooperative effect involves an increased affinity of a second site for a substrate when a first site is occupied. As in the case of hemoglobin this interaction involves a structural change. According to the Monod–Changeaux–Wyman model the multisite enzyme can exist in at least two states; this is probably an over-simplification but two is the minimum number of states required to explain the observed effects. In each of the two states the conformations of all the subunits are assumed to be the same. The effector molecule, which may be a substrate, drives the equilibrium toward one or the other of these two states. If the effector drives the equilibrium in the direction that produces an enhanced rate of reaction, the effector is called an *activator*. If it causes a reduction in rate it is called an inhibitor. As we have seen in the case of hemoglobin the effect is multiplied by the fact that one effector molecule affects several catalytic sites on the molecule. The fact that enzymatic activities may be affected by various substances present in the cell provides a mechanism for the control of the rates of reactions in living things so that intermediates do not accumulate.

10.27   HETEROGENEOUS REACTIONS

A homogeneous reaction is one occurring in a single phase and a heterogeneous reaction is one occurring in more than one phase, for example, in the gas phase and on the surface of a solid catalyst. Heterogeneous reactions are carried out on a large scale in industry. The development of catalysts is especially important for slow reactions that evolve heat. Although the equilibrium constant may be favorable at room temperature, the rate may be so low that the reaction cannot be used practically. When the temperature is raised the rate is increased, but at high temperatures the equilibrium constant may be unfavorable, as is readily surmised from Le Châtelier's principle. A reaction of this type is $N_2 + 3H_2 = 2NH_3$, which is the basis for the Haber process for the manufacture of ammonia. To carry out this reaction at temperatures where the equilibrium constant is sufficiently favorable special iron catalysts are used. Vanadium and other metals are added to make the catalyst more effective and to protect it from too rapid inactivation by traces of impurities in the gases. Substances that enhance the activity of catalysts are called *promoters*, and substances that inhibit catalytic activity are called *poisons*. Since only a fraction of the surface of the catalyst may be involved, it is easy to see how relatively small amounts of promoters and poisons may be effective. In the manufacture of sulfuric acid by the contact process, the presence of a very minute amount of arsenic completely destroys the catalytic activity of the platinum catalyst by forming platinum arsenide at the surface.

Enormous quantities of solid catalysts are used in the petroleum industry for "cracking," which increases the yield of gasoline from petroleum, and for "reforming," which causes the rearrangement of the molecular structures and raises the octane rating of gasoline. To facilitate the regeneration of these catalysts they may be circulated as fine particles in the gas stream.

The catalytic activity of the walls of the containing vessel is a factor in many gaseous reactions. For surface reactions the temperature coefficient is usually small, because the activation energies are low and the slowest process is the diffusion of the products away from the walls. A $10°$ rise at about 300 K increases the diffusion rate by only about 3 %, whereas the rate of a chemical reaction often increases about 300 %, as already explained. Since homogeneous reactions generally have higher activation energies than the corresponding heterogeneous reactions, homogeneous reactions are favored at higher temperatures and heterogeneous reactions at lower temperatures.

The mechanisms of surface-catalyzed reactions involve a sequence of steps: (a) adsorption of the reacting gases on the surface, (b) reaction on the surface, and (c) desorption of the products. Under the influence of forces at the surface, reactions occur at much higher rates than in the gas phase. The nature of the surface determines the products which are obtained, different catalysts yielding different products with the same reacting gases.

Surface-catalyzed reactions often obey a rate equation that may be derived from the Langmuir adsorption isotherm (Section 8.9). The rate of appearance of product dx/dt formed by the surface reaction is proportional to the fraction of the surface occupied by reactant molecules as given by equation 8.30.

$$\frac{dx}{dt} = k'\theta = \frac{k'(k/r)P}{1 + (k/r)P}$$ (10.156)

It can be seen from equation 10.156 that at low pressures, where $1 \gg (k/r)P$, the rate is directly proportional to the pressure.

$$\frac{dx}{dt} = k'\left(\frac{k}{r}\right)P$$ (10.157)

It can also be seen from equation 10.156 that at high pressures, where $1 \ll (k/r)P$, the rate is independent of the pressure of the reactant: $(dx/dt) = k'$. Under these conditions the surface is saturated, and further increasing the pressure cannot increase the number of reactant molecules adsorbed. When ammonia is decomposed on a tungsten filament, for example, the rate of decomposition is independent of the pressure, over a wide range of pressure.

It is often found that surface-catalyzed reactions are slowed down by an accumulation of product molecules. This is a result of the adsorption of product molecules on the active sites, which may be taken into account quantitatively in the theory.

References

I. Amdur and G. G. Hammes, *Chemical Kinetics*, McGraw-Hill Book Co., New York, 1966.

R. P. Bell, *The Proton in Chemistry*, Cornell University Press, Ithaca, 1959.

S. W. Benson, *The Foundations of Chemical Kinetics*, McGraw-Hill Book Co., New York, 1960.

S. W. Benson, *Thermochemical Kinetics*, Wiley, New York, 1968.

S. W. Benson and H. E. O'Neal, *Kinetic Data on Gas Phase Unimolecular Reactions*, NSRDS-NBS-21, National Bureau of Standards, U.S. Dept. of Commerce, Washington D.C., 1970.

P. D. Boyer, *The Enzyme*, Vol. II, Kinetics and Mechanism, 3rd ed., Academic Press, New York, 1970.

J. N. Bradley, *Shock Waves in Chemistry and Physics*, Wiley, New York, 1962.

F. S. Dainton, *Chain Reactions; An Introduction*, 2nd ed., Barnes and Noble, New York, 1966.

A. A. Frost and R. G. Pearson, *Kinetics and Mechanism*, Wiley, New York, 1961.

W. C. Gardiner, Jr., *Rates and Mechanisms of Chemical Reactions*, W. A. Benjamin, Inc., New York, 1969.

S. Glasstone, K. J. Laidler, and H. Eyring, *The Theory of Rate Processes*, McGraw-Hill Book Co., New York, 1941.

E. F. Green and J. P. Toennies, *Chemical Reactions in Shock Waves*, Academic Press, New York, 1964.

H. Gutfreund, *An Introduction to Study of Enzymes*, Wiley, New York, 1965.

H. S. Johnston, *Gas Phase Reaction Rate Theory*, The Ronald Press Co., New York, 1966.

E. L. King, *How Chemical Reactions Occur*, W. A. Benjamin, Inc., New York, 1963.

K. J. Laidler, *Chemical Kinetics*, McGraw-Hill Book Co., New York, 1965.

N. M. Rodiguin and E. N. Rodiguina, *Consecutive Chemical Reactions*, D. Van Nostrand, Princeton, 1964.

A. F. Trotman-Dickenson, *Gas Kinetics*, Butterworths Scientific Publications, London, 1955.

A. F. Trotman-Dickenson and G. S. Milne, *Tables of Bimolecular Gas Phase Reactions*, NSRDS-NBS-9, National Bureau of Standards, U.S. Dept. of Commerce, Washington D.C., 1967.

R. E. Weston, Jr., and H. A. Schwartz, *Chemical Kinetics*, Prentice-Hall, Inc., Englewood Cliffs, N.J., 1972.

Problems

10.1 The half life of a first-order chemical reaction A → B is 10 min. What percent of A remains after 1 hr?   *Ans.* 1.56%.

10.2 The following data were obtained on the rate of hydrolysis of 17% sucrose in 0.099 M HCl aqueous solution at 35°:

Time, min	9.82	59.60	93.18	142.9	294.8	589.4
Sucrose remaining, %	96.5	80.3	71.0	59.1	32.8	11.1

What is the order of the reaction with respect to sucrose and the value of the rate constant k?
Ans. First order, 6.20×10^{-5} s^{-1}.

10.3 Methyl acetate is hydrolyzed in approximately 1 M HCl at 25°. Aliquots of equal volume are removed at intervals and titrated with a solution of NaOH. Calculate the first-order rate constant from the following experimental data:

Time, s	339	1242	2745	4546	∞
Volume, cm^3	26.34	27.80	29.70	31.81	39.81

Ans. 1.26×10^{-4} s^{-1}.

10.4 The reaction between propionaldehyde and hydrocyanic acid has been studied at 25° by W. J. Svirbely and J. F. Roth [*J. Am. Chem. Soc.*, **75,** 3106 (1953)]. In a certain aqueous solution at 25° the concentrations at various times were as follows:

Time, min	2.78	5.33	8.17	15.23	19.80	∞
HCN, mol l^{-1}	0.0990	0.0906	0.0830	0.0706	0.0653	0.0424
C_3H_7CHO, mol l^{-1}	0.0566	0.0482	0.0406	0.0282	0.0229	0.0000

What is the order of the reaction and the value of the rate constant k?

Ans. Second order, 0.676 1 mol^{-1} min^{-1}.

10.5 The reaction $CH_3CH_2NO_2 + OH^- \rightarrow H_2O + CH_3CHNO_2^-$ is of second order, and k at 0° is 39.1 M^{-1} min^{-1}. An aqueous solution is made 0.004 molar in nitroethane and 0.005 molar in NaOH. How long will it take for 90% of the nitroethane to react?

Ans. 26.3 min.

10.6 Hydrogen peroxide reacts with thiosulfate ion in slightly acid solution as follows:

$$H_2O_2 + 2S_2O_3^{2-} + 2H^+ \rightarrow 2H_2O + S_4O_6^{2-}$$

This reaction rate is independent of the hydrogen-ion concentration in the pH range 4–6. The following data were obtained at 25° and pH 5.0:

Initial concentrations: $(H_2O_2) = 0.03680$ M; $(S_2O_3^{2-}) = 0.02040$ M

Time, min	16	36	43	52
$(S_2O_3^{2-}) \times 10^3$	10.30	5.18	4.16	3.13

(*a*) What is the order of the reaction?
(*b*) What is the rate constant?

Ans. (*a*) Second order, (*b*) 0.572 liter mol^{-1} s^{-1}.

10.7 A solution of A is mixed with an equal volume of a solution of B containing the same number of moles, and the reaction A + B = C occurs. At the end of 1 hr A is 75% reacted. How much of A will be left unreacted at the end of 2 hr if the reaction is (*a*) first order in A and zero order in B; (*b*) first order in both A and B; (*c*) zero order in both A and B?

Ans. (*a*) 6.25, (*b*) 14.3, (*c*) 0%.

10.8 Derive the integrated rate equation for a reaction of $\frac{1}{2}$ order. Derive the expression for the half-life of such a reaction.

$$Ans. \ (A)_0^{1/2} - (A)^{1/2} = \frac{k}{2}t$$

$$t_{1/2} = \frac{\sqrt{2}}{k}(\sqrt{2}-1)(A)_0^{1/2}$$

10.9 For the reaction 2A = B + C the rate law for the forward reaction is

$$-\frac{d(A)}{dt} = k(A)$$

Give two possible rate laws for the reverse reaction.

Ans. $k(B)(C)/(A)$; $k(B)^{1/2}(C)^{1/2}$.

10.10 The reaction $2NO + O_2 \rightarrow 2NO_2$ is third order. Assuming that a small amount of NO_3 exists in rapid reversible equilibrium with NO and O_2 and that the rate-determining step is the slow bimolecular reaction $NO_3 + NO \rightarrow 2NO_2$, derive the rate equation for this mechanism.

Ans. $d(NO_2)/dt = k(NO)^2(O_2)$.

10.11   The hydrolysis of $(CH_2)_6C\!\!\begin{smallmatrix}\diagup Cl\\ \diagdown CH_3\end{smallmatrix}$ in 80% ethanol follows the first-order rate equation.

The values of the specific reaction-rate constants, as determined by H. C. Brown and M. Borkowski [*J. Am. Chem. Soc.*, **74**, 1896 (1952)], are as follows:

Temp., °C	0	25	35	45
k, s^{-1}	1.06×10^{-5}	3.19×10^{-4}	9.86×10^{-4}	2.92×10^{-3}

(a) Plot $\log k$ against $1/T$; (b) calculate the activation energy; (c) calculate the preexponential factor.   *Ans.* (b) 21,000 cal mol^{-1}, (c) 7.8×10^{11} s^{-1}.

10.12   (a) The viscosity of water changes about 2% per degree at room temperature. What is the activation energy for this process? (b) The activation energy for a reaction is 15 kcal mol^{-1}. Calculate $k_{35°}/k_{25°}$.   *Ans.* (a) 3.6 kcal mol^{-1}. (b) 2.27.

10.13   If a first-order reaction has an activation energy of 25,000 cal mol^{-1}, and, in the equation $k = Ae^{-E_a/RT}$, A has a value of 5×10^{13} s^{-1}, at what temperature will the reaction have a half-life of (a) 1 min; (b) 30 days?   *Ans.* (a) 76°, (b) −4°.

10.14   Isopropenyl allyl ether in the vapor state isomerizes to allyl acetone according to a first-order rate equation. The following equation gives the influence of temperature on the rate constant (in s^{-1}):

$$k = 5.4 \times 10^{11}e^{-29,300/RT}$$

At 150° how long will it take to build up a partial pressure of 300 Torr of allyl acetone. starting with 760 Torr of isopropenyl allyl ether [L. Stein and G. W. Murphy, *J. Am. Chem, Soc.*, **74**, 1041 (1952)]?   *Ans.* 1210 s.

10.15   F. W. Schuler and G. W. Murphy [*J. Am. Chem. Soc.*, **72**, 3155 (1950)] studied the thermal rearrangement of vinyl allyl ether to allyl acetaldehyde in the range 150–200° and found that

$$k = 5 \times 10^{11}e^{-30,600/RT}$$

Calculate (a) the enthalpy of activation and (b) the entropy of activation, and (c) give an interpretation of the latter.

Ans. (a) 29.7 kcal mol^{-1}, (b) −7.7 cal K^{-1} mol^{-1},

(c) The activated complex may be an improbable ring structure.

10.16   For the two parallel reactions $A \xrightarrow{k_1} B$ and $A \xrightarrow{k_2} C$ show that the activation energy E' for the disappearance of A is given in terms of the activation energies E_1 and E_2 for the two paths by

$$E' = \frac{k_1E_1 + k_2E_2}{k_1 + k_2}$$

10.17   The thermal decomposition of gaseous acetaldehyde is a second-order reaction. The value of E_a is 45,500 cal mol^{-1}, and the molecular diameter of the acetaldehyde molecule is 5×10^{-8} cm. (a) Calculate the number of molecules colliding per milliliter per second at 800 K and 760 Torr pressure. (b) Calculate k in liters mol^{-1} s^{-1}.

Ans. (a) 2.9×10^{28} molecules ml^{-1} s^{-1}, (b) 0.077 l mol^{-1} s^{-1}.

10.18   Set up the rate expressions for the following mechanism:

$$A \underset{k_2}{\overset{k_1}{\rightleftharpoons}} B$$

$$B + C \xrightarrow{k_3} D$$

If the concentration of B is small compared with the concentrations of A, C, and D, the steady-state approximation may be used to derive the rate law. Show that this reaction may follow the first-order equation at high pressures and the second-order equation at low pressures. *Ans.* $d(D)/dt = k_1 k_3 (A)(C)/[k_2 + k_3(C)]$.

10.19 The reaction $NO_2Cl = NO_2 + \frac{1}{2}Cl_2$ is first order and appears to follow the mechanism

$$NO_2Cl \xrightarrow{k_1} NO_2 + Cl$$

$$NO_2Cl + Cl \xrightarrow{k_2} NO_2 + Cl_2$$

(*a*) Assuming a steady state for the chlorine atom concentration, show that the empirical first-order rate constant can be identified with $2k_1$. (*b*) The following data were obtained by H. F. Cordes and H. S. Johnston [*J. Am. Chem. Soc.*, **76**, 4264 (1954)] at 180°. In a single experiment the reaction is first order, and the empirical rate constant is represented by k. Show that the reaction is second order at these low gas pressures and calculate the second-order rate constant:

$c \times 10^8$, mol cm^{-3}	5	10	15	20
$k \times 10^4$, s^{-1}	1.7	3.4	5.2	6.9

Ans. 3.4×10^3 cm^3 mol^{-1} s^{-1}.

10.20 The following table gives kinetic data [Y. T. Chia and R. E. Connick, *J. Phys. Chem.*, **63**, 1518 (1959)] for the following reaction at 25°

$$OCl^- + I^- = OI^- + Cl^-$$

(OCl$^-$)	(I$^-$)	(OH$^-$)	$\dfrac{d(IO^-)}{dt} \times 10^4$
	mol liter^{-1}		mol liter^{-1} s^{-1}
0.0017	0.0017	1.00	1.75
0.0034	0.0017	1.00	3.50
0.0017	0.0034	1.00	3.50
0.0017	0.0017	0.5	3.50

What is the rate law for the reaction and what is the value of the rate constant?

$$Ans. \quad \frac{d(OI^-)}{dt} = \frac{(60 \text{ s}^{-1})(I^-)(OCl^-)}{(OH^-)}.$$

10.21 Calculate the first-order rate constants for the dissociation of the following weak acids: acetic acid, acid form of imidazole $C_3N_2H_5^+$, NH_4^+. The corresponding acid dissociation constants are 1.75×10^{-5}, 1.2×10^{-7}, 5.71×10^{-10}, respectively. The second-order rate constants for the formation of the acid forms from a proton plus the base are 4.5×10^{10}, 1.5×10^{10}, and 4.3×10^{10} M^{-1} s^{-1}, respectively. *Ans.* 7.9×10^5, 1.8×10^3, 25 s^{-1}.

10.22 Derive the relation between the relaxation time τ and the rate constants for the reaction $A \underset{k_2}{\overset{k_1}{\rightleftharpoons}} B$, which is subjected to a small displacement from equilibrium.

Ans. $\tau^{-1} = k_1 + k_2$.

10.23 For acetic acid at 25°

$$CH_3CO_2H \underset{4.5 \times 10^{10} \text{ M}^{-1} \text{ s}^{-1}}{\overset{7.8 \times 10^5 \text{ s}^{-1}}{\rightleftharpoons}} CH_3CO_2^- + H^+$$

what is the relaxation time τ for a 0.1 M solution?   *Ans.* 0.51 μs.

10.24 The mutarotation of glucose is first order in glucose concentration and is catalyzed by acids (A) and bases (B). The first-order rate constant may be expressed by an equation of the type that is encountered in reactions with parallel paths.

$$k = k_0 + k_{H^+}(H^+) + k_A(A) + k_B(B)$$

where k_0 is the first-order rate constant in the absence of acids and bases other than water. The following data were obtained by J. N. Brönsted and E. A. Guggenheim [*J. Am. Chem. Soc.*, **49**, 2554 (1927)] at 18° in a medium containing 0.02 M sodium acetate and various concentrations of acetic acid:

(CH_3CO_2H), mol l^{-1}	0.020	0.105	0.199
$k \times 10^4$, min^{-1}	1.36	1.40	1.46

Calculate k_0 and k_A. The term involving k_{H^+} is negligible under these conditions.

Ans. $k_0 = 1.34 \times 10^{-4}$ min^{-1}, $k_A = 5 \times 10^{-5}$ M^{-1} min^{-1}.

10.25 At pH 7 the measured Michaelis constant and maximum velocity for the enzymatic conversion of fumarate to L-malate

$$\text{fumarate} + H_2O = \text{L-malate}$$

are 4.0×10^{-3} M and $(1.3 \times 10^3 \text{ s}^{-1})(E)_0$, where $(E)_0$ is the total molar concentration of the enzyme. The Michaelis constant and maximum velocity for the reverse reaction are 10×10^{-6} M and $(0.80 \times 10^3 \text{ s}^{-1})(E)_0$. What is the equilibrium constant for this hydration reaction? (The activity of water is set equal to unity since it is in excess.) *Ans.* 4.1.

10.26 At 25° and pH 8 the maximum initial velocities V and Michaelis constants K of the fumarase reaction $F + H_2O = M$ are

$$V_F = (0.2 \times 10^3 \text{ s}^{-1})(E)_0 \qquad V_M = (0.6 \times 10^3 \text{ s}^{-1})(E)_0$$
$$K_F = 7 \times 10^{-6} \text{ M} \qquad\qquad K_M = 100 \times 10^{-6} \text{ M}$$

where $(E)_0$ is the total molar concentration of enzymatic sites. Calculate the values of the four rate constants in the mechanism

$$E + F \underset{k_2}{\overset{k_1}{\rightleftharpoons}} EX \underset{k_4}{\overset{k_3}{\rightleftharpoons}} E + M$$

and the equilibrium constant $(M)_{eq}/(F)_{eq}$.

Ans. $k_1 = 0.1 \times 10^9$ M^{-1} s^{-1}, $k_2 = 0.6 \times 10^3$ s^{-1},
$k_3 = 0.2 \times 10^3$ s^{-1}, $k_4 = 8 \times 10^6$ M^{-1} s^{-1}, 4.2.

10.27 The following initial velocities were determined spectrophotometrically for solutions of sodium succinate to which a constant amount of succinoxidase was added. The velocities are given as the change in absorbancy at 2500 Å in 10 s. Calculate V, K_S and K_I for malonate.

(succ.) (mM)	$\dfrac{A \times 10^3}{10 \text{ s}}$ No. Inhib.	15×10^{-6} M malonate
10	16.7	14.9
2	14.2	10.0
1	11.3	7.7
0.5	8.8	4.9
0.33	7.1	—

Ans 15.1, 0.38 mM, 8.6×10^{-3} mM.

10.28 The maximum initial velocities for an enzymatic reaction are determined at a series of pH values

pH	V
6.0	11
6.5	30
7.0	74
7.5	129
8.0	147
8.5	108
9.0	53

Calculate the values of the parameters V', K_a and K_b in

$$V = \frac{V'}{1 + (H^+)/K_a + K_b/(H^+)}$$

Hint: A plot of V versus pH may be constructed and the hydrogen ion concentration at the midpoint on the acid side referred to as $(H^+)_a$ and the hydrogen ion concentration at the midpoint on the basic side is referred to as $(H^+)_b$. Then

$$K_a = (H^+)_a + (H^+)_b - 4\sqrt{(H^+)_a(H^+)_b}$$

$$K_b = \frac{(H^+)_a(H^+)_b}{K_a}$$

Ans 234, 4.41×10^{-8}, 3.67×10^{-9}.

10.29 The reaction

$$SO_2Cl_2 = SO_2 + Cl_2$$

is first order with a rate constant of 2.2×10^{-5} s^{-1} at 320°. What percentage of SO_2Cl_2 is decomposed after being heated at 320° for 2hr?

10.30 The kinetics of the hydrolysis of an ester is studied by titrating the acid produced. A sample is withdrawn and titrated with alkali. The volumes required at various times are

Time, min	0	27	60	∞
Volume, ml	0	18.1	26.0	29.7

(a) Prove this reaction is first order. (b) Calculate the half-life.

10.31 A solution of ethyl acetate and sodium hydroxide was prepared that contained (at $t = 0$) 5×10^{-3} M ethyl acetate and 8×10^{-3} M sodium hydroxide. After 400 sec at 25° a 25-ml aliquot was found to neutralize 33.3 ml of 5×10^{-3} M hydrochloric acid. (a) Calculate the rate constant for this second-order reaction. (b) At what time would you expect 20.0 ml of hydrochloric acid to be required?

10.32 The second-order rate constant for an alkaline hydrolysis of ethyl formate in 85% ethanol (aqueous) at 29.86° is 4.53 liters mol^{-1} s^{-1} [H. M. Humphreys and L. P. Hammett, *J. Am. Chem. Soc.*, **78**, 521 (1956)]. (a) If the reactants are both present at 0.001 M, what will be the half-life of the reaction? (b) If the concentration of one of the reactants is doubled and of the other is cut in half, how long will it take for half the reactant present at the lower concentration to react?

10.33 It is found that the decomposition of HI to $H_2 + I_2$ at 508° has a half-life of 135 min when the initial pressure of HI is 0.1 atm and 13.5 min when the pressure is 1 atm. (a) Show that this proves that the reaction is second order. (b) What is the value of the rate constant in liters mol^{-1} s^{-1}? (c) What is the value of the rate constant in atm^{-1} s^{-1}?

10.34   The reaction $2NO + O_2 \rightarrow 2NO_2$ is third order and $dc_{NO_2}/dt = kc_{NO}^2 c_{O_2}$. The rate constant k has a value of 7.1×10^9 mol^{-2} ml^2 s^{-1} at $25°$. Air blown through a certain hot chamber and cooled quickly to $25°$ and 760 Torr contains 1% by volume of nitric oxide, NO, and 20% of oxygen. (a) How long will it take for 90% of this NO to be converted into nitrogen dioxide, NO_2 (or N_2O_4)? (b) If the gases are blown through at the rate of 5000 ft^3 min^{-1}, how large a chamber must be constructed to obtain this 90% conversion?

10.35   Derive the integrated rate equation for an nth order reaction

$$-\frac{d(A)}{dt} = k(A)^n, \qquad n > 1$$

Derive the expression for the half-life.

10.36   For the consecutive first-order reactions

$$A \xrightarrow[k=0.15\ s^{-1}]{} B \xrightarrow[k=1.1\ s^{-1}]{} C$$

plot curves which give the concentrations of A, B, and C as a function of time.

10.37   Parallel first-order reactions that compete for the reactant are often encountered because many products may be thermodynamically possible. For the reactions

$$A \xrightarrow{k_1} B$$

$$A \xrightarrow{k_2} C$$

show that B and C appear with the same half-life even though k_1 and k_2 have different values.

10.38   The initial rate of the reaction

$$BrO_3^- + 3SO_3^{2-} = Br^- + 3SO_4^{2-}$$

was found by F. S. Williamson and E. L. King [*J. Am. Chem. Soc.*, **79,** 5397 (1953)] to be given by $k(BrO_3^-)(SO_3^{2-})(H^+)$. Give one of the thermodynamically possible rate laws for the reverse reaction.

10.39   For the mechanism

setting

$$\frac{d(A)}{dt} = 0$$

$$\frac{d(B)}{dt} = 0$$

$$\frac{d(C)}{dt} = 0$$

yields

$$\frac{(B)_{eq}}{(A)_{eq}} = \frac{k_1 k_4 + k_1 k_5 + k_4 k_6}{k_2 k_4 + k_2 k_5 + k_3 k_5}$$

which looks strange to a chemist. If the three reactions are *each* at equilibrium, what is the relation between the six rate constants? Show that this relation reduces the above expression for $(B)_{eq}/(A)_{eq}$ to the correct result.

10.40 For the reaction

$$CO_2(+H_2O) \underset{k_{-1}}{\overset{k_1}{\rightleftharpoons}} H_2CO_3$$

(The parentheses indicate that H_2O is not included in the equilibrium constant expression or in the rate equation.)

$$\Delta H° = 1130 \text{ cal mol}^{-1}$$

$$\Delta S° = -8 \text{ cal K}^{-1} \text{ mol}^{-1}$$

At 25° $\qquad k_1 = 0.0375 \text{ s}^{-1}$

At 0° $\qquad k_1 = 0.0021 \text{ s}^{-1}$

Calculate: (a) the activation energy for the forward reaction, (b) the activation energy for the reverse reaction, (c) k_{-1} at 25°, and (d) k_{-1} at 0°, assuming that $\Delta H°$ is independent of temperature in this range.

10.41 The following rate constants were obtained by Wiig for the first-order decomposition of acetone dicarboxylic acid in aqueous solution:

Temp., °C	0	20	40	60
$k \times 10^5$, s^{-1}	2.46	47.5	576	5480

(a) Calculate the energy of activation. (b) Calculate the preexponential factor A. (c) What is the half-life of this reaction at 80°?

10.42 Although the thermal decomposition of ethyl bromide is complex, the overall rate is first order and the rate constant is given by the expression $k = 3.8 \times 10^{14} e^{-54,800/RT}$. Estimate the temperature at which (a) ethyl bromide decomposes at the rate of 1% per sec and (b) the decomposition is 70% complete in 1 hr.

10.43 Given that the first-order rate constant for the over-all decomposition of N_2O_5 is $k = 4.3 \times 10^{13} e^{-24,700/RT}$ s^{-1}, calculate (a) the half-life at $-10°$; (b) the time required for 90% reaction at 50°.

10.44 At 700 K what is the half-life of SiH_4? The equation for the first-order rate constant for decomposition is $k = 2 \times 10^{13} e^{-51,700/RT}$.

10.45 The first-order rate constant for the thermal decomposition of $C_2H_5Br(g)$ is given by

$$k = 3.8 \times 10^{14} e^{-55,000/RT}$$

where k is in s^{-1}. Calculate (a) $\Delta H^{\ddagger°}$ and (b) $\Delta S^{\ddagger°}$ at 500°.

10.46 A reaction $A + B + C \rightarrow D$ follows the mechanism

$$A + B \rightleftharpoons AB$$

$$AB + C \rightarrow D$$

in which the first step remains essentially in equilibrium. Show that the dependence of rate on temperature is given by

$$k = Ae^{-(E_a + \Delta H)/RT}$$

where ΔH is the enthalpy change for the first reaction.

10.47 For the gas reaction

$$O + O_2 + M \underset{k'}{\overset{k}{\rightleftharpoons}} O_3 + M$$

When $M = O_2$ Benson and Axworthy [*J. Chem. Phys.*, **26**, 1718 (1957)] obtained

$$k = 6.0 \times 10^7 e^{0.6/RT} \, l^2 \, mol^{-2} \, s^{-1}$$

where the activation energy is in kcal mol^{-1}. Calculate the values of the parameters in the Arrhenius equation for the reverse reaction.

10.48 (*a*) Calculate the second-order rate constant for collisions of dimethyl ether molecules with each other at 777 K. It is assumed that the molecules are spherical and have a radius of 2.5 Å. If every collision was effective in producing decomposition, what would be the half-life of the reaction (*b*) at 1 atm pressure, (*c*) at a pressure of 10^{-3} Torr?

10.49 The solution reaction

$$I^- + OCl^- = OI^- + Cl^-$$

is believed to go by the mechanism

$$OCl^- + H_2O \underset{}{\overset{K_1}{\rightleftharpoons}} HOCl + OH^- \qquad \text{(fast)}$$

$$I^- + HOCl \xrightarrow{k} HOI + Cl^- \qquad \text{(slow)}$$

$$HOI + OH^- \underset{}{\overset{K_2}{\rightleftharpoons}} H_2O + OI^- \qquad \text{(fast)}$$

Derive the rate equation for the forward rate of this reaction which shows the effect of the concentration of OH^-.

10.50 Ozone is decomposed by the catalytic chain

$$NO + O_3 \xrightarrow{k_1} NO_2 + O_2$$

$$NO_2 + O \xrightarrow{k_2} NO + O_2$$

What is the steady-state rate law for the formation of O_2?

10.51 From the fact that the bimolecular rate constant for the reaction of H^+ with NH_3 is $4.3 \times 10^{10} \, M^{-1} \, s^{-1}$, calculate the reaction radius for this reaction using equation 10.120. The diffusion coefficient of H^+ may be calculated from its mobility (Table 11.6) using equations 11.21 and 11.34. The diffusion coefficient of NH_3 may be neglected because it is quite a bit smaller.

10.52 Derive the relation between the relaxation time τ and the rate constants for the reaction $A + B \underset{k_2}{\overset{k_1}{\rightleftharpoons}} C + D$, which is subjected to a small displacement from equilibrium.

10.53 The reaction between selenious acid and iodide ion in acid solution is:

$$H_2SeO_3 + 6I^- + 4H^+ = Se(s) + 2I_3^- + 3H_2O$$

The initial reaction rates were measured at $0°$ at a variety of concentrations, as indicated in the following table, in moles per liter. These initial rates were evaluated from plots of H_2SeO_3 versus time. Determine the form of the rate law. (*Note:* This rate law holds only as long as insignificant quantities of I_3^- are present.)

$(H_2SeO_3) \times 10^4$	$(H^+) \times 10^2$	$(I^-) \times 10^2$	Initial Rate $\times 10^7$
0.712	2.06	3.0	4.05
2.40	2.06	3.0	14.6
7.20	2.06	3.0	44.6
0.712	2.06	1.8	0.93
0.712	2.06	3.0	4.05
0.712	2.06	9.0	102
0.712	2.06	15.0	508
0.712	2.06	3.0	4.05
0.712	5.18	3.0	28.0
0.712	12.5	3.0	173.0

10.54   Calculate the first-order rate constants for the following reactions at 25°:
H^+ production

$$HOAc \rightarrow H^+ + OAc^-$$

$$ImH^+ \rightarrow H^+ + Im$$

$$NH_4^+ \rightarrow H^+ + NH_3$$

OH^- production

$$OAc^- + H_2O \rightarrow HOAc + OH^-$$

$$Im + H_2O \rightarrow ImH^+ + OH^-$$

$$NH_3 + H_2O \rightarrow NH_4^+ + OH^-$$

where HOAc is acetic acid and Im is imidazole ($C_3N_2H_4$). The reverse reactions given above may all be assumed to be diffusion controlled with $k = 10^{10}$ $M^{-1}s^{-1}$. Acid dissociation constants at 25°:

HOAc	1.75×10^{-5}
ImH^+	1.2×10^{-7}
NH_4^+	5.71×10^{-10}

Which conjugate acid-base pair can play both H^+ and OH^- production roles about equally effectively?

10.55   The initial rate v of oxidation of sodium succinate to form sodium fumarate by dissolved oxygen in the presence of the enzyme succinoxidase may be represented by

$$v = \frac{V}{1 + K_S/(S)}$$

where V is the maximum initial velocity obtainable with a given amount of enzyme, K_S is the Michaelis constant, and (S) is the concentration of sodium succinate. Calculate V and K_S from the following data. [For this calculation it is convenient to plot v^{-1} versus $(S)^{-1}$.]

$(S) \times 10^3$, M	10	2	1	0.5	0.33
$v \times 10^6$, M/s	1.17	0.99	0.79	0.62	0.50

10.56   Derive the steady-state rate law for the mechanism

$$E + S \underset{k_{-1}}{\overset{k_1}{\rightleftharpoons}} EP_2 + P_1 \underset{k_{-2}}{\overset{k_2}{\rightleftharpoons}} E + P_2$$

in the form including both the forward and reverse reactions. Give the steady-state rate equations for the initial velocities of the forward and reverse reactions. Can all four rate constants be determined from steady-state velocities of the forward and reverse reactions?

10.57   For the fumarase reaction

$$fumarate + H_2O = L\text{-malate}$$

at pH 7, 25° and 0.01 ionic strength, the Michaelis-Menten parameters have the following values

$$V_F(M\ s^{-1}) = 1.3 \times 10^3(E)_0 \qquad V_M(M\ s^{-1}) = 0.8 \times 10^3(E)_0$$

$$K_F = 4 \times 10^{-6}\ M \qquad\qquad K_M = 10 \times 10^{-6}\ M$$

where $(E)_0$ is the molar concentration of the enzyme, which has four catalytic sites per molecule. Calculate (a) the four rate constants in the mechanism

$$E + F \underset{k_{-1}}{\overset{k_1}{\rightleftharpoons}} EX \underset{k_{-2}}{\overset{k_2}{\rightleftharpoons}} E + M$$

and (b) ΔG°_{obs} for the overall reaction.

10.58   At $25°$ and pH 7.8 the following values are obtained for the Michaelis constant and maximum initial velocity for the forward reaction catalyzed by fumarase

$$F + H_2O = M \qquad K = (M)_{eq}/(F)_{eq} = 4.4$$
$$V_F = (0.8 \times 10^3 \text{ s}^{-1})(E)_0$$
$$K_F = 7 \times 10^{-6} \text{ M}$$

where the enzyme concentration is in moles of enzyme per liter. The enzyme has four catalytic sites per molecule. In some experiments L-malate was added and was found to be inhibitory with a constant

$$K_M = 100 \times 10^{-6} \text{ M}$$

Calculate the values of the four rate constants in the mechanism

$$E + F \underset{k_{-1}}{\overset{k_1}{\rightleftharpoons}} X \underset{k_{-2}}{\overset{k_2}{\rightleftharpoons}} E + M$$

where E represents an enzymatic *site*.

10.59   The Michaelis constant for succinate on succinoxidase is 0.5×10^{-3} M and the competitive inhibition constant for malonate is 10×10^{-3} M. In an experiment with 10^{-3} M succinate and 15×10^{-3} M malonate, what is the percent inhibition?

10.60   Alcohol dehydrogenase catalyzes the reaction

$$\text{alcohol} + \text{NAD}^+ = \text{aldehyde} + \text{NADH} + \text{H}^+$$

where NAD^+ is nicotinamide adenine dinucleotide in the oxidized form and NADH in the reduced form. For enzyme isolated from yeast the maximum velocity for ethyl alcohol is given by

$$V(\text{M min}^{-1}) = 2.7 \times 10^4 (E)_0$$

where $(E)_0$ is the molar concentration of enzyme of molecular weight 150,000. The reaction is conveniently followed spectrophotometrically at 3400 Å because NADH has a molar absorbancy coefficient of 6200 M^{-1} cm^{-1}

$$\text{Absorbancy} = \log \frac{I_0}{I} = 6200(\text{NADH})d$$

where d is the thickness of the spectrophotometer cuvette. How many grams of enzyme must be placed in a 3 ml reaction mixture to produce a rate of change of absorbancy of 0.001 per minute when a 1.0 cm cuvette is used?

10.61   The following values of percent transmission are obtained with a spectrophotometer at a series of times during the decomposition of a substance absorbing light at a particular wavelength. Calculate k, $t_{1/2}$ and τ assuming the reaction is first order.

t	Percent Transmission
5 min	14.1
10 min	57.1
∞	100

Beer's law: $\log 100/T = abc$ where T = percent transmission, a = absorbancy index, b = cell thickness, and c = concentration.

10.62 Prove that in a first-order reaction, where $dn/dt = -kn$, the average life, that is, the average life expectancy of the molecules, is equal to $1/k$.

10.63 The hydrolysis of 1-chloro-1-methylcycloundecane in 80% ethanol has been studied by H. C. Brown and M. Borkowiski [*J. Am. Chem. Soc.*, **74**, 1894 (1952)] at 25°. The extent of hydrolysis was measured by titrating the acid formed after measured intervals of time with a solution of NaOH. The data are as follows:

Time, hr	0	1.0	3.0	5.0	9.0	12	∞
x, cm^3	0.035	0.295	0.715	1.055	1.505	1.725	2.197

(*a*) What is the order of the reaction? (*b*) What is the value of the rate constant? (*c*) What fraction of the 1-chloro-1-methylcycloundecane will be left unhydrolyzed after 8 hr?

10.64 It is found that 30% of a compound decomposes in 10 hr at a certain temperature. Assuming a first-order reaction, how long will be required for 99% of the compound to decompose?

10.65 In studying a first-order reaction, observations of concentration of reactant are made at equally spaced time intervals. Show that it is not satisfactory to average all the constants obtained in successive time intervals using equation 10.11. [See W. E. Roseveare, *J. Am. Chem. Soc.*, **53**, 1651 (1931) and L. J. Reed and E. J. Theriault, *J. Phys. Chem.*, **35**, 673 (1931).]

10.66 From the following data on the rate of the rearrangement of 1-cyclohexenyl allyl malonitrile at 135.7°, determine graphically the first-order rate constant. Check the graphical evaluation by calculating k from data at two different times.

Time, min	0	5	10	20	30	45
% rearranged	19.8	34.2	46.7	64.7	77.0	86.3

10.67 A second-order reaction A + B = C, where $a = b$, is 20% completed in 500 s. How long will it take for the reaction to go to 60% completion?

10.68 The second-order rate constant for a gas reaction is 1×10^3 liters mol^{-1} s^{-1} at 25°. Calculate the value of the rate constant when the rate equation is written in terms of pressure in atmospheres.

10.69 Equal molar quantities of A and B are added to a liter of a suitable solvent. At the end of 500 s half of A has reacted according to the reaction $A + B = C$. How much of A will be reacted at the end of 800 s if the reaction is (*a*) zero order with respect to both A and B; (*b*) first order with respect to A and zero order with respect to B; (*c*) first order with respect to both A and B?

10.70 It is more difficult than commonly realized to differentiate between a first- and a second-order reaction by the shape of the plot of percentage reaction versus time. Make these plots for first- and second-order reactions ($a = b$) having the same half-life. By what percent do they differ at (*a*) $t = t_{1/2}/2$ and (*b*) $t = 2t_{1/2}$?

10.71 Show that the interconversion of ortho- and parahydrogen will be $\frac{3}{2}$ order, as obtained experimentally in the range 600–750°, if the rate-determining step is that between atoms and molecules of hydrogen

$$H + \text{para-}H_2 \underset{k_2}{\overset{k_1}{\rightleftharpoons}} \text{ortho-}H_2 + H$$

where the arrows represent the directions of the nuclear spins (cf. Section 16.2).

10.72 The equations for (B) and (C) in Section 10.11 give an indeterminate result if $k_1 = k_2$. Rederive the equations, giving (B) and (C) as functions of time for the special case that
$$A \xrightarrow{k_1} B \xrightarrow{k_1} C.$$

10.73 For a reversible second-order reaction $A + B \underset{k_2}{\overset{k_1}{\rightleftharpoons}} C$ the rate equation is $dx/dt = k_1(a - x)(b - x) - k_2x^2$, where a and b are the initial concentrations of A and B and the concentration of C at any time is x. Integrate this equation to find the equation obeyed by the overall reaction.

10.74 When an optically active substance is isomerized, the optical rotation decreases from that of the original isomer to zero in a first-order manner. In a given case the half-time for this process is found to be 10 min. Calculate the rate constant for the conversion of one isomer to another.

10.75 The rate of rearrangement of 1-ethyl propenyl allyl malonitrile to 1-ethyl-2-methyl-4-pentenylidene malonitrile, can be followed by measuring the refractive index. The following first-order constants were obtained:

Temp., °C	120.0	130.0	140.0
$k \times 10^4$, s^{-1}	4.02	9.12	19.83

(a) What is the energy of activation? (b) What is the frequency factor?

10.76 Suppose that a substance X decomposes into A and B in parallel paths with rate constants given by
$$k_A = 10^{15}e^{-30,000/RT}$$
$$k_B = 10^{13}e^{-20,000/RT}$$

(a) At what temperature will the two products be formed at the same rate? (b) At what temperature will A be formed 10 times as fast as B? (c) At what temperature will A be formed 0.1 as fast as B? (d) State a generalization concerning the effect of temperature on the relative rates of reactions with different activation energies.

10.77 For the reaction
$$O + NO + M \rightarrow NO_2 + M$$

$k_{1000\ K} = 6 \times 10^9\ l^2\ mol^{-2}\ s^{-1}$ and $k_{300\ K} = 3 \times 10^{10}\ l^2\ mol^{-2}\ s^{-1}$. Calculate the parameters in the Arrhenius equation.

10.78 The vapor-phase decomposition of di-*t*-butyl peroxide is first order in the range 110–280° and follows the equation
$$k = 3.2 \times 10^{16}e^{-39,100/RT}$$

where k is in s^{-1}. Calculate (a) $\Delta H^{\ddagger\circ}$ and (b) $\Delta S^{\ddagger\circ}$.

10.79 In the reaction
$$N_2 + O_2 \underset{k_2}{\overset{k_1}{\rightleftharpoons}} 2NO$$
$$-\frac{d(N_2)}{dt} = k_1\left(p'_{N_2} - \frac{p_{NO}}{2}\right)\left(p'_{O_2} - \frac{p_{NO}}{2}\right) - k_2p_{NO_2}$$

where p'_{N_2} is the original pressure of N_2, p'_{O_2} is the original pressure of O_2, and p_{NO} is the pressure of NO formed. Values of the equilibrium constant K are given in Table 5.4. The rate constant for the reverse reaction is given in $atm^{-1}\ s^{-1}$ by
$$k_2 = 1 \times 10^9e^{-70,000/RT}$$

(a) Calculate k_1 at 2400 K and 1900 K. (b) Calculate the time required for NO at 0.02 atm to undergo 10% decomposition at 2400 K and at 1900 K, using k_2 and neglecting k_1.

10.80 For the first order gaseous decomposition of propylene oxide $\Delta H^{\ddagger\circ} = 56.9$ kcal mol^{-1} and $\Delta S^{\ddagger\circ} = 6$ cal K^{-1} mol^{-1} in the neighborhood of 285°. Calculate (a) the frequency factor A and (b) the first-order rate constant at 285°.

10.81 For the mechanism

$$A + B \underset{k_2}{\overset{k_1}{\rightleftharpoons}} C$$

$$C \xrightarrow{k_3} D$$

(a) Derive the rate law using the steady state approximation to eliminate the concentration of C. (b) Assuming that $k_3 \ll k_2$, express the preexponential factor A and E_a for the apparent second-order rate constant in terms of A_1, A_2, and A_3 and E_{a1}, E_{a2}, and E_{a3} for the three steps.

10.82 For the decomposition of ozone

$$2O_3 = 3O_2$$

the rate law is

$$-\frac{d(O_3)}{dt} = k\frac{(O_3)^2}{(O_2)}$$

Devise a mechanism to explain this rate law. (*Hint:* the first step might be expected to be the production of an oxygen atom.)

10.83 The following mechanism has been proposed for the thermal decomposition of ethyleneoxide [*J. Chem. Phys.*, **31**, 506 (1959)]:

Initiation: $CH_2{-}CH_2 \overset{k_1}{\longrightarrow} CH_2{-}CH\cdot + H$ (with O bridges)

Propagation: $CH_2{-}CH\cdot \overset{k_2}{\longrightarrow} CH_3\cdot + CO$ (with O bridge)

$CH_2{-}CH_2 + CH_3\cdot \overset{k_3}{\longrightarrow} CH_2{-}CH\cdot + CH_4$ (with O bridges)

Termination: $CH_2{-}CH\cdot + CH_3\cdot \overset{k_4}{\longrightarrow}$ stable products (with O bridge)

Assuming the radicals are in a steady state, show that the decomposition is first order in ethyleneoxide concentration.

10.84 Derive the relation between the relaxation time τ and the rate constants for the mechanism

$$A \underset{k_{-1}}{\overset{k_1}{\rightleftharpoons}} B$$
$$K_A \updownarrow \qquad \updownarrow K_B$$
$$A' \underset{k_{-1}'}{\overset{k_1'}{\rightleftharpoons}} B'$$

which is subjected to a small displacement from equilibrium. It is assumed that the equilibria, $A \rightleftharpoons A'$, $K_A = (A')/(A)$, and $B \rightleftharpoons B'$, $K_B = (B')/(B)$, are adjusted very rapidly so that these steps remain in equilibrium.

10.85 The mutarotation of glucose is catalyzed by acids and bases and is first order in the concentration of glucose. When perchloric acid is used as a catalyst, the concentration of hydrogen ions may be taken to be equal to the concentration of perchloric acid, and the catalysis by perchlorate ion may be ignored since it is such a weak base. The following first-order constants were obtained by J. N. Brönsted and E. A. Guggenheim [*J. Am. Chem. Soc.*, **49**, 2554 (1927)] at 18°:

$(HClO_4)$, M	0.0010	0.0048	0.0099	0.0192	0.0300	0.0400
$k \times 10^4$, min^{-1}	1.25	1.38	1.53	1.90	2.15	2.59

Calculate the values of the constants in the equation $k = k_0 + k_{H^+}(H^+)$.

10.86 Derive the steady-state rate equation for the mechanism.

$$E + S \xrightarrow{k_3} E' + P$$
$$k_1 \Big\updownarrow k_2$$
$$E'$$

The Michaelis-Menten form is obtained, but, nevertheless, why is this an unsuitable mechanism for explaining enzymatic action?

10.87 Suppose an enzyme has a turnover number of 10^4 min^{-1} and a molecular weight of 60,000. How many moles of substrate can be turned over per hour per gram of enzyme if the substrate concentration is twice the Michaelis constant? It is assumed that the substrate concentration is maintained constant by a preceding enzymatic reaction and that products do not accumulate and inhibit the reaction.

10.88 The kinetics of the fumarase reaction

$$\text{fumarate} + H_2O = \text{L-malate}$$

is studied at 25° C using a 0.01 ionic strength buffer of pH 7. The rate of the reaction is obtained using a recording ultraviolet spectrometer to measure the fumarate concentration. The following rates of the forward reaction are obtained using a fumarase concentration of 5×10^{-10} M:

(F)	v_F
2×10^{-6} M	2.2×10^{-7} M s^{-1}
40×10^{-6} M	5.9×10^{-7} M s^{-1}

The following rates of the reverse reaction are obtained using a fumarase concentration of 5×10^{-10} M:

(M)	v_M
5×10^{-6} M	1.3×10^{-7} M s^{-1}
100×10^{-6} M	3.6×10^{-7} M s^{-1}

(*a*) Calculate the Michaelis constants and turnover numbers for the two substrates. In practice many more concentrations would be studied. (*b*) Calculate the four rate constants in the mechanism

$$E + F \underset{k_{-1}}{\overset{k_1}{\rightleftharpoons}} X \underset{k_{-2}}{\overset{k_2}{\rightleftharpoons}} E + M$$

where E represents the catalytic site. There are four catalytic sites per fumarase molecule. (c) Calculate K_{eq} for the reaction catalyzed. The concentration of H_2O is omitted in the expression for the equilibrium constant because its concentration cannot be varied in dilute aqueous solutions.

10.89 A certain enzyme E catalyzes an essentially irreversible reaction S → P. Addition of small amounts of E to a solution of pure S results in a rapid initial production of P, but then the rate of production of P gradually slows down and comes to almost a complete halt, *before* all of the S is converted to P. If a large amount of additional S is now added, there is again a rapid initial production of P that again slows down to almost a complete halt before all the S is converted to P. Explain these findings.

10.90 Derive the steady-state rate equation for the mechanism

$$E + S \underset{k_2}{\overset{k_1}{\rightleftharpoons}} X \overset{k_3}{\longrightarrow} E + P$$

$$E + I \underset{k_5}{\overset{k_4}{\rightleftharpoons}} EI$$

for the case that $(S) \gg (E)_0$ and $(I) \gg (E)_0$.

10.91 Derive the steady-state rate equation for the mechanism

$$\begin{array}{ccccc} E & + S \underset{k_{-1}}{\overset{k_1}{\rightleftharpoons}} & ES & \overset{k_2}{\longrightarrow} & E + P \\ \updownarrow K_{EH} & & \updownarrow K_{EHS} & & \\ EH & & EHS & & \end{array}$$

Sketch the shape of the plots of V_S and K_S versus pH.

10.92 As a first approximation the kinetics of the alcohol dehydrogenase reaction

$$\text{alcohol} + NAD^+ = \text{aldehyde} + NADH + H^+$$

may be represented by the mechanism

$$E + NAD^+ \overset{k_1}{\longrightarrow} ENAD^+$$

$$ENAD^+ + \text{alcohol} \overset{k_2}{\longrightarrow} E + NADH + \text{aldehyde} + H^+$$

Derive the steady-state rate equation for the initial velocity.

10.93 Pure solutions of the α and β chains of hemoglobin can be prepared. Assuming that α and β exist only as monomers in these solutions, and that they react on the first collision, estimate the half-life for the reaction

$$\alpha + \beta \rightarrow \alpha\beta$$

in water at 25°. The viscosity of water at this temperature is 0.00895 $g\,cm^{-1}\,s^{-1}$. Calculate the half-life if equal volumes of 10^{-6} M solutions of α and β are mixed.

CHAPTER 11

IRREVERSIBLE PROCESSES IN SOLUTION

In this chapter we are concerned with the *rates* with which certain processes occur. The examples to be considered are viscosity, electrical conductivity, and diffusion. Determinations of coefficients of viscosity, diffusion, and sedimentation are useful for calculating the molecular weight and shape of macromolecules; these applications are discussed in Chapter 20. The theory for the rates of irreversible processes in solution is handicapped by the lack of an adequate theory for the liquid state. The study of electrical conductivity has told us a lot about ionic solutions.

11.1 VISCOSITY

The coefficient of viscosity was defined in the discussion of the kinetic theory of gases in Section 9.12. This definition applies to laminar flow, that is, flow in which one layer (lamina) slides smoothly relative to another. When the flow velocity is great enough, turbulence develops. The coefficient of viscosity of a liquid may be measured by a number of methods that are illustrated in Fig. 11.1; these methods include the determination of the rate of flow through a capillary, the rate of settling of a sphere in a liquid, and the force required to turn one of two concentric cylinders at a certain angular velocity.

The SI unit of viscosity is the pascal second (Pa s) which is the viscosity of a fluid in which the velocity under a shear stress of 1 Pa has a gradient of 1 m s^{-1} per meter perpendicular to the plane of shear.*

An estimate as to whether the flow will be turbulent may be obtained by calculating a dimensionless quantity referred to as the Reynolds number. The Reynolds number for flow in a tube is defined by $d\bar{v}\rho/\eta$, where d is the diameter of the tube, $\bar{v}$ is the average velocity of the fluid along the tube, ρ is the density of the fluid, and η is its coefficient of viscosity. At flow velocities corresponding with values of the Reynolds number of greater than 2000, turbulence is encountered.

For certain colloidal suspensions and solutions of macromolecules the coefficient of viscosity depends on the rate of shear, and this is referred to as non-Newtonian behavior. If the shear stress orients or distorts the suspended particles, the coefficient of viscosity may decrease as the shear rate is increased.

---

* The cgs unit of viscosity is the poise that is the viscosity of a fluid in which the velocity under a shear stress of 1 dyne cm^{-2} has a gradient of 1 cm s^{-1} per centimeter perpendicular to the plane of shear. A poise is equal to 0.1 Pa s.

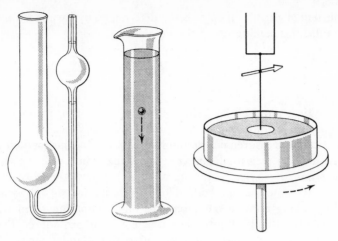

Fig. 11.1 Various types of viscometers.

The force causing a particle to settle in a fluid is equal to its effective mass times the acceleration of gravity; the effective mass is the mass of the particle minus the mass of the fluid it displaces. The force retarding motion is the frictional coefficient times the velocity. The frictional coefficient f is the frictional force per unit of velocity. Stokes showed that for spheres and nonturbulent flow*

$$f = 6\pi\eta r \qquad (11.1)$$

where r is the radius and η the coefficient of viscosity. If the density of the sphere is ρ and the density of the medium is ρ_0, the force causing motion is $\frac{4}{3}\pi r^3(\rho - \rho_0)g$, where g is the acceleration of gravity. When the rate of settling of the sphere in the liquid is constant, the retarding force $f(dx/dt) = 6\pi r\eta\ (dx/dt)$ is equal to the force due to gravity, and so

$$\tfrac{4}{3}\pi r^3(\rho - \rho_0)g = 6\pi\eta r\left(\frac{dx}{dt}\right) \qquad (11.2)$$

$$\frac{dx}{dt} = \frac{2r^2(\rho - \rho_0)g}{9\eta} \qquad (11.3)$$

Thus by measuring the velocity dx/dt of settling of a sphere of known r and ρ in a liquid of known density ρ_0, the coefficient of viscosity η may be obtained. This method is especially valuable for solutions of high viscosity, such as concentrated solutions of high polymers. Conversely, the determination of the rate of settling of colloidal particles of known density in a liquid of known viscosity provides a means for determining the effective particle radius.

---

* For a derivation see R. B. Bird, W. E. Stewart, and E. N. Lightfoot, *Transport Phenomena*, Wiley, New York, 1960.

The coefficient of viscosity η may also be determined by passing a liquid through a capillary tube and making use of the Poiseuille equation

$$\eta = \frac{P\pi r^4 t}{8Vl} \tag{11.4}$$

where t is the time required for volume V of liquid to flow through a capillary tube of length l and radius r under an applied pressure P.

The quantitative measurement of absolute viscosity by this method is difficult. Accordingly, indirect measurements are usually made in which the viscosity of a liquid is determined relative to that of another liquid whose viscosity has been obtained by an absolute method. This determination of physical-chemical constants by making relative measurements and comparing them with similar measurements of a standard substance is a common procedure. The absolute measurements of the standard substance may well require years of research, whereas the relative measurements can often be made quickly and easily with a high degree of accuracy.

In a simple viscometer such as that illustrated in Fig. 11.1, the pressure that causes flow through the capillary is proportional to the difference in height h of liquid levels in the two tubes, the density of the liquid ρ_1, and the acceleration of gravity g. Thus, it is readily shown that if the same volume of two liquids is allowed to flow through the same capillary

$$\frac{\eta_1}{\eta_2} = \frac{\rho_1 t_1}{\rho_2 t_2} \tag{11.5}$$

where 1 and 2 represent the two liquids. This method is useful for determining the viscosity of a second liquid when that of the first is known.

The viscosities of most liquids decrease with increasing temperature. According to the "hole theory" there are vacancies in a liquid, and molecules are continually moving into these vacancies so that the vacancies move around. This process permits flow, but requires energy because there is an activation energy that a molecule has to have to move into a vacancy. The activation energy is more readily available at higher temperatures and so the liquid can flow more easily at higher temperatures. The viscosities of several liquids at different temperatures are shown in Table 11.1. The variation of the coefficient of viscosity with temperature may be represented quite well by

$$\frac{1}{\eta} = Ae^{-E_a/RT} \tag{11.6}$$

where E_a is the activation energy for the fluidity, $1/\eta$.

The viscosity of a liquid increases as the pressure is increased because the number of holes is reduced, and it is therefore more difficult for molecules to move around each other.

Table 11.1 Viscosity of Liquids in Pa s (kg m^{-1} s^{-1})

Liquid	0°	25°	50°	75°
Water	0.001793	0.000895	0.000549	0.000380
Ethanol	0.00179	0.00109	0.000698	—
Benzene	0.00090	0.00061	0.00044	—

In contrast with liquids, the viscosity of a gas increases as the temperature increases. The viscosity of an ideal gas is independent of pressure.

11.2 ELECTRIC CONDUCTIVITY

Electric conductivity may be classified according to four types:

1. *Metallic conductivity*, which results from the mobility of electrons. Metallic conductors become poorer conductors at higher temperatures because it is more difficult for electrons to pass through the crystal lattice when the atoms of the lattice are in more active thermal motion.

2. *Electrolytic conductivity of liquids*, which results from the mobility of ions. Electrolytic conductors become better conductors as the temperature is raised because ions can move through the solution more readily at higher temperatures, where the viscosity is smaller and there is less solvation of the ions.

3. *Semiconductivity of solids*, which results from electrons populating an unfilled band separated from the highest filled band by an energy of the order of kT at room temperature (Section 19.23). The conductivity of semiconductors increases exponentially with the absolute temperature.

4. *Electric conductivity in gases* by gas ions and electrons.

The SI unit of electric resistance R is the ohm, represented by Ω. The ohm is defined by VA^{-1} where V is the volt and A is the ampere. (The definitions of A and other base units are given in the Appendix.)

The electric resistance R of a uniform conductor is directly proportional to its length l and inversely proportional to its cross-sectional area A.

$$R = \frac{rl}{A} = \frac{l}{\kappa A} \tag{11.7}$$

where the proportionality constant r is called the *resistivity* and the proportionality constant κ is called the *conductivity*. As may be seen from equation 11.7, the conductivity has the units of Ω^{-1} m^{-1} in the SI system.

The conductivities of various materials are given in Table 11.2. In the neighborhood of absolute zero the resistance of metals becomes extremely low. For superconductors the resistance becomes zero at a critical temperature.

Table 11.2   Conductivities κ of Typical Conductors

Material	Temperature, °C	Conductivity Ω^{-1} m^{-1}
Silver	0	6.812×10^7
Copper	0	6.406×10^7
Aluminum	0	3.900×10^7
Iron	0	1.102×10^7
Lead	0	4.882×10^6
Mercury	0	1.043×10^6
Fused sodium nitrate	500	1.760×10^2
Fused zine chloride	500	8.38
1 M potassium chloride	25	11.93
0.001 M potassium chloride	25	1.468×10^{-2}
1 M acetic acid	18	1.320×10^{-1}
0.001 M acetic acid	18	4.09×10^{-3}
Diamond	25	10^{-5}
Water	18	4.0×10^{-6}
SiO$_2$	25	10^{-15}
Xylene	25	1.429×10^{-17}

Resistances may be determined by use of a Wheatstone bridge in which an unknown resistance is balanced against a known resistance. Alternating current is used in measuring the resistances of cells containing electrolytic solutions. When alternating current is used, and the electrodes are platinized, the electrolysis that occurs when the current passes in one direction is reversed when the current passes in the other direction. The coating of platinum black, produced by electrolytic deposition, adsorbs gases and catalyzes their reaction. In this way the formation of a nonconducting gas film is prevented.

The conductivity κ of an electrolyte is inversely proportional to the measured resistance of the cell.

$$\kappa = \frac{K_{\text{cell}}}{R} \tag{11.8}$$

The cell constant K_{cell} is determined by measuring the resistance of the cell when it contains a solution of known conductivity. The conductivities of certain standard solutions have been carefully measured.

Example 11.1   When a certain conductance cell was filled with 0.0200 M potassium chloride, which has a conductivity of 0.2768 Ω^{-1} m^{-1}, it had a resistance of 82.40 Ω at 25° as measured with a Wheatstone bridge; when filled with 0.0025 M potassium sulfate, it had a resistance of 326.0 Ω.

(a) What is the cell constant?

$$K_{\text{cell}} = (0.2768 \ \Omega^{-1} \ \text{m}^{-1})(82.40 \ \Omega) = 22.81 \ \text{m}^{-1}$$

(b) What is the conductivity κ of the K$_2$SO$_4$ solution?

$$\kappa = \frac{K_{\text{cell}}}{R} = \frac{22.81 \ \text{m}^{-1}}{326.0 \ \Omega} = 6.997 \times 10^{-2} \ \Omega^{-1} \ \text{m}^{-1}$$

11.3 MOLAR CONDUCTIVITY

The molar conductivity Λ is defined by

$$\Lambda = \frac{\kappa}{c} \qquad (11.9)$$

where c is concentration in moles per unit volume, and so it is the product of the conductivity and the volume containing 1 mol of electrolyte. In the SI system concentrations are in moles per m^3 so that the molar conductivity Λ has the units $\Omega^{-1}\ m^2\ mol^{-1}$. The formula unit used in expressing the concentration must be specified; examples are $\Lambda(KCl)$, $\Lambda(MgCl_2)$, $\Lambda(\frac{1}{2}MgCl_2)$, $\Lambda(\frac{2}{3}AlCl_3 + \frac{1}{3}KCl)$.*

Example 11.2 Calculate the molar conductivity of 0.0025 M K_2SO_4 in water from the data of Example 11.1.

$$\Lambda = \frac{\kappa}{c} = \frac{6.997 \times 10^{-2}\ \Omega^{-1}\ m^{-1}}{2.5\ mol\ m^{-3}}$$

$$= 2.799 \times 10^{-2}\ \Omega^{-1}\ m^2\ mol^{-1}$$

Table 11.3 gives the conductivities and molar conductivities of aqueous solutions of potassium chloride at 25°. It may be observed that, when the volume is increased tenfold and the number of moles per liter is consequently decreased to one-tenth, the conductivity decreases nearly but not quite to one-tenth its value. The molar conductivity on the other hand, changes only slightly but does increase and approaches a limiting value at infinite dilution.

Table 11.3 Molar Conductivity of Aqueous Solutions of Potassium Chloride at 25°

Moles per liter	Moles per m^3, c	Conductivity κ, $\Omega^{-1}\ m^{-1}$	Molar Conductivity Λ, $\Omega^{-1}\ m^2\ mol^{-1}$
1	10^3	11.19	0.01119
0.1	10^2	1.289	0.01289
0.01	10	0.1413	0.01413
0.001	1	0.01469	0.01469
0.0001	10^{-1}	0.001489	0.01489

The molar conductivity of potassium chloride and other electrolytes at 25° is plotted versus the square root of concentration in Fig. 11.2. Extrapolation of this plot to infinite dilution gives Λ_0, the limiting value of the molar conductivity. This is the value when the ions are so far apart that they do not interact and retard

---

* In the older literature, data are tabulated in terms of equivalent conductance defined by 1000 κ/C, where C is the concentration of the electrolyte in equivalents per liter. The equivalent conductance has the units $\Omega^{-1}\ cm^2\ equiv^{-1}$.

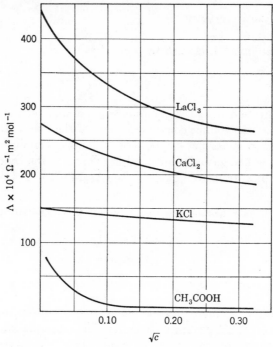

Fig. 11.2 Molar conductivity of typical electrolytes plotted against the square root of the concentration.

each other as they move towards the electrodes. In accurate work at low concentrations it is necessary to subtract the conductivity of the pure solvent from that of the solution to obtain the conductivity due to the electrolyte. Thus,

$$\kappa_{solute} = \kappa_{solution} - \kappa_{solvent} \tag{11.10}$$

The purest water has a conductivity of about $5 \times 10^{-6}\,\Omega^{-1}\,m^{-1}$, but it is difficult to obtain and store such pure water because of the absorption of carbon dioxide and other gases from the atmosphere and of alkali and other electrolytes from the glass containing vessel. Ordinary distilled water in equilibrium with air has a conductivity of about $10^{-3}\,\Omega^{-1}\,m^{-1}$.

The reason for plotting Λ versus $\sqrt{c}$ rather than c would be made evident by making a Λ vs c plot for the potassium chloride data. It would be found that the slope of the plot increases rapidly as zero concentration is approached so that it is not possible to extrapolate to Λ^0.

The dependence of the molar conductivity of acetic acid on concentration is quite different, as shown in Fig. 11.2. Acetic acid dissociates more at the greater dilution, giving a considerable increase in the total number of ions. For a substance like acetic acid it is not possible to extrapolate to infinite dilution to obtain the molar conductivity of the completely dissociated substance. However, the value of Λ_0 for acetic acid may be determined indirectly, using equation 11.6.

Electrolytes may be divided into two general classes: *strong electrolytes*, such as potassium chloride, with high conductances and slight increases in conductance on dilution; and *weak electrolytes*, such as acetic acid, with low conductances and larger increases in conductance on dilution.

Table 11.4 gives the molar conductivities of a number of electrolytes.

Table 11.4   Molar Conductivities of Electrolytes at 25° in Water $\Lambda \times 10^4$, $\Omega^{-1}\,m^2\,mol^{-1}$

Moles per liter	NaCl	KCl	NaI	KI	HCl	AgNO$_3$	CaCl$_2$	NaC$_2$H$_3$O$_2$	HC$_2$H$_3$O$_2$[1]
0.0000	126.5	149.9	126.9	150.3	426.1	133.4	271.6	91.0	(390.6)
0.0005	124.5	147.8	125.4	—	422.7	141.4	—	89.2	67.7
0.001	123.7	146.9	124.3	—	421.4	130.5	—	88.5	49.2
0.005	120.6	143.5	121.3	144.4	415.8	127.2	240.8	85.7	22.9
0.01	118.5	141.3	119.2	142.2	412.0	124.8	231.1	83.8	16.3
0.02	115.8	138.3	116.7	139.5	407.2	121.4	—	81.2	11.6
0.05	111.1	133.4	112.8	135.0	399.1	115.7	205.0	76.9	7.4
0.10	106.7	129.0	108.8	131.1	391.3	109.1	—	72.8	5.2

[1] Acetic acid is a weak electrolyte, and the conductance at infinite dilution is obtained by indirect methods. All the other electrolytes given in the table are strong electrolytes.

The molar conductivity of a strong electrolyte is the sum of contributions of the individual ions. This relationship, which was discovered by Kohlrausch, is especially useful at infinite dilution since the motion of the various ions is completely independent, so that the same ionic conductivities are obtained using different electrolytes. For a strong electrolyte $M_{\nu_+}X_{\nu_-}$ Kohlrausch's law is

$$\Lambda^0 = \nu_+\lambda_+^0 + \nu_-\lambda_-^0 \qquad (11.11)$$

where Λ^0 is the molar conductivity at infinite dilution, λ_+^0 is the ionic conductivity of the cation at infinite dilution, and λ_-^0 is the ionic conductivity of the anion at infinite dilution. We have not yet shown how to determine λ_+^0 and λ_-^0, but the correctness of equation 11.11 is indicated by the fact that the following differences from Table 11.4 are equal:

$$\Lambda_{KCl}^0 - \Lambda_{NaCl}^0 = 23.4 \times 10^{-4}\Omega^{-1}\,m^2\,mol^{-1}$$

$$\Lambda_{KI}^0 - \Lambda_{NaI}^0 = 23.4 \times 10^{-4}\Omega^{-1}\,m^2\,mol^{-1}$$

11.4 CONDUCTIVITY OF NONAQUEOUS SOLUTIONS

When electrolytes that are completely dissociated in water are dissolved in solvents of low dielectric constant, the coulombic attraction is sufficient to cause ionic association at extremely low concentrations of ions. The force between ions is inversely proportional to the dielectric constant (Section 6.1) of the medium.

Thus all electrolytes are "weak electrolytes" in solvents of low dielectric constant. Among the solvents that have been important in studies of nonaqueous solutions of electrolytes are alcohols, liquid ammonia, dioxane, acetone and other ketones, anhydrous formic acid and acetic acid, pyridine, and several amines and nitro compounds.

A few miscellaneous examples of conductivities in nonaqueous solutions are given in Table 11.5.

Table 11.5   Conductivity of Nonaqueous Solutions

Solvent	Temperature, °C	Electrolyte	Concentration, moles per liter	$\Lambda \times 10^4$, $\Omega^{-1} \, m^2 \, mol^{-1}$
Ammonia	−33	NaI	1.14	245
Ethanol	18	NaI	1.00	35.2
Acetone	25	NaI	1.0	26.4
			0.1	64.1
			0.01	109.7
			0.0	(176.2)
Ammonia	−33	Na	0.8	2017

In water there is no detectable difference in the ionization of potassium chloride, potassium bromide, and potassium iodide, but there is a marked difference in liquid ammonia, which has a dielectric constant of 22 and a lesser tendency to solvate. At $-34°$ in ammonia, the dissociation constants for potassium iodide and chloride are, respectively, 4.2×10^{-4} and 8.7×10^{-4}.

The conductivity of metallic sodium in liquid ammonia is interesting. The neutral sodium atoms dissociate into positive Na^+ ions and electrons.

11.5   ELECTRIC MOBILITY

The electric mobility u of a species is the magnitude of the velocity in an electric field divided by the magnitude of the strength of the electric field E.

The magnitude of an electric field is defined as the force on a test charge due to the electric field divided by the magnitude of the test charge. In the SI system a unit electric field E is one that exerts a force of a newton on a charge of a coulomb. Thus the unit of E is NC^{-1}. The electric potential, which is measured in volts, is the energy per unit charge. Thus $1 \, V = 1 \, JC^{-1} = 1 \, N \, m \, C^{-1}$. This shows that $1 \, V \, m^{-1} = 1 \, N \, C^{-1}$; therefore the SI unit for the magnitude of an electric field is $V \, m^{-1}$. In the SI system the velocity is expressed in $m \, s^{-1}$, and so the electric mobility has the units $m \, s^{-1}/V \, m^{-1} = m^2 \, V^{-1} \, s^{-1}$.

The most direct method for determining electric mobilities is to measure the velocity of a boundary between two electrolytic solutions in a tube of uniform cross section through which a current is flowing. For example, if a 0.1 M solution

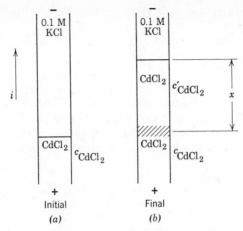

Fig. 11.3 Determining the electric mobility of the potassium ion with the moving-boundary method.

of potassium chloride is layered over a solution of cadmium chloride in a tube, as illustrated in Fig. 11.3a, and a current i is caused to flow through the tube, the potassium ions will move upward toward the negative electrode away from the position of the initial boundary. They will be followed by the slower moving cadmium ions, so that there will be no gap in the column of electrolyte. Since the concentration of cadmium chloride above the initial boundary position, (c'_{CdCl_2}) will, in general, be different from that initially placed below the potassium chloride solution, there will be a change in cadmium chloride concentration at the initial boundary position, which is represented by $//////$ in Fig. 11.3b.

In order to calculate the electric mobility of potassium ions from the velocity with which they move in the KCl solution it is necessary to know the electric field strength E in the KCl solution. The electric field strength E is the negative gradient of the electric potential ϕ. When the electric potential varies only in the x direction

$$E = -\frac{d\phi}{dx} \tag{11.12}$$

For a uniform conductor the difference in the potential per unit distance may be calculated using Ohm's law. For a conductor of unit cross section the difference in potential between two points is equal to the current density I/A, where I is the current and A the area, multiplied by the resistance $1/\kappa$.

$$E = \frac{I}{A\kappa} \tag{11.13}$$

In the experiment illustrated in Fig. 11.3 the boundary between the KCl and $CdCl_2$ solutions moves with the velocity of the potassium ions in the potassium chloride solution, and so the electric mobility u of the potassium ions may be calculated from the distance x moved by the boundary in the time t in an electric field of strength E.

$$u = \frac{dx/dt}{E} \tag{11.14}$$

Example 11.3 In the experiment with 0.1 M potassium chloride illustrated in Fig. 11.3 the boundary moved 4.64 cm during 67 min when a current of 5.21×10^{-3} A was used. The cross-sectional area of the tube was 0.230 cm², and $\kappa = 1.29 \ \Omega^{-1} \ m^{-1}$ at 25°. Calculate the electric field strength and the electric mobility of the potassium ion.

$$E = \frac{(5.21 \times 10^{-3} \ \text{A})}{(0.230 \times 10^{-4} \ \text{m}^2)(1.29 \ \Omega^{-1} \ \text{m}^{-1})}$$
$$= 176 \ \text{V m}^{-1}$$

$$u = \frac{(0.0464 \ \text{m})}{(67 \times 60 \ \text{s})(176 \ \text{V m}^{-1})}$$
$$= 6.56 \times 10^{-8} \ \text{m}^2 \ \text{V}^{-1} \ \text{s}^{-1}$$

To obtain a sharp moving boundary in an experiment such as that illustrated in Fig. 11.3 it is necessary that the leading ion (in this case, potassium) have a higher mobility than the indicator ion (in this case, cadmium). The cadmium chloride solution below the moving boundary (c'_{CdCl_2}) has a lower conductivity than the potassium chloride solution above the boundary. By reference to equation 11.13 it can be seen that the electric field strength is therefore greater in the cadmium chloride solution below the moving boundary than in the potassium chloride solution. Therefore, if potassium ions diffuse down into the cadmium chloride solution, they will be in a stronger electric field and will catch up with the boundary. On the other hand, if cadmium ions diffuse ahead of the boundary, they will have a lower velocity than the potassium ions because of their lower mobility and will soon be overtaken by the boundary. This so-called "adjusting effect" keeps the boundary sharp.

The moving-boundary method may be used for the study of mixtures of ions, which may be macromolecular ions such as proteins. The study of colloids by this method is referred to as *electrophoresis* (see Section 20.2). The electric mobilities of a number of small ions at infinite dilution in water at 25° are given in Table 11.6.

Table 11.6 Electric Mobilities at 25° at Infinite Dilution (in m² V⁻¹ s⁻¹)

Ion	$u \times 10^8$	Ion	$u \times 10^8$	Ion	$u \times 10^8$
Li^+	4.01	Cl^-	7.90	Ca^{2+}	6.16
Na^+	5.20	Br^-	8.12	Ba^{2+}	6.60
K^+	7.62	I^-	7.96	SO_4^{2-}	8.27
H^+	36.3	OH^-	20.5	NO_3^-	7.40

It is interesting that in the alkali metal and halogen families of the periodic table the lighter ions have lower mobilities. This is a result of the stronger electric field in the neighborhood of the ions with smaller radii, which causes them to be more highly hydrated, by the ion-dipole interaction, than the larger ions. A hydrated ion drags along a shell of water when it moves in a solution between two electrodes, and so it moves more slowly than an unhydrated ion.

11.6 ELECTRIC MOBILITIES OF HYDROGEN AND HYDROXYL IONS

There is a variety of kinds of evidence* that hydrogen ions in aqueous solutions are hydrated to form $H_9O_4^+$, that is a trihydrate of hydronium ion H_3O^+ with the following structure.

$$
\begin{array}{c}
\text{H} \qquad \text{H} \\
\diagdown \quad \diagup \\
\text{O} \\
\vdots \\
\text{H} \\
| \quad + \\
\text{O} \\
\diagup \quad \diagdown \\
\text{H} \qquad \text{H} \\
\vdots \qquad \qquad \vdots \\
\text{H—O} \qquad \qquad \text{O—H} \\
| \qquad\qquad\qquad | \\
\text{H} \qquad\qquad\quad \text{H}
\end{array}
$$

According to this structure the electric mobility of the hydrogen ion might be expected to be rather low, but as a matter of fact its mobility is about $10\times$ that of other ions, except the hydroxyl ion. The high electric mobility of the hydrogen ion is due to the fact that a proton may *in effect* be transferred along a series of hydrogen-bonded (Section 14.8) water molecules by rearrangement of the hydrogen bonds. In the following figure (*a*) shows the initial bonding in a group of oriented water molecules and (*b*) shows the final bonding.

$$
\overset{+}{\text{H}}\text{—O—H}\cdots\text{O—H}\cdots\text{O—H}\cdots\text{O—H} \qquad \text{H—O}\cdots\text{H—O}\cdots\text{H—O}\cdots\text{H}\overset{+}{\text{—O}}\text{—H}
$$

$$
\begin{array}{cccc}
| & | & | & | \\
\text{H} & \text{H} & \text{H} & \text{H}
\end{array}
\qquad\qquad
\begin{array}{cccc}
| & | & | & | \\
\text{H} & \text{H} & \text{H} & \text{H}
\end{array}
$$

$$(a) \qquad\qquad\qquad\qquad (b)$$

In order for another hydrogen ion to be transferred to the right through this group of water molecules molecular rotations must occur to again produce a favorable orientation for charge transfer.

---

* H. L. Clever, *J. Chem. Ed.*, **40**, 637 (1963).

This model for hydrogen ion mobility helps us to understand the remarkable fact that hydrogen ions move about $50 \times$ more rapidly through ice than through liquid water. In ice each oxygen atom is surrounded by four oxygen atoms at a distance of 2.76 Å in a tetrahedral arrangement. Each hydrogen is near the line through the centers of the oxygen atoms and is about 1 Å from one oxygen and 1.76 Å from the other. Hydrogen ions may be conducted rapidly through this structure by the above mechanism because the water molecules are oriented correctly.

The transfer of hydroxyl ion in the opposite direction is illustrated by

O⁻ H—O····H—O····H—O   O—H····O—H····O—H O⁻
 | | | | | | | |
H   H   H   H   H   H   H   H
      (a)   *(b)*

11.7 RELATION BETWEEN CONDUCTIVITY AND ELECTRIC MOBILITY

The conductivity κ of an electrolyte is the sum of the contributions of all the ionic species in the electrolyte. The electric current contributed by an ion depends upon its charge number z_i as well as on its electric mobility. The concentration of ion i expressed in faradays of electric charge is $|z_i| c_i$. If the ions all move with a velocity of u_i the transport of electric charge through a plane perpendicular to the direction of motion is $F |z_i| c_i u_i$. Thus for the electrolyte as a whole

$$\kappa = F \sum |z_i| c_i u_i$$
$$= Fc \sum |z_i| \nu_i u_i \qquad (11.15)$$

since equation 11.13 shows that κ is the current density divided by the electric field strength. The number of ions of type i in a molecule of electrolyte is represented by ν_i.

Example 11.4 Calculate the conductivity of 0.100 M sodium chloride at 25° from the mobilities of sodium and chloride ions at this concentration, which are 4.26×10^{-8} and 6.80×10^{-8} m² V⁻¹ s⁻¹, respectively.

$$\kappa = (96{,}485 \text{ C mol}^{-1})(0.100 \times 10^3 \text{ mol m}^{-3})[(4.26 + 6.80) \times 10^{-8} \text{ m}^2 \text{ V}^{-1} \text{s}^{-1}]$$
$$= 1.067 \ \Omega^{-1} \text{ m}^{-1}$$

The relation between the molar conductivity Λ and the mobilities of the ions in that electrolyte may be obtained by substituting equation 11.15 into equation 11.9.

$$\Lambda = F \sum |z_i| \nu_i u_i \qquad (11.16)$$

A comparison with equation 11.11 shows that

$$\lambda_i^\circ = F |z_i| u_i^\circ \qquad (11.17)$$

Mobilities and ionic conductivities increase with the temperature, and the temperature coefficients are very nearly the same for all ions in a given solvent and are approximately equal to the temperature coefficient of the viscosity; for water this is 2 % per degree in the neighborhood of 25°.

Another quantity frequently used in discussing electrolytic solutions is the transference number. With strong electrolytes the *transference number T* is the fraction of the current carried by a given ion. The ions that move faster carry the larger fraction of electricity through the solution. The fraction of the current carried by the cation is

$$T_c = \frac{|z_c|\, c_c u_c}{|z_c|\, c_c u_c + |z_a|\, c_a u_a} = \frac{u_c}{u_c + u_a} \tag{11.18}$$

since $|z_c|\, c_c = |z_a|\, c_a$.

Likewise, the transference number of the anion T_a is equal to $u_a/(u_c + u_a)$. It is obvious that $T_c + T_a = 1$. The transference number can be determined directly from the velocity in a moving boundary experiment or from changes in electrolyte concentration in electrolysis experiments.

11.8 DETERMINATION OF THE ION PRODUCT OF WATER

For pure water or aqueous solutions of low ionic strength, concentrations may be substituted for activities in writing the ion product.

$$K_w = (\text{H}^+)(\text{OH}^-) \tag{11.19}$$

The concentrations of these ions in pure water may be calculated from its electrical conductivity. At 25° the conductivity κ of the purest water* is $5.5 \times 10^{-6}\ \Omega^{-1}\,\text{m}^{-1}$. The concentrations of hydrogen and hydroxyl ions are, of course, equal and may be calculated using equation 11.15, and the values of the limiting ion mobilities in Table 11.6.

$$\kappa = Fc(u_{\text{H}^+} + u_{\text{OH}^-})$$

$$c = \frac{\kappa}{F(u_{\text{H}^+} + u_{\text{OH}^-})}$$

$$= \frac{5.5 \times 10^{-6}\ \Omega^{-1}\,\text{m}^{-1}}{(96{,}485\ C\ \text{mol}^{-1})(5.68 \times 10^{-7}\ \text{m}^2\ V^{-1}\ \text{s}^{-1})}$$

$$= 1.00 \times 10^{-4}\,\text{mol m}^{-3}$$

$$= 1.00 \times 10^{-7}\,\text{M}$$

Thus $K_w = (1.00 \times 10^{-7})^2 = 1.00 \times 10^{-14}$ at 25°.

---

* F. Kohlrausch and A. Heydweiller, Z. *physik. Chem.*, **14, 317** (1894).

The ion product of water is given at various temperatures in Table 11.7. The ion product for D_2O at 25° is 1.54×10^{-15} showing that the properties of isotopic compounds are very much alike but not identical.

Table 11.7 Ion Product of Pure Water[1]

Temperature	0°	10°	25°	40°	50°
$K_w \times 10^{14}$	0.113	0.292	1.008	2.917	5.474

[1] H. S. Harned and W. J. Hamer, *J. Am. Chem. Soc.*, **55,** 2194 (1933).

Example 11.5 Using the values of the ion product at 10° and 40° calculate the enthalpy of ionization of water. Using equation 5.71

$$\log \frac{K_2}{K_1} = \frac{\Delta H°(T_2 - T_1)}{2.3RT_1T_2}$$

$$\log \frac{2.917}{0.292} = \frac{\Delta H°(30)}{2.3(1.987)(283.1)(313.1)}$$

$$\Delta H° = 13,500 \text{ cal mol}^{-1}$$

for the reaction $H_2O(l) = H^+ + OH^-$. This is in agreement with the enthalpy of neutralization of a strong acid by a strong base obtained calorimetrically at 25° (see Section 1.19).

11.9 RELATION BETWEEN MOBILITY AND ION FRICTIONAL COEFFICIENT

The mobilities of a number of familiar ions at infinite dilution, given in Table 11.6, may be interpreted in terms of the frictional coefficient f (equation 11.1) of the ion. When an electric field is applied to an electrolytic solution, an ion is accelerated until it reaches a velocity v where the product of the frictional coefficient f and the velocity v is equal to the force of the field on the charge of the ion. Thus

$$f_iv_i = |z_i| eE \tag{11.20}$$

where z_i is the charge number of ion i. Since the electric mobility is $u_i = v_i/E$, the frictional coefficient of an ion may be calculated from

$$f_i = \frac{|z_i| e}{u_i} \tag{11.21}$$

Apparent radii of the hydrated ions may be calculated using Stokes' law (equation 11.1).

11.10 DIFFUSION

Fick's first law of diffusion has been introduced in connection with the kinetic theory of gases (Section 9.11). Although the driving force for diffusion is the gradient of the chemical potential ($d\mu/dx$ if the chemical potential changes only in the x direction), the phenomenological equation for diffusion is written in terms of the concentration gradient dc/dx. According to Fick's first law, the flux J of a substance through a plane perpendicular to the direction of diffusion is directly proportional to the concentration gradient dc/dx.

$$J = -D\left(\frac{dc}{dx}\right) \tag{11.22}$$

where D is the diffusion coefficient. The flux J is the quantity of substance diffusing per unit time through a unit area. This equation is analogous to those for the flux of electric charge (which is proportional to the potential gradient according to Ohm's law), the flux of heat (which is proportional to the temperature gradient), and the flux of momentum (which is proportional to the velocity gradient) in viscous flow.

If the flux J is expressed in kg m^{-2} s^{-1} and the concentration gradient in kg m^{-4}, it is seen that the diffusion coefficient will have the units of m^2 s^{-1}. The diffusion coefficient for potassium chloride in very dilute aqueous solution at 25° is 1.99×10^{-10} m^2 s^{-1}, and for sucrose it is 5.23×10^{-10} m^2 s^{-1}. These values may be compared with those for several proteins given in Section 20.8.

The interpretation of most types of diffusion experiments is based on Fick's second law which is obtained from Fick's first law by introducing the idea of conservation of mass to eliminate the flow J. In the hypothetical diffusion experiment illustrated in Fig. 11.4, the quantity of material crossing the plane at x in time δt is $J\,\delta t$ whereas the quantity leaving through the plane at $x + \delta x$ in the same time is

$$\left(J + \frac{\partial J}{\partial x}\,\delta x\right)\delta t$$

The net gain in the quantity of diffusing material between these hypothetical planes may be expressed in terms of the change of concentration in the volume

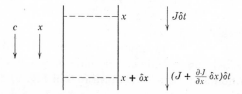

Fig. 11.4 Tube of unit-cross sectional area in which there is transport by diffusion, electrical migration, or sedimentation.

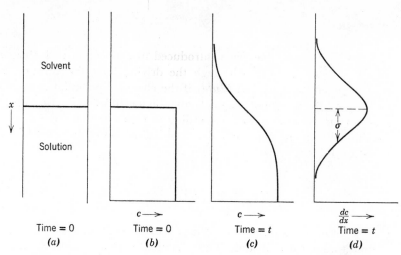

Fig. 11.5 Diffusion of an initially sharp boundary in a cell of uniform cross section.

δx and in terms of the difference between these two quantities of material transported.

$$\delta c \, \delta x = J \, \delta t - \left(J + \frac{\partial J}{\partial x} \delta x \right) \delta t$$

$$= - \frac{\partial J}{\partial x} \delta x \, \delta t \tag{11.23}$$

In the limit as the distances and times are made smaller

$$\frac{\partial c}{\partial t} = - \frac{\partial J}{\partial x} \tag{11.24}$$

This expression is referred to as the equation of continuity. Substitution of Fick's first law yields

$$\frac{\partial c}{\partial t} = \frac{\partial}{\partial x} D \frac{\partial c}{\partial x} \tag{11.25}$$

If the diffusion coefficient D is independent of the concentration and therefore of distance, then

$$\frac{\partial c}{\partial t} = D \frac{\partial^2 c}{\partial x^2} \tag{11.26}$$

which is known as Fick's second law.

Diffusion coefficients may be measured by a number of different methods.* Probably the most widely used methods are those in which a sharp boundary is formed between solution and solvent, as illustrated in Fig. 11.5. Initially the plot

---

* L. J. Gosting, *Advances in Protein Chemistry*, Vol. XI, Academic Press, New York, 1956; W. Jost, *Diffusion in Solids, Liquids, and Gases*, Academic Press, New York, 1952.

of concentration on the horizontal axis versus height in the cell on the vertical axis has the shape indicated in (b). At a later time this boundary will have become diffuse, and the concentration will vary with height as illustrated in (c). Instead of an abrupt change in concentration there is a more gradual one. When the solute is a colored substance, its concentration may be determined photometrically as a function of height. One of the most generally useful methods for determining the diffuseness of the boundary depends on measuring the deflection of light by the refractive-index gradient associated with the concentration gradient.[*] Since the bending of a light ray in such a boundary is proportional to the refractive index gradient dn/dx, the curve obtained with such a schlieren optical system has the shape indicated in (d). This curve has the shape of a normal probability curve (sometimes referred to as a Gaussian curve) if D is independent of concentration.

To derive the expression for concentration as a function of distance and time for the experiment illustrated in Fig. 11.5, equation 11.26 is integrated with the following boundary conditions: when $t = 0$, $c = c_0$ for $x > 0$, and $c = 0$ for $x < 0$; and when $t > 0$, c approaches c_0 as x approaches ∞ and c approaches 0 as x approaches $-\infty$. The result is

$$c = \frac{c_0}{2} \left[1 + \frac{2}{\sqrt{\pi}} \int_0^{x/2\sqrt{Dt}} e^{-\beta^2} \, d\beta \right] \tag{11.27}$$

where the second term in the square brackets is referred to as the Gaussian error function. The equation for the derivative curve in Fig. 11.5d is

$$\frac{\partial c}{\partial x} = \frac{c_0}{2\sqrt{\pi Dt}} e^{-x^2/4Dt} \tag{11.28}$$

The function $(2\pi)^{-1/2} e^{-y^2/2}$ is referred to as the normal probability function of y. This bell-shaped probability curve is referred to as a Gaussian curve.

The square of the standard deviation of the experimental bell-shaped curve is

$$\sigma^2 = \frac{\displaystyle\int_{-\infty}^{\infty} x^2 \left(\frac{\partial c}{\partial x}\right) dx}{\displaystyle\int_{-\infty}^{\infty} \left(\frac{\partial c}{\partial x}\right) dx} \tag{11.29}$$

Substituting equation 11.28

$$\sigma^2 = \frac{\displaystyle\int_{-\infty}^{\infty} x^2 e^{-x^2/4Dt} \, dx}{\displaystyle\int_{-\infty}^{\infty} e^{-x^2/4Dt} \, dx} \tag{11.30}$$

$$= 2Dt$$

---

[*] L. G. Longsworth, *Ind. Eng. Chem., Anal. Ed.*, **18,** 219 (1946).

where the last form is obtained by using the values of the definite integrals.*
Since the standard deviation σ of a Gaussian curve is the half width at the inflection
point, and the inflection points are at a height of 0.606 of the maximum ordinate,
σ is readily obtained from the experimental curve, and D may be calculated using
equation 11.30.

11.11   RELATION BETWEEN THE DIFFUSION COEFFICIENT AND THE FRICTIONAL COEFFICIENT

The driving force of diffusion is the negative gradient of the chemical potential,
and the retarding force is the frictional coefficient f times the velocity v. Thus, if
the chemical potential changes only in the x direction, then

$$N_A f v = - \frac{d\mu}{dx} \tag{11.31}$$

At any level in a diffusion cell the flux J is equal to the velocity v times the concentration c. Using the preceding equation, we have

$$J = vc = \frac{-c}{N_A f} \frac{d\mu}{dx} \tag{11.32}$$

If the solution is ideal $\mu = \mu^* + RT \ln c$ so that

$$J = - \frac{RT}{N_A f} \frac{dc}{dx} \tag{11.33}$$

Comparing this with Fick's first law (equation 11.22), we see that

$$D = \frac{RT}{N_A f} \tag{11.34}$$

This relation was first obtained by Einstein. The radius of spherical particles
may be calculated from the measured diffusion coefficient using this relation and
Stokes' law (equation 11.1).

---

*

$$\int_0^\infty e^{-a^2 x^2}\, dx = \frac{\sqrt{\pi}}{2a}$$

$$\int_0^\infty x^{2n} e^{-ax^2}\, dx = \frac{1 \cdot 3 \cdot 5 \cdots (2n-1)}{2^{n+1} a^n} \sqrt{\frac{\pi}{a}}$$

References

R. B. Bird, W. E. Stewart, and E. N. Lightfoot, *Transport Phenomena*, Wiley, New York, 1960.

J. O'M. Bockris and D. M. Drazic, *Electrochemical Science*, Taylor and Francis Ltd., London, 1972.

H. L. Friedman, *Ionic Solution Theory*, Wiley-Interscience, New York, 1962.

R. M. Fuoss and F. Accascina, *Electrolytic Conductance*, Wiley-Interscience, New York, 1959.

S. R. de Groot and P. Mazur, *Nonequilibrium Thermodynamics*, Wiley-Interscience, New York, 1962.

H. S. Harned and B. B. Owen, *The Physical Chemistry of Electrolytic Solutions*, Reinhold Publishing Corp., New York, 1958.

W. Jost, *Diffusion in Solids, Liquids, and Gases*, Academic Press, New York, 1952.

I. Prigogine, *Introduction to the Thermodynamics of Irreversible Processes*, Wiley, New York, 1967.

R. A. Robinson and R. H. Stokes, *Electrolyte Solutions*, Academic Press, New York, 1959.

Problems

11.1 Ten ml of water at $25°$ is forced through 20 cm of 2 mm diameter capillary in 4 s. Calculate the pressure required and the Reynolds number.

Ans. 1.14×10^{-5} N m^{-2}. 1780.

11.2 A steel ball ($\rho = 7.86$ g cm^{-3}) 0.2 cm in diameter falls 10 cm through a viscous liquid ($\rho_0 = 1.50$ g cm^{-3}) in 25 s. What is the absolute viscosity at this temperature?

Ans. 3.46 Pa s.

11.3 How long will it take a spherical air bubble 0.5 mm in diameter to rise 10 cm through water at $25°$?   *Ans.* 0.66 s.

11.4 Calculate the time necessary for a quartz particle 10 μm in diameter to sediment 50 cm in distilled water at $25°$. The density of quartz is 2.6 g cm^{-3}. The coefficient of viscosity of water may be taken to be 8.95×10^{-4} kg m^{-1} s^{-1}.   *Ans.* 95.6 min.

11.5 A conductance cell was calibrated by filling it with a 0.02 M solution of potassium chloride ($\kappa = 0.2768$ Ω^{-1} m^{-1}) and measuring the resistance at $25°$, which was found to be 457.3 Ω. The cell was then filled with a calcium chloride solution containing 0.555 gram of CaCl$_2$ per liter. The measured resistance was 1050 Ω. Calculate (a) the cell constant for the cell, (b) the conductivity of the CaCl$_2$ solution, and (c) the molar conductivity of CaCl$_2$ at this concentration.

Ans. (a) 126.6 m^{-1}, (b) 0.1206 Ω^{-1} m^{-1}, (c) 0.0241 Ω^{-1} m^2 mol^{-2}.

11.6 It is desired to use a conductance apparatus to measure the concentration of dilute solutions of sodium chloride. If the electrodes in the cell are each 1 cm^2 in area and are 0.2 cm apart, calculate the resistance which will be obtained for 1, 10, and 100 ppm NaCl at $25°$.

Ans. 92,700, 9320, and 950 Ω.

11.7 Determine the value of Λ^0 for lithium chloride from the following data at $25°$.

Moles per liter	0.05	0.01	0.005	0.001	0.0005
Molar conductivity	0.010011	0.010732	0.010940	0.011240	0.011315

Ans. 0.01143 Ω^{-1} m^2 mol^{-1}.

11.8 A moving-boundary experiment is carried out with a 0.1 M solution of hydrochloric acid at $25°$ ($\kappa = 4.24$ Ω^{-1} m^{-1}). Sodium ions are caused to follow the hydrogen ions. Three milliamperes is passed through the tube of 0.3 cm² cross-sectional area, and it is observed that the boundary moves 3.08 cm in 1 hr. Calculate (a) the hydrogen-ion mobility, (b) the hydrogen-ion transference number, (c) the chloride-ion mobility, and (d) the electric field strength.

> *Ans.* (a) 3.63×10^{-7} m² V⁻¹ s⁻¹.
> (b) 0.826.
> (c) 7.64×10^{-8} m² V⁻¹ s⁻¹.
> (d) 23.6 V m⁻¹.

11.9 Calculate the conductivity of 0.001 M HCl at 25°. The limiting ion mobilities may be used for this problem. *Ans.* 0.0426 Ω^{-1} m⁻¹

11.10 Using Stokes' law calculate the effective radius of a nitrate ion from its mobility (74.0×10^{-9} m² V⁻¹ s⁻¹ at 25°). *Ans.* 1.29 Å.

11.11 Using a table of the probability integral, calculate enough points on a plot of *c* versus *x* (like Fig. 11.6c) to draw in the smooth curve for diffusion of 0.1 M sucrose into water at 25° after 3 hr ($D = 5.23 \times 10^{-10}$ m² s⁻¹).

11.12 A sharp boundary is formed between a dilute aqueous solution of sucrose and water at 25°. After 5 hr the standard deviation of the concentration gradient is 0.434 cm. (a) What is the diffusion coefficient for sucrose under these conditions? (b) What will be the standard deviation after 10 hr? *Ans.* (a) 5.23×10^{-10} m² s⁻¹. (b) 0.614 cm

---

11.13 The water flow time for a viscometer of the type illustrated on the left in Fig. 11.1 is 59.2 s at 25°. If 46.2 s is required for the same volume of ethyl benzene ($\rho = 0.867$ g cm⁻³) to flow through the capillary, calculate its absolute viscosity in Pa s at 25°.

11.14 Ten cm³ of water at 25° is forced through 20 cm of 2 mm diameter capillary in 4 s. Calculate the pressure required and the Reynolds number. Will the flow be laminar?

11.15 Using data in Table 11.1 and equation 11.6, estimate the activation energy for water molecules to move into a vacancy at 25° C.

11.16 One hundred grams of sodium chloride is dissolved in 10,000 liters of water at 25°, giving a solution which may be regarded in these calculations as infinitely dilute. (a) What is the molar conductivity of the solution? (b) What is the conductivity of the solution? (c) This dilute solution is placed in a glass tube of 4-cm diameter provided with electrodes filling the tube and placed 20 cm apart. How much current will flow if the potential drop between the electrodes is 80 V?

11.17 Estimate the conductivity at 25° of water which contains 70 ppm (parts per million) by weight of magnesium sulfate.

11.18 The conductivity of a saturated solution of thallous bromide at 20° is 2.158×10^{-2}, and the conductivity of the water used is 0.444×10^{-4} Ω^{-1} m⁻¹. The molar conductivity at infinite dilution is 0.0138 Ω^{-1} m² mol⁻¹. Calculate the solubility of thallous bromide in grams per liter.

11.19 In 0.1 M hydrochloric acid at 0° the mobilities of hydrogen and chloride ions are 365×10^{-9} and 79×10^{-9} m² V⁻¹ s⁻¹, respectively. (a) Calculate the conductivity for this solution at 0°. (b) A moving-boundary experiment is carried out in a tube with a uniform cross-sectional area of 0.200 cm², and sodium ions are caused to follow the hydrogen ions. If a current of 5 ma is passed for 1 hr, how far will the hydrogen ions move? (c) What is the field strength in this experiment?

11.20 A study of conductivities at high electric field strengths reveals that the molar conductivity increases slightly with increasing electric field strength. A microsecond pulse at 10^3 V m⁻¹ may be used. Approximately how far will a sodium ion move during such a pulse at room temperature?

11.21   Calculate the electric mobility of NO_3^- in a very dilute solution, given the following molar conductivities at infinite dilution (in Ω^{-1} m^2 mol^{-1}, at 25°): KCl, 0.01499; KNO$_3$, 0.01449; HCl, 0.04261; and the transference number of H$^+$ in HCl as 0.821.

11.22   Estimate the electric mobility of a $(CH_3)_4N^+$ ion assuming it is not hydrated in solution at 25°. The effective radius may be taken as 3 Å.

11.23   Using a table of the normal probability function, calculate enough points on a plot of dc/dx versus x (like Fig. 11.5) to draw in the smooth curve for diffusion of 0.01 M sucrose into water at 25° after 3 hr.

11.24   (a) Calculate the time required for the half-width of a freely diffusing boundary of dilute potassium chloride in water to become 0.5 cm at 25° ($D = 1.77 \times 10^{-9}$ m^2 s^{-1}) (b) Calculate the corresponding time for serum albumin ($D = 6.15 \times 10^{-11}$ m^2 s^{-1}).

11.25   Estimate the rate of sedimentation of water droplets of 1 μm diameter in air at 20°. The viscosity of air at this temperature is 1.808×10^{-5} Pa s.

11.26   Plot log of viscosity of mercury against the reciprocal of the absolute temperature from the following data, and estimate the viscosity of mercury at 50°:

t, °C	0	20	35	98	203
$\eta \times 10^3$, Pa s	1.661	1.547	1.476	1.263	1.079

11.27   A sample of water from a large pool had a resistance of 9200 Ω at 25° when placed in a certain conductance cell. When filled with 0.020 M potassium chloride the cell had a resistance of 85 Ω at 25°. Five hundred grams of sodium chloride was dissolved in the pool, which was then thoroughly stirred. A sample of this solution gave a resistance of 7600 Ω. With the help of graphical interpolation calculate the number of liters of water in the pool.

11.28   Show that if a potential difference of 1 V is applied across opposite faces of a 1 m cube of electrolyte, the current in amperes is numerically equal to the conductivity κ.

11.29   A glass tube 4 cm in diameter and 30 cm long is closed at each end with a sheet silver electrode and filled with 0.01 M silver nitrate. Sixty volts is applied. (a) How much current flows? (b) How many degrees will the temperature of the solution rise in 10 min, if it is assumed that the heat capacity of the solution is nearly 1 cal K^{-1} cm^{-3} and that all the heat is retained by the solution?

11.30   Calculate the conductivity at 25° of a solution containing 0.001 M hydrochloric acid and 0.005 M sodium chloride. The limiting ionic mobilities at infinite dilution may be used to obtain a sufficiently good approximation.

11.31   The molar conductivity of a solution of NH$_4$Cl is 0.01497 Ω^{-1} m^2 mol^{-1}, and the ionic conductivities of the ions OH$^-$ and Cl$^-$ are 0.01980 and 0.00763, respectively. Calculate the molar conductivity of NH$_4$OH at infinite dilution.

11.32   The conductivity of a saturated solution of AgCl in water at 25° was found to be 2.28×10^{-4} Ω^{-1} m^{-1}. The conductivity of the water used was 1.16×10^{-4} Ω^{-1} m^{-1}. With the information that the ionic conductivity of Ag$^+$ is 0.00619 Ω^{-1} m^2 mol^{-1} and of Cl$^-$ is 0.00763 Ω^{-1} m^2 mol^{-1} at infinite dilution, calculate: (a) the solubility of AgCl in grams per liter; (b) the solubility product of AgCl.

11.33   Calculate the apparent radii of Li$^+$ and Na$^+$ in infinitely dilute aqueous solution from their mobilities in Table 11.6. How do you explain the fact that the ion with the smaller crystal radius (see Table 19.1) has the larger apparent radius in solution?

11.34   If only a small amount of material q is allowed to diffuse through a porous plate from a solution of concentration c'' into a solution of concentration c', Fick's law (equation 11.22) may be written

$$D = \frac{-q}{Kt(c' - c'')}$$

where K is the cell constant which must be determined in an experiment with a substance of known diffusion coefficient. If a 0.10 M aqueous solution of potassium chloride is allowed to diffuse into water for 12 hr and 38 min at 25°, it is found that 1.25×10^{-4} mole of salt diffuses through the porous plate. Calculate D if K has previously been found to be 1.5 cm.
11.35   An initially sharp boundary is formed between aqueous solutions containing 0.3 M glycine and 0.1 M glycine at 25°. $D = 1.022 \times 10^{-9}$ m^2 s^{-1}. Plot c versus x and dc/dx versus x for a diffusion time of 15 hr.

PART THREE

QUANTUM CHEMISTRY

Quantum theory, which revolutionized physics in the early part of this century, is required for an understanding of chemistry. For example, spectra, the structure of the periodic table, the configurations of molecules and their properties cannot be understood without quantum theory. In the first chapter of this section we consider the historical development of quantum theory and its application to simple idealized systems including the particle in a box, the simple harmonic oscillator, and the hydrogen atom. Because of mathematical complexities, it is not possible to give complete treatments. Simple ideas about symmetry are presented next because of the symmetry of simple molecules and of their wave functions.

The nature of chemical bonding and the structures of molecules are explained by quantum theory. Because of quantum mechanics this last half century has been a period of very rapid increase in our understanding of molecules. Certain properties of very simple molecules can be calculated more accurately than they have been measured in the laboratory.

One of the most important sources of information about molecular structure and molecular energy levels is spectroscopy. For example, spectra yield quantitative information about bond lengths and angles, vibration frequencies, dissociation energies, dipole moments, and the shapes of potential-energy curves. Nuclear magnetic resonance spectroscopy and electron spin resonance spectroscopy have become so important in the practice of chemistry that they are discussed in a separate chapter.

Information about molecules obtained from spectroscopy and the theoretical concepts of statistical mechanics provide the means for calculating thermodynamic properties from first principles. Thermodynamic properties of ideal gases are readily calculated, but for liquids and solids the theory is more difficult because of molecular interactions.

All the concepts described here are involved in photochemistry—the study of chemical reactions produced by the absorption of electromagnetic radiation.

CHAPTER 12

QUANTUM THEORY

In the late nineteenth century it became apparent that classical mechanics was unable to account for many experimental facts concerning the behavior of systems of atomic size. We have already referred to the heat capacities of gases in Chapter 9. In 1900 Planck assumed that electromagnetic radiation was quantized in his derivation of the equation for the intensity of radiation of different frequencies from a cavity. Planck's idea of quantization was used by Einstein to interpret the photoelectric effect in 1905 and by de Broglie to predict the wave properties of particles in 1924. Bohr developed his theory of the hydrogen atom in 1913. Heisenberg and Schrödinger developed quantum mechanics in 1926. Quantum mechanics has been of tremendous importance for the understanding of chemistry.

In this chapter we will consider the application of quantum mechanics to simple systems, including the hydrogen atom, which can be treated exactly, and to many electron atoms for which analytic solutions cannot be obtained. Applications of quantum mechanics to molecules are considered in Chapter 14, after an analysis of symmetry of molecules in the next chapter.

12.1 PHYSICAL CONSTANTS

The basic physical constants that will be encountered in quantum mechanics are summarized in Table 12.1.

Table 12.1 Physical Constants for Quantum Mechanics (more accurate values of these constants are given in the Appendix)

Electronic charge	e	1.602×10^{-19} C
Electron mass	m_e	9.11×10^{-31} kg
Proton mass	m_p	1.673×10^{-27} kg
Planck's constant	h	6.63×10^{-34} J s
First Bohr radius	a_0	5.29×10^{-11} m
Ionization potential H[1]		13.61 eV
Rydberg constant[1]	R_∞/hc	109737 cm^{-1}
Speed of light in a vacuum	c	2.9979×10^8 m s^{-1}

[1] For infinite nuclear mass.

The charge on the electron was measured by Millikan in the classical oil-drop experiment. A fine mist of oil was blown into a thermostated air space between two electrodes, and the radius of one particular drop was calculated from its rate of fall by means of Stokes's equation (Section 11.1). When the electric field was turned on, the velocity of the drop increased or decreased, depending on the sign of the charge and the direction of the field. If the surrounding air was ionized by an X-ray beam, the oil drop frequently picked up a gas ion, and hence its velocity in the electric field changed. The charge on the electron calculated from the minimum change in velocity was 1.602×10^{-19} C.

The mass of the electron was obtained from e/m_e determined by J. J. Thomson from the deflection of an electron beam.

12.2   WAVE MOTION

Since quantum mechanics is based on ideas about waves, it is useful to review certain aspects of wave motion. An electromagnetic wave consists of oscillating electric and magnetic fields directed perpendicular to each other and perpendicular to the direction of propagation, as shown in Fig. 12.1. The speed of propagation in a vacuum is represented by c.

The frequency ν of radiation is the number of cycles per second and is therefore equal to the distance traversed by the radiation in 1 s, c, divided by the wavelength.

$$\nu = \frac{c}{\lambda} \tag{12.1}$$

The unit of frequency is the Hertz Hz, which is one cycle per second.

The wave number $\tilde{\nu}$ is the number of cycles per centimeter.

$$\tilde{\nu} = \frac{1}{\lambda} \tag{12.2}$$

where the wavelength is given in centimeters. The wave number is thus given

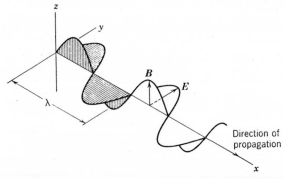

Fig. 12.1   Electric and magnetic field strengths in electromagnetic radiation.

in cm^{-1}. From equations 12.1 and 12.2, $v = c\tilde{v}$. Wave numbers are more useful in discussing spectroscopy than wavelengths because wave numbers, like frequencies, are proportional to the energy of a photon. The magnitudes of the electric field and the magnetic flux density at time t for a wave moving in the x direction are given by

$$E_y = E_{y0} \sin \frac{2\pi}{\lambda} (x - ct) = E_{y0} \sin 2\pi(\tilde{v}x - vt) \qquad (12.3)$$

$$B_z = B_{z0} \sin \frac{2\pi}{\lambda} (x - ct) = B_{z0} \sin 2\pi(\tilde{v}x - vt) \qquad (12.4)$$

where E_y is the magnitude of the electric field in the direction perpendicular to the direction of propagation and B_z is the magnitude of the magnetic field perpendicular to both of these directions. The symbols E_{y0} and B_{z0} represent the magnitudes of these fields when they have their maximum values.

Since the wavelengths of electromagnetic radiation vary over such a tremendous range, many different units are used. In the X ray, ultraviolet and visible ranges, the angstrom unit Å (10^{-10} m = 10^{-8} cm) is the most frequently used unit. The visible range extends from about 4,000 Å to about 8,000 Å (400–800 nm). The angstrom is not an SI unit, and so the nanometer nm (10^{-9} m) is receiving increasing use; 1 nm = 10 Å. Since the accuracy of measuring wavelengths is greater than that of measuring the length of the meter bar, the length of the meter is now defined as equal to 1,650,763.73 wavelengths of a certain line of ^{86}Kr in vacuum.

12.3 CAVITY RADIATION

Hot gases produce line spectra, but hot solids produce continuous radiation. The electromagnetic radiation from different solids at the same temperature may show a rather different spectral distribution. It is found, however that if the radiation inside an isothermal hollow body is viewed through a small hole in its wall, the intensity of radiation and distribution of wavelengths are independent of the material and of the size and shape of the cavity. Such radiation is often called black-body radiation, but we will refer to it as cavity radiation.

The spectral distribution of cavity radiation is shown at three temperatures in Fig. 12.2. The ordinate is the intensity I_λ, which is defined so that the rate of emission of energy in the wavelength range λ to $\lambda + d\lambda$ per unit area of surface is $I_\lambda\, d\lambda$.

Typical units for I_λ are watts cm^{-2} μm^{-1} and the corresponding units of $I_\lambda\, d\lambda$ are watts cm^{-2}.

The total intensity I (energy radiated per unit area of surface per unit time) is

$$I = \int_0^\infty I_\lambda\, d\lambda \qquad (12.5)$$

When the experimentally determined form of I_λ for a cavity radiator is inserted into equation 12.5 integration yields $I = \sigma T^4$ where $\sigma = 5.69 \times 10^{-8}$ W m^{-2} K^{-4}. This law for the total intensity was discovered experimentally by Stefan and was derived later by Boltzmann,

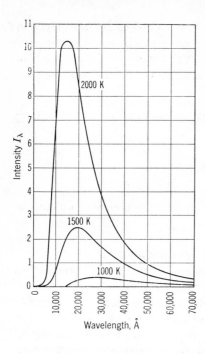

Fig. 12.2 Emission of radiation from a cavity at different temperatures. The area under the curve between specified wavelengths, divided by 10^4, gives the energy in calories per second radiated from 1 cm² of a cavity in the given range of wavelengths.

using the principles of thermodynamics. However, all attempts to derive the dependence of intensity on wavelength on the basis of classical physics were unsuccessful.

12.4 PLANCK'S THEORY

Planck was able to formulate a successful theory for the intensity of cavity radiation on the basis of a bold assumption: The oscillators in the solid that emit and absorb radiation can only have energies that are some multiple of $h\nu$ where h is a constant (later termed Planck's constant) and ν is the frequency of radiation emitted and absorbed. Thus, the oscillators cannot radiate or absorb any amount of energy, but only quanta of size $h\nu$.

On the basis of this assumption Planck was able to derive

$$I_\lambda = \frac{2\pi c^2 h}{\lambda^5 (e^{hc/kT\lambda} - 1)} = \frac{2\pi c^2 h}{\lambda^5 (e^{h\nu/kT} - 1)} \tag{12.6}$$

which represented the data accurately. In this equation c is the velocity of light $(2.99792458 \times 10^8 \text{ m s}^{-1})$ and k is the Boltzmann constant $(1.380662 \times 10^{-23} \text{ J K}^{-1})$. From the data he was able to calculate the value of what is now known as Planck's constant. The best value is

$$h = 6.626176 \times 10^{-34} \text{ J s}$$

Planck presented his theory to the Berlin Physical Society on December 14, 1900. Since the assumption did not agree with everyday experience Planck himself worked hard, but unsuccessfully, to find other assumptions that would allow him to derive equation 12.6. In 1905, however, the new idea of quantization received further support from Einstein, who used it to explain the photoelectric effect.

12.5 THE PHOTOELECTRIC EFFECT

The ejection of electrons from a metal surface by light is referred to as the photoelectric effect. The energy of the ejected electrons may be determined with the circuit shown in Fig. 12.3.

A receiver K is coated with a film of potassium or other metal in a highly evacuated tube. A wire screen W placed in the tube is connected with the receiver through a battery B and a sensitive galvanometer G. When the receiver is exposed to light, electrons are ejected from the receiver and attracted to the positively charged screen, thus completing the circuit. The current registered by the galvanometer is directly proportional to the number of electrons ejected per second, which, in turn, is proportional to the intensity of the light.

It is found that for a given surface the frequency of the incident light must be greater than a certain value, called the *threshold frequency*, in order for any electrons to be emitted from the surface. At frequencies above the threshold frequency the electrons are ejected with excess kinetic energy. The maximum energy of the electrons ejected perpendicular to the surface may be determined by reversing the polarity of the battery shown in Fig. 12.3 and determining the potential difference required to stop the current completely while the metal surface is being illuminated. It is found that the *maximum speed* of the ejected electrons is independent of the intensity of the light and depends only on its *frequency*.

$$\tfrac{1}{2}mv^2 = h(v - v_0) \tag{12.7}$$

It was not possible to understand these results in terms of classical physics. According to the classical view the square of the amplitude of the electric field

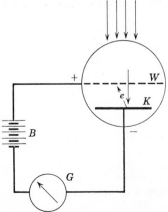

Fig. 12.3 Photoelectric cell and circuit.

(i.e., the intensity) is directly proportional to the energy of the radiation so that more intense radiation might be expected to eject electrons at higher velocities. Instead, for monochromatic light the velocity stays constant, but the number of electrons ejected goes up with the intensity.

Einstein was able to explain these effects by making the assumption that the energy of a light beam travels through space in quanta of energy $h\nu$, now called *photons*. When light is absorbed by a metal the total energy of a photon $h\nu$ is given to a single electron within the metal. If this quantity of energy is sufficiently large, the electron may penetrate the potential barrier at the surface of the metal and still retain some energy as kinetic energy. The kinetic energy retained by the electron depends on the energy, and therefore the frequency, of the photon that ejected it. The number of electrons ejected depends on the number of incident photons, and therefore the intensity of light.

From an analysis of data on the photoelectric effect, Millikan calculated a value for the constant of proportionality between the photon's energy and its frequency and found it to be in good agreement with the one calculated from Planck's radiation equation. In view of the nature of the photoelectric effect it is necessary to regard light as having a dual character; under some circumstances it behaves like a wave and under others like a particle.

12.6 LINE SPECTRA

Whereas cavity radiation is continuous, various other types of spectra contain lines. The nature of line spectra could not be explained by classical theories. Attempts to find regularity in spectra had shown that the frequencies of spectral lines could be calculated by taking differences between quantities called "terms." The various lines in a spectrum are accounted for by taking various differences between a relatively small number of term values.

The simplest spectrum is that of hydrogen atoms, and Fig. 12.4 illustrates a small region of it. In 1885 Balmer discovered that the wavelengths λ (or wave numbers, $\tilde{\nu}$) of the lines in the visible region of the hydrogen-atom spectrum

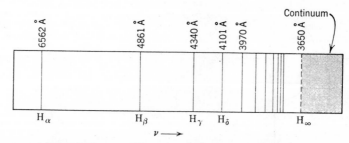

Fig. 12.4 Balmer series of lines in the spectrum of atomic hydrogen. At shorter wavelengths than the series limit there is continuous emission. (For the energy level diagram see Fig. 2.15.)

could be expressed by a simple relation that may be written as

$$\frac{1}{\lambda} = \tilde{\nu} = R\left(\frac{1}{2^2} - \frac{1}{n_2^2}\right) \tag{12.8}$$

where n_2 is an integer greater than 2, and R is the Rydberg constant, 109,677.58 cm^{-1}. The value of R may be determined very accurately because of the high precision with which the wavelengths of spectral lines can be measured.*

It will be noticed in equation 12.8 that n_2 cannot be less than 2, for then $\tilde{\nu}$ would be a meaningless, negative number, and it cannot be 2, for then $\tilde{\nu}$ becomes zero. As n_2 becomes larger than 2, the corresponding value of $\tilde{\nu}$ becomes larger. When n_2 is already large, however, further increases cause $\tilde{\nu}$ to increase only very slightly, and, as n_2 approaches infinity, $\tilde{\nu}$ approaches $\frac{1}{4}R$ as a limit. As indicated in Fig. 12.4, there is continuous radiation at wavelengths shorter than this series limit.

The success of the Balmer formula led to further exploration, and other series of lines were discovered in the atomic hydrogen spectrum which could be represented by the equation

$$\tilde{\nu} = R\left(\frac{1}{n_1^2} - \frac{1}{n_2^2}\right) \tag{12.9}$$

where n_1 is also an integer. The series for which $n_1 = 1$ (Lyman series) is in the ultraviolet; the series for which $n_1 = 3$ (Paschen series), 4 (Brackett series), or 5 (Pfund series) are in the infrared region. It is important to note that every line in the spectrum can be represented as a difference of two terms, R/n_1^2 and R/n_2^2. The spectra of other atoms are more complicated, but in general it is found possible to represent the lines of the spectrum as differences between term values. This concept may be readily understood by the application of the principle of conservation of energy which requires that

$$h\nu = E_2 - E_1 \tag{12.10}$$

where E_2 is the energy of the atom or molecule before emission of a photon $h\nu$ and E_1 is the energy after emission. This equation is basic to all types of spectroscopy.

12.7 BOHR'S THEORY OF THE HYDROGEN ATOM

In 1911 Rutherford had deduced from the deflection of alpha particles by thin metal foils that an atom contains a small positively charged nucleus. The number of positive charges on the nucleus is defined as the *atomic number*; the number of electrons surrounding the nucleus in the neutral atom is also equal to the atomic number.

---

* The value of the Rydberg constant R_∞ for infinite nuclear mass is 109,737.3177 cm^{-1}.

A successful theory of the spectrum of the hydrogen atom was developed in 1913 by Bohr on the basis of the quantum theory. Bohr made a complete break with classical mechanics by assuming that in the hydrogen atom the angular momentum of the orbital electron can only have values that are integral multiples of a quantum of angular momentum of magnitude $\hbar$, referred to as h-bar, which is $h/2\pi$. Bohr assumed that an electron moves in a circular orbit around the positively charged nucleus. We now know that orbital electrons do not behave in this way, but nevertheless Bohr was able to derive a correct expression for the energy levels of hydrogenlike atoms (i.e., atoms with one electron). He was also able to calculate the sizes of hydrogenlike atoms; he calculated the radius of the inner orbit of the hydrogen atom to be 0.529×10^{-10} m.

The electronic energy levels in the hydrogen atom, as calculated from the Bohr theory, are summarized in Fig. 12.5. The Lyman series of lines is produced by electrons jumping from orbits with quantum numbers 2, 3, 4, ... into the lowest permitted orbit $(n_1 = 1)$. The Balmer series of lines is produced by electrons falling from larger orbits into the second orbit $(n_1 = 2)$, etc. The energies of the various orbits may be expressed in several ways. The energies in wave numbers given at the right in Fig. 12.5 are the wave numbers for radiation produced when an electron falls from an infinite distance into a given orbit with no initial kinetic energy. The wave number $\tilde{\nu}$ of any line in the spectrum may be obtained by subtracting the values at the right for the two energy levels involved. Thus the

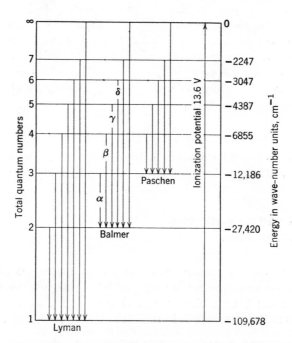

Fig. 12.5 Energy levels for the hydrogen atom as calculated from the Bohr theory. Ionization potentials are discussed in Section 12.26.

second line in the Balmer series is due to an electron falling from the fourth orbit into the second, and its wave number is $27,420 - 6855 = 20,565$ cm^{-1}.

The continuous emission at wavelengths shorter than 356 nm shown in Fig. 12.5 is due to electrons having positive total energies (i.e., "ionized electrons") falling into an orbit of the hydrogen atom. The energy is not quantized for positive values, and hence the loss in energy of the electron may have a continuous range of values, leading to the emission of continuous radiation. The "convergence limit" of emission lines in a spectral series corresponds then to complete separation of the electron, that is, to ionization yielding a positively charged ion and a free electron. In the absorption of light the electron makes a transition to a level of higher energy or to dissociation by the absorption of a quantum of energy.

In spite of the great success of the Bohr theory in accounting for the spectrum of atomic hydrogen and hydrogenlike atoms, attempts to extend it to spectra of many-electron atoms were unsuccessful. The Bohr theory was also regarded as unsatisfactory in that apparently arbitrary postulates had been introduced into classical mechanics. The search for a more general theory led to quantum mechanics, which provides a rather different picture of the hydrogen atom.

12.8 THE DE BROGLIE RELATION

Different experiments involving light seem to call for two contradictory views of the nature of light: diffraction phenomena seem to require a wave picture, yet photoelectric emission is suggestive of particlelike nature. These two aspects of light can be reconciled as follows: the energy in a radiation field occurs only in multiples of a basic unit ("photon") that depends on the frequency or wavelength according to

$$E = h\nu = \frac{hc}{\lambda} \qquad (12.11)$$

A photon's momentum $p = mc$, where m is its mass and c is the velocity of light, can be calculated from the Einstein expression for the energy $E = mc^2$.

$$p = mc = \frac{E}{c} \qquad (12.12)$$

Substituting equation 12.11

$$p = \frac{h}{\lambda} \qquad (12.13)$$

In 1924 de Broglie suggested that this relation is also applicable to material particles. That is, a particle with momentum p has wave properties, with the wavelength given by

$$p = mv = \frac{h}{\lambda} \qquad (12.14)$$

where v is the velocity of the particle. The wavelengths of particles calculated in this way are called de Broglie wavelengths. The wavelike nature of matter suggested by de Broglie was verified in 1928 by Davisson and Germer, who obtained a diffraction pattern from electrons impinging on the face of a nickel crystal.

12.9 HEISENBERG UNCERTAINTY PRINCIPLE

In 1927 Heisenberg* showed that there is a fundamental limit to the accuracy with which certain physical measurements may be made simultaneously. This limitation applies to many combinations of dynamical variables (coordinates, velocities, angular momenta, energy, time) that have the physical dimensions of *action*, that is mass·length^2·time^{-1}, and can be expressed by the relations

$$\Delta q \, \Delta p \geqslant h/4\pi \tag{12.15}$$

$$\Delta E \, \Delta t \geqslant h/4\pi \tag{12.16}$$

Where Δq is the root-mean-square uncertainty in position, Δp is the root-mean-square uncertainty in momentum, Δt is the root mean square uncertainty in time, and ΔE is the root mean square uncertainty in energy.

Because of the small value of h, this uncertainty is not detectable for macroscopic objects, but for electrons, atoms, and molecules Heisenberg's relations are significant. The Heisenberg uncertainty relation indicates that it is meaningless to ask about the exact position and exact velocity of an electron in an atom.

It is important to realize that the uncertainties in equations 12.15 and 12.16 are not "experimental errors" dependent, say, on the quality of one's laboratory equipment, but are inherent in any measurement process. For example, suppose the position of a moving electron is measured by shining light on it, that is, by bombarding it with photons. It would be necessary to use short wavelengths to locate the electron precisely, but this would make the momenta h/λ of the photons large, so that those that struck the electron would knock it significantly off course, and its original velocity would become very uncertain. Thus increasing certainty in the position of the electron implies decreasing certainty in its momentum, in agreement with equation 12.15.

The motion of electrons when struck by photons is known as the *Compton effect*. Compton irradiated carbon and other light elements with X rays having an energy of about 20 keV. He found that the X rays scattered by the atomic electrons at a particular angle have a sharply defined frequency lower than that of the incident rays. The observed wavelengths were in exact agreement with an equation derived on the assumption that this is a collision of two particles, a photon and an electron. The laws of conservation of energy and momentum lead to an expression which correctly describes the experimental results, with the momentum of a photon given by $p = h/\lambda$, and its energy is $E = h\nu$.

12.10 SCHRÖDINGER EQUATION

Quantum mechanics was developed independently in 1926 by W. Heisenberg and Erwin Schrödinger. Heisenberg's approach is referred to as matrix mechanics

---

* W. Heisenberg, Z. *Physik* **43,** 172 (1927).

and Schrödinger's approach is referred to as wave mechanics. Although the two methods appear different they can be shown to be mathematically equivalent. We will consider only the Schrödinger formulation that uses ideas about wave motion.

Schrödinger's equation for the one-dimensional motion of a particle of mass m in a potential V, which is a function of x, is

$$\frac{d^2\psi}{dx^2} + \frac{2m}{\hbar^2}(E - V)\psi = 0 \qquad (12.17)$$

The solution ψ of this equation is called the wave function and expresses all that can be known about the properties of the system when it is in a steady state, that is not changing with time.

Every particle or collection of particles (for example, a hydrogen atom or even a mole of gas molecules) has a quantum mechanical wave function which describes the state of the system. The wave function ψ depends on the coordinates of the particles ($3N$ coordinates are required for N particles) and may depend on time. The wave function itself does not have a simple physical meaning, and as a matter of fact it may involve an imaginary part. The product of the wave function ψ with its complex conjugate ψ^* is proportional to the *probability density* ρ associated with the particle. (The complex conjugate of a function is obtained by substituting $-i$ for i, where $i = \sqrt{-1}$.) The probability density ρ is defined so that the probability that a particle is located in a small volume $dx\,dy\,dz$ is represented by $\rho\,dx\,dy\,dz$. Probability densities have been discussed earlier in connection with kinetic theory (Section 9.2). The integral of the probability over the whole volume containing a particle is unity

$$\int \rho\,d\tau = 1 \qquad (12.18)$$

since the probability is unity that the particle is somewhere. The symbol $d\tau$ represents a differential volume. Equation 12.18 is referred to as the normalization condition on the probability density ρ.

The interpretation of wave functions in terms of probabilities is in accord with the Heisenberg uncertainty principle. Since it is impossible to know both the position and velocity of a particle at the same time, we can deal only with the probability that a particle will be in a certain element of volume. The probability density ρ is given by

$$\rho = \frac{\psi^*\psi}{\displaystyle\int \psi^*\psi\,d\tau} \qquad (12.19)$$

If the integral in the denominator has the value unity, the wave function is said to be normalized.

Although two independent solutions of the Schrödinger equation exist (since it is a second order equation) for any value of the energy E, it turns out that most of these are unacceptable on physical grounds. For $\psi^*\psi$ to be interpreted as a probability density it must be single valued and have a finite integral. These boundary conditions can usually be met only for certain discrete values of the energy.

12.11 OPERATORS

In quantum mechanics, mechanical quantities are represented by operators. An operator is a description of a mathematical operation to be applied to a function to obtain a new function. Simple operators are c, x, d/dx, and d^2/dx^2; that is, multiplication by a constant c, multiplication by a variable x, differentiation with respect to x, and two successive differentiations with respect to x. In quantum mechanics each observable quantity, like the x coordinate, momentum in the x direction, energy, and angular momentum, has a corresponding operator.

To express the Schrödinger equation in terms of operators it is convenient to introduce the classical expression for the energy of the system in terms of momenta and coordinates. This function is known as the Hamiltonian in classical mechanics, and it is represented by H. If the potential energy V of the system depends only on coordinates,

$$H = T + V \qquad (12.20)$$

where T is kinetic energy. Rather than using Cartesian coordinates to express T and V it may be more convenient to use other coordinates; for example, in discussing the vibrations of a molecule, the displacements of each atom from its equilibrium position might be used.

Quantum mechanical operators are obtained from the classical expressions for the observables according to definite rules. For Cartesian coordinates the transformation is illustrated by

$$x \rightarrow x \qquad (12.21)$$

$$p_x \rightarrow \frac{\hbar}{i}\frac{\partial}{\partial x} \qquad (12.22)$$

For example, let us consider a particle of mass m moving in the x direction in a potential V that depends on x only. The classical Hamiltonian H is

$$H = \frac{p_x^2}{2m} + V \qquad (12.23)$$

The corresponding quantum mechanical operator for a particle constrained to move in the x direction only is, therefore,

$$\mathscr{H} = -\frac{\hbar^2}{2m}\frac{\partial^2}{\partial x^2} + V \qquad (12.24)$$

This operator $\mathscr{H}$ is called the *Hamiltonian operator*.

If the particle can move in three dimensions, then

$$H = \frac{1}{2m}(p_x^2 + p_y^2 + p_z^2) + V \tag{12.25}$$

Thus the Hamiltonian operator is given by

$$\mathscr{H} = \frac{-\hbar^2}{2m}\left(\frac{\partial^2}{\partial x^2} + \frac{\partial^2}{\partial y^2} + \frac{\partial^2}{\partial z^2}\right) + V \tag{12.26}$$

An operator has to have a function on which to operate, and in quantum mechanics this is the wave function. Application of the Hamiltonian operator $\mathscr{H}$ to the wave function for a system yields its energy E according to

$$\mathscr{H}\psi = E\psi \tag{12.27}$$

Substituting equation 12.24 into equation 12.27 shows it to be the same as equation 12.17, the time-independent Schrödinger equation for a particle moving in the x direction.

$$\frac{-\hbar^2}{2m}\frac{\partial^2\psi}{\partial x^2} + V\psi = E\psi \tag{12.28}$$

For a particle moving in three dimensions,

$$\frac{\partial^2\psi}{\partial x^2} + \frac{\partial^2\psi}{\partial y^2} + \frac{\partial^2\psi}{\partial z^2} + \frac{2m}{\hbar^2}(E - V)\psi = 0 \tag{12.29}$$

$$\nabla^2\psi + \frac{2m}{\hbar^2}(E - V)\psi = 0 \tag{12.30}$$

where the second form has been expressed using the Laplacian operator ∇^2.

$$\nabla^2 = \frac{\partial^2}{\partial x^2} + \frac{\partial^2}{\partial y^2} + \frac{\partial^2}{\partial z^2} \tag{12.31}$$

As previously stated, wave functions ψ that are physically reasonable (single valued, continuous, and having a finite value for the integral of the square of the function) exist in these equations only for certain values of E. These values of E are called *eigenvalues* and the corresponding wave functions are called *eigenfunctions*. The eigenvalues are the stationary energy states of the system under consideration.

In general, if α is an operator, a is a constant and f is a function, and if $\alpha f = af$, then it is said that f is an eigenfunction of α with eigenvalue a.

Example 12.1   Show that the function $\psi = ce^{ax}$, where a is a constant, is an eigenfunction of the operator d/dx and calculate the eigenvalue.

$$\frac{d}{dx}(ce^{ax}) = cae^{ax} = a\psi$$

Thus the eigenvalue is a.

Quantum mechanics provides the means for calculating the average values of observables. The average value of an observable is the average of the series of values that would be obtained if a given experiment were repeated many times, the initial state of the system being the same in each experiment. The average value is called the expectation value and is denoted by $\langle B \rangle$, which is given by

$$\langle B \rangle = \int \psi^* \mathcal{B} \psi \, d\tau \qquad (12.32)$$

where the integration extends over all space, ψ is a normalized wave function, and $\mathcal{B}$ is the quantum mechanical operator which corresponds to observable B. For example, if we were interested in the average value of the x coordinate of a particle we would use

$$\langle x \rangle = \int \psi^* x \psi \, dx \qquad (12.33)$$

If ψ in equation 12.32 is an eigenfunction of the operator $\mathcal{B}$ with an eigenvalue b, then

$$\mathcal{B}\psi = b\psi \qquad (12.34)$$

Substituting this relation in equation 12.32 we have

$$\langle B \rangle = b \int \psi^* \psi \, d\tau = b \qquad (12.35)$$

where b has been taken out of the integral because it is a constant. The integral is equal to unity because the wave function is normalized. Hence, the expectation value is simply the eigenvalue.

We will now discuss the solutions of the Schrödinger equation for four simple systems: (1) the particle in a box, (2) the simple harmonic oscillator, (3) the rigid rotor, and (4) the hydrogen atom. These illustrations show how the predictions of quantum mechanics differ from those of classical mechanics.

12.12 QUANTUM-MECHANICAL TREATMENT OF A PARTICLE IN A ONE-DIMENSIONAL BOX

The simplest problem related to that of an electron in an atom involves the calculation of the wave function for an electron constrained to move within a distance of length a in the x direction. The potential energy V is taken as zero within this length a and infinite for other x values. This model is related to an atom in that the electron is bound in a small space.

To determine the wave function for the particle between $x = 0$ and $x = a$ where $V = 0$, equation 12.17 may be written

$$\frac{d^2\psi}{dx^2} + \frac{2m}{\hbar^2} E\psi = 0 \qquad (12.36)$$

The solution of this equation is

$$\psi = A \sin\left(\frac{2\,mE}{\hbar^2}\right)^{\!\frac{1}{2}}\!x + A' \cos\left(\frac{2\,mE}{\hbar^2}\right)^{\!\frac{1}{2}}\!x \qquad (12.37)$$

as may be shown by substitution into equation 12.36. The values of the constants A and A' are calculated later.

Since the potential outside the box is infinite, the probability of finding the particle outside must be zero.* To avoid a discontinuity at $x = 0$ and $x = a$ the wave function ψ must have a value of zero at these points. To satisfy the boundary condition at $x = 0$ the constant A' in equation 12.37 must be taken to be zero. The boundary condition at $x = a$ is satisfied† only if

$$\left(\frac{2\,mE}{\hbar^2}\right)^{\!\frac{1}{2}}\!a = n\pi \qquad (12.38)$$

where n is an integer. According to this equation, the energy of the particle in the box is given by

$$E = \frac{h^2 n^2}{8\,ma^2} \qquad (12.39)$$

A particle moving between two points on a line can have only the energies given by this equation for integer values of n, whereas a perfectly free particle can have any energy. Such discrete energy levels are characteristic of solutions of the Schrödinger equation for bound particles. No such discrete energy levels are expected on the basis of classical mechanics.

The lowest energy level $(n = 1)$ is $E = h^2/8ma^2$, and the particle would necessarily have at least this much energy. A "zero-point energy" is found whenever a particle is constrained to a finite region; if this were not so the uncertainty principle would be violated.‡ The next higher energy levels are at 4 times $(n = 2)$ and 9 times $(n = 3)$ this energy as shown in Fig. 12.6a. The wave functions are superimposed on this plot, and we can see that the wavelength is equal to $2a/n$. It is apparent from equation 12.39 that the bigger the box or the heavier the particle the lower and more closely spaced are the energy levels.

The value of the constant A in equation 12.37 is calculated by *normalizing* the wave function. Equation 12.37 may be rewritten in terms of the quantum number

---

* If the wave function did not have a value of zero outside of the box, equation 12.17 could not be satisfied except with infinite energy, which is impossible.

† There is a discontinuity in the slope of the plot of ψ versus x at $x = 0$ and $x = a$ resulting from the infinite jump in potential at these points. Infinite potential barriers do not exist in nature, and wave functions for real systems decrease to zero as distance is increased without discontinuities in the first derivative of the wave function.

‡ The zero point energy of the particle in a box is in agreement with the requirements of the Heisenberg uncertainty principle. Since $\Delta x \approx a$, then $\Delta p \approx h/a$, where $\approx$ means approximately equal to. Thus $\Delta E = (\Delta p)^2/2m \approx h^2/2ma^2$, which is of the correct order of magnitude.

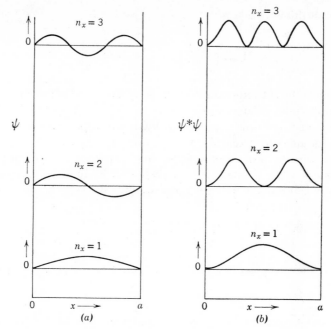

Fig. 12.6   (a) Wave functions ψ and (b) probability density functions $\psi^*\psi$ for the lowest three energy levels for a particle in a box. The plots are placed at vertical heights which correspond to the energies of the levels. As the number of nodes goes up, the energy goes up.

by introducing equation 12.38.

$$\psi = A \sin \frac{\pi x n}{a} \tag{12.40}$$

The probability that the particle is somewhere between $x = 0$ and $x = a$ is, of course, unity, and this is expressed mathematically by integrating $\psi^*\psi$ over this distance:

$$1 = \int_0^a \psi^*\psi \, dx = A^2 \int_0^a \sin^2 \frac{\pi x n}{a} \, dx = \frac{A^2 a}{\pi} \int_0^\pi \sin^2 (n\alpha) \, d\alpha \tag{12.41}$$

where $\alpha = \pi x/a$. Since ψ is a real function, $\psi^*\psi$ is simply ψ^2. Since

$$\int_0^\pi \sin^2 (n\alpha) \, d\alpha = \frac{\pi}{2} \tag{12.42}$$

we find that $A = (2/a)^{1/2}$, so that the final wave function for a particle in a one-dimensional box is

$$\psi = \left(\frac{2}{a}\right)^{1/2} \sin \frac{\pi x n}{a} \tag{12.43}$$

The probability density given by $\psi^*\psi$, as calculated by squaring equation 12.43 is given in Fig. 12.6b for various values of the quantum number n.

If two different wave functions are used in equation 12.41 then

$$\int_0^a \psi^*\psi' \, dx = A^2 \int_0^a \sin \frac{\pi x n}{a} \sin \frac{\pi x n'}{a} \, dx = \frac{A^2 a}{\pi} \int_0^\pi \sin (n\alpha) \sin (n'\alpha) \, d\alpha$$

$$= 0 \quad \text{if} \quad n \neq n' \tag{12.44}$$

When such an integral is equal to zero, the functions ψ and ψ' are said to be orthogonal. Solutions of the Schrödinger equation corresponding to different energy eigenvalues are always orthogonal.

12.13   AVERAGE VALUES AND THE CORRESPONDENCE PRINCIPLE

According to Bohr's correspondence principle, quantum mechanics must yield results identical with those of classical physics in the limit that the quantum numbers involved are large. This may be illustrated with the particle in a box. For a particle in a box of quite large dimensions the energy levels become so close together that they appear continuous in agreement with classical mechanics.

Let us compare the calculation of the mean position $\langle x \rangle$ of a particle in a box and the mean square position $\langle x^2 \rangle$ using quantum mechanics, and using classical mechanics. According to equation 12.32 these average values may be calculated as follows:

$$\langle x \rangle = \int_0^a \psi^* x \psi \, dx = \int_0^a x \frac{2}{a} \sin^2 \frac{\pi x n}{a} \, dx = \frac{a}{2} \tag{12.45}$$

$$\langle x^2 \rangle = \int_0^a \psi^* x^2 \psi \, dx = \int_0^a x^2 \frac{2}{a} \sin^2 \frac{\pi x n}{a} \, dx = \frac{a^2}{3}\left(1 - \frac{3}{2n^2\pi^2}\right) \tag{12.46}$$

For the classical case of a particle of fixed energy all positions in the box are equally probable, and we say that the probability density $\rho(x)$ is equal to $1/a$ so that

$$\int_0^a \rho(x) \, dx = 1 \tag{12.47}$$

The classical mean and mean square positions are

$$\langle x \rangle = \int_0^a \frac{x \, dx}{a} = \frac{a}{2} \tag{12.48}$$

$$\langle x^2 \rangle = \int_0^a \frac{x^2}{a} \, dx = \frac{a^2}{3} \tag{12.49}$$

The mean position is the same classically and quantum mechanically, and as the quantum number approaches infinity, the mean square position calculated from quantum mechanics by equation 12.46 approaches the classical limit.

12.14   PARTICLE IN A THREE-DIMENSIONAL BOX

The Schrödinger equation may readily be solved for a three-dimensional rectangular potential well with infinite potential everywhere outside. The Schrödinger equation in the form of equation 12.29 may be solved by writing the wave function as the product of three functions, each depending on just one coordinate.

$$\psi(x, y, z) = X(x)Y(y)Z(z) \tag{12.50}$$

By substituting this for ψ in equation 12.29 and then dividing by $X(x)Y(y)Z(z)$ we obtain

$$-\frac{\hbar^2}{2m}\left[\frac{1}{X(x)}\frac{d^2X(x)}{dx^2} + \frac{1}{Y(y)}\frac{d^2Y(y)}{dy^2} + \frac{1}{Z(z)}\frac{d^2Z(z)}{dz^2}\right] = E \tag{12.51}$$

since V is everywhere zero inside the box. If the energy is written as the sum of three contributions associated with the three coordinates, then

$$E = E_x + E_y + E_z \tag{12.52}$$

and equation 12.51 can be separated into three equations because the functions $X(x)$, $Y(y)$, and $Z(z)$ are each functions of variables that can change independently of each other. For example, if y and z are held constant the second and third terms on the left-hand side of equation 12.51 will be zero. Since E is constant, the first term must be constant. We call this constant value E_x and find equation 12.53a. Similar arguments yield equations 12.53b and 12.53c. The important mathematical result here is that a partial differential equation has been converted into three ordinary differential equations that can be easily solved.

$$-\frac{\hbar^2}{2m}\left[\frac{1}{X(x)}\frac{d^2X(x)}{dx^2}\right] = E_x \tag{12.53a}$$

$$-\frac{\hbar^2}{2m}\left[\frac{1}{Y(y)}\frac{d^2Y(y)}{dy^2}\right] = E_y \tag{12.53b}$$

$$-\frac{\hbar^2}{2m}\left[\frac{1}{Z(z)}\frac{d^2Z(z)}{dz^2}\right] = E_z \tag{12.53c}$$

These equations are just like equation 12.36 and may be solved in the same way to obtain

$$X(x) = A_x \sin\frac{n_x\pi x}{a} = A_x \sin\left(\frac{2mE_x}{\hbar^2}\right)^{1/2}x \tag{12.54a}$$

$$Y(y) = A_y \sin\frac{n_y\pi y}{b} = A_y \sin\left(\frac{2mE_y}{\hbar^2}\right)^{1/2}y \tag{12.54b}$$

$$Z(z) = A_z \sin\frac{n_z\pi z}{c} = A_z \sin\left(\frac{2mE_z}{\hbar^2}\right)^{1/2}z \tag{12.54c}$$

where a, b, and c are the lengths of the sides in the x, y, and z directions, respectively, and n_x, n_y, and n_z are quantum numbers. Thus there is a quantum number for each coordinate.

The allowed energy levels are

$$E = \frac{n_x^2 h^2}{8ma^2} + \frac{n_y^2 h^2}{8mb^2} + \frac{n_z^2 h^2}{8mc^2} \tag{12.55}$$

If the lengths of no two edges of the box are in the ratio of integers, the energy levels will be different for all possible sets of the quantum numbers n_x, n_y, and n_z. If, however, some of the ratios are integers, several distinct combinations of the three quantum numbers give rise to the same total energy. Such an energy level is said to be *degenerate*, and the degeneracy is equal to the number of independent wave functions associated with a given energy level. An understanding of degenerate levels is important for later calculations.

12.15 THE SIMPLE HARMONIC OSCILLATOR

The quantum mechanical treatment of a simple harmonic oscillator is required for the understanding of the vibration of molecules. To provide a background, the simple harmonic oscillator will be considered from a classical point of view before it is considered from a quantum mechanical point of view. In a simple harmonic oscillator the force tending to restore the particle to its equilibrium position is directly proportional to the displacement from the equilibrium position. Thus

$$\text{Force} = -kx \tag{12.56}$$

where x is the distance measured from the equilibrium position and k is the force constant. The minus sign is required because the force is in the negative x direction if the displacement x from the equilibrium position is positive. When the particle is at the equilibrium distance the force is zero.

A force of this type may be represented as the negative gradient of the potential energy V.

$$\text{Force} = -\frac{\partial V}{\partial x} = -kx \tag{12.57}$$

Integrating, we have

$$V = \tfrac{1}{2}kx^2 \tag{12.58}$$

if the integration constant is taken equal to zero, so that $V = 0$ when $x = 0$. Thus a plot of potential energy V versus displacement x is parabolic for a simple harmonic oscillator, as shown in Fig. 12.7a. A small object sliding without friction in a parabolic well would execute simple harmonic motion. As it slides through the minimum, the velocity is a maximum and the potential energy is a minimum. As it slides up the other side, kinetic energy is converted to potential energy. When the particle is at either one of its two highest points in the oscillation, the velocity is zero, and the total energy is potential energy.

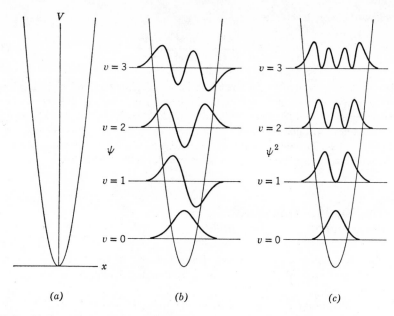

Fig. 12.7   (*a*) Potential well for a classical harmonic oscillator. (*b*) Allowed energy levels and wave functions for a quantum mechanical harmonic oscillator. (*c*) Probability density functions for a quantum mechanical harmonic oscillator.

The force in equation 12.57 may be equated to the mass of the particle times its acceleration. Thus

$$m \frac{d^2 x}{dt^2} = -kx \tag{12.59}$$

The solution of this differential equation is

$$x = a \sin \left(\frac{k}{m} \right)^{1/2} t$$

$$= a \sin 2\pi \nu_0 t \tag{12.60}$$

where

$$\nu_0 = \frac{1}{2\pi} \left(\frac{k}{m} \right)^{1/2} \tag{12.61}$$

is the fundamental vibration frequency. The frequency of vibration of a harmonic oscillator is independent of the amplitude of the vibration.

The energy (classical Hamiltonian) of a classical harmonic oscillator is given by the sum of the kinetic and potential energies.

$$H = \frac{p_x^2}{2m} + \frac{k}{2} x^2 \tag{12.62}$$

In order to treat the harmonic oscillator from a quantum mechanical point of view, the Hamiltonian operator is obtained by substituting $(\hbar/i)(d/dx)$ for p_x in equation 12.62. Thus the Hamiltonian operator $\mathscr{H}$ for a simple harmonic oscillator is

$$\mathscr{H} = -\frac{\hbar^2}{2m}\frac{d^2}{dx^2} + \frac{k}{2}x^2 \tag{12.63}$$

To obtain the allowed energy levels we use this operator in equation 12.27 to obtain

$$-\frac{\hbar^2}{2m}\frac{d^2\psi}{dx^2} + \frac{1}{2}kx^2\psi = E\psi \tag{12.64}$$

This equation is now solved to find the wave functions and corresponding eigenvalues E, which are the allowed energy levels.

The following eigenfunctions and eigenvalues are obtained from equation 12.64

$$\psi_0 = \left(\frac{2a}{\pi}\right)^{1/2} e^{-ax^2} \qquad\qquad E_0 = \tfrac{1}{2}h\nu_0$$

$$\psi_1 = \left(\frac{2a}{\pi}\right)^{1/2} 2a^{1/2}xe^{-ax^2} \qquad\qquad E_1 = \tfrac{3}{2}h\nu_0$$

$$\psi_2 = \left(\frac{2a}{\pi}\right)^{1/2}(4ax^2 - 1)e^{-ax^2} \qquad E_2 = \tfrac{5}{2}h\nu_0 \tag{12.65}$$

$$\cdot \qquad\qquad\qquad\qquad \cdot$$
$$\cdot \qquad\qquad\qquad\qquad \cdot$$
$$\cdot \qquad\qquad\qquad\qquad \cdot$$

$$\psi_v \qquad\qquad\qquad\qquad E_v = (v + \tfrac{1}{2})h\nu_0$$

where $\nu_0 = (1/2\pi)\sqrt{k/m}$, the classical frequency for the harmonic oscillator, and $a = (\pi/h)\sqrt{km}$.

Exercise 1   Verify that ψ_0 and ψ_1 are solutions of the Schrödinger equation by insertion into equation 12.64. Also verify that these eigenfunctions are orthogonal.

The energy levels are equally spaced as shown in Fig. 12.7b, where the wave functions are also shown. Thus the quantum mechanical treatment of the harmonic oscillator yields rather different results from the classical treatment. According to classical mechanics the oscillator may have any energy, but according to quantum mechanics the possible energy levels are given by $E = (v + \tfrac{1}{2})h\nu_0$ where v is 0, 1, 2, ... According to classical mechanics the oscillator may be at rest and have zero energy, but according to quantum mechanics the lowest energy level permitted is $E = \tfrac{1}{2}h\nu_0$, called the zero point energy. This is an illustration of the Heisenberg uncertainty principle according to which $\Delta p_x\,\Delta x \approx h$. If the particle were at rest at the origin the uncertainties in p_x and x would each be zero, and the Heisenberg uncertainty principle would be violated (Section 12.9).

The wave functions given above have been normalized. The probability that the x coordinate of the harmonic oscillator is between x and $x + dx$ is given by $\psi^2 \, dx$, since the wave functions are real. If a large number of identically prepared systems are examined the fraction having coordinates between x and $x + dx$ is equal to this probability. The probability *densities* ψ^2 are plotted in Fig. 12.7c, versus x for the first three energy levels. In the ground state ($v = 0$) the most probable internuclear distance occurs at the position of the minimum in the potential well. This is in distinct contrast with that for a classical simple harmonic oscillator which would spend the longest times at the turning points. As the quantum number increases, however, the quantum mechanical probability density function approaches that for the classical harmonic oscillator. This is an example of the correspondence principle (Section 12.13) according to which the quantum mechanical result must approach the classical result in the limit of infinite quantum number. As the quantum number increases, the range of distance over which there is an appreciable probability of finding the particle increases. This is like the increased amplitude of the motion of a classical harmonic oscillator at higher energy. It may be verified that as the quantum number increases the probability density function approaches that expected from classical mechanics.

At any temperature the number of molecules in a given vibrational state can be computed from the Boltzmann distribution (Section 17.3). For example, for $H^{35}Cl$, $\tilde{\nu}_0 = 2890 \text{ cm}^{-1}$ which corresponds to $8.25 \text{ kcal mol}^{-1}$. At 298 K, the ratio of the number of molecules in state 1 to the number in the ground state is

$$\frac{N_{v=1}}{N_{v=0}} = e^{-E/RT} = \exp\left[\frac{-(8250 \text{ cal mol}^{-1})}{(1.987 \text{ cal K}^{-1} \text{ mol}^{-1})(298 \text{ K})}\right] \qquad (12.66)$$

$$= 8.9 \times 10^{-7}$$

Thus virtually all HCl molecules are in the lowest vibrational state at room temperature. Other common diatomic molecules, however, may have a significant fraction of excited vibrational states. For example, for $^{127}I_2$, $\tilde{\nu}_0 = 213 \text{ cm}^{-1}$ which corresponds to $0.609 \text{ kcal mol}^{-1}$. Then

$$\frac{N_{v=1}}{N_{v=0}} = e^{-E/RT} = \exp\left[\frac{-(609 \text{ cal mol}^{-1})}{(1.987 \text{ cal K}^{-1} \text{ mol}^{-1})(298 \text{ K})}\right]$$

$$= 0.358 \qquad (12.67)$$

12.16 THE RIGID ROTOR

The quantum mechanical treatment of the rotation of two masses m_1 and m_2 held a fixed distance r apart is of interest since this approximates the rotational behavior of diatomic molecules, and because the mathematical problem is identical with that for the angular part of the wave equation for the motion of an electron in the field of an atom. The particles rotate about their mutual center of mass with a moment of inertia I of

$$I = m_1 r_1^2 + m_2 r_2^2 = \mu r^2 \qquad (12.68)$$

where $r = r_1 + r_2$ and μ is the reduced mass

$$\mu = \frac{1}{\dfrac{1}{m_1} + \dfrac{1}{m_2}} \tag{12.69}$$

It may be shown that the rigid rotor behaves like a single particle of mass I, and the potential energy is zero, so that the Schrödinger equation (12.30) may be written

$$\nabla^2\psi + \frac{2I}{\hbar^2} E\psi = 0 \tag{12.70}$$

The eigenvalues are

$$E_{\text{rot}} = \frac{\hbar^2}{2I} J(J+1) \qquad J = 0, 1, 2, \ldots \tag{12.71}$$

These eigenvalues are $(2J + 1)$-fold degenerate. The wave functions for the rigid rotor are the same as those we will encounter later in the angular dependence of the wave function for the hydrogen atom.

12.17 THE HYDROGEN ATOM

The simplest atomic systems are those consisting of a nucleus of mass M and charge Ze, where Z is the atomic number, and one electron with mass m_e and charge $-e$. The Schrödinger equation may be solved exactly for these hydrogen-like atoms, and these solutions are of tremendous importance for the treatment of atoms with two or more electrons, for which closed mathematical solutions cannot be obtained. The complete treatment of the hydrogen atom is too complicated to give here, and so only the main features will be outlined.

The starting point for the quantum mechanical treatment is the classical Hamiltonian for the system consisting of an electron and nucleus of charge Ze. To simplify the expression for the Hamiltonian we will assume that the electron moves around the stationary nucleus. Actually the electron and nucleus move around their mutual center of mass. The nucleus of charge Ze is taken as the origin of the coordinate system, so the potential energy is $-Ze^2/4\pi\epsilon_0 r$, where r is the distance between nucleus and electron. The classical Hamiltonian is

$$H = \frac{1}{2m_e} (p_x^2 + p_y^2 + p_z^2) - \frac{Ze^2}{4\pi\epsilon_0 r} \tag{12.72}$$

where m is the mass of the electron. Converting this to the quantum mechanical Hamiltonian (Section 12.11) and applying this operator to ψ yields

$$\frac{\partial^2\psi}{\partial x^2} + \frac{\partial^2\psi}{\partial y^2} + \frac{\partial^2\psi}{\partial z^2} + \frac{2m_e}{\hbar^2} \left(E + \frac{Ze^2}{4\pi\epsilon_0 r} \right)\psi = 0 \tag{12.73}$$

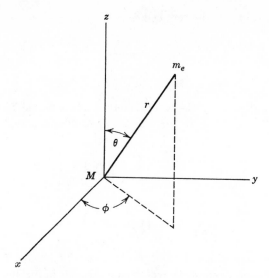

Fig. 12.8 Spherical coordinates used in describing the hydrogenlike atom.

In order to solve this equation it is convenient to transform the Cartesian co-ordinates to spherical coordinates r, θ, ϕ which are defined in Fig. 12.8 since with $r = \sqrt{x^2 + y^2 + z^2}$ it is not possible to separate variables. This transformation yields

$$\frac{1}{r^2}\frac{\partial}{\partial r}\left(r^2\frac{\partial \psi}{\partial r}\right) + \frac{1}{r^2 \sin \theta}\frac{\partial}{\partial \theta}\left(\sin \theta \frac{\partial \psi}{\partial \theta}\right)$$

$$+ \frac{1}{r^2 \sin^2 \theta}\frac{\partial^2 \psi}{\partial \phi^2} + \frac{8\pi^2 m_e}{h^2}\left(E + \frac{Ze^2}{4\pi\epsilon_0 r}\right)\psi = 0 \quad (12.74)$$

If the fact that the electron and the nucleus rotate about their mutual center of mass had been taken into account, the m_e in this equation would be replaced by the reduced mass μ (equation 12.69), which is given by $m_e M/(M + m_e)$, where m_e is the mass of the electron and M is the mass of the nucleus.

The ψ of equation 12.74 may be written as a product of three functions, one dependent on r, one dependent on θ, and one dependent on ϕ.

$$\psi(r, \theta, \phi) = R(r)\Theta(\theta)\Phi(\phi) \quad (12.75)$$

when this is substituted in equation 12.74, three ordinary differential equations result. Each of the resulting three differential equations may be solved, and each leads to the introduction of a quantum number that must be an integer. An analogous situation was encountered earlier where it was found that for a particle in a three-dimensional box, the wave function is the product of three functions, one for each coordinate. For the hydrogen atom the quantum numbers are the principal quantum number n, the angular momentum quantum number l, and the magnetic quantum number m. There is one quantum number for each degree of freedom. The values of these quantum numbers are restricted in the following way:

n = principal quantum number. Permissible solutions of the Schrödinger equation correspond to $n = 1, 2, 3, \ldots$

l = angular momentum quantum number. This quantum number can have only values of $0, 1, \ldots, (n-1)$, where n is the principal quantum number.

When this quantum number is $l = 0\ \ 1\ \ 2\ \ 3$

The corresponding symbols are $s\ \ p\ \ d\ \ f$

m = magnetic quantum number. The permissible values are

$$-l, -(l-1), \ldots, -1, 0, +1, \ldots, (l-1), l$$

The eigenvalues for the energy of the hydrogenlike atom are given by

$$E = -\frac{m_e e^4 Z^2}{2(4\pi\epsilon_0)^2 \hbar^2 n^2} \qquad n = 1, 2, 3, \ldots \qquad (12.76)$$

Thus the energies of the various states of the hydrogen atom are inversely proportional to the square of the principal quantum number n. The energy is negative because the electron in a hydrogenlike atom has *less* energy than when it is free. By convention the energy of a separated electron and nucleus is taken as zero.

Example 12.2 Calculate the energy of a hydrogen atom in its ground state ($n = 1$) using the mass of the electron rather than the reduced mass of the hydrogen atom.

$$E = -\frac{2\pi^2 m_e e^4}{(4\pi\epsilon_0)^2 \hbar^2}$$

$$= -\frac{2\pi^2 (9.1095 \times 10^{-31}\ \text{kg})(1.6022 \times 10^{-19}\ \text{C})^4 (0.8988 \times 10^{10}\ \text{N m}^2\ \text{C}^{-2})^2}{(6.6262 \times 10^{-34}\ \text{J s})^2}$$

$$= -2.1802 \times 10^{-18}\ \text{J}$$

$$= -\frac{2.1802 \times 10^{-18}\ \text{J}}{1.6021 \times 10^{-19}\ \text{J eV}^{-1}}$$

$$= -13.61\ \text{eV}$$

Thus the ionization energy of a hydrogen atom in the ground state is 13.61 electron volts.*

In the absence of magnetic or electric fields the energy of a hydrogenlike atom depends only on the principal quantum number n. Equation 12.76 applies to He^+, Li^{2+}, Be^{3+}, and all the other single-electron species. As Z increases the orbitals become smaller, and the electrons become more tightly bound. The

---

* The electron volt is the work done by moving an electronic charge through a potential difference of 1 V or, in other words, the energy acquired by an electron being accelerated by a potential difference of 1 V. The energy in joules may be calculated as follows:

$$eE = (1.602 \times 10^{-19}\ \text{C})(1\ \text{V}) = 1.602 \times 10^{-19}\ \text{J}$$

Table 12.2 Hydrogenlike Wave Functions[1]

n	l	m	Wave Function
1	0	0	$\psi_{1s} = \dfrac{1}{\sqrt{\pi}} \left(\dfrac{Z}{a_0}\right)^{3/2} e^{-\sigma}$
2	0	0	$\psi_{2s} = \dfrac{1}{4\sqrt{2\pi}} \left(\dfrac{Z}{a_0}\right)^{3/2} (2 - \sigma)e^{-\sigma/2}$
2	1	0	$\psi_{2p_z} = \dfrac{1}{4\sqrt{2\pi}} \left(\dfrac{Z}{a_0}\right)^{3/2} \sigma e^{-\sigma/2} \cos\theta$
2	1	± 1	$\psi_{2p_x} = \dfrac{1}{4\sqrt{2\pi}} \left(\dfrac{Z}{a_0}\right)^{3/2} \sigma e^{-\sigma/2} \sin\theta \cos\phi$
			$\psi_{2p_y} = \dfrac{1}{4\sqrt{2\pi}} \left(\dfrac{Z}{a_0}\right)^{3/2} \sigma e^{-\sigma/2} \sin\theta \sin\phi$
3	0	0	$\psi_{3s} = \dfrac{1}{81\sqrt{3\pi}} \left(\dfrac{Z}{a_0}\right)^{3/2} (27 - 18\sigma + 2\sigma^2)e^{-\sigma/3}$
3	1	0	$\psi_{3p_z} = \dfrac{\sqrt{2}}{81\sqrt{\pi}} \left(\dfrac{Z}{a_0}\right)^{3/2} (6 - \sigma)\sigma e^{-\sigma/3} \cos\theta$
3	1	± 1	$\psi_{3p_x} = \dfrac{\sqrt{2}}{81\sqrt{\pi}} \left(\dfrac{Z}{a_0}\right)^{3/2} (6 - \sigma)\sigma e^{-\sigma/3} \sin\theta \cos\phi$
			$\psi_{3p_y} = \dfrac{\sqrt{2}}{81\sqrt{\pi}} \left(\dfrac{Z}{a_0}\right)^{3/2} (6 - \sigma)\sigma e^{-\sigma/3} \sin\theta \sin\phi$
3	2	0	$\psi_{3d_{z^2}} = \dfrac{1}{81\sqrt{6\pi}} \left(\dfrac{Z}{a_0}\right)^{3/2} \sigma^2 e^{-\sigma/3} (3\cos^2\theta - 1)$
3	2	± 1	$\psi_{3d_{xz}} = \dfrac{\sqrt{2}}{81\sqrt{\pi}} \left(\dfrac{Z}{a_0}\right)^{3/2} \sigma^2 e^{-\sigma/3} \sin\theta \cos\theta \cos\phi$
			$\psi_{3d_{yz}} = \dfrac{\sqrt{2}}{81\sqrt{\pi}} \left(\dfrac{Z}{a_0}\right)^{3/2} \sigma^2 e^{-\sigma/3} \sin\theta \cos\theta \sin\phi$
3	2	± 2	$\psi_{3d_{x^2-y^2}} = \dfrac{1}{81\sqrt{2\pi}} \left(\dfrac{Z}{a_0}\right)^{3/2} \sigma^2 e^{-\sigma/3} \sin^2\theta \cos 2\phi$
			$\psi_{3d_{xy}} = \dfrac{1}{81\sqrt{2\pi}} \left(\dfrac{Z}{a_0}\right)^{3/2} \sigma^2 e^{-\sigma/3} \sin^2\theta \sin 2\phi$

[1] $\sigma = \dfrac{Z}{a_0} r.$

ionization of the $1s$ orbital of hydrogen is 13.6 eV, of He^+ is $2^2 \cdot 13.6 = 54.4$ eV, and of Li^{+2} is $3^2 \cdot 13.6 = 122.4$ eV. Electronic orbitals in atoms are generally described by giving the principal quantum number and the symbol representing the angular momentum quantum number. Thus we speak of the $1s$, $2s$, $2p$, $3s$, $3p$, $3d$, etc., orbitals.

The orbital angular momentum is given by $\sqrt{l(l+1)}\hbar$. Thus s electrons have no orbital angular momentum, and p electrons have an angular momentum of $\sqrt{2}\hbar$. The angular momentum in a specified direction is given by $m\hbar$. The angular momentum in a particular direction cannot be greater than the total angular momentum and so $|m| \leqslant |l|$. Thus m may have any integer value between $+l$ and $-l$, including zero. Therefore, $2l + 1$ values of m are possible. For $l = 3$, $m = -3, -2, -1, 0, 1, 2, 3$.

The wave functions for the hydrogenlike atoms through $n = 3$ are given in Table 12.2. It is convenient to write these equations in terms of the Bohr radius a_0, which, as we will see, is the most probable distance between the electron and proton in a hydrogen atom in the $1s$ state.

$$a_0 = \frac{\hbar^2(4\pi\epsilon_0)}{m_e e^2} \tag{12.77}$$

Example 12.3   Calculate the most probable distance between the electron and proton in a hydrogen atom in the $1s$ state using the mass of the electron rather than the reduced mass of the hydrogen atom.

$$a_0 = \frac{h^2(4\pi\epsilon_0)}{4\pi^2 m_e e^2}$$

$$= \frac{(6.6262 \times 10^{-34} \text{ J s})^2}{4\pi^2(0.8988 \times 10^{10} \text{ N m}^2 \text{ C}^{-2})(9.1095 \times 10^{-31} \text{ kg})(1.6022 \times 10^{-19} \text{ C})^2}$$

$$= 0.529 \times 10^{-10} \text{ m}$$

$$= 0.529 \text{ Å}$$

In atomic units this is taken as the unit of distance and is referred to as 1 Bohr.

To visualize the nature of these functions it is helpful to consider $R(r)$ and $\Theta(\theta)\Phi(\phi)$ separately.

The radial functions $R(r)$ are plotted in Fig. 12.9 for the wave functions in Table 12.2, with $Z = 1$. The radial function always contains the factor e^{-Zr/na_0}, where n is the principal quantum number. As Z is increased the amplitude of the wave function falls off more rapidly with increasing r indicating that the electron is attracted more closely to the higher positively charged nuclei.

The radial wave functions have $n - l$ nodes ($R(r) = 0$), where n is the principal quantum number. The wave function changes sign at a node, but the square of the wave function does not. The existence of nodes is required so that the $1s$ and $2s$ and other orbitals will be orthogonal (Section 12.12); that is,

$$\int \psi_{1s} \psi_{2s} \, d\tau = 0 \tag{12.78}$$

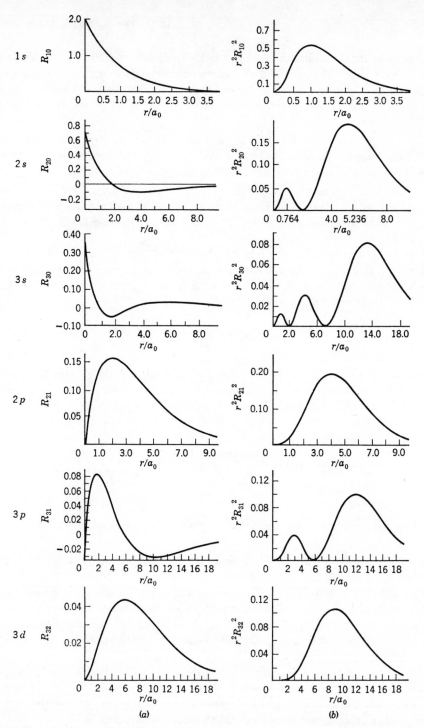

Fig. 12.9 (a) Electronic radial-wave functions $R(r)$ for the hydrogen atom. (b) Probability density for finding an electron at a distance between r and $r + dr$ (From M. Karplus and R. N. Porter, *Atoms and Molecules*, copyright © 1970, by W. A. Benjamin, Inc., Menlo Park, California).

Figure 12.9*a* shows that the radial part $R(r)$ of the hydrogen atom wave functions for *s* orbitals make the probability density for the electron greatest at the nucleus. However, we may be interested in another question, and that is how does the probability of finding the electron at a distance between r and $r + dr$ depend on r? To calculate this so-called radial distribution for *s* orbitals we multiply $R(r)^2$ by $4\pi r^2$ since $4\pi r^2 \, dr$ is the volume of the spherical shell centered at the origin. The radial distribution function for the 1*s* orbital has a maximum at a_0, as shown in Fig. 12.9*b*. This most probable radius for the electron agrees with the radius of the first Bohr orbit. The more diffuse cloud of probability density provided by quantum mechanics is very different from the Bohr theory and is in accord with the Heisenberg uncertainty principle.

In looking at Fig. 12.9, it is important to remember that this represents just the radial part of the ψ^2, and that when $l = 1, 2, \ldots$, this is modified by the angular dependence.

It is rather difficult to present electron density as a function of three spatial coordinates for various orbitals. One way to do this is to use the density of dots to represent the probability of finding an electron in a region of space and use stereo plots with a stereo viewer to see probability densities in three dimensions.* These plots show the decrease in electron density radially as well as the angular part of the wave function. A more commonly used method is to use a surface to represent a constant value of $\Theta(\theta)\Phi(\phi)$. For *s* orbitals these surfaces are all spheres, as shown in Fig. 12.10 because the orbitals are spherically symmetric. For the 2*p* orbitals the contour surfaces are two separated and somewhat distorted ellipsoids, as shown in Fig. 12.10. One of these surfaces comes from the wave function with a positive sign, and the other comes from the wave function with the negative sign. These signs are indicated because they will be of interest later when we discuss molecular orbitals. The probability density is of course always positive. The orientations of the *p* orbitals can be checked by considering the magnitudes and signs of the trigonometric functions at several angles. In the absence of an electric or magnetic field electrons in p_x, p_y, and p_z orbitals all have the same energy; that is, the energy depends only on the total quantum number n. In the presence of a magnetic field, however, electrons in the *p* orbital in the direction of the field have different energies, which is why m is called the magnetic quantum number. This is an example of the distinction between a quantum state and an energy level. For the hydrogen atom with quantum number $n = 2$ there are four states, all having the same energy in the absence of a magnetic or electric field. Such an energy level is said to be degenerate, and the degeneracy is the number of wave functions that have that particular energy associated with them.

The *p* orbitals do not have to point along the x, y, and z directions. Linear combinations of p_x, p_y, and p_z may be formed to point in any three mutually perpendicular directions. It will be seen in Chapter 14 that the directional character of certain chemical bonds results from the directed orientation of these and other orbitals.

---

* D. T. Cromer, *J. Chem. Ed.*, **45**, 626 (1968).

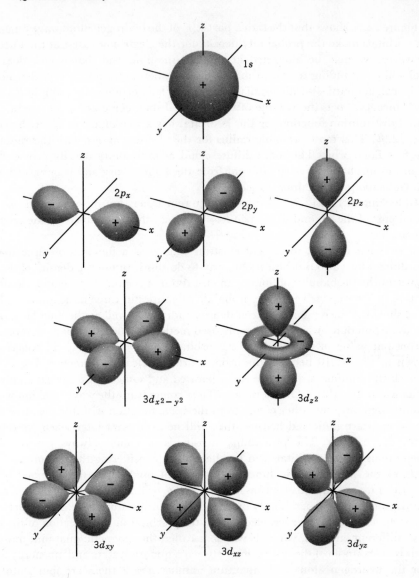

Fig. 12.10 Plots showing the angular dependence of $1s$, $2p$, and $3d$ functions for a fixed value of r. The indicated signs are those of the wave functions.

There are five independent d orbitals. The $3d_{z^2}$ orbital has two large regions of electron density along one axis, by convention the z axis, and a small donut-shaped orbital in the xy plane. The other four d orbitals have four equivalent lobes of electron density with two nodal planes separating them. Note that lobes that are opposite each other come from wave functions with the same signs.

One of the deficiencies in the diagrams in Fig. 12.10 is that the nodal surfaces resulting from the radial functions $R(r)$ are not shown. Although the hydrogen atom has an infinite number of orbitals, most questions of chemical significance involve only the lowest energy orbitals.

12.18 ANGULAR MOMENTUM

The magnitude of the square of the electronic angular momentum in the hydrogen atom is given by

$$M^2 = l(l + 1)\hbar^2 \qquad l = 0, 1, 2, \ldots, n - 1 \tag{12.79}$$

and the component of the angular momentum along one axis (conventionally taken as the z axis) is given by

$$M_z = m\hbar \qquad m = 0, \pm 1, \pm 2, \ldots, \pm l \tag{12.80}$$

Thus the hydrogen atom, and the other quantum mechanical systems, are quite different from classical rotating objects. The angular momentum of a classical object may have any value, and the angular momentum vector may point in any direction. In quantum mechanics M^2 is limited to certain values, and the component in the z-direction is limited to certain values.

The possible orientations of angular momentum vectors are shown in Fig. 12.11 for p orbitals ($l = 1$) and d orbitals ($l = 2$). The angular momentum vector cannot point in the z direction in either case; to do so would violate the Heisenberg uncertainty principle because it would mean that the electronic motion is confined to a plane. Note that the component of angular momentum in a particular direction is less than the total angular momentum. The number of possible orientations of the angular momentum vector is $2l + 1$. It does not make sense to speak of definite components of angular momentum in the x and y directions; the vector of total

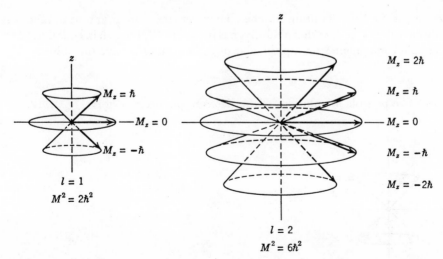

Fig. 12.11 Possible orientations for angular momentum vectors for $l = 1$ (p orbitals) and $l = 2$ (d orbitals).

angular momentum can point in an infinite number of directions that yield a cone-shaped surface of possibilities. In a magnetic field the energy of an electron in an orbital depends on the magnetic quantum number m. The angle θ between the total angular momentum vector M and the component M_z in the direction of the field (equation 12.80) is given by

$$\cos \theta = \pm \frac{m}{\sqrt{l(l+1)}} \tag{12.81}$$

12.19   SPIN

The spectra of atoms show a fine structure that is not explained by the theory we have just discussed. For example some of the lines may be resolved into closely spaced multiplets when a magnetic field is applied (Zeeman effect) or when an electric field is applied (Stark effect). This fine structure was interpreted in 1925 by Goudsmit and Uhlenbeck as being due to an intrinsic magnetic moment of the electron that is independent of its orbital angular momentum. Later Dirac applied relativity theory to the quantum mechanical formulation and demonstrated a really satisfactory theoretical basis was provided for the intrinsic angular momentum of an electron. The term electron spin is used, but it is not really correct to think of the intrinsic magnetic effects of the electron as being due to a spinning motion of the electron mass on its axis. The intrinsic angular momentum of an electron can be treated in a way that is analogous to that for orbital angular momentum. The magnitude S of the total spin angular momentum can be written as

$$S^2 = s(s+1)\hbar^2 = \tfrac{3}{4}\hbar^2 \tag{12.82}$$

where s is the spin quantum number. However, the spin quantum number can only have the value $\tfrac{1}{2}$, so that S^2 always has the value $3\hbar^2/4$ shown in equation 12.82. The spin component S_z in a particular direction can only have the values

$$S_z = \pm\tfrac{1}{2}\hbar \tag{12.83}$$

These two possible orientations of the electron spin are shown in Fig. 12.12.

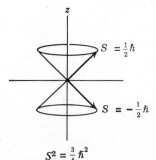

$$S^2 = \frac{3}{4}\hbar^2$$

Fig. 12.12   Possible orientations of the angular momentum vectors for an electron for which $s = \tfrac{1}{2}$.

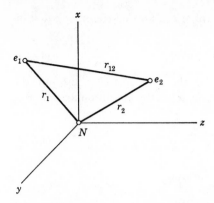

Fig. 12.13 Coordinates in the helium atom.

Since the wave functions for the hydrogen atom discussed earlier do not include spin they need to be multiplied by functions representing the possible spin states of the system. It is customary to use α to represent one spin wave function and β to represent the other. Thus the wave function of the hydrogen atom depends upon four quantum numbers n, l, m and m_s.

12.20 HELIUM ATOM

The helium atom has two electrons and the coordinates used in writing the Hamiltonians are shown in Fig. 12.13. The two electrons repel each other with a potential energy $e^2/4\pi\epsilon_0 r_{12}$. The Hamiltonian operator is

$$\mathcal{H} = -\frac{h^2}{8\pi^2 m_e}(\nabla_1^2 + \nabla_2^2) - \frac{1}{4\pi\epsilon_0}\left(\frac{2e^2}{r_1} + \frac{2e^2}{r_2} - \frac{e^2}{r_{12}}\right) \qquad (12.84)$$

In writing such equations it is convenient to introduce new units that are more appropriate for atomic dimensions and eliminate some constants. They are referred to as atomic units (a.u.). The unit of mass is taken as the mass of the electron m_e. The unit of charge is taken as the charge of the electron e. The unit of distance is taken as the Bohr radius a_0 of the hydrogen atom in its ground state (equation 12.77). The unit of energy is the potential energy of two unit charges a unit distance apart.

$$\frac{e^2}{4\pi\epsilon_0 a_0} = 27.2 \text{ eV} = 4.35942 \times 10^{-18} \text{ J} = 1 \text{ H} \qquad (12.85)$$

This unit is referred to as a Hartree H. In atomic units the energy of a hydrogen atom in its ground state is

$$E = -\frac{m_e e^4}{2(4\pi\epsilon_0)\hbar^2} = -\frac{1}{2}\frac{e^2}{a_0} = -\tfrac{1}{2}\text{H} \qquad (12.86)$$

Thus the Hartree is twice the energy corresponding with the Rydberg frequency.

In atomic units Planck's constant h has the value 2π. The Schrödinger equation becomes

$$\nabla^2\psi + 2(E - V)\psi = 0 \qquad (12.87)$$

In writing the Hamiltonian in atomic units, $\hbar$, the electronic charge e, and the electron mass m_e are all taken as unity. In atomic units the Hamiltonian operator for the hydrogenlike atom is

$$\mathscr{H} = -\tfrac{1}{2}\nabla^2 - \frac{Z}{r} \qquad (12.88)$$

and the Hamiltonian operator for the helium atom is

$$\mathscr{H} = -\tfrac{1}{2}[\nabla_1^2 + \nabla_2^2] - \frac{2}{r_1} - \frac{2}{r_2} + \frac{1}{r_{12}} \qquad (12.89)$$

Unfortunately the presence of the interelectronic distance r_{12} makes the Schrödinger equation inseparable, and so an exact solution, such as was obtained for the hydrogenlike atoms, cannot be obtained. Approximation methods have been developed to deal with such calculations. The two most used are the variation method and the perturbation method.

12.21 THE VARIATION METHOD

In the variation method an approximate wave function ψ is used in

$$E' = \frac{\displaystyle\int \psi^* \mathscr{H} \psi \, d\tau}{\displaystyle\int \psi^*\psi \, d\tau} \geq E \qquad (12.90)$$

to obtain an estimate E' of the energy. In this equation $\mathscr{H}$ is the complete Hamiltonian operator for the system of interest. If ψ were the correct wave function then the true energy eigenvalue E would be obtained. When ψ is any arbitrary function, it can be shown* that E' is greater (more positive, less negative) than the true energy E. Thus if the wave function ψ has parameters that may be varied, the best wave function of this form will be obtained by varying the parameters so as to obtain the lowest energy. As more variable terms are introduced a closer approximation may be obtained to the true energy, but at the price of increased cost of computation.

As a first approximation we can solve the Schrödinger equation for the helium atom by ignoring the electron-electron repulsion term. The wave function ψ is taken as the product of the $1s$ wave function for electron number 1 in a He$^+$ ion

---

* J. C. Davis, *Advanced Physical Chemistry*, The Ronald Press Co., New York, 1965.

and the $1s$ wave function for electron number 2 in a He^+ ion

$$\psi = 1s(1)1s(2) = \frac{8}{\pi} \exp\left[-2(r_1 + r_2)\right] \qquad (12.91)$$

The value of E' computed using this wave function is $2E_{He^+} = -108.3$ eV. The experimental value of $2E_{He^+}$ is -78.6 eV.

A better value can be obtained from the variation method by introducing a single parameter into the "trial" wave function. Each electron tends to shield the other from the full nuclear charge. Therefore $Z'e$ is used to represent an effective nuclear charge, and the wave function is written

$$\psi = N \exp\left[-Z'(r_1 + r_2)\right] \qquad (12.92)$$

where N is the normalization factor. The value of Z' is obtained by using equation 12.90. This leads to an energy that is within 1.7% of the experimental value. Better agreement can be obtained by adding further terms to the trial-wave function.

12.22 PAULI EXCLUSION PRINCIPLE

Although the wave function for the helium atom given in equation 12.92 is a useful approximation it fails in two ways to form a satisfactory basis for an accurate calculation of the properties of helium and to provide a satisfactory theoretical basis for extension to atoms with more electrons.

There is an additional principle that is required as a supplement to the Schrödinger equation for systems containing more than one electron. That is the Pauli principle (1925) that no two electrons in an atom may have the same four quantum numbers n, l, m_l, and m_s. Spin functions may be included in equation 12.92 by use of α to represent the spin function for $m_s = \frac{1}{2}$ and β to represent the spin function for $m_s = -\frac{1}{2}$. Therefore, the following wave function for the helium atom satisfies the Pauli principle because the two electrons have different spin quantum numbers.

$$\psi = 1s\alpha(1)1s\beta(2) \qquad (12.93)$$

However, this wave function is still not satisfactory because it implies that it is possible to distinguish between electron 1 and electron 2. Since all electrons are identical they are indistinguishable. A wave function for helium that does not have this problem is

$$\begin{aligned}\psi &= 1s\alpha(1)1s\beta(2) + 1s\alpha(2)1s\beta(1) \\ &= 1s(1)1s(2)[\alpha(1)\beta(2) + \alpha(2)\beta(1)]\end{aligned} \qquad (12.94)$$

Using this wave function we obtain a probability density ψ^2, which is unchanged by interchanging the labels of electrons 1 and 2. This objective is also achieved by the wave function

$$\begin{aligned}\psi &= 1s\alpha(1)1s\beta(2) - 1s\alpha(2)1s\beta(1) \\ &= 1s(1)1s(2)[\alpha(1)\beta(2) - \alpha(2)\beta(1)]\end{aligned} \qquad (12.95)$$

where the second form shows that the wave function may be written as a spatial part $1s(1)1s(2)$ and a spin part $[\alpha(1)\beta(2) - \alpha(2)\beta(1)]$. Since wave functions 12.94 and 12.95 give different energy eigenvalues and different electron densities, while there is only one form of helium atoms, they cannot both be correct; and it is found that function 12.95 is of the correct form. We note that function 12.94 is symmetric to electron interchange while function 12.95 is antisymmetric. The generalization of this experience is an alternative statement of the Pauli exclusion principle: *to represent a system of two or more electrons a wave function must be antisymmetric with respect to the interchange of the labels of any two electrons.*

Antisymmetric wave functions may conveniently be written in determinantal form. For example, equation 12.95 may be written

$$\psi = \left(\tfrac{1}{2}\right)^{\frac{1}{2}} \begin{vmatrix} 1s\alpha(1) & 1s\beta(1) \\ 1s\alpha(2) & 1s\beta(2) \end{vmatrix} \tag{12.96}$$

where the factor $\left(\tfrac{1}{2}\right)^{\frac{1}{2}}$ has been introduced so that the wave function is normalized. This is referred to as a Slater determinant after J. C. Slater. The rows are labeled with the indices of the different electrons, and the columns are the different one-electron wave functions.

The Pauli exclusion principle applies to all particles of half integral spin, but not particles of integral spin. Thus the Pauli principle applies to protons and neutrons in addition to electrons, but not to deuterons, alpha particles, and photons, which have integral spin and symmetric wave functions.

12.23   FIRST EXCITED STATE OF HELIUM

Discussion of the first excited state of helium provides the opportunity to learn more about spin functions of atoms. As a first approximation we may consider that one electron is in the $1s$ orbital, and the other electron is in the $2s$ orbital. In order to provide for the indistinguishability of the two electrons we can write two wave functions.

$$\psi_+ = 1s(1)2s(2) + 1s(2)2s(1) \tag{12.97}$$

$$\psi_- = 1s(1)2s(2) - 1s(2)2s(1) \tag{12.98}$$

These spatial wave functions need to be considered with spin functions. There are four spin functions for two electrons when they are in different orbitals so that they can have the same spin.

$$\alpha(1)\alpha(2) \qquad \beta(1)\beta(2) \qquad \alpha(1)\beta(2) \qquad \alpha(2)\beta(1) \tag{12.99}$$

However, to provide for the indistinguishability of electrons the spin functions need to be written

$$\alpha(1)\alpha(2)$$
$$\beta(1)\beta(2)$$
$$\alpha(1)\beta(2) + \alpha(2)\beta(1)$$
$$\alpha(1)\beta(2) - \alpha(2)\beta(1)$$
$$\tag{12.100}$$

Each of these four spin functions may be used to multiply each of the two spatia functions (12.97 and 12.98), but only the following four are antisymmetric and useful in representing the excited helium atom, according to the Pauli principle:

$$\psi_1 = (\tfrac{1}{2})[1s(1)2s(2) + 1s(2)2s(1)][\alpha(1)\beta(2) - \alpha(2)\beta(1)] \text{ singlet} \qquad (12.101)$$

$$\psi_2 = (\tfrac{1}{2})[1s(1)2s(2) - 1s(2)2s(1)][\alpha(1)\beta(2) + \alpha(2)\beta(1)] \qquad (12.102)$$

$$\psi_3 = (\tfrac{1}{2})[1s(1)2s(2) - 1s(2)2s(1)]\alpha(1)\alpha(2) \qquad \text{triplet} \quad (12.103)$$

$$\psi_4 = (\tfrac{1}{2})[1s(1)2s(2) - 1s(2)2s(1)]\beta(1)\beta(2) \qquad (12.104)$$

Since to a good approximation the energy of each state depends on the spatial part of the wave function, ψ_2, ψ_3, and ψ_4 are degenerate and form a spin *triplet* (multiplicity of three). ψ_1 has a different energy and is not degenerate; it forms a spin *singlet* (multiplicity of one).

In the ground state of helium the electrons are paired, and the electron spin is zero. However, when one of the electrons is excited to the $2s$ level, the electrons may be paired, as in the singlet state represented by ψ_1, or unpaired as in the triplet state represented by ψ_2, ψ_3, and ψ_4. The three components of the triplet have spins in the z-direction of 0, +1, and −1, respectively. In the presence of an external magnetic field the singlet level is not split into components, but the triplet level is split into three components.

12.24 ELECTRONIC STRUCTURE OF ATOMS

The exact calculation of wave functions for many electron atoms becomes difficult because of the many electron-electron repulsions that we have ignored for simplicity up to now. In 1927 Hartree suggested what is now known as the self-consistent field (SCF) method for coping with this problem in the calculation of wave functions for atoms, which was later modified by Fock to include the Pauli principle. In this method it is assumed that each electron moves in a spherically symmetrical potential due to the nucleus and the averaged fields of all the other electrons except the one being considered. The calculation is started with approximate wave functions for all the electrons but one. The average potential attributable to the other electrons is calculated, and then the Schrödinger equation is solved for this one electron using the average potential due to the other electrons and the nucleus. The resulting wave function is incorporated into an improved calculation of the average field, and an approximate wave function is then obtained from the Schrödinger equation for a second electron. This process is continued until the calculated set of wave functions differs insignificantly from the previous set. This set of wave functions is then said to be self-consistent. A considerable amount of computing is required to calculate the wave functions for a polyelectron atom. The self-consistent field theory treatment of a given atom yields a series of atomic orbitals each characterized by four quantum numbers

Atom	Electronic Configuration	Total

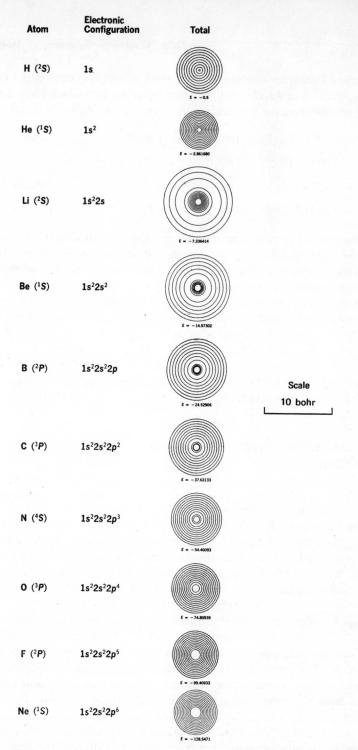

H (^2S)   1s $E = -0.5$

He (^1S)   1s^2 $E = -2.861680$

Li (^2S)   1s^{2}2s $E = -7.236414$

Be (^1S)   1s^{2}2s^2 $E = -14.57302$

B (^2P)   1s^{2}2s^{2}2p $E = -24.52906$

Scale

10 bohr

C (^3P)   1s^{2}2s^{2}2p^2 $E = -37.63133$

N (^4S)   1s^{2}2s^{2}2p^3 $E = -54.40093$

O (^3P)   1s^{2}2s^{2}2p^4 $E = -74.80939$

F (^2P)   1s^{2}2s^{2}2p^5 $E = -99.40933$

Ne (^1S)   1s^{2}2s^{2}2p^6 $E = -128.5471$

Fig. 12.14 Contour diagrams of the electron densities of the first 10 elements (*Atomic and Molecular Structure: 4 Wall Charts* by Arnold C. Wahl, Copyright © 1970 by McGraw-Hill, Inc. Used with permission of McGraw-Hill Book Co.). The outer contour line is at 4.9×10^{-4} electron/(bohr)^3 for each atom, and the inner contour line is at 1 electron/(bohr)^3, except for the hydrogen atom, where it represents 0.25 electron/(bohr)^3.

and a characteristic energy. In contrast with the case for hydrogenic atoms, the orbital energies depend both on the principal quantum number n and the orbital quantum number l.

The Hartree-Fock calculated energies agree with experimental values to about 1%. Their method provides for the interactions between electrons in an average way but does not provide for their instantaneous interactions. Since electrons tend to stay away from each other we may speak of electron correlation. The *correlation energy* is the difference between the exact energy and the Hartree-Fock energy. This energy, which is of the order of an electron volt, is large enough to be a serious problem in the calculation of chemical properties where we are interested in differences in energies.

Contour diagrams of the electron densities of the first 10 elements calculated by the Hartree-Fock method are shown in Fig. 12.14. These diagrams show that in spite of the large differences in numbers of orbital electrons these atoms are roughly all of the same size in their ground states because the increasing nuclear charge results in the electrons being held more tightly.

12.25 THE PERIODIC TABLE AND THE AUFBAU PRINCIPLE

The ground state of an atom is that in which the electrons are in the lowest possible energy level consistent with the Pauli principle. Thus the electron configurations of successive elements in the periodic table are obtained by putting electrons in the levels shown in Fig. 12.15 starting with the lowest level. Each level may hold two electrons of opposite spin. When there are several equivalent orbitals of the

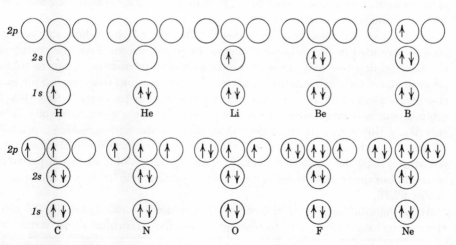

Fig. 12.15 Assignment of electrons to orbitals according to Hund's rules.

same energy the following rules, due to Hund, may be used to decide how the electrons are distributed between orbitals:

1. If the number of electrons is equal to or smaller than the number of equivalent orbitals, the electrons are assigned to different orbitals.

2. If two electrons singly occupy two equivalent orbitals, their spins will be parallel in the ground state.

Assignment of electrons to orbitals in accord with Hund's rules tends to keep them as far apart, on the average, as possible and thus to minimize the contribution of interelectron repulsion to the energy.

The chemical properties of the elements may be better understood in terms of these assignments of electrons to orbitals. The two electrons of helium both go into the $1s$ level and have opposite spins ($m_s = +\frac{1}{2}$ and $-\frac{1}{2}$). No more electrons are allowed in this orbital. Since considerable energy is required to remove an electron from this completed orbital, helium is a very inert substance.

Lithium has three electrons, two of which are in the $1s$ level and the third in the $2s$ level ($n = 2$, $l = 0$). Since the $2s$ electron is much farther from the nucleus and is partially shielded from the $+3$ charge of the nucleus by the two inner electrons, the outer electron is easily removed, producing an ion with the electronic structure of helium. In going from lithium to neon, there are eight elements ending with neon, which again has a stable structure with eight electrons with $n = 2$. The next element, sodium, has one $3s$ electron ($n = 3$, $l = 0$). This electron is shielded from the $+11$ nuclear charge by ten inner electrons, so that it is loosely bound.

A description of the orbitals of an atom which are occupied is called the electron configuration. The electron configuration is given by using exponents to indicate the number of electrons in the $1s$, $2s$, $2p$, etc., orbitals. For example, hydrogen in the ground state is represented by $1s$, helium by $1s^2$, lithium by $1s^2\,2s$, boron by $1s^2\,2s^2\,2p$, and sodium by $1s^2\,2s^2\,2p^6\,3s$.

The form of the periodic table shown in Fig. 12.16* shows how the various orbitals are filled in forming the elements. In the hydrogen atom the energy of the electron depends only upon its principal quantum number n, but in many electron atoms the inner electrons screen the outer electrons from the full nuclear charge. As a result of the screening effect s orbitals have a lower energy than p orbitals, which have a lower energy than d orbitals, for a given value of n. The energies of the p orbitals are higher than those of the corresponding s orbitals because the p electrons spend most of their time further from the nucleus and are not attracted by the nucleus as strongly. This separation of levels is not perfectly clear cut, but the relative energies are shown approximately by the vertical heights in Fig. 12.16.

At lanthanum La(57) an electron goes into the $5d$ orbital, but in the next element cerium Ce(58) an electron goes into the $4f$ orbital. This starts the

---

* H. C. Longuet-Higgins, *J. Chem. Ed.*, **34**, 30 (1957).

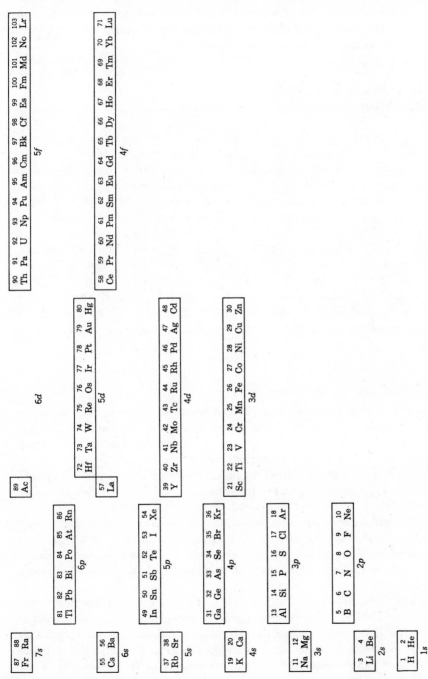

Fig. 12.16 Periodic table arranged to show electron configurations of the atoms. The relative energies of the orbitals are shown by their vertical heights. We can think of the electrons being fed successively into higher and higher energy orbitals.

427

lanthanide series (elements 57–71) in which 14 electrons are put into the $4f$ orbital. Because the chemical properties are largely determined by the outer valence electrons, these elements are much alike. At actinium Ac(89) a similar thing happens; the $6d$ orbital is started but is not completed. Elements 89 to 103 form the actinide series in which the $5f$ orbital is filled up.

12.26 IONIZATION POTENTIAL AND ELECTRON AFFINITY

The ionization potential is the voltage corresponding to the energy required to remove an electron completely from a gaseous atom or molecule without giving the free electron any kinetic energy. The ionization potential may be determined by bombarding a gas with electrons that have been accelerated by a difference in electric potential between a grid and the hot filament that emits the electrons. If the accelerated electrons have insufficient kinetic energy to cause a shift from one energy level to another in the atoms or molecules they strike, the collisions are said to be elastic. As the potential is increased, the accelerated electrons gain sufficient energy to excite an orbital electron from one energy level to the next higher level. Light is emitted when the electron returns to an empty lower level. As the potential is further increased, new spectral lines appear. The potentials required to cause emission of light are called *resonance potentials*. The relation between the accelerating potential E and the frequency of the light emitted is

$$Ee = h\nu \qquad (12.105)$$

where e is the charge on the electron.

If the accelerating potential is sufficiently great an electron may be driven from the atom or molecule, and this potential is called an *ionization potential*. The ionization potential of an atom or ion may be calculated from spectroscopic data, since this potential is given by the convergence limit (Section 12.6). The singly charged positive ion produced may be ionized further by bombarding with electrons of still higher energy, that is, the second, third, . . . , ionization potentials correspond with the ejection of the second, third, . . . , electrons.

A plot of the first ionization potentials of gaseous atoms versus atomic number is given in Fig. 12.17. The ionization potentials change in a periodic way because of the progressive filling up of shells with electrons. The principal maxima in this plot are given by the inert gases, and the principal minima by the alkali metal atoms.

The alkali metal atoms are easily ionized, since they have a single electron in the outer orbital and the effective nuclear charge is low. The attraction of the nucleus for the outermost electron of the alkali metal atoms is quite effectively shielded by the electrons of the inner orbits. In the series lithium, sodium, potassium, rubidium, and cesium the ionization potential decreases because of the increase in size of the outer orbit containing a single electron.

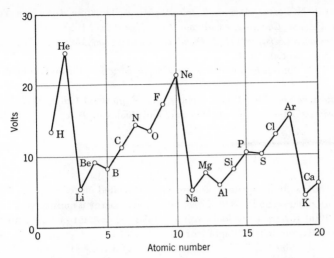

Fig. 12.17   First ionization potentials of gaseous atoms early in the periodic table.

In contrast, the ionization potentials of the halogens are almost as great as those of the inert gases. The electrons in the outer orbits of the halogen atoms are shielded from the nuclear charge mainly by the electrons in inner orbits, since the electrons in the outer orbits are all approximately the same distance from the nucleus. A direct result of this incomplete shielding of the nuclear charge, so far as electrons in the outer orbit are concerned, is the fact that the halogen atoms readily take on an additional electron to form negative ions.

The electron affinity EA is defined as the energy released in the process

$$A + e \rightarrow A^- \tag{12.106}$$

If we consider the reverse of this process we can see that the electron affinity of A is the ionization potential of A^-. The electron affinity increases with increasing atomic number in a row of the periodic table; the electron affinity of Li is 0.6 eV, and the electron affinity of F is 3.45 eV. The affinities of chlorine, bromide, and iodine for an additional electron are 3.71, 3.49, and 3.19 eV. Oxygen atoms and sulfur atoms also have affinities for an additional electron (3.07 and 2.8 eV, respectively).

References

J. M. Anderson, *Introduction to Quantum Chemistry*, W. A. Benjamin, Inc., New York, 1969.
J. C. Davis, Jr., *Advanced Physical Chemistry*, The Ronald Press Co., New York, 1965.
M. W. Hanna, *Quantum Mechanics in Chemistry*, W. A. Benjamin, Inc., New York, 1969.
G. Herzberg, *Atomic Spectra and Atomic Structure*, Prentice-Hall, Inc., Englewood Cliffs, N.J., 1937.
R. M. Hochstrasser, *Behavior of Electrons in Atoms*, W. A. Benjamin, Inc., New York, 1965.

M. Karplus and R. N. Porter, *Atoms and Molecules*, W. A. Benjamin, Inc., New York, 1970.

W. Kauzmann, *Quantum Chemistry*, Academic Press, New York, 1957.

I. N. Levine, *Quantum Chemistry Vol. I, Quantum Mechanics and Molecular Electonic Structure*, Allyn and Bacon, Boston, 1970.

L. Pauling and E. B. Wilson, *Introduction to Quantum Mechanics*, McGraw-Hill Book Co., New York, 1935.

K. S. Pitzer, *Quantum Chemistry*, Prentice-Hall, Inc., Englewood Cliffs, N.J., 1953.

J. C. Slater, *Quantum Theory of Matter*, McGraw-Hill Book Co., New York, 1968.

Problems

12.1   A hollow box with an opening of 1 cm² area is heated electrically. (*a*) What is the total energy emitted per second at 800 K? (*b*) How much energy is emitted per second if the temperature is 1600 K? (*c*) How long would it take the radiant energy emitted at this temperature, 1600 K, to melt 1000 g of ice?

Ans. (*a*) 2.33×10^4 J m^{-2} s^{-1}, (*b*) 3.73×10^5 J m^{-2} s^{-1}, (*c*) 8950 s.

12.2   Assuming that the sun radiates as a black body, calculate its surface temperature from the fact that the wavelength of maximum emission is 5000 Å. The wavelength of maximum emission is given by $\lambda_{max} = ch/4.97kT$. *Ans.* 5800 K.

12.3   The yellow doublet of the sodium lamp has an average wavelength of 5890 Å. Calculate the energy in (*a*) electron volts and (*b*) kilocalories per mole.   *Ans.* (*a*) 2.11, (*b*) 48.7.

12.4   Most chemical reactions require activation energies ranging between 10 and 100 kcal mol^{-1}. What are the equivalents of 10 and 100 kcal mol^{-1} in terms of (*a*) angstroms, (*b*) wave numbers, (*c*) electron volts?

Ans. (*a*) 28500, 2850 Å, (*b*) 3.50×10^3, 3.50×10^4 cm^{-1}, (*c*) 0.434, 4.34 eV.

12.5   An experiment on the emission of photoelectrons from a sodium surface by light of different wavelengths gave the following values for the potentials at which the photoelectric current was reduced to zero. Plot voltage against frequency, and calculate (*a*) the threshold frequency and (*b*) Planck's constant.

λ, Å	E, V
3651	-0.950
3125	-0.382

Ans. (*a*) 10.5×10^{14} s^{-1}, (*b*) 6.58×10^{-34} J s.

12.6   What potential difference is required to accelerate a singly charged gas ion in a vacuum so that it has (*a*) a kinetic energy equal to that of an average gas molecule at 25°, (*b*) an energy equivalent to 20 kcal mol^{-1}?

Ans. (*a*) 0.0385, (*b*) 0.867 V.

12.7   Calculate the velocity of an electron that has been accelerated by a potential difference of 1000 V.   *Ans.* 1.87×10^7 m s^{-1}.

12.8   Calculate the wavelengths (in micrometers) of the first three lines of the Paschen series for atomic hydrogen.   *Ans.* 1.8756, 1.2822, 1.0941 μm.

12.9   In the Balmer series for atomic hydrogen what is the wavelength of the series limit?   *Ans.* 3647 Å.

12.10   Calculate the radius for the first Bohr orbit with the reduced mass of the hydrogen atom.   *Ans.* 0.529 465 Å.

12.11 How many kcal mol^{-1} are required to excite a hydrogen atom from $n = 1$ to $n = 2$? How many times larger than the translational energy of a hydrogen atom at room temperature is this?   *Ans.* 236 kcal mol^{-1}, 266.

12.12 Calculate the de Broglie wavelength of electrons accelerated by 40,000 V.
Ans. 0.0614 Å.

12.13 Electrons are accelerated by a 1000 V potential drop. (*a*) Calculate the de Broglie wavelength. (*b*) Calculate the wavelength of the X rays that could be produced when these electrons strike a solid.   *Ans.* (*a*) 0.387, (*b*) 12.4 Å.

12.14 Show that the function $\psi = 8e^{5x}$ is an eigenfunction of the operator d/dx. What is the eigenvalue?   *Ans.* 5.

12.15 Calculate the approximate quantum number corresponding with the translational energy of a 1 g bullet fired with a velocity of 300 m at a target 100 m away. What is the de Broglie wavelength?   *Ans.* 9.06×10^{34}, 2.21×10^{-33} m.

12.16 (*a*) Calculate the energy levels for $n = 1$ and $n = 2$ for an electron in a potential well of width 5 Å with infinite barriers on either side. The energies should be expressed in kcal mol^{-1}. (*b*) If an electron makes a transition from $n = 2$ to $n = 1$ what will be the wavelength of the radiation emitted?   *Ans.* (*a*) 34.6 kcal mol^{-1}; 138.4 kcal mol^{-1}, (*b*) 2747 Å.

12.17 Calculate the average distance between the electron and nucleus of a hydrogenlike atom in the 1s state. See the discussion of Fig. 12.9.
Ans. $\frac{3}{2}\frac{a_0}{Z}$.

12.18 Calculate the degeneracies of the first three levels for a particle in a cubical box.
Ans. 1, 3, 3.

12.19 Show that for a 1s orbital of a hydrogenlike atom the most probable distance from proton to electron is a_0/Z.

12.20 What are the electron configurations for H$^-$, Li$^+$, O^{2-}, F$^-$, Na$^+$, and Mg^{2+}?
Ans. $1s^2$, $1s^2$, $1s^2$, $2s^2\,2p^6$, $1s^2\,2s^2\,2p^6$, $1s^2\,2s^2\,2p^6$, $1s^2\,2s^2\,2p^6$.

12.21 Apply Hund's rules to obtain the electron configurations for Si, P, S, Cl, Ar.
Ans. Si $1s^2\,2s^2\,2p^6\,3s^2\,3p^2$
P $1s^2\,2s^2\,2p^6\,3s^2\,3p^3$
S $1s^2\,2s^2\,2p^6\,3s^2\,3p^4$
Cl $1s^2\,2s^2\,2p^6\,3s^2\,3p^5$
Ar $1s^2\,2s^2\,2p^6\,3s^2\,3p^6$

12.22 The first ionization potential for atomic lithium is 5.39 V (Li $=$ Li$^+$ $+$ e). The second ionization potential is 75.62 V (Li$^+$ $=$ Li^{+2} $+$ e). Calculate the wavelengths for the convergence limits indicated by these potentials.   *Ans.* 2300, 164 Å.

12.23 Absorption by the fundamental vibration of the HCl molecule occurs at 3.64 μm. (*a*) What is the frequency of light of this wavelength? (*b*) What is the energy in joules per quantum? (*c*) What is the energy in terms of electron volts?

12.24 (*a*) How many calories per mole are equivalent in energy to 2 eV? A lead storage battery gives about 2 V. (*b*) How many calories per mole are equivalent to 100,000 V? X rays have energies of about 100,000 eV. (*c*) How many calories per mole are equivalent in energy to 5 million eV? Some alpha particles have energies of about 5 million eV.

12.25 If 10,000 cal is lost per min by radiation from the door of an electrically heated furnace at 900 K, how many more watts of electricity must be applied to offset the losses due to radiation if the furnace is heated to 110 K?

12.26 Plot the intensity of cavity radiation versus wavelength for a temperature of 10,000 K. What is the wavelength of maximum emission?

12.27 The threshold frequency for photoelectric emission from lithium is approximately

5200 Å. Calculate the velocity of electrons emitted as the result of absorption of light at 3600 Å.

12.28 Calculate the velocity of an electron that has been accelerated by a potential difference of 1.00 volt.

12.29 Calculate the frequency and the wavelength in angstroms for the line in the Paschen series of the hydrogen spectrum that is due to a transition from the sixth quantum level to the third.

12.30 Calculate the fraction of hydrogen atoms which at equilibrium at 1000° C would have $n = 2$.

12.31 The first ionization potential of atomic hydrogen is 13.54 V. Calculate the wavelength of the light produced when a free electron without kinetic energy returns to the inner orbit.

12.32 Positronium consists of an electron and a positron (the mass of a positron is the same as that of an electron). (a) Calculate the wavelength of the radiation emitted when the electron falls from the orbital $n = 2$ to $n = 1$. (b) Calculate the ionization potential.

12.33 Calculate the value of the Rydberg constant R for hydrogen and compare it with the experimentally determined value.

12.34 Calculate the de Broglie wavelength for thermal neutrons at a temperature of 100° C.

12.35 Calculate the de Broglie wavelength of electrons that have been accelerated by 50,000 V.

12.36 Show that the function $\psi = xe^{-ax^2}$ is an eigenfunction of the operator $d^2/dx^2 - 4a^2x^2$ What is the eigenvalue?

12.37 Calculate the difference in energy in ergs for $n_x = 1$ and $n_x = 2$ for a 100 gram ball in a 10 cm box.

12.38 For a hydrogen atom in a one-dimensional box calculate the value of the quantum number for the energy level for which the energy is equal to $\frac{3}{2}kT$ at 25° (a) for a box 10 Å long and (b) for a box 1 cm long.

12.39 For a simple harmonic oscillator in its ground state, show that one obtains the classical results for $\langle x \rangle$ and $\langle p \rangle$.

12.40 Calculate $\langle r \rangle$ for a 2s electron in a hydrogen atom.

Given :
$$\int_0^\infty x^n e^{-ax}\, dx = \frac{n!}{a^{n+1}}$$

for $n > -1, a > 0$

12.41 For the H atom show that ψ_{1s} and ψ_{2s} are orthogonal. (See the integral in the preceding problem.)

12.42 Show that the wave function for a 1s hydrogenlike orbital is normalized. (see the integral in problem 12.40.)

12.43 Sketch the possible orientations for angular momentum vectors for $l = 3/2$.

12.44 Considering only the first 18 elements of the periodic table, list those whose outer electrons are in spherically symmetrical orbits and those whose outer electrons are not.

12.45 Calculate the ionization potentials for He^+, Li^{2+}, Be^{3+}, B^{4+}, and C^{5+}.

12.46 Calculate the ionization potential for He^+ and the wavelengths of the first two lines of the Balmer series.

12.47 Since it is frequently necessary to interconvert energies in the units centimeters^{-1}, electron volts, and kilocalories per mole, calculate the factors for converting (a) electron volts to cm^{-1}, and (b) cm^{-1} to kilocalories per mole.

12.48 Calculate the mean kinetic energy in electron volts of a molecule in a gas at 300 K.

12.49 An electric heater of 10 cm² has a temperature of 800 K. How many calories of radiant heat are emitted per minute, if it is assumed that the heater is a cavity?

12.50 Calculate the ratio of the intensities of light of 5000 Å wavelength from cavities of 1000 K and 5000 K.

12.51 What is the wavelength of light which has energy equal to the energy of electrons which have been accelerated by a potential of (a) 400 V, (b) 3 V?

12.52 The work to get an electron through the surface of tungsten is 4.58 V. (a) What is the wavelength of the photoelectric threshold? (b) When light of 2000 Å is used, what potential must be applied to keep the most energetic electrons from reaching the collector? (c) What is the maximum velocity of electrons emitted by 2000 Å light?

12.53 There is a Brackett series in the hydrogen spectrum where $n_1 = 4$. Calculate the wavelengths, in angstroms, of the first two lines of this series.

12.54 Calculate the shift of wavelength of the first two lines of the Balmer series for hydrogen when one neutron is added to the nucleus, giving deuterium.

12.55 After Bohr's theory of the hydrogen atom Sommerfeld suggested a quantization rule that, when applied to the hydrogen atom, reconciled the Bohr model with the wave nature of the electron suggested by de Broglie. Using Sommerfeld's suggestion that the allowed electron orbits must have a circumference equal to an exact multiple of the wavelength of the electron, derive the expression for the energy levels of the hydrogen atom.

12.56 As pointed out in the text, cavity radiation, emitted from the interior of a hollow object through an infinitesimally small hole, is often referred to as black-body radiation. A black body is an object that absorbs all radiation that falls on it. What similarity does a cavity have to this?

12.57 Calculate the de Broglie wavelength of a hydrogen atom with a translational energy corresponding to room temperature.

12.58 What is the energy in electron volts of a neutron which has a de Broglie wavelength the same as 20,000 eV X rays?

12.59 Show by a simple calculation why a beam of thermal neutrons is a suitable source from which to obtain a monoenergetic beam of suitable energy for the study of crystal structures.

12.60 Show that the function $\psi = Ke^{r/k}$ is an eigenfunction of the operator d/dr. What is the eigenvalue?

12.61 Calculate the first three energy levels in kilocalories per mole for an electron in a potential well 5 Å in width with infinitely high potential outside.

12.62 Show that a simple harmonic oscillator in its ground state obeys the uncertainty principle by computing

$$\Delta x = \sqrt{\langle x^2 \rangle - \langle x \rangle^2} \quad \text{and} \quad \Delta p = \sqrt{\langle p^2 \rangle - \langle p \rangle^2}$$

12.63 Calculate the reduced mass of a hydrogen atom.

12.64 Show that the following wave function for the hydrogen atom is antisymmetric to the interchange of the two electrons

$$\begin{vmatrix} 1s\alpha(1) & 1s\beta(2) \\ 1s\alpha(2) & 1s\beta(2) \end{vmatrix}$$

12.65 Using the Bohr theory, calculate the ionization potential for $Be^{3+} = Be^{4+} + e$.

CHAPTER 13

SYMMETRY

Ideas about symmetry are of great importance in connection with both theoretical and experimental studies of atomic and molecular structure. The basic principles of symmetry are applied in quantum mechanics, spectroscopy, and structural determinations by X-ray, neutron, and electron diffraction. Nature exhibits a great deal of symmetry, and this is especially evident when we examine molecules in their equilibrium configurations. By equilibrium configuration we refer to that with the atoms fixed in their mean positions. When symmetry is present certain calculations are simplified if this symmetry is taken into account. Symmetry also determines whether a molecule can be optically active or whether it may have a dipole moment. Single molecules, unlike crystalline solids (see Chapter 19), are not restricted in the symmetry that they may possess.

13.1 SYMMETRY ELEMENTS AND SYMMETRY OPERATIONS

Five types of symmetry elements are used to describe the symmetry of molecules: center of symmetry, proper rotation axis, mirror plane, improper rotation axis, and the identity element. Each of these symmetry elements has associated with it a symmetry operation. The symmetry elements and operations are described in Table 13.1. After a symmetry operation has been applied to a molecule, the appearance of a molecule may be changed. But if it is not, the molecule is said to "possess" the symmetry operation and the corresponding symmetry element.

13.2 THE CENTER OF SYMMETRY AND THE INVERSION OPERATION

A molecule has a center of symmetry i if a straight line from any atom projected through the center of the molecule encounters an equivalent atom equidistant from the center. The center of symmetry (or inversion center) is the symmetry element, and the operation is inversion through the center by which one-half of

Table 13.1   Symmetry Elements and Associated Operations

Symbol	Element	Operation
i	*Center of symmetry* (or inversion center).	A projection through the center of symmetry to an equal distance on the other side from the center.
C_n	*Proper rotation axis*	A counterclockwise rotation about the C_n axis by $2\pi/n$ (or $360°/n$).
σ_h	*Horizontal mirror plane* perpendicular to principal C_n axis (*i.e.*, proper axis of highest symmetry).	Reflection across the plane of symmetry.
σ_v	*Vertical mirror plane* containing the principal C_n axis.	Reflection across the plane of symmetry.
σ_d	*Diagonal mirror*, containing the principal C_n axis; the plane bisects the angle formed by two horizontal C_2 axes which are perpendicular to the principal axis C_n of highest symmetry.	Reflection across the plane of symmetry.
S_n	*Improper rotation axis* (also referred to as a *rotation-reflection axis or alternating axis*).	A counterclockwise rotation about the S_n axis by $2\pi/n$ followed by reflection in a plane perpendicular to the axis (*viz.*, the *combined operation* of a C_n rotation followed by reflection across a σ_h mirror plane).
E	*Identity element*	A C_1 operation corresponding to a rotation of $360°/1$ about any axis.

the molecule can be generated from the other half. The action of the inversion operation is to transform the coordinates (x, y, z) into their respective negatives $(-x, -y, -z)$. The inversion operation may be represented by

$$i \cdot \begin{bmatrix} x \\ y \\ z \end{bmatrix} = \begin{bmatrix} -x \\ -y \\ -z \end{bmatrix} \tag{13.1}$$

If the inversion operation is applied twice we obtain the original configuration as shown by

$$i \cdot i \cdot \begin{bmatrix} x \\ y \\ z \end{bmatrix} = i \cdot \begin{bmatrix} -x \\ -y \\ -z \end{bmatrix} = \begin{bmatrix} x \\ y \\ z \end{bmatrix} \tag{13.2}$$

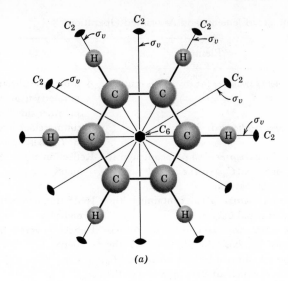

(a)

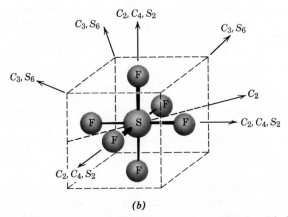

(b)

Fig. 13.1 Molecules with a center of symmetry. (a) C_6H_6 molecule with D_{6h} point-group symmetry. (b) SF_6 molecule with O_h point-group symmetry. This centrosymmetric molecule possesses three equivalent C_4 axes at right angles to one another. Three S_4 and three C_2 axes are coincident with the three C_4 axes. Four C_3 axes (and four coincident S_6 axes) are located along the four body-diagonals of the cube.

Thus the successive application of i an even number of times produces the identity operation E. It is convenient to include E as an operation even though it does not change the configuration of a molecule. The identity operation corresponds with a rotation of 360° about any axis.

Since $i^k = i$ when k is odd, and $i^k = E$ when k is even, the center of symmetry generates only *one distinct* operation. In molecules with a center of symmetry, the atoms may be thought of as occurring in centrosymmetric pairs with the exception of an unshifted atom if one lies at the center of symmetry. Centrosymmetric molecules include C_6H_6 (Fig. 13.1a), SF_6 (Fig. 13.1b), the *staggered* conformation of C_2H_6 (Fig. 13.3b), CO_2, and C_2H_4.

13.3   THE SYMMETRY AXIS AND THE ROTATION OPERATION

A symmetry axis is a line about which rotation through an angle $2\pi/n$ radians brings a structure into coincidence with itself. The rotation operation is represented by C_n, where n is referred to as the order of the rotation, and the rotation is conventionally taken as positive in the counterclockwise direction. If the z axis is the rotation axis, the action of the rotation operation is to transform the coordinate (x, y, z) to $(-x, -y, z)$. The C_2 rotation operation may be represented by

$$C_2^1 \cdot \begin{bmatrix} x \\ y \\ z \end{bmatrix} = \begin{bmatrix} -x \\ -y \\ z \end{bmatrix} \tag{13.3}$$

As shown in Fig. 13.2a a water molecule has a C_2 axis passing through the oxygen and bisecting the angle between the O—H bonds. The ammonia molecule NH_3 (Fig. 13.2b) has a C_3 axis passing through the nitrogen. The benzene molecule (Fig. 13.1a) has a C_6 axis perpendicular to the plane of the ring. Any linear molecule, such as HCl, has a C_∞ axis (Fig. 13.2c) because the appearance of the molecule is not changed by a rotation of any angle (infinity in number) about the internuclear axis.

A symmetry axis may generate several operations. The symmetry operation involving counterclockwise rotation by $2\pi/n$ carried out successively k times, is denoted by the symbol C_n^k. The operations generated by a twofold axis are

$$C_2^1$$
$$C_2^2 = E$$

The first is rotation through 180°, and the second is rotation through 360°, which produces the same effect as the identity operation E. Thus a C_2 axis has only one distinct operation C_2^1, in that C_2^1 cannot be represented in any other way, whereas the C_2^2 operation is usually represented as E. Inasmuch as $C_2^3 = C_2^1$ and $C_2^4 = E$, it follows for rotation axes (C_n) that operations with k greater than n do not yield separate operations.

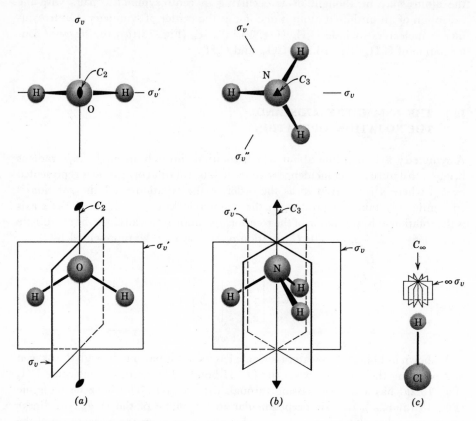

Fig. 13.2 (a) H_2O molecule with C_{2v} point-group symmetry. (b) NH_3 molecule with C_{3v} point-group symmetry. (c) HCl molecule with $C_{\infty v}$ point-group symmetry.

The operations generated by a fourfold axis are

$$C_4^1$$

$$C_4^2 = C_2^1$$

$$C_4^3$$

$$C_4^4 = E$$

Since a C_2 axis is always coincident with a C_4 axis, only the operations C_4^1 and C_4^3 are *distinct* operations of the C_4 axis.

Example 13.1 How many distinct operation are implied by a C_6 axis?
  Ans.

$$C_6^1$$
$$C_6^2 = C_3^1$$
$$C_6^3 = C_2^1$$
$$C_6^4 = C_3^2$$
$$C_6^5$$
$$C_6^6 = E$$

Thus two operations (C_6^1 and C_6^5) are characteristic only of a C_6 axis and cannot be represented in any other way.

In discussing symmetry operations, it is convenient to orient the molecules in a right-hand Cartesian coordinate system. The thumb, index, and middle fingers of the right hand are pointed in three mutually perpendicular directions, and these are taken as the x, y, and z directions, respectively. The center of mass of the molecule under consideration is located at the origin of the Cartesian coordinate system, and its *principal axis* is aligned with the z axis. The *principal axis* is defined as the C_n axis with highest order n; if there are several rotational axes of the same highest symmetry (e.g., three twofold axes at right angles to one another), the z axis is taken along the one passing through the greatest number of atoms.

13.4 THE SYMMETRY PLANE AND
THE REFLECTION OPERATION

A plane of symmetry is a plane that bisects a molecule in such a way that the part of the molecule on one side of the plane is the mirror image of the part of the other side. The symbol σ represents both the symmetry element, the plane, and the symmetry operation, reflection through the plane. Since the operation σ gives a configuration equivalent to the original and since the application of the same σ twice to a molecule produces its original configuration, it follows that a mirror plane generates only *one distinct* operation in that $\sigma^k = \sigma$ when k is odd, and $\sigma^k = E$ when k is even.

If the xz plane is a mirror plane the reflection operation σ may be represented by

$$\sigma \cdot \begin{bmatrix} x \\ y \\ z \end{bmatrix} = \begin{bmatrix} x \\ -y \\ z \end{bmatrix} \tag{13.4}$$

A mirror plane (and corresponding operation of reflection) perpendicular to the direction of the principal C_n axis (i.e., the normal of the mirror plane is coincident with the C_n axis of highest order n) is called a *horizontal* mirror plane

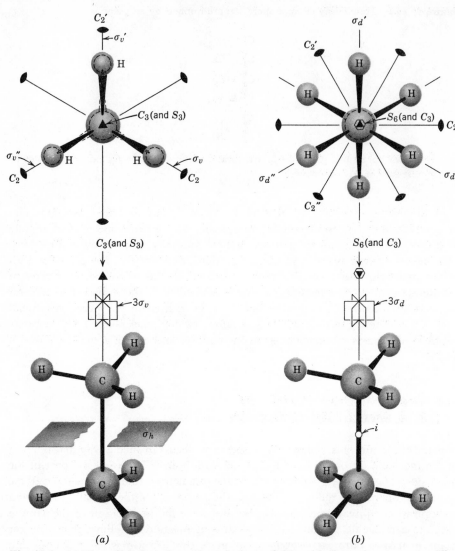

Fig. 13.3 (a) The eclipsed conformation of ethane, C_2H_6, with D_{3h} point symmetry. The view along the C_3 (and S_3) axis of the C—C bond is shown at the top. (b) The staggered conformation of ethane, C_2H_6 with D_{3d} point symmetry. The view along S_6 (and C_3) axis of the C—C bond is shown at the top.

and denoted as σ_h. Molecules with a horizontal mirror plane include C_6H_6 (Fig. 13.1a) (for which σ_h is perpendicular to the C_6 axis and contains all the atoms of this planar molecule) and the *eclipsed* conformation of ethane (Fig. 13.3a), which has a σ_h perpendicular to a C_3 principle axis.

Mirror planes which contain the principal C_n axis, are called *vertical* mirror planes and *generally* are symbolized as σ_v. The particular mirror planes that bisect the angles formed by pairs of horizontal C_2 axes are designated as σ_d. As is evident from Fig. 13.2a, a H_2O molecule has two *vertical* mirror planes, σ_v and σ_v', which are perpendicular to each other. One of these mirror planes (σ_v) comprises the plane of the molecule, and the other (σ_v') is perpendicular to it. The twofold axis lies in the intersection of the two mirror planes. Ammonia has three σ_v containing the C_3 axis, and benzene has six σ_v containing the C_6 principal axis. The linear molecule HCl has an *infinite* number of *vertical* mirror planes of type σ_v, all of which include the C_∞ rotational axis. Homonuclear diatomic molecules such as H_2 or Cl_2 have a σ_h in addition. Figure 13.3b shows the *staggered* conformation of ethane to possess three vertical σ_d which contain the principal C_3 axis and which bisect the three horizontal C_2 axes. In the *eclipsed* conformation of ethane (Fig. 13.3a), however, each of the three vertical mirror planes contains one of the three horizontal C_2 axes (as well as the C_3 principal axis). Consequently, these mirror planes are called σ_v.

13.5   THE IMPROPER AXIS AND THE OPERATION OF IMPROPER ROTATION

The improper axis is a compound element consisting of a symmetry axis perpendicular to a plane, neither of which necessarily generates its own symmetry operation independently. The operation of improper rotation consists of a rotation followed by a reflection. The nth order improper rotation is represented by S_n and is the product of two operations.

$$S_n = \sigma C_n \tag{13.5}$$

This means that the operations σ and C_n are applied successively. The molecule is always oriented so that the symmetry plane is a horizontal plane.

An S_1 axis is equivalent to a *plane of symmetry* (σ) as the operation involving rotation of 360° followed by reflection across the mirror plane perpendicular to the axis of rotation can be simply represented in terms of reflection across the mirror plane.

An S_2 axis is equivalent to a *center of symmetry* (i), since the S_2 operation consisting of a counterclockwise rotation about an axis of $2\pi/2$ or 180° followed by reflection across a horizontal mirror plane perpendicular to this axis yields the same configuration as an inversion through a center of symmetry located at the intersection of the axis of rotation and reflection plane.

An S_2 axis has six operations:

$$S_3^1$$
$$S_3^2 = C_3^2$$
$$S_3^3 = \sigma_h$$
$$S_3^4 = C_3^1$$
$$S_3^5$$
$$S_3^6 = E$$

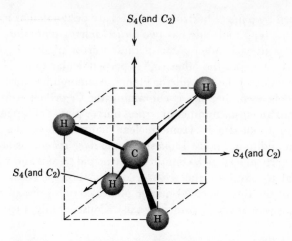

Fig. 13.4   Three equivalent S_4 rotation-reflection axes at right angles to one another are shown for the tetrahedral methane molecule, CH_4, which possesses T_d point group symmetry. Any symmetrical tetrahedral molecule like methane also has four C_3 axes (at angles of 109° 28′ with one another) along the C—H bonds, three C_2 axes coincident with the S_4 axes, and six vertical planes (σ_d), each of which passes through the carbon and two hydrogens and relates the other two hydrogens to each other.

Thus an S_3 axis implies a C_3 axis and a horizontal mirror plane σ_h. The *eclipsed* conformation of ethane has an S_3 axis, as shown in Fig. 13.3a.

An S_4 axis has four associated operations:

$$S_4^1$$
$$S_4^2 = C_2^1$$
$$S_4^3$$
$$S_4^4 = E$$

Thus an S_4 axis implies a C_2 axis. This is illustrated by methane that has three equivalent S_4 axes at right angles to each other as shown in Fig. 13.4.

An S_6 axis has six associated operations:

$$S_6^1$$
$$S_6^2 = C_3^1$$
$$S_6^3 = i$$
$$S_6^4 = C_3^2$$
$$S_6^5$$
$$S_6^6 = E$$

Thus an S_6 axis implies the existence of both a C_3 axis coincident with the S_6 axis and a center of symmetry (i). Therefore only the S_6^1 and S_6^5 operations are *distinct* operations characteristic of only the S_6 axis. The *staggered* configuration of ethane has an S_6 axis, as shown in Fig. 13.3b.

In general, S_n axes with n even contain n operations with neither a σ_h nor a C_n axis (but with a $C_{n/2}$ axis). In general S_n axes with n odd contain a total of $2n$ operations, including σ_h and the operations generated by C_n.

13.6 COMBINATION OF SYMMETRY OPERATIONS—POINT GROUPS

In general, molecules possess more than one of the possible elements of symmetry. For example, the water molecule H_2O (Fig. 13.2a) has two mirror planes (σ_v and σ'_v) at right angles and a twofold axis (C_2) along the line of their intersection; this C_2 axis is an automatic consequence of the mirror planes being perpendicular to each other. Since certain combinations of symmetry operations give rise to other symmetry operations, it is not necessary to specify every symmetry operation to describe completely the symmetry of molecules.

The set of symmetry operations that a molecule possesses can be collected into a *group*. This group obeys the following four conditions:

1. If A represents a symmetry operation of the group and B represents another symmetry operation of the same group, the product $A \times B = F$ also is an operation of the group. The product $A \times B = F$ means that the operation B followed by the operation A is equivalent to an operation F. In general $A \times B \neq B \times A$, which means that the operation B followed by operation A is *not* necessarily equivalent to the operation A followed by operation B. In other words, A *does not commute* with B. If $A \times B = B \times A$, the multiplication is commutative. For a given molecule the various products can be summarized in a group multiplication table.

2. In each group there exists an identity operation E (corresponding to the C_1 rotation of 360° about any given axis) such that for any other operation of the group (e.g., A).

$$A \times E = E \times A = A \qquad (13.6)$$

3. For each operation A there exists in the group an inverse operation A^{-1} such that $A^{-1} \times A = A \times A^{-1} = E$. The inverse operation A^{-1} is that which returns the object to its original position. Hence, the inverse of a C_2, σ, or i operation is itself (viz., $C_2 \times C_2 = E$; $\sigma \times \sigma = E$; $i \times i = E$). The inverse of a C_3^1 operation is a C_3^2, while the inverse of an S_4^1 operation is an S_4^3.

4. The Associative Law of multiplication holds:

$$A \times (B \times C) = (A \times B) \times C \qquad (13.7)$$

The group of symmetry operations that leaves a molecule (or other finite objects such as a crystal) congruent with itself is called a *point group*. For a point group all symmetry elements pass through a fixed point or origin at the center of the object. Each point group is represented by a *Schoenflies symbol* that designates sufficient symmetry elements to obtain the other symmetry elements and all *distinct* operations of the point group.

Consider H_2O (Fig. 13.2a), which has four symmetry operations: E, C_2^1, σ_v, and σ_v'. The operation of one vertical mirror plane (σ_v) followed by the operation of the other vertical mirror plane (σ_v') is equivalent to the twofold operation, that is, $\sigma_v' \times \sigma_v = C_2^1$. Similarly, the successive operations of C_2^1 followed by σ_v yield the same result as the σ_v' operation (viz., $\sigma_v \times C_2^1 = \sigma_v'$). Each of the four operations is its own inverse (e.g., $\sigma_v \times \sigma_v = E$). These product operations for H_2O are summarized in Table 13.2 as a group multiplication table, which shows (*since no additional operations are generated*) that these four symmetry operations form a group and that the operations are commutative. The point group is designated by the Schoenflies symbol C_{2v}. The subscript "2" of this symbol signifies not only that the principal proper rotation axis (C_n) is a C_2, but also that there are *two vertical* mirror planes, at right angles to each other, which contain the C_2 axis.

Table 13.2   Multiplication Table for the Group C_{2v}

		Operation B		
	E	C_2^1	σ_v	σ_v'
E	E	C_2^1	σ_v	σ_v'
C_2^1	C_2^1	E	σ_v'	σ_v
σ_v	σ_v	σ_v'	E	C_2^1
σ_v'	σ_v'	σ_v	C_2^1	E

(Operation A labels the rows)

The table contains the products $A \times B$ for the indicated operations. Note that each column and each row has each symmetry operation represented once only.

13.7   CLASSIFICATION OF SCHOENFLIES POINT GROUPS

In Table 13.3 a number of Schoenflies point groups are listed together with examples. Although Schoenflies symbols are shown in boldface type they are not vectors. There is, in principle, an infinite number of point groups. The external symmetries of crystals fall into only 32 point groups (Section 19.4).

13.8   DETERMINATION OF SCHOENFLIES POINT GROUP NOTATION

In determining the Schoenflies point group of any molecule, it is convenient to carry out a systematic examination according to the following sequence of steps:

1. Determine whether a molecule possesses T_d point group symmetry, that of a regular tetrahedron, or O_h point group symmetry, that of a regular octahedron. In both cubic

Table 13.3   Common Schoenflies Point Groups with Examples

Schoenflies Symbol	Symmetry Elements	Molecular Configuration	Schoenflies Symbol	Symmetry Elements	Molecular Configuration
C_1	E		D_{3h}	$E, C_3(S_3), 3C_2, \sigma_h, 3\sigma_v$	
C_s	E, σ		D_{4h}	$E, C_4(C_2, S_4), 4C_2, \sigma_h, 2\sigma_v, 2\sigma_d, i$	
C_i	E, i		D_{5h}	$E, C_5(S_5), 5C_2, \sigma_h, 5\sigma_v$	
C_2	E, C_2		D_{6h}	$E, C_6(C_3, C_2, S_6, S_3), 6C_3, \sigma_h, 3\sigma_v, 3\sigma_d, i$	
C_{2v}	$E, C_2, 2\sigma_v$		$D_{\infty h}$	$E, C_\infty(S_\infty), \infty C_3, \sigma_h, \infty\sigma, i$	
C_{3v}	$E, C_3, 3\sigma_v$		D_{2d}	$E, C_2(S_4), 2C_2, 2\sigma_d$	
C_{4v}	$E, C_4(C_2), 4\sigma_v$		D_{3d}	$E, C_3(S_6), 3C_2, 3\sigma_d, i$	
$C_{\infty v}$	$E, C_\infty, \infty\sigma_v$		D_{4d}	$E, C_4(S_8, C_2), 4C_2, 4\sigma_d$	
C_{2h}	E, C_2, σ_h, i				

445

Table 13.3—(*Continued*)

Schoenflies Symbol	Symmetry Elements	Molecular Configuration	Schoenflies Symbol	Symmetry Elements	Molecular Configuration
C_{3h}	$E, C_3(S_3), \sigma_h$	(boric acid $B(OH)_3$ structure)	D_{5d}	$E, C_5(S_{10}), 5C_3, 5\sigma_d, i$	(ferrocene, Fe)
			T_d	$E, 3C_2(3S_2), 4C_3, 6\sigma_d$	(CH$_4$)
D_{2h}	$E, C_2, 2C_2, \sigma_h, 2\sigma_v, i$	(ethylene $H_2C=CH_2$)	O_h	$E, 3C_4(3C_2, 3S_4), 4C_2(4S_6), 3\sigma_h 6C_2, 6\sigma_d, i$	(SF$_6$)

point groups there are *four* C_3 axes along the body diagonals of a cube. If the molecule does not belong to either of these point groups but contains more than one threefold axis (i.e., more than one axis of symmetry greater than a twofold), it is of a point group not given in this chapter.

2. Determine whether or not the molecule possesses any C_n proper axes of symmetry. If so, proceed to step 3. If not, look for a mirror plane, in which case it belongs to the point group C_s, or a center of symmetry, in which case it has the point symmetry C_i. If no symmetry elements are located, the molecule then belongs to the point group C_1, which possesses only the identity element (and corresponding operation).

3. If at least one proper axis C_n is present, locate the principal axis C_n of highest symmetry if one exists (e.g., for the point group D_2 with three C_2 axes perpendicular to one another, there is no unique principal axis). Next determine if there is an improper axis S_{2n} collinear with the principal (or other) axis C_n. If an S_{2n} axis exists but no other elements of symmetry are present except possibly i, the molecule belongs to one of the S_n (n even) point groups. Otherwise, if no S_{2n} axes are found or if an S_{2n} axis is observed along with other symmetry elements, proceed to step 4.

4. Determine if the point group is dicyclic (i.e., D_n, D_{nh}, D_{nd}) by observing whether n twofold equally spaced axes lie in a plane perpendicular to the principal (or other) axis C_n. If so, proceed to step 5; if not, proceed to step 6.

5. To differentiate among the point groups D_n, D_{nh}, and D_{nd}, determine whether a horizontal mirror plane (σ_h) is perpendicular to the C_n axis. If a σ_h is present, the point group is D_{nh}. If not, determine whether there are *vertical* mirror planes (σ_d) containing the C_n axis that bisect pairs of horizontal C_2 axes. If σ_d planes are present, the point group is D_{nd}. If the molecule contains no mirror planes of any kind, the point group then is D_n.

6. If a molecule does not possess n horizontal C_2 axes perpendicular to the C_n axis, it must belong to a cyclic point group (viz., C_n, C_{nh}, or C_{nv}) containing only one proper rotation axis. If the C_n axis contains a horizontal mirror plane (σ_h), the point group is C_{nh}. If the

C_n axis possesses n *vertical* mirror planes intersecting the C_n axis, the molecule belongs to the point group C_{nv}. If the molecule contains neither horizontal nor vertical mirror planes, the point group is C_n.

13.9 SYMMETRY PROPERTIES OF WAVE FUNCTIONS

The presence of symmetry in a molecule indicates a simplification of the mathematical description of the molecule. The wave functions for the molecule must have all of the symmetry of the molecule.

An s orbital, being spherically symmetrical, has all possible point symmetry elements. A p_x orbital of Fig. 12.12 is symmetric with respect to reflection in the xz plane or the xy plane. However, when a p_x orbital is reflected in the yz plane the wave function an equal distance on the other side of the plane has the same magnitude, *but the opposite sign*. Thus we say that the p_x wave function is *antisymmetric* with reflection in the yz plane. The p_x orbital is antisymmetric with respect to the C_2^z operation, the C_2^y operation, and the inversion operation i, but it is symmetric with respect to any rotation about the x axis, a C_∞ axis. The d_{xz} orbital of Fig. 12.10 is symmetric with respect to σ^{xz}, i, and C_2^z, and antisymmetric with respect to σ^{xy}, σ^{yz}, C_2^z, C_2^x, and C_4^y.

13.10 SYMMETRY AND DIPOLE MOMENT

A dipole moment (Section 14.10) is a vector quantity that is not affected either in direction or in magnitude by any symmetry operation of the molecule. Therefore the vector must be contained in each of the symmetry elements. Consequently, molecules that possess nonzero dipole moments belong only to the point groups C_n, C_s, and C_{nv}. The presence or absence of a dipole moment therefore tells something about the symmetry of a molecule. For example, carbon dioxide and water might have structures corresponding to a symmetrical linear molecule, to an unsymmetrical linear molecule, or to a bent molecule. The dipole moments recorded in Table 14.4 show that carbon dioxide has zero moment; therefore, the molecule must be symmetrical and linear. If it were unsymmetrical or bent, there would have been a permanent dipole moment. On the other hand, water has a pronounced dipole moment and cannot have the symmetrical linear structure.

The symmetry of the structure of benzene prohibits the existence of a permanent electric dipole moment. Symmetry requires that the dipole moment of chlorobenzene be along the direction of the carbon-chloride bond.

13.11 SYMMETRY AND OPTICAL ACTIVITY

If a molecule and its mirror image cannot be superimposed, it is potentially optically active. Since a rotation followed by a reflection always converts a right-handed object to a left-handed object, an S_n axis guarantees that a molecule cannot exist in separate left and right-handed forms.

All *improper* rotation axes (S_n) including a mirror plane $(\sigma = S_1)$ and center of symmetry $(i = S_2)$ convert a right-handed object into a left-handed object (i.e., produce a mirror image of the original object), whereas all *proper* rotation axes (C_n) leave a right-handed object unchanged in this respect. Hence only molecules having only proper rotation elements of symmetry can be optically active.

In a molecule in which internal rotation can take place (e.g., ethane or H_2O_2) it is possible to have optically active conformations, but in a gas or solution these conformers would be expected to occur in essentially equal numbers of right- and left-handed forms so that the gas or solution is rendered optically inactive.

References

I. Bernal, W. C. Hamilton, and J. S. Ricci, *Symmetry*, W. H. Freeman and Co., San Francisco, 1972.

F. A. Cotton, *Chemical Applications of Group Theory*, Wiley, New York, 1971.

L. H. Hall, *Group Theory and Symmetry in Chemistry*, McGraw-Hill Book Co., New York, 1969.

G. Herzberg, *Molecular Spectra and Molecular Structure, Vol. 2, Infrared and Raman Spectra of Polyatomic Molecules*, D. Van Nostrand Co., Princeton, N.J., 1945, pp. 1–12.

H. H. Jaffé and M. Orchin, *Symmetry in Chemistry*, Wiley, New York, 1965.

Problems

List the Schoenflies symbols and symmetry elements for each of the following molecules.

13.1 H_2S *Ans.* C_{2v}.

13.2 PCl_3 *Ans.* C_{3v}.

13.3   trans-$[CrBr_2(H_2O)_4]^+$
(ignore the H's)   *Ans.* D_{4h}.

13.4 FeF_6^{3-} *Ans.* O_h.

13.5 *gauche*-CH_2ClCH_2Cl

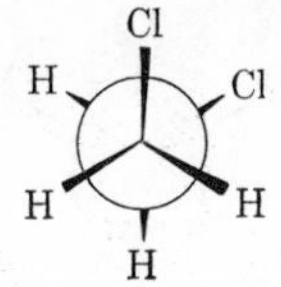

Ans. C_2.

13.6 $C_6H_3Br_3$
(1,3,5-tribromobenzene)

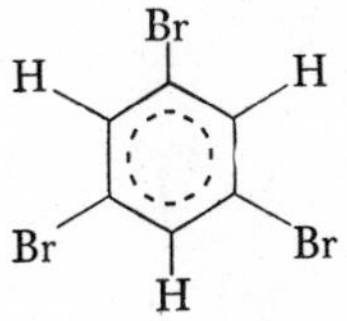

Ans. D_{3h}.

13.7 $CHClBr(CH_3)$

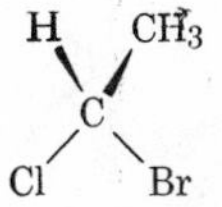

Ans. C_1.

13.8 IF_5

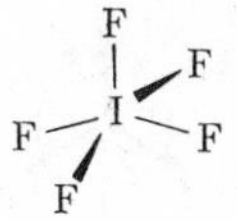

Ans. C_{4v}.

13.9 C_6H_{12}
(cyclohexane)

Ans. D_{3d}.

13.10 B_2H_6

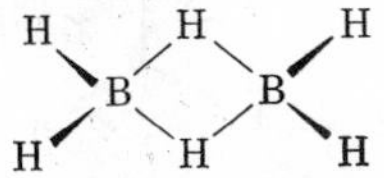

Ans. D_{2h}.

13.11 N_2

N≡N

Ans. $D_{\infty h}$.

13.12 $C_{10}H_8$
(naphthalene)

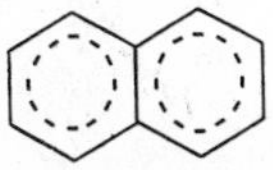

(planar)

Ans. D_{2h}.

13.13 C_5H_8
(spiropentane)

(triangles ⊥
to each other)

Ans. D_{2d}.

13.14 H_4C_4S
(thiophene)

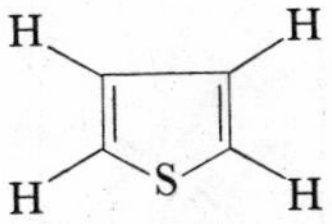

(planar)

Ans. C_{2v}.

13.15 $C_6H_4Cl_2$
(*p*-dichlorobenzene)

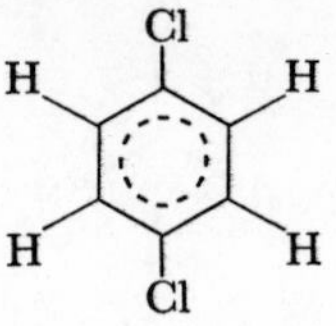

Ans. D_{2h}.

13.16 CCl_4

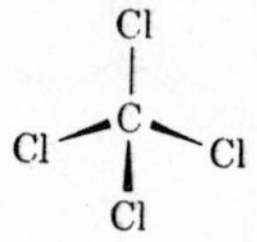

Ans. T_d.

13.17 PF_5

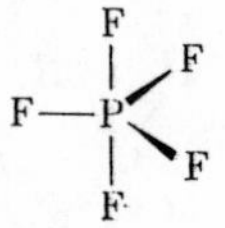

Ans. D_{3h}.

13.18 trans-CFClBrCFClBr

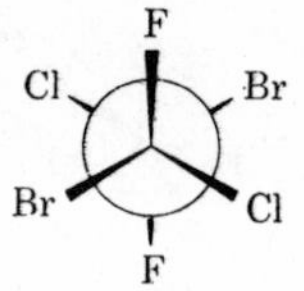

Ans. C_i.

13.19 HCN

H—C≡N

Ans. $C_{\infty v}$.

13.20 C_2F_6

Ans. D_{3d}.

13.21 $Cr(CO)_6$

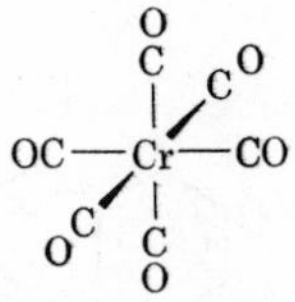

13.22 C_8H_8
(cyclooctatetraene)

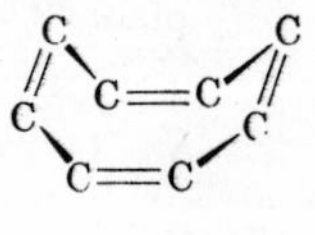

13.23 HCO_2H

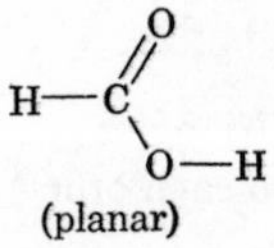

(planar)

13.24 $UO_2F_5^{3-}$

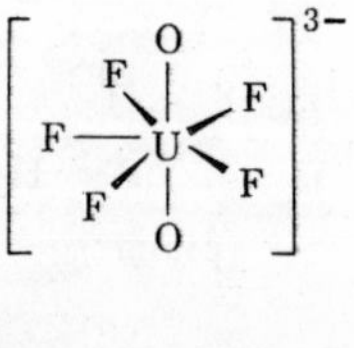

13.25 $C_{14}H_{10}$
(phenanthrene)

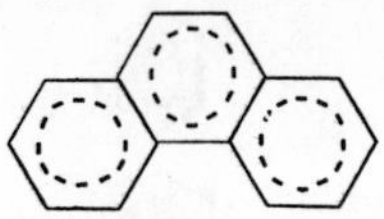

13.26 C_6Cl_6

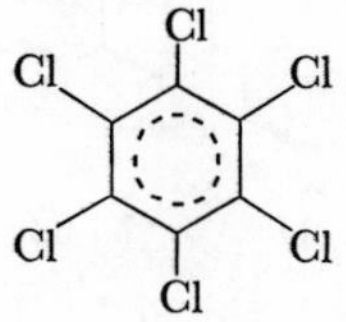

13.27 S_8

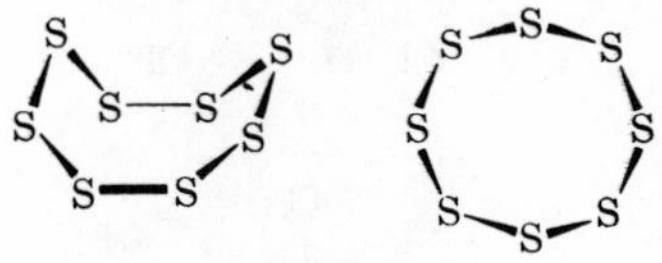

13.28 $Fe(CN)_6^{3-}$

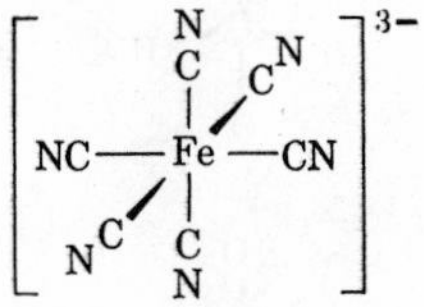

13.29 $[HNBCl]_3$ 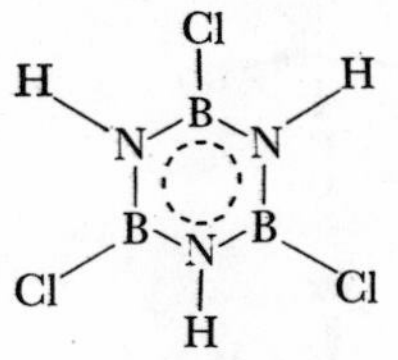   (planar)

13.30 $C_{10}H_{16}$
(adamantane)   (ignoring the H's)

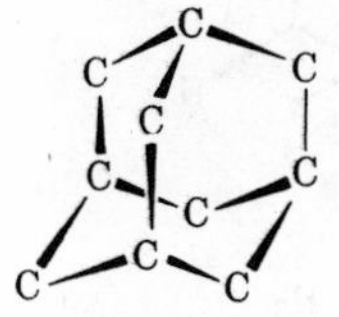

13.31 $C_6H_2O_2Cl_2$
(2,5-Dichloroquinone)   (planar)

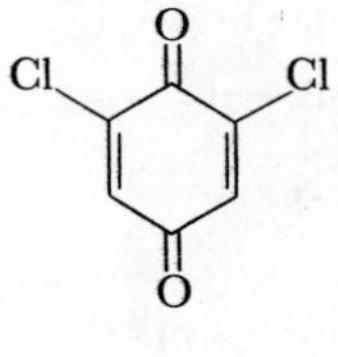

13.32 HOCl

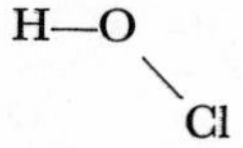

13.33 $[HNBH]_3$ 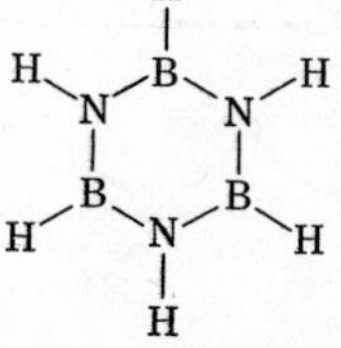   (planar)

13.34 $C_6H_3(C_6H_5)_3$
(1,3,5-Triphenylbenzene)
(not planar)

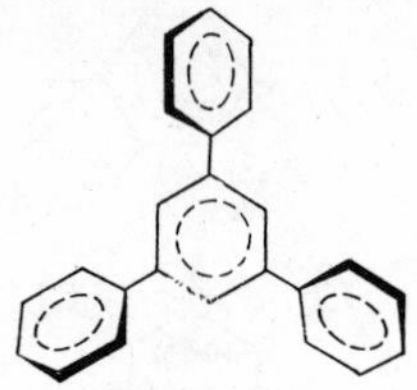

13.35 cis-CHCl=CHCl

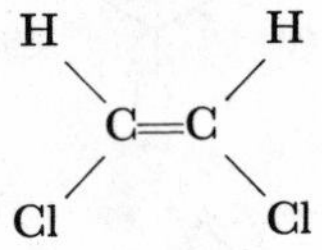

13.36 CH_3Cl
(methyl chloride)

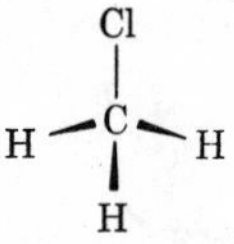

13.37 $Ni(CO)_4$

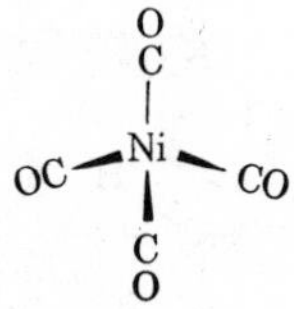

13.38 H_3CCH_2Br

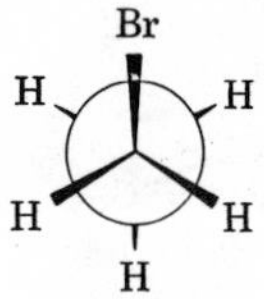

13.39 $PtBr_4(NH_3)_2$
(tetrabromodiammine-
platinum (IV))
(ignore the H's)

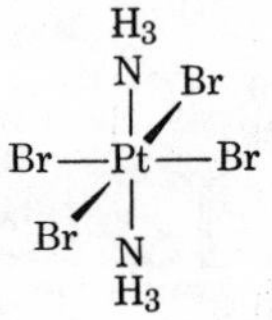

13.40 B_2Cl_4

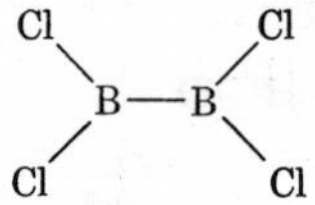

13.41 $(CH_2)_2O$
(ethylene oxide)

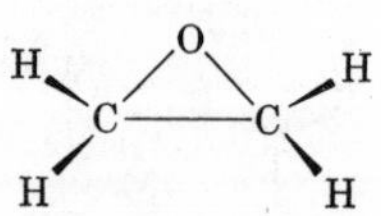

13.42 $C_4H_4N_2$
(pyrazine)   (planar)

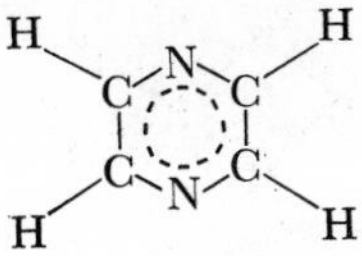

13.43 $Hg_3Cl_3O^+$
(planar)

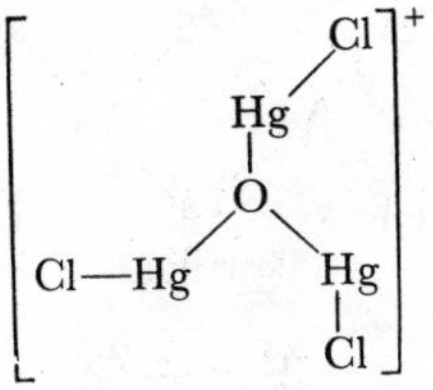

13.44 $CH_2BrCHBr_2$

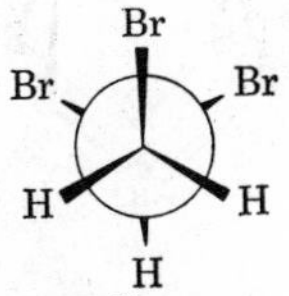

13.45 *trans*-CH_2ClCH_2Cl

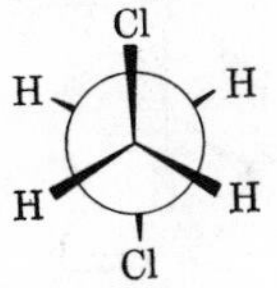

13.46 $TeCl_4$

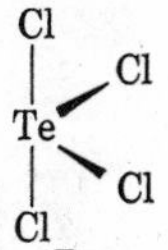

13.47 Al_2Br_6

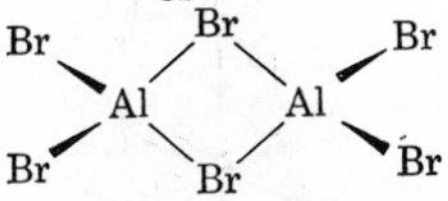

13.48 P_4S_3

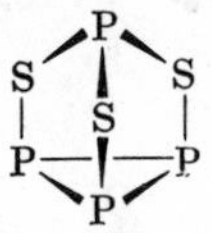

13.49 P_4O_6

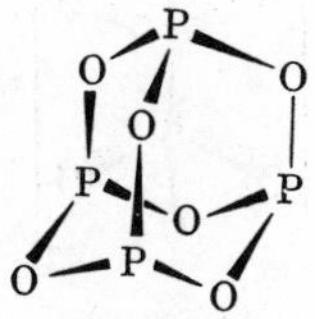

13.50 $[Tl_2Cl_9]^{3-}$

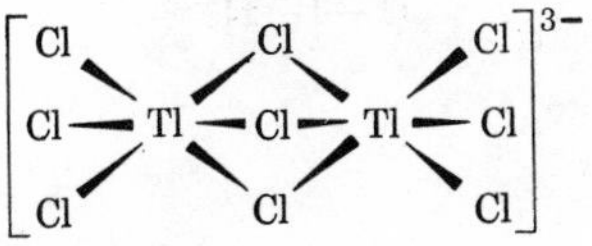

13.51 $[B_3O_6]^{3-}$
(planar)

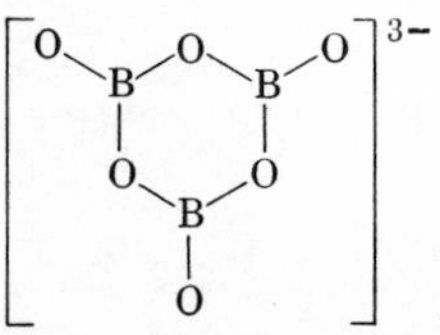

13.52 B_2H_5Br

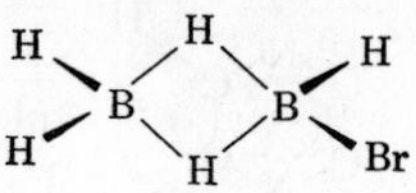

13.53 Au_2Cl_6

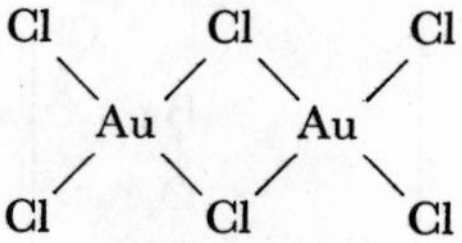

(planar)

13.54 As_4S_4

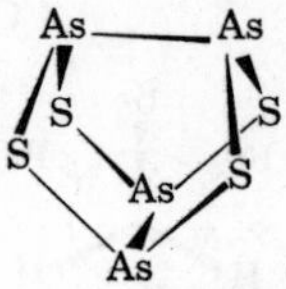

13.55 IOF_5

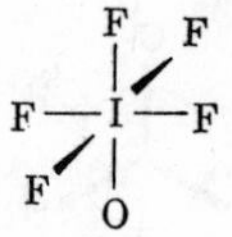

13.56 OF_2

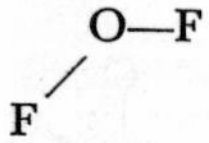

13.57 S_2F_{10}

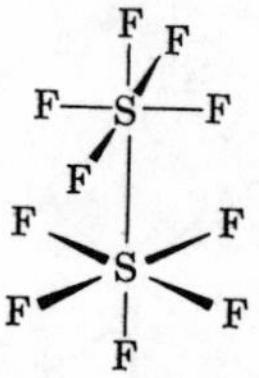

13.58 $[S_2O_6]^{2-}$

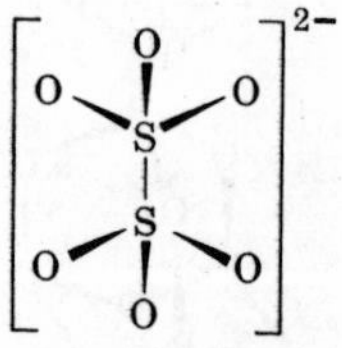

13.59 $[I_3]^-$

$[I-I-I]^-$

CHAPTER 14

MOLECULAR ELECTRONIC STRUCTURE

Quantum mechanics has made it possible to understand the nature of chemical bonding and to predict the structures and properties of simple molecules. Our ideas about covalent bonds go back to 1916 when Lewis described the sharing of electron pairs between atoms. The pairs of electrons held jointly by two atoms were considered to be effective in completing a stable electronic configuration for each atom. This approach provided only a qualitative picture of chemical bonding. Today the electronic structures of simple molecules may be calculated with a high degree of accuracy, and energy levels, bond angles, bond distances, dipole moments, and other properties may be calculated. For molecules with a larger number of electrons approximations have to be introduced. However, even approximate calculations are very helpful in understanding molecular structure, chemical properties, and molecular spectra.

14.1 MOLECULAR WAVE FUNCTIONS

The methods of quantum mechanics may be applied to molecules in much the same way as we have discussed for atoms in Chapter 12 to obtain energy levels and electron densities. A wave function with variable parameters is improved by use of the variation method (Section 12.21), and then this wave function is used to calculate various properties of the molecule.

In making quantum mechanical calculations on molecules it is assumed that the total wave function ψ may be factored into an electronic wave function ψ_e and a nuclear wave function ψ_N.

$$\psi = \psi_e \psi_N \tag{14.1}$$

This approximation, introduced by Born and Oppenheimer, comes from the fact that nuclei are much heavier than electrons and move much more slowly. Therefore the Schrödinger equation may be solved for the electronic wave function with fixed nuclei.

$$\mathscr{H}_e \psi_e = E_e \psi_e \tag{14.2}$$

This process yields the energy eigenvalues E_e for the molecule with the nuclei at fixed distances apart. The electronic energies E_e obtained for different possible positions of the N nuclei form a potential energy surface, and the wave function

455

ψ_{N} describing the motion of the nuclei is obtained from the Schrödinger equation for the nuclear motion.

$$\mathscr{H}_{\mathrm{N}}\psi_{\mathrm{N}} = E_{\mathrm{N}}\psi_{\mathrm{N}} \tag{14.3}$$

In the remainder of this chapter we will be concerned only with electronic wave functions, and so the subscript e will be dropped.

14.2 THE HYDROGEN MOLECULE ION H_2^+

Although the role of electrostatic attraction in bonding of ions, as in gaseous $\mathrm{Na^+Cl^-}$, was understood prior to the development of quantum mechanics, the origin of covalent bonding was obscure. Basically the quantum mechanical explanation of covalent bonding is that the wave nature of the electron leads to a buildup of electronic charge in the region between the nuclei.

The hydrogen molecule ion H_2^+ is the simplest example of a molecule since it has a single bonding electron. This molecular ion is formed in electrical discharges through hydrogen, and its properties are well known from spectroscopic measurements. The equilibrium internuclear distance is 1.06 Å, and the dissociation energy* is 2.78 eV or 64.10 kcal mol^{-1}. The energy levels and electron densities of the hydrogen molecule ion can be calculated to any desired degree of accuracy because the Schrödinger equation may be solved exactly. The coordinates of the three particles involved are defined in Fig. 14.1. Using the Born-Oppenheimer approximation the Hamiltonian is

$$\mathscr{H} = -\frac{\hbar^2}{2\,m_e}\nabla^2 + \frac{1}{4\pi\epsilon_0}\left(-\frac{e^2}{r_a} - \frac{e^2}{r_b} + \frac{e^2}{r_{ab}}\right) \tag{14.4}$$

After a change to more convenient coordinates that we will not go into, it is possible to solve the Schrödinger equation using this Hamiltonian for a series of values of the internuclear distance r_{ab}. For each internuclear distance r_{ab} there is a series of energy eigenvalues; the lowest is that of the ground state and, in principle, there is an infinite number of excited states. The potential energy curves for the ground state and a number of excited states of H_2^+ are shown in Fig. 14.2. The lowest curve (labeled $\sigma_g 1s$) shows that a molecule can be formed with internuclear distance 1.06 Å and a dissociation energy (into a proton and a hydrogen atom) of 2.78 eV in exact agreement with the experiment. The potential energy curve labeled $\sigma_u^* 1s$ does not lead to a stable molecule. A hydrogen molecule ion in this state immediately dissociates into a proton and a hydrogen atom.

Although the Schrödinger equation may be solved exactly for H_2^+ the addition of more electrons or nuclei requires the introduction of approximations. It is, therefore, of interest to see how a simple molecular orbital method works for H_2^+.

---

* The energy of an electron accelerated by a potential difference of one volt is 1.60219×10^{-19} J. Multiplying by the Avogadro constant yields the energy of a mole of electrons, which is 96,485 J mol^{-1}, and dividing by 4.184 J cal^{-1} yields 23,060 cal mol^{-1} eV^{-1}.

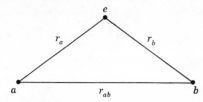

Fig. 14.1   Coordinates in H$^+$. Two protons are represented by a and b.

In the molecular orbital approach to H$_2^+$ we start out with the idea that if the two protons were far apart each could have an electron in the $1s$ orbital. Therefore, we assume that as a first approximation the electronic wave function for H$_2^+$ is

$$\psi = c_1(1s_a) + c_2(1s_b) \tag{14.5}$$

where $1s_a$ and $1s_b$ represent the $1s$ wave functions associated with protons a and b and the constants c_1 and c_2 are to be evaluated by the variation method (Section 12.21). Such a wave function is referred to as a LCAO function because it is a linear combination of atomic orbitals.

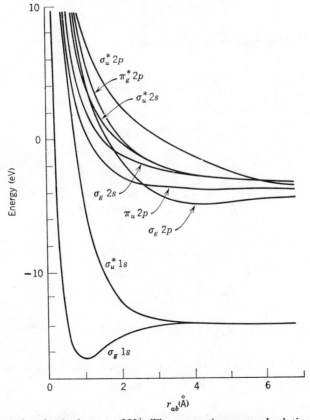

Fig. 14.2   Ground and exited states of H$_2^+$. The energy is measured relative to completely separated protons and one electron. (From J. C. Davis, *Advanced Physical Chemistry*, The Ronald Press Co., New York, 1965.)

The variation energy E is given by

$$E = \frac{\int \psi^* \mathscr{H} \psi \, d\tau}{\int \psi^* \psi \, d\tau}$$

$$= \frac{\int [c_1(1s_a) + c_2(1s_b)] \mathscr{H} [c_1(1s_a) + c_2(1s_b)] \, d\tau}{\int [c_1(1s_a) + c_2(1s_b)]^2 \, d\tau}$$

$$= \frac{c_1^2 H_{aa} + 2c_1 c_2 H_{ab} + c_2^2 H_{bb}}{c_1^2 S_{aa} + 2c_1 c_2 S_{ab} + c_2^2 S_{bb}}$$

$$= \frac{c_1^2 H_{aa} + 2c_1 c_2 H_{ab} + c_2^2 H_{aa}}{c_1^2 + 2c_1 c_2 S + c_2^2} \tag{14.6}$$

where the following symbols have been used

$$H_{aa} = \int (1s_a) \mathscr{H} (1s_a) \, d\tau = H_{bb} \tag{14.7}$$

$$H_{ab} = \int (1s_a) \mathscr{H} (1s_b) \, d\tau \tag{14.8}$$

$$H_{ba} = \int (1s_b) \mathscr{H} (1s_a) \, d\tau = H_{ab} \tag{14.9}$$

$$H_{bb} = \int (1s_b) \mathscr{H} (1s_b) \, d\tau = H_{aa} \tag{14.10}$$

$$S_{aa} = \int (1s_a)(1s_a) \, d\tau = 1 \tag{14.11}$$

$$S_{ab} = \int (1s_a)(1s_b) \, d\tau = S_{ba} = S \tag{14.12}$$

$$S_{ba} = \int (1s_b)(1s_a) \, d\tau = S_{ab} = S \tag{14.13}$$

$$S_{bb} = \int (1s_b)(1s_b) \, d\tau = 1 \tag{14.14}$$

The integrals represented by S are referred to as overlap integrals. According to the variation method the best values of c_1 and c_2 in equation 14.5 are the ones that give the lowest value of E. Equation 14.6 is multiplied out and the derivative taken with respect to c_1 and c_2. This yields

$$\frac{\partial E}{\partial c_1} = \frac{2c_1(H_{aa} - E) + 2c_2(H_{ab} - SE)}{c_1^2 + 2c_1c_2S + c_2^2} = 0 \tag{14.15}$$

$$\frac{\partial E}{\partial c_2} = \frac{2c_1(H_{ab} - SE) + 2c_2(H_{bb} - E)}{c_1^2 + 2c_1c_2S + c_2^2} = 0 \tag{14.16}$$

since these derivatives equal zero at the minimum energy. The values of c_1 and c_2 are obtained by solving the two simultaneous equations

$$c_1(H_{aa} - E) + c_2(H_{ab} - SE) = 0 \tag{14.17}$$

$$c_1(H_{ab} - SE) + c_2(H_{bb} - E) = 0 \tag{14.18}$$

There is a nontrivial solution for c_1 and c_2 from these simultaneous equations only if the determinant of the coefficients of c_1 and c_2 is equal to zero.

$$\begin{vmatrix} H_{aa} - E & H_{ab} - SE \\ H_{ab} - SE & H_{aa} - E \end{vmatrix} = 0 \tag{14.19}$$

This equation is referred to as a secular determinant. In this case it is a quadratic with two solutions E_g and E_u.

$$E_g = \frac{H_{aa} + H_{ab}}{1 + S} \tag{14.20}$$

$$E_u = \frac{H_{aa} - H_{ab}}{1 - S} \tag{14.21}$$

The integral H_{aa} is called the coulomb integral because the difference between H_{aa} and the energy of a single hydrogen atom is just that of the coulomb interaction of $(1s_a)^2$ with nucleus b. Since $(1s_a)^2$ represents a distribution of negative charge while the nucleus is positive, H_{aa} will be a negative number. The integral H_{ab} is referred to as the resonance integral. When equations 14.20 and 14.21 are substituted back into equation 14.17 it is found that

$$c_1 = \pm c_2 \tag{14.22}$$

Substituting this relation into equation 14.5 and normalizing yields

$$\psi_g = \frac{1}{[2(1 + S)]^{\frac{1}{2}}} [(1s_a) + (1s_b)] \tag{14.23}$$

$$\psi_u = \frac{1}{[2(1 - S)]^{\frac{1}{2}}} [(1s_a) - (1s_b)] \tag{14.24}$$

Because of the symmetry of H_2^+ it is not surprising that the orbitals of the two protons are weighted equally. The first wave function ψ_g is symmetric, that is, inverting through the midpoint of the two nuclei does not change the magnitude or the sign of the wave function. Therefore this wave function is designated with a g, which stands for *gerade* (German for even). The second wave function ψ_u is antisymmetric and is designated by a u for *ungerade* (German for uneven).

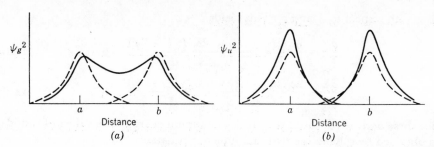

Fig. 14.3 Hydrogen molecule ion. (a) Probability density ψ_g^2 for the even wavefunction along a line passing through the two nuclei. (b) Probability density ψ_u^2 for the uneven wave function. In both parts the dashed lines give the electron densities for the two protons separately.

The electron probability densities along the line of the two nuclei are given for ψ_g and ψ_u in Fig. 14.3 compared with the probability densities expected for two separate hydrogen atoms in the $1s$ state. The squares of the wave functions

$$\psi_g^2 = \frac{1}{2(1 + S)} \left[(1s_a)^2 + 2(1s_a)(1s_b) + (1s_b)^2 \right] \tag{14.25}$$

$$\psi_u^2 = \frac{1}{2(1 - S)} \left[(1s_a)^2 - 2(1s_a)(1s_b) + (1s_b)^2 \right] \tag{14.26}$$

show that the electron density is increased between the nuclei with ψ_g and decreased with ψ_u. In fact, with ψ_u^2 the electron density is zero in a plane bisecting the line between the protons, that is, there is a node midway between the protons.

It is the buildup of electron density between the protons with ψ_g^2 that is responsible for the bonding. Since this orbital is symmetrical around the internuclear axis it is referred to as a sigma orbital, and since it is even (gerade) and is made up of two $1s$ orbitals it is designated $\sigma_g 1s$.

For the wave function ψ_u the nuclei repel each other, and so this molecular orbital is referred to as an *antibonding orbital*. This antibonding orbital is symmetrical about the internuclear axis and is referred to as a σ^*1s orbital, where the * reminds us that it is an antibonding orbital. The energy for ψ_u is evidently *higher* than that for ψ_g, and so H_{ab} must be negative.

Further molecular orbitals may be constructed using $2s$, $2p$, and other orbitals. The angular momentum about the internuclear axis is given by $\lambda \hbar$ and the following symbols are used to represent λ:

$$
\begin{array}{cccc}
\lambda & 0 & 1 & 2 \\
\text{orbital} & \sigma & \pi & \delta
\end{array}
$$

The value of E_g calculated from equation 14.20 by carrying out the indicated integrations for different values of r_{ab} does show a minimum, and so this very simple molecular orbital theory does account for the bonding in H_2^+. However it yields an internuclear distance of 1.32 Å (actual 1.06 Å) and a dissociation energy of 1.77 eV or 40.82 kcal mol^{-1} (actual 2.78 eV or 64.10 kcal mol^{-1}). The calculations can be improved by adding further terms to the wave functions and using the variational method to evaluate further adjustable parameters. In this way the potential energy curve for H_2^+ can be calculated to any desired degree of accuracy.

14.3   THE HYDROGEN MOLECULE

When the Born-Oppenheimer approximation is used the Hamiltonian for the hydrogen molecule may be written

$$\mathscr{H} = -\frac{\hbar^2}{2 m_e}(\nabla_1^2 + \nabla_2^2) + \frac{1}{4\pi\epsilon_0}\left(-\frac{e^2}{r_{a1}} - \frac{e^2}{r_{a2}} - \frac{e^2}{r_{b1}} - \frac{e^2}{r_{b2}} + \frac{e^2}{r_{ab}} + \frac{e^2}{r_{12}}\right) \quad (14.27)$$

where the coordinates are defined in Fig. 14.4. The Schrödinger equation with this Hamiltonian cannot be solved exactly, but it may be solved to as high a degree of approximation as desired by use of the variation method. The potential energy curves for the ground and excited states are given in Fig. 14.5. The dissociation energy of H_2(4.7451 eV) is greater than that of H_2^+ (2.78 eV), and the bond length in H_2 (0.740 Å) is shorter than that in H_2^+ (1.060 Å).

In order to illustrate how simple molecular orbital theory is used with more than one electron, we will extend the treatment of H_2^+ given in the preceding section to H_2.

If electron-electron repulsion is ignored, the probability $\psi(1)^2 d\tau$ of finding electron 1 in a volume $d\tau$ is independent of the probability $\psi(2)^2 d\tau$ of finding electron 2 in a volume $d\tau$. The probability of simultaneously finding electron 1 and electron 2 in volume $d\tau$ is the product of the individual probabilities or $\psi(1)^2\psi(2)^2 d\tau^2$. Therefore it seems reasonable that the wave function that would describe the electron distribution of a hydrogen molecule without electron-electron repulsion is the product of the wave functions for the two

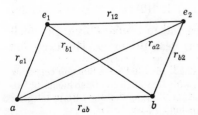

Fig. 14.4   Coordinates in the hydrogen molecule. The two protons are represented by a and b.

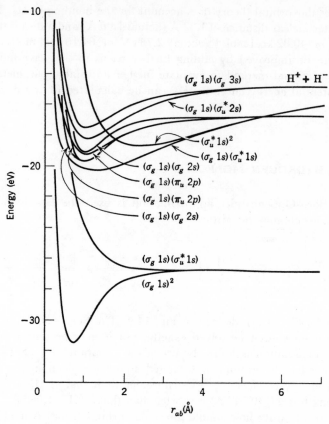

Fig. 14.5   Potential energy curves for ground and excited states of H_2. (From J. C. Davis, *Advanced Physical Chemistry*, The Ronald Press Co., New York, 1965.)

electrons; $\psi = \psi(1)\psi(2)$. As a first approximation we can put two electrons into the H_2^+ LCAO-MO orbital (equation 14.23).*

$$\psi_{\mathrm{MO}} = \psi_g(1)\psi_g(2) \tag{14.28}$$

$$\psi_{\mathrm{MO}} = \frac{1}{2(1+S)} \left[(1s_a) + (1s_b)\right](1)\left[(1s_a) + (1s_b)\right](2)$$

$$= \frac{1}{2(1+S)} \left[(1s_a)(1)(1s_b)(2) + (1s_a)(2)(1s_b)(1)\right.$$

$$\left. + (1s_a)(1)(1s_a)(2) + (1s_b)(1)(1s_b)(2)\right] \tag{14.29}$$

The first two terms in this last expression correspond to configurations in which one electron is near one proton and the other is near the other. However, the

---

* As discussed in the case of the helium atom (Section 12.23) the total wave function consisting of a spatial part times a spin part must be antisymmetric with respect to the interchange of two electrons. However, the spin function is relatively unimportant in evaluating the various integrals encountered in the variation method, and so we will not include it here.

third and fourth terms correspond to configurations in which both electrons 1 and 2 are near the same proton; that is, they correspond to the ionic structures $H_a^- H_b^+$ and $H_a^+ H_b^-$. The wave function ψ_{MO} leads to a dissociation energy for H_2 of 2.681 eV and an internuclear distance of 0.85 Å. Although this agreement with experiment is not very good, the simple LCAO-MO wave function given by equation 14.29 can be improved by introducing an effective nuclear charge and including excited H_2^+-like MOs in the trial function. When properties are calculated in this way it is not possible to make simple interpretations of the terms of the wave function, but it does become possible to calculate the bond length and binding energy more accurately than they can be measured.

Another simple theory for chemical bonding is the valence-bond (VB) method. In this method complete atoms are brought together and allowed to interact. Whereas in the molecular-orbital theory all electrons are considered to belong to the whole molecule, in the valence-bond theory a pair of electrons is considered to belong to the pair of atoms that is bonded together.

The valence-bond theory leads to a different trial wave function for H_2. If two hydrogen atoms a and b are infinitely far apart, the system is described by the wave function

$$\psi = (1s_a)(1)(1s_b)(2) \tag{14.30}$$

where 1 designates the electron near proton a, and 2 designates the electron near proton b. The square of the wave function ψ is the probability density function for the system which is equal to the product of two probability densities $[(1s_a)(1)]^2$ and $[(1s_b)(2)]^2$. Thus the probability of finding electron 1 in a given volume and electron 2 in another specified volume is equal to the product of two probabilities.

As the two atoms are brought closer together the system might continue to be described by wave function 14.30 except that the electrons are indistinguishable. This indistinguishability of the electrons is taken into account by using as a wave function the sum of the two wave functions which are equally likely to be correct.

$$\psi_{VB} = (1s_a)(1)(1s_b)(2) + (1s_a)(2)(1s_b)(1) \tag{14.31}$$

This wave function is the basis of the Heitler-London treatment of 1927, which gave the first reasonable theoretical values for the dissociation energy of H_2. This wave function given by simple valence-bond theory differs from that for the simple molecular-orbital theory by lack of the "ionic terms." The simple molecular-orbital theory overemphasizes the ionic terms, and the simple-valence theory lacks these terms. The VB wave function can be improved by adding contributions from these ionic structures, and the MO wave function can be improved by providing for a smaller contribution by the ionic structures. Thus better results can be obtained with

$$\psi = c_1 \psi_{VB} + c_2 \psi_{ionic} \tag{14.32}$$

where

$$\psi_{ionic} = (1s_a)(1)(1s_a)(2) + (1s_b)(1)(1s_b)(2) \tag{14.33}$$

Thus with further improvement the distinction between the two methods disappears, and the same result is obtained from either starting point. Another way of improving the MO wave function, which again leads to the same result, is to "mix in" the excited-state wave function $\sigma_u^* 1s$, which is given by

$$\psi'_{\text{MO}} = C[(1s_a)(1) - (1s_b)(1)][(1s_a)(2) - (1s_b)(2)] \qquad (14.34)$$

to obtain the wave function

$$\psi = c_3 \psi_{\text{MO}} + c_4 \psi'_{\text{MO}} \qquad (14.35)$$

By bringing in other functions and changing the effective nuclear charge in the wave functions it is possible to get closer and closer to the experimental values for the dissociation energy and the equilibrium internuclear distance.

In the ground state of the hydrogen molecule the net electron spin for the molecule is zero. This is a *singlet* state and the energy level is not split in an electric or magnetic field. The ground states of most molecules containing an even number of electrons are singlet states. A hydrogen molecule with one electron in the bonding orbital and another with the same spin in the antibonding orbital is in a *triplet* state (Section 12.23).

The bond between two hydrogen atoms is exceptional in that the principal source of repulsion energy is due to the electrostatic repulsion of the protons. For a covalent bond not involving a hydrogen atom, each nucleus is shielded by a core of inner electrons. A repulsion arises when these cores begin to interpenetrate. I1 a third electron is added to the bond giving H_2^-, it must be added to the antibonding orbital, as required by the Pauli exclusion principle. Since the antibonding orbital has a much higher energy, this leads to a high energy for H_2^- and instability with respect to $H_2 + e$.

In order to compare the dissociation energy of H_2 calculated quantum mechanically with values obtained from spectroscopic measurements, it is necessary to understand the relation between the potential energy curve for the molecule and the zeroth vibrational level.

14.4 POTENTIAL ENERGY CURVE OF THE HYDROGEN MOLECULE

Figure 14.6 gives the potential energy of the ground state of the hydrogen molecule as a function of the distance between the protons. The potential energy is taken as zero at the minimum of the potential energy curve, but the distance between protons in a hydrogen molecule would not remain constant at this value even at absolute zero. The Heisenberg uncertainty principle (Section 12.9) would be violated if the vibrational momentum was zero and the internuclear distance was constant. In the treatment of the harmonic oscillator (Section 12.15) we found that the lowest energy eigenvalue is $h\nu_0/2$ where ν_0 is the vibrational frequency. At absolute zero a harmonic oscillator has this "zero point energy."

Because of the zero point energy we have to be careful to distinguish between three dissociation energies: (1) the dissociation energy D' referred to earlier in this chapter, which is measured from the minimum of the potential energy curve,

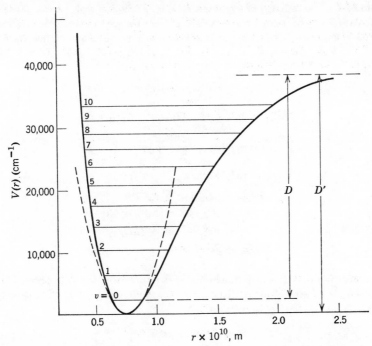

Fig. 14.6 Potential-energy diagram of the ground state of the hydrogen molecule with the first 10 vibrational levels shown. The dashed curve shows the parabola that fits the potential-energy curve at the minimum.

(2) the spectroscopic dissociation energy D measured from the zero point vibrational level, and (3) the thermodynamic dissociation energy measured at some particular temperature where molecules are in various vibrational, rotational, and perhaps even electronic states. As seen from Fig. 14.6 the spectroscopic dissociation energy D is smaller than D' by the zero point energy.

$$D' = D + \tfrac{1}{2}h\nu_0 \tag{14.36}$$

Example 14.1 The spectroscopic dissociation energy D for the hydrogen molecule is 103.24 kcal mol^{-1}. Calculate the dissociation energy D' measured from the minimum of the potential energy curve, given that the fundamental vibration frequency is 4395 cm^{-1}.

$D' = 103.24$ kcal mol^{-1} +

$$\frac{(6.626 \times 10^{-34} \text{ J s})(3 \times 10^8 \text{m s}^{-1})(4395 \text{ cm}^{-1})(100 \text{ cm m}^{-1})(6.022 \times 10^{23} \text{ mol}^{-1})}{2(4.184 \text{ J cal}^{-1})(10^3 \text{ cal kcal}^{-1})}$$

$= (103.24 + 6.29)$ kcal mol^{-1}

$= 109.53$ kcal mol^{-1}

$$\frac{109.53 \text{ kcal mol}^{-1}}{23.060 \text{ kcal mol}^{-1} \text{ eV}^{-1}} = 4.750 \text{ eV}$$

Example 14.2   The enthalpy of formation H(g) is 52.089 kcal mol^{-1} at 25° (Section 1.18). Calculate the enthalpy of formation of H(g) from the spectroscopic dissociation energy of 103.24 kcal mol^{-1}. To simplify the calculation it may be assumed that C_V for H(g) is 3 cal K^{-1} mol^{-1} from 0–298 K and C_V for H$_2$(g) is 3 cal K^{-1} mol^{-1} from 0–100 K and 5 cal K^{-1} mol^{-1} from 100–298 K (See Section 17.15).

$$H_2(g) = 2H(g)$$

$$\Delta U_{298}^\circ = \Delta U_0^\circ + \int_0^{298} (2C_{V,H} - C_{V,H_2})\, dT$$

$$= 103{,}240 + \int_0^{100} 3\, dT + \int_{100}^{298} dT$$

$$= 103{,}740 \text{ cal}$$

$$\Delta H_{298}^\circ = \Delta U_{298}^\circ + \sum \nu_i RT$$

$$= 103{,}740 + (1.987)(298)$$

$$= 104{,}330 \text{ cal} = 2\Delta H_{f,H}^\circ$$

$$\Delta H_{f,H}^\circ = 52.17 \text{ kcal}$$

It is convenient to have a mathematical expression for the potential energy V as a function of internuclear distance r. An approximate function that is useful because it is simple is that due to Morse.

$$V = D'[1 - e^{-\beta(r-r_e)}]^2 \tag{14.37}$$

where r_e is the equilibrium internuclear distance, that is, the distance corresponding with the minimum of the potential curve.

The constant β, which determines the steepness of the potential-energy curve, is related to the fundamental vibrational frequency ν_0 and the reduced mass μ of the diatomic molecule by

$$\beta = \nu_0\left(\frac{2\pi^2\mu}{D'hc}\right)^{1/2} \tag{14.38}$$

when D' is given in cm^{-1}. Thus the potential-energy curve for a diatomic molecule may be estimated if experimental values of ν_0, r_e, and D' are available. For hydrogen, $\nu_0 = 4395$ cm^{-1}, $r_e = 0.7412 \times 10^{-8}$ cm, and $D' = 38{,}300$ cm^{-1}, and these parameters have been used to calculate Fig. 14.6.

The Morse function is an empirical one which has qualitatively the right shape. It does not, however, accurately reproduce the potential energy curve for most molecules.

14.5 ELECTRON CONFIGURATIONS OF HOMONUCLEAR DIATOMIC MOLECULES

From the quantum mechanical treatments of the hydrogen molecule ion and the hydrogen molecule we conclude that bonding occurs where there is overlap of atomic orbitals. In order to treat the bonding of diatomic molecules with more electrons we need to consider the possibility of forming bonds with other atomic orbitals. Not all combinations lead to bonding. First it can be shown that only atomic orbitals of approximately the same energies can be combined. Second the greater the overlap the greater the strength of the bond. Overlap increases as atoms are brought closer together, but at close distances repulsion arises because of the electrostatic repulsions of the inner electron clouds and of the two positively-charged nuclei.

As we discovered earlier when two atomic orbitals are combined two molecular orbitals are formed—a bonding orbital and an antibonding orbital. Antibonding orbitals are designated with a *. Figure 14.7 shows the various possibilities for combining s and p orbitals. As we have seen the combination of two $1s$ orbitals leads to a $\sigma_g 1s$ bonding orbital and a $\sigma_u^* 1s$ antibonding orbital. The $2s$ atomic orbitals combine in a similar way. The spherical node surrounding each nucleus in this case may be neglected because the orbital overlap is not significant this close to the nucleus.

An s orbital does not form a molecular orbital with a p_y or p_z orbital, that is, with a p orbital that is perpendicular to the internuclear axis. The overlap integral is equal to zero because the positive contribution due to one-half of the p orbital is exactly compensated by a negative contribution due to the other half of the p orbital. However, a p_x orbital can combine with an s orbital as shown in Fig. 14.7.

There are two different ways of forming bonding orbitals from p orbitals. If the lobes of the p orbitals are pointed at each other along the internuclear axis two σ orbitals are formed. In contrast with the other LCAO-MOs the wave function for the bonding orbital has a negative sign. If the lobes of the p orbitals are perpendicular to the internuclear axis (p_y and p_z) they can overlap sideways to form π orbitals. The $\pi_u 2p$ orbital produces bonding because there is electron density that tends to draw the two nuclei together, even though it is not on the internuclear axis. There are two $\pi_u 2p$ bonding orbitals and two $\pi_u^* 2p$ antibonding orbitals because there are two p_y and two p_z atomic orbitals on the two nuclei. Thus six molecular orbitals are formed from the six $2p$ orbitals on the two nuclei, three bonding and three antibonding.

Orbitals of still higher energy can be formed from $2s$, $3p$, $3d$, and higher atomic orbitals, but we have carried the process far enough for the discussion of possible homonuclear diatomic molecules from He_2 to Ne_2. This simple molecular orbital theory helps us understand whether stable molecules are formed and tells us something about the relative values of binding energies and bond lengths.

The electronic structure of molecules can be discussed in terms of the aufbau principle (Section 12.25) that is used to explain the periodic table. Following the Pauli principle that there may be only two electrons in an orbital, electrons are placed in orbitals starting with the lowest. In order to estimate the relative energies of various molecular orbitals of homonuclear diatomic molecules it is convenient to use the correlation diagram given in Fig. 14.8. In constructing this diagram two separated atoms in the indicated electronic state are imagined to be pushed together until their nuclei coincide, in other words until a *united*

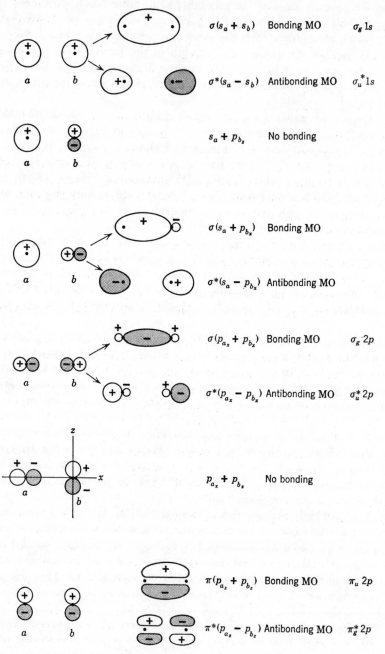

Fig. 14.7 Combinations of orbitals on nuclei a and b to form bonding and antibonding (indicated by *) orbitals. The MO wave functions and designations are given for the molecular orbitals. The x axis goes through the two nuclei. In the shaded region the wavefunction is negative.

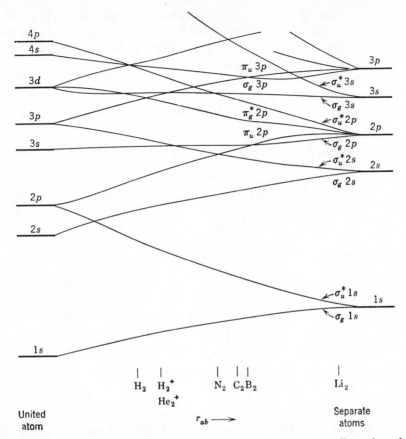

Fig. 14.8 Correlation diagram for molecular orbitals of homonuclear diatomic molecules. (From J. C. Davis, *Advanced Physical Chemistry*, The Ronald Press Co., New York, 1965.)

atom of twice the nuclear charge is formed. Figure 14.8 is based on the idea that the energies of the orbitals change smoothly in going from the separated atoms to the united atoms. The abscissa represents bond distances of homonuclear diatomic molecules. In carrying lines through to the united atom the principle is followed that molecular orbitals with a given angular momentum connect with atomic orbitals of the united atom with the same angular momentum $|m|$. The lines are also drawn so that the orbitals are either symmetric or antisymmetric with respect to a plane halfway between the nuclei or through the nucleus of the united atom. The molecular orbitals $(1s_a + 1s_b)$, $(2s_a + 2s_b)$, $(2p_{za} - 2p_{zb})$, $(2p_{xa} + 2p_{xb})$, and $(2p_{ya} + 2p_{yb})$ are all symmetric and connect in the united atom with the symmetric atomic orbitals $1s$, $2s$, $3s$, $2p_x$, and $2p_y$, respectively. The molecular orbitals $(1s_a - 1s_b)$, $(2s_a - 2s_b)$, $(2p_{za} + 2p_{zb})$, $(2p_{xa} - 2p_{xb})$, and $(2p_{ya} - 2p_{yb})$ are antisymmetric and connect in the united atom with the antisymmetric atomic orbitals $2p_z$, $3p_z$, $4p_z$, $3d_{xz}$, and $3d_{yz}$, respectively. Such a correlation diagram is only approximate and is not a completely reliable guide to the order of energies of molecular orbitals. The net number of bonding electrons in a molecule equals the number of electrons in bonding orbitals minus the number of electrons in antibonding orbitals. The covalent bond order is the net number of bonding electrons divided by two. The bond orders and other information about homonuclear diatomic molecules from H_2^+ to F_2 are given in Table 14.1.

Table 14.1 Ground States of Homonuclear Diatomic Molecules and Ions

Molecule	No. of Electrons	Configuration	Bond Order	r_{ab} Å	D' eV	D' kcal mol^{-1}
H_2^+	1	$(\sigma 1s)$	$\frac{1}{2}$	1.060	2.793	64.41
H_2	2	$(\sigma 1s)^2$	1	0.7412	4.7476	109.484
He_2^+	3	$(\sigma 1s)^2(\sigma^*1s)$	$\frac{1}{2}$	1.080	2.5	57
He_2	4	$(\sigma_g 1s)^2(\sigma^*1s)^2$	0	—	—	—
Li_2	6	$[He_2](\sigma 2s)^2$	1	2.673	1.14	26.3
Be_2	8	$[He_2](\sigma 2s)^2(\sigma^*2s)^2$	0	—	—	—
B_2	10	$[Be_2](\pi 2p)^2$	1	1.589	~ 3.0	~ 70
C_2	12	$[Be_2](\pi 2p)^4$	2	1.242	6.36	146.7
N_2^+	13	$[Be_2](\pi 2p)^4(\sigma 2p)$	$2\frac{1}{2}$	1.116	8.86	204.3
N_2	14	$[Be_2](\sigma 2p)^2(\pi 2p)^4$	3	1.094	9.902	228.35
O_2^+	15	$[N_2](\pi^*2p)$	$2\frac{1}{2}$	1.1227	6.77	156.1
O_2	16	$[N_2](\pi^*2p)^2$	2	1.2074	5.213	120.2
F_2	18	$[N_2](\pi^*2p)^4$	1	1.435	1.34	28.4
Ne_2	20	$[N_2](\pi^*2p)^4(\sigma^*2p)^2$	0	—	—	—

He_2^+

The third electron in this molecular ion is an antibonding orbital, but there is net bonding because there are two electrons in bonding orbitals. This ion is observed in electric arcs.

He_2

According to the simple molecular orbital theory a pair of electrons would occupy the $\sigma_g 1s$ orbital, and a pair would occupy the $\sigma_u^* 1s$ orbital. Since both the bonding and antibonding orbitals are filled there is no decrease in energy as compared to two isolated helium atoms, and a stable He_2 molecule is not formed.

Li_2

The additional pair of electrons beyond those for He_2 go into the $\sigma_g 2s$ orbital. Thus there is a single bond between the two nuclei.

Be_2

The additional pair of electrons beyond Li_2 goes into the $\sigma_u^* 2s$ antibonding orbital. Thus as in the case of He_2 there is no net stabilization as compared with isolated Be atoms.

B_2

According to Fig. 14.8 the additional pair of electrons beyond Be_2 would be expected to go into the $\sigma_g 2p$ orbital producing a net stabilization and the formation of a molecule in which all the electrons are paired (a singlet state). A stable B_2 molecule is formed, as indicated in Table 14.1, but spectroscopic measurements show that its ground state is a triplet (Section 12.23) so that the outer two electrons are in different $\pi 2p$ states. If there are several orbitals with the same energy, Hund's rules (Section 12.25) require that the electrons spread themselves between several orbitals. Since there are two unpaired electrons the ground state is a triplet.

C_2

The additional two electrons beyond B_2 fill the two half-vacant $\pi_u 2p$ bonding orbitals. Thus C_2 has four electrons in bonding orbitals that are not compensated by electrons in corresponding antibonding orbitals. Thus we would expect C_2 to have tighter binding and a smaller internuclear distance than B_2.

N_2

The additional pair of electrons beyond C_2 fill the remaining p bonding orbitals. Thus N_2 has a singlet ground state and a triple bond.

O_2

According to the simple molecular orbital theory the additional pair of electrons beyond N_2 would go into the $\pi_u^* 2p$ orbitals. According to Hund's rule one electron would go into each of these degenerate orbits, and their spins would be parallel. This is in accord with the facts since molecular oxygen is paramagnetic (see Section 16.1), and its ground state is a triplet.

F_2

The additional two electrons beyond O_2 fill the $\pi_u^* 2p$ orbitals so that the ground state is a singlet. Since the electron pairs in two π_u^* antibonding orbitals approximately cancel the bonding due to the electron pairs in the two π_g orbitals, the bonding is weaker than in O_2, and the internuclear distance is greater.

Ne_2

The $\sigma_u^* 2p$ orbital would be filled, and so the antibonding effects cancel the bonding effects, and there is no tendency to form a stable molecule.

Figure 14.9 displays contour diagrams for the electron densities of the stable diatomic molecules from H_2 to F_2. These electron densities were calculated by the Hartree-Fock method.

The treatment of diatomic molecules with different nuclei is complicated by the fact that the same atomic orbitals will not be involved on both atoms because of size, energy, and directional differences. In order for atomic orbitals on two atoms to be involved in bonding they must have the same σ, π, ... properties, and energies that are not too different.

In heteronuclear diatomic molecules, bonding electrons are not shared equally between the two atoms. In an extreme case like Na^+Cl^- an electron is transferred completely, and the electrostatic attraction of the two atoms makes the major contribution to the bonding, as discussed later in this chapter under ionic bonding.

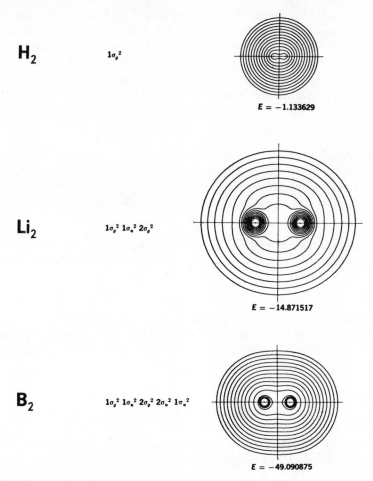

H$_2$ $1\sigma_g^2$

$E = -1.133629$

Li$_2$ $1\sigma_g^2 \, 1\sigma_u^2 \, 2\sigma_g^2$

$E = -14.871517$

B$_2$ $1\sigma_g^2 \, 1\sigma_u^2 \, 2\sigma_g^2 \, 2\sigma_u^2 \, 1\pi_u^2$

$E = -49.090875$

Fig. 14.9 Electron densities for covalent diatomic molecules. The charge density is plotted in electrons per cubic bohr. Adjacent contour lines differ by a factor of 2. The innermost contour line for each diagram is 1 electron (bohr)$^{-3}$, except for H$_2$ where it is 0.25 electron (bohr)$^{-3}$. The outermost contour line is 6.1×10^{-5} electron (bohr)$^{-3}$ in all cases. A bohr is 5.29×10^{-11} m. (*Atomic and Molecular Structure: 4 Wall Charts* by Arnold C. Wahl. Copyright © 1970 by McGraw-Hill, Inc. Used with permission of McGraw-Hill Book Co.).

The difference in affinity of different atoms in a molecule for electrons is discussed in terms of electronegativity.

14.6   ELECTRONEGATIVITY

There are several ways of estimating the tendency of an atom in a molecule to attract electrons to itself and become more negative. For example, this tendency in a molecule AB is measured by the relative stabilities of A$^+$B$^-$ and A$^-$B$^+$. The

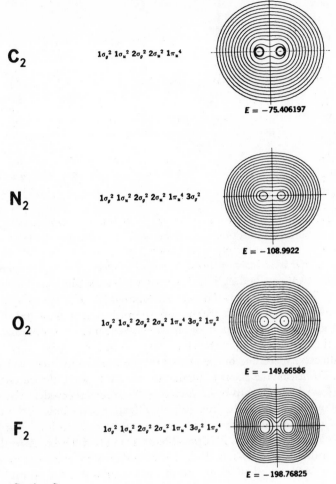

C_2 $1\sigma_g{}^2\ 1\sigma_u{}^2\ 2\sigma_g{}^2\ 2\sigma_u{}^2\ 1\pi_u{}^4$

$E = -75.406197$

N_2 $1\sigma_g{}^2\ 1\sigma_u{}^2\ 2\sigma_g{}^2\ 2\sigma_u{}^2\ 1\pi_u{}^4\ 3\sigma_g{}^2$

$E = -108.9922$

O_2 $1\sigma_g{}^2\ 1\sigma_u{}^2\ 2\sigma_g{}^2\ 2\sigma_u{}^2\ 1\pi_u{}^4\ 3\sigma_g{}^2\ 1\pi_g{}^2$

$E = -149.66586$

F_2 $1\sigma_g{}^2\ 1\sigma_u{}^2\ 2\sigma_g{}^2\ 2\sigma_u{}^2\ 1\pi_u{}^4\ 3\sigma_g{}^2\ 1\pi_g{}^4$

$E = -198.76825$

Fig. 14.9   (*Continued*)

energy difference between these structures is given by

$$E(A^+B^-) - E(A^-B^+) = (IP_A - EA_B) - (IP_B - EA_A)$$
$$= (IP_A + EA_A) - (IP_B + EA_B) \qquad (14.39)$$

where IP is ionization potential and EA is electron affinity (Section 12.26). Mulliken defined the electronegativity difference between atoms A and B as $\frac{1}{2}[E(A^+B^-) - E(A^-B^+)]$. Thus he defined the electronegativity of an atom as $\frac{1}{2}(IP + EA)$. The electronegativity of an atom depends upon its valence state and so the ionization potential and electron affinity used are not simply those of the ground state of the atom.

The most frequently used scale of electronegativities is that calculated by Pauling from thermochemical data. Values on the Mulliken scale in eV can be converted to the Pauling scale by dividing by 3.17. Exact agreement is not obtained,

		H 2.1				
Li 1.0	Be 1.5	B 2.0	C 2.5	N 3.0	O 3.5	F 4.0
Na 0.9	Mg 1.2	Al 1.5	Si 1.8	P 2.1	S 2.5	Cl 3.0
K 0.8	Ca 1.0		Ge 1.7	As 2.0	Se 2.4	Br 2.8
Rb 0.8	Sr 1.0		Sn 1.7	Sb 1.8	Te 2.1	I 2.4
Cs 0.7	Ba 0.9					

Fig. 14.10   Dependence of electronegativity on position in the periodic table.

but the scales are in relatively good agreement. Fluorine is the most electronegative atom (4.0 on Pauling's scale), and cesium is the least electronegative
atom (0.7 on Pauling's scale). The electronegativities of a number of elements
are given in Fig. 14.10, which shows that the electronegativity depends on the
position of the element in the periodic table. As we go down the halogen column
of the periodic table, the atoms become less electronegative because of the increasingly effective screening of the charge on the nucleus by inner electrons. The
alkali-metal atoms have a great tendency to lose their outer electrons and therefore have a low electronegativity. Again the electronegativity decreases as we go
down a column because of the increasingly effective screening of the charge on
the nucleus by inner electrons.

By use of electronegativities it is possible to predict which bonds will be ionic,
and which bonds will be covalent. Two elements of very different electronegativity,
like a halogen and an alkali metal, form an ionic bond because an electron is
almost completely transferred to the atom of higher electronegativity. Two
elements with nearly equal electronegativities form covalent bonds. For instance,
carbon, which occupies an intermediate position in the electronegativity scale,
forms covalent bonds with elements near it in the periodic table. If there is a
considerable difference between the electronegativities of the two elements, the
bond is polar (that is, possessed of a high degree of ionic character), as in the
case of sodium chloride. In the majority of chemical bonds the sharing of the
electron pair is not exactly equal, so that the bond has some ionic character which
results in a dipole moment (Section 14.10) for the bond.

14.7   BOND RADII

It has been found that the distance between two kinds of atoms connected by a
covalent bond of a given type (single, double, etc.) is nearly the same in different

molecules. The distance between two atoms is taken to be equal to the sum of the bond radii of the two atoms. Since the C—C bond distance is 1.54 Å in many compounds, the radius for a carbon single bond is taken to be 0.77 Å. Since the C≡C distance in acetylene is 1.20 Å, the radius for a carbon triple bond is taken to be 0.60 Å. By consideration of the bond distances in many compounds it has been possible to build up tables of bond radii, such as Table 14.2, which are useful

Table 14.2 Covalent Radii for Atoms[1] (radii in Å)

	H	C	N	O	F
Single-bond radius	0.30	0.772	0.70	0.66	0.64
Double-bond radius		0.667	0.60	0.56	
Triple-bond radius		0.603			
		Si	P	S	Cl
Single-bond radius		1.17	1.10	1.04	0.99
Double-bond radius		1.07	1.00	0.94	0.89
Triple-bond radius		1.00	0.93	0.87	
		Ge	As	Se	Br
Single-bond radius		1.22	1.21	1.17	1.14
Double-bond radius		1.12	1.11	1.07	1.04
		Sn	Sb	Te	I
Single-bond radius		1.40	1.41	1.37	1.33
Double-bond radius		1.30	1.31	1.27	1.23

[1] Taken from L. Pauling, *The Nature of the Chemical Bond*, Cornell University Press, Ithaca, N.Y., 1960, which should be consulted for details concerning the source and constancy of these radii.

in describing the structure of molecules. It must be realized, however, that the effective radius of an atom depends in part also on its structure and environment and on the nature of bonds that it forms with other atoms in the molecule.

The interatomic distances and configurations of many molecules and ions have been summarized.*

14.8 HYDROGEN BONDS

A number of unusual structures such as HF_2^- and acetic acid dimer in the gas phase (see Fig. 14.11) are evidence for the formation of hydrogen bonds. The unusually high acid dissociation constant of orthosalicylic acid, as compared with the meta and para isomers is also evidence for a hydrogen bond. A hydrogen bond results when a proton may be shared between two electronegative atoms, such as F, O, or N, which are the right distance apart. The proton of the hydrogen bond is attracted by the high concentration of negative charge in

* *Tables of Interatomic Distances and Configurations in Molecules and Ions*, The Chemical Society, London, 1958.

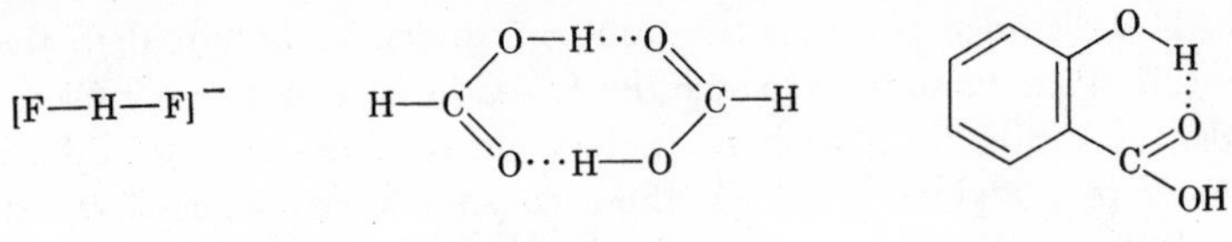

$[F{-}H{-}F]^-$

Formic acid dimer

Salicylic acid

Fig. 14.11   Examples of hydrogen bonding.

the vicinity of these electronegative atoms. Fluorine forms very strong hydrogen bonds; oxygen, weaker ones; and nitrogen still weaker ones. The unusual properties of water are due to a large extent to the formation of hydrogen bonds involving the four lone pair electrons on oxygen, as described in Section 11.6. In ice there is a tetrahedral arrangement with each oxygen atom bonded to four hydrogen atoms. Hydrogen bonds are formed along the axis of each lone pair in ice, and their existence in liquid water is responsible for the high boiling point of water as compared with the boiling points of hydrides of other elements in the same column of the periodic table (H_2S, $-62°$; H_2Se, $-42°$; H_2Te, $-4°$). When water is vaporized, these hydrogen bonds are broken, but in formic and acetic acids the hydrogen bonds are strong enough for double molecules of the type illustrated in Fig. 14.11 to exist in the vapor. Benzoic and other carboxylic acids form dimers in certain nonpolar solvents, such as benzene and carbon tetrachloride.

Hydrogen bonds between N and O are responsible for the stability of the α helix formed by polypeptides. This structure is shown in Fig. 20.3. Helices of this structure are found in protein molecules.

14.9   IONIC BONDING

When there is a large difference in the electronegativities of the atoms in a diatomic molecule, an electron may be almost completely transferred from the atom with the lower electronegativity to the atom with the higher electronegativity. This is the case for gaseous potassium chloride. The dipole moment indicates that there is essentially one positive charge on the potassium atom and one negative charge on the chlorine atom. The potassium and chloride ions are held together by an ionic bond. Ionic bonds can be treated quantitatively fairly readily. To obtain the potential energy curve for the molecule let us consider an isolated potassium atom and an isolated chlorine atom. The ionization potential of potassium is 4.34 V and so 4.34 eV is required to form $K^+ + e$. The electron affinity of a chlorine atom is 3.82 eV so that this energy is obtained when Cl^- is formed. The net energy required is $4.34 - 3.82 = 0.52$ eV. Thus the separated ions have this much more energy than the separated atoms (Fig. 14.12). When the ions are brought closer together, however, there is an electrostatic attraction. The electron distributions around K^+ and Cl^- are both spherically symmetrical, since each has the argon structure. Provided that the charge clouds due to the filled orbitals do not overlap appreciably, the energy of electrostatic attraction

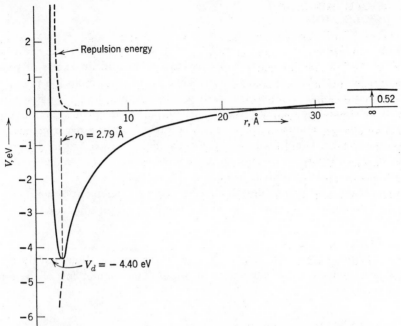

Fig. 14.12 Potential energy of the system K^+, Cl^- as a function of interatomic distance r. The zero of potential energy is taken as the energy of isolated K and Cl atoms.

measured relative to the energy at infinite separation is $-e^2/4\pi\epsilon_0 r$, where r is the distance between the centers of the ions. The electrostatic energy plus the energy required to form the two ions from the two atoms is plotted in Fig. 14.12.

As the ions approach each other a repulsion arises as soon as the closed electron shells of the ions begin to overlap appreciably. This may simply be regarded as resulting from the natural repulsion of the orbital electrons. The repulsion energy is plotted at the top of Fig. 14.12. The sum of the energy of electrostatic attraction and the energy of repulsion gives a potential-energy curve with a minimum, as shown by the solid line. The minimum occurs at the average internuclear distance for the gaseous potassium chloride molecule, 2.79 Å. The potential energy V may be represented by

$$V = -\frac{e^2}{4\pi\epsilon_0 r} + be^{-ar} + 0.52 \qquad (14.40)$$

where $-e^2/4\pi\epsilon_0 r$ represents the energy of electrostatic attraction and be^{-ar} the repulsive energy. The constants b and a are determined empirically. The second term becomes more important than the first at sufficiently small distances.

Electrostatic bonds may also be formed between an ion and a molecule having a dipole moment. Complex ions such as $[Ni(NH_3)_4]^{2+}$ contain electrostatic bonds of this *ion-dipole* type. An ion may also induce a dipole moment in a neutral molecule to form an *ion-induced dipole* bond. Such a bond is formed in the reaction $I^- + I_2 = I_3^-$.

14.10 DIPOLE MOMENT

When atoms of different electronegativity are bonded together there is an excess of electron charge on the more-electronegative atom and an excess of positive charge on the less-electronegative atom. Thus a diatomic molecule may be an electric dipole and have a dipole moment $\boldsymbol{\mu}$. If a positive charge $+q$ is separated from a negative charge $-q$ by distance r, the dipole moment has the magnitude $\mu = qr$. The dipole moment $\boldsymbol{\mu}$ is a vector pointing from the negative charge to the positive charge. Dipole moments are conveniently expressed in debye units D after Peter Debye, who made so many contributions to the understanding of polar molecules. A debye unit D is 3.336×10^{-30} C m*.

The dipole moment $\boldsymbol{\mu}$ may be calculated for any arrangement of point charges q_i having a net charge of zero. If the vector from the origin to q_i is $\boldsymbol{r}_i$

$$\boldsymbol{\mu} = \sum_i q_i \boldsymbol{r}_i \tag{14.41}$$

If the net charge of the system is zero ($\sum q_i = 0$) it is easy to show that the dipole moment $\boldsymbol{\mu}$ is independent of the choice of origin. The components of the dipole moment vector are

$$\mu_x = \sum_i q_i x_i \qquad \mu_y = \sum_i q_i y_i \qquad \mu_z = \sum_i q_i z_i \tag{14.42}$$

where x_i, y_i, and z_i are the coordinates of charge q_i.

14.11 DIELECTRIC CONSTANT

If a nonconducting substance is placed in an electric field E, say in a capacitor, negative charges in the sample tend to be drawn toward the positive plate and positive charges toward the negative plate. Under the influence of the field the sample is said to become polarized. The polarization of a sample containing dipolar molecules is illustrated in Fig. 14.13. The polarization of a sample is represented by a vector P, called the polarization. The magnitude of the polarization is equal to the resultant dipole moment per unit volume. For an isotropic substance (i.e., one with properties independent of direction) the vector P lies in the same direction as the vector E. The magnitude of the polarization of a sample containing a large number of molecules is proportional to the magnitude of the electric field strength

$$P = \chi E \tag{14.43}$$

where χ is the electric susceptibility.

---

* When dipole moments are expressed in cgs units it is convenient to use the unit 10^{-18} esu cm, which is the original debye unit. The magnitude of the dipole moment for a dipole consisting of an electron separated from a unit positive charge by 1 Å is $(4.803 \times 10^{-10}$ esu$)(10^{-8}$ cm$) = 4.803 \times 10^{-18}$ esu cm. The debye may be expressed in SI units as follows:

$$\frac{(10^{-18} \text{ esu cm})(1.602 \times 10^{-19} \text{ C})(10^{-2} \text{ m cm}^{-1})}{(4.803 \times 10^{-10} \text{ esu})} = 3.336 \times 10^{-30} \text{ C m}$$

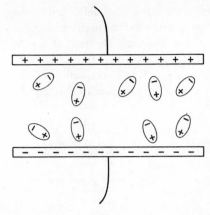

Fig. 14.13 Diagrammatic representation of a capacitor containing dipolar molecules. Although the molecules are all shown as partially oriented with the field, this is far from the case in reality because of the disorienting effect of thermal agitation.

Rather than using the electric susceptibility it is more common to use the dielectric constant ϵ, which is defined for the isotropic case by

$$\epsilon = \frac{D}{E} \tag{14.44}$$

where E is the magnitude of the electric field strength and D is the magnitude of the electric displacement $\boldsymbol{D}$ which is defined by

$$\boldsymbol{D} = \boldsymbol{E} + 4\pi\boldsymbol{P} \tag{14.45}$$

For isotropic media these vectors all have the same direction. Substitution of equation 14.43 into 14.45 and comparison with equation 14.44 shows that

$$\epsilon = 1 + 4\pi\chi \tag{14.46}$$

so that equation 14.43 may be written

$$\boldsymbol{P} = \frac{\epsilon - 1}{4\pi} \boldsymbol{E} \tag{14.47}$$

When an electric field is applied to a sample, time is required for the establishment of the equilibrium polarization. Thus the dielectric constant is a frequency dependent quantity. We will first consider a static field or one that is changing so slowly that the polarization has its equilibrium value at all times. The static dielectric constant may be obtained experimentally from measurements of capacitance.

The capacitance C of a capacitor is equal to the charge on one plate divided by the potential difference between the plates. The capacitance is increased if a vacuum between the plates is replaced by some nonconducting medium. The ratio of the capacitance C_x of a condenser filled with medium x to the capacitance when the space between the plates is evacuated, C_{vac}, is the dielectric constant ϵ of the medium.

$$\epsilon = \frac{C_x}{C_{\text{vac}}} \tag{14.48}$$

Thus the dielectric constant reduces to unity for a vacuum. This definition of the dielectric constant applies only at frequencies low enough so that many molecular collisions occur in one cycle. The magnitude of the dielectric constant of a substance depends on the temperature and the frequency at which the alternating electric field varies. A few values are given in Table 14.3. These values of the dielectric constant apply at frequencies which are sufficiently low so that equilibrium is maintained as the electric field varies.

Table 14.3 Dielectric Constants ϵ of Gases and Liquids

Gas (1 atm)	At 0°	Liquid	At 20°
Hydrogen	1.000272	Hexane	1.874
Argon	1.000545	Benzene	2.283
Air (CO_2 free)	1.000567	Toluene	2.387
Carbon dioxide	1.00098	Chlorobenzene	5.94
Hydrogen chloride	1.0046	Ammonia	15.5
Ammonia	1.0072	Acetone	21.4
Water (steam at 110°)	1.0126	Methanol	33.1
		Water	80

14.12 POLARIZATION

Three contributions to the polarization must be considered: orientation polarization, electronic polarization, and vibrational polarization. Orientation polarization is due to the partial alignment of permanent dipoles. The extent to which dipoles can be oriented by an applied field was calculated by Debye* using the Boltzmann distribution law. The electric field acting on the molecule is represented by E_i and referred to as the internal field. The energy of a dipole in the field E_i is $-\boldsymbol{\mu}\cdot E_i$ where $\boldsymbol{\mu}$ is the permanent dipole moment vector of the molecule and the dot refers to the dot product: $-\boldsymbol{\mu}\cdot E_i = -\mu E_i \cos\theta$, where θ is the angle between the two vectors. If the energy of the dipole in the field is small compared with kT, it may be shown that the contribution per molecule in a gas to the mean moment in the direction of the field is given by $\mu^2 E_i/3kT$, where E_i is the magnitude of the internal field. As the temperature is increased, the thermal agitation becomes more vigorous and fewer of the permanent dipoles are oriented in the direction of the field.

The second contribution is from the field-induced displacement of the average positions of the electrons relative to the nuclei of the molecule. This electronic contribution $\alpha_e E_i$ is practically independent of the temperature and is proportional to the electric field strength. The proportionality constant α_e is called the *mean molecular electronic polarizability*.

* P. Debye, *Polar Molecules*, Chemical Catalog Co., New York, 1929; or Dover Publications, New York.

The ease with which different molecules become polarized by distortion of the electron cloud differs greatly. The magnitude of the induced moment depends on the orientation of the molecule in the electric field. Because of thermal agitation, however, all possible orientations are encountered; for a macroscopic sample the experimentally determined induced moment is an average over all orientations.

The third contribution $\alpha_v E_i$ is from the deformation of the nuclear skeleton of the molecule by the electric field. The proportionality constant α_v is called the *mean molecular vibrational polarizability*. To a first approximation α_v is independent of temperature.

The magnitude P of the total polarization, which includes all three contributions, is equal to the mean moment in the direction of the field multiplied by the number N of molecules per unit volume ($N = N_A/\bar{V}$).

$$P = N\left(\frac{\mu^2}{3kT} + \alpha_e + \alpha_v\right)E_i \qquad (14.49)$$

There is a problem in knowing the average internal or local electric field strength E_i. If the concentration of polar molecules is low it can be shown that $E_i = E(\epsilon + 2)/3$, where E_i is the internal field and E the applied field. Eliminating E_i from equation 14.49 using this relation, eliminating P between this equation and equation 14.47, and introducing $N = N_A\rho/M$, we obtain

$$\mathscr{P} = \frac{\epsilon - 1}{\epsilon + 2}\frac{M}{\rho} = \frac{4\pi N_A}{3}\left(\frac{\mu^2}{3kT} + \alpha_e + \alpha_v\right) \qquad (14.50)$$

The quantity $\mathscr{P}$ is called the molar polarization, and ρ is the density. This equation is accurate for dilute gases and holds approximately for slightly polar liquids, but it does not apply to liquids of high dielectric constant because of the approximation of taking $E_i = E(\epsilon + 2)/3$. Equation 14.50 neglects short-range molecular interactions; more complicated theories have been developed by Onsager and Kirkwood to take those effects into account.

14.13 DIPOLE MOMENTS OF GASEOUS MOLECULES

Since the orientation polarization depends on the absolute temperature, it is possible to calculate the polarizability $\alpha_e + \alpha_v$ and the dipole moment μ of molecules of a gas from measurements of the dielectric constant ϵ at a series of temperatures. In the gaseous state the molecules are so far apart that they do not induce dipole moments in each other.

To obtain the dipole moment of a gaseous molecule the dielectric constants ϵ and gas densities are measured at several temperatures, and the corresponding values of the total molar polarization $\mathscr{P}$ are calculated. According to equation 14.50 a plot of $\mathscr{P}$ versus $1/T$ will have an intercept of $\frac{4}{3}\pi N_A(\alpha_e + \alpha_v)$ and a slope of $4\pi N_A\mu^2/9k$. Plots of molar polarization $\mathscr{P}$ versus reciprocal absolute temperature are illustrated in Fig. 14.14 for the chlorine-substituted compounds of methane as

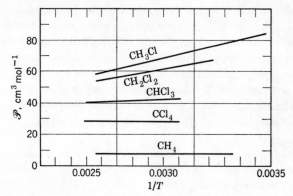

Fig. 14.14 Variation of molar polarization with temperature.

measured by Sanger.* It is seen that zero slopes are obtained for the symmetrical molecules CH_4 and CCl_4, indicating that these molecules have no permanent dipole moment. Because of the electronegative character of chlorine atoms, CH_3Cl, CH_2Cl_2, and $CHCl_3$ have permanent dipole moments with the chlorine atom the negative end of the dipole. The dipole moment of CH_3Cl is the largest of the three.

The dependence of the dielectric constant of a typical polar substance on frequency is shown in Fig. 14.15. At low frequencies all the terms on the right-hand side of equation 14.50 contribute. As the frequency is increased above the radio-frequency range the dielectric constant decreases and the orientation polarization eventually becomes negligible because there is insufficient time for molecular orientation to occur before the field is reversed. In the region where the dielectric constant is labeled ϵ_∞, the polarizability is made up of vibrational, α_v, and electronic, α_e, terms. The dielectric constant shows dispersion (Section 15.15) in the neighborhood of absorption lines in the infrared and ultraviolet.

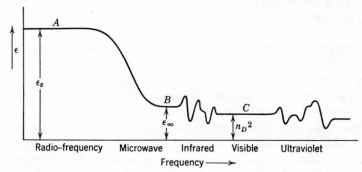

Fig. 14.15 Variation of dielectric constant with frequency for a typical polar substance.

* R. Sanger, *Physik. Z.*, **27**, 562 (1926).

The dipole moment of a solute in a nonpolar solvent may be determined from dielectric constant and density measurements on dilute solutions.* Refractive index measurements are used to evaluate the molecular electronic polarizability α_e. Dipole moments of a variety of substances are given in Table 14.4.

Table 14.4 Dipole Moments (in debye units)

$AgClO_4$	4.7	C_6H_5Cl	1.55	SO_2	1.61
$C_6H_5NO_2$	3.96	C_6H_5OH	1.70	HCl	1.03
$(CH_3)_2CO$	2.8	C_2H_5OH	1.70	HBr	0.79
H_2O	1.84	$C_6H_5NH_2$	1.56	HI	0.30
CH_4	0	NH_3	1.46	N_2O	0.14
CH_3Cl	1.85	H_2S	1.10	CO	0.12
CH_3Br	1.45	H_2	0	CS_2	0
CH_3I	1.35	Cl_2	0	C_2H_4	0
CH_2Cl_2	1.59	CO_2	0	C_2H_6	0
$CHCl_3$	1.15	$C_6H_5CH_3$	0.40	C_6H_6	0
CCl_4	0				

14.14 MOLAR REFRACTION

The ratio of the velocity of light in a vacuum to that in a medium, referred to as refractive index n, is readily determined by measuring the angles of incidence and refraction at an interface of a medium and a vacuum. The refractive index depends on the wavelength of the light and on the temperature, which are usually specified with subscripts and superscripts, respectively. Thus n_D^{25} indicates a refractive index at 25° taken with monochromatic yellow D light of the sodium arc.

Maxwell showed that the dielectric effect measured with a capacitor and the refraction of light are both due to the dielectric constant. If the dielectric constant ϵ and refractive index n are measured at the same frequency, Maxwell showed that

$$\epsilon = n^2 \qquad (14.51)$$

In the visible and ultraviolet regions only α_e contributes to the polarizability because only the electrons can move fast enough to follow the high-frequency electric field.

When $\epsilon = n^2$ is substituted in equation 14.50 the Lorentz-Lorenz equation is obtained.

$$\frac{n^2 - 1}{n^2 + 2} \frac{M}{\rho} = \tfrac{4}{3}\pi N_A \alpha_e = R_m \qquad (14.52)$$

The quantity R_m is referred to as the molar refraction. It is proportional to the molecular electronic polarizability α_e due to electron displacements.

---

* F. Daniels, J. W. Williams, P. Bender, R. A. Alberty, C. D. Cornwell, and J. E. Harriman, *Experimental Physical Chemistry*, McGraw-Hill Book Co., New York, 1970.

The molar refraction is nearly independent of temperature, pressure, and state of aggregation. The molar refractions of a large number of organic and inorganic compounds have been determined, and it has been found that atoms and groups of atoms always contribute the same amount to the molar refraction of any compound. To obtain the dielectric constant at very high frequencies it is more practical to measure the refractive index and calculate ϵ. The electronic polarizability is the sum of atomic and bond contributions. This is the reason that the molar refraction R_m, defined in equation 14.52, is an additive property.

14.15   VAN DER WAALS FORCES

The short range attractive forces between molecules are named after van der Waals who discussed these forces in gases and liquids (Section 3.4). These forces contribute to the nonideal behavior of gases, and at sufficiently high pressures and low temperatures, to condensation to the liquid state. The origin of these forces was explained in 1930 by London. The intermolecular attraction arises from the fluctuations of charge in the two atoms or molecules that are close together. Since the electrons are moving, each molecule has an instantaneous dipole moment that is not zero. If the electron density fluctuations in the two atoms or molecules were unrelated there would be no net attraction between the molecules since there would be repulsion as often as attraction. However, an instantaneous dipole in one atom or molecule induces an oppositely oriented dipole in the neighboring atom or molecule, and these instantaneous dipoles attract each other. This leads to a force of attraction that is referred to as a *dispersion force*.

The interaction energy is proportional to the square of the polarizability α (Section 14.12) and inversely proportional to r^6 where r is the distance between the two atoms or molecules. Thus the interaction is strong only at very short distances, and polarizable molecules attract each other more strongly than nonpolarizable molecules.

References

C. J. Ballhausen and H. B. Gray, *Molecular Orbital Theory*, W. A. Benjamin, Inc., New York, 1965.

F. A. Cotton and G. Wilkinson, *Advanced Inorganic Chemistry*, Wiley-Interscience, New York, 1972.

J. C. Davis, *Advanced Physical Chemistry*, The Ronald Press Co., New York, 1965.

H. B. Gray, *Electrons and Chemical Bonding*, W. A. Benjamin, Inc., New York, 1964.

R. M. Hochstrasser, *Behavior of Electrons in Atoms*, W. A. Benjamin, Inc., New York, 1964.

M. Karplus and R. N. Porter, *Atoms and Molecules*, W. A. Benjamin, Inc., New York, 1970.

J. W. Linnett, *The Electronic Structure of Molecules*, Barnes and Noble, New York, 1966.

J. W. Linnett, *Wave Mechanics and Valency*, Barnes and Noble, New York, 1966.

M. Orchin and H. H. Jaffe, *The Importance of Antibonding Orbitals*, Houghton Mifflin, Boston, 1967.

R. G. Parr, *Quantum Theory of Molecular Electronic Structure*, W. A. Benjamin, Inc., New York, 1963.

F. L. Pilar, *Elementary Quantum Chemistry*, McGraw-Hill Book Co., New York, 1968.

L. Pauling, *The Nature of the Chemical Bond*, Cornell University Press, Ithaca, 1960.

G. C. Pimental and A. L. McClellan, *The Hydrogen Bond*, W. H. Freeman and Co., San Francisco, 1960.

J. A. Pople and D. L. Beveridge, *Approximate Molecular Orbital Theory*, McGraw-Hill Book Co., New York, 1970.

W. G. Richards and J. A. Horsley, *Ab Initio Molecular Orbital Calculations for Chemists*, Clarendon Press, Oxford, 1970.

H. F. Schaefer, III, *The Electronic Structure of Atoms and Molecules*, Addison-Wesley Publ. Co., Reading, Mass., 1972.

O. Sinanoglu and K. B. Wiberg, *Sigma Molecular Orbital Theory*, Yale University Press, New Haven, Conn., 1970.

C. P. Smyth, *Dielectric Behavior and Structure*, McGraw-Hill Book Co., New York, 1955.

Problems

14.1 Derive the expression for the normalization constant N in

$$\psi = N[(1s_a) + (1s_b)]$$

Ans. $N = \pm 1/\sqrt{2(1 + S)}$, where S is the overlap integral.

14.2 Given the dissociation energy D' for N_2 in Table 14.1 and the fundamental vibration frequency 2331 cm^{-1}, calculate the spectroscopic dissociation energy D in kcal mol^{-1}.

Ans. 225 kcal mol^{-1}.

14.3 If the following compounds have ionic bonds only, which noble gases have the same electronic structures as the ions? HF, NaCl, MgO. *Ans.* Ne; Ne—Ar; Ne—Ne.

14.4 The dipole moments of HCl, HBr, and HI are given in Table 14.4. Explain the relative magnitudes of these moments in terms of the electronegativity.

Ans. The electronegativities increase in going from I to Cl and so the negative charge on the halogen is greater in HCl than in HI and the dipole moment is larger.

14.5 Calculate the dipole moment HCl would have if it consisted of a proton and a chloride ion (considered to be a point charge) separated by 1.27 Å (the internuclear distance obtained from the infrared spectrum). The experimental value is 1.03 debye units. How do you explain the difference?

Ans. 6.10 debye units. The charges are not completely separated in HCl.

14.6 The molar polarization $\mathscr{P}$ of ammonia varies with temperature as follows:

t, °C	19.1	35.9	59.9	113.9	139.9	172.9
P, cm^3 mol^{-1}	57.57	55.01	51.22	44.99	42.51	39.59

Calculate the dipole moment of ammonia. *Ans.* 5.28×10^{-30} C m.

14.7 By extrapolation the molar polarization of pyridine oxide, C_5H_5NO, in an infinitely dilute solution in dioxane is found to be 411 cm^3 mol^{-1} at 25°. The molar refraction of C_5H_5NO is 28 cm^3 mol^{-1}, and this is approximately equal to $4\pi N_A(\alpha_e + \alpha_v)/3$ Calculate the dipole moment. *Ans.* 14.4×10^{-30} C m.

14.8 The refractive index n_D^{20} of carbon tetrachloride is 1.4573, the density at 20° is 1.595 g cm^{-3}, and the molecular weight is 153.84. Calculate the molar refraction.

Ans. 26.28 cm^3 mol^{-1}

14.9   The molar refraction for liquid chloroform for the D line of sodium at 20° is 21.25 cm³ mol⁻¹. Calculate the refractive index of chloroform vapor at a hypothetical pressure of 1 atm at 20°.   *Ans.* 1.00133.

14.10   Calculate the normalization factors shown in equations 14.23 and 14.24.

14.11   Carry out the differentiations required to get from equation 14.6 to 14.15 and 14.16.

14.12   By multiplying out terms show that the wave functions given in equation 14.32 and 14.35 are the same.

14.13   Explain the relative enthalpies of formation (Table 1.3) of HCl, HBr, and HI in terms of electronegativity concept.

14.14   The dipole moments of CH_3Cl, CH_3Br, and CH_3I are given in Table 14.4. Explain the relative magnitudes of these moments in terms of the electronegativity.

14.15   Calculate the dipole moment of fluorobenzene, using a plot of $\mathscr{P}$ versus $1/T$, from the following molar polarizations for gaseous samples measured by K. B. McAlpine and C. P. Smyth [*J. Chem. Phys.*, **3**, 55 (1935)]:

T, K	343.6	371.4	414.1	453.2	507.0
$\mathscr{P}$, cm³ mol⁻¹	69.9	66.8	62.5	59.3	55.8

14.16   Show from the units of the various quantities involved that the polarization has the units of cm³ mol⁻¹.

14.17   Calculate the molar refraction of water vapor and compare it with the molar refraction of liquid water. The refractive index of water vapor at 20° for the sodium D line is 1.000241 for a (hypothetical) pressure of 1 atm. The density of water vapor under these conditions would be 0.749 g liter⁻¹. The molar refraction of liquid water at 20° for the sodium D line is 3.75 cm³ mol⁻¹.

14.18   Show that wave equations ψ_g and ψ_u in equations 14.23 and 14.24 are orthogonal.

14.19   Given the dissociation energy D' for O_2 in Table 14.1 and the fundamental vibration frequency 1556 cm⁻¹, calculate the spectroscopic dissociation energy D in eV.

14.20   Alternating positive and negative charges are arranged in a linear array with distances between successive charges of 1, 2, and 3 Å. Calculate the magnitude of the dipole moment. Show that the magnitude of the moment is independent of the choice of origin of the coordinate system.

14.21   The dipole moment of water is 1.84×10^{-18} esu cm. What is the value of the dipole moment in coulomb meters?

14.22   The molar polarization of silicobromoform, $SiHBr_3$, is 46 cm³ mol⁻¹ at 25°. The induced polarizability α may be taken as 1.308×10^{-23}. Calculate the dipole moment of this molecule.

14.23   The refractive index of benzene vapor for the D line of sodium at 20° is 1.00170, corrected to a pressure of 1 atm. Calculate the molar refraction of benzene vapor, and compare it with the value for the liquid, which is 26.18 cm³ mol⁻¹.

14.24   Derive equation 14.38 by approximating the Morse potential function (equation 14.37) to a simple harmonic potential.

CHAPTER 15

MOLECULAR SPECTROSCOPY

Spectroscopy is the study of the interactions between matter and electromagnetic radiation. The spectra of atoms have been discussed briefly in connection with quantum mechanics. Atomic spectra involve only transitions of electrons from one electronic energy level to another. Molecular spectra involve transitions between rotational and vibrational energy levels in addition to electronic transitions. As a result, the spectra of molecules are much more complicated than those of atoms. But this means that information may be obtained about molecular vibrations and rotations that reveal a great deal about molecular structure.

In this chapter we will first discuss the spectroscopy of diatomic molecules in the order rotational spectra, vibrational spectra, and electronic spectra. Then we will consider polyatomic molecules, optical rotatory dispersion, and circular dichroism.

Additional topics in spectroscopy are discussed in other chapters including nuclear magnetic resonance and electron spin resonance in Chapter 16 and fluorescence and phosphorescence in Chapter 18.

15.1 THE ELECTROMAGNETIC SPECTRUM

The spectrum of electromagnetic radiation is continuous, but it is convenient to divide it arbitrarily into regions according to the sources and detectors required. Typical limits of the various spectral regions are indicated in Table 15.1.

The chemical and physical effects of various types of radiation are quite different, and these differences can be understood in terms of the differing energies of the photons. The energy ϵ of a single photon is given by

$$\epsilon = h\nu \tag{15.1}$$

where h is Planck's constant (Section 12.4), 6.626×10^{-34} J s.

The energy of a mole of electrons that have been accelerated by 1 V is of the magnitude commonly encountered for chemical reactions.

$$\frac{(1.60219 \times 10^{-19}\,\text{J})(6.02205 \times 10^{23}\,\text{mol}^{-1})}{4.184\,\text{J cal}^{-1}} = 23{,}060\,\text{cal mol}^{-1}$$

Table 15.1   Energy of Electromagnetic Radiation

Description	Wavelength Range	Wave Number, cm^{-1}	Frequency, Hz	Energy	
				kcal mol^{-1}	electron volts
Radio	$\begin{cases} 3 \times 10^3 \text{ m} \end{cases}$	3.33×10^{-6}	10^5	9.51×10^{-9}	4.12×10^{-10}
frequency	0.30 m	0.0333	10^9	9.51×10^{-5}	4.12×10^{-6}
Microwave	0.0006 m	16.6	4.98×10^{11}	0.0457	2.07×10^{-3}
Far infrared	$(600 \ \mu\text{m})$				
	$30 \ \mu\text{m}$	333	10^{13}	0.951	0.0412
Near infrared	$0.8 \ \mu\text{m}$	1.25×10^4	3.75×10^{14}	35.8	1.55
Visible	(800 nm)				
	400 nm	2.5×10^4	7.5×10^{14}	71.5	3.10
Ultraviolet					
	150 nm	6.06×10^4	19.98×10^{14}	190	8.25
Vacuum ultraviolet X rays and	5 nm	2×10^6	6×10^{16}	5.72×10^5	247.8
γ rays	10^{-4} nm	10^{11}	3×10^{21}	2.86×10^8	12.4×10^6

This is a useful conversion factor between two commonly used energy units. Whereas kcal mol^{-1} is the most frequently used energy unit in chemistry, wave numbers are used in spectroscopy, and electron volts are used in physics. Useful conversion factors are given in the front cover.

Example 15.1   Calculate the energy in joules per quantum, calories per mole, and electron volts of photons of wavelength 3000 Å.

$$h\nu = hc/\lambda = (6.62 \times 10^{-34} \text{ J s})(3 \times 10^8 \text{ m s}^{-1})/(3000 \times 10^{-10} \text{ m})$$

$$= 6.62 \times 10^{-19} \text{ J}$$

$$N_A h\nu = (6.02 \times 10^{23} \text{ mol}^{-1})(6.62 \times 10^{-19} \text{ J})$$

$$= 3.98 \times 10^5 \text{ J mol}^{-1}$$

$$= \frac{3.98 \times 10^5 \text{ J mol}^{-1}}{4.184 \text{ J cal}^{-1}} = 95,300 \text{ cal mol}^{-1}$$

$$= \frac{95,300 \text{ cal mol}^{-1}}{23,060 \text{ cal mol}^{-1} \text{ eV}^{-1}} = 4.13 \text{ eV}$$

15.2 THE FUNDAMENTAL EQUATION

In absorption a photon of electromagnetic radiation is absorbed by a molecule that is in a particular quantum state characterized by a certain wave function and a definite energy E''. There is a transition to another quantum state, characterized by another wave function and a higher energy E'. In emission, a photon is emitted by a molecule which makes a transition from a higher energy state to a lower energy state. For either the absorption or emission process, the total energy must be conserved. This leads to the relation

$$h\nu = E' - E'' = \Delta E \qquad (15.2)$$

where E' is the energy of the higher energy state, E'' is the energy of the lower energy state, and ν is the frequency of the radiation.

The types of molecular change that accompany the absorption of electromagnetic radiation may be described as follows: in the radio-frequency range the energy of one photon is very low, and the only changes are those of nuclear spin states of substances in a magnetic field (discussed in Chapter 16). In the microwave range there are changes in electron spin states for substances with unpaired electrons in a magnetic field (discussed in the next chapter) and transitions between rotational energy levels of gas molecules. In the infrared region absorption causes changes in vibrational energy accompanied by changes in rotational energy. In the visible and ultraviolet regions absorption causes changes in the electronic energy of the molecule accompanied by changes in its vibrational and rotational energy. The changes in electronic energy involve the most loosely held, or so-called outer, electrons. In the far ultraviolet and X-ray regions the energies are sufficiently great to move inner electrons to higher levels and to dissociate molecules.

In spectroscopy we will not be concerned with translational energy, but only with the "internal" energy of a molecule. The internal energy E of a molecule in a certain state may be considered to be made up of contributions from rotational energy E_{rot}, vibrational energy E_{vib}, and electronic energy E_{el},

$$E = E_{rot} + E_{vib} + E_{el} \qquad (15.3)$$

Actually this division into separate contributions is not exact; for example, the rotational energy depends somewhat on the amount of vibrational energy because the mean internuclear distance, and hence the moment of inertia, depends on the vibrational energy.

The state of an isolated molecule may be specified by giving a series of quantum numbers to describe the rotational state, the vibrational state and the electronic state. In the process of absorption or emission there is a change in this set of quantum numbers. For example, if a diatomic molecule in the ground electronic state and the zero point vibrational state has some particular rotational energy, absorption of a photon may raise it to an excited electronic state, with several quanta of vibrational energy, and a different rotational energy. There are many transitions that might give rise to absorption or emission, but only those satisfying certain selection rules are allowed.

Electronic, vibrational, and rotational changes may all occur upon the absorption or emission of a single photon, and so equation 15.2 may be written

$$h\nu = (E'_{\text{el}} - E''_{\text{el}}) + (E'_{\text{vib}} - E''_{\text{vib}}) + (E'_{\text{rot}} - E''_{\text{rot}}) \tag{15.4}$$

where the upper state is represented by a single prime, and the lower state by a double prime. A fact that considerably simplifies the interpretation of spectra is that usually,

$$(E'_{\text{el}} - E''_{\text{el}}) \gg (E'_{\text{vib}} - E''_{\text{vib}}) \gg (E'_{\text{rot}} - E''_{\text{rot}}) \tag{15.5}$$

Consequently, rotational, vibrational, and electronic spectra occur in different regions of the spectrum.

15.3 ROTATIONAL SPECTRA OF DIATOMIC MOLECULES

As a first approximation we may consider a rotating diatomic molecule to be a rigid rotor with an internuclear distance r equal to the equilibrium value. As discussed in Section 12.16 the eigenvalues for a rigid rotor are

$$E_{\text{rot}} = \frac{\hbar^2}{2I}J(J+1) \qquad J = 0, 1, 2, \ldots \tag{15.6}$$

where I is the moment of inertia

$$I = \mu r^2 \tag{15.7}$$

and μ is the reduced mass (Section 15.4). The rotational levels are $(2J + 1)$-fold degenerate. The rotational energy levels for a rigid diatomic molecule, as calculated using equation 15.6, are shown in Fig. 15.1.

The quantum mechanical treatment of rotational transitions shows that a molecule has a pure rotational spectrum only if it possesses a permanent electric dipole moment. This is expected from the fact that a rotating dipole produces an oscillating electric field that can interact with the oscillating electric field of a light wave. Since homonuclear diatomic molecules like H_2 and N_2 do not have dipole moments, they do not interact with the electromagnetic field of light and thus do not show pure rotational spectra as do molecules like HCl and CH_3Cl, which have dipole moments. Since the orientation of the dipole moment vector in a molecule must be unchanged by any symmetry operation of the molecule, the vector must be contained in each of the symmetry elements. Only molecules belonging to point groups C_n, C_s, and C_{nv} can possess dipole moments (Section 13.10).

In addition the quantum mechanical treatment shows that the only transitions between rotational levels that are allowed are those for which $\Delta J = \pm 1$. In absorption $\Delta J = +1$, and in emission $\Delta J = -1$.

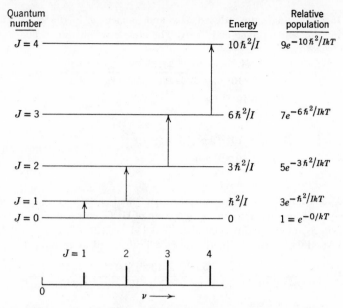

Fig. 15.1 Rotational levels for a rigid diatomic molecule and the absorption spectrum that results from $\Delta J = 1$. The energies and populations of the levels are indicated on the right. The transitions are labeled by the upper of the two J values involved.

We are interested in expressing the frequencies of lines in the rotational spectrum in reciprocal centimeters. Since $\nu = \Delta E / hc$, it is convenient to rewrite equation 15.6 as

$$\frac{E}{hc} = \frac{h}{8\pi^2 cI} J(J+1) = BJ(J+1) \tag{15.8}$$

where $B = h/8\pi^2 cI$ is referred to as the rotational constant. The frequency in cm^{-1} for a transition from a level with quantum number $J - 1$ to the level J is given by

$$\tilde{\nu} = \frac{E'}{hc} - \frac{E''}{hc} = B[J(J+1) - (J-1)J] = 2BJ \tag{15.9}$$

where J, the quantum number for the upper level, can only have values of 1, 2, 3, ... Thus the frequency of the radiation absorbed is directly proportional to the quantum number of the rotational level to which the molecule is excited by the absorption of a quantum of radiation. As a result we find a series of nearly equally spaced lines in the far-infrared and microwave spectra of diatomic molecules. The moments of inertia of isotopically substituted molecules are appreciably different, and therefore a separate series of lines is found for each isotopically different species of a given molecule.

Example 15.2 Calculate the reduced mass and moment of inertia of $H^{35}Cl$, using the mean internuclear distance of 1.275Å. The atomic weights are 1.008 and 34.98 for H and ^{35}Cl.

$$\mu = \frac{(1.008 \times 10^{-3}\ kg)(34.98 \times 10^{-3}\ kg)}{(1.008 \times 10^{-3}\ kg + 34.98 \times 10^{-3}\ kg)(6.022 \times 10^{23})}$$

$$= 1.628 \times 10^{-27}\ kg$$

$$I = \mu r^2$$

$$= (1.628 \times 10^{-27}\ kg)(1.275 \times 10^{-10}\ m)^2$$

$$= 2.647 \times 10^{-47}\ kg\ m^2$$

Example 15.3 Calculate the frequencies in cm^{-1} and the wavelengths in micrometers for the pure rotational lines in the spectrum of $H^{35}Cl$ corresponding to the following changes in rotational quantum number: $0 \to 1$, $1 \to 2$, $2 \to 3$, and $8 \to 9$. Using equation 15.9 and the moment of inertia for $H^{35}Cl$ calculated in example 15.2, we obtain:

$$\tilde{\nu} = \frac{(6.626 \times 10^{-34}\ J\ s)\ J}{4\pi^2(2.647 \times 10^{-47}\ kg\ m^2)(3 \times 10^8\ m\ s^{-1})(100\ cm\ m^{-1})}$$

$$= 21.1\ J\ cm^{-1}$$

Thus

J	$\tilde{\nu} = 21.1\ J$	$\lambda = 10^4/\tilde{\nu}$
1	21.1 cm^{-1}	474 μm
2	42.2	237
3	63.3	158
9	189.9	53

The usual experimental situation is the reverse of this example; that is, the wavelengths of the rotational lines are measured in the far-infrared or microwave region and the moment of inertia and finally the mean internuclear distance are calculated.

The relative intensities of rotational lines are dependent on the relative populations of the levels in the initial state. The populations of the levels at thermal equilibrium are given by the Boltzmann distribution

$$N_i = Kg_i e^{-\epsilon_i/kT} \tag{15.10}$$

where g_i is the degeneracy (Section 12.14) of the ith level. As discussed earlier the component of the angular momentum in a particular direction is equal to $m\hbar$, where m may have values of J, $(J-1) \ldots, 0, \ldots, -J$, where J is the rotational quantum number. Thus there are in all $2J + 1$ different possible states with quantum number J. In the absence of an external electric field the energies are identical for these various sublevels, and so the Jth energy level is said to have a degeneracy of $2J + 1$.

The rotational energy in the absence of an electric field is given by $\epsilon_i = hcJ(J + 1)B$ so that using equation 15.10 the population of the Jth rotational level is given by

$$N_J = K(2J + 1)e^{-[hcJ(J+1)B]/kT} \tag{15.11}$$

According to this equation the number of molecules in level J increases with J at low J values, goes through a maximum and then, because of the exponential term, decreases as J is further increased. The lines in the spectrum at the bottom of Fig. 15.1 have been labeled with the rotational quantum number J of the upper of the two states involved, and the relative lengths of these lines have been used to show the relative populations of these states.

For molecules with larger moments of inertia I the rotational energies are smaller, in fact small compared with kT. The quantum numbers may become quite large before $e^{-\epsilon_J/kT}$ becomes appreciably different from unity. For small quantum numbers populations are given by the degeneracies since $e^{-\epsilon_J/kT} \approx 1$ for $\epsilon_J \ll kT$.

The lines in the rotational spectrum of a diatomic molecule are split when the molecules being studied are placed in an external electric field. This rotational Stark effect results from the interaction of the molecular dipole moment with the electric field.

Although homonuclear diatomic molecules do not have permanent electric dipole moments and do not exhibit pure rotational spectra, they do show rotational Raman spectra (Section 15.10), and their electronic spectra show rotational fine structure.

15.4   VIBRATION SPECTRA OF DIATOMIC MOLECULES

As a first approximation we may consider a vibrating diatomic molecule to be a simple harmonic oscillator. As discussed in Section 12.15 the vibrational energy E_{vib} of a simple harmonic oscillator with a classical frequency of ν is

$$E_{\text{vib}} = (v + \tfrac{1}{2})h\nu \qquad v = 0, 1, 2, \ldots \tag{15.12}$$

where v is the vibrational quantum number. The classical vibrational frequency of a simple harmonic oscillator is given by

$$\nu = \frac{1}{2\pi}\left(\frac{k}{m}\right)^{\frac{1}{2}} \tag{15.13}$$

where k is the force constant and m is the mass. The vibration of a diatomic molecule involves the relative motion of two atoms, but we will now show that the solution to this problem is of the same functional form as that for an oscillating mass point with m in equation 15.13 replaced by the reduced mass μ.

In a diatomic molecule the masses of the two atoms are represented by m_1 and m_2, and their distances from the center of mass are represented by r_1 and r_2. The internuclear distance r is equal to $r_1 + r_2$. The center of mass is located so that

$$r_1 m_1 = r_2 m_2 \tag{15.14}$$

Combining this equation with $r = r_1 + r_2$ yields

$$r_1 = \frac{m_2}{m_1 + m_2} r \qquad (15.15a)$$

$$r_2 = \frac{m_1}{m_1 + m_2} r \qquad (15.15b)$$

The force tending to restore the diatomic molecule to its equilibrium internuclear distance is equal to $m_1 a_1 = m_2 a_2$, where a_1 is the acceleration of nucleus 1 and a_2 is the acceleration of nucleus 2. By use of equations 15.15a and 15.15b we have

$$a_1 = \frac{d^2 r_1}{dt^2} = \frac{m_2}{m_1 + m_2} \frac{d^2 r}{dt^2} \qquad (15.16a)$$

$$a_2 = \frac{d^2 r_2}{dt^2} = \frac{m_1}{m_1 + m_2} \frac{d^2 r}{dt^2} \qquad (15.16b)$$

Thus

$$\text{Force} = \frac{m_1 m_2}{m_1 + m_2} \frac{d^2 r}{dt^2} = \mu \frac{d^2 r}{dt^2} = \mu \frac{d^2 (r - r_e)}{dt^2} \qquad (15.17)$$

where r_e is the equilibrium internuclear distance, and μ is the reduced mass. Equating the force of attraction or repulsion $-k(r - r_e)$ to that given in equation 15.17, we obtain

$$\frac{d^2 (r - r_e)}{dt^2} = -\frac{k}{\mu} (r - r_e) \qquad (15.18)$$

If the nuclei are closer together than r_e, there is repulsion. If the nuclei are further apart than r_e, there is attraction. The attraction is due to electronic bonding and the repulsion is due to the overlapping of inner electronic shells.

By comparing equations 15.18 and 12.59 we see that we can write

$$\nu_0 = \frac{1}{2\pi} \left(\frac{k}{\mu} \right)^{1/2} \qquad (15.19)$$

Thus the fundamental vibration frequency for a diatomic molecule depends upon the force constant and the reduced mass of the molecule.

Example 15.4   Calculate the force constant for $H^{35}Cl$ from the fact that the fundamental vibration frequency is $8.667 \times 10^{13} \text{ s}^{-1}$ and the reduced mass is 1.628×10^{-27} kg.

$$k = (2\pi 8.667 \times 10^{13} \text{ s}^{-1})^2 (1.628 \times 10^{-27} \text{ kg})$$

$$= 483 \text{ kg s}^{-1}$$

$$= 483 \text{ N m}^{-1}$$

The possible energy levels for a simple harmonic oscillator were shown in Section 12.15.

Vibrational spectra result from changes in vibrational energy levels. Absorption and emission spectra of this type, however, are shown only by diatomic molecules for which a *change* in dipole moment accompanies the vibrational motion. A fluctuating dipole moment offers a mechanism for interaction between the molecule and electromagnetic radiation. Homonuclear diatomic molecules such as H_2, N_2, etc., have zero dipole moment for all bond lengths and therefore do not show vibrational spectra. In general, heteronuclear diatomic molecules do have dipole moments that depend on internuclear distance, and so they exhibit vibrational spectra.

The quantum mechanical treatment of transitions between vibrational levels shows that for the simple harmonic oscillator with a dipole moment proportional to the internuclear distance the selection rule is

$$\Delta v = \pm 1$$

where the $+$ applies to absorption and the $-$ to emission.

Since for real molecules the dipole moment is not proportional to the internuclear distance, the overtones $\Delta v = \pm 2, \pm 3, \ldots$ are allowed. Anharmonicity in the potential also allows these overtones.

Using equation 15.12 we see that the difference in vibrational energy between two levels with vibrational quantum numbers v' and v'' is given by

$$\Delta E = E'_v - E''_v = (v' + \tfrac{1}{2})h\nu_0 - (v'' + \tfrac{1}{2})h\nu_0 = (v' - v'')h\nu_0 \quad (15.20)$$

Since most molecules are in the ground vibrational state at room temperature, absorption will result in the transition $v = 0 \to 1$. This is the only transition expected on absorption for a quantum mechanical *harmonic* oscillator in its ground state. Since the equilibrium ratio of HCl molecules having $v = 1$ to those with $v = 0$ is 8.9×10^{-7} at room temperature (see equation 12.66) the molecules with $v = 1$ do not produce a significant absorption.

The vibrational absorption spectrum of HCl is represented schematically in Fig. 15.2. The strongest absorption band is at 3.46 μm; there is a much weaker band at 1.76 μm and a very much weaker one at 1.197 μm. In order to explain the existence of these higher "overtones" it is necessary to include more terms in the expression for the potential energy of the oscillating molecule. The potential-energy function for a molecule (cf, the plot for H_2 in Fig. 14.6) may be expressed

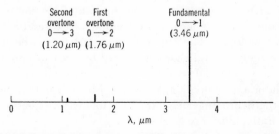

Fig. 15.2 Vibrational absorption spectrum of HCl. The relative intensities of the lines fall off five times as fast as indicated.

by a power series about the equilibrium internuclear distance r_e.

$$E(r) = E(r_e) + \left(\frac{dE}{dr}\right)_{r_e} (r - r_e) + \frac{1}{2!}\left(\frac{d^2E}{dr^2}\right)_{r_e} (r - r_e)^2$$

$$+ \frac{1}{3!}\left(\frac{d^3E}{dr^3}\right)_{r_e} (r - r_e)^3 + \frac{1}{4!}\left(\frac{d^4E}{dr^4}\right)_{r_e} (r - r_e)^4 + \cdots \qquad (15.21)$$

If we arbitrarily choose the zero of energy at the minimum of the curve, $E(r_e) = 0$,

$$E(r) = \frac{1}{2!}\left(\frac{d^2E}{dr^2}\right)_{r_e} (r - r_e)^2 + \frac{1}{3!}\left(\frac{d^3E}{dr^3}\right)_{r_e} (r - r_e)^3 + \frac{1}{4!}\left(\frac{d^4E}{dr^4}\right)_{r_e} (r - r_e)^4 + \cdots \quad (15.22)$$

since the first derivative is zero at the bottom of the curve. Taking only the first term gives the harmonic oscillator approximation. Taking the higher terms into account changes the selection rule to include $\Delta v = \pm 2, \pm 3, \ldots$, but the expected intensities of the lines drop sharply as Δv increases, as illustrated by Fig. 15.2.

The simple harmonic oscillator approximation is useful because the potential energy curve is nearly parabolic in the region of the minimum. If the potential were actually parabolic, the vibrational levels would be equally spaced in energy. Accordingly we would expect to find overtones of the fundamental vibration frequency at exactly integral multiples of the fundamental. Real molecules are not simple harmonic oscillators, however, because the potential wells are non-parabolic. This is also illustrated by Fig. 14.6, and it can be seen that for larger displacements the potential is not parabolic and the vibrational levels lie closer together as the quantum number (or energy) increases.

For HCl the strongest absorption in the vibrational spectrum is due to the transition $v = 0 \rightarrow 1$ and occurs at 3.46 μm (8.66 $\times$ 10^{13} s^{-1} or 2889 cm^{-1}). The next strongest absorption is the first overtone which is due to the transition $v = 0 \rightarrow 2$ and occurs at 1.76 μm (17.0 $\times$ 10^{13} s^{-1} or 5668 cm^{-1}). The first overtone is shown later in Fig. 15.4. The complicated shape results from changes in rotational energy which accompany the changes in vibrational energy. The second overtone band is due to $v = 0 \rightarrow 3$ and occurs at 1.197 μm (25.1 $\times$ 10^{13} s^{-1} or 8348 cm^{-1}). If HCl were a harmonic oscillator the overtones would appear at multiples of the fundamental frequency, that is at 5772 cm^{-1} and 8657 cm^{-1}, rather than the actual frequencies of 5668 cm^{-1} and 8348 cm^{-1}.

It is often convenient to have a mathematical expression for the potential energy curve of a molecule. The Morse function has already been given in Section 14.4. It has the right general shape and may be used to obtain a much better expression for the vibrational energy levels of a molecule than simply using the harmonic oscillator approximation. When the Morse equation (equation 14.37) is substituted in the Schrödinger equation the following expression is obtained for the first few vibrational energy levels.

$$E = (v + \tfrac{1}{2})h\nu_0 - x_e h\nu_0 (v + \tfrac{1}{2})^2 \qquad (15.23)$$

The second term is a small correction term, and

$$x_e = \frac{h\nu_0}{4D'}$$ (15.24)

The second term represents the decrease in energy of successive vibrational levels below that expected for a simple harmonic oscillator.

15.5 VIBRATION-ROTATION SPECTRUM

A diatomic molecule in the ground vibrational level ($v = 0$) may be in one of several rotational levels. The populations of the various rotational levels at equilibrium are determined by the Boltzmann distribution as discussed in Section 15.3.

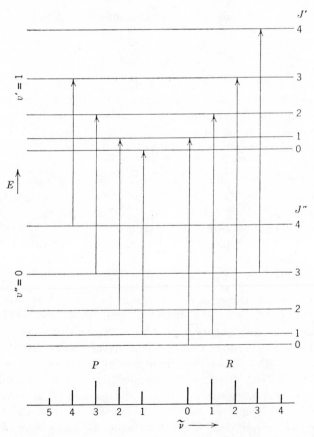

Fig. 15.3 Energy levels for a diatomic molecule and the transitions observed in the vibration-rotation spectrum. The lines are indexed according to the rotational quantum number of the initial state. The relative heights of the spectral lines indicate relative intensities of absorption.

When a diatomic molecule absorbs a quantum of radiation which is sufficiently large to raise it to a higher vibrational level, the molecule has to go to a different rotational sublevel of the higher vibrational state because the selection rule for the rotational quantum number J is $\Delta J = \pm 1$. The possible transitions are shown in Fig. 15.3 where the lower group of rotational levels all belong to the $v'' = 0$ vibrational level and the upper group of rotational levels all belong to the $v' = 1$ vibrational level. The energy before absorption of a quantum is given by

$$\frac{E''}{hc} = \tilde{v}_0(v'' + \tfrac{1}{2}) + BJ''(J'' + 1) \tag{15.25}$$

The energy after absorption of a quantum is given by

$$\frac{E'}{hc} = \tilde{v}_0(v' + \tfrac{1}{2}) + BJ'(J' + 1) \tag{15.26}$$

The frequency of an absorption line in wave numbers is given by

$$\tilde{v} = \frac{E'}{hc} - \frac{E''}{hc} \tag{15.27}$$

The transitions for which $\Delta J = -1$ produce the lines designated as the P branch of the spectrum. Substituting equations 15.25 and 15.26 into equation 15.27 with $v'' = 0$, $v' = 1$, and $J' = J'' - 1$, we obtain the wave numbers of the lines in the P branch of the spectrum, indicated at the bottom left of Fig. 15.3.

$$\tilde{v}_P = \tilde{v}_0 - 2BJ'' \tag{15.28}$$

where J'' is the rotational quantum number of the initial state and may have values of $1, 2, 3, \ldots$.

The transitions for which $\Delta J = +1$ produce the lines designated as the R branch of the spectrum. Substituting equations 15.25 and 15.26 into equation 15.27 with $v'' = 0$, $v' = 1$, and $J' = J'' + 1$, we obtain the wave numbers of the lines in the R branch.

$$\tilde{v}_R = \tilde{v}_0 + 2B + 2BJ'' \tag{15.29}$$

where J'' is the rotational quantum number of the initial state and may have values of $0, 1, 2, \ldots$. It is evident from equations 15.28 and 15.29 that there will not be a line with a frequency of v_0. This is illustrated by the HCl spectrum in Fig. 15.4. These are the lines for the first overtone band of HCl which is due to the transition $v = 0 \rightarrow 2$. The spectrum is made up of double peaks because the isotopically different molecules $H^{35}Cl$ (75% abundance) and $H^{37}Cl$ (25% abundance) have different moments of inertia and different fundamental vibration frequencies. The $H^{37}Cl$ lines occur at lower frequencies than the $H^{35}Cl$ lines, and the spacing of the rotational lines is smaller for $H^{37}Cl$ than for $H^{35}Cl$.

According to the foregoing oversimplified theory, the lines would be equally spaced on a wave number scale and adjacent lines in either branch would be separated by $2B$. A more accurate theory has to make allowance for the fact that the rotational constant B has different values in the two vibrational states. For a

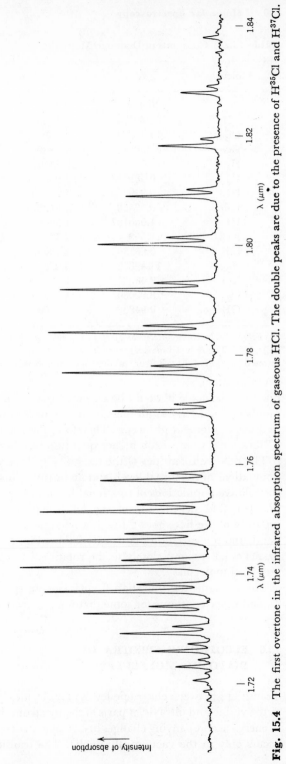

Fig. 15.4 The first overtone in the infrared absorption spectrum of gaseous HCl. The double peaks are due to the presence of H³⁵Cl and H³⁷Cl.

Table 15.2[1] Constants of Diatomic Molecules

Molecule	$N_A\mu$	$r \times 10^{10}$, m	$\tilde{\nu}_0$, cm^{-1}	D, eV
Br$_2$	39.958	2.283$_6$	323.2	1.971
CH	0.930024	1.1198	2861.6	3.47
Cl$_2$	17.48942	1.988	564.9	2.475
CO	6.85841	1.1282	2170.21	11.108
H$_2$	0.504066	0.7416$_6$	4395.24	4.476$_3$
H$_2^+$	0.503928	1.06	2297	2.648$_1$
HCl	0.979889	1.27460	2989.74	4.430
HBr	0.99558	1.4138	2649.67	3.75$_4$
HI	1.000187	1.604$_1$	2309.53	3.056$_4$
KCl	18.599	2.79	280	4.42
LiH	0.881506	1.5953$_5$	1405.649	2.5
Na$_2$	11.49822	3.078$_6$	159.23	0.73
NO	7.46881	1.1508	1904.03	6.487
O$_2$	8.00000	1.20739$_8$	1580.361	5.080
OH	0.94838	0.9706	3735.21	4.35

[1] From G. Herzberg, *Molecular Spectra and Molecular Structure*, D. Van Nostrand Co., Princeton N.J., 1950. The reduced masses are for the molecules with the most abundant isotopic species. The lowered position of the last figure of some of the numbers indicates that this figure is uncertain.

diatomic molecule it would be expected that an increase in vibrational quantum number would lead to an increase in the mean internuclear distance, and hence to a larger moment of inertia. Therefore, the rotational constant B is smaller for a vibrational state with a higher quantum number.

The relative intensities of the lines in Fig. 15.4 are proportional to the relative populations of the initial levels and an intrinsic line intensity. As explained earlier the relative populations of rotational levels are proportional to the product of the $(2J + 1)$-fold degeneracy of the levels and the Boltzmann factor. Thus the intensity of the lines goes through a maximum on either side of the center of the band, and the variation in intensity follows Fig. 15.1. The maxima in the P and R branches correspond closely to the rotational levels that have the greatest populations of molecules.

The fundamental vibration frequencies $\tilde{\nu}_0$, spectroscopic dissociation energies D, and other constants for some diatomic molecules are given in Table 15.2.

15.6 ELECTRONIC SPECTRA OF DIATOMIC MOLECULES

Transitions between electronic levels of molecules lead to absorption and emission in the visible and ultraviolet parts of the spectrum. For some molecules the changes in energy accompanying changes in elctronic structures are so great that absorption occurs only in the vacuum ultraviolet. The transitions between electronic states may be accompanied by transitions between rotational and vibrational states, and so the spectra may be very complicated.

All molecules show electronic spectra because the change from one electronic structure to another always provides the opportunity for interaction with electromagnetic radiation. Homonuclear diatomic molecules, which do not have rotation or vibration-rotation spectra, do show vibrational and rotational structure in their electronic spectra.

An important principle for the interpretation of electronic spectra was stated by Franck and Condon. They pointed out that since electrons move much more rapidly than nuclei it is a good approximation to assume that, in an electronic transition, the nuclei do not change their positions. Therefore an electronic transition may be represented by a vertical line in the figures we will be considering.

Figure 15.5a shows the potential energy curves for the ground state and first excited state of a diatomic molecule XY. The molecule has a different electronic structure in the excited state, and the potential energy of the excited state is higher. The equilibrium internuclear separation is greater in the excited state that has an antibonding electron and therefore weaker bonding, as shown by the lower dissociation energy for the excited state. When the molecule in the ground state absorbs radiation it is most likely to be at its equilibrium internuclear separation, as discussed in Section 12.15 for the ground state of a harmonic oscillator. This is not true for the higher vibrational levels where the classical situation in which the highest probability is at the turning points is approached more and more closely as the vibrational energy is increased. According to the Franck-Condon principle the most probable transition is $0 \rightarrow 2$ because the excited molecule in the $v = 2$ state does not have much internuclear kinetic energy and has a high probability at the equilibrium internuclear distance of the vibrational zero point

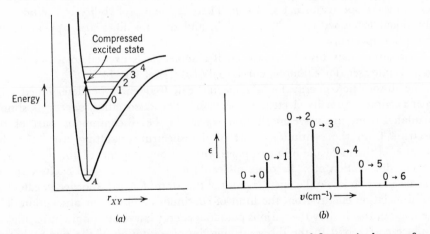

Fig. 15.5   (a) Potential energy curves for the ground state and first excited state of a diatomic molecule XY. (b) Absorption spectrum of diatomic molecule XY. (From N. J. Turro, *Molecular Photochemistry*, copyright (c) 1965 by W. A. Benjamin, Inc., Menlo Park, California.)

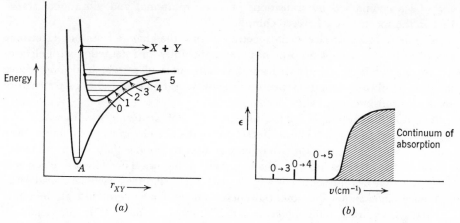

Fig. 15.6 (*a*) Potential energy curves for the ground state and first excited state of a diatomic molecule *XY*. The excited state has a larger equilibrium internuclear distance than the one in the preceding figure. (*b*) Absorption spectrum of diatomic molecule *XY*. (From N. J. Turro, *Molecular Photochemistry*, copyright (c) 1965 by W. A. Benjamin, Inc., Menlo Park, California.)

level of the ground electronic state. Transitions to other vibrational levels of the excited state occur with lower probabilities as indicated by the absorption spectrum shown in Fig. 15.5*b*.

If the minimum potential energy of the excited state is at a greater internuclear distance, the absorption of radiation by the ground state will produce excited molecules with enough energy to dissociate, as shown in Fig. 15.6*a*. Since variable amounts of energy may lead to dissociation, continuous absorption is produced as shown by the spectrum in Fig. 15.6*b*. There is some probability of excitation to vibrational levels with $v = 3$, 4, and 5, and so some lines also appear in the absorption spectrum.

Dissociation may also be produced if a molecule is excited to a nonbonded excited state (see, for example, curve $(\sigma_g 1s)(\sigma_u^* 1s)$ of Fig. 14.5).

The dissociation energy for a molecule can frequently be determined with great accuracy from its electronic spectrum. The dissociation energy is simply calculated from the wavelength that separates the discontinuous part of the spectrum from the continuous part of the spectrum resulting from the dissociation of the molecule into two parts with various amounts of kinetic energy.

Since dissociation of a diatomic molecule may take place into excited atoms, it is necessary to know the energy states of the atoms formed in order to calculate the dissociation energy from the limit of continuous emission or absorption. The energies approached by the various potential energy curves at infinite internuclear distance correspond to the different energies of excitation of the two separated atoms.

The dissociation energy of O_2 has been determined very accurately from its ultraviolet spectrum. Continuous absorption begins at 1759 Å (56,877 cm^{-1}).

One of the oxygen atoms produced is in an excited state with an energy of $15,868$ cm^{-1} above the ground state, and the other is in the ground electronic state. Therefore the dissociation energy into two atoms in their ground states is $56,877 - 15,868 = 41,009$ cm^{-1} or 5.08 eV.

The dissociation generally takes place from the lowest vibrational level of the ground electronic state, but for heavy diatomic molecules such as I_2 the separation of the vibrational levels is small, and so at room temperature the molecules occupy several of the lower vibrational levels of the ground electronic state.

The fluorescence and phosphorescence spectra of diatomic molecules are discussed in Section 18.3.

15.7 ROTATIONAL SPECTRA OF POLYATOMIC MOLECULES

For the treatment of the pure rotational spectrum we may consider the polyatomic molecule to be a rigid framework with fixed bond lengths and angles equal to their mean values. For a polyatomic molecule the moment of inertia about a particular axis is simply the sum of the moments due to the various nuclei about that axis

$$I = \sum_i m_i r_i^2 \tag{15.30}$$

where r_i is the perpendicular distance of the nucleus of mass m_i from the axis.

Molecules may be classified according to their ellipsoid of revolution, constructed as follows: lines are drawn from the center of mass of the molecule in various directions with lengths proportional to the moment of inertia of the molecule about that line as an axis. The principal axes x, y, and z of the ellipsoid formed by the ends of these lines are used to calculate the principal moments of inertia I_x, I_y, and I_z. These principal moments of inertia are used to classify molecules as shown in Table 15.3. For molecules with one or more axes of symmetry, one principal axis is the axis of highest symmetry. The second is perpendicular to the first and to a "vertical" plane of symmetry if one exists. The third is perpendicular to the other two. Since a molecule must have a permanent dipole moment in order to

Table 15.3 Classification of Polyatomic Molecules According to Their Moments of Inertia

Moments of Inertia	Type of Rotor	Examples
$I_x = I_y \quad I_z = 0$	Linear	HCN
$I_x = I_y = I_z$	Spherical top	CH_4, SF_6, UF_6
$I_x = I_y > I_z$	Prolate $\Big\}$ symmetric top	NH_3, CH_3Cl
$I_x = I_y < I_z$	Oblate	
$I_x \neq I_y \neq I_z$	Asymmetric top	CH_2Cl_2, CH_3OH

have a pure rotational spectrum, spherical top molecules have no observable rotational spectrum. Some linear molecules, symmetric tops, and asymmetric tops also have zero dipole moment by symmetry and so no rotational spectrum.

The classical rotational energy of a rigid molecule is given by

$$E_{rot} = \frac{M_x^2}{2I_x} + \frac{M_y^2}{2I_y} + \frac{M_z^2}{2I_z} \tag{15.31}$$

where

$$M_x = I_x \omega_x \tag{15.32}$$

$$M_y = I_y \omega_y \tag{15.33}$$

$$M_z = I_z \omega_z \tag{15.34}$$

and ω_x, ω_y, and ω_z are the angular velocities about the three axes. The square of the total angular momentum M^2 is given by

$$M^2 = M_x^2 + M_y^2 + M_z^2 = I_x^2 \omega_x^2 + I_y^2 \omega_y^2 + I_z^2 \omega_z^2 \tag{15.35}$$

For linear molecules the principal axis is taken as the z axis and so $I_z = 0$. The other two moments are equal so that

$$E_{rot} = \tfrac{1}{2} I(\omega_x^2 + \omega_y^2) = \tfrac{1}{2} I \omega^2 = \frac{M^2}{2I} \tag{15.36}$$

Quantum mechanically the square of the angular momentum M^2 is given by $J(J+1)\hbar^2$, and so we obtain from equation 15.36 the equation used for diatomic molecules (equation 15.8).

A symmetric top molecule is one for which two of the moments of inertia are equal. It can be shown that this will be true for molecules with an axis of threefold or higher symmetry. The threefold or higher axis is taken to be the z axis. Letting $I_z = I_A$ and $I_x = I_y = I_B$

$$
\begin{aligned}
E_{rot} &= \frac{1}{2I_B}(M_x^2 + M_y^2) + \frac{M_z^2}{2I_A} \\
&= \frac{M^2}{2I_B} + \frac{1}{2}\left(\frac{1}{I_A} - \frac{1}{I_B}\right)M_z^2
\end{aligned} \tag{15.37}
$$

Quantum mechanically the allowed values for M^2 and M_z are

$$M^2 = J(J+1)\hbar^2 \qquad J = K, K+1, K+2, \cdots \tag{15.38}$$

$$M_z = K\hbar \qquad\qquad K = 0, \pm1, \pm2, \cdots, \pm J \tag{15.39}$$

Thus the rotational levels for a symmetric top are given by

$$E = \frac{\hbar^2}{2I_B} J(J+1) + \frac{\hbar^2}{2}\left(\frac{1}{I_A} - \frac{1}{I_B}\right)K^2 \tag{15.40}$$

Asymmetric top molecules have three different moments of inertia. There is no simple formula for the rotational levels in terms of J and K. However, the rotational spectrum may be calculated for any set of three moments of inertia using a computer or the three moments may be calculated from an observed spectrum.

15.8 MICROWAVE SPECTROSCOPY

Microwave radiation is produced by special electronic oscillators. Monochromatic radiation is produced, and the frequency may be varied continuously over wide ranges. The usual experimental arrangement is described in Fig. 15.7. Microwave radiation is transmitted down a wave guide that contains the gas being studied. The intensity of the radiation at the other end of the wave guide is measured by use of a crystal diode detector and amplifier. The oscillator frequency is swept over a range, and the intensity at each frequency is presented on an oscilloscope.

According to the Heisenberg uncertainty principle the accuracy with which an energy level may be determined is inversely proportional to the time the molecule is in this level. Hence, to obtain sharp rotational lines of a gas the pressure must be maintained sufficiently low so that the average time between collisions is long compared to the period of a rotation. Usually it is necessary to determine micro-wave spectra at pressures below 0.1 Torr to reduce the broadening effects of collisions.

Microwave lines at low pressures are very sharp, and their frequencies may often be determined to the order of 1 ppm. Therefore, the moments of inertia of a molecule may be determined with great accuracy. As a matter of fact, this method has made possible the most precise evaluations of bond lengths and bond angles. To obtain these values for polyatomic molecules it is necessary to study the

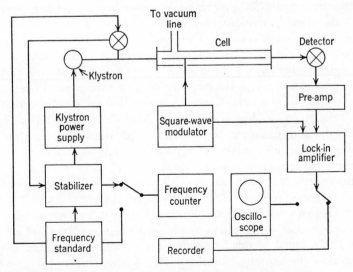

Fig. 15.7 Block diagram of a Stark-modulated microwave spectrometer.

spectra of isotopically different molecules. In effect a number of simultaneous equations are solved for the internuclear distances and angles.

The lines in the microwave spectrum are split if the molecules being studied are in an electrostatic field. This so-called Stark effect is due to the interaction of the dipole moment of the gaseous molecule and the electric field. Since the splitting is proportional to the permanent dipole moment, the magnitude of the dipole moment may be calculated.

15.9 VIBRATIONAL SPECTRA OF POLYATOMIC MOLECULES

In order to describe the positions of N nuclei in a molecule it is necessary to specify $3N$ coordinates. Three of these may be used to specify the position, and therefore the translational motion, of the center of mass of the molecule. Another three may be used to specify the angular position of a nonlinear molecule with respect to some coordinate system. For example, the orientation of a water molecule may be described by giving two angles to locate the direction of the C_2 axis (Section 13.3) and the angle of the plane of the molecule. For a linear molecule only two angles need to be specified. Thus $3N$-6 coordinates remain for a nonlinear molecule to describe vibrations, and $3N$-5 coordinates remain for a linear molecule. These are commonly called internal degrees of freedom, and they include vibrations and internal rotations. If there is an appreciable potential energy for an internal rotation about some bond there will be an oscillation about a mean position. In ethylene, $CH_2{=}CH_2$, there is a large potential barrier for internal rotation so that there are only small oscillations about the $C{=}C$ bond. In some cases, like $CH_3{-}CH_3$, the potential energy barrier is small enough that the internal rotation is said to be "free" at room temperature.

The vibratory motion of a polyatomic molecule may be very complicated, but all of the possible motions may be described in terms of simple motions, called the *normal modes of vibration*. The number of normal modes is equal to the number of internal degrees of freedom that we have just calculated.

In a normal mode of vibration, the nuclei move in straight lines and in phase (i.e., the nuclei pass through the extremes of their motion simultaneously). The motions of the nuclei in a normal mode are such that the center of mass does not move, and the molecule does not rotate as a whole. This means that different atoms move different distances. Each normal mode has a characteristic vibration frequency. Any vibratory motion of a molecule may be expressed as a sum of displacements in the various normal modes.

For a linear triatomic molecule like CO_2, $3N - 5 = 4$, and so there are four vibrational degrees of freedom. Two are symmetrical and asymmetrical stretching vibrations. These normal modes are shown in Fig. 15.8. Two are bending motions that differ only in that they are in mutually perpendicular planes and are therefore degenerate. It is noted in this figure that the stretching vibrations have higher frequencies than the bending vibrations. This is often true and is a result of the fact that a larger force is required to stretch a molecule than to bend it.

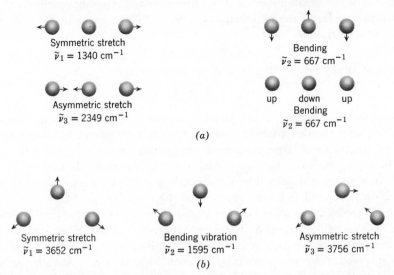

Fig. 15.8   (a) Normal modes of vibration of symmetrical linear triatomic molecule CO_2. (b) Normal modes of vibration of nonlinear triatomic molecule H_2O. The frequencies of these vibrations are given in wave numbers.

Of the three normal vibrations for CO_2 the symmetric stretch is not active in the infrared but the other vibrations are. Since CO_2 is symmetrical in its equilibrium state it does not have a dipole moment, and the symmetrical stretching vibration does not create one. The unsymmetrical normal vibrations produce a dipole moment.

For a nonlinear triatomic molecule, like H_2O, $3N - 6 = 3$, and so there are 3 normal modes of vibration. These normal modes are shown in Fig 15.8b.

The frequencies, in cm^{-1}, of the strongest bands for H_2O vapor are summarized in Table 15.4. The weaker bands in the spectrum are the overtones and combinations shown in the table. The vibrations are not strictly harmonic, and so the overtones are not exact multiples, and the combinations are not exact sums as shown in the table.

The use of normal coordinates in describing molecular vibrations allows the

Table 15.4   Infrared Bands of H_2O Vapor

$\tilde{\nu}$, cm^{-1}	Intensity	Interpretation
1595.0	very strong	$\tilde{\nu}_2$
3151.4	medium	$2\tilde{\nu}_2$
3651.7	strong	$\tilde{\nu}_1$
3755.8	very strong	$\tilde{\nu}_3$
5332.0	medium	$\tilde{\nu}_2 + \tilde{\nu}_3$
6874	weak	$2\tilde{\nu}_2 + \tilde{\nu}_3$

separation of the Schrödinger equation into as many separate equations as there are normal modes, if the potential is quadratic. The eigenvalue for each of the separated equations is

$$E_i = (v_i + \tfrac{1}{2})h\nu_i \tag{15.41}$$

and so the total vibrational energy of the molecule is

$$E_{\mathrm{vib}} = \sum_i (v_i + \tfrac{1}{2})h\nu_i \tag{15.42}$$

The vibrational spectra of polyatomic molecules are very useful in identifying them and serve also as a criterion of purity. For such practical applications the infrared spectra for a large number of compounds have been cataloged and are used like "fingerprints". Groups of atoms within the molecule have quite characteristic absorption bands. The wavelengths at which a certain group absorbs vary slightly, depending on the structure of the rest of the molecule.

The infrared spectrum may be considered to be made up of several regions.*

$3700-2500$ cm^{-1}—hydrogen stretching vibrations. These vibrations occur at high frequencies because of the low mass of the hydrogen atom. If an OH group is not involved in hydrogen bonding (Section 14.8) it usually has a frequency in the vicinity of $3600-3700$ cm^{-1}. Hydrogen bonding causes this frequency to drop by 300 to 1000 cm^{-1} or more. The NH absorption falls in the $3300-3400$ cm^{-1} range, and the CH absorption falls in the $2850-3000$ cm^{-1} range. As the atom to which hydrogen is attached becomes heavier, the frequencies decrease; for SH, PH, and SiH they are approximately 2500, 2400 and 2200 cm^{-1}.

$2500-2000$ cm^{-1}—triple-bond region. These bonds have high frequencies because of the large force constants. The C≡C group usually causes absorption between 2050 and 2300 cm^{-1}, but this absorption may be weak or absent because of the symmetry of the molecule. The C≡N group absorbs near $2200-2300$ cm^{-1}.

$2000-1600$ cm^{-1}—double-bond region. Absorption bands of substituted aromatic compounds fall in this range and are a good indicator of the position of the substitution. Carbonyl groups, C=O, of ketones, aldehydes, acids, amides, and carbonates usually show strong absorption in the vicinity of 1700 cm^{-1}. Olefins, C=C, may show absorption in the vicinity of 1650 cm^{-1}. The bending of the C—N—H bond also occurs in this region.

$1500-1700$ cm^{-1}—single-bond stretch and bend region. This region is not diagnostic for particular functional groups, but it is a useful "fingerprint" region since it shows differences between similar molecules. Organic compounds usually show peaks in the region between 1300 and 1475 cm^{-1} due to the bending motions of hydrogens. Out-of-plane bending motions of olefinic and aromatic CH groups usually occur between 700 and 1000 cm^{-1}.

15.10   RAMAN SPECTRA

When light passes through a liquid or gas a small fraction of the light is scattered. A perfect crystalline solid would not scatter light because the light scattered by one

---

* R. P. Bauman, *Absorption Spectroscopy*, Wiley, New York, 1962, pp. 336–347.

unit of the crystal would be destroyed by interference with light scattered from another unit. The mechanism of the scattering of light involves the polarization of the molecules or atoms by the electric field of the light. Thus the electric field of the light induces a rapidly fluctuating dipole in the atoms or molecules in its path. As discussed in Section 20.12 the fluctuating dipole leads to the emission of electromagnetic waves in various directions at the same frequency as the incident radiation, and this radiation is seen as the scattered light. This scattering, referred to as Rayleigh scattering, may be viewed as the elastic scattering of a photon by a molecule.

In 1928 Raman discovered experimentally that there was sometimes also present in the scattered light weak radiation of frequencies not present in the incident light. The frequency differences between the weak lines and the exciting line are characteristic of the scattering substance and are independent of the frequency of the exciting line. Raman scattering may be viewed as inelastic scattering of a photon by a molecule. It is different from fluorescence or phosphorescence (Section 18.3) in that the sample does not have an absorption band at the wavelength of the incident light. In other words, any wavelength may be used to study the Raman effect. Since it is a scattering phenomenon the scattered intensity is proportional to the fourth power of the frequency.

Since Raman lines are weak, long exposures have been required with conventional light sources, but the development of lasers has made available powerful monochromatic light sources for the study of Raman scattering. The advantages of these sources are their high intensity of light collimated in a particular direction, the small line width and the polarization of the beam. The most commonly used laser exciting line is 6328 Å from a He-Ne continuous laser.

The experimental arrangement for observing Raman spectra is indicated in Fig. 15.9. Light scattered by the sample is focussed on the slit of the spectrophotometer. Intensity is measured as a function of wave length by use of a photomultiplier tube. The Raman spectrum of liquid CCl_4 obtained in this way is shown in Fig. 15.10.

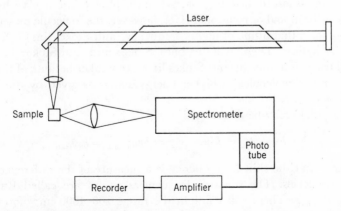

Fig. 15.9   Apparatus for obtaining Raman spectra.

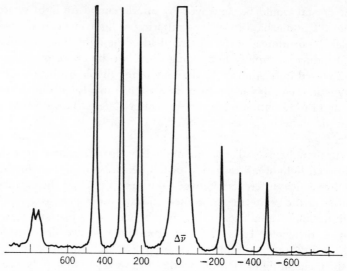

Fig. 15.10 Raman spectrum of liquid CCl_4 obtained using a He-Ne laser. The Raman frequencies of the lines on either side of the parent line ($\Delta\bar{\nu} = 0$) are indicated in cm^{-1}.

The change in frequency between the incident and scattered light results from an exchange of energy between the incident photon and the scattering molecule. The photon, of energy $h\nu$ insufficient to cause a transition to an excited electronic state and be absorbed, induces a forced oscillation in the molecule, which is in its ground electronic state and in a low vibration and rotational level. If the molecule is shifted from a level with energy E'' to another energy E', the scattered light has a frequency of ν', and from conservation of energy

$$h\nu + E'' = h\nu' + E' \tag{15.43}$$

Generally, the scattered light has a lower frequency (called a Stokes line) because energy is lost to the molecule. When the photon interacts with a molecule in a higher vibrational or rotational level, however, the molecule may give energy to the photon so that a line of higher frequency (anti-Stokes line) is found in the scattered radiation. There will be both Stokes and anti-Stokes lines for any permitted transition, but the anti-Stokes lines are weaker because of the relatively small number of molecules in higher energy states, as given by the Boltzmann distribution.

Equation 15.43 may be rearranged to

$$E' - E'' = h(\nu - \nu') = h\nu_{\text{Raman}} = hc\tilde{\nu}_{\text{Raman}} \tag{15.44}$$

which shows that the shift in frequency is a measure of the difference in energy between two levels. The shifts in frequencies $\nu - \nu'$ are called Raman shifts ν_{Raman} or $\tilde{\nu}_{\text{Raman}}$. These shifts fall in the range 100–4000 cm^{-1} for vibrational changes.

The selection rule for the vibrational Raman effect is

$$\Delta v = \pm 1 \tag{15.45}$$

In order for a vibrational motion to be Raman active it is necessary for the particular vibration to change the polarizability α (Section 14.12) of the molecule. The reason for this selection rule is readily understood because if there is a change in polarizability attending the vibration there will be a variation in the induced moment at the frequency of the vibration. Since the polarizability of a homonuclear diatomic molecule does change during a vibration, these molecules have vibrational Raman lines.

The data obtained from Raman and infrared spectra are complementary. A detailed consideration of the selection rules for the Raman effect shows that if a molecule has a center of symmetry (Section 13.2), then any vibration that is active in the infrared is inactive in the Raman effect, and vice versa.

Rotational transitions may also be observed by use of the Raman effect if the polarizability of a molecule depends upon its orientation. This is true for linear molecules, even those that because of symmetry do not show infrared or microwave spectra. The rotational selection rule for the Raman effect of linear molecules is

$$\Delta J = 0, \pm 2 \tag{15.46}$$

The 2 arises from the fact that the polarizability ellipsoid of a molecule, being the same at either end, presents the same appearance in a given direction twice in every complete rotation.

15.11 SPECTROPHOTOMETERS

When polychromatic radiation is passed into a substance, some of the radiation may be absorbed and the rest is either transmitted or scattered. The fraction transmitted may be determined as a function of frequency by use of a spectrophotometer. The design of a spectrophotometer is shown schematically in Fig. 15.11.

The principal parts of a spectrophotometer are the source of electromagnetic radiation, the monochromator, the cell compartment, the photoelectric detector, and a device for indicating the output from the detector (electrical meter, potentiometer, or recording potentiometer). The cell compartment contains one

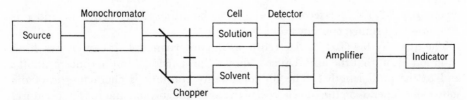

Fig. 15.11 A spectrophotometer.

optical absorption cell filled with the solution to be studied and another absorption cell filled with a reference solution, usually pure solvent. The ratio of the intensity I of transmitted light for the solution to the intensity I_0 for the solvent is called the *transmittancy*.

The transmittancy I/I_0 can be determined at different wavelengths, and the absorption spectrum can be mapped. With some spectrophotometers such a plot is recorded automatically. The positions and intensities of the absorption bands and lines serve for identification and for criteria of purity; the percentage transmissions serve for quantitative analyses of the concentration of material present.

15.12 LAMBERT-BEER LAW

The probability that a photon will be absorbed is usually directly proportional to the concentration of absorbing molecules and to the thickness of the sample for a very thin sample. This probability is expressed mathematically by the equation

$$\frac{dI}{I} = -kc \, dx \tag{15.47}$$

where I is the intensity of light of a particular wavelength, that is, the number of photons per unit area per unit time, and dI is the change in light intensity produced by absorption in a thin layer of thickness dx and concentration c. Distance x is measured through the cell in the direction of the beam of light which is being absorbed.

The intensity of a beam of light after passing through length l of solution is related to the incident intensity I_0 by equation 15.49, which is obtained by integrating equation 15.47 between the limits I_0 when $x = 0$ and I when $x = l$.

$$\int_{I_0}^{I} \frac{dI}{I} = -kc \int_{0}^{l} dx \tag{15.48}$$

$$\ln \frac{I}{I_0} = 2.303 \log \frac{I}{I_0} = -kcl \tag{15.49}$$

Since it is convenient to use logarithms to the base 10, the Lambert-Beer law is used in the form

$$\log \frac{I_0}{I} = A = \epsilon cl \tag{15.50}$$

where $\epsilon = k/2.303$ is referred to as the molar absorption coefficient and c is amount-of-substance concentration.

The quantity $\log (I_0/I)$ is referred to as the absorbance A. It can be seen from equation 15.50 that the absorbance is directly proportional to the concentration c and to the path length l. The proportionality constant is characteristic of the solute and depends on the wavelength of the light, the solvent, and the temperature. The Lambert-Beer law will not be obeyed unless the radiation is monochromatic.

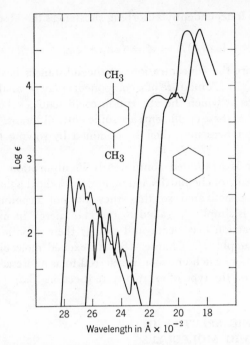

Fig. 15.12 Absorption spectra of benzene and paraxylene.

If the radiation is not monochromatic the absorption coefficient may vary significantly over the band of wavelengths used. The apparent absorption coefficient for a substance that associates or dissociates will change with concentration because of the changing ratio of concentrations of the absorbing species.*

Figure 15.12 shows the absorption spectra of benzene and paraxylene.*

Example 15.5 The percentage transmittancy of an aqueous solution of disodium fumarate at 250 nm and 25° is 19.2 % for a 5 $\times$ 10^{-4} M solution in a 1-cm cell. Calculate the absorbance A and the molar absorption coefficient ϵ.

$$A = \log (I_0/I) = \log (100/19.2) = 0.716$$

$$\epsilon = \frac{A}{lc} = \frac{0.716}{(1 \text{ cm})(5 \times 10^{-4} \text{ M})} = 1.43 \times 10^3 \text{ M}^{-1} \text{ cm}^{-1}$$

What will be the percentage transmittancy of a 1.75 $\times$ 10^{-5} M solution in a 10-cm cell?

$$\log (I_0/I) = (1.43 \times 10^3 \text{ M}^{-1} \text{ cm}^{-1})(10 \text{ cm})(1.75 \times 10^{-5} \text{ M})$$

$$= 0.250$$

$$I_0/I = 1.778 \quad \text{and} \quad 100 I/I_0 = 56.2 \%$$

---

* P. E. Stevenson, *J. Chem. Ed.*, **41**, 234 (1964).

For mixtures of independently absorbing substances the absorbance is given by the equation

$$\log (I_0/I) = A = (\epsilon_1 c_1 + \epsilon_2 c_2 + \cdots)l \qquad (15.51)$$

where c_1, c_2, ... are the concentrations of the substances having absorption coefficients of ϵ_1, ϵ_2, ... A mixture of n components may be analyzed by measuring A at n wavelengths at which the absorption coefficients are known for each substance, provided that these coefficients are sufficiently different. The concentrations of the several substances may then be obtained by solving the n simultaneous linear equations.

When a sample is irradiated continuously its absorption coefficient remains constant in the absence of chemical reaction, and this indicates that excited molecules are continuously deactivated so that they do not accumulate. Usually the excitation energy is simply degraded to thermal energy in molecular collisions, but a chemical reaction may occur and change the composition and absorption spectrum of the sample (cf., Chapter 18). An excited molecule may also emit a quantum of radiation. Such emission is referred to as fluorescence or phosphorescence, depending on the type of excited state (Section 18.3)

15.13 ELECTRONIC SPECTRA OF POLYATOMIC MOLECULES

Electronic transitions lead to absorption spectra in the visible and ultraviolet regions. For organic compounds this involves promotion of electrons in n, σ and π orbitals in the ground state to σ^* and π^* antibonding orbitals in the excited state. Since n electrons do not form bonds there are no antibonding orbitals associated with them. In the excited state the density of electrons between nuclei is lower than in the ground state and so the molecule is in a higher energy state.

The energy required for a $\sigma^* \leftarrow \sigma$ transition is generally so high that the absorption occurs in the vacuum ultraviolet. Thus saturated hydrocarbons and other compounds in which all valence shell electrons are involved in single bonds do not absorb in the ordinary ultraviolet or visible regions.

Ultraviolet absorptions due to $\sigma^* \leftarrow n$ transitions are shown by compounds that contain nonbonding electrons on oxygen, nitrogen, sulfur, or halogen atoms. For example, methyl alcohol vapor shows an absorption maximum at 1830 Å with a molar absorption coefficient of 150 M^{-1} cm^{-1}.

Groups like C=C, C=O, —N=N—, and —N=O, which cause absorption above 1800 Å, are called *chromophores*. The positions and intensities of the absorptions caused by these groups are rather characteristic. For example, double bonds C=C generally produce an absorption maximum at 1800–1900 Å with a molar absorption coefficient of about 10^4 M^{-1} cm^{-1}. Ketone and aldehyde C=O groups usually have absorption maxima at 2700–2900 Å with molar-absorption coefficients of 10–30 M^{-1} cm^{-1}. The most intense electronic-absorption spectra are produced by certain dyes that may have molar absorption coefficients as high as 10^5 M^{-1} cm^{-1} in solution.

Unsaturated molecules can show $\pi^* \leftarrow n$ and $\pi^* \leftarrow \pi$ transitions. One of the best understood cases is the carbonyl absorption of aldehydes and ketones. The stronger absorption band around 180 nm is due to the $\pi^* \leftarrow \pi$ transition, and the weaker absorption band around 285 nm is due to the $\pi^* \leftarrow n$ transition. The structure of the rest of the molecule affects the strength of the absorption and the wavelength of the maximum absorption, but a series of compounds with the same chromophore will generally show about the same ultraviolet absorption spectrum. When chromophoric groups are separated by two or more single bonds their effects are usually additive, but if they are a part of a conjugated* system large effects are seen because the π electron system is spread over at least four atomic centers. The absorption band is generally shifted 15–45 nm to longer wavelength. As further unsaturated groups are added to the conjugated system, there are further shifts to a longer wavelength and an increase in molar absorption coefficent because the energy required for the $\pi^* \leftarrow \pi$ transition is less, and the transition probability is higher. Single-ring aromatics absorb in the vicinity of 250 nm, naphthalenes in the vicinity of 300 nm, and anthracenes and phenanthrenes in the vicinity of 360 nm.

15.14 FREE-ELECTRON MODEL

For molecules with conjugated systems of double bonds (that is, $R(CH=CH)_n R'$) it is found that the electronic absorption bands shift to longer wavelengths as the number of conjugated double bonds is increased. Approximate quantitative calculations of the absorption frequencies may be made on the basis of the free-electron model for the π electrons of these molecules. The energy for the lowest electronic transition is that required to raise an electron from the highest filled level to the lowest unfilled level. In a system of conjugated double bonds each carbon atom has three σ bonds that lie in a plane and each sigma bond involves one outer electron of that carbon atom. Above and below this plane are the π orbital systems (Fig. 14.7). Each carbon atom contributes one electron to this π system, but these electrons are free to move the entire length of the series of π orbitals and are not localized at a given carbon atom. In the free-electron model it is assumed that the π system is a region of uniform potential and that the potential energy rises sharply to infinity at the ends of the system (that is, a square-well potential). Thus the energy levels E available to the π electrons would be expected to be those calculated for the particle restricted to movement in one direction (Section 12.12).

$$E = \frac{n^2 h^2}{8 m_e a^2} \tag{15.52}$$

The length of the box a is usually taken to be the length of the chain between terminal carbon atoms plus a bond length or two.

The π electrons (one for each carbon atom) are assigned to energy levels so that there are two (one with spin $+\frac{1}{2}$, and the other with spin $-\frac{1}{2}$) in each level starting with the lowest. For a completely conjugated hydrocarbon the number of π electrons is even, and the quantum number of the highest filled level will be $n = N/2$, where N is the number of π electrons

---

* In conjugated molecules double and single bonds alternate.

(the number of carbon atoms involved). In absorption an electron from the highest filled level is excited to the next higher level with quantum number $n' = N/2 + 1$. The difference in energy of these two levels is

$$\Delta E = \frac{h^2}{8m_e a^2} (n'^2 - n^2)$$

$$= \frac{h^2}{8m_e a^2} [(N/2 + 1)^2 - (N/2)^2]$$

$$= \frac{h^2}{8m_e a^2} (N + 1) \tag{15.53}$$

The absorption frequency in wave numbers is given by

$$\tilde{\nu} = \frac{\Delta E}{hc} = \frac{h(N + 1)}{8cm_e a^2} \tag{15.54}$$

Example 15.6 Calculate the lowest absorption frequency for octatetraene (C_8H_{10}) that contains a series of four conjugated double bonds. The length of the π bond system is about 9.5 Å.

$$\tilde{\nu} = \frac{h(N + 1)}{8cm_e a^2}$$

$$= \frac{(6.62 \times 10^{-34} \text{ J s})(9)(10^{-2} \text{ m cm}^{-1})}{8(3 \times 10^8 \text{ m s}^{-1})(9.110 \times 10^{-31} \text{ kg})(9.5 \times 10^{-10} \text{ m})^2}$$

$$= 30{,}200 \text{ cm}^{-1}$$

The observed absorption band is at 33,100 cm^{-1}.

15.15 DISPERSION AND ABSORPTION

When electromagnetic radiation passes through a medium there are two inter-related effects; dispersion and absorption. Dispersion refers to the dependence of the refractive index upon wavelength. The term dispersion is used because this dependence indicates how the light of various wavelengths would be dispersed by a prism of the medium.

When a medium has an absorption band the dispersion curve has a characteristic shape in the vicinity of the absorption band. Figure 15.13 shows the dependence of refractive index n and molar absorption coefficient ϵ on wavelength for the ideal case that there is a single absorption band. The refractive index normally increases on going to shorter wavelengths, and so when it decreases as an absorption band is approached from longer λ, the dispersion is said to be anomalous. A strong absorption band will give rise to a large effect and a weak absorption band will give rise to a small effect. As seen in Fig. 15.13 the effect on the refractive index extends over a much greater range of wavelength than the absorption.

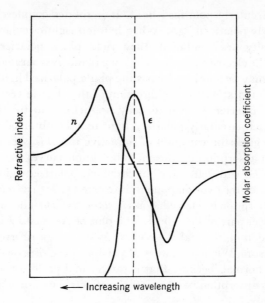

Fig. 15.13 Variation of refractive index n and molar absorption coefficient ϵ in the neighborhood of a single absorption line. (From J. G. Foss, *J. Chem. Ed.*, **40**, 592 (1963).)

These effects are encountered with all kinds of radiation, including absorption and dispersion of sound, and, since Chapter 16 is on nuclear magnetic resonance and electron spin resonance, we may note that they exhibit absorption and dispersion phenomena analogous to the ones we have discussed. These phenomena are explained classically in terms of damped harmonic oscillators. When atomic or molecular oscillators are set in motion by a light wave they absorb and absorption is a maximum at their resonant frequency. Since oscillating electrons radiate light, the interaction of this scattered light with the original light beam leads to dispersion.

15.16   CIRCULAR BIREFRINGENCE
AND CIRCULAR DICHROISM*

The dependence of refractive index and absorption coefficient on wavelength can also be measured with circularly polarized light. Circularly polarized light is light in which the electric vector rotates as the light beam advances. If the electric vector rotates clockwise as observed facing the light source the light is said to be

---

* D. J. Caldwell and H. Eyring, *The Theory of Optical Activity*, Wiley, New York, 1971; C. Djerassi, *Optical Rotatory Dispersion*, McGraw-Hill Book Co., New York, 1959; J. G. Foss, *J. Chem. Ed.*, **40**, 592 (1963); G. Snatzke, *Optical Rotatory Dispersion and Circular Dichroism in Organic Chemistry*, Sadtler Research Labs., Inc., Philadelphia, Pa., 1967; L. Velluz, M. Legrand, and M. Grosjean, *Optical Circular Dichroism*, Academic Press, New York, 1965.

right-handed circularly polarized light. If it rotates counterclockwise, it is left-handed circularly polarized light. When left and right circularly polarized beams of equal intensity are combined, they yield plane polarized light. In plane polarized light the electric vector remains in a plane. A separated beam of circularly polarized light may be obtained by passing plane polarized light through a quarter wave plate oriented at 45° to the direction of the electric vector of the polarized light. Since the quarter wave plate may be inclined to the right or to the left, both right and left circularly polarized light may be obtained in this way.*

Figure 15.14 gives the variation of refractive index (dispersion curve) and the variation of the absorption coefficient (absorption curve) for an optically active material measured with left and right circularly polarized light. The difference in refractive index for the two components is referred to as *circular birefringence*, and the difference in absorption is referred to as *circular dichroism*. The difference curves are shown in the lower part of Fig. 15.14. The plot of Δn versus λ is referred to as the rotatory dispersion curve, and the plot of $\Delta \epsilon$ versus λ is referred to as the circular dichroism spectrum. When a rotatory dispersion curve rises on going to shorter λ it is referred to as normal, because that is the normal pattern for a refractive index curve. When an absorption band causes the effects shown in Fig. 15.14 the whole

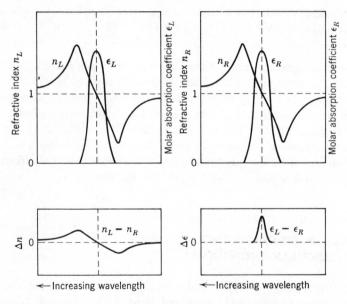

Fig. 15.14   Variation of refractive index n and molar absorption coefficient ϵ in the neighborhood of a single absorption line of an optically active substance, as measured with left- and right-handed circularly polarized light. The difference curves are referred to as the rotatory dispersion curve (Δn) and the circular dichroism spectrum ($\Delta \epsilon$). (From J. G. Foss, *J. Chem. Ed.*, **40,** 592 (1963).)

---

* R. P. Feynman, R. B. Leighton, and M. Sands, *The Feynman Lectures on Physics*, Addison-Wesley Publ. Co., Reading, Mass., 1963.

phenomenon is referred to as a Cotton effect. In contrast with ordinary dispersion a strong absorption bond may or may not give a large effect on the rotatory dispersion and a weak absorption band may give a large effect on the rotatory dispersion.

Substances that show circular birefringence and circular dichroism are said to be optically active. They can be divided into two classes: one in which optical activity is found only in the crystal form, for example, quartz, and the other in which it is found in the solid, gaseous, and liquid states of the pure substance or in solutions. Optical activity arises in the former group due to a right- or left-hand spiral structure in the crystal and disappears when this structure is melted. Substances in the latter category are optically active because of the asymmetry of the molecule itself. For a molecule whose mirror image is not superimposable on the molecule, left and right circularly polarized light have different refractive indices and correspondingly different absorption coefficients. This may happen for any molecule having only proper rotation elements of symmetry (Section 13.11). A molecule possessing any improper rotation axis (S_n), including a mirror plane or a center of symmetry, cannot be optically active.

In order to clarify what is observed experimentally let us consider Fig. 15.15. This figure shows the vector sum of the intensities of left and right circularly polarized light at one instant only. If plane polarized light (Fig. 15.15a) is passed through a sample having different refractive indices n_L and n_R for left- and right-handed circularly polarized light, there will be a rotation of the plane of polarization due to the different velocities of the left and right circularly polarized components, as shown in Fig. 15.15b. If the refractive indices of the medium are different for the two circularly polarized components, the angles of the two electric vectors measured with respect to the plane of the plane polarized light will be different after passage through a layer of the medium. It can be seen from the figure that the sum of the two electric vectors does not lie in the original plane, so that the plane has been rotated. The angle δ_n (in radians) through which the resultant electric vector is rotated is given by

$$\delta_n = \frac{\pi}{\lambda} (n_R - n_L) l \qquad (15.55)$$

where n_R and n_L are the refractive indices for the right and left circularly polarized rays, l is the length of the medium, and λ is the wavelength in vacuum. Thus optical rotation may arise from a difference in the refractive indices for the two circularly polarized components.

The rotation of plane-polarized light is measured with a polarimeter that consists of a light source, linear polarizer, sample, and analyzer (another linear polarizer). The rotation of the plane of polarization by the sample is measured by rotating the analyzer. If a substance rotates the plane of polarized light to the right, or clockwise, as viewed looking toward the light source, it is said to be dextrorotatory; if the rotation is counter-clockwise the substance is levorotatory. The magnitude of the rotation α is directly proportional to the length l of the sample and the concentration c of the optically active molecules, and so it is convenient to calculate

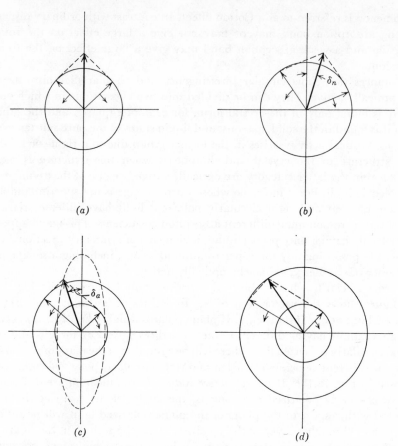

Fig. 15.15 Circular birefringence and circular dichroism. (*a*) Representation of plane polarized light as the sum of two components of circularly polarized light rotating in opposite directions. (*b*) Rotation of plane polarized light by an angle δ_n due to different velocities of propagation of left and right circularly polarized light. (*c*) Production of elliptically polarized light from plane polarized light by different absorption coefficients for left and right circularly polarized light. (*d*) Effects on plane polarized light of both differences in refractive indices and absorption coefficients for the two circularly polarized components. In (*c*) and (*d*) the emerging beam is elliptically polarized.

a *specific rotation* [α] that is defined by

$$[\alpha] = \frac{\alpha}{cl} \tag{15.56}$$

where l is the path length in decimeters and c is the concentration in grams per cubic centimeter of solution. For a pure substance, c equals ρ, the density of the pure substance. The specific rotation varies with the wavelength, temperature, and solvent so that these variables must be specified.

In Fig. 15.15c we consider the effect of greater absorption of one of the circularly polarized components than the other. In this figure the electric vectors of the left and right circularly polarized components are shown as having rotated through equal angles in opposite directions, but the intensity of the right circularly polarized component has been reduced by absorption. The locus traced out by by the tip of the resultant vector is an ellipse for this case, and so the resultant beam is elliptically polarized light, in contrast with the situation in (b), which produces plane-polarized light. Although partial absorption of one of the components produces elliptically polarized light, the major axis of the ellipse still lies along the vertical direction, that is, it is not rotated.

In general the effects described in Fig. 15.15b and c occur together, and the resultant is described in Fig. 15.15d. The emerging beam is elliptically polarized, and the major axis is rotated. The ellipticity is determined by the difference in absorbance, and the optical rotation is determined by the difference in refractive indices.

A molecule may have several optically active absorption bands and therefore several Cotton effects. Both the classical electromagnetic and quantum mechanical theories for optical rotation predict a variation of $[\alpha]$ with wavelength of the form:

$$[\alpha] = \sum_{i=1}^{n} \frac{k_i}{\lambda^2 - \lambda_{oi}^2} \tag{15.57}$$

where the k_i and λ_{oi} are constants, and λ is the wavelength of the incident monochromatic light. Formally the λ_{oi} are the wavelengths of certain absorption bands in the molecule—those that are optically active. This equation applies only at some distance from an absorption band; the rotation does not become infinite. Not all absorption bands in an optically active molecule are necessarily optically active. If the chromophore is far removed from the asymmetric center the difference in absorption of left and right circularly polarized light may be negligible.

The rotatory dispersion curve and circular dichroism spectrum are used in determining the structure, configuration, and conformation of complicated optically active molecules such as steroids. Another region of extensive interest is that of proteins and synthetic polypeptides. Here information about gross conformational changes can be obtained since optical rotation is very sensitive to the configuration and conformation of molecules.

The Faraday effect is the rotation of plane polarized light by an otherwise optically inactive medium placed in a magnetic field. The Kerr electric-optic effect is the rotation of plane-polarized light by an otherwise optically isotropic substance placed in an electric field.

References

C. N. Banwell, *Fundamentals of Molecular Spectroscopy*, McGraw-Hill Book Co., New York, 2nd ed., 1972.

G. M. Barrow, *Introduction to Molecular Spectroscopy*, McGraw-Hill Book Co., New York, 1962.

G. M. Barrow, *The Structure of Molecules*, W. A. Benjamin, Inc., New York, 1963.

R. P. Bauman, *Absorption Spectroscopy*, Wiley, New York, 1962.

R. Chang, *Basic Principles of Spectroscopy*, McGraw-Hill Book Co., New York, 1971.

R. T. Conley, *Infrared Spectroscopy*, Allyn and Bacon, Boston, 1966.

J. C. Davis, Jr., *Advanced Physical Chemistry*, Ronald Press, New York, 1965.

H. B. Dunford, *Elements of Diatomic Molecular Spectra*, Addison-Wesley Publ. Co., Reading, Mass., 1968.

J. R. Dyer, *Applications of Absorption Spectroscopy of Organic Compounds*, Prentice-Hall, Inc., Englewood Cliffs, N.J., 1965.

T. R. Gilson and P. J. Hendra, *Laser Raman Spectroscopy*, Wiley, New York, 1970.

W. Gordy, W. V. Smith, and R. F. Trambarulo, *Microwave Spectroscopy*, Wiley, New York, 1963.

M. D. Harmony, *Introduction to Molecular Energies and Spectra*, Holt, Rinehart and Winston, Inc., New York, 1972.

G. Herzberg, *Infrared and Raman Spectra of Polyatomic Molecules*, D. Van Nostrand Co., Princeton, N.J., 1950.

G. Herzberg, *Molecular Spectra and Molecular Structure*, D. Van Nostrand Co., Princeton, N.J., 1950.

G. Herzberg, *The Spectra and Structures of Simple Free Radicals*, Cornell University Press, Ithaca, 1971.

H. H. Jaffe and M. Orchin, *Theory and Application of Ultraviolet Spectroscopy*, Wiley, New York, 1962.

G. W. King, *Spectroscopy and Molecular Structure*, Holt, Rinehart and Winston, Inc., New York, 1964.

H. A. Szymanski, *Raman Spectroscopy*, Plenum Press, New York, 1967.

C. H. Townes and A. L. Schawlow, *Microwave Spectroscopy*, McGraw-Hill Book Co., New York, 1955.

E. B. Wilson, J. C. Decius, and P. C. Cross, *Molecular Vibrations*, McGraw-Hill Book Co., New York, 1955.

Problems

15.1 Calculate the reduced mass and the moment of inertia of $D^{35}Cl$.

Ans. 3.162×10^{-27} kg, 5.140×10^{-47} kg m².

15.2 Calculate the frequency in wave numbers and the wavelength in cm of the first rotational transition $(J = 0 \to 1)$ for $D^{35}Cl$.   *Ans.* 10.89 cm⁻¹, 0.09183 cm.

15.3 Calculate (*a*) the reduced mass and (*b*) the moment of inertia for gaseous $Na^{35}Cl$, which has a mean internuclear distance of 2.36 Å.

Ans. (*a*) 2.31×10^{-23} g, (*b*) 12.9×10^{-46} kg m².

15.4 The moment of inertia of $^{12}C^{16}O$ is 18.75×10^{-47} kg m². Calculate the frequencies in wave numbers and the wavelengths in centimeters for the first four lines in the pure rotational spectrum.   *Ans.* 2.98, 5.96, 9.00, 11.92 cm⁻¹; 0.335, 0.167, 0.1118, 0.0839 cm.

15.5 Calculate the zero-point energies of (*a*) H_2 and (*b*) Cl_2 in kilocalories per mole. The fundamental vibration frequencies are to be found in Table 15.2.

Ans. (*a*) 6.3, (*b*) 0.809 kcal mol⁻¹.

15.6 (*a*) What vibrational frequency in wave numbers corresponds to a thermal energy of kT at 25°? (*b*) What is the wavelength of this radiation?

Ans. (*a*) 207 cm⁻¹, (*b*) 4.83×10^{-3} cm.

15.7 The fundamental vibration frequency of Br_2 given in Table 15.2 does not appear in the simple infrared absorption spectrum because the molecule does not have a dipole moment. Calculate the energy difference between the ground and first vibrational levels in (a) joules per molecule, (b) kilocalories per mole, and (c) electron volts.

Ans. (a) 6.41×10^{-21} J per molecule, (b) 0.924 kcal mol^{-1}, (c) 4.00×10^{-2} eV.

15.8 (a) Calculate the wavelength in μm which corresponds with the fundamental vibration frequency of CO. The necessary data are in Table 15.2. What wavelengths correspond with the (b) first and (c) second overtones of the fundamental vibration frequency?

Ans. (a) 4.61, (b) 2.305, (c) 1.54 μm.

15.9 Calculate the wavelengths in (a) wave numbers and (b) micrometers of the center two lines in the vibration-rotation spectrum of HBr for the fundamental vibration. The necessary data are to be found in Table 15.2.

Ans. (a) 2632.8, 2666.6 cm^{-1}, (b) 3.80, 3.75 μm.

15.10 The fundamental vibration frequency of $H^{35}Cl$ is 8.667×10^{13} s^{-1}. What would be the separation in angstroms between the infrared absorption lines for $H^{35}Cl$ and $H^{37}Cl$ if the force constants of the bonds are assumed to be the same?   *Ans.* 26 Å.

15.11 Plot the Morse curve for H_2 with energy units of kilocalories per mole from the fact that the fundamental vibration frequency is 4395.2 cm^{-1}, the equilibrium internuclear distance is 0.7416×10^{-8} cm, and the dissociation energy is 4.476 eV.

Ans. See Fig. 14.6.

15.12 According to the hypothesis of Franck, the molecules of the halogens dissociate into one normal atom and one excited atom. The wavelength of the convergence limit in the spectrum of iodine is 4995 Å. (a) What is the energy of dissociation of iodine into one normal and one excited atom? (b) The lowest excitation energy of the iodine atom is 0.94 eV. What is the energy corresponding to this excitation? (c) Compute the heat of dissociation of the iodine molecule into two normal atoms, and compare it with the value obtained from thermochemical data, 34.5 kcal mol^{-1}.

Ans. (a) 57.22, (b) 21.67, (c) 35.55 kcal mol^{-1}.

15.13 List the numbers of translational, rotational, and vibrational degrees of freedom for (a) Ne, (b) N_2, (c) CO_2, and (d) CH_2O.

Ans. (a) 3, 0, 0, (b) 3, 2, 1, (c) 3, 2, 4, (d) 3, 3, 6.

15.14 When CCl_4 is irradiated with the 4358 Å mercury line, Raman lines are obtained at 4399, 4418, 4446, and 4507 Å. Calculate the Raman frequencies of CCl_4 (expressed in wave numbers). Also calculate the wavelengths (expressed in microns) in the infrared at which absorption might be expected. To calculate the Raman frequencies accurately with a slide rule the equation must be rearranged as follows:

$$\tilde{\nu}_{Raman} = \frac{\nu_{incid} - \nu_{scatt}}{c} = \left(\frac{1}{\lambda_{incid}} - \frac{1}{\lambda_{scatt}} \right)$$

$$= \frac{(\lambda_{scatt} - \lambda_{incid})}{\lambda_{incid}\lambda_{scatt}}$$

Ans. $\tilde{\nu}_{Raman}$	213	312	454	759
λ, μm	46.8	32.0	22.0	13.2

15.15 A solution of a dye containing 1 g per 100 ml transmits 80% of the light at 4356 Å in a glass cell 1 cm thick. (a) What percent of light will be absorbed by a solution containing 2 g per 100 cm^3 in a cell 1 cm thick? (b) What concentration will be required to absorb 50% of the light? (c) What percent of the light will be transmitted by a solution of the dye containing 1 g per 100 cm³ in a cell 5 cm thick? (d) What thickness should the cell be in order to absorb 90% of the light with solution of this concentration?

Ans. (a) 36.0%, (b) 3.10 g/100 cm³, (c) 32.7%, (d) 10.3 cm.

15.16   The absorption coefficient α for a solid is defined by $I = I_0 e^{-\alpha x}$, where x is the thickness of the sample. The absorption coefficients for NaCl and KBr at a wavelength of 28 μm are 14 cm⁻¹ and 0.25 cm⁻¹. Calculate the percentage of this infrared radiation transmitted by 0.5 cm thicknesses of these crystals.   *Ans.* NaCl, 0.09%; KBr, 88.3%.

15.17   The protein metmyoglobin and imidazole form a complex in solution. The molar absorption coefficients in M⁻¹ cm⁻¹ of the metmyoglobin (Mb) and the complex (C) are as follows:

λ	$\epsilon_{Mb} \times 10^{-3}$	$\epsilon_C \times 10^{-3}$
5000 Å	9.42	6.88
6300 Å	3.58	1.30

An equilibrium mixture in a cell of 1 cm path length has an absorbance of 0.435 at 5000 Å and 0.121 at 6500 Å. What are the concentrations of metmyoglobin and complex?

Ans. 2.17×10^{-5}, 3.37×10^{-5} M.

15.18   (a) Calculate the energy levels for $n = 1$ and $n = 2$ for an electron in a potential well of width 5 Å with infinite barriers on either side. The energies should be expressed in J and kcal mol⁻¹. (b) If an electron makes a transition from $n = 2$ to $n = 1$ what will be the wavelength of the radiation emitted?

Ans. (a) 2.41×10^{-19}, 9.64×10^{-19} J; 34.7, 138.8 kcal mol⁻¹, (b) 2750 Å.

15.19   The specific rotation $[\alpha]_D$ of a solution of *d*-ethoxysuccinic acid in water is 33.02° at 17°. Calculate the concentration of this compound in grams per liter in a solution which has a rotation of 2.02° when measured in a polarimeter at 17° in which the tube of solution is 20 cm long.   *Ans.* 30.6 g/l.

15.20   When α-D-mannose ($[\alpha]_D^{20} = +29.3°$) is dissolved in water, the optical rotation decreases as β-D-mannose is formed until at equilibrium $[\alpha]_D^{20} = +14.2°$. This process is referred to as mutarotation. As expected, when β-D-mannose ($[\alpha]_D^{20} = -17.0°$) is dissolved in water, the optical rotation increases until $[\alpha]_D^{20} = +14.2°$ is obtained. Calculate the percentage of α form in the equilibrium mixture.   *Ans.* 86%.

15.21   The specific rotation for sucrose in water at 20° for the D line of sodium is $66.412 + 0.01267c$, where c is the concentration in grams per 100 cm³ of solution. Calculate the rotation for a solution containing 20 g per 100 cm³ in a 2-decimeter tube.   *Ans.* 26.67°.

15.22   The internuclear distance in CO is 1.282 Å. Calculate (a) the reduced mass, and (b) the moment of inertia.

15.23   Calculate the wavelengths in centimeters of the first four lines in the pure rotational spectrum of Na³⁵Cl. The moment of inertia is given in problem 15.3.

15.24   Assuming that the internuclear distance is 0.742 Å for (a) H₂, (b) HD, (c) HT, and (d) D₂, calculate the moments of inertia of these molecules.

15.25   Calculate the energy difference between the $J = 0$ and $J = 1$ rotational levels for OH radicals, in kilocalories per mole. Calculate the wavelength in centimeters at which this transition will appear. The equilibrium internuclear distance and reduced mass are given in Table 15.2.

15.26   The far infrared spectrum of HI consists of a series of equally spaced lines with $\Delta \tilde{\nu} = 12.8$ cm⁻¹. What is (a) the moment of inertia and (b) the internuclear distance?

15.27   Calculate the zero-point energies of gaseous HCl and KCl in electron volts. The fundamental vibration frequencies are to be found in Table 15.2.

15.28 As explained in Section 15.5, there is no absorption at exactly the fundamental vibration frequency of a diatomic molecule consisting of two different kinds of atoms or exactly at the first, second, etc., harmonics. However, these frequencies are the approximate centers of vibrational-rotational bands. Calculate the wavelengths in μm of the centers of bands for HBr and HI, including first and second harmonics. The necessary data are in Table 15.2.

15.29 The fundamental vibration frequency of $H^{35}Cl$ is 8.667×10^{13} s^{-1}. Calculate the fundamental vibration frequency of $D^{35}Cl$ on the assumption that the force constants of the bonds are the same. About what wavelength in micrometers will the fundamental infrared absorption of $D^{35}Cl$ be grouped? The observed value is 4.8 μm.

15.30 The fundamental ($v = 0 \rightarrow 1$) absorption of gaseous $H^{81}Br$ occurs at 2650 cm^{-1}. Calculate the force constant.

15.31 The fundamental vibration frequency of $H^{35}Cl$ is 2990 cm^{-1}. (a) To what wavelength (in Å units) does this correspond? (b) What is the zero point energy in ergs per molecule? (c) What is the ratio of the populations in the ground state and in the first vibrational level at 25°.

15.32 Plot the potential-energy curve for $H^{35}Cl$, using information in Table 15.2.

15.33 The dissociation energies of $HCl(g)$, $H_2(g)$, and $Cl_2(g)$ into normal atoms have been determined spectroscopically and are 4.431, 4.476, and 2.476 eV, respectively. Calculate the enthalpy of formation of $HCl(g)$ at 0 K in kilocalories per mole from these data.

15.34 List the numbers of translational, rotational, and vibrational degrees of freedom of Cl_2, H_2O, and C_2H_2.

15.35 Acetylene has two C—H stretching vibrations, a symmetrical one at 3374 cm^{-1}, and an unsymmetrical one at 3287 cm^{-1}.

$$\overset{\longleftarrow\quad\longrightarrow}{H-C\equiv C-H} \qquad \overset{\longrightarrow\quad\longrightarrow}{H-C\equiv C-H}$$

Which vibration will be active in the Raman effect? Calculate the Raman wavelengths for a 4358Å exciting line. Which vibration will be active in the infrared? Calculate the infrared wavelength.

15.36 Commercial chlorine from electrolysis contains small amounts of chlorinated organic impurities. The concentrations of impurities may be calculated from infrared absorption spectra of liquid Cl_2. Calculate the concentration of $CHCl_3$ in grams per milliliter in a sample of liquid Cl_2 if the transmittancy at $\tilde{\nu} = 1216$ cm^{-1} is 45% for a 5-cm cell. At this wavelength liquid Cl_2 does not absorb, and the absorption coefficient for $CHCl_3$ dissolved in liquid Cl_2 is 900 ± 80 cm^{-1} $(g\ cm^{-3})^{-1}$.

5.37 To test the validity of Beer's law in the determination of vitamin A, solutions of known concentration were prepared and treated by a standard procedure with antimony trichloride in chloroform to produce a blue color. The percent transmission of the incident filtered light for each concentration expressed in micrograms per milliliter, was as follows:

Concentration, μg ml^{-1}	Transmission, %
1.0	66.8
2.0	44.7
3.0	29.2
4.0	19.9
5.0	13.3

Plot these data so as to test Beer's law. A solution, when treated in the standard manner with antimony chloride, transmitted 35% of the incident light in the same cell. What was the concentration of vitamin A in the solution?

15.38   The protein metmyoglobin and the azide ion (N_3^-) form a complex. The molar absorption coefficients (M^{-1} cm^{-1}) of the metmyoglobin (Mb) and of the complex (C) in a buffer are as follows:

λ	$\epsilon_{Mb} \times 10^{-4}$	$\epsilon_C \times 10^{-4}$
4900 Å	0.850	0.744
5400 Å	0.586	1.028

An equilibrium mixture in a 1-cm cell gave an absorbance of 0.656 at 4900 Å and of 0.716 at 5400 Å. (a) What are the concentrations of metmyoglobin and complex? (b) Since the total azide concentration is 1.048×10^{-4} M, what is the equilibrium constant for

$$Mb + N_3^- = C$$

15.39   The specific rotation of l-leucine is $[\alpha]_D^{25} = -14.0°$. If the specific rotation of a mixture of d and l forms is $[\alpha]_D^{25} = +2.3°$, calculate the fraction of l form in the mixture.

15.40   When α-D-glucose ($[\alpha]_D^{20} = +112.2°$) is dissolved in water, the optical rotation decreases as β-D-glucose is formed until at equilibrium $[\alpha]_D^{20} = +52.7°$. As expected, when β-D-glucose ($[\alpha]_D^{20} = +18.7°$) is dissolved in water, the optical rotation increases until $[\alpha]_D^{20} = +52.7°$ is obtained. Calculate the percentage of the β form in the equilibrium mixture.

15.41   Since the energy of a molecular quantum state is divided by kT in the Boltzmann equation, it is of interest to calculate the temperature at which kT is equal to the energy of photons of different wavelength. Calculate the temperature at which kT is equal to the energy of photons of wavelength 10^3 cm, 10^{-1} cm, 10^{-3} cm, 10^{-5} cm.

15.42   The separation of the pure rotation lines in the spectrum of CO is 3.86 cm^{-1}. Calculate the equilibrium internuclear separation.

15.43   Calculate the frequencies (in kilomegacycles) of the microwave absorption lines that you would expect for NaCl vapor up to $J = 3$. The isotopes ^{35}Cl and ^{37}Cl are present to the extent of 75.4 and 24.6% respectively.

15.44   For a linear triatomic molecule XYZ with internuclear distances r_{xy} and r_{yz}, derive the expression for the moment of inertia in terms of r_{xy} and r_{yz}.

15.45   Show that for large J the frequency of radiation absorbed in exciting a rotational transition is approximately equal to the classical frequency of rotation of the molecule in its initial or final state.

15.46   The moment of inertia of $^{16}O{=}^{12}C{=}^{16}O$ is 71.67×10^{-47} kg m^2. (a) Calculate the CO bond length, r_{CO}, in CO_2. (b) Assuming that isotopic substitution does not alter r_{CO}, calculate the moments of inertia of

$$1. \ ^{18}O{=}^{12}C{=}^{18}O \text{ and}$$

$$2. \ ^{16}O{=}^{13}C{=}^{16}O.$$

15.47   How many cm^{-1} correspond to 1 kcal mol^{-1}?

15.48   Calculate the zero-point energy of CO in kilocalories per mole and the energy difference between the zero and first vibrational-energy levels in kilocalories per mole. The fundamental vibration frequency is 2170 cm^{-1}.

15.49   Calculate the energy difference between the ground and first vibrational levels for H_2 in (a) joules per molecule (b) kilocalories per mole, and (c) electron volts. The fundamental vibration frequency is given in Table 15.2.

15.50 Calculate the wavelengths in (*a*) wave numbers and (*b*) μm of the center four lines in the infrared spectrum of HI at the first harmonic of the fundamental vibration frequency. The necessary data are to be found in Table 15.2.

15.51 Gaseous HBr has an absorption band centered at about 2645 cm^{-1} consisting of a series of lines approximately equally spaced with an interval of 16.9 cm^{-1}. For gaseous DBr estimate the frequency in wave numbers of the band center and the interval between lines.

15.52 The fundamental vibrational frequency of HCl is 2890 cm^{-1}. At 25° what is the ratio of the number of molecules with $v = 1$ to the number with $v = 0$?

15.53 At what temperature would the population of the first ($v = 1$) vibrational level of H_2 (fundamental vibration frequency, $\tilde{\nu}_0$, is 4395 cm^{-1}) be one-half the population of the lowest ($v = 0$) vibrational level?

15.54 Plot the potential-energy curve for Br_2 with energy units of kilocalories per mole according to the Morse function, using $D' = 45.7$ kcal mol^{-1}, $\tilde{\nu}_0 = 323$ cm^{-1}, and $r_e = 2.28$ Å.

15.55 Given the following fundamental vibrational frequencies:

H^{35}Cl	2989 cm^{-1}	H^{2}D	3817 cm^{-1}
^{2}D^{35}Cl	2144 cm^{-1}	^{2}D^{2}D	2990 cm^{-1}

Calculate $\Delta H°$ for the reaction

$$H^{35}Cl(v = 0) + {}^2D^2D(v = 0) = {}^2D^{35}Cl(v = 0) + H^2D(v = 0)$$

15.56 The following table gives the wave numbers for the lines of the H^{35}Cl band at 3.46 μm [C. F. Meyer and A. A. Levin, *Phys. Rev.*, **34**, 44 (1929)]. Only a few of the lines near the center of the band are included.

	J	$\tilde{\nu}$
R branch	3	2963.24
	2	2944.89
	1	2925.78
	0	2906.25
P branch	1	2865.09
	2	2843.56
	3	2821.49
	4	2798.78

The wave number of a line in the vibration-rotation spectrum may be represented by $\tilde{\nu} = \tilde{\nu}_0 + BJ'(J' + 1) - BJ''(J'' + 1)$ when the interaction between rotation and vibration is neglected. The rotational quantum number is represented by J' in the upper state and J'' in the lower state. The *R* branch in the spectrum is due to transitions from $J'' = J$ to $J' = J + 1$ (i.e., $\Delta J = +1$), and the *P* branch in the spectrum is due to transitions from $J'' = J$ to $J' = J - 1$ (i.e., $\Delta J = -1$). Derive the expressions for $\tilde{\nu}_R$ and $\tilde{\nu}_P$ in terms of J and calculate (*a*) ν_0, (*b*) the rotational constant B, and (*c*) the internuclear distance. The B value obtained by extrapolation to the center of the band should be used. In an exact calculation it would be necessary to take account of the fact that the B value is different for the excited state.

15.57 The limit of continuous absorption for Br_2 gas occurs at a wave number of 19,750 cm^{-1}. The dissociation which occurs is

$$Br_2(\text{normal}) = Br(\text{normal}) + Br(\text{excited})$$

The transition of a normal bromine atom to an excited one corresponds to a wave number of 3685 cm^{-1}.

$$Br(\text{normal}) = Br(\text{excited})$$

Calculate the energy increase for the process

$$Br_2(\text{normal}) = 2Br(\text{normal})$$

in (a) ergs per molecule and (b) electron volts.

15.58   List the numbers of translational, rotational, and vibrational degrees of freedom of NNO (a linear molecule) and NH_3.

15.59   How many normal modes of vibration are there for (a) SO_2(bent), (b) H_2O_2(bent), (c) HC≡CH(linear), and (d) C_6H_6?

15.60   Alkenes show a Raman frequency of 1642 cm^{-1}, which is due to the C=C stretching motion. Calculate the corresponding wavelengths in the Raman spectrum if the 4358 Å exciting line is used.

15.61   The following absorption data are obtained for solutions of oxyhemoglobin in pH 7 buffer at 575 nm in a 1-cm cell:

Grams per 100 cm^3	Transmission, %
0.03	53.5
0.05	35.1
0.10	12.3

(a) Is Beer's law obeyed? What is the molar absorption coefficient? (b) Calculate the per cent transmission for a solution containing 0.01 gram per 100 ml.

15.62   When a 1.9-cm absorption cell was used, the transmittancy of 4360 Å light by bromine in carbon tetrachloride solution was found to be as follows:

c, M	0.00546	0.00350	0.00210	0.00125	0.00066
Transmittancy, I/I_0	0.010	0.050	0.160	0.343	0.570

Calculate the molar absorption coefficient. What percentage of the incident light would be transmitted by 2 cm of solution containing 1.55×10^{-3} M bromine in carbon tetrachloride?

15.63   The following equilibrium in solution is to be studied spectrophotometrically.

$$B + C = D$$

Assuming that only B and D absorb light at the wavelength used, express the absorbancy of the solution in terms of the initial concentrations of B and C (no D is added initially), the absorbancy indices of these substances and the equilibrium constant for the reaction.

15.64   A dye having a conjugated series of double bonds has an electronic absorption band at 3000 Å. Calculate the length of the π bond system assuming that the transition is from $n = 1$ to $n = 2$ for an electron in a one-dimensional box.

15.65   The electronic spectrum of a molecule

$$R-(CH=CH)_k-R'$$

can be described by considering the transitions of $2k\pi$ electrons among the levels in a one-dimensional box of length kl where l is the length of the unit —CH=CH—. (a) Ignore electron-electron interactions and assign electrons to orbitals in accord with the Pauli principle. Obtain an expression for the lowest energy transition as a function of k. (b) If $l = 2.8$ Å, what is the wavelength of the lowest energy transition for $k = 2, 5$, and 10?

15.66   What is the concentration, in grams per 1000 cm^3, of a solution of lactose that gives a rotation of 7.24° in a 10-cm cell at 20° with sodium D light? The value of $[\alpha]_D^{20}$ for lactose is 55.4.

CHAPTER 16

MAGNETIC RESONANCE SPECTROSCOPY

Magnetic resonance spectroscopy is different from other kinds of spectroscopy in that the separation of energy levels exists only in the presence of a magnetic field. For magnetic fields that can be routinely produced in the laboratory the transitions between energy levels for nuclei that are magnetic dipoles occur in the radiofrequency range, and the transitions between energy levels for the spins of unpaired electrons occur in the microwave range. These new types of spectroscopy— nuclear magnetic resonance (NMR) and electron spin resonance (ESR)—yield such valuable structural information that they have become indispensable in chemistry.

In order to relate these magnetic phenomena to the bulk magnetic properties of matter we begin the chapter with a discussion of magnetic susceptibility.

16.1 MAGNETIC SUSCEPTIBILITY*

When a substance is placed in magnetic field of strength H the magnetic flux density B inside the medium is given by

$$B = H + 4\pi M \tag{16.1}$$

where M is the magnetization of the substance. The simplest magnetic unit is the magnetic dipole, and magnetic dipole moments being vectors are represented by $\boldsymbol{\mu}$; their scalar magnitudes are represented by μ. A magnetic dipole is a little magnet; it can be visualized as a little current loop, but this is an oversimplification. The electron and many nuclei are magnetic dipoles.

The magnetization M of a sample is the magnetic dipole moment per unit volume and may be written as

$$M = n\overline{\boldsymbol{\mu}} \tag{16.2}$$

where n is the number of magnetic dipoles per unit volume and $\overline{\boldsymbol{\mu}}$ is the average value of the individual magnetic moments in the direction of the magnetic field H.

For isotropic substances at equilibrium the magnetization M is proportional to H and the proportionality constant χ is the magnetic susceptibility per unit volume.

$$M = \chi H \tag{16.3}$$

---

* E. M. Purcell, *Electricity and Magnetism*, Berkeley Physics Course Vol. 2, McGraw-Hill Book Co. New York, 1963, Chapter 10.

The magnetic susceptibility χ is a dimensionless quantity. The magnetization does not appear instantly when a field is applied (cf. Section 16.4).

In contrast with polarization by an electric field the magnetic moment may be in the direction of the applied field (χ is positive) or in the opposite direction (χ is negative). Substances for which the magnetization is in the direction of the field are *paramagnetic*, and substances for which the magnetization is opposed to the field are *diamagnetic*. For *ferromagnetic* substances the magnetic susceptibility is large and positive, but the susceptibility depends on the history of the sample, and the magnetization does not drop to zero when the field is removed.

Since magnetic fields are produced by currents it is possible to connect the strength of the field of any magnet, including a single magnetic dipole, with basic mechanical and electrical units. The SI unit of magnetic field strength is a weber m^{-2}, which is named a tesla and represented by T. The earlier and more widely used unit of magnetic field strength is the gauss; we will use gauss, represented by G, throughout this book. The conversion is $10,000 \text{ G} = 1 \text{ T}$.

The field strength in a sample is represented by B. In a good deal of the literature the symbol H is used for the magnetic field in a sample, but we will use B.

Paramagnetism results from the orientation of permanent magnetic dipoles in a substance. These permanent magnetic dipoles are due to the spin of unpaired electrons or to the angular momentum of electrons in orbitals of atoms or molecules. Electrons in orbitals with $l = 1, 2, 3, \ldots$ have angular momentum and therefore produce a magnetic dipole moment. Nuclei with magnetic moments produce a paramagnetic effect, but this effect is only about a millionth as large as paramagnetism due to orbital moments or unpaired electrons. Nuclear magnetic properties are studied by nuclear magnetic resonance.

In the absence of a magnetic field the magnetic dipoles that produce paramagnetism are not oriented. In the presence of a field the orientation of permanent magnetic dipoles by the field is opposed by the disorganizing effect of thermal motion, as in the alignment of electric dipoles by an electric field (Section 14.12). In most substances the magnetic effects of electron spin and electron orbital motions cancel because electrons are paired in filled shells. Many rare earth ions are paramagnetic because they have unpaired electrons. Free radicals have an odd number of electrons and are therefore paramagnetic. The most familiar substance that is paramagnetic is molecular oxygen. As we have seen earlier (Section 14.5) it has two unpaired electrons. This property of oxygen gas makes it possible to determine its partial pressure in a gas stream with a little torsion balance in the field of a magnet.

The magnetic moment μ of a paramagnetic molecule or ion is commonly measured in units of Bohr magnetons μ_B.

$$\mu_B = \frac{e\hbar}{2m_e} = \frac{(1.60219 \times 10^{-19} \text{ C})(6.6262 \times 10^{-34} \text{ J s})}{4\pi(9.1095 \times 10^{-31} \text{ kg})}$$

$$= 9.2742 \times 10^{-24} \text{ J T}^{-1}$$

$$= 9.2742 \times 10^{-28} \text{ J G}^{-1} \tag{16.4}$$

This is the magnetic moment in the direction of the magnetic field for an electron in the first Bohr orbit.

Diamagnetism results from the induction of microscopic currents in a sample by the external magnetic field. The magnetic dipoles produced in this way are aligned in a direction opposite of that of the external field. Since the induced magnetic dipoles oppose the field, a diamagnetic substance experiences a force in the direction of the weaker part of an inhomogeneous magnetic field. A paramagnetic substance is attracted by the poles of a magnet with an inhomogeneous magnetic field. The diamagnetic effect is produced in all substances, but it is only about a hundredth or a thousandth as strong as the paramagnetic effect when the latter exists.

The diamagnetic effect for an anisotropic molecule depends on its orientation with respect to the applied magnetic field. As an example the delocalized electrons in π orbitals of aromatic substances can be induced to circulate about the whole ring when the field is perpendicular to the plane of the ring but not when the field is directed along the plane of the ring.

The metals iron, cobalt, nickel, gadolinium, dysprosium, and certain of their alloys and compounds are ferromagnetic below a certain critical temperature for each substance. The origin of ferromagnetism was a puzzle which was explained by quantum mechanics. The question is: Why do so many electrons in incomplete shells have their spins aligned, and why do they remain aligned even after the applied magnetic field is removed? The explanation is that the lowest energy state for certain solids is one in which spins are parallel, rather than opposed as they are for the two electrons in a hydrogen molecule, for example. The requirements of certain atomic distances and certain radii of the orbitals for d electrons limit this phenomena to just a few elements. Ferromagnetic substances show hysteresis in their magnetic properties. This means that the magnetic moment depends on the magnetic history of the sample; the increase in moment with increasing field is not retraced when the field is reduced.

The atoms in an antiferromagnetic substance are arranged so that the magnetic moments of nearest neighbors are opposed.

16.2 PRINCIPLES OF MAGNETIC RESONANCE

In the last section it was pointed out that since electrons and certain nuclei have spin angular momentum they behave like little magnets and tend to be oriented in an applied magnetic field. However, in contrast with macroscopic magnets only certain energy levels are possible because of quantization. In nuclear magnetic resonance (NMR) and electron spin resonance (ESR) transitions between these energy levels are studied. Because of the similarities of these methods we will discuss the basic equations for both in this section.

All nuclei with odd mass numbers have the property of spin, and those nuclei with even mass number that have an odd number of protons also have spin. The

Table 16.1   Nuclear Magnetic Properties

Nucleus	Spin I	g_N	γ_N, s^{-1} G^{-1}
^{1}H	$\frac{1}{2}$	5.585	26,753
^{2}D	1	0.857	4,107
^{7}Li	$\frac{3}{2}$	2.171	10,398
^{13}C	$\frac{1}{2}$	1.405	6,728
^{14}N	1	0.403	1,934
^{15}N	$\frac{1}{2}$	-0.567	-2,712
^{17}O	$\frac{5}{2}$	-0.757	-3,628
^{19}F	$\frac{1}{2}$	5.257	25,179
^{23}Na	$\frac{3}{2}$	1.478	7,081

spin I of a nucleus with an odd mass number is always an odd integral multiple of $\frac{1}{2}$, and the spin of a nucleus with an even mass number, but an odd number of protons, is 1, 2, 3, ... The spins of a number of nuclei are summarized in Table 16.1. The commonly occurring nuclei ^{12}C and ^{16}O do not have spin.

The spin angular momentum vector is represented by $I\hbar$ since the spin angular momentum is measured in units of $\hbar = h/2\pi$.

A nucleus that has both spin and charge has a magnetic moment $\boldsymbol{\mu}_N$ that is proportional to the spin angular momentum vector.

$$\boldsymbol{\mu}_N = \gamma_N \hbar \boldsymbol{I} \tag{16.5}$$

where γ_N is the gyromagnetic ratio of the nucleus. Since the spin angular momentum $I\hbar$ is commonly measured in units of J s and the magnetic moment $\boldsymbol{\mu}_N$ in J G^{-1}, the gyromagnetic ratio has the units of s^{-1} G^{-1}.

The magnetic moment vector $\boldsymbol{\mu}_N$ may also be expressed in terms of a dimensionless constant g_N, called the nuclear g value, and the magnitude of the nuclear magneton μ_N given by

$$\mu_N = \frac{e\hbar}{2m_p} = \frac{(1.60219 \times 10^{-19}\text{ C})(6.6262 \times 10^{-34}\text{ J s})}{4\pi(1.67265 \times 10^{-27}\text{ kg})}$$

$$= 5.0508 \times 10^{-27}\text{ J T}^{-1}$$

$$= 5.0508 \times 10^{-31}\text{ J G}^{-1} \tag{16.6}$$

where e and m_p are the charge and mass of the proton. Thus

$$\boldsymbol{\mu}_N = g_N \mu_N \boldsymbol{I} \tag{16.7}$$

The values of I, g_N and γ_N for a number of nuclei are summarized in Table 16.1.

The proton and neutron each have spin $\frac{1}{2}$, and the spin of a nucleus (which is always integral or half integral) may be thought of as the resultant of the spins of the protons and neutrons that comprise the nucleus. A deuterium nucleus, which contains one proton and one neutron, would be expected to have a spin of 1 or 0, depending on whether the spins of the proton and neutron were aligned parallel (↑↑) or antiparallel (↑↓). The deuteron is found to have a spin of 1 in its nuclear ground state, and so the proton and neutron are in parallel alignment.

So far we have not said anything directly about quantum numbers or magnetic field strengths. If a magnetic field is applied to a nucleus, the nucleus gains an energy E_{mag} that is proportional to the magnetic dipole moment μ_z of the nucleus in the direction of the field.

$$E_{mag} = -\mu_z B \tag{16.8}$$

Here B is the scalar magnitude of the field. Taking the z component of both sides of equation 16.7 and substituting in equation 16.8 yields

$$E_{mag} = -g_N \mu_N I_z B \tag{16.9}$$

where I_z is the allowed component of the nuclear spin in the z-direction. According to quantum mechanics the angular momentum in any particular direction is quantized. If the nuclear spin quantum number m_I is $\frac{1}{2}$, as it is for the proton, the allowable component of nuclear spin I_z in the z-direction is $+\frac{1}{2}$ or $-\frac{1}{2}$. If the nuclear spin quantum number is 1, as it is for deuterium, the allowable component of nuclear spin in the z-direction is $+1, 0,$ or -1. Figure 16.1a shows how the energy levels of a proton depend on the magnetic field strength. The proton is in the lower spin level when its magnetic moment is parallel with the field; the component of

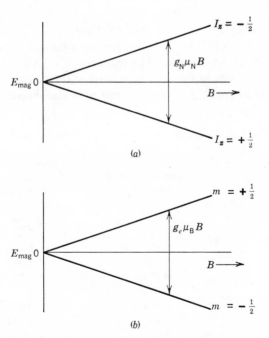

(a)

(b)

Fig. 16.1 (a) Energy levels of a proton in a magnetic field. (b) Energy levels of an electron in a magnetic field.

spin in the z-direction is $+\frac{1}{2}$. In the upper level the magnetic moment is anti-parallel, and $I_z = -\frac{1}{2}$. The selection rule for a magnetic dipole transition is

$$\Delta I_z = \pm 1 \qquad (16.10)$$

Figure 16.1a shows that the change in energy for protons and the frequency that corresponds to the transition from one spin level to the other (which is called the precessional frequency) are directly proportional to the field strength.

$$\Delta E = h\nu = g_N \mu_N B \qquad (16.11)$$

The frequencies for a 14,094 G field are shown for various nuclei in Table 16.2 together with the relative sensitivity for equal numbers of nuclei.

Example 16.1 Calculate the magnetic field strength required to give a precessional frequency for protons of 60 MHz.

$$B = \frac{h\nu}{g_N \mu_N} = \frac{(6.6262 \times 10^{-34} \text{ J s})(60 \times 10^6 \text{ s}^{-1})}{(5.585)(5.0508 \times 10^{-31} \text{ J G}^{-1})}$$

$$= 14,094 \text{ G}$$

The basic equations for electron spin resonance (ESR) follow the same pattern as for nuclear magnetic resonance. The magnetic energy of an electron in a magnetic field is

$$E_{\text{mag}} = g_e \mu_B m_s B \qquad (16.12)$$

where g_e is the g-factor for the electron (2.002322 for a free electron), μ_B is the Bohr magneton (Section 16.1), and m_s is the quantum number corresponding to to the z-component of the electron spin. The two energy levels of a single electron in a magnetic field are shown in Fig. 16.1b. Because of the negative charge of the electron the magnetic moment $\boldsymbol{\mu}_e$ of an electron is in the direction opposite to

Table 16.2 Nuclear Magnetic Resonance Frequencies and Relative Sensitivities

Nucleus	Frequency in MHz at 14,094 G	Sensitivity*
^{1}H	60.000	1.00
^{2}D	9.2104	9.65×10^{-3}
^{7}Li	23.317	0.293
^{13}C	15.087	1.59×10^{-2}
^{14}N	4.3343	1.01×10^{-3}
^{15}N	6.0798	1.04×10^{-3}
^{17}O	8.134	2.91×10^{-2}
^{19}F	56.446	0.833
^{23}Na	15.871	9.25×10^{-2}

* Relative sensitivity for equal number of nuclei at constant field.

its spin angular momentum (rather than in the same direction as for a nucleus with spin), and the electron spin quantum number is $-\frac{1}{2}$ in the lower level in contrast with the situation with nuclei. For a transition from $m_s = -\frac{1}{2}$ to $m_s = +\frac{1}{2}$

$$\Delta E = h\nu = g_e \mu_{\mathrm{B}} B \tag{16.13}$$

Example 16.2 Calculate the magnetic field strength required to give a precessional frequency of 9500 MHz for a free electron.

$$\begin{aligned} B &= \frac{h\nu}{g_e \mu_{\mathrm{B}}} \\ &= \frac{(6.6262 \times 10^{-34} \text{ J s})(9500 \times 10^6 \text{ s}^{-1})}{(2.0023)(9.2741 \times 10^{-28} \text{ J G}^{-1})} \\ &= 3390 \text{ G} \end{aligned}$$

16.3 HIGH RESOLUTION NMR SPECTROMETER

An NMR spectrometer provides the magnetic field to produce the energy levels, the radio frequency to excite transitions, and a radiofrequency receiver to detect emitted radiation. The resolution of an NMR spectrometer depends on the strength and homogeneity of the magnetic field and the constancy of the radiofrequency radiation. High resolution is required for most chemical applications.

The high resolution spectra are usually obtained with magnetic field strengths of the order of 10,000 G. A field of 14,094 G corresponds with a precession frequency of 60 MHz for protons as shown in Example 16.1. For several reasons it is advantageous to use the highest possible fields, but there are great practical difficulties in obtaining sufficiently stable and homogeneous magnetic fields for high resolution NMR work above about 70,000 G. Since chemical shifts (Section 16.5) are in the range of parts per million, the field needs to be constant to a part in 10^8 or 10^9. The fixed frequency radiofrequency sources that are used are crystal controlled and are stable to a part in 10^9.

Different methods for studying nuclear magnetic resonance were developed independently by Purcell and Bloch in 1946. A diagram for a crossed-coil NMR spectrometer is shown in Fig. 16.2. The sample to be investigated is placed in a tube between the poles of the electromagnet. The direction of the constant magnetic field B_0 is taken as the z direction. The strength of this magnetic field is set at a value just below that required for resonance, and then a secondary field is applied by use of coils around the faces of the pole pieces. The current for these coils is supplied by a sweep generator so that the magnetic field strength at the sample can be swept through the resonance condition. A sweep generator produces a steadily increasing voltage that drops to zero when a certain voltage is reached, and then repeats the process. The sweep generator also controls the sweep of the oscilloscope or chart recorder.

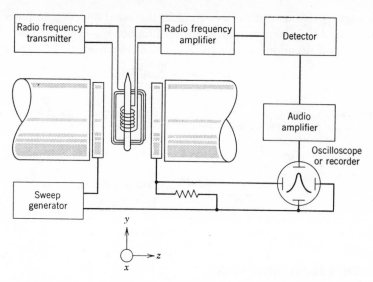

Fig. 16.2 Block diagram of a high resolution nuclear magnetic resonance spectrometer. The axis of the transmitter coil is the x axis, and that of the receiver coil is the y axis.

The magnetic nuclei in the sample precess about the z direction (the axis of the electromagnet). Transitions between the possible orientations of the nuclear spin are produced by exposing the sample to radio waves of a constant frequency, which are transmitted by a coil around the sample tube with its axis perpendicular to the z direction. The axis of the transmitter coil is the x axis, and that of the receiver coil is the y axis, which is the axis of the sample tube. When the radiofrequency and the magnetic field strength correspond with a resonance for a certain set of nuclei, energy is absorbed from the transmitter. This energy is radiated by the precessing nuclei and picked up by the detector coil, which is oriented so that it does not otherwise receive energy from the transmitter. If the signal from the radiofrequency receiver is applied to the vertical amplifier of the oscilloscope the absorption line is traced out as shown. Very frequently the limiting factor in high resolution NMR spectroscopy is the inhomogeneity of the magnetic field. An improvement in resolution can be obtained by spinning the sample tube so that the nuclei experience a field that is averaged over the volume of the sample. Even if the field were perfect an NMR absorption line would have a finite width because of the interaction of a magnetic moment with its surroundings.

In Fourier transform NMR spectroscopy the sample is irradiated with a whole range of radiofrequencies (so called "white" radiation). To avoid saturating the system the irradiation is carried out in very short pulses. After a pulse the nuclei reemit their energy. This reemitted energy is a composite of the resonance frequencies of all the nuclei in the sample. If there are two nuclei and no coupling, two frequencies ν_A and ν_X will be emitted. The two frequencies will produce a "beat" pattern in the detector, and from the beat pattern it is possible to calculate

ν_A and ν_X. This process is referred to as a Fourier transform, and a small computer is required to analyze the beat pattern if there are several frequencies. This method is important because of the increase in sensitivity in essentially looking at all the resonances after each pulse, rather than one at a time in the conventional type of NMR experiment described above. Thus smaller samples can be used and less abundant isotopic species (like ^{13}C) may be studied.

16.4 THERMAL EQUILIBRIUM AND SPIN RELAXATION

The ratio of the populations of two spin levels at thermal equilibrium may be calculated using Boltzmann's law (Section 17.3). For nuclei that are distributed between two levels α and β, as shown in Fig. 16.1, with a difference in energy of $g_N \mu_N B$,

$$\frac{N_\alpha^0}{N_\beta^0} = e^{g_N \mu_N B/kT} \tag{16.14}$$

where N_α^0 is the number of spins in the lower state at equilibrium and N_β^0 is the number of spins in the upper state at equilibrium. At ordinary temperatures $g_N \mu_N B \ll kT$ so that the exponential may be approximated by a power series, and to a sufficiently good approximation

$$\frac{N_\alpha^0}{N_\beta^0} = 1 + \frac{g_N \mu_N B}{kT} \tag{16.15}$$

For nuclei in the vicinity of room temperature the difference in population of the two levels is very small. For protons at 10,000 G, the magnetic energy is about 10^{-3} cm^{-1} compared to $kT \approx 200$ cm^{-1}, so that the population of the two spin states differ by only one part in 10^5.

Example 16.3 What is the ratio of the number of proton spins in the lower state to the number in the upper state in a magnetic field of 10,000 G at room temperature?

$$\frac{N_\alpha^0}{N_\beta^0} = 1 + \frac{g_N \mu_N B}{kT}$$

$$= 1 + \frac{(5.585)(5.05 \times 10^{-31} \text{ J G}^{-1})(10^4 \text{ G})}{(1.38 \times 10^{-23} \text{ J K}^{-1})(298 \text{ K})}$$

$$= 1 + 6.86 \times 10^{-6}$$

Because of the exchange of spin with the lattice the lower state is more highly populated, and the sample may absorb energy if the frequency of the applied field corresponds with the transition frequency. If the system is irradiated with too high a power the excess spins at the lower energy will be used up and no resonance signal will be obtained. This phenomenon is referred to as saturation.

As suggested earlier it takes a little time for spins to become oriented in a magnetic field or to lose their orientation once the field is turned off. These rate processes are referred to as relaxation processes, and they are characterized by the relaxation time defined in Section 10.2. Two different processes are involved in relaxing nuclear spins. In the spin-lattice relaxation process (relaxation time T_1) the excess spin energy equilibrates with the lattice, which is a general term used to describe the surroundings. The vibrational, rotational, and translational motions of atoms and molecules in the neighborhood cause a time-varying magnetic field at a nucleus or unpaired electron. This field, which results from the magnetic moments of the neighboring atoms and molecules, has components of the frequency required to produce transitions between α and β. The value of T_1 may be measured by starting with a spin system with a population difference between the two states, removing the radiation field, and watching the exponential decay. The values of T_1 range from 10^{-2} to 10^4 s for solids and 10^{-4} to 10 s for liquids.

In the spin-spin relaxation process (relaxation time T_2) the excess spin energy equilibrates directly between nuclei with spin. For solids T_2 is of the order of 10^{-4} s and for liquids $T_2 \approx T_1$.

The NMR absorption lines of solids are rather broad, but they give important information about the distance between nuclei with magnetic moments and about their orientations in the crystal. The nuclei interact by a direct dipole-dipole coupling between their magnetic moments, and this causes very broad absorption lines.

The NMR absorption lines in liquids are very much narrower than in solids because the dipole-dipole coupling with nuclei in other molecules is averaged to zero by the rapid random rotations of molecules in a liquid.

16.5 THE CHEMICAL SHIFT

If protons in all molecules had the same NMR frequency chemists would not be very interested in NMR, but fortunately different absorption lines are obtained for protons in different chemical environments in a molecule. For example, Fig. 16.3 shows a sketch of the proton magnetic resonance spectrum of CH_3CH_2OH

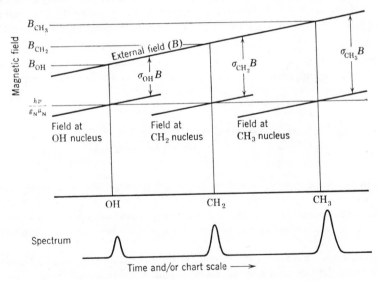

Fig. 16.3 Low-resolution proton-magnetic-resonance spectrum of CH_3CH_2OH. As the magnetic field is increased protons in different chemical environments come into resonance. As indicated the magnetic field strengths at the various protons are different from the strength of the external field. (Adapted from J. C. Davis, *Advanced Physical Chemistry*, The Ronald Press, New York, 1965.)

under conditions of low resolution. The areas under the three peaks are proportional to the numbers of protons of the three different types. Several absorption lines are obtained for such a molecule because a nucleus in a molecule experiences a magnetic field that is slightly different from the applied field. This difference is a secondary field arising from the magnetic effect of induced electronic currents set up in a molecule by the applied magnetic field. As explained in connection with diamagnetism (Section 16.1), and as illustrated in Fig. 16.3, the secondary field opposes the applied field so that the effective field at a proton is less than the applied field. The more a proton is shielded by the surrounding electrons the higher the applied field required to produce resonance at a given frequency. The local field depends on the chemical environment and so the effect is referred to as a *chemical shift*. In ethyl alcohol the protons in CH_3 are shielded to a greater extent than those in CH_2, which are shielded to a greater extent than the proton in OH.

The secondary field is proportional to the applied field but in the opposite direction, and so the effective magnetic field B_{eff} at the nucleus is expressed by

$$B_{eff} = (1 - \sigma)B \tag{16.16}$$

where the screening constant σ

$$\sigma = \frac{B - B_{eff}}{B} \approx \frac{B - B_{eff}}{B_{eff}} \tag{16.17}$$

is of the order of 10^{-6}. Thus a very slightly higher field is required to produce resonance of a proton when it is in a molecule as compared to when it is isolated. Although chemical shifts are not large the nuclear resonance lines obtained from liquids are so narrow that very small changes in local fields may be detected.

Table 16.3 Chemical Shifts of ^1H

Compound	Shift (τ) (ppm)	Compound	Shift (τ) (ppm)
Methyl protons		Olefinic protons	
$(CH_3)_4Si$	10.00	$(CH_3)_2C{=}CH_2$	5.4
$(CH_3)_4C$	9.08	Cyclohexene	4.43
CH_3CH_2OH	8.83		
CH_3COCH_3	7.93	Acetylenic protons	
CH_3OH	6.62	$HOCH_2C{\equiv}CH$	7.67
Methylene protons		Aromatic protons	
Cyclopropane	9.78	Benzene	2.73
Cyclohexane	8.56	Naphthalene	2.27
CH_3CH_2OH	6.41		
		Aldehydic protons	
Methine protons		CH_3CHO	0.28
$(CH_3)_2CHOH$	6.05	C_6H_5CHO	0.04

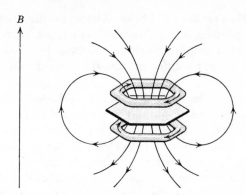

Fig. 16.4 Magnetic field produced by electron circulation in the π orbitals of the benzene ring. At the protons the induced field is in the same direction as the applied field.

Since it is not convenient to use an isolated nucleus as a reference, chemical shifts are measured relative to σ for some standard compound in the same solvent. If there are nuclei in two chemical environments the chemical shift difference can be calculated using equation 16.17.

$$\sigma_1 - \sigma_2 = \frac{B_2 - B_1}{B_{\text{eff}}} \tag{16.18}$$

Chemical shifts may be expressed in frequency units by multiplying $\sigma_1 - \sigma_2$ by the resonance frequency, but it is generally more convenient to use $\sigma_1 - \sigma_2$ in parts per million because this applies to experiments at any frequency.

The usual standard for measuring proton chemical shifts is tetramethylsilane, $Si(CH_3)_4$, which has a large chemical shift—implying that the protons are effectively screened by surrounding electrons. On the τ scale the protons of tetramethylsilane in the solvent used are assigned a chemical shift of $+10$ ppm.

Table 16.3 gives chemical shifts on the τ scale. Protons bonded to aromatic rings have resonances at lower magnetic field strengths than aliphatic protons. When a benzene ring is oriented perpendicular to the magnetic field the circulation of electrons in the π orbitals induces a field that is in the same direction as the applied field at the protons, as shown in Fig. 16.4. Therefore aromatic protons resonate at a lower field than they otherwise would. This effect is reduced by molecular tumbling because when the benzene ring is oriented parallel to the field there is no such effect.

Chemical shifts for ^{19}F and ^{11}B are much larger than those for protons because they are surrounded by a larger number of electrons capable of having currents induced in them.

16.6 SPIN-SPIN SPLITTING

Figure 16.5, which shows the high-resolution proton-magnetic-resonance spectrum for CH_3CH_2OH, shows that the spectrum is more complicated than indicated

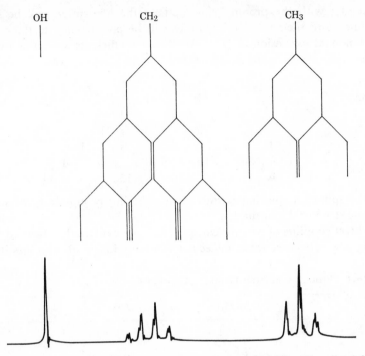

Fig. 16.5 Proton resonance spectrum of ethyl alcohol at 40 MHz. The signal from the radiofrequency receiver is plotted vertically, and the magnetic field strength is plotted horizontally. The field increases linearly from left to right.

in Fig. 16.3. With higher resolution the lines in the spectrum are split into multiplets. This splitting is called spin-spin splitting, and it results from the effects of nuclear spins in the same molecule. In the pattern for ethyl alcohol, the absorption line due to methyl (CH_3) protons is split into three components because the neighboring methylene group (CH_2) contains two protons, each with spin $\frac{1}{2}$. We can think of the first methylene proton splitting the methyl proton resonance into a doublet, as shown in the figure, and then the second methylene proton splitting the doublet into a triplet with the center line twice as intense as the other two. The two methylene protons produce the same spin-spin splitting because there is rapid internal rotation around the C—C bond.

The absorption line due to methylene protons is split into four components by the three protons of the neighboring CH_3 group. The reason for the relative intensities of $1:3:3:1$ is evident from the diagram.

In this spectrum the proton in the hydroxyl group does not cause further splitting because it undergoes chemical exchange so rapidly with protons in other molecules that it does not produce a splitting effect. This rapid chemical exchange occurs only when acid catalyzed, but only traces of acid are required to eliminate the splitting.

We can generalize these results by saying that if there are n equivalent protons that interact with the proton being studied, the absorption will be split into $n + 1$ lines, and their relative intensities will be proportional to the coefficients of the binomial expansion of $(1 + x)^n$. These coefficients are given by Pascal's triangle.

0 proton						1						
1 proton					1		1					
2				1		2		1				
3			1		3		3		1			
4		1		4		6		4		1		
5	1		5		10		10		5		1	
6	1	6		15		20		15		6		1

Since the spin-spin splitting is due to neighboring groups it gives important additional structural information.

The direct coupling of two nuclear spins does not explain the observed spin-spin coupling effects because rapid molecular tumbling in liquids averages it to zero.

Table 16.4 Proton Spin–Spin Coupling Constants

Structure	$J(\text{Hz})$
	-20 to $+6$
	5.5 to 7.5
	7 to 10
	12 to 19
	o 6 to 9 m 0.5 to 4 p 0 to 2.5
	ax, ax 9 to 14 ax, eq 2 to 4 eq, eq 2.5 to 4

Nuclear spins in molecules are coupled via intervening electrons. The major effect results from the coupling of the spin of the first nucleus with electrons via the so-called contact interaction, first proposed by Fermi. This effect, which tends to align the spin of orbital electrons antiparallel to the nuclear spin, can only occur when the electron has a significant probability density at the nucleus. The electrons that become partially aligned in this way affect the magnetic field at the second nucleus. In addition, nuclear spin induces a current in the valence electrons by the action of the magnetic field of the nucleus on the orbital magnetic moment of the electrons. There is also a direct dipolar coupling between the nuclear spin and the electron spin.

Nuclear magnetic resonance is so useful for determining molecular structures that it has become an indispendable tool of the organic chemist.*

The strength of spin-spin splitting is measured by the coupling constant J which in the case we have been discussing is simply the separation of peaks in a multiplet, usually expressed as a frequency by multiplying the σ value for the splitting by the resonance frequency. A few spin-spin coupling constants are summarized in Table 16.4.

The interpretation of NMR spectra is often more complicated than we have indicated as shown by the structure of the lines in Fig. 16.5. The complete quantitative interpretation of such a spectrum depends upon consideration of the Hamiltonian (Section 12.11) for the spin system and the spin wave functions.

16.7 ANALYSIS OF NMR SPECTRA
OF LIQUIDS

If the chemical shifts are large compared with the coupling constants, the interpretation of the spectrum is relatively simple. The chemical shifts indicate the chemical environments of the protons, and the areas of the absorption peaks indicate the numbers of protons of various types. If the chemical shifts are large compared with the spin-spin coupling constants, the type of first order analysis described for CH_3CH_2OH suffices. However, if the differences in chemical shifts are of the same order of magnitude as the coupling constants, the spectrum cannot be understood in terms of a first-order analysis because there are additional lines and a different pattern.

For the purpose of discussing different types of spin systems it is convenient to use letters of the alphabet to represent nuclei. A spin system with two nuclei may be represented by AX, AB, or A_2 depending on whether the difference in chemical shifts is large compared with the coupling constant, whether they are of the same order of magnitude, or whether the nuclei are equivalent. The changes in the spectrum of the two proton system as we go from $J/\delta = 0$ (AX case) to $J \approx \delta$ (AB case) to $\delta \ll J$ (A_2 case) are illustrated in Fig. 16.6. If there is no coupling the two protons give separate lines that are separated by the difference in chemical

---

* J. B. Henderson, D. J. Cram, and G. S. Hammond, *Organic Chemistry*, McGraw-Hill Book Co, New York, 1970, Chapter 7.

shifts. If the coupling constant J is small enough compared with the chemical shift δ so that first-order analysis is sufficient, each line is split into a doublet with splitting equal to J. As the coupling becomes larger compared with the chemical shift the center two lines become stronger with respect to the outer lines until the spectrum is almost a single close doublet. If the chemical shifts of the two protons are identical the spectrum is a single line. In other words no spin-spin couplings are observed within a group of magnetically equivalent nuclei.

It is important to make a distinction between nuclei that are chemically equivalent, and nuclei that are magnetically equivalent. This distinction is illustrated by CH_2F_2 and $CH_2{=}CF_2$, which at first both look like A_2X_2 systems. Fluorine, represented by X, has a spin of $\frac{1}{2}$. In both compounds the protons are chemically equivalent and have the same chemical shift; they are equivalent by symmetry operations. The proton resonance spectra of these compounds are shown in Fig. 16.7. The proton spectrum of CH_2F_2 is the expected $1:2:1$ triplet.

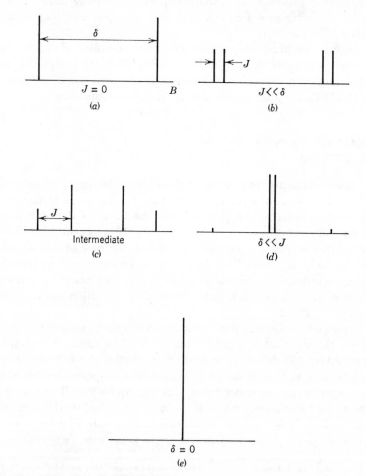

Fig. 16.6 Spectrum of a two-proton system.

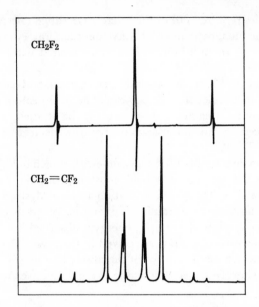

Fig. 16.7 Proton-resonance spectra of CH_2F_2 and $CH_2{=}CF_2$ at 60 MHz (From E. D. Becker, *J. Chem. Ed.*, **42,** 591 (1965)).

The proton spectrum of $CH_2{=}CF_2$ is more complicated because the coupling of H and F, which are cis (on the same side) is different from the coupling of H and F, which are trans (on opposite sides). Thus the two protons in $CH_2{=}CF_2$ are not magnetically equivalent. In order for nuclei to be magnetically equivalent, it is necessary for them to be chemically equivalent *and* to couple to all other nuclei in the system in exactly the same way. If the ethanol molecule CH_3CH_2OH were perfectly rigid the methyl protons would not be equivalent. However, because of rapid rotation about the C—C bond the electronic environments of the three methyl protons are magnetically equivalent.

The separation of peaks due to the chemical shift is directly proportional to the field strength, but the separation due to spin-spin splitting J is independent of field strength. Thus at higher field strengths the spin-spin splittings do not cause as much complication in the interpretation of the NMR pattern.

16.8 ELECTRON SPIN RESONANCE

The basic theory of electron spin resonance (ESR) has been described in Section 16.1. An electron resonance experiment is similar to an NMR experiment, but since the gyromagnetic ratio of the electron is about 10^3-fold larger than for nuclei the frequencies required fall in the microwave range rather than the radio frequency range when magnetic fields of convenient laboratory strength are used. Usually a frequency of about 10 GHz ($\lambda = 3$ cm) is used with a magnetic field of 3–4 kG.

Substances that show ESR spectra include free radicals, odd electron molecules, triplet states of organic molecules, and paramagnetic transition metal ions and their complexes. Any paramagnetic substance can be studied, but, as we have seen, most substances are not paramagnetic because electron spins are usually paired.

The block diagram for a simple ESR spectrometer is shown in Fig. 16.8. As in the NMR spectrometer the frequency is held constant, and the magnetic field is swept through resonance. Microwave radiation from a klystron passes down a waveguide to a resonant cavity containing the sample. When there are transitions between electron spin levels in the sample, energy is absorbed from the microwave radiation, and less microwave energy is received at the crystal detector. By use of a phase sensitive detector the derivative of the absorption line is recorded on the oscilloscope or strip chart recorder. The shape of the derivative curve is illustrated in Fig. 16.8.

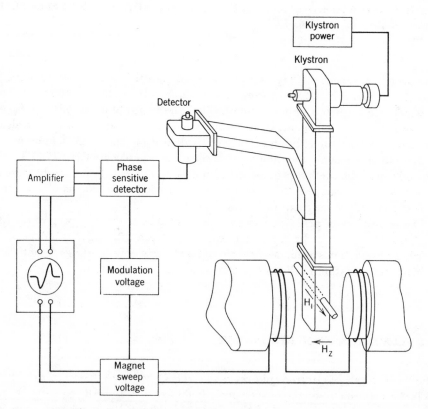

Fig. 16.8 Block diagram of simple ESR spectrometer.

16.9 HYPERFINE COUPLING

ESR spectroscopy is especially useful for chemistry because the electron magnetic moment interacts with other magnetic moments in the molecule including protons and other nuclei listed in Table 16.1. The splitting of absorption lines that results is called hyperfine splitting, rather than spin-spin splitting as in NMR.

As a simple example of hyperfine splitting let us consider a free radical with two protons that affect the electron energy levels in a magnetic field to different extents. Figure 16.9a shows the effects of the two protons on the possible energy levels for the electron. In the presence of a magnetic field an unpaired electron has two energy levels $m_s = \frac{1}{2}$ and $m_s = -\frac{1}{2}$. The two protons each split these levels so that the unpaired electron has eight energy levels. In electron spin resonance transitions the electron spin flips, but the nuclear spins do not. Thus in absorbing energy in an ESR transition the electron goes from an energy level in the lower group $(m_s = -\frac{1}{2})$ to the corresponding level in the upper group $(m_s = +\frac{1}{2})$. As the magnetic field strength is increased the four possible transitions one-by-one come into resonance, and four ESR lines are obtained. Since the four nuclear-spin states $(\alpha_1\alpha_2, \alpha_1\beta_2, \beta_1\alpha_2, \text{and } \beta_1\beta_2)$ are equally probable the four lines are of equal intensity. The two hyperfine splittings a_1 and a_2 may be calculated from the spectrum as shown.

If the two protons have the same coupling constant to the unpaired electron the spectrum has three lines, and the center line is twice as intense because it represents two possible transitions, as shown in Fig. 16.9b. Two or more protons having the same coupling to an unpaired electron are said to be equivalent, and they usually occupy symmetrically equivalent positions in the molecule.

Figure 16.10 shows a simple way of looking at these two spectra. We can describe the effect of two proton spins in the molecule by saying that the first proton splits the original single line due to electron spin into a doublet, and the second proton further splits the two lines into a quadruplet. If the two hyperfine splittings are different, four absorption lines are obtained. If the two hyperfine splittings are the same (the protons are equivalent), three lines are obtained.

Thus in general if n equivalent protons interact with an unpaired electron there will be $n + 1$ lines, and their relative intensities are given by Pascal's triangle (Section 16.6).

The ESR spectrum of benzene negative ion ($C_6H_6^-$) in Fig. 16.11a illustrates this very nicely. The ESR spectrum of naphthalene negative ion ($C_{10}H_8^-$) shown in Fig. 16.11b is more complicated because there are two types of protons, labeled A and B in the following structural formula

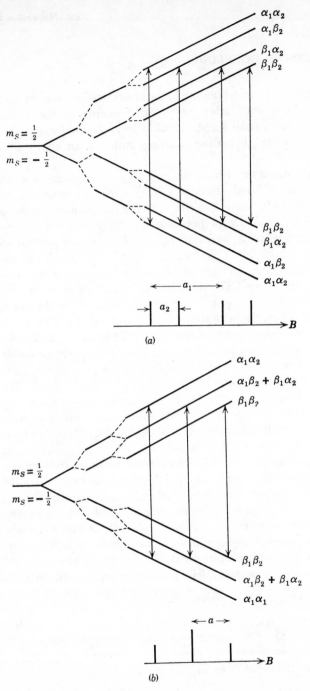

Fig. 16.9 (a) Hyperfine energy levels and ESR spectrum for an unpaired electron in the presence of two nonequivalent protons. (b) Hyperfine energy levels and ESR spectrum for an unpaired electron in the presence of two equivalent protons. The spin states of the protons are represented by $\alpha_1\alpha_2$, $\alpha_1\beta_2$, $\beta_1\alpha_2$, and $\beta_1\beta_2$.

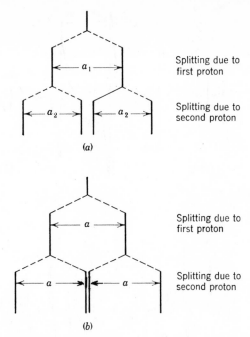

Splitting due to
first proton

Splitting due to
second proton

(a)

Splitting due to
first proton

Splitting due to
second proton

(b)

Fig. 16.10 (a) Hyperfine splitting due to two nonequivalent protons. (b) Hyperfine splitting due to two equivalent protons.

The four A protons alone would give rise to a five-line spectrum. Each of these lines is further split into five lines by the B protons, to give the 25-line spectrum shown in the figure.

The splitting by other magnetic nuclei is similar, except that of course nuclei of spin 1, like ^{14}N, cause splitting into triplets of equal intensity since the quantum number for the ^{14}N nucleus may be $+1$, 0, or -1.

The hyperfine coupling between the proton and electron of the hydrogen atom is 506.8 G. This coupling (due largely to Fermi's contact interaction discussed in Section 16.6) is proportional to the probability of finding the $1s$ electron at the nucleus. The coupling between unpaired electrons and protons in organic radicals are much less than this (often in the range 2–100 G) because the unpaired electron occupies orbitals that extend over the whole molecule, and the probability that the unpaired electron is at a particular hydrogen nucleus is quite small.

16.10 ESR APPLICATIONS

ESR spectroscopy has been very useful in determining the structure of organic and inorganic free radicals. Free radicals may be produced chemically, photochemically, or by use of high energy radiation. If a radical has a very short life a flow system or continuous radiation may have to be used to maintain a sufficiently

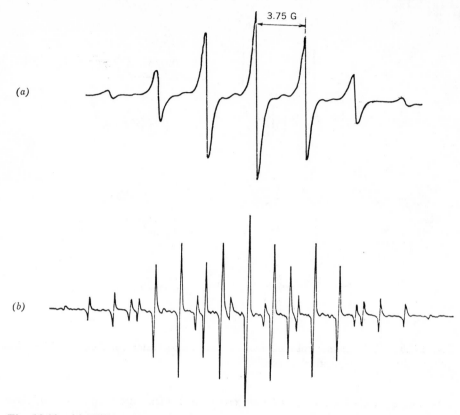

Fig. 16.11   (a) ESR spectrum of benzene negative ion. (b) ESR spectrum of naphthalene negative ion. (From A. Carrington and A. D. McLachlan, *Introduction to Magnetic Resonance*, Harper and Row, New York, 1967.)

high concentration for detection. Actually a concentration of only about 10^{-10} M is required to obtain a spectrum.

ESR spectra may be determined for unstable radicals trapped in glasses, frozen rare gas matrices, or in crystals. If the radicals are regularly oriented with respect to crystal axes the directional effects of magnetic interactions may be studied.

A molecule in a triplet state has a total electron spin of one $(S = 1)$. In this case there are three sublevels that have spin angular momentum S_z about a chosen axis of $+1$, 0, or -1. A triplet molecule has an even number of electrons, two of them unpaired, while a radical with spin $\frac{1}{2}$ has an odd number of electrons. In order for a molecule to have a triplet state the unpaired electrons must interact; a molecule with two unpaired electrons a great distance apart is a diradical, not a triplet.

The information about electron densities in molecules that is obtained from ESR spectra have been useful for checking molecular-orbital and valence-bond calculations.

Groups with unpaired electrons (spin labels) may be attached to proteins to provide a very sensitive detector of changes in protein structures.

References

C. N. Banwell, *Fundamentals of Molecular Spectroscopy*, McGraw-Hill Book Co., New York, 1972.

E. Becker, *High Resolution NMR*, Academic Press, New York, 1969.

M. Bersohn and J. C. Baird, *An Introduction to Electron Paramagnetic Resonance*, W. A. Benjamin, Inc., New York, 1966.

F. Bovey, *Nuclear Magnetic Resonance Spectroscopy*, Academic Press, New York, 1969.

E. F. H. Brittain, W. O. George, and C. H. J. Wells, *Introduction to Molecular Spectroscopy*, Academic Press, New York, 1970.

A. Carrington and A. D. McLachlan, *Introduction to Magnetic Resonance*, Harper and Row, New York, 1967.

R. Chang, *Basic Principles of Spectroscopy*, McGraw-Hill Book Co., New York, 1971.

J. C. Davis, *Advanced Physical Chemistry*, The Ronald Press Co., New York, 1965.

F. Daniels, J. W. Williams, P. Bender, R. A. Alberty, C. D. Cornwell, and J. E. Harriman, *Experimental Physical Chemistry*, McGraw-Hill Book Co., New York, 1970.

H. G. Hecht, *Magnetic Resonance Spectroscopy*, Wiley, New York, 1967.

W. W. Paudler, *Nuclear Magnetic Resonance*, Allyn and Bacon, Inc., Boston, 1971.

J. A. Pople, W. G. Schneider, and H. J. Bernstein, *High Resolution Nuclear Magnetic Resonance*, McGraw-Hill Book Co., New York, 1959.

J. D. Roberts, *Nuclear Magnetic Resonance*, McGraw-Hill Book Co., New York, 1959.

Problems

16.1 Calculate the magnetic field strength to give a precessional frequency for fluorine of 60 MHz.   *Ans.* 14,973 G.

16.2 Frequencies used in nuclear magnetic resonance spectra are of the order of 60 MHz. Calculate the corresponding energy in kilocalories per mole.

Ans. 5.72×10^{-6} kcal mol^{-1}.

16.3 What magnetic field strength is required for proton magnetic resonance at 220 MHz?

Ans. 51,670 G.

16.4 What is the ratio of the number of ^{31}P spins in the lower state to the number in the upper state in a magnetic field of 17,240 G at room temperature?

Ans. $1 + 4.79 \times 10^{-6}$.

16.5 The proton resonance pattern of 2,3-dibromothiopene shows an *AB*-type spectrum with lines at 403.83, 411.64, 425.46, and 431.47 Hz measured from tetramethylsilane at 14,100 G (K. F. Kuhlmann and C. L. Braun, *J. Chem. Ed.*, **46**, 750 (1969)). (*a*) What is the coupling constant *J*? (*b*) What is the difference in the chemical shifts of the *A* and *B* hydrogens? (*c*) At what frequencies would the lines be found at 20,000 G?

Ans. (*a*) 6.91, (*b*) 20.73, (*c*) 574.89, 581.80, 604.30, 611.21 Hz.

16.6 Are protons a and b magnetically equivalent in

> *Ans.* They are not magnetically equivalent (although they are chemically equivalent) because the coupling $H_a - H_c$ is different from the coupling $H_a - H_d$.

16.7 Calculate the precessional frequency of electrons in a 15,000 G field.

Ans. 42,000 MHz.

16.8 Line separations in ESR may be expressed in G or MHz. Show how the conversion factor 1 G = 2.80 MHz is obtained.

16.9 Sketch the ESR spectrum expected for p-benzosemiquinone radical ion

The four hydrogens are magnetically equivalent.

Ans. Five lines with relative intensities of 1, 4, 6, 4, 1.

16.10 An unpaired electron in the presence of two protons gives the following four-line ESR spectrum: $\Delta B = 0, 1, 3, 4$ G. What are the two coupling constants in G and in MHz?

Ans. $a_1 = 3$ G or 8.40 MHz,

$a_2 = 1$ G or 2.80 MHz.

16.11 At room temperature the chemical shift of cyclohexane protons is an average of the chemical shifts of the axial and equatorial protons. Explain.

16.12 Calculate the magnetic field strength B required for ^{19}F resonance at 20 MHz.

16.13 What is the ^{13}C resonance frequency at 10,000 G?

16.14 Sketch the proton-resonance spectrum for a compound, represented by AMX, with three protons with rather different chemical shifts ($\delta_A = 100$, $\delta_B = 200$, and $\delta_C = 700$ Hz). The three protons are coupled with $J_{AB} = 9$, $J_{AX} = 7$, and $J_{BX} = 3$ Hz.

16.15 Sketch the expected nuclear-magnetic-resonance spectra for the protons in n-propyl, isopropyl, and *tert*-butyl alcohols, showing the number of components in each band. (Note that only protons on adjacent carbon atoms need to be considered in predicting the splitting.)

16.16 What is the magnitude of the magnetic moment of the proton?

16.17 Calculate the frequency required for an electron spin resonance experiment using a magnetic field of 10,000 G.

16.18 Sketch the ESR spectrum of $CD_3\cdot$, giving the relative intensities of the various lines.

16.19 Sketch the ESR spectrum for an unpaired electron in the presence of three protons for the following cases: (a) the protons are not equivalent, (b) the protons are equivalent, (c) two protons are equivalent and the third is different.

16.20 Using data from Table 16.1, calculate the magnetic field strengths at which (*a*) ^{6}Li and (*b*) ^{10}B will precess at 5 MHz.

16.21 A sample containing protons is placed in a magnetic field with a strength of 10,000 G. Calculate the fraction of the protons that at 25° will have their spins lined up with the field.

16.22 Using information from Tables 16.3 and 16.4 sketch the spectrum you would expect for ethyl acetate $(CH_3CO_2CH_2CH_3)$.

16.23 Describe the protons and deuteron NMR spectra of HD.

16.24 On the basis of the spin-spin coupling constants in Table 16.4 describe the spectrum expected for

The protons are labeled to assist with labeling the spectrum.

16.25 Calculate the resonance frequency for electrons at 0.33 T (3,300 G).

16.26 Sketch the ESR spectrum for the ethyl radical with the indicated hyperfine splittings.

$$\cdot CH_2 - CH_3$$

$$22.38 \quad 26.87$$

$$G \qquad G$$

16.28 Sketch the spectrum for an unpaired electron in the presence of ^{14}N and ^{1}H.

16.28 Sketch the ESR spectrum of the ethyl radical $CH_3CH_2\cdot$ on the basis of the known facts that the coupling with the more-distant protons (hyperfine coupling of 26.9 G) is larger than for the closer protons (-22.4 G).

16.29 Sketch the ESR spectrum of $CH_3\cdot$.

CHAPTER 17

STATISTICAL MECHANICS

The equilibrium properties of matter may be considered from two points of view: the macroscopic and the microscopic. From the macroscopic point of view represented by thermodynamics, the behavior of large numbers of molecules is described in terms of pressure, volume, temperature, composition, and exchanges of heat and work. The relationships between various types of equilibrium measurements derived in thermodynamics are not based on any theory of the structure of matter or of the mechanisms by which changes occur.

From the microscopic point of view represented by the applications of classical mechanics in kinetic theory and by quantum mechanics, the behavior of individual molecules is described in terms of their structures and the mechanisms by which they interact with other molecules. We have already shown how kinetic theory provides a means for understanding the pressure, energy, and heat capacities of ideal gases with no internal degrees of freedom. But we have also seen that kinetic theory fails to provide an explanation for the change of the heat capacities of molecules or crystals with temperature. Ideally we would like to be able to predict the thermodynamic behavior of substances using our knowledge about individual molecules obtained from spectroscopic measurements and from theoretical calculations such as those discussed in Chapter 14.

Statistical mechanics provides the needed bridge between microscopic mechanics (classical and quantum) and macroscopic thermodynamics. The classical aspects of this science were developed during the latter part of the nineteenth century by Boltzmann in Germany, Maxwell in England, and Willard Gibbs in the United States. For simple systems such as ideal gases, the calculations are not very difficult, and values of thermodynamic properties so obtained are often more accurate than the ones measured directly. For more complicated systems, especially those involving strong interactions between molecules, the theory is much more difficult and is the subject of current research.

Statistical mechanics provides insight into the laws of thermodynamics, and through it we will see heat, work, temperature, reversible processes, and state functions in a new light.

17.1 MICROSTATES AND RANDOMNESS

A microstate of a system is a particular spatial configuration and distribution of energy between particles that is described in full detail. For a given thermodynamic state (that is, a certain number of particles at a specified temperature, pressure, and volume) there are many possible microstates.

Although it will take a whole chapter to give an introduction to statistical mechanics, the basic idea is quite simple, and we will illustrate it with two examples, expansion of a gas into a vacuum and the distribution of energy among energy levels of atoms in small idealized crystals. The basic idea is that large groups of molecules can exist in many, many microstates and will tend to move toward equilibrium because that is the thermodynamic state with the largest number of microstates.

Consider a gas in a container in a thermostat connected through a stopcock with a second container that is evacuated. From a quantum mechanical point of view a complete description of the microstate of a gas in a container at a specified pressure and temperature requires a large number of quantum numbers.

When the stopcock is opened we know that the gas will expand into the second container. From the microscopic point of view, this happens because there are more microstates available to the gas in the expanded condition. If a single molecule can move into the new volume, the number of microstates is essentially doubled because with the molecule in the new position the number of new microstates equals the number of microstates previously available to all the other molecules in the first container. Thus as the gas explores all of the accessible microstates in the new total volume, most of them are states with gas in the new volume made available by opening the stopcock. Consequently, at equilibrium we find the gas distributed throughout the volume.

In classical mechanics and quantum mechanics everything is reversible in time, that is, an observer of a movie of an event cannot tell whether the projector is being run forwards or backwards. But the example of the expansion of a gas shows how irreversible processes occur; the gas expands because there are many more microstates available to the system when the stopcock is opened and there is a very large number of microstates with gas in both containers, compared with the number with gas in only one container. It is helpful to think of the equilibrium state as the one of the greatest randomness—in this example, spatial randomness.

As a second example let us consider the distribution of energy between atoms in a crystal. By considering crystals with three, five, and seven atoms we can actually count the possible microstates and see the origin of the Boltzmann distribution. It is assumed that the atoms in the crystal are perfect one-dimensional harmonic oscillators so that an atom may have energies ϵ_v given by $vh\nu$, where v is the vibrational quantum number, h is Planck's constant, and ν is the vibrational frequency. This amounts to taking the zero-point energy as the reference and measuring energy with respect to the zero point level.

Figure 17.1 shows the different ways of distributing three quanta of energy between the three atoms of the crystal. The atoms may be distinguished because they have different locations, and they are labeled a, b, and c. Each little diagram represents a microstate of the system. Thus our crystal can be in any one of 10 different microstates, each of which has exactly the same energy, three quanta of

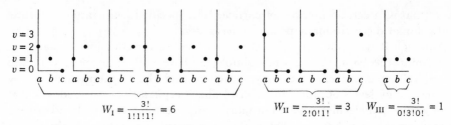

Fig. 17.1 Different ways of distributing three quanta of energy between atoms a, b, and c in an idealized crystal. When there are three quanta of energy there are 10 possible ways of distributing them, and so we say there are 10 microstates.

a particular frequency. There doesn't appear to be any reason why any of these 10 states should be favored over others. Thus it is assumed that they are all equally probable, and that at any given instant the crystal would be as likely to be in one microstate as another. We assume that there is a mechanism for quanta to move from one atom to another, and therefore we expect that over a long period of time the crystal spends equal amounts of time in each of the 10 microstates consistent with an overall energy of $3h\nu$ for the entire three-atom crystal.

The microstates of the three-atom crystal are of three different types: I, II, and III. We will refer to them as three different distributions. In the first distribution one atom has two quanta, another has one, and the third is in the ground state: there are six of these microstates. In the second distribution one atom has three quanta and the other two are in the ground state: there are three of these microstates. In the third distribution each atom has one quantum of energy: there is only one microstate with this distribution.

The number W of microstates in a distribution can be calculated using the relation

$$W = \frac{N!}{N_1! N_2! N_3! \cdots N_n!} \tag{17.1}$$

where $N! = 1 \cdot 2 \cdot 3 \ldots N$, N_1 is the number of atoms in state 1 (e.g., having zero quanta), N_2 is the number of atoms in state 2 (e.g., having one quantum), N_3 is the number of atoms in state 3 (e.g., having two quanta), etc., and N is the total number of atoms. Remember that $0! = 1$. We do not really need equation 17.1 at this point because we can write out all the possible microstates, as in Fig. 17.1, but this relation will be useful later when the numbers are larger.

With Fig. 17.1 we can answer the question: "On the average how many atoms of the crystal will be in the $v = 0$ level, how many in the $v = 1$ level, etc.?"

In considering crystals we are not interested in *which* atoms have a certain number of quanta, but *how many* atoms have 1, 2, 3, . . . quanta. The numbers of atoms in various levels are shown in Fig. 17.1 and are given in the second column of Table 17.1. The average occupation numbers N_i are obtained by dividing by the total number of microstates. We can compare the average occupation numbers N_i with what we would expect from the Boltzmann equation, which

Table 17.1 Average Occupation N_i of Energy Levels in the Three-Atom Crystal

Level (v)	Number of Atoms in Level	N_i^*	N_i (Boltzmann equation 17.5)
0	12	1.2	1.896
1	9	0.9	0.698
2	6	0.6	0.257
3	3	0.3	0.094
4	0	0	0.035
.			.
.			.
.			.
Total		3	3

* Number of atoms in level $\div 10$.

gives the correct answer when the number of atoms is very large. We will derive the Boltzmann equation in Section 17.3, but we will use it here to calculate the occupation numbers N_i assuming an infinite sequence of levels and a temperature for which $h\nu = kT$. The Boltzmann relation is

$$N_i = N \frac{e^{-\epsilon_i/kT}}{\sum e^{-\epsilon_i/kT}} \qquad (17.2)$$

where N_i is the average occupation number of level i, $N = \sum N_i$, and ϵ_i is the energy of level i. Since for a crystal of harmonic oscillators $\epsilon_i = v_i h\nu$,

$$N_i = N \frac{e^{-v_i h\nu/kT}}{\sum e^{-v_i h\nu/kT}} \qquad (17.3)$$

If the temperature is such that $h\nu/kT = 1$,

$$N_i = N \frac{e^{-v_i}}{\sum_i e^{-v_i}} \qquad (17.4)$$

For the three-atom crystal

$$N_i = \frac{3}{1.582} e^{-v_i} \qquad (17.5)$$

since $\sum_i e^{-v_i} = 1.582$.

We have to realize that when we analyzed the crystal by counting microstates, we implicitly assumed something about the temperature. Since the energy of the crystal is $Nh\nu$, and since we know from thermodynamics that the energy of the crystal is NkT (see Section 9.5), we are assuming in these examples that $kT = h\nu$. We would not expect the Boltzmann equation to give the correct answer for this small a crystal, but it should get better as we consider larger crystals.

If we enumerate the microstates of an idealized crystal with five atoms and five quanta of energy, we find that there are 126, as summarized in Fig. 17.2b. Rather

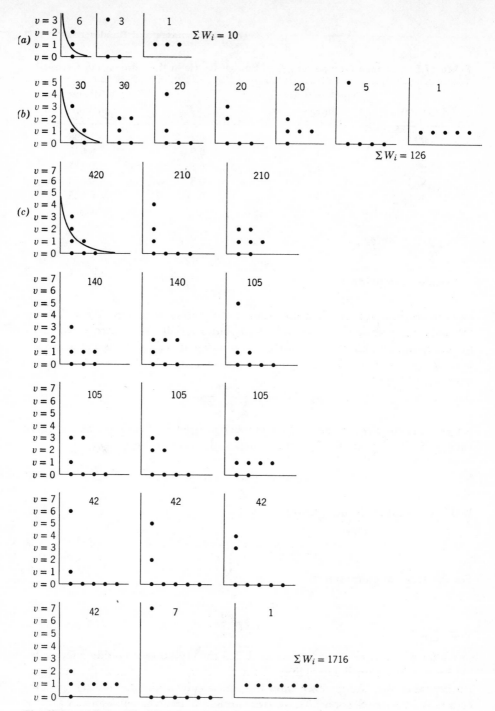

Fig. 17.2 (a) Ways W_i of distributing three quanta of energy between three distinguishable oscillators with equally spaced vibrational levels. The numbers are the numbers of micro-states for each distribution, and $\sum W_i$ is the total number of microstates. (b) Ways of distributing five quanta of energy between five distinguishable oscillators with equally spaced vibrational levels. (c) Ways of distributing seven quanta of energy between seven distinguishable oscillators with equally spaced vibrational levels. The line on the first diagram in (a), (b), and (c) gives the Boltzmann distribution.

than writing out all these microstates we simply summarize the numbers W_i in the various distributions. These numbers are calculated using equation 17.1. The average occupation numbers N_i are compared with the Boltzmann distribution for $h\nu = kT$

$$N_i = \frac{5}{1.582}\, e^{-v_i} \qquad (17.6)$$

in Table 17.2. The Boltzmann equation gives a little better answer than for the three-atom crystal.

Finally the calculation is repeated for a seven-atom crystal with seven quanta, and the results are compared with the Boltzmann distribution for $h\nu = kT$

$$N_i = \frac{7}{1.582}\, e^{-v_i} \qquad (17.7)$$

in Table 17.3. The point of this calculation is not so much to show that the Boltzmann relation is giving a closer approximation, as to point out that one of the distributions (the first with 420 arrangements) alone represents over one-fifth of the microstates and therefore has a major effect in the calculation of N_i shown in Table 17.3.

Let us think for a moment about the characteristics of the predominant distribution in terms of equation 17.1. In order to get the largest value of W the quanta should ideally be distributed so that each atom is in a different energy state. Then the denominator would simply be a product of 1!'s and $W = N!$, the maximum possible value. However, if the number of quanta is only equal to the number of atoms there is not enough energy to put all the atoms in different states. The best compromise involves distributing the atoms between the lower energy levels with as many levels represented as possible.

In going from the three-atom crystal to the seven-atom crystal we see two trends that continue—and become even more significant—as the size of the

Table 17.2 Average Occupation N_i of Energy Levels in the Five-Atom Crystal

Level (v)	Number of Atoms in Level	N_i^*	N_i (Boltzmann) equation 17.6)
0	280	2.222	3.161
1	175	1.389	1.163
2	100	0.794	0.428
3	50	0.397	0.157
4	20	0.159	0.058
5	5	0.040	0.021
6	0	0	0.008
.			.
.			.
.			.
Total		5	5

* Number of atoms in level ÷ 126.

Table 17.3 Average Occupation N_i of Energy Levels in Seven-Atom Crystal

Level (v)	Number of Atoms in Level	N_i^*	N_i (Boltzmann) equation 17.7)
0	5544	3.231	4.425
1	3234	1.885	1.628
2	1764	1.028	0.599
3	1022	0.596	0.220
4	252	0.147	0.081
5	147	0.086	0.030
6	42	0.024	0.011
7	7	0.004	0.004
8	0	0	0.001
.			.
.			.
.			.
Total	12,012	7	7

* Number of atoms in level ÷ 1716.

crystal is increased. The first trend is that the number of microstates $\sum W_i$ increases rapidly with the size of the crystal. The second trend is that the fraction of the microstates that are in the predominant distribution increases as the size of the crystal is increased. If this type of analysis is applied to larger and larger crystals it is found that one distribution dominates more and more so that the system spends almost all of its time in a group of microstates having this particular distribution.

The Boltzmann distribution curve is shown in the first diagram for each crystal. Since the number of quanta is equal to the number of atoms in each crystal they are at the same temperature since their vibrational frequencies are the same. Since the crystals are at the same temperature, the Boltzmann curve has the same shape in each case. These small idealized crystals are accurately described by the microstates as shown in Fig. 17.2 and not by the Boltzmann distribution, but as the number of particles and the number of quanta increase the Boltzmann distribution becomes a progressively more accurate description of the distribution of energy in the crystal.

17.2 ENSEMBLES

Before quantum mechanics was developed, Gibbs, Boltzmann, and others developed classical statistical mechanics. Since molecules follow quantum mechanics it was not possible to calculate certain thermodynamic properties using classical statistical mechanics. However, classical statistical mechanics is still useful for treating situations where particles behave classically. We will restrict our attention to quantum statistical mechanics. This approach is not only more general but is simpler.

A typical example of a system we might be interested in is a mole of identical gas molecules. Such a system is completely described by a wave function Ψ, which is a function of n generalized coordinates and time. For equilibrium systems time is not a variable, and so we may consider the system to be described by its stationary states ψ, which are a function of n generalized coordinates. For a collection of N independent atoms without electronic excitation, $n = 3N$. For a collection of N molecules the number of degrees of freedom is larger because of vibrational and rotational motions. (If electronic excitation is a possibility, n is still larger for both atoms and molecules.) Obviously a very large number of quantum numbers would be required to describe a mole of ideal gas, but we will find that it is not actually necessary to write down all of this information. This idea of talking about the quantum state of a whole system of molecules is very useful because it allows us, in principle, to treat systems of interacting molecules as well as independent molecules.

In order to calculate the thermodynamic properties of a mole of gas we have to know the probabilities of the various microstates (quantum states) and the contribution each makes to the property being calculated.

Gibbs showed how the averaging required to calculate thermodynamic quantities could be carried out using the concept of an ensemble. To develop this concept let us consider the calculation of thermodynamic properties of a system of N identical molecules in a volume V and in contact with a thermostat at temperature T. The specification of V, N, and T determines the thermodynamic state of the system and fixes the values of the other thermodynamic properties. Although we have fixed the macroscopic state of the system, we have not fixed its microstate. In fact there is a tremendously large number of microstates (from either a classical or quantum mechanical point of view) that are all compatible with a particular thermodynamic state. According to quantum mechanics each of these microstates is described by a time-independent wave function ψ.

An ensemble is a hypothetical collection of η systems each having the same characteristics from a thermodynamic or macroscopic point of view but each in a different microstate (that is, having a different wave function ψ). In the case we are considering, each of the systems has the same V, N, and T. The correct average values of properties are obtained by letting $\eta \to \infty$.

The ensemble we are talking about is referred to as a *canonical ensemble*,* and it is represented schematically in Fig. 17.3. Each square represents a system with N particles and volume V that has been brought into contact with a thermostat of temperature T. Since each system can exchange energy with the thermostat it may, in principle, have any energy, and all quantum states of the collection of molecules are, in principle, accessible. The systems in the ensemble exist in different quantum states, and so the energy of E_i of each system is specified in the figure.

---

* The other principal types of ensembles are the microcanonical and grand canonical. In the microcanonical ensemble the systems are isolated systems each having the same N, V, and E. In the grand canonical ensemble the systems are open isothermal systems each having the same V, T, and average number of particles. The constraint of constant average number of particles in the ensemble is equivalent to a constant μ (chemical potential).

V, N, E_1	V, N, E_2	V, N, E_3
V, N, E_4	V, N, E_5	V, N, E_6
V, N, E_7	V, N, E_8	V, N, E_9

Fig. 17.3 Canonical ensemble of η systems showing only nine of them.

We will first consider the calculation of the average energy and pressure for a system in the ensemble. In order to do this we will need the two postulates of statistical mechanics.

First Postulate. The time average of the energy or pressure (or other mechanical property) is equal to the average of these properties for the ensemble as $\eta \rightarrow \infty$.

According to this postulate we can calculate the energy or pressure (or other mechanical property) if we know the values in the various quantum states and the fractions of the systems in the ensemble in the various quantum states.

The systems in a canonical ensemble may be distributed in many different ways between the many possible quantum mechanical states that are consistent with V, N, and T. Let us assume that we can observe the individual systems of the ensemble and find that they are distributed between energy states as summarized in the following table.

Energy	E_1	E_2	E_3	$\cdots$
Number of systems with that energy	N_1	N_2	N_3	$\cdots$

In this table we arrange possible energy states in order of increasing energy eigenvalue E_i. This table is less than a complete description of the ensemble in the sense that it does not tell which systems have which microstates, only the total numbers of systems that have each allowable energy. The complete description of an ensemble is more than is actually required to determine the thermodynamic properties of the system. These properties, which are averages over the ensemble, depend on the *numbers* of members of the ensemble in each quantum state and not on *which* members of the ensemble are in *which* state. A description of an ensemble in terms of the *numbers* N_i of members of the ensemble that have a certain property is said to define a *class of states* of the ensemble. The class of states that occurs most often will make the largest contribution to the average or thermodynamic properties of the system. A particular system in the ensemble may have any one of the possible energies, but for the ensemble as a whole the distribution must satisfy the relations

$$\sum_i N_i = \mathcal{N} \qquad (17.8)$$

$$\sum_i N_i E_i = \mathcal{E} \qquad (17.9)$$

where $\mathcal{N}$ is the total number of systems in the ensemble and $\mathcal{E}$ is the total energy of the ensemble.

In order to calculate the number of systems with a particular value of the energy or pressure (or other mechanical property) we need another postulate.

Second Postulate. In an isolated system all of the possible quantum states consistent with specified values of the number of particles, the volume, and the energy are equally probable. This is the *principle of equal a priori probabilities.*

The canonical ensemble as a whole is an isolated system with volume $\mathcal{N}V$, $\mathcal{N}N$ particles, and total energy $\mathcal{E}$. The application of the second postulate to this isolated system means that the many quantum mechanical states of the *ensemble as a whole* are equally probable. We now want to use this idea to obtain the most probable values of the numbers N_i of systems with various energies, assuming that $E_1, E_2, \ldots$ are known.

Looking at the state of the ensemble as a whole there are many possible arrangements of numbers of systems in possible energy levels that are permitted by equations 17.8 and 17.9. For a particular distribution $N_1, N_2, \ldots$ we can choose systems in many different ways. We are concerned with the number of ways the systems can be assigned to the energy levels. The number of ways $\mathcal{N}$ systems can be distributed with N_1 systems having E_1, N_2 having E_2, etc. is given by equation 17.1 with $\mathcal{N}$ replacing N in the numerator. *We will assume that there is one distribution with so many more microstates than any other that the thermodynamic properties of the system may be calculated from this distribution alone.* In a more detailed analysis than we can go into here, it can be shown that this is an excellent assumption. In the next section we will see how we can calculate the values of $N_1, N_2, \ldots$ that will give the maximum number of arrangements W subject to the constraints of equations 17.8 and 17.9.

17.3 DERIVATION OF THE BOLTZMANN DISTRIBUTION

Since the number of microstates is very large it is possible to treat W as a quantity that varies continuously and to use the methods of differential calculus. Actually it is more convenient to maximize the natural logarithm of W. Since the N_i are all large we first use Stirling's approximation to eliminate the factorials in equation 17.1. Stirling's approximation

$$\ln N! \cong N \ln N - N \qquad (17.10)$$

gets more and more accurate as N increases. Substituting equation 17.10 into equation 17.1 yields

$$\ln W = \mathcal{N} \ln \mathcal{N} - \mathcal{N} - \sum_i N_i \ln N_i + \sum N_i$$

$$= \mathcal{N} \ln \mathcal{N} - \sum_i N_i \ln N_i \tag{17.11}$$

since $\mathcal{N} = \sum N_i$. In order to find the maximum we let the populations of the states be changed by small amounts $dN_1, dN_2, dN_3, \ldots$. The resulting change in $\ln W$ is obtained by taking the differential of equation 17.11.

$$d \ln W = - \sum (1 + \ln N_i) \, dN_i = 0 \tag{17.12}$$

If we are at a maximum $d \ln W = 0$.* Because of the constraints expressed in equation 17.8 and 17.9 the values of $dN_1, dN_2, dN_3, \ldots$ must also satisfy the relations

$$\sum dN_i = 0 \tag{17.13}$$

$$\sum E_i \, dN_i = 0 \tag{17.14}$$

As a result of these two relations all but two of the N_i can be treated as independent variables; these last two could be calculated from relations 17.8 and 17.9 if the others were known.

There are several ways to proceed, but the best approach to this problem of a conditional maximum was discovered by Lagrange and is referred to as the method of undetermined multipliers. In this method the conditions are combined by multiplying the condition expressed by 17.13 by α and the condition expressed by 17.14 by β and subtracting them from condition 17.12.

$$- \sum (1 + \ln N_i) \, dN_i - \alpha \sum dN_i - \beta \sum E_i \, dN_i = 0 \tag{17.15}$$

where α and β are two constants whose values are to be determined later. This equation can be rewritten

$$\sum (1 + \ln N_i + \alpha + \beta E_i) \, dN_i = 0 \tag{17.16}$$

The various dN_1 may now all be considered to be independent because the two constants α and β may be given values later to satisfy equations 17.8 and 17.9. Since equation 17.16 must be satisfied for any set of values of dN_i, each term must be individually be equal to zero.

$$\ln N_i = -(\alpha + 1) - \beta E_i \tag{17.17}$$

$$N_i = e^{-(\alpha+1)} e^{-\beta E_i} \tag{17.18}$$

Since $\mathcal{N} = \sum N_i$

$$\mathcal{N} = e^{-(\alpha+1)} \sum e^{-\beta E_i}$$

$$= e^{-(\alpha+1)} Q \tag{17.19}$$

where

$$Q = \sum e^{-\beta E_i} \tag{17.20}$$

---

* Strictly, the set of N_i that satisfies equation 17.12 just determines an *extremum* of $\ln W$, that is, $\ln W$ may be a maximum or a minimum. It may be shown that the solution to equation 17.12 corresponds to a maximum of $\ln W$.

is the canonical ensemble partition function. Thus equation 17.18 becomes

$$N_i = \frac{\mathcal{N}}{Q} e^{-\beta E_i} \qquad (17.21)$$

which is the Boltzmann equation we have been seeking.*

It is sometimes convenient to write the partition function in terms of allowed energy levels E_j rather than the energies E_i of quantum states.

$$Q = \sum_{j(\text{levels})} \Omega_j e^{-\beta E_j} \qquad (17.22)$$

where Ω_j is the degeneracy (Section 12.14) of the jth level.

The exponential factor $e^{-\beta E_i}$ is called the Boltzmann factor. β has to be a positive quantity so that it will be possible to normalize as shown above. The reason behind the exponential decrease with E_i is that as one system gets more energy the number of states available to the other $\mathcal{N} - 1$ systems is substantially reduced because of the constraint of constant energy.

We will be able to evaluate β in the next section.

17.4 INTERNAL ENERGY, ENTROPY, AND PRESSURE

The value of any thermodynamic quantity for a system can be obtained by averaging over the systems in an ensemble. As an example we will first consider the internal energy U. Let p_i represent the probability that system i will have energy E_i. The average value of the energy may be identified as the internal energy U.

$$U = \sum p_i E_i \qquad (17.23)$$

In the symbols of the preceding section $p_i = N_i/\mathcal{N}$. Therefore substituting from equation 17.21 we obtain

$$U = \frac{1}{Q} \sum E_i e^{-\beta E_i} \qquad (17.24)$$

Since the numerator is the negative derivative of the denominator with respect to β

$$\sum E_i e^{-\beta E_i} = -\frac{\partial}{\partial \beta} \sum e^{-\beta E_i} \qquad (17.25)$$

---

* In Section 17.1 we have already used a Boltzmann distribution

$$N_i = \frac{N e^{-\epsilon_i/kT}}{\sum_i e^{-\epsilon_i/kT}}$$

which is a more specialized version of equation 17.21.

we obtain

$$-\left(\frac{\partial Q}{\partial \beta}\right)_{V,N} = UQ \qquad (17.26)$$

$$U = -\left(\frac{\partial \ln Q}{\partial \beta}\right)_{V,N} \qquad (17.27)$$

In order to evaluate β we will compare the relations of statistical mechanics with the combined first and second laws of thermodynamics for a closed system with only pressure-volume work, that is, $dU = T\,dS - P\,dV$.

In statistical mechanical terms if V and T are changed reversibly at constant N for our ensemble, the differential of U may be calculated from equation 17.23.

$$dU = \sum E_i\,dp_i + \sum p_i\,dE_i \qquad (17.28)$$

The first term is due to changes that do not affect the energy levels but involves a transfer of energy between energy levels; thus it comes from thermal changes. The second term does involve a change in energy levels, and this suggests that it corresponds with work done on the system.

Inserting the logarithmic form of the Boltzmann equation 17.21 where $p_i = N_i/\mathcal{N}$

$$E_i = -\frac{1}{\beta}\,(\ln p_i + \ln Q) \qquad (17.29)$$

in the first term, and recognizing in the second that the energy states E_i at constant N_i can only be changed by changing the volume, equation 17.28 becomes

$$dU = -\frac{1}{\beta}\sum_i (\ln p_i + \ln Q)\,dp_i + \sum_i p_i\left(\frac{\partial E_i}{\partial V}\right)_N dV \qquad (17.30)$$

This equation may be simplified by noting that $\ln Q \sum dp_i = 0$ since $\sum dp_i = 0$ from $\sum p_i = 1$. The first term may be rewritten using

$$d\left(\sum_i p_i \ln p_i\right) = \sum_i dp_i + \sum_i \ln p_i\,dp_i$$

$$= \sum_i \ln p_i\,dp_i \qquad (17.31)$$

so that equation 17.30 becomes

$$dU = -\frac{1}{\beta}\,d(\sum \ln p_i\,dp_i) - P\,dV \qquad (17.32)$$

where the pressure P_i of state i is

$$P_i = -\left(\frac{\partial E_i}{\partial V}\right)_N \qquad (17.33)$$

This relation comes from the fact that $-P_i\, dV = dE_i$ is the work that has to be done on the system to change the energy of the ith state by dE_i and

$$P = \sum p_i P_i \qquad (17.34)$$

Comparing equation 17.32 with the first law of thermodynamics indicates that

$$dq = -\frac{1}{\beta}\, d(\sum p_i \ln p_i) \qquad (17.35)$$

This equation is the statistical mechanical analog of the statement $dq = T\, dS$ in the second law of thermodynamics, which says that there is an integrating factor for heat in a reversible process. Thus β is inversely proportional to the temperature T introduced in the second law.

$$\beta = \frac{1}{kT} \qquad (17.36)$$

where the proportionality constant k depends on the size of the unit of temperature. The Boltzmann constant $(k = R/N_A)$ is used because this brings the second law temperature scale into agreement with the ideal gas temperature scale.

Since for a reversible process $dS = dq/T$

$$S = -k \sum_i p_i \ln p_i \qquad (17.37)*$$

Thus the entropy depends only upon the probabilities p_i of the various states accessible to a system. Since the probabilities are fractions, the entropy is necessarily a positive quantity. If there is only one possible state (for example, a perfect crystal at absolute zero), $S = 0$.

17.5 EXPRESSION OF THERMO-DYNAMIC QUANTITIES IN TERMS OF THE PARTITION FUNCTION Q

With the identification $\beta = 1/kT$, equation 17.27 for the internal energy U may be written

$$U = kT^2 \left(\frac{\partial \ln Q}{\partial T}\right)_{V,N} \qquad (17.38)$$

---

* This expression may be used to derive the relation $S = k \ln \Omega$ used in Section 2.9 as follows: if all of the members of the ensemble have the same energy E_i, they all have the same degeneracy, $\Omega_j = \Omega$. Since every state of a member of the ensemble has equal a priori probability $p_i = 1/\Omega$. Therefore the summation of equation 17.37 consists of Ω identical terms

$$S = -k\Omega \left(\frac{1}{\Omega} \ln \frac{1}{\Omega}\right)$$

$$= k \ln \Omega$$

There should be a constant of integration in equation 17.37, which turns out to be equal to zero.

On a molar basis

$$\bar{U} = RT^2\left(\frac{\partial \ln Q}{\partial T}\right)_{V,N} \tag{17.39}$$

where N is the number of moles. This equation gives the internal energy measured from a ground state energy $\bar{U}_0$ of zero. If the ground state energy is not zero the left side of the equation becomes $\bar{U} - \bar{U}_0$.*

The expression for the entropy in terms of the partition function may be derived from the expression for the entropy in terms of the probabilities of the various states (equation 17.37). Substituting $\beta = 1/kT$ in equation 17.29 and substituting it in equation 17.37 yields

$$S = -k \sum_i p_i\left(-\frac{E_i}{kT} - \ln Q\right)$$

$$= \frac{1}{T}\sum p_i E_i + k \ln Q \sum p_i \tag{17.40}$$

Since $\sum p_i E_i = U$ (equation 17.23) and $\sum p_i = 1$,

$$S = k \ln Q + \frac{U}{T} \tag{17.41}$$

Since $A = U - TS$

$$A = -kT \ln Q \tag{17.42}$$

Since pressure $P = -(\partial A/\partial V)_{T,N}$

$$P = kT\left(\frac{\partial \ln Q}{\partial V}\right)_{T,N} \tag{17.43}$$

Since $G = H - TS$ and $H = U + PV$ it is a simple matter to also express G and H in terms of the canonical ensemble partition function Q.

Now we need to discuss the calculation of Q so that these expressions for thermodynamic quantities can be used.

17.6 EVALUATION OF THE PARTITION FUNCTION

If there are only weak interactions between molecules, the energy of the system is given to a high degree of approximation by the sum of the energies of the individual parts. The evaluation of the partition function is very much more complicated for systems in which the molecules interact strongly. If the independent particles are labeled a, b, c, ... the possible energy eigenvalues for the whole system are given by

$$E_i = \epsilon_{ai} + \epsilon_{bi} + \cdots \tag{17.44}$$

---

* It should be noted that, at absolute zero,

$$U_0 = H_0 = A_0 = G_0$$

because U and H differ by RT and U and A differ by TS.

where ϵ_{ai} is the energy of molecule a in one of its possible states. In calculating the canonical ensemble partition function all of the possible energy eigenvalues for the systems are to be used. The following example shows that all possible combinations of energies of individual molecules are included if the canonical ensemble partition function is written as the product of the molecular partition functions q of the individual particles.

$$Q = q_a q_b \cdots \tag{17.45}$$

where

$$q_a = \sum_i e^{-\epsilon_{ai}/kT} \tag{17.46}*$$

$$q_b = \sum_i e^{-\epsilon_{bi}/kT} \tag{17.47}$$

Example 17.1 Verify that the product of summations indicated in equation 17.45 does indeed yield the indicated summation. Taking the first two terms in the molecular partition functions

$$q_a q_b = (e^{-\epsilon_{a1}/kT} + e^{-\epsilon_{a2}/kT})(e^{-\epsilon_{b1}/kT} + e^{-\epsilon_{b2}/kT})$$
$$= e^{-(\epsilon_{a1}+\epsilon_{b1})/kT} + e^{-(\epsilon_{a1}+\epsilon_{b2})/kT} + e^{-(\epsilon_{a2}+\epsilon_{b1})/kT} + e^{-(\epsilon_{a2}+\epsilon_{b2})/kT}$$

Thus all possible combinations of molecular energy levels are obtained.

Molecular partition functions may also be written in terms of summations over levels j by including the degeneracies g_j.

$$q = \sum_j g_j e^{-\epsilon_j/kT} \tag{17.48}$$

If molecules $a, b, \ldots$ are identical their molecular partition functions are identical so that for N independent molecules the canonical ensemble partition function is given by

$$Q = q^N \qquad \text{(localized subsystems)} \tag{17.49}$$

This relation is limited to localized subsystems (molecules) because we have assumed that molecules can be identified as $a, b, \ldots$. If molecules are in a crystal they are localized and may be identified by position. If molecules are in a gas they are not localized and according to quantum mechanics they are indistinguishable. The modification of the calculation of the canonical ensemble partition function for nonlocalized subsystems is described in the next section.

17.7 CANONICAL ENSEMBLE PARTITION FUNCTION FOR NONLOCALIZED SYSTEMS

In a gas the number of microstates is far greater than the number of molecules. Therefore most of the possible states are unoccupied. If N molecules in a gas

---

* Molecular partition functions were used in Section 5.17.

could be identified, they could be arranged $N!$ ways between the N different microstates that they happen to occupy. Therefore the canonical ensemble partition function calculated using equation 17.49, which assumes that the molecules are distinguishable, yields too large a value by a factor of $N!$. Therefore, for nonlocalized subsystems

$$Q = \frac{q^N}{N!} \quad \text{(nonlocalized subsystems)} \tag{17.50}$$

The molecular partition function q may be calculated from molecular properties calculated from quantum mechanics or deduced from spectroscopy.

17.8 THERMODYNAMIC PROPERTIES OF IDEAL GASES

As discussed in Chapter 15 on spectroscopy, a molecule in a gas has various types of energy: translational, rotational, vibrational, and electronic. To a fairly high degree of approximation it may be considered that the total energy of a molecule in a particular state is the sum of these various types of energy:

$$\epsilon = \epsilon_{tr} + \epsilon_{rot} + \epsilon_{vib} + \epsilon_{el} \tag{17.51}$$

In general the rotational- and vibrational-energy level schemes depend on the electronic state. At normal temperatures, however, only the ground electronic state, or a group of electronic states with nearly the same energy, need be considered in calculating the molecular partition function. Substituting equation 17.51 into equation 17.46 and using $e^{a+b+c} = e^a e^b e^c$, we obtain

$$q = \sum e^{-\epsilon_{tr}/kT} \sum e^{-\epsilon_{rot}/kT} \sum e^{-\epsilon_{vib}/kT} \sum e^{-\epsilon_{el}/kT} \tag{17.52}$$

or

$$q = q_{tr}q_{rot}q_{vib}q_{el} \tag{17.53}$$

where

$$q_{tr} = \sum e^{-\epsilon_{tr}/kT} \tag{17.54}$$

etc. To this degree of approximation the canonical ensemble partition function is given by

$$Q = \frac{1}{N!}(q_{tr}q_{rot}q_{vib}q_{el})^N$$

$$= Q_{tr}Q_{rot}Q_{vib}Q_{el} \tag{17.55}$$

or

$$\ln Q = \ln Q_{tr} + \ln Q_{rot} + \ln Q_{vib} + \ln Q_{el} \tag{17.56}$$

where

$$Q_{tr} = \frac{q_{tr}^N}{N!} \tag{17.57}$$

$$Q_{rot} = q_{rot}^N \tag{17.58}$$

$$Q_{vib} = q_{vib}^N \tag{17.59}$$

$$Q_{el} = q_{el}^N \tag{17.60}$$

The $N!$ is included in the translational factor for the reasons described in the preceding section.

Since the thermodynamic quantities are obtained from $\ln Q$ and $\ln Q$ is made up of a sum of contributions according to equation 17.56, we can see that the thermodynamic quantities are made up from sums of contributions from the translational, rotational, vibrational, and electronic degrees of freedom. If the vibrational or rotational constants depend on electronic state (which they do in general), then the partition function does not separate so nicely.

17.9 TRANSLATIONAL PARTITION FUNCTION

The quantum mechanical energy levels for the molecules of an ideal gas in a container with dimensions $a \times b \times c$ are those given in equation 12.55 for a particle in a box. Substituting the equation for the energies of the allowed quantum states of a molecule of mass m in equation 17.46,

$$
\begin{aligned}
q_{\text{tr}} &= \sum_{n_1=1}^{\infty} \sum_{n_2=1}^{\infty} \sum_{n_3=1}^{\infty} \exp\left[-\frac{h^2}{8mkT}\left(\frac{n_1^2}{a^2} + \frac{n_2^2}{b^2} + \frac{n_3^2}{c^2}\right)\right] \\
&= \sum_{n=1}^{\infty} \exp\left[-\frac{h^2 n^2}{8ma^2kT}\right] \sum_{n=1}^{\infty} \exp\left[-\frac{h^2 n^2}{8mb^2kT}\right] \sum_{n=1}^{\infty} \exp\left[-\frac{h^2 n^2}{8mc^2kT}\right]
\end{aligned} \tag{17.61}
$$

In order to evaluate these sums we may note that for macroscopic containers the exponents are very small, unless T is very small. Since the successive terms in the summations differ by only small amounts, we may replace the summations by integrations, considering the quantum number n as a continuous variable that is not restricted to integer values.

$$
\int_0^{\infty} e^{-h^2 x^2/8ma^2kT}\, dx = (2\pi mkT)^{1/2} a/h \tag{17.62}
$$

since

$$
\int_0^{\infty} e^{-\alpha x^2}\, dx = \frac{1}{2}\left(\frac{\pi}{\alpha}\right)^{1/2} \tag{17.63}
$$

Substituting integrals obtained in this way in equation 17.61 and noting that $abc = V$, we obtain the following expression for the translational contribution to the molecular partition function.

$$
q_{\text{tr}} = \frac{(2\pi mkT)^{3/2} V}{h^3} \tag{17.64}
$$

For a monatomic gas without electronic excitation this is the only contribution to the molecular partition function. The canonical ensemble partition function is given by $q^N/N!$ so that

$$
Q_{\text{tr}} = \frac{(2\pi mkT)^{3N/2} V^N}{h^{3N} N!} = \frac{(2\pi mkT)^{3N/2} V^N e^N}{h^{3N} N^N} \tag{17.65}
$$

where the last form is obtained using Stirling's approximation $N! = N^N/e^N$.

17.10   THERMODYNAMIC PROPERTIES OF AN IDEAL MONATOMIC GAS

Using equation 17.65 for the molecular partition function, all of the thermodynamic properties of an ideal monatomic gas without electronic excitation may be calculated.

$$P = \frac{NkT}{V} \tag{17.66}$$

$$U = \tfrac{3}{2}NkT \tag{17.67}$$

$$C_V = \tfrac{3}{2}Nk \tag{17.68}$$

$$H = \tfrac{5}{2}NkT \tag{17.69}$$

$$C_P = \tfrac{5}{2}Nk \tag{17.70}$$

$$S = Nk \ln\left[\frac{(2\pi mkT)^{3/2} V e^{5/2}}{h^3 N}\right] \tag{17.71}$$

$$A = -NkT \ln\left[\frac{(2\pi mkT)^{3/2} V e}{h^3 N}\right] \tag{17.72}$$

$$G = -NkT \ln\left[\frac{(2\pi mkT)^{3/2} V}{h^3 N}\right] \tag{17.73}$$

The molar quantities may be obtained by letting N equal Avogadro's constant N_A. For U, H, A, and G the above relations give the corresponding property with respect to the ground state for which $U_0 = H_0 = A_0 = G_0$.

Example 17.2   Calculate the entropy of $H(g)$ at $25°C$.

$$m = \frac{1.0079 \times 10^{-3}\ \text{kg mol}^{-1}}{6.02205 \times 10^{23}\ \text{mol}^{-1}} = 1.67368 \times 10^{-27}\ \text{kg}$$

$$\bar{V} = 22.4138 \times 10^{-3}\ \text{m}^3\ \text{mol}^{-1}\,\frac{298.15}{273.15} = 24.4652 \times 10^{-3}\ \text{m}^3\ \text{mol}^{-1}$$

Substituting these and other values in

$$\bar{S}° = R \ln\left[\frac{(2\pi mkT)^{3/2} \bar{V} e^{5/2}}{h^3 N_A}\right]$$

yields 26.0141 cal K^{-1} mol^{-1}. Since the hydrogen atom has one electron, the spin direction may be either parallel to or antiparallel to an arbitrary direction. Thus the degeneracy is 2 and this contributes $R \ln 2 = 1.3774$ cal K^{-1} mol^{-1} to the entropy so that we obtain $\bar{S}° = 27.3915$ cal K^{-1} mol^{-1}, compared with 27.3927 cal K^{-1} mol^{-1} in Table 2.4.

17.11 THE ROTATIONAL PARTITION FUNCTION FOR A DIATOMIC MOLECULE

The expression for the rotational energy of a rigid diatomic molecule has already been discussed in Section 15.3.

$$\epsilon_{\text{rot}} = \frac{J(J+1)h^2}{8\pi^2 I} \tag{17.74}$$

Here J is the rotational quantum number, and I is the moment of inertia of the diatomic molecule. Although the rotational energy depends only on J, the state of a rigid rotator is specified by the quantum number J and an additional quantum number M, where M can have integral values between $-J$ and $+J$. Thus there are $2J + 1$ values of M for each value of J; in other words the rotational levels are $(2J + 1)$-fold degenerate. The rotational partition function is thus

$$q_{\text{rot}} = \sum_{J=0}^{\infty} (2J + 1)e^{-J(J+1)h^2/8\pi^2 IkT} \tag{17.75}$$

For molecules with large moments of inertia, the energy levels are ordinarily so so close together that they may be considered to be continuous, above 10–100 K, so that the summation may be replaced by an integration:

$$q_{\text{rot}} = \int_0^\infty (2J + 1)e^{-J(J+1)h^2/8\pi^2 IkT} dJ \tag{17.76}$$

Since $(2J + 1)\,dJ$ is the differential of $J(J + 1) = J^2 + J$, this equation may be integrated by multiplying and dividing through by $8\pi^2 IkT/h^2$ to obtain

$$q_{\text{rot}} = \frac{8\pi^2 IkT}{h^2} \int_0^\infty e^{-u}\,du$$

$$= \frac{8\pi^2 IkT}{h^2} \tag{17.77}$$

If the two atoms in the diatomic molecule are identical, twice too many terms have been included in the summation, and to a good approximation the correct result is obtained by dividing by 2. In general

$$q_{\text{rot}} = \frac{8\pi^2 IkT}{\sigma h^2} \tag{17.78}$$

where the so-called symmetry number σ is 2 if the two atoms in the molecule are identical and is 1 if they are different.

17.12 VIBRATIONAL PARTITION FUNCTION

The vibrational levels of diatomic molecules have been discussed in Section 15.4 on spectroscopy. If there is excitation to very high vibrational levels the anharmonicity has to be taken into account in calculating the partition function, but for most purposes it is sufficient to use the harmonic oscillator approximation because only lower levels are occupied. For the purpose of calculating the partition function vibrational energy is usually measured from the ground state ($v = 0$) rather than from the bottom of the potential energy curve. This simplifies the equations because the *zero* point energy ($h\nu_0/2$ where ν_0 is the fundamental vibration frequency) does not appear. We will say more about the origin for measuring energy in connection with the calculation of equilibrium constants.

In the harmonic oscillator approximation

$$\epsilon_{\text{vib}} = vh\nu \qquad v = 0, 1, 2, \cdots$$

so that

$$q_{\text{vib}} = \sum_{v=0}^{\infty} e^{-vh\nu/kT} = 1 + x + x^2 + x^3 + \cdots \tag{17.79}$$

where $x = e^{-h\nu/kT}$. When $x < 1$

$$\frac{1}{1-x} = 1 + x + x^2 + x^3 + \cdots \tag{17.80}$$

Since $e^{-h\nu/kT} < 1$

$$q_{\text{vib}} = \frac{1}{1 - e^{-h\nu/kT}}$$

$$= \frac{1}{1 - e^{-\Theta_v/T}} \tag{17.81}$$

where Θ_v is the vibrational characteristic temperature; $\Theta_v = h\nu/k = hc\tilde{\nu}/k$. The characteristic temperature of a vibration Θ_v is the temperature at which the energy of the vibrational quantum is equal to kT. From the vibrational characteristic temperatures listed in Table 17.4 it is evident that the vibration of

Table 17.4[1] Parameters for Diatomic Molecules

	Θ_v, K	Θ_r, K	r_e, Å	D_0, eV
H_2	6210	85.4	0.740	4.454
N_2	3340	2.86	1.095	9.76
O_2	2230	2.07	1.204	5.08
CO	3070	2.77	1.128	9.14
NO	2690	2.42	1.150	5.29
HCl	4140	15.2	1.275	4.43
HBr	3700	12.1	1.414	3.60
HI	3200	9.0	1.604	2.75
Cl_2	810	0.346	1.989	2.48
Br_2	470	0.116	2.284	1.97
I_2	310	0.054	2.667	1.54

[1] T. Hill, *Statistical Thermodynamics*, Addison-Wesley Publishing Co., Reading, Mass. (1960).

lighter diatomic molecules is relatively unexcited at room temperature. Molecules with heavier atoms have lower vibrational frequencies and therefore vibrational motion contributes to thermodynamic quantities at lower temperatures.

The vibrations of polyatomic molecules can be discussed in terms of normal modes of vibration, as discussed in Section 15.9 on spectroscopy. There are $3N - 6$ normal modes of vibration for a nonlinear molecule with N atoms and $3N - 5$ normal modes for a linear molecule. For harmonic motion the vibrational partition function for a polyatomic molecule may be written

$$q_{\text{vib}} = q_1 q_2 \cdots q_{3N-6} \tag{17.82}$$

where each factor represents a different normal mode, since the normal modes are independent. Thus the various normal modes contribute additively to thermodynamic quantities.

These equations explain why vibrational motion is not fully excited at ordinary temperatures (cf. Section 17.15). As mentioned in Section 9.5 classical mechanics was not able to explain this fact.

17.13 ELECTRONIC PARTITION FUNCTION

In the electronic partition function q_{el}, the summation must be carried out by writing out the individual terms. Usually only the first term of the electronic partition function

$$q_{\text{el}} = g_0 e^{-\epsilon_0/kT} + g_1 e^{-\epsilon_1/kT} + \cdots \tag{17.83}$$

has to be considered because the first excited level is so far above the ground state: $(\epsilon_1 - \epsilon_0) \gg kT$. However, a few molecules such as NO, NO_2, and O_2 do have low-lying electronic states that have to be taken into consideration.

For atoms the energy of the ground state is set equal to zero. Thus $\epsilon_0 = 0$ and $q_{\text{el}} = g_0$, the degeneracy of the ground state.*

For diatomic molecules it is convenient to take the origin as the energy of the separated atoms in their ground states. Thus $\epsilon_0 = -D_0$, the experimentally determined dissociation energy (Section 15.6), and equation 17.83 becomes

$$q_{\text{el}} = g_0 e^{D_0/kT} \tag{17.84}$$

---

* S. J. Strickler, *J. Chem. Ed.*, **43**, 364 (1966), discusses the fact that the electronic partition function q_{el} for the hydrogen atom at 25° calculated from

$$q_{\text{el}} = \sum_{n=1}^{\infty} n^2 e^{-\epsilon_n/kT}$$

is infinite. In this equation n^2 is the degeneracy, and ϵ_n is the energy of the orbital with quantum number n. This is contrary to experience since it would require the population of the ground state to be zero. The problem is that there is an infinite number of terms in the partition function with an energy less than 13.60 eV. The paradox is resolved by considering the size of the orbital of a hydrogen atom for a very large quantum number n. For a finite container the maximum n is finite, and the value of q_{el} is not detectably different from unity. In fact the cutoff probably comes from neighboring molecules that limit the size of a hydrogen atom to about $(V/N)^{1/3}$.

17.14 SUMMARY OF STATISTICAL CALCULATIONS OF THERMODYNAMIC QUANTITIES

Since the partition function of an ideal gas molecule is a product of the partition functions for translation, rotation, vibration, and electronic excitation (to the extent these energies are separable according to equation 17.51), the thermodynamic quantities are made up of additive terms from these various sources. The contributions of translation, rotation, vibration, and electronic excitation to the various thermodynamic quantities for diatomic molecules are summarized in Table 17.5. For atoms without electronic excitation translation makes the only contribution and the formulas given in the first column may be used.

17.15 HEAT CAPACITIES OF GASES

In the chapter on kinetic theory we discussed the heat capacities of gases (Section 9.5) and how the principle of equipartition led to a correct value for the rotational contribution to C_V of diatomic molecules at room temperature but not to the vibrational contribution. We can now see how this failure of classical mechanics was corrected by quantum mechanics. Vibrational motion is quantized, and the occupation of vibrational levels depends on the temperature. It is not until the absolute temperature is in the range of $\Theta_v = h\nu/k$ that a vibration with frequency

Table 17.5   Molar Thermodynamic Quantities for Diatomic Ideal Gases[1]

	Translation	Rotation	Vibration	Electronic
$\bar{U} - \bar{U}^\circ$	$\frac{3}{2}RT$	RT	$RT\,\dfrac{x}{e^x - 1}$	$-N_A D_0$
$\bar{H} - \bar{H}^\circ$	$\frac{5}{2}RT$	RT	$RT\,\dfrac{x}{e^x - 1}$	$-N_A D_0$
$\bar{C}_V$	$\frac{3}{2}R$	R	$R\,\dfrac{x^2 e^x}{(e^x - 1)^2}$	0
$\bar{C}_P$	$\frac{5}{2}R$	R	$R\,\dfrac{x^2 e^x}{(e^x - 1)^2}$	0
$\bar{S}$	$R \ln\left[\dfrac{(2\pi mkT)^{3/2}\bar{V}e^{5/2}}{h^3 N_A}\right]$	$R \ln\left(\dfrac{eT}{\sigma\Theta_{\text{rot}}}\right)$	$R\left[\dfrac{x}{e^x - 1} - \ln(1 - e^{-x})\right]$	$R \ln g_0$
$\bar{A} - \bar{A}^\circ$	$-RT \ln\left[\dfrac{(2\pi mkT)^{3/2}\bar{V}e}{h^3 N_A}\right]$	$-RT \ln\left(\dfrac{T}{\sigma\Theta_{\text{rot}}}\right)$	$RT \ln(1 - e^x)$	$-N_A D_0 - RT \ln g_0$
$\bar{G} - \bar{G}^\circ$	$-RT \ln\left[\dfrac{(2\pi mkT)^{3/2}\bar{V}}{h^3 N_A}\right]$	$-RT \ln\left(\dfrac{T}{\sigma\Theta_{\text{rot}}}\right)$	$RT \ln(1 - e^x)$	$-N_A D_0 - RT \ln g_0$

[1] $\Theta_{\text{rot}} = h^2/2Ik$ and $x = h\nu/kT$. D_0 is the experimental dissociation energy. $\bar{U}^\circ = \bar{H}^\circ = \bar{A}^\circ = \bar{G}^\circ$.

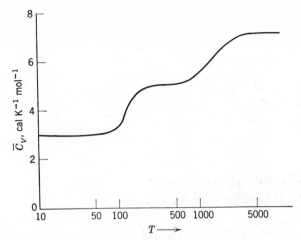

Fig. 17.4 Dependence of $\bar{C}_v$ for hydrogen on temperature. The temperature axis is on a logarithmic scale.

v will make a significant contribution to the heat capacity. This is illustrated by Fig. 17.4 for hydrogen. In addition this graph illustrates the relatively large Θ_r for hydrogen.

At very low temperatures H_2 behaves like a monatomic gas and the molar heat capacity at constant volume is $\frac{3}{2}R$. Starting at about 100 K rotation is excited and from 1000 K vibration is excited as shown by the increases in C_V. Energy is absorbed into translational motions at very low temperatures because the translational energy levels are very closely spaced. The rotational levels have wider separations and they do not contribute to the heat capacity until kT is of the order of the energy of separation between rotational energy levels, as may be seen from the Boltzmann equation. The temperatures Θ_r at which kT is equal to the separation between rotational levels are given for a variety of gases in Table 17.4. Usually Θ_r is below the boiling point for the gas. Molecular hydrogen is an exception because of the very small value of the moment of inertia, and the resulting wide spacing of rotational energy levels. For gases with heavier atoms the separation of rotational levels is smaller so that rotations are excited at lower temperatures.

Example 17.3 Calculate the molar heat capacity $\bar{C}_P$ of water vapor at 1000 K, given the frequencies of the three normal modes of vibration, 1595, 3652, and 3756 cm^{-1}.

Translation contributes $\frac{5}{2}R$ to $\bar{C}_P$, and assuming ideal gas behavior $\bar{C}_P = \bar{C}_V + R$. Since three angles are required to describe the orientation of a water molecule, there are three degrees of rotational freedom that contribute $\frac{3}{2}R$ (Section 9.5). The contribution of the first normal mode is

$$R \frac{x^2 e^x}{(e^x - 1)^2} = \frac{(1.987 \text{ cal K}^{-1} \text{ mol}^{-1})(2.294)^2 e^{2.294}}{(e^{2.294} - 1)^2}$$

$$= 1.305 \text{ cal K}^{-1} \text{ mol}^{-1}$$

where

$$x = \frac{hc\tilde{\nu}}{kT} = \frac{(6.626 \times 10^{-34} \text{ J s})(2.998 \times 10^8 \text{ m s}^{-1})(1595 \text{ cm}^{-1})(100 \text{ cm m}^{-1})}{(1.381 \times 10^{-23} \text{ J K}^{-1})(1000 \text{ K})}$$

The other two normal modes contribute 0.290 and 0.264 cal K^{-1} mol^{-1}. The sum of these contributions is 9.807 cal K^{-1} mol^{-1}, compared with 9.769 cal K^{-1} mol^{-1} obtained with experimental parameters from Table 1.2. In the statistical mechanical calculation we have assumed that water vapor is an ideal gas, and this is not quite true.

17.16 THEORETICAL CALCULATION OF EQUILIBRIUM CONSTANTS

A very simple introduction to the calculation of the equilibrium constant for a reaction $A(g) = B(g)$, where A and B are ideal gases, has been given in Section 5.17. This simple derivation may be extended to obtain[*] the expression for a general chemical reaction of ideal gases.

$$0 = \sum \nu_i A_i \tag{17.85}$$

The result is

$$K_p = \left(\prod_i p_i^{\nu_i}\right)_{eq} = e^{-\Delta\epsilon_0/kT} \prod_i \left(\frac{q_i^0}{N_A}\right)^{\nu_i} \tag{17.86}$$

where ν_i is positive for a product and negative for a reactant. The superscript zero indicates that the molecular partition coefficient is calculated for the gas in its standard state of one atmosphere pressure. The difference $\Delta\epsilon_0$ between the energies of the products and reactants at absolute zero is given by

$$\Delta\epsilon_0 = \sum_i \nu_i \epsilon_{0i} \tag{17.87}$$

and is related to the change in internal energy ΔU_0 at absolute zero by $\Delta U_0 = N_A \Delta\epsilon_0$. In equation 17.87 it is important that the various energies ϵ_i be relative to the same arbitrary zero of energy. If the dissociation energies of the various reactants and products are known, this is conveniently taken as that of the state of the separated atoms. The dissociation of reactants into atoms yields, of course, the same atoms as the products.

Since the molecular partition function q_i is the product of the translational, rotational, vibrational, and electronic partition functions (equation 17.53), the partition function for gas i in its standard state may be written

$$q_i^0 = \frac{RT}{P}\left(\frac{2\pi m_i kT}{h^2}\right)^{3/2} q_{\text{rot }i}^0 q_{\text{vib }i}^0 q_{\text{ele }i}^0 \tag{17.88}$$

---

[*] N. Davidson, *Statistical Mechanics*, McGraw-Hill Book Co., New York, 1962, p. 124; J. H. Knox *Molecular Thermodynamics*, Wiley-Interscience, New York, 1971.

where V in the expression for the translational partition function has been replaced with RT/P since we are considering an ideal gas. In calculations P is set equal to 101,325 N m^{-2} if the standard state used in defining the equilibrium constant is one atmosphere. The volume must be expressed in SI units if the rest of the quantities are expressed in SI units.

Substituting equation 17.88 in equation 17.86 yields

$$K_p = \left(\frac{kT}{P}\right)^{\Sigma \nu_i} \left(\frac{2\pi kT}{h^2}\right)^{\frac{3}{2}\Sigma \nu_i} (\prod_i m_i^{\nu_i})^{3/2} \prod_i (q_{\text{rot}\,i})^{\nu_i} \prod_i (q_{\text{vib}\,i})^{\nu_i} \prod_i (q_{\text{ele}\,i})^{\nu_i} e^{-\Delta \epsilon_0/kT}$$

(17.89)

The superscript zeros on the rotational, vibrational, and electronic partition functions have been omitted because their calculation does not depend upon pressure for ideal gases. Using equation 17.89 the equilibrium constants may be calculated for reactions of ideal gases if the following quantities are known for the reactants and products:

1. molecular masses m_i
2. moments of inertia I_i
3. vibration frequencies
4. electronic energies and degeneracies
5. change in energy $\Delta \epsilon_0$ for the reaction at absolute zero.

The value of K_p for any reaction involving only atoms and diatomic molecules may be calculated with equations given in this chapter.

If nonideal gases or liquids are involved, the value of the thermodynamic equilibrium constant calculated in this way cannot be used to predict accurately the pressure or mole fraction of a reactant in the equilibrium mixture. The thermodynamic equilibrium constant is expressed in terms of the *activities* of the reactants and products at equilibrium. In nonideal systems the activities depend on interactions of all components of the mixture, and such systems are very difficult to treat theoretically.

References

F. C. Andrews, *Equilibrium Statistical Mechanics*, Wiley, New York, 1963.

S. M. Blinder, *Advanced Physical Chemistry*, The Macmillan Co., London, 1969.

N. Davidson, *Statistical Mechanics*, McGraw-Hill Book Co., New York, 1962.

J. C. Davis, Jr., *Advanced Physical Chemistry*, The Ronald Press, New York, 1965, Chapter 5.

T. L. Hill, *An Introduction to Statistical Mechanics*, Addison-Wesley Publishing Co., Reading, Mass., 1960.

E. A. Jackson, *Equilibrium Statistical Mechanics*, Prentice-Hall, Englewood Cliffs, N. J., 1968.

G. J. Janz, *Estimation of Thermodynamic Properties of Organic Compounds*, Academic Press, New York, 1958.

W. Kauzmann, *Kinetic Theory of Gases*, W. A. Benjamin, Inc., New York, 1966.

W. Kauzmann, *Thermodynamics and Statistics: With Applications to Gases*, W. A. Benjamin, Inc., New York, 1967.

J. H. Knox, *Molecular Thermodynamics*, Wiley-Interscience, New York, 1971.

J. E. Mayer and M. G. Mayer, *Statistical Mechanics*, Wiley, New York, 1940.

B. J. McClelland, *Statistical Thermodynamics*, Wiley, New York, 1973.

L. K. Nash, *Chemthermo: A Statistical Approach to Chemical Thermodynamics*, Addison-Wesley Publishing Co., Reading, Mass., 1971.

D. Rapp, *Statistical Mechanics*, Holt, Rinehart and Winston, Inc., New York, 1972.

F. Reif, *Fundamentals of Statistical and Thermal Physics*, McGraw-Hill Book Co., New York, 1965.

F. Reif, *Statistical Physics*, Berkeley Physics Course. Vol. 5, McGraw-Hill Book Co., New York, 1965.

Problems

17.1 (*a*) How many different ways can two distinguishable balls be placed in two boxes? (*b*) How many different ways can two distinguishable balls be placed in three boxes? (*c*) What are the answers to (*a*) and (*b*) if the balls are indistinguishable?

Ans. (*a*) 4, (*b*) 9, (*c*) 3, 6.

17.2 For an idealized crystal consisting of nine simple harmonic oscillators with nine quanta of energy, what is the most probable distribution of quanta between oscillators? The answer may be obtained by analogy with Fig. 17.2 with a few trials. What probabilities does the Boltzmann distribution give for the first 5 levels?

Ans. $v_i = 0$	1	2	3	4	5
Most probable 4	2	2	1	0	0
Boltzmann 5.70	2.10	0.77	0.28	0.10	0.04

17.3 The difference of 1.10 cal K^{-1} mol^{-1} between the third law entropy of CO and the statistical mechanical value can be attributed to the randomness of orientation of CO molecules in its crystals at absolute zero. If half of the molecules are oriented CO and half OC, calculate the entropy of a crystal at absolute zero using equation 17.1 and Stirling's approximation.   *Ans.* 1.38 cal K^{-1} mol^{-1}.

17.4 Using the Boltzmann equation, calculate the ratio of populations at 25° of energy levels separated by (*a*) 2 kcal mol^{-1}, (*b*) 10 kJ mol^{-1}.   *Ans.* (*a*) 0.0342, (*b*) 0.0177.

17.5 Calculate the temperature at which 10% of the molecules in a system will be in the first excited electronic state if this state is 400 kJ mol^{-1} above the ground state.

Ans. 22,000 K.

17.6 Calculate the entropy of one mole of H-atom gas at 1000 K and (*a*) 1 atm and (*b*) 1000 atm.   *Ans.* (*a*) 33.404 cal K^{-1} mol^{-1},

(*b*) 19.677 cal K^{-1} mol^{-1}.

17.7 Calculate the entropy of neon at 25° and 1 atm.

Ans. 146.217 J K^{-1} mol^{-1} or 34.947 cal K^{-1} mol^{-1}.

17.8 Calculate $\bar{S}°$ and $\bar{C}_P^0$ for argon ($M = 39.948$) at 25° and 1 atm.

Ans. 154.733 J K^{-1} mol^{-1} or 36.982 cal K^{-1} mol^{-1}.

20.786 J K^{-1} mol^{-1} or 4.968 cal K^{-1} mol^{-1}.

17.9 Calculate the temperature at which the (a) rotational and (b) vibrational energy contributions to the partition function of Cl_2 become important. The contribution may be taken as important when the energy of the first excited state becomes equal to kT.

Ans. (a) 0.691 K, (b) 813 K.

17.10 Calculate the ratio of the number of HBr molecules in state $v = 2$, $J = 5$ to the number in state $v = 1$, $J = 2$ at 1000 K. Assume that all of the molecules are in their electronic ground states. *Ans.* 0.0407.

17.11 Derive the expression for the vibrational contribution to the internal energy

$$\bar{U} - \bar{U}_0^\circ = \frac{RTx}{e^x - 1}$$

where $x = h\nu/kT$.

17.12 Calculate $(H° - H_0°)/T$ and the entropy of nitrogen gas at 25° and 1 atm pressure. The equilibrium separation of atoms is 1.095 Å, and the vibrational wave number is 2330.7 cm^{-1}. *Ans.* 6.9555, 45.7489 cal K^{-1} mol^{-1}.

17.13 The ground state of $Cl(g)$ is fourfold degenerate. The first excited state is 881 cm^{-1} higher in energy and is twofold degenerate. What is the value of the electronic partition function at 25°C? At 1000 K? *Ans.* 4.0285, 4.563.

17.14 Calculate the value of the equilibrium constant K_p for the dissociation of CO at 2000 K.

$$CO(g) = C(g) + O(g)$$

Spectroscopic data on CO are available in Table 15.2. The degeneracies of the electronic ground states of C and O are both 9. (The value of K_p at 2000 K calculated from data in Table 5.5 is 7.427×10^{-22} atm.) *Ans.* 7.790×10^{-22} atm.

17.15 What fraction of oxygen molecules is dissociated into atoms at 4000 K and 1 atm ignoring electronic contributions? $O_2 = 2O$. *Ans.* 3.70×10^{-2}.

17.16 Calculate the equilibrium constant for the isotope exchange reaction

$$D + H_2 = H + DH$$

at 25°. Assume that the equilibrium distance and force constants of H_2 and DH are the same. *Ans.* 7.17.

17.17 (a) How many different ways can three distinguishable balls be placed in two boxes? (b) How many different ways are there if the balls are indistinguishable?

17.18 How many microstates are there for a three-atom crystal of one-dimensional harmonic oscillators if only two quanta of energy are available? On the average how many atoms will have zero, one, or two quanta?

17.19 Calculate the ratio of populations at 25° of energy levels separated by (a) 1 eV and (b) 10 eV. (c) Calculate the ratios at 1000°.

17.20 Calculate the entropies of one mole of each of the atomic gases H, N, and C at 25° and 1 atm. Compare these values with those in Table 2.3.

17.21 Calculate the molar entropy of helium in the ideal gas state at 25° and 1 atm pressure.

17.22 At what temperatures do the rotational and vibrational energy contributions become important in the calculation of the partition function of H_2? The contribution may be taken as important when the energy of the first excited state becomes equal to kT.

17.23 What fraction of HCl molecules is in the state $v = 2$, $J = 7$ at 500°C?

17.24 Calculate $(\bar{H}° - \bar{H}_0°)/T$, and the entropy of hydrogen gas at 25° and 1 atm. Table 15.2 gives the values of μ, $\tilde{\nu}_0$, and r_e.

17.25 A molecule exists in singlet and triplet forms with the singlet having a higher energy by 4.11×10^{-21} J per molecule. The singlet level has a degeneracy of 1 and the triplet level has a degeneracy of 3. (*a*) Ignoring higher levels, what is the electronic partition function? (*b*) What is the ratio of the concentration of triplets to singlet molecules at 298 K?

17.26 What is the value of the electronic partition function for I(*g*) at 10,000 K? The ground state is doubly degenerate, and the first excited state at 7603 cm^{-1} is also doubly degenerate.

17.27 Calculate the equilibrium constant and degree of dissociation for $H_2(g)$ at 3000 K and 1 atm.

$$H_2(g) = 2H(g)$$

In studying the conduction of heat from tungsten filaments by gases Langmuir obtained a degree of dissociation of 0.072 under these conditions. The degeneracy of hydrogen atoms is 2.

17.28 Calculate the equilibrium constant at 25° for the reaction

$$H_2 + D_2 = 2HD$$

It may be assumed that the equilibrium distance and the force constant k are the same for all three molecular species, so that the additional vibrational frequencies required may be calculated from $2\pi\nu = (k/\mu)^{1/2}$. Because of the zero-point vibration, $\Delta \bar{U}_0$ for this reaction is given by

$$\Delta \bar{U}_0 = \tfrac{1}{2} N_A h [2\nu_{HD} - \nu_{H_2} - \nu_{D_2}]$$

17.29 (*a*) How many different ways can four distinguishable balls be placed in two boxes? (*b*) How many·different ways are there if the balls are indistinguishable?

17.30 In the text small crystals having numbers of vibrational quanta equal to the number of atoms were considered. Calculate the average occupation N_i of energy levels of a five-atom crystal having three vibrational quanta.

17.31 A crystal contains N atoms, each of which can have energy $+u$ or energy $-u$ (this is the type of situation one finds when a paramagnetic substance is placed in a magnetic field). (*a*) Suppose $N = 3$. Write down all the possible microstates that the system could be in. (*b*) Determine the values of $\mathcal{N}$, E_i, N_i, and p_i for the ensemble of microstates in the answer to (*a*). (*c*) Show that $q^N = Q$ for this system. (*d*) Find S and U for this system. (*e*) Find S and U for the system for arbitrary $N \gg 1$.

17.32 The energies of the $n = 1$ and $n = 2$ orbits in the hydrogen atom are 27,420 and 109,678 cm^{-1}, respectively. What are the relative populations in these levels (*a*) at 25°, (*b*) at 2000°?

17.33 Using relations from statistical mechanics, show that the change in entropy for an ideal monatomic gas is given by equation 2.15 for a process in which the temperature and volume change. Thermodynamics shows that this is true for any ideal gas provided $\bar{C}_V$ is independent of temperature.

17.34 Calculate the entropy of argon at (*a*) 25° and (*b*) 727° and 1 atm pressure.

17.35 Calculate the rotational energy level of $H^{35}Cl$ that has the highest occupation number at (*a*) 25°C and (*b*) 500 K.

17.36 Calculate the partition function q for $I_2(g)$ at 500 K and 1 atm pressure. Given: $M = 256$, $I = 7.64 \times 10^{-45}$ kg cm^2, $\sigma = 2$, and $\nu = 6.396 \times 10^{12}$ s^{-1}.

17.37 Calculate $(\bar{H}^\circ - \bar{H}_0^\circ)/T$, and the entropy of nitrogen gas at 2000 K and 1 atm pressure. (Compare problem 17.12.)

17.38 Calculate the entropy for chlorine gas at 25° and 1 atm pressure. The equilibrium separation of the nuclei is 1.989 Å and the vibrational wave number is 556.9 cm^{-1}.

17.39 Calculate $\bar{C}_P$ for hydrogen gas at 300 and 2000 K. This calculation is discussed in some detail by C. Marzzacco and M. Waldman, *J. Chem. Ed.*, **50**, 444 (1973).

17.40 A hypothetical molecule can exist in only three states with energies of 0, ϵ, and ϵ. Derive the molar partition function Q and the expression for $\bar{U}$ and $\bar{C}_V$ in terms of ϵ.

17.41 Calculate $\bar{C}_P$ for CO_2 at 1000°. Compare the actual contributions to $\bar{C}_P$ from the various normal modes with the classical expectations.

17.42 Calculate the equilibrium constant K_p for

$$I_2(g) = 2I(g)$$

at 1000 K. Information about $I_2(g)$ is given in problem 17.36. The ground state of $I(g)$ is fourfold degenerate. The first excited state, which is 7603 cm^{-1} higher, is doubly degenerate. The dissociation energy at absolute zero is 1.544 eV.

17.43 It may be shown (W. Kauzmann, *Thermodynamics and Statistics: With Application to Gases*, W. A. Benjamin, Inc., New York, 1967, p. 249) that the second virial coefficient in

$$\frac{P\bar{V}}{RT} = 1 + \frac{B}{\bar{V}} + \frac{C}{\bar{V}^2} + \cdots$$

is given by

$$B = 2\pi N_A \int_0^\infty (1 - e^{-U/kT}) r^2 \, dr$$

where r is intermolecular distance, k is the Boltzmann constant (R/N_A), and U is the potential energy of interaction. Derive the expression for B for a gas of rigid spheres of diameter d. The molecules do not interact unless they touch, and so $U = 0$ if $r > d$. The molecules cannot penetrate each other, and so $U = \infty$ if $r < d$.

17.44 The measurement of the equilibrium constant for

$$H_2(g) + I_2(g) = 2HI(g) \qquad K = \frac{(HI)^2}{(H_2)(I_2)} = 54.5$$

at 698.2 K has been discussed in Section 5.8. Calculate the value expected from statistical mechanics.

	M	I	σ	$\bar{\nu}$
	g mol^{-1}	kg m^2		cm^{-1}
$H_2(g)$	2.016	4.59×10^{-48}	2	4405
$I_2(g)$	256	7.43×10^{-45}	2	214
$HI(g)$	129	4.31×10^{-47}	1	2309

$\Delta U_0^\circ = -9.728$ kJ mol^{-1}

CHAPTER 18

PHOTOCHEMISTRY

Photochemistry comprises the study of chemical reactions produced directly or indirectly by light. Ordinary thermal reactions proceeding in the dark acquire their activation energy through random, successive collisions between molecules. Photochemical reactions receive their activation energy by the absorption of photons of light by molecules, and so they offer the possibility of a high degree of selectivity in that the energy of a quantum of light may be just sufficient for a particular reaction or may be absorbed only by a particular component of a mixture. Thus the activation step of a photochemical reaction is quite different, and more selective, than the activation of an ordinary (thermal) reaction. Two important biological processes involving photochemical reactions are photosynthesis and vision, which are just partially understood.

18.1 LAWS OF PHOTOCHEMISTRY

There are two basic laws of photochemistry. According to the first which is due to Grotthus (1817) and Draper (1843) only light that is absorbed can produce photochemical change. This law now seems rather self evident, but we must also be aware that there is another effect not visualized by Grotthus and Draper according to which radiation that is not absorbed may stimulate an excited molecule to radiate (Section 18.10).

The second law of photochemistry, proposed by Stark and Einstein (1908–1912), is that a molecule absorbs a single quantum of light in becoming activated.

$$A + h\nu \rightarrow A^* \tag{18.1}$$

Avogadro's number of photons is called an "einstein," just as Avogadro's number of electrons is called a "faraday." The calculation of the energy carried by an einstein of photons has been discussed in Section 15.1. Rather recently exceptions to this law have been found in simultaneous two quantum absorptions in systems illuminated with the intense and coherent radiation from a laser.

$$A + 2h\nu \rightarrow A^{**} \tag{18.2}$$

The second law of photochemistry provides the basis for the calculation of quantum yield Φ of a particular process.

$$\Phi = \frac{\text{number of molecules undergoing the process}}{\text{number of quanta absorbed}} \tag{18.3}$$

The quantum yield, which is an efficiency rating, provides a convenient means for describing the experimental facts and is useful for drawing conclusions about the mechanism of the reaction.

It is useful to distinguish between the *primary* process and the subsequent *secondary* reactions. In the primary process one photon is absorbed by one molecule, and the molecule is activated to a higher state. The activated molecule may lose energy in several ways, including luminescence or internal conversion, or before it loses all this energy, it may undergo chemical change. Sometimes the primary process is followed by a simple stoichiometric reaction that gives some integer number or fraction of moles of product. On the other hand, dissociated fragments resulting from the absorption of light may recombine so as to give low apparent quantum yields. Again each activated molecule can give rise to a chain reaction in which a reaction is repeated many times thus giving large quantum yields.

Example 18.1 In the photobromination of cinnamic acid to dibromocinnamic acid, using blue light of 435.8 nm at 30.6°, a light intensity of 1.4×10^{-3} J s^{-1} produced a decrease of 0.075 millimole of Br_2 during an exposure of 1105 s. The solution absorbed 80.1% of the light passing through it. Calculate the quantum yield.

$$E = N_A hc/\lambda = \frac{(6.02 \times 10^{23} \text{ mol}^{-1})(6.62 \times 10^{-34} \text{ J s})(3 \times 10^8 \text{ m s}^{-1})}{(4.358 \times 10^{-7} \text{ m})}$$

$$= 2.74 \times 10^5 \text{ J mol}^{-1} = 65.6 \text{ kcal mol}^{-1}$$

$$\text{Number of einsteins absorbed} = \frac{(1.4 \times 10^{-3} \text{ J s}^{-1})(0.801)(1105 \text{ s})}{(2.74 \times 10^5 \text{ J})}$$

$$= 4.52 \times 10^{-6}$$

Number of moles of Br_2 reacting $= 7.5 \times 10^{-5}$

$$\phi = \frac{7.5 \times 10^{-5}}{4.52 \times 10^{-6}} = 16.6$$

18.2 MOLECULAR EXCITATION

The absorption of light by a diamagnetic molecule generally raises it from a singlet ground state to a singlet excited state. To remind you of this terminology that was used in Section 12.23, the excited electron of a singlet excited state has a spin antiparallel to the one with which it is paired. There are many singlet excited states, but usually only several of those of the lowest energy can be produced with visible or near ultraviolet light. The singlet ground state is usually represented by S_0 and singlet excited states by $S_1, S_2, \ldots$, as shown in Fig. 18.1.

For each excited singlet state (S_1, S_2, S_3, etc.) there is a corresponding triplet state (T_1, T_2, T_3, etc.). The difference between the singlet and triplet states is that the electrons are not paired in the triplet states. Since the spin angular momentum of an electron is one-half, the spin for a triplet state is 1. The spin angular momentum vector for a triplet state may have one of three orientations in a magnetic field, hence the designation triplet.

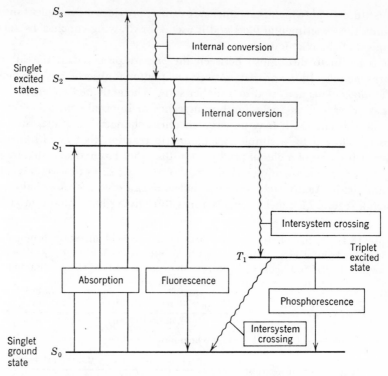

Fig. 18.1 Schematic representation of changes in molecular energy levels which may occur upon absorption of radiation. Nonradiative processes are represented by ⤳ and radiative processes are represented by →.

A triplet state always has a lower energy than the corresponding singlet state. The explanation is roughly that since the two outer electrons have the same spin they cannot come very close to each other without violating the Pauli principle. Since the electrons are further apart there is a decrease in electronic repulsion and the energy of the molecule is lower.

After a molecule has been excited by the absorption of radiation two types of radiationless processes occur so rapidly that they precede fluorescence, phosphorescence, and chemical reaction. These fast processes are (a) internal conversion and (b) intersystem crossing.

Internal conversion refers to radiationless loss of energy from excited states without a change in multiplicity (that is, singlet-singlet transitions or triplet-triplet transitions). When a molecule absorbs radiation it may be excited to one of several singlet excited states S_1, S_2, S_3, etc. The higher excited states rapidly (10^{-11} s) lose energy so that the molecules arrive very quickly in the first excited singlet state S_1. Excited molecules may also exist in a series of excited triplet states T_1, T_2, T_3, etc., and internal conversion rapidly converts these to T_1. In the process of internal conversion energy is degraded to heat.

Intersystem crossing refers to radiationless transitions between states of different multiplicity. Since for the reason mentioned above the first triplet excited state usually has a lower energy than the first singlet excited state, the most familiar example of intersystem crossing is $S_1 \rightsquigarrow T_1$.

A molecule in the S_1 state can return to the ground state S_0 by fluorescence, provided that internal conversion to S_0 does not occur more rapidly, and a molecule in the T_1 state can return to the ground state S_0 by phosphorescence, provided that intersystem crossing to S_0 does not occur more rapidly. Since S_1 and T_1 states are the starting points for most photochemical reactions, and since photochemical reactions have to be faster than fluorescence or phosphorescence so that these processes do not drain away the energy, we will start our consideration of photochemistry with fluorescence and phosphorescence.

18.3 FLUORESCENCE AND PHOSPHORESCENCE

The radiation emitted in a transition between states of the same multiplicity (that is, singlet-singlet transitions or triplet-triplet transitions) is called *fluorescence*, and the radiation emitted in a transition between states of different multiplicity is called *phosphorescence*. The fluorescence lifetimes of organic molecules are generally about 10^{-9} s. Phosphorescence lifetimes are 10^{-3} s up to minutes or longer because transitions between states of different multiplicity have very low probabilities.

The absorption, fluorescence, and phosphorescence spectra of a molecule have some features in common, as illustrated by Fig. 18.2. Since internal conversion occurs so rapidly, fluorescence arises from radiative transitions from the lowest vibrational level of S_1 to the various vibrational levels of S_0. Therefore, if the spacings of the vibrational levels in the ground state and excited state are similar there will be an approximate "mirror image" relationship between the absorption and fluorescence spectra. The mirror-image relationship is not expected to be exact because the vibrational spacings are not the same in the ground and excited states. The small difference in the 0—0 bands for absorption and fluorescence is due to the difference between the solvation of the initial excited state in absorption and the equilibrated excited state from which fluorescence occurs.

Phosphorescence arises from radiative transitions from the lowest vibrational level of the triplet state T_1 to the various vibrational levels of S_0. Therefore, the phosphorescence spectrum resembles the fluorescence spectrum. Since the T_1 state has a lower energy than the S_1 state the phosphorescence spectrum is observed at longer wavelengths than the fluorescence spectrum. The phosphorescence spectrum is weaker than the fluorescence spectrum because there is more time for energy to be lost by intersystem crossing.

The origin of the fluorescence spectrum may be discussed in greater detail if we restrict our attention to diatomic molecules. Figure 18.3a shows the potential energy curves and vibrational levels for the ground state S_0 and first excited singlet state S_1 of a diatomic molecule. Internal conversion rapidly brings the excited

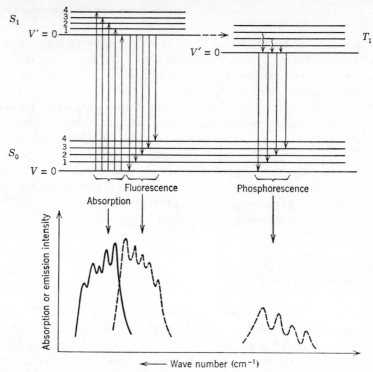

Fig. 18.2 Schematic diagram showing the origin of absorption, fluorescence, and phosphorescence spectra. (From Brittain, George and Wells, *Introduction to Molecular Spectroscopy* Academic Press, New York, 1970.)

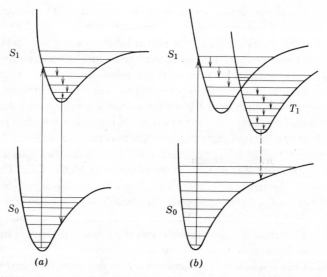

Fig. 18.3 (*a*) Absorption and fluorescence by a diatomic molecule. (*b*) Absorption and phosphorescence by a diatomic molecule. (From J. C. Davis, *Advanced Physical Chemistry*, The Ronald Press Co., New York, 1965.)

588

molecule to its lowest vibrational level. When a transition occurs to the S_0 level the molecule is most likely to end up in a higher vibrational level because the potential energy curve for the excited state is displaced to a greater internuclear distance than the ground state curve, and the transition is represented by a vertical line according to the Franck-Condon principle (Section 15.6). In the case illustrated the transition $0 \rightarrow 3$ is the most intense.

The origin of the phosphorescence spectrum for a diatomic molecule is shown in Fig. 18.3b. If during the loss of vibrational energy from S_1 a molecule happens to have the vibrational energy and internuclear separation that corresponds with the intersection of the potential energy curves for S_1 and T_1, then loss of vibrational energy may continue in the T_1 state. At the intersection of the two potential energy curves the nuclei are not moving relative to one another, and the energy is the same in the two states. Under these conditions there is a high probability that a transition will occur in spite of the selection rule that transitions between states of different multiplicity are forbidden. Loss of vibrational energy in the T_1 state continues by internal conversion as shown. The triplet state has a long lifetime, and so the phosphorescent radiation is emitted slowly. Since the triplet state T_1 has a much longer lifetime than S_1, it is much more likely to be involved in a photochemical reaction.

Certain fluorescent dyes emit visible light when activated by light in the near ultraviolet or short visible range present in daylight, and this fluorescent light added to the light reflected from the colored paint or cloth gives an appearance of unusual brightness.

If an excited state that might fluoresce or phosphoresce is prevented from doing so by interaction with another component of the solution this substance is called a quencher.

18.4 QUENCHING OF EXCITED STATES

A substance added to a reaction mixture may interact with an excited molecule to remove its excitation energy. This is referred to as quenching, and the rate constant for the quenching reaction may be determined by measuring the dependence of the quantum yield for fluorescence on the concentration of the quencher. According to the following mechanism a molecule in the singlet ground state S_0 absorbs a photon and the excited singlet S_1 fluoresces, is quenched by an added molecule Q, or is deactivated in a radiationless process.

$$S_0 + h\nu \rightarrow S_1 \qquad\qquad I \qquad\qquad\qquad (18.4)$$

$$S_1 \rightarrow S_0 + h\nu' \qquad k_1(S_1) \qquad\qquad (18.5)$$

$$S_1 + Q \rightarrow S_0 + Q^* \qquad k_2(S_1)(Q) \qquad (18.6)$$

$$S_1 \rightarrow S_0 \qquad\qquad k_3(S_1) \qquad\qquad (18.7)$$

These reactions come to a steady state if the radiation is constant and no ir-reversible photochemical reactions occur. The steady-state rate equation is

$$\frac{d(S_1)}{dt} = 0 = I - [k_1 + k_2(Q) + k_3](S_1) \tag{18.8}$$

The quantum yield Φ_0 for the emission from S_1 in the absence of Q is given by

$$\Phi_0 = \frac{k_1(S_1)}{I} = \frac{k_1}{k_1 + k_3} \tag{18.9}$$

where I has been eliminated by use of equation 18.8 with $(Q) = 0$. The quantum yield Φ_Q when Q is present is given by

$$\Phi_Q = \frac{k_1(S_1)}{I} = \frac{k_1}{k_1 + k_3 + k_2(Q)} \tag{18.10}$$

The ratio of these quantum yields is

$$\frac{\Phi_0}{\Phi_Q} = 1 + \frac{k_2}{k_1 + k_3}(Q)$$
$$= 1 + k_2\tau(Q) \tag{18.11}$$

where the second form has been obtained by substituting the expression for the relaxation time τ of the fluorescence of S_1 in the absence of Q

$$\tau = \frac{1}{k_1 + k_3} \tag{18.12}$$

Since τ can be measured in the absence of Q, the bimolecular quenching constant k_2 can be calculated from the slope of a plot of Φ_0/Φ_Q against quencher concentration. For very efficient quenchers k_2 approaches the value calculated for diffusion controlled reactions (Section 10.22).

Very often the molecules that absorb light take part in the photochemical reaction only in an indirect manner and act merely as carriers of energy. One of the outstanding examples is mercury vapor activated by the absorption of ultra-violet light of 253.67 nm, which is emitted by a mercury-vapor lamp. The energy corresponding to this radiation is very large ($112 \text{ kcal mol}^{-1}$), and it is more than the $102.4 \text{ kcal mol}^{-1}$ necessary to dissociate hydrogen molecules into atoms. When mercury vapor is mixed with hydrogen and exposed to light from a mercury-vapor lamp, the chief reactions are

$$\text{Hg} + h\nu \rightarrow \text{Hg}^* \tag{18.13}$$
$$\text{Hg}^* + \text{H}_2 \rightarrow \text{Hg} + 2\text{H} \tag{18.14}$$

where Hg^* represents an activated mercury atom. Hydrogen is transparent to this radiation. Mercury acts as a photosensitizer. The hydrogen atoms readily reduce metallic oxides, nitrous oxide, ethylene, carbon monoxide, and other materials. The excited mercury atoms decompose not only hydrogen but also ammonia and various organic compounds.

Excitation energy may be transferred from one molecule to another without a collison. The transfer process

$$D^* + A \rightarrow A^* + D \tag{18.15}$$

may occur over distances of 5–10 nm if the emission spectrum of D^* overlaps strongly with the absorption spectrum of A and the corresponding dipole transitions in both D^* emission and A absorption are singlet-singlet or triplet-triplet transitions. Although this intermolecular energy transfer might appear to be due to emission and reabsorption, it occurs more rapidly than emission from D^* and only over restricted distances. The rate of this so-called resonance energy transfer is inversely proportional to the sixth power of the distance between D^* and A molecules.†

18.5 QUANTUM YIELDS

The quantum yields of a few photochemical reactions are summarized in Table 18.1.

Reaction 1 has the same value of Φ from 280 to 300 nm, at low pressures and high pressures, in the liquid state or in solution in hexane. The primary process $HI + h\nu = H + I$ is followed by the reactions $H + HI = H_2 + I$ and $I + I = I_2$, thus giving two molecules of HI decomposed for each photon absorbed. Reaction 2

Table 18.1[1] Quantum Yields in Photochemical Reactions at Room Temperature

Reaction	Approximate Wavelength Region, nm	Approximate Φ
1. $2HI \rightarrow H_2 + I_2$	300–280	2
2. $C_{14}H_{10} \rightleftharpoons \frac{1}{2}(C_{14}H_{10})_2$	<360	1–0
3. $2NO_2 \rightarrow 2NO + O_2$	>435	0
	366	2
4. $CH_3CHO \rightarrow CO + CH_4(+C_2H_6 + H_2)$	310	0.5
	253.7	1
5. $(CH_3)_2CO \rightarrow CO + C_2H_6(+CH_4)$	<330	0.2
6. $NH_3 \rightarrow \frac{1}{2}N_2 + \frac{3}{2}H_2$	210	0.2
7. $H_2C_2O_4(+UO_2^{2+}) \rightarrow CO + CO_2 + H_2O(+UO_2^{2+})$	430–250	0.5–0.6
8. $Cl_2 + H_2 \rightarrow 2HCl$	400	10^5

[1] W. A. Noyes, Jr., and P. A. Leighton, *Photochemistry of Gases*, Reinhold Publishing Corp., New York, 1941, Appendixes, pp. 415–465; F. Daniels, *J. Phys. Chem.*, **41**, 713 (1938).

† N. J. Turro, *Molecular Photochemistry*, W. A. Benjamin, Inc., New York, 1967, p. 96.

has a quantum yield of unity initially, but the reverse thermal reaction reduces it as the product accumulates. In reaction 3 at 366 nm the quantum yield is 2 if correction is made for internal screening by the accompanying N_2O_4, which absorbs some light at 366 nm. The reactions are $NO_2 + h\nu = NO_2^*$, $NO_2^* + NO_2 = 2NO + O_2$, where the asterisk indicates an excited molecule. At 435 nm and longer wavelengths no reaction occurs when the radiation is absorbed.

Reaction 4 similarly shows a greater quantum yield at shorter wavelengths. This reaction is interesting because at 300° Φ has a value of more than 300, indicating that the free radicals that are first produced by the absorption of light are able to propagate a chain reaction at the higher temperatures. At room temperature the reactions involved in the chain do not go fast enough to be detected. The products given in parentheses are present also but in small amounts.

The experimental determination of the quantum yield constitutes an excellent method for detecting *chain reactions* (Section 10.21). If several molecules of products are formed for each photon of light absorbed, the reaction is obviously a chain reaction in which the products of the reaction are able to activate additional molecules of reactants.

Reaction 5 is an example of the fact that the absorption of light in a particular bond does not necessarily cause the rupture of that bond. Acetone, like most molecules which contain the carbonyl group C=O, absorbs ultraviolet light at about 280 nm. The C=O bond, however, is very strong and does not break to give atomic oxygen. Instead, the absorption energy leads to the cleavage of an adjacent C—C bond that is weaker, thus:

$$\begin{matrix} CH_3 \\ \diagdown \\ \quad C{=}O + h\nu \longrightarrow CH_3{\cdot} + CH_3\dot{C}{=}O \qquad (18.16) \\ \diagup \\ CH_3 \end{matrix}$$

giving a methyl radical and an acetyl radical. The acetyl radical can then decarboxylate giving CO and $CH_3\cdot$, or it can react with $CH_3\cdot$ to give back acetone. The methyl radicals can couple to form ethane.

In the photolysis of ammonia, reaction 6, hydrogen atoms are split off, and the low yield is probably due to partial recombination of the fragments. The quantum yield varies with pressure and reaches a maximum at 80–90 Torr.

Reaction 7 illustrates a photosensitized reaction. The photodecomposition of oxalic acid, sensitized by uranyl ion, is so reproducible that it is suitable for use as an actinometer. The light is absorbed by the colored uranyl ion, and the energy is transferred to the colorless oxalic acid, which then decomposes. The uranyl ion remains unchanged and can be used indefinitely as a sensitizer. The fact that the molar absorption coefficient of uranyl nitrate is increased by the addition of colorless oxalic acid indicates that a complex is formed. The formation of a loose chemical complex is often necessary for photosensitization.

Reaction 8 is the best known example of a chain reaction. About a million molecules react for each quantum absorbed. The molecules of hydrogen chloride formed undergo further reaction with the hydrogen and chlorine atoms produced (Section 10.21). The measurement of the number of molecules per photon gives a measure of the average number of molecules involved in the chain.

To determine a quantum yield it is necessary to measure the intensity of the light. This may be done by use of a thermopile, which is a series of thermocouples with one set of junctions blackened to absorb all the radiation, which is then converted into heat. The other set of junctions is protected from radiation. The temperature difference between the two sets of junctions is measured by the galvanometer deflection. The galvanometer readings may be converted into joules of radiation per second per square meter striking the thermopile, by calibrating with a standard carbon-filament lamp from the National Bureau of Standards.

The amount of radiation may also be measured with a chemical *actinometer*, in which the amount of chemical change is determined. The yield of the photochemical reaction in the actinometer was determined originally by use of a thermopile.

18.6 FLASH PHOTOLYSIS

The unstable intermediates in a photochemical reaction are usually present at such low concentrations that they cannot be studied directly. One way to increase their concentrations is to use a very powerful flash of light. The high-energy flash of short duration is obtained by discharging a bank of capacitors through a gas-discharge tube. The flashes are so intense that in some cases practically all the molecules in the reaction tube are dissociated into free radicals and atoms. Powers of 50 MW may be obtained for a few microseconds. By this technique it has been possible to determine the absorption spectra of radicals such as NH_2, ClO, and CH_3.

In addition the flash-photolysis method is useful for studying the kinetics of the unstable intermediates. The rate of the disappearance of the free radicals can be followed by the rate of increase in the monochromatic transmitted light as measured with an oscilloscope and photomultiplier. The spectrophotometer is set so that the light passing through the illuminated cell is of the wavelength which is absorbed by the free radical. Another method for obtaining unstable intermediates of photochemical reactions at a concentration at which they can be studied spectroscopically is to form them in a rigid, unreactive medium such as a frozen rare gas at a temperature so low that they have a long lifetime.

18.7 CHEMILUMINESCENCE AND THERMOLUMINESCENCE

Chemiluminescence is the emission of light resulting from certain chemical reactions. For example, the oxidation of ether solutions of magnesium *p*-bromophenyl bromide gives rise to marked chemiluminescence, the greenish blue glow that accompanies the exposure of the solution to air being visible in daylight. The oxidations of decaying wood containing certain forms of bacteria, of luciferin in fireflies, and of yellow phosphorus are further examples.

The effect of radiation on crystals is interesting. X rays give rise to characteristic colors when absorbed by alkali halides and other crystals. Sodium chloride becomes yellow and potassium chloride blue, the coloration being due to the absorption of light by electrons that have been released by X rays and are trapped in negative-ion "vacancies" in the crystal lattice. When an irradiated crystal is heated, the trapped electrons are released, and in returning to a lower energy level they give off light, a phenomenon known as *thermoluminescence*. If the crystal is heated slowly, a series of light emissions occurs at definite temperatures. The nature of these curves, in which the intensity of light emitted is plotted against the temperature, depends on the extent of the radiation exposure, the impurities present, and other factors. Certain minerals, such as limestones and fluorites, exhibit thermoluminescence even without laboratory exposure to radiation, because they contain traces of uranium in parts per million which have been giving off radioactive radiations for geological ages.

18.8 PHOTOGRAPHY

If silver chloride or bromide is mixed with gelatin and exposed very briefly to light, no change is observed; but, when this emulsion is immersed in a solution of a mild reducing agent, such as, for example, pyrogallic acid, the parts that have been exposed to light are reduced to metallic silver much more rapidly than the unexposed parts. The photographic plate consists of a large number of minute grains of crystalline silver halide; some of these grains are completely reduced by the developer to black metallic silver, and others are unaffected. The unaffected grains are dissolved out with sodium thiosulfate ("hypo").

It is necessary for a quantum of energy to strike a sensitive spot in the crystal lattice so that the silver halide may be reduced to give a nucleus of silver, which then spreads, on further reduction, to include the whole grain. These sensitive spots seem to be identified with minute traces of silver sulfide in the crystal lattice which are particularly responsive to the action of light. By increasing the number of these sensitive spots and by other means, the speed of photographic films and plates has been greatly increased.

The silver halides respond only to the ultraviolet and to the shorter wavelengths of the visible spectrum; but if certain red dyes, such as dicyanin, are mixed in the emulsion, the plate becomes sensitive also to red. This phenomenon constitutes another example of photo-sensitization.

18.9 PHOTOSYNTHESIS

The most important photochemical reaction in the world is the union of carbon dioxide and water in plants through the agency of sunlight and chlorophyll. Chlorophyll, which gives plants their green color, is a complex organic compound containing magnesium. It absorbs red, blue, and, to a lesser extent, green light. The excited chlorophyll thus formed is responsible for the production from carbon dioxide and water of the starting material of all plant growth.

$$CO_2 + H_2O + \text{light} \rightarrow (1/n)(CH_2O)_n + O_2 \qquad (18.17)$$

where $(CH_2O)_n$ represents a carbohydrate such as cellulose or sugar.

When cellulose is burned, the reaction is

$$(1/n)(CH_2O)_n + O_2 \rightarrow CO_2 + H_2O \qquad \Delta H^\circ = -112{,}000 \, \text{cal mol}^{-1} \qquad (18.18)$$

The endothermic formation of carbohydrates from carbon dioxide and water must require, therefore, the absorption of 112 kcal mol^{-1}, and the activation energy must be at least as great as this. It is probably more than this because of the requirement of extra energy to put the reacting molecules into an activated state.

This activation energy is equivalent to radiation of 230 nm or less. There is no radiation this short in the sunlight that reaches the earth's surface. Chlorophyll, however, acts as a photosensitizer, absorbing visible light and making it available for photosynthesis in the plant. But there is something unique about the reaction. Red light will cause photosynthesis, but red light corresponds to only 40 kcal mol^{-1}; and more than 112 kcal mol^{-1} are required to cause the reaction. Apparently the reaction takes place in steps. Laboratory experiments with algae have shown that normally about eight photons are required for each carbon dioxide molecule utilized and each oxygen molecule evolved in photosynthesis under favorable conditions with low light intensity.

Exercise 18.1   Show that, in photosynthesis, if eight photons of absorbed light at 600 nm produce one molecule of a product that has a heat of combustion of 112 kcal mol^{-1}, the efficiency of conversion of absorbed light into stored chemical energy is 30%.

18.10   EINSTEIN COEFFICIENTS

When a molecule is exposed to electromagnetic radiation it may absorb radiation that corresponds with the energy difference between the state in which it exists and some higher energy state. Einstein showed that the absorbed energy may be lost by *spontaneous* emission or *stimulated* emission. The number of molecules excited from state n to state m (See Fig. 18.4) by the absorption of a quantum $h\nu_{nm}$ is proportional to the number in state n and to the density $\rho(\nu_{nm})$ of radiation of frequency ν_{nm}.

$$\text{Rate of excitation} = B_{nm} N_n \rho(\nu_{nm}) \qquad (18.19)$$

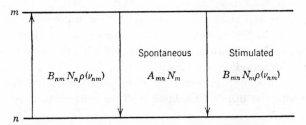

Fig. 18.4   Spontaneous and stimulated emission.

where B_{nm} is the Einstein transition probability for absorption. The excited molecules have a certain probability of spontaneously emitting $h\nu_{nm}$ and returning to state n.

$$\text{Rate of spontaneous emission} = A_{mn}N_m \tag{18.20}$$

In addition Einstein suggested that the irradiation of excited molecules with radiation of frequency ν_{nm} would stimulate emission according to

$$\text{Rate of stimulated emission} = B_{mn}N_m\rho(\nu_{nm}) \tag{18.21}$$

where B_{mn}, the Einstein transition probability for stimulated emission is equal to B_{nm}, the Einstein transition probability for absorption. In a steady state

$$N_m[A_{mn} + \rho(\nu_{nm})B_{mn}] = N_nB_{nm}\rho(\nu_{nm}) \tag{18.22}$$

We will not pursue the relation between A_{mn} and B_{mn} ($=B_{nm}$), but simply need to point out that stimulated emission is the basis for the operation of lasers and is important for NMR and other kinds of spectroscopy.

18.11   LASERS*

Normally a beam of light loses intensity as it passes through an absorbing material. However, if there are molecules in an excited state, stimulated emission can occur, and the light beam can gain in intensity. A laser achieves this condition and produces an intense and coherent beam. By coherent we mean that the light waves are in phase. The name "laser" comes from *light amplification* by *stimulated emission* of *radiation*.

In order to get laser action it is necessary to get a population inversion in the absorber. In a population inversion there are more molecules in the *upper* state than in the *lower* state. This cannot be achieved by simply using a high-intensity light source of the proper frequency to excite molecules from a lower level to a higher level. The incident radiation stimulates emission of photons from the upper state at the same rate as they are absorbed by the lower state, and therefore no more than 50% of the molecules can be forced into the upper state by use of an intense source. However, if an upper state (e.g., S_1) can be converted to another excited state (e.g., T_1) by a radiationless process, the population of T_1 may become greater than 50%. This process is referred to as optical pumping. If after pumping, radiation corresponding with the transition $T_1 \rightarrow S_0$ is passed into the material, stimulated emission occurs.

Population inversions may also be achieved in chemical reactions in which products are in excited states. In a chemical laser the energy producing laser radiation comes from the energy of the chemical reaction. For example, a gas stream containing fluorine atoms may be mixed with deuterium (or hydrogen) and carbon dioxide, producing a chain reaction that yields vibrationally excited deuterium (or hydrogen) fluoride. The vibrational-rotational energy from the excited DF pumps the upper CO_2 laser level by intermolecular energy transfer

---

* A. L. Schawlow, *Lasers and Light*, W. H. Freeman Co., San Francisco, 1969.

processes. The resulting continuous-wave laser action of CO_2 at 10.6 μm can be provided entirely from chemical sources if the fluorine atoms are produced by a reaction such as

$$F_2 + NO = NOF + F \qquad (18.23)$$

Lasers have many important uses because of their high peak power, coherence, and high degree of monochromaticity. Flux densities as high as 5×10^{14} W cm^{-2} have been attained. This corresponds with fluctuating electrical fields of the order of 3×10^8 V cm^{-1} and fluctuating magnetic fields of 10^6 G.

Lasers are important in chemistry because they can be used to initiate photochemical reactions, and the short pulses permit the study of very fast reactions. Lasers have revolutionized Raman spectroscopy because exposures are greatly reduced and even weak lines may be detected in experiments of short duration.

References

J. G. Calvert and J. N. Pitts, Jr., *Photochemistry*, Wiley, New York, 1966.
R. B. Cundall and A. Gilbert, *Photochemistry*, Appleton-Century-Crofts, New York, 1970.
N. J. Turro, *Molecular Photochemistry*, W. A. Benjamin, Inc., New York, 1967.
C. H. J. Wells, *Introduction to Molecular Photochemistry*, Chapman and Hall, London, 1972.

Problems

18.1 A certain photochemical reaction requires an activation energy of 30 kcal mol^{-1}. To what values does this correspond in the following units: (*a*) J per mole, (*b*) frequency of light, (*c*) wave number, (*d*) wavelength in angstroms, (*e*) electron volts?

Ans. (*a*) 125 kJ mol^{-1}, (*b*) 3.16×10^{14} s^{-1}, (*c*) 10,500 cm^{-1}
(*d*) 9520 Å, (*e*) 1.30 eV.

18.2 A sample of gaseous acetone is irradiated with monochromatic light having a wavelength of 313 nm. Light of this wavelength decomposes the acetone according to the equation.

$$(CH_3)_2CO \rightarrow C_2H_6 + CO$$

The reaction cell used has a volume of 59 cm^3. The acetone vapor absorbs 91.5% of the incident energy. During the experiment the following data are obtained:

Temperature of reaction $= 56.7°$

Initial pressure $= 766.3$ Torr

Final pressure $= 783.2$ Torr

Time of radiation $= 7$ hr

Incident energy $= 48.1 \times 10^{-4}$ J s^{-1}

What is the quantum yield? *Ans.* 0.17.

18.3 A 100-cm^3 vessel containing hydrogen and chlorine was irradiated with light of 400 nm. Measurements with a thermopile showed that 11×10^{-7} J of light energy was absorbed by the chlorine per second. During an irradiation of 1 min the partial pressure of chlorine, as determined by the absorption of light and the application of Beer's law, decreased from 205 to 156 Torr (corrected to 0°). What is the quantum yield?

Ans. 2.6×10^6 mol HCl einstein^{-1}.

18.4 Discuss the economic possibilities of using photochemical reactions to produce valuable products with electricity at 1 cent per kilowatt-hour. Assume that 5 % of the electric energy consumed by a quartz-mercury-vapor lamp goes into light, and 30 % of this is photochemically effective. (*a*) How much will it cost to produce 1 lb (453.6 g) of an organic compound having a molecular weight of 100, if the average effective wavelength is assumed to be 4000 Å and the reaction has a quantum yield of 0.8 molecule per photon? (*b*) How much will it cost if the reaction involves a chain reaction with a quantum yield of 100?

Ans. (*a*) 31.4 cents, (*b*) 0.251 cent.

18.5 Chloroform is often kept in dark bottles to prevent photooxidation by air, giving phosgene, which is poisonous. The quantum yield of the photooxidation has been reported to be about 100 molecules per photon with light of 436 nm. How many calories of this light will be required to oxidize 1 mg of chloroform when the chloroform containing dissolved air is placed in a transparent bottle?

Ans. 5.50×10^{-3} cal

18.6 The quantum yield is 2 for the photolysis of gaseous HI to $H_2 + I_2$ by light of 253.7 nm wavelength. Calculate the number of moles of HI that will be decomposed if 300 J of light of this wavelength is absorbed.

Ans. 1.27×10^{-3} mol.

18.7 The following calculations are made on a uranyl oxalate actinometer, on the assumption that the energy of all wavelengths between 254 and 435 nm is completely absorbed. The actinometer contains 20 cm^3 of 0.05 M oxalic acid, which also is 0.01 M with respect to uranyl sulfate. After 2 hr of exposure to ultraviolet light, the solution required 34 cm^3 of potassium permanganate, $KMnO_4$, solution to titrate the undecomposed oxalic acid. The same volume, 20 cm^3, of unilluminated solution required 40 cm^3 of the $KMnO_4$ solution. If the average energy of the quanta in this range may be taken as corresponding to a wavelength of 350 nm how many joules were absorbed per second in this experiment? ($\Phi = 0.57$.)

Ans. 124×10^{-4} J.

18.8 A solution absorbs 300 nm radiation at the rate of 1 W. What does this correspond to in einsteins per second?

Ans. 2.5×10^{-6} einsteins s^{-1}.

18.9 Ketone A dissolved in *t*-butyl alcohol is excited to a triplet state A^* by light of 320–380 nm. The triplet may then return to the ground state A or rearrange to give the isomer B, thus

$$A \qquad\qquad\qquad A^* \qquad\qquad\qquad B$$

The unimolecular rate constant k_d determines the rate of deactivation of excited molecules and the unimolecular rate constant k_r determines the rate of rearrangement. The symbol I represents the number of einsteins per second and it is assumed that each photon absorbed produces one excited molecule of A^*. The quantum yield Φ for the formation of B is given by

$$\Phi = \frac{d(B)/dt}{I} = \frac{k_r(A^*)}{I}$$

Zimmerman, McCullough, Staley and Padwa measured the quenching effect of dissolved naphthalene and report the following data:

Moles naphthalene						
liter^{-1}	0.0099	0.0330	0.0620	0.0680	0.0775	0.0960
Φ	0.0049	0.0023	0.0020	0.0017	0.0014	0.0012

The quenching rate by naphthalene is controlled by the bimolecular quenching constant of A which is equal to the diffusion-controlled constant $k_q = 1.2 \times 10^9$ M^{-1} s^{-1}.

Assuming a steady state

$$d(A^*)/dt = I - k_r(A^*) - k_d(A^*) - k_q(A^*)(N) = 0$$

where (N) is the concentration of naphthalene.

Calculate k_r and k_d by plotting $1/\Phi$ versus (N) and determining the slope and intercept of the line. *Ans.* $k_r = 1.74 \times 10^5$ s^{-1}, $k_d = 2.36 \times 10^7$ s^{-1}.

18.10 If a good agricultural crop yields about 2 tons acre^{-1} of dry organic material per year with a heat of combustion of about 4000 cal g^{-1}, what fraction of a year's solar energy is stored in an agricultural crop if the solar energy is about 1000 cal min^{-1} ft^{-2} and the sun is shining about 500 min day^{-1} on the average? 1 acre $= 43,560$ ft^2, and 1 ton $= 907,000$ g.
Ans. 10^{-3}.

18.11 What intensities of light in 10^{-7} J s^{-1} are required to produce a microeinstein s^{-1} at (a) 700 nm and (b) 300 nm. (c) and (d) What are these powers in watts?

18.12 When CH$_3$I molecules in the vapor state absorb 253.7 nm light, they dissociate into methyl radicals and iodine atoms. Assuming that the energy required to rupture the C—I bond is 50 kcal mol^{-1}, what is the kinetic energy of each of the fragments if they are produced in their ground states?

18.13 Assuming a hypothetical photochemical reaction in which one-tenth of the solar radiation is absorbed and utilized with a quantum yield of unity, how many tons of product can be produced per acre per day, if the molecular weight of the product is 100, the average effective light is 510 nm, and the solar radiation is 1 cal min^{-1} cm^{-2} for 500 min during the day?

18.14 The oxidation of rubrene, C$_{42}$H$_{28}$, is effected by oxygen at a wavelength of 436 nm with a quantum yield of unity. How many calories of this light will be required to photooxidize 1 g of C$_{42}$H$_{28}$?

18.15 Sunlight between 290 and 313 nm can produce sunburn (erythema) in 30 min. The intensity of radiation between these wavelengths in summer and at 45° latitude is about 50 μW cm^{-2}. Assuming that 1 photon produces chemical change in 1 molecule, how many molecules in a square centimeter of human skin must be photochemically affected to produce evidence of sunburn?

18.16 The quantum yield for the photolysis of acetone

$$(CH_3)_2CO = C_2H_6 + CO$$

at 300 nm is 0.2. How many moles per second of CO are formed if the intensity of the 300 nm radiation absorbed is 10^{-2} J s^{-1}?

18.17 A uranyl oxalate actinometer is exposed to light of wavelength 390 nm for 1980 s, and it is found that 24.6 cm^3 of 0.00430 M potassium permanganate is required to titrate an aliquot of the uranyl oxalate solution after illumination, in comparison with 41.8 cm^3 before illumination. Using the known quantum yield of 0.57, calculate the number of joules absorbed per second. The chemical reaction for the titration is $2MnO_4^- + 5H_2C_2O_4 + 6H^+$ $= 2Mn^{2+} + 10CO_2 + 8H_2O$.

18.18   Nitrogen dioxide is decomposed photochemically by light of 366 nm with a quantum yield of 2.0 molecules per photon, according to the reaction

$$2NO_2 \rightarrow 2NO + O_2$$

The thermal reaction runs in the reverse direction. When an enclosed sample of nitrogen dioxide is illuminated for a long period of time, the quantum yield decreases and approaches zero. Suggest a mechanism to explain these facts, and write the chemical equations.

18.19   Calculate the maximum possible theoretical yield in tons of carbohydrate material $(H_2CO)_n$ that can be produced on an acre of land by green plants or trees during a 100-day growing season. Similar calculations apply to algae growing in a lake or ocean. Assume that the sun's radiation approximately averages 1.0 cal cm^{-2} min^{-1} for 8 hr day^{-1} and that one-half the area is covered by green leaves. Assume that one-third of the radiation lies between 400 and 650 nm, which is the range of the light absorbed by chlorophyll, and that the average wavelength is 550 nm. Assume that the leaves are thick enough to absorb practically all the light that strikes them. Assume that the quantum yield is 0.12 molecule per photon; that is, 8 photons with chlorophyll can produce 1 H_2CO unit from 1 molecule of CO_2 and 1 molecule of H_2O. Criticize these several assumptions.

18.20   A photochemical reaction of biological importance is the production of vitamin D, which prevents rickets and brings about the normal deposition of calcium in growing bones. Steenbock found that rickets could be prevented by subjecting the food as well as the patient to ultraviolet light below 310 nm. When ergosterol is irradiated with ultraviolet light below 310 nm, vitamin D is produced. When irradiated ergosterol was included in a diet otherwise devoid of vitamin D, it was found that absorbed radiant energy of about 7.5×10^{-5} J was necessary to prevent rickets in a rat when fed over a period of 2 weeks. The light used has a wavelength of 265 nm (a) How many quanta are necessary to give 7.5×10^{-5} J? (b) If vitamin D has a molecular weight of the same order of magnitude as ergosterol (382), how many grams of vitamin D per day are necessary to prevent rickets in a rat? It is assumed that the quantum yield is unity.

18.21   If a reaction responds to both red and violet light, 700 and 400 nm, with an equal quantum efficiency, will there be more photochemical reaction per 100 cal of light in the red or in the blue? How much more?

18.22   For 900 s, light of 4360 Å was passed into a carbon tetrachloride solution containing bromine and cinnamic acid. The average absorbed was 19.2×10^{-4} J s^{-1}. Some of the bromine reacted to give cinnamic acid dibromide, and in this experiment the total bromine content decreased by 3.83×10^{19} molecules. (a) What was the quantum yield? (b) State whether or not a chain reaction was involved. (c) If a chain mechanism was involved, suggest suitable reactions which might explain the observed quantum yield.

18.23   Ammonia is decomposed by ultraviolet light of 200 nm with a quantum yield of 0.14. (a) How many calories of this light would be necessary to decompose 1 g of ammonia? (b) Offer a suggestion to explain this comparatively low quantum yield.

18.24   A cold high-voltage mercury lamp is to be used for a certain photochemical reaction which responds to ultraviolet light of 253.7 nm. The chemical analysis of the product is sensitive to only 10^{-4} mol. The lamp consumes 150 W and converts 5% of the electric energy into radiation of which 80% is at 253.7 nm. The amount of the light that gets into the monochromator and passes out the exit slit is only 5% of the total radiation of the lamp. Fifty percent of this 253.7 nm radiation from the monochromator is absorbed in the reacting system. The quantum yield is 0.4 molecule of product per quantum of light absorbed. How long an exposure must be given in this experiment if it is desired to measure the photochemical change with an accuracy of 1%?

18.25 The photochemical oxidation of phosgene, sensitized by chlorine, has been studied by G. K. Rollefson and C. W. Montgomery [*J. Am. Chem. Soc.*, **55**, 142, 4025 (1932)]. The overall reaction is

$$2COCl_2 + O_2 = 2CO_2 + 2Cl_2$$

and the rate expression which gives the effect of the several variables is

$$\frac{dc_{CO_2}}{dt} = \frac{kI_0 c_{COCl_2}}{1 + k' c_{Cl_2}/c_{O_2}}$$

where I_0 is the intensity of the light. The quantum yield is about two molecules per quantum. Devise a series of chemical equations involving the existence of the free radicals ClO and COCl which will give a mechanism consistent with the rate expression.

18.26 The photopolymerization of anthracene reaches a stationary state, owing to the thermal decomposition of the dianthracene. For the photoreaction the temperature coefficient r is 1.1, and for the thermal reaction r is 2.8, where r is defined as k_{t+10}/k_t. Calculate the effect of a $5°$ rise in temperature on the amount of dianthracene formed when the photostationary state is reached.

18.27 The following reactions describe the photochemical decomposition of hydrogen bromide with light of 253 nm at $25°$. The primary process dissociates the molecule into atoms of hydrogen and bromine, which can then undergo further reactions. The quantum yield for the primary process is designated by ϕ; the quantum yield for the overall reaction is designated by Φ. The intensity of light absorbed is designated by I.

(1) $HBr + h\nu \rightarrow H + Br$ Rate $= \phi I$
(2) $H + HBr \rightarrow H_2 + Br$ Rate $= k_2(c_H)(c_{HBr})$
(3) $Br + Br + M \rightarrow Br_2 + M$ Rate $= k_3(c_{Br})^2(c_M)$
(4) $H + Br_2 \rightarrow HBr + Br$ Rate $= k_4(c_H)(c_{Br_2})$
(5) $H + H + M \rightarrow H_2 + M$ Rate $= k_5(c_H)^2(c_M)$ (negligible)

Derive the expression for the quantum yield for the overall reaction and show that early in the reaction $\Phi = 2\phi$.

18.28 Biacetyl triplets have a quantum yield of 0.25 for fluorescence and a measured lifetime of the triplet state of 10^{-3} s. If its phosphorescence is quenched by a compound Q with a diffusion-controlled rate (10^{10} M^{-1} s^{-1}), what concentration of Q is required to cut the phosphorescence yield in half?

STRUCTURES

Structure is so important in chemistry that we have not been able to delay all structural considerations to Part Four; but we have left to this Part structural information about crystals and macromolecules. In both cases the relation between structure and function is of special interest.

In order to understand how the structures of crystals are determined we need to understand the basic ideas of crystal geometry and the way in which the symmetry in solids is different from the symmetry of individual molecules discussed in Chapter 13. The three-dimensional structure of simple solids is determined very much by the sizes of atoms and the ways that they can be packed.

Theories of the solid state have the problem of explaining a very wide range of physical properties; for example, the electrical conductivities of solids range over 10^{30}. Semiconductors, which are in the middle of the range of electrical conductivity, are especially important because of their use in electronic circuits.

The study of macromolecules produced by living things (proteins, polynucleotides, and polysaccharides) and synthetic high polymers requires special methods for determining very high molecular weights. These high molecular weights are required to provide the special properties shown by these substances.

CHAPTER 19

CRYSTAL STRUCTURE AND SOLID STATE

In this chapter we will first consider crystal geometry, then X-ray diffraction and the structures of specific crystals, and finally solid-state chemistry. The ideas of point-group symmetry developed in Chapter 13 are basic to the understanding of crystals, but the introduction of translations produces new kinds of symmetry and requires space groups in addition to point groups.

In 1912 Laue suggested that the wavelength of X rays might be about the same as the distance between atoms in a crystal so that a crystal could serve as a diffraction grating for X rays. This experiment was carried out by Friederick and Knipping, and they observed the expected diffraction. Almost immediately afterward, W. L. Bragg (1913) improved on the Laue experiment, mainly by substituting monochromatic for polychromatic radiation and by providing a more physical interpretation to the Laue theory of the scattering experiment. Bragg also determined the structures of a number of simple crystals, including those of NaCl, CsCl, and ZnS. Since the birth of X-ray crystallography as a science at that time, this tool of structural analysis by single crystal X-ray diffraction has developed into the most powerful method known for obtaining the atomic arrangement in the solid state. Since the 1950s, with the advent of large high-speed computers capable of handling the X-ray data, it has been possible to analyze in even greater detail the structures of compounds as complex as proteins.

19.1 LATTICES*

A crystal may be described as a three-dimensional pattern in which a structural motif is repeated in such a way that the environment of every motif is the same throughout the crystal. The motif is often a molecule, but it may also be a group of molecules or in inorganic crystals a cluster of atoms bonded with atoms in adjacent structural motifs.

A linear pattern may be described by saying that there is a set of parallel motifs at the end of a vector $\boldsymbol{a}$ and its multiples. These vectors are described by

$$\boldsymbol{T} = u\boldsymbol{a} \tag{19.1}$$

---

* In discussing crystal geometry we will follow M. J. Buerger, *Introduction to Crystal Geometry*, McGraw-Hill Book Co., New York, 1971, which may be referred to for more detailed information.

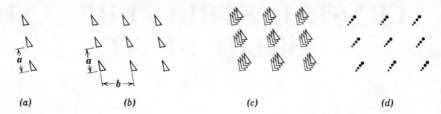

Fig. 19.1   (*a*) One-dimensional pattern with vector *a*. (*b*) Two-dimensional pattern with vectors *a* and *b*. (*c*) Three-dimensional pattern with vectors *a*, *b*, and *c*. (*d*) Lattice of pattern *c*.

where u is an integer. An example of a linear pattern is shown in Fig. 19.1*a*. The linear pattern formed in this way can be repeated in a second direction represented by the vector *b* to form a two-dimensional pattern in which every structural pattern has the same environment as every other, as shown in Fig. 19.1*b*. Finally the whole two-dimensional pattern can be repeated in a third direction *c* to produce a translationally ordered pattern in three dimensions, as shown in Fig. 19.1*c*. The structural pattern is repeated at the end of every vector of the form

$$T = ua + vb + wc \qquad u, v, w, \text{ integers} \qquad (19.2)$$

For many purposes it is convenient to concentrate on the geometry of the repetition and replace the structural pattern by a point, as shown in Fig. 19.1*d*, to obtain a *lattice*. The lattice may be generated from a single starting point by the infinite repetition of a set of fundamental translations that characterize the lattice. The lattice is described by equation 19.2, but it has no true origin and can be shifted around parallel with itself. Any three noncoplanar vectors, *a*, *b*, and *c* describe a lattice, but a given lattice can be described by an infinite number of sets of three vectors. This is illustrated in two dimensions in Fig. 19.2. Any one of the *b* vectors may be chosen to generate the pattern.

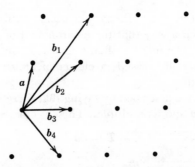

Fig. 19.2   Alternate choices for the second translational vector in a two-dimensional lattice.

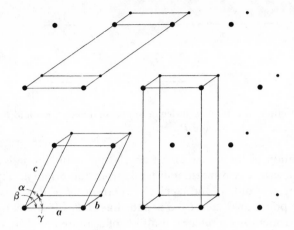

Fig. 19.3 Cells in a lattice.

19.2 UNIT CELLS

The space occupied by a lattice may be divided into unit cells. The repetition of a cell (with everything in it) in three dimensions generates the entire pattern of a crystal. A given lattice can be blocked out in cells in different ways as shown in Fig. 19.3. If the corners of the cells include all of the lattice points in the crystal the cell is called a *primitive unit cell*. Primitive cells have one lattice point per cell because each of the corner lattice points is shared by eight cells. A lattice can also be blocked out in cells that do not include all lattice points as corners. This is illustrated by one of the cells in Fig. 19.3. Such cells, referred to as *multiple unit cells*, are useful in simplifying the geometry of crystals for which the primitive cell is oblique, but the multiple cell has two or more edges that are at 90°.

The most general kind of lattice is composed of unit cells with three unequal edges and three unequal angles: this is referred to as a triclinic lattice. This is the only lattice that does not have some kind of rotational symmetry or a mirror plane. All lattices have centres of inversion.

19.3 ROTATIONAL SYMMETRY IN CRYSTALS

A lattice may have various types of rotational symmetry of the type discussed for molecules in Section 13.3; in other words, rotation through an angle of $360°/n$ may bring the lattice into an equivalent position. However, in contrast with individual molecules where rotational axes of order unity to infinity are in principle possible, in crystals only axes of 1-, 2-, 3-, 4-, and 6-fold rotational symmetry are possible. This statement can be proved by use of Fig. 19.4. The lattice shown in this figure has an axis of n-fold symmetry. The lattice points A_1, A_2, A_3, and A_4 are each separated by distance a. Because of the assumed

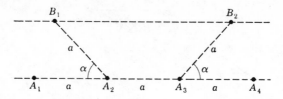

Fig. 19.4   Restriction on rotational order in a crystal. There are n-fold rotational axes at lattice points A_1, A_2, A_3, A_4, B_1, and B_2.

symmetry, rotation of the lattice about any lattice point through an angle $\alpha = 2\pi/n$ will produce a lattice indistinguishable from the original. Therefore clockwise rotation by α about A_3 and clockwise rotation by α about A_2 requires that there be lattice points at B_1 and B_2. Since the line $B_1 B_2$ is parallel to $A_1 A_4$, B_1 and B_2 must be separated by an integral multiple of a, represented by ma. Thus

$$a + 2a \cos \alpha = ma$$

$$\cos \alpha = \frac{N}{2} \qquad \text{where } N = m - 1.$$

The only values of α that satisfy this equation are $0°$, $60°$, $90°$, $120°$, $180°$, and $360°$, which means that the rotational symmetry of the lattice can be only 1, 2, 3, 4, or 6-fold. In Chapter 13 on point-group symmetry, these axes were represented by C_1, C_2, C_3, C_4, and C_6, but crystallographers use the Hermann-Mauguin or international symbols* and in this system the rotational axes are simply referred to by the numbers 1, 2, 3, 4, and 6.

In discussing improper rotations in Chapter 13 we used rotary reflections, but in crystallography the same purpose is accomplished with a rotatory inversion. The crystallographic rotary inversion axes are represented by $\bar{1}$, $\bar{2}$, $\bar{3}$, $\bar{4}$, and $\bar{6}$ where the number represents the number of equivalent positions in a $360°$ rotation. The $\bar{1}$ axis is equivalent to an inversion i. A $\bar{2}$ axis is equivalent to a mirror plane. A $\bar{3}$ axis is equivalent to a three-fold rotation plus an inversion. A $\bar{6}$ axis is equivalent to a three-fold axis and a mirror plane. It is important to note that a rotatory-inversion operation converts an object into its mirror image. Therefore an object that cannot be superimposed on its mirror image cannot possess any element of rotatory-inversion symmetry. In the Hermann-Mauguin system mirror planes are presented by m. A mirror plane perpendicular to an n-fold axis is represented by n/m.

19.4   32 CRYSTALLOGRAPHIC POINT GROUPS

In discussing the symmetry of molecules in Chapter 13 it was remarked that there is in principle an infinite number of point groups. Perfect crystals (crystals grown

---

* N. F. M. Henry and K. Lonsdale, ed., *International Tables for X-Ray Crystallography*, Vol. I, Symmetry Groups, Kynoch Press, Birmingham, England, 1952.

in a symmetrical environment) can be classified according to the point groups, but, because of the limitation of crystal lattices to rotational axes 1, 2, 3, 4, and 6 discussed in the preceding section, a crystal must belong to one of 32 crystallographic point groups. In other words only 32 point groups result from combinations of proper and improper rotations of one-, two-, three-, four-, and sixfold. Although as we will see the symmetries of the arrangements of atoms in crystals are more complicated than the 32 crystallographic point groups, the symmetry of crystals that is visible to the eye is the point-group symmetry. Crystallographers have recognized this since the nineteenth century, and so the 32 crystallographic point groups are often referred to as the 32 crystal classes. Certain crystal forms can arise from a single one of the crystal classes, and so these characteristic shapes immediately identify the point group.

19.5 SIX CRYSTAL SYSTEMS

It is useful to devise coordinate systems for describing the planes and directions in a lattice in such a way that as many of the three coordinate axes a, b, and c as possible are taken along the directions of symmetry, even if it requires a multiple cell. The 32 crystallographic point groups require 14 space lattices (often called Bravais lattices) and their geometries can be referred to six crystal systems shown in Fig. 19.5. The angles α, β, and γ are defined in Fig. 19.3. Symmetry imposes certain restrictions on the lengths of the cell edges (a, b, and c) and the angles between them, and these restrictions are shown. However, it is important to remember that the crystal classes are defined by symmetry and not by the restrictions shown in the last column. The sign $\neq$ is used to mean "not generally equal to" but the angles or edges indicated may be equal accidentally within the error of measurement.

The symmetry of a crystal may be very different from that of the molecule in a crystal. If there are two unsymmetrical molecules in a unit cell, the crystal may have one or more twofold axes. With more molecules in a unit cell, higher symmetries may also be obtained, even though the molecule itself is unsymmetrical. Conversely, symmetrical molecules may crystallize in lattices of lower symmetry.

The position of an atom in a unit cell is designated by giving its coordinates as fractions x, y, z of the unit cell edges a, b, c. (Note that here x, y, and z are not Cartesian coordinates but are measured in the directions of the edges of the unit cell.) An atom at the corner of the unit cell has coordinates designated by 0, 0, 0 and the body-centered position is $\frac{1}{2}, \frac{1}{2}, \frac{1}{2}$. If an atom is at x, y, z, it must also be at every other point that is equivalent to this point by space-group symmetry operations. For example, in the body-centered space lattice if there is an atom at x, y, z there is an equivalent atom at $x + \frac{1}{2}, y + \frac{1}{2}, z + \frac{1}{2}$. In a crystal a lattice point is not necessarily occupied by an atom or molecule, but the lattice point represents a repeating unit.

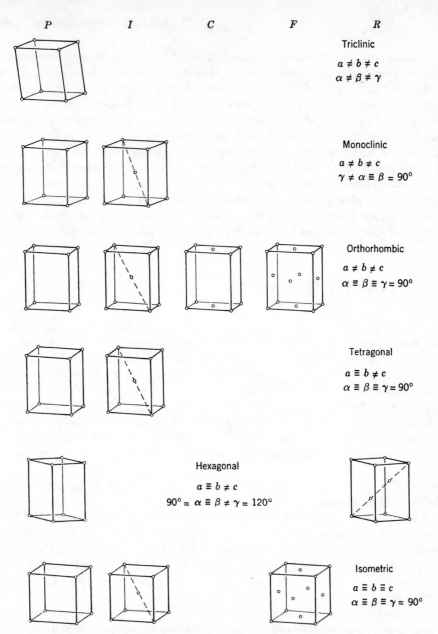

Fig. 19.5 The 14 space lattices (often called Bravais lattices) and the six crystal systems. *P* refers to primitive, *I* to body centered, *C* to end centered, *F* to face centered, and *R* to rhombohedral. (From M. J. Buerger, *Elementary Crystallography*, Wiley, New York, 1963. p. 101.)

19.6 230 SPACE GROUPS

The introduction of the operation of translation in crystals leads to more symmetry operations and their combinations than in the 32 crystallographic groups. To describe the patterns of crystals two new kinds of symmetry operations are required; glide planes and screw axes. A glide plane is the combination of a reflection in a plane with a translation by one half of a lattice translation. A screw axis is a combination of a rotation and a fraction of a lattice translation parallel to the axis. The translation accompanying the screw motion must be n/p times the unit translation, where p is an integer and the angle d between successive motifs is $360°/n$. There are 11 possible screw axes represented by the symbol n_p: 2_1, 3_1, 3_2, 4_1, 4_2, 4_3, 6_1, 6_2, 6_3, 6_4, 6_5.

The fact that there are 230 ways in which these symmetry operations may be combined in the three-dimensional patterns of crystals was derived independently by three men: Fedorow, a Russian crystallographer, in 1890; Schoenflies, a German mathematician, in 1891; and Barlow, a British amateur, in 1895. Space groups are labeled by first giving the Bravais lattice symbol followed by the point-group symbol with appropriate changes for the replacement of rotational axes by screw axes and mirror planes by glide planes. The actual determination of space groups of crystals did not become possible until diffraction techniques were utilized to determine the internal symmetry of crystals. Knowledge of the space group of a crystal·simplifies the determination of the structure because only the asymmetric portion of the unit cell needs to be studied, the rest of the contents may be obtained from symmetry operations.

19.7 DESIGNATION OF CRYSTAL PLANES

The lattice points in one plane of a crystal are illustrated in Fig. 19.6. This figure is an end view of the crystal along the c axis, which is perpendicular to the plane of the a and b axes. A number of different planes may be drawn through the lattice points parallel to the c axis, as illustrated in Fig. 19.6, in which the planes are seen edgewise.

The orientation of a stack of planes in a crystal is designated by its indices,* hkl, which are defined as the number of parts into which these three integers divide the a, b, and c axes, which the particular plane nearest the origin makes with the crystallographic axes. A plane that intercepts the a axis at a/h, the b axis at b/k, and the c axis at c/l is referred to as an hkl plane. The indices of a stack of planes through a lattice may be obtained by counting the number of planes

---

* Often called "Miller indices" because this designation, invented by Whewell in 1825 and Grossman in 1829, was popularized by Miller's textbook of crystallography in 1829. They showed that faces of crystals could be designated by three integers although nothing was known about the internal structures of crystals.

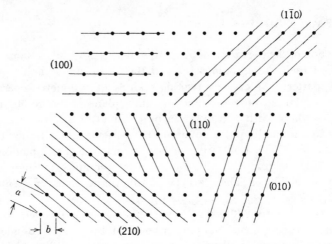

Fig. 19.6 Stacks of planes through lattice points as seen along the c axis of a crystal.

crossed in moving one lattice spacing in the **a**, **b**, and **c** directions, respectively. For the set of planes in the lower left-hand corner of Fig. 19.6, two planes are crossed in going one lattice distance in the direction of the a axis, and one plane is crossed in going one lattice distance horizontally in the direction of the b axis whereas no plane would be crossed in going one lattice distance into the paper, since the planes are parallel to the c axis. Thus, this set of planes is designated by the numbers 210. The indices of the stack of planes in the upper right-hand corner are 1$\bar{1}$0. The negative sign indicates that, if a particular plane is intercepted by going in the positive direction along a, it is necessary to go in the negative direction along b in order to intercept the same plane. This representation for the exterior faces of a crystal and for the internal planes within the crystal will specify the orientation, but not the spatial position. The spacing, d_{hkl}, represents the perpendicular interplanar distance between planes of the stack hkl.

The faces of crystals are usually planes with high densities of atoms or molecules and so they are planes with low indices. Another result of the fact that the external crystal faces are planes with high densities of atoms or molecules is that a given crystalline form of a substance has a constancy of angle between two given faces at a given temperature, as discovered by Steno in 1669. Crystal faces with the closest packing of atoms tend to have the lowest Gibbs free energy and are therefore the most stable at constant temperature and pressure.

19.8 DIFFRACTION METHODS

In a crystal it is the electrons that scatter X rays; therefore they behave as sources of X rays in the diffraction process. Bragg pointed out that it is convenient to consider that the X rays are "reflected" from a stack of planes in the crystal. For

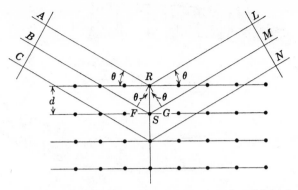

Fig. 19.7   Diagram used in proving the $n\lambda = 2d \sin \theta$.

a given stack of planes (*hkl*) the reflected beam of monochromatic radiation occurs only at a certain angle that is determined by the wavelength of the X rays and the interplanar spacing in the crystal. The relationship correlating these variables is given by the *Bragg equation*, which may be derived by reference to Fig. 19.7, in which the horizontal lines represent layers in the crystal separated by the distance *d*. The plane *ABC* is perpendicular to the incident beam of parallel monochromatic X rays, and the plane *LMN* is perpendicular to the reflected ray. As the angle of incidence θ is changed, a reflection will be obtained only when the waves are in phase at plane *LMN*, that is, when the difference in distance between planes *ABC* and *LMN*, measured along rays reflected from different planes, is a whole-number multiple of the wavelength. This occurs when

$$FS + SG = n\lambda \tag{19.3}$$

Since $\sin \theta = \dfrac{FS}{d} = \dfrac{SG}{d}$, then

$$n\lambda = 2d \sin \theta \tag{19.4}$$

This important equation gives the relationship of the distance between planes in a crystal and the angle at which the reflected radiation has a maximum intensity for a given wavelength λ; that is, all the X-ray waves are in phase. If λ is longer than $2d$, there is no solution for *n* and no diffraction. Thus light waves pass through crystals without being diffracted by the atomic planes. If $\lambda \ll d$ the X rays are diffracted through inconveniently small angles. The Bragg equation does not indicate the intensities of the various diffracted beams. The intensities depend on the nature and arrangement of the atoms within each unit cell.

The reflection corresponding to $n = 1$ for a given family of planes is called the first-order reflection; the reflection corresponding to $n = 2$ is the second-order reflection; and so on. Each successive order exhibits a wider angle. In discussing X-ray reflections it is customary to set $n = 1$ in equation 19.4 and consider that the second-order reflection is from a different stack of planes separated by half

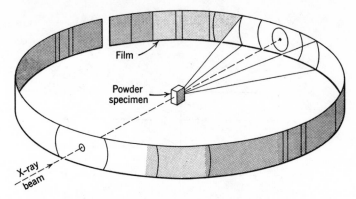

Fig. 19.8   X-ray powder camera.

the lattice distance, etc. Equation 19.4 may be written

$$\lambda = 2\left(\frac{d}{n}\right) \sin \theta$$

$$= 2d_{nh,nk,nl} \sin \theta \qquad (19.5)$$

where $d_{nh,nk,nl}$ is the perpendicular distance between adjacent planes having the indices nh, nk, nl. The planes nh, nk, nl, are parallel to the hkl planes, and the perpendicular interplanar distance is $d_{nh,nk,nl} = d/n$.

To determine the angles at which X rays are diffracted, an oriented single crystal may be rotated in an X-ray beam and the intensity of X rays at the reflection angle determined with a counter. Various types of X-ray cameras have been developed in which the photographic film is moved as the crystal is rotated, so that the problem of getting spots superimposed on other spots in the pattern is avoided.

Instead of scattering X rays from a single large crystal, it is convenient (and sometimes necessary when suitable single crystals are not available) in some types of work to pass a collimated beam of X rays through a powdered sample containing microcrystals presumably oriented in random directions. This method was discovered originally by Debye and Scherrer and later by Hull. The reflections may be recorded on a circular photographic film as illustrated in Fig. 19.8. If coarse crystals are used, the powder pattern is seen to be made up of rings of spots, each spot being produced by a suitably oriented small crystal. If the crystals are very fine, a large number of spots of reflected beams are produced by the different crystal planes and continuous arcs are obtained on the film.

For cubic crystals, X-ray powder patterns are all that is required to differentiate between primitive, body-centered, and face-centered cubic crystals.

19.9   CUBIC LATTICES

Since the cubic system is the simplest, it is explored in some detail here. There are three independent Bravais lattices that have all the symmetry of a cube:

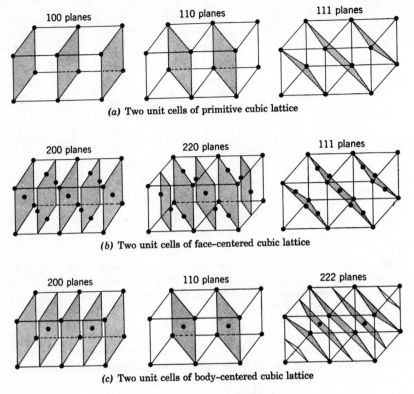

(a) Two unit cells of primitive cubic lattice

(b) Two unit cells of face–centered cubic lattice

(c) Two unit cells of body–centered cubic lattice

Fig. 19.9 Planes through cubic lattices.

primitive, body-centered, and face-centered. These are illustrated in Fig. 19.9. Since these lattices are based upon microscopic translations, they cannot be distinguished by macroscopic examination of the crystals.

In the *primitive cubic lattice in* Fig. 19.9a the lattice points shown as black dots at the corners of a cube may represent atoms, ions, molecules, or any repeated structure unit. Three types of reflecting planes are illustrated: 100, 110, and 111. The reflections from these planes occur at the smallest angles of all the possible planes that can be imagined in a cubic crystal, since planes with higher indices are closer together and θ will be larger, according to equation 19.4.

In a cubic crystal the perpendicular distance between a stack of planes with indices h, k, and l is given by

$$d_{hkl} = \frac{a}{\sqrt{h^2 + k^2 + l^2}} \tag{19.6}$$

where a is the length of the side of the unit cell. The spacings in a cubic crystal are obtained by substituting $0, 1, 2, 3, \ldots$ for h, k, and l in this equation.

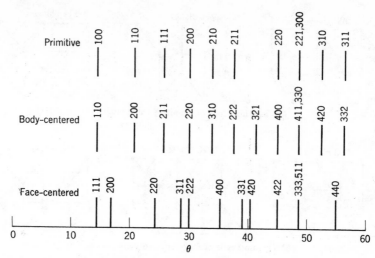

Fig. 19.10   Angles of incidence θ and reflection indices from cubic crystals. The values of λ/a have been chosen arbitrarily to cause the first reflection to fall at the same angle for each type of crystal. For primitive cubic, $\lambda/a = 0.500$; for body-centered cubic, $\lambda/a = 0.353$; for face-centered cubic, $\lambda/a = 0.289$.

Exercise I   Prove equation 19.6, making use of the fact that the sum of the squares of the sides of a right triangle is equal to the square of the hypotenuse. Test this relation for planes with indices (100), (110), and (111).

For *primitive cubic crystals* d_{hkl} may have the following values: a, $a/\sqrt{2}$, $a/\sqrt{3}$, $a/\sqrt{4}$, $a/\sqrt{5}$, $a/\sqrt{6}$, $a/\sqrt{8}$, etc., where $a/\sqrt{7}$ is missing because 7 cannot be obtained from $h^2 + k^2 + l^2$, where h, k, and l have values of 0, 1, 2, 3, ...

A face centered lattice has four equivalent positions per unit cell related by pure translational symmetry. In the *face-centered cubic lattice* illustrated in Fig. 19.9*b* there are lattice points in the center of each face of the unit cell in addition to the lattice points at the corners. One-half of the face-centered lattice points and one-eighth of the corner lattice points belong to the unit cell, making a total of four per cell. Hence, the equivalent positions are xyz; $\frac{1}{2} + x$, $\frac{1}{2} + y$, z; $\frac{1}{2} + x$, y, $\frac{1}{2} + z$; $x, \frac{1}{2} + y$, $\frac{1}{2} + z$. These four equivalent positions also can be given as $000 + xyz$; $\frac{1}{2}\frac{1}{2}0 + xyz$; $\frac{1}{2}0\frac{1}{2} + xyz$; $0\frac{1}{2}\frac{1}{2} + xyz$. NaCl has a face-centered cubic lattice (Fig. 19.12). Since all of the lattice points can be considered to be occupied by Na$^+$ there are four Na$^+$ per unit cell. Since the Cl$^-$ ions on the 12 cell edges (in which each edge is shared by 4 unit cells) and at the center of the unit cell also are related to one another by face-centering, there are four Cl$^-$ per unit cell. The positions of the four Na$^+$ and four Cl$^-$ ions are as follows:

Four Na$^+$ $x = 0$, $y = 0$, $z = 0$ + face-centering $\rightarrow 000$; $\frac{1}{2}\frac{1}{2}0$; $\frac{1}{2}0\frac{1}{2}$; $0\frac{1}{2}\frac{1}{2}$.

Four Cl$^-$ $x = \frac{1}{2}$, $y = 0$, $z = 0$ + face-centering $\rightarrow \frac{1}{2}00$; $0\frac{1}{2}0$; $00\frac{1}{2}$; $\frac{1}{2}\frac{1}{2}\frac{1}{2}$.

The diffraction patterns of NaCl and any other face-centered crystal (regardless of the crystal system) show an *absence* of all reflections for which the indices *hkl* are *not all even* or *all odd*. Hence, only the reflections 111, 200, 220, 311, 222, 400, 331, 420, etc., may be observed. It can be shown mathematically that the X-ray scattering from the face-centered symmetry-related atoms due to the equivalent points at 000; $\frac{1}{2}\frac{1}{2}0$; $\frac{1}{2}0\frac{1}{2}$; $0\frac{1}{2}\frac{1}{2}$ is completely in phase for reflections with indices all even or all odd but completely out of phase for all other *hkl* reflections. Consequently, the spacings corresponding to the *hkl* reflections which may appear on a powder diagram of a face-centered cubic crystal are (from equation 19.6) $a/\sqrt{3}$, $a/\sqrt{4}$, $a/\sqrt{8}$, $a/\sqrt{11}$, $a/\sqrt{12}$, $a/\sqrt{16}$, $a/\sqrt{19}$, $a/\sqrt{20}$, etc.

A body-centered lattice has two positions per unit cell related by translational symmetry, namely xyz; $\frac{1}{2} + x$, $\frac{1}{2} + y$, $\frac{1}{2} + z$. Hence there are two lattice points per unit cell at 000 and $\frac{1}{2},\frac{1}{2},\frac{1}{2}$ that have identical environments as shown in Fig. 19.9c. An examination of the diffraction data for body-centered crystals shows that *hkl* reflections for which the sum $h + k + l$ is odd are not observed. This means that the scattering of each atom in a body-centered unit cell is completely in phase with that of its corresponding body-centered equivalent atom if the sum of the indices is even, but 180° out of phase if the sum is odd. Accordingly, the interplanar spacings found for a *body-centered cubic lattice* are $a/\sqrt{2}$, $a/\sqrt{4}$, $a/\sqrt{6}$, $a/\sqrt{8}$, etc., which are the distances between the (110), (200), (211), and (220) planes.

The reflections for the three types of cubic crystals are summarized in Fig. 19.10, in which the presence of a reflection is indicated by a line at the corresponding angle of incidence. The angle of reflection for a real crystal depends on the length of the side of the unit cell *a* and the wavelength λ of the X rays used. For the purposes of this illustration, the ratio λ/a is arbitrarily taken as 0.500 for primitive cubic, 0.353 for body-centered cubic, and 0.289 for face-centered cubic. It may be seen that the various types of cubic crystals may be distinguished by their diffraction patterns, since the lines are qualitatively different. In the powder pattern for primitive cubic the successive lines are closer together, and there is a gap after the sixth line. In the powder pattern for body-centered cubic this gap is filled in. In the powder pattern for face-centered cubic the first two lines are relatively close together, the second line is by itself, and the next two lines are close together. Thus by use of powder patterns it is possible to determine whether a cubic crystal is primitive, body centered, or face centered.

19.10   POWDER PATTERNS FOR CUBIC CRYSTALS

The powder patterns of three substances forming cubic crystals are shown in Fig. 19.11. As may be seen by comparison with Fig. 19.10 the reflections for sodium chloride are found to correspond to those expected for a face-centered cubic lattice. The reflection indices have been assigned on this basis. The 100 reflection is missing, and so none of the spacings calculated using the **Bragg** equation is equal to the length of the side of the unit cell a. However, a may be calculated from the angle of any reflection by use of equations 19.5 and 19.6. The value of a for sodium chloride is 5.64 Å.

The structural units of the face-centered lattice that have thus been found could be sodium chloride molecules, or there could be a lattice of equal numbers of sodium and chloride ions. If the lattice were made up of sodium chloride molecules, all units of the lattice would be the same, and a more detailed consideration of the theory shows that the intensities of the X-ray reflections would always decrease progressively from first- to second- to third-order reflections. This is true for the 200, 400, and 600 reflections, but not for the 111, 222, 333, . . . reflections. It may be noted in Fig. 19.11 that the 111 reflection is very weak, whereas 222 is strong and 333 is apparently missing. This fact leads to the requirement of a lattice with ions at the lattice points.

The square connecting sodium ions is drawn in Fig. 19.12, showing that there is a face-centered sodium ion. Similar squares could be drawn on any part of the faces. A square is drawn with dashed lines for the chloride ions; it is evident that the chloride ions are displaced half a cell edge from the sodium ions.

A further examination of Fig. 19.12 shows that the (111) planes that cut diagonally through the sodium chloride crystal include only sodium ions or only chloride ions. Thus the (222) planes are alternately planes of sodium and chloride

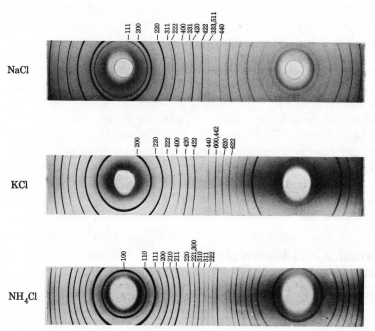

Fig. 19.11   X-ray powder patterns for cubic crystals. The X-ray beam enters through the hole at the right and leaves through the hole at the left (see Fig. 19.8). (Courtesy Prof. S. Bailey of the University of Wisconsin.)

Fig. 19.12 Interpenetrating face-centered lattice of sodium chloride; small spheres, sodium ions; large spheres, chloride ions. Each Na^+ is octahedrally surrounded by six Cl^- and each Cl^- octahedrally surrounded by six Na^+.

ions. It may be remembered that the maximum in X-ray reflections occurs when the angle is such that the paths between successive layers of ions are equal to one wavelength of the reflected radiation. If the rays are reflected from these planes at such an angle that the rays from successive planes of chloride ions differ in path length by one wavelength, the rays coming from successive sodium-ion planes, which are spaced equally between them, will then differ by half a wavelength and cause interference. The interference would be complete except for the fact that the chloride ions have more electrons and scatter X rays more efficiently than the sodium ions. The reflections from the (222) planes, however, have a difference of a whole wavelength between the rays from the chloride and from the sodium planes, so that there is no interference, and the 222 reflection is intense. The 333 reflection again corresponds to a difference of one-half wavelength between the two sets of reflecting planes, and this interference, combined with the fact that the third-order spectrum is naturally weaker, leads to a very weak reflection.

The powder pattern for potassium chloride given in Fig. 19.11 is superficially that of a primitive cubic lattice. This appears surprising, since the structure would be expected to be face centered like that for sodium chloride. The reflections look very much like those from a simple cubic structure because the scattering power of the potassium ion is almost exactly equal to that of the chloride ion, since both ions have the argon electronic structure. A more accurate determination of the intensities of reflection shows that potassium chloride does indeed form a face-centered lattice.

The powder pattern for ammonium chloride given in Fig. 19.11 shows that it has a primitive cubic lattice. If the center of a chloride ion is taken as the corner of the unit cell, the ammonium ion lies at the center of the cell, but the crystal is not body-centered cubic, since the ions are not equivalent.

Although the powder camera technique has been widely utilized for finger-printing compounds, its use to determine the arrangement of atoms is mainly limited to the simple crystal structures comprising the cubic, hexagonal, and tetragonal systems. In general, it is much more convenient to utilize single crystal X-ray techniques to determine the arrangement of atoms. The main advantage of X-ray diffraction over other structural methods is that X-ray diffraction in practically all cases provides a *direct, unique* solution of the structure.

19.11   UNIT-CELL DIMENSIONS

It has been shown that the spacings and intensities of the X-ray reflections reveal the type of crystal lattice. After it is known whether a cubic crystal is of the primitive, face-centered, or body-centered type, the size of the unit cell may be calculated by use of Bragg's law from the angles at which X rays are reflected. For example, when X rays from a palladium target having a wavelength of 0.581 Å are used, the 200 reflection of sodium chloride occurs at an angle of 5.9°. According to Bragg's law,

$$d_{200} = \frac{\lambda}{2 \sin \theta} = \frac{0.581 \times 10^{-8} \text{ cm}}{2 \sin 5.9°} = 2.82 \times 10^{-8} \text{ cm} \qquad (19.7)$$

Since the distance between (200) planes is one-half the length of the side of this unit cell, $a = 5.64$ Å.

If the density and the size of the unit cell are known, the number of atoms, ions or molecules per unit cell may be calculated. Since the density of sodium chloride is 2.163 g cm^{-3} at 25°, the formula weight is 58.443 g mol^{-1}, and the length of the side of the unit cell is 5.64 Å,

$$2.163 \text{ g cm}^{-3} = \frac{n(58.443 \text{ g mol}^{-1})}{(6.02205 \times 10^{23} \text{ mol}^{-1})(5.64 \times 10^{-8} \text{ cm})^3} \qquad (19.8)$$

$$n = 3.999$$

Thus the unit cell contains four sodium ions and four chloride ions.

The wavelength of X rays may be determined from the diffraction of X rays by finely ruled gratings or by crystals for which the interplanar spacings are known. If the wavelength of the X rays and the interplanar spacings are known, Avogadro's constant may be calculated from the measured angles of reflection.

Example 19.1   Potassium crystallizes with a body-centered cubic lattice and has a density of 0.856 g cm^{-3}. Calculate the length of the side of the unit cell a and the distance between (200), (110), and (222) planes.

$$0.856 \text{ g cm}^{-3} = \frac{(2)(39.102 \text{ g mol}^{-1})}{(6.02205 \times 10^{23} \text{ mol}^{-1})a^3}$$

$$a = 5.333 \times 10^{-8} \text{ cm} = 5.333 \text{ Å}$$

$$d_{hkl} = \frac{5.333 \text{ Å}}{\sqrt{h^2 + k^2 + e^2}}$$

For (200) planes, $d_{200} = 5.333/\sqrt{4} = 2.667$ Å

For (110) planes, $d_{110} = 5.333/\sqrt{2} = 3.771$ Å

For (222) planes, $d_{222} = 5.333/\sqrt{12} = 1.540$ Å

19.12   IONIC RADII IN CRYSTALS

In a number of inorganic crystals the binding is primarily due to electrostatic attractions between positive and negative ions. Since the coulombic force is undirected, the relative sizes of the ions largely determine how the ions are packed to form a three-dimensional array. The radius of an ion is nearly the same in different crystals because the repulsive force increases very sharply as the internuclear distance becomes smaller than a certain value. The ionic radii of the halide ions and the alkali metal ions may be calculated fairly simply from the unit-cell distances of the alkali halide crystals, since these all belong to the face-centered cubic system, except the cesium salts, which are primitive cubic.

Ionic radii in crystals for a number of anions and cations are summarized in Table 19.1. It is seen that in each column of the periodic table the ionic radius increases with the number of orbital electrons.

Table 19.1[1]   Ionic Radii in Crystals (in Å)

Li^+	0.60	Be^{2+}	0.31	O^{2-}	1.40	F^-	1.36
Na^+	0.95	Mg^{2+}	0.65	S^{2-}	1.84	Cl^-	1.81
K^+	1.33	Ca^{2+}	0.99	Se^{2-}	1.98	Br^-	1.95
Rb^+	1.48	Sr^{2+}	1.13	Te^{2-}	2.21	I^-	2.16
Cs^+	1.69	Ba^{2+}	1.35				

[1] L. Pauling, *The Nature of the Chemical Bond*, Cornell University Press, Ithaca, 1960.

19.13   CLOSE PACKING OF SPHERES

When bonding is not highly directional it is often found that the lowest energy structure is that in which each particle is surrounded by the greatest possible number of neighbors. It is, therefore, of interest to consider the ways in which uniform spheres can be stacked to form close-packed structures. When spheres are packed in a plane they arrange themselves so that each sphere is surrounded hexagonally by six others. When the second layer is formed by placing spheres in the hollows on top of the first layer, it is evident that all of the hollows in the first layer are not occupied, as may be seen from Fig. 19.13. When a third layer is added there is a choice as to whether the spheres in this layer are stacked so that they are above the spheres in the first layer, as in Fig. 19.13a, or not, as in Fig. 19.13b. The layers of the first type of packing, which is referred to as *hexagonal close packing*, may be described as *ABABAB* . . . . The layers of the second type of packing, which is referred to as *cubic close packing*, may be described by *ABCABC* . . . . In both kinds of close packing the number of nearest neighbors is the same, 12, and the fraction of the volume occupied by the spheres is the same.

The fact that cubic close packing is really face-centered cubic may be seen from Fig. 19.14. Since the length of the diagonal of the face of a unit cell is $\sqrt{2}a$

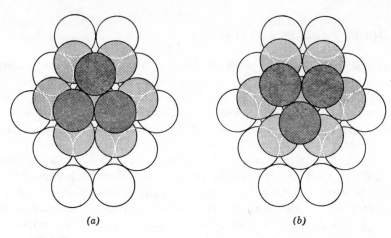

Fig. 19.13 (*a*) Hexagonal close packing. (*b*) Cubic close packing.

the radii of the spheres that just touch are given by $(\sqrt{2}/4)a$. Since there are four spheres per unit cell the fraction of the volume occupied by spheres is

$$\frac{4(\tfrac{4}{3}\pi)\left(\dfrac{\sqrt{2}}{4}a\right)^3}{a^3} = 0.740 \tag{19.9}$$

Hexagonal close packing and cubic close packing are the only two ways of close packing identical spheres so that the environment of each sphere is identical with the environment of all the other spheres, but there are other ways of close packing spheres so that the environment of each sphere is not identical; for example *ABCABABCAB* . . . . In principle there is an infinity of these other ways.

There are two types of interstitial sites, or holes, in close-packed structures: tetrahedral holes and octahedral holes. A tetrahedral hole is surrounded by four

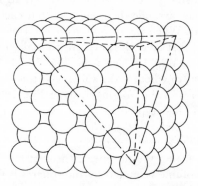

Fig. 19.14 Cubic close packing (face-centered cubic packing). Some atoms have been omitted to show that the close packed planes are (111) planes.

spheres and will accommodate a smaller sphere with a radius of 0.23 that of the larger spheres. An octahedral hole is surrounded by six spheres. It will accommodate a smaller sphere with a radius of 0.41 that of the larger spheres. Most of the carbides, nitrides, borides, and hydrides of the transition metals form crystals with the smaller nonmetallic atoms in the interstitial sites of the close-packed structures formed by the metal atoms.

Although the majority of metallic elements crystallize with hexagonal close packing or cubic close packing, some crystallize with the body-centered cubic arrangement, which is not a close-packed structure.

19.14 BODY-CENTERED CUBIC STRUCTURE OF SPHERES

In this structure, which is illustrated in Fig. 19.15, each atom has eight nearest neighbors and six other next nearest neighbors slightly further away at the body-centered positions of neighboring cells. By use of the Pythagorean theorem it is readily shown that the distance from the body-centered point to one of the corners of the cubic unit cell is $(\sqrt{3}/2)a$. If the structure is made up of spheres that touch they must have a radius of $(\sqrt{3}/4)a$. The fraction of the volume of the unit cell (and hence of the entire crystal) occupied by spheres is

$$\frac{2(\tfrac{4}{3}\pi)\left(\dfrac{\sqrt{3}}{4}a\right)^3}{a^3} = 0.680 \tag{19.10}$$

A number of metallic elements crystallize in the body-centered cubic structure even though it is not a close-packed structure.

Copper, silver, and lead crystallize in cubic close packing (face-centered cubic), and zinc and magnesium in hexagonal close packing. The alkali metals and tungsten crystallize in a body-centered cubic structure.

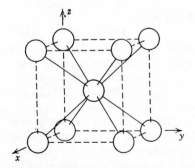

Fig. 19.15 Body-centered cubic structure.

19.15 RESULTS FROM X-RAY
DIFFRACTION STUDIES

X-ray analyses of thousands of crystal structures have led to detailed knowledge of the geometrical properties of different groups of atoms, including well-established values of bond lengths and angles. The resulting stereochemical principles have been of great help in the determination of new crystal structures, particularly for large molecules of biological origin that are composed of small basic units. Modern crystallographic analyses using data-collecting diffractometers, and high-speed computers have enabled the molecular architecture of proteins to be determined. X-ray diffraction data were used in the determination of the structure of deoxyribonucleic acid and in learning about the hydrogen bonding that makes that structure stable.

In certain cases X-ray diffraction may be used to determine the absolute configuration of an optically active substance. In 1951 Bijroet, Peerdeman, and van Bommel studied sodium rubidium $(+)$-tartaric acid by X-ray diffraction and found that the absolute configuration was the one arbitrarily chosen from the two possible enantiomorphic structures by Fischer 100 years earlier. X-ray diffraction also has been widely used in inorganic chemistry to determine *both* the *structure* and correct *formulas* of many boron hydride and metal carbonyl complexes where erroneous assignments of formulas have been previously made. In many cases, it is the only practical method of ascertaining the correct composition of compounds. In the study of the synthetic elements neptunium, plutonium, curium, and americium it was possible to establish quickly the purity of compounds and the chemical composition with exceedingly small amounts of material and without destroying the samples.

The X-ray pattern for an unknown substance may be used to identify it, or the presence of impurities may be detected, and even measured quantitatively, in a known substance. This application is greatly facilitated by such tables as those of the American Society for Testing Materials, which give the spacings calculated from the three strongest powder lines for a very large number of substances.

19.16 NEUTRON DIFFRACTION

The average de Broglie wavelength (Section 12.8) of thermal neutrons is 1.4 Å at room temperature. An essentially monochromatic beam may be obtained by diffraction from a crystal monochromator that selects a small band of wavelengths from the incident beam obtained from a nuclear reactor. Neutron diffraction also can be used to study the structures of crystals in the form of powders or single crystals. Although the principles of neutron diffraction are similar to those of X-ray diffraction, several fundamental differences between them result in neutron diffraction being a complementary technique to that of X-ray diffraction. Whereas X rays are scattered by electrons, neutrons are scattered primarily by the nuclei in a crystal. Hence, the atomic scattering factors for neutrons do not vary directly with atomic number as do the scattering factors for X rays, but instead have roughly the same scattering factors (with no dependence on the Bragg scattering angle). This means in contrast to X-ray diffraction that neutron diffraction is especially useful for accurately locating hydrogen atoms in a crystalline structure. For example, in a compound such as uranium hydride X-ray diffraction was utilized to determine the uranium coordinates and neutron diffraction the hydrogen coordinates.

Since neutrons possess a magnetic moment by virtue of having a nuclear spin of $\frac{1}{2}$, there is an additional scattering if the compound contains paramagnetic atoms or ions with unpaired electrons. Thus, neutron diffraction has been widely utilized to investigate structures of magnetic materials in order to differentiate the arrangement of the atomic magnetic moments in solids such as MnO and Fe_3O_4.

19.17 STRUCTURE OF LIQUIDS

In a perfect crystal the atoms, ions, or molecules occur at definite distances from any individual atom, ion, or molecule that is taken as the center of a coordinate system. In a gas the molecules have random positions at a given time. Liquids are intermediate between crystals and gases in that the molecules are not arranged in a definite lattice, but there is some order. By a detailed analysis of the intensity of the scattered X-rays it is possible to calculate the distribution of atoms or molecules in the liquid and obtain a plot such as Fig. 19.16. The ordinate of this plot gives the probability of finding other atoms at a distance r from a certain

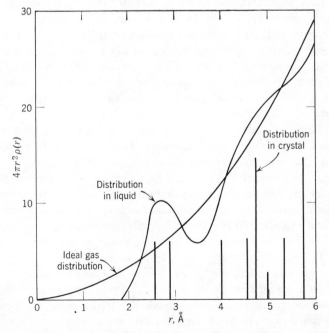

Fig. 19.16 Atomic distribution for liquid Zn at 460°. (Adapted from L. S. Darken and R. W. Gurry, *Physical Chemistry of Metals*, McGraw-Hill Book Co., New York, 1953, p. 112.)

atom. This probability is given by $4\pi r^2 \rho$, where ρ is the local density of atoms (number of atoms per unit volume). The area under a plot of the radial distribution function $4\pi r^2 \rho$ against r between two values of r is equal to the number of atoms contained in the corresponding spherical shell. In Fig. 19.16 the smooth parabolic curve represents the purely random distribution of an ideal mon-atomic gas, and the vertical lines represent the positions and numbers of the atoms in the crystal. As the temperature of a liquid is raised, the maxima and minima in the distribution curve become less pronounced and the distribution function becomes more like that for a gas. The X-ray diffraction method of investigating liquids is useful for determining the nature of the molecules in the liquid state. Studies of liquid phosphorus show the existence of P_4 molecules.

The theory of liquids is in a much less satisfactory state than the theories of gases and crystals, but important progress is being made in our understanding of the structure of liquids. The thermodynamic properties of a liquid may be expressed in terms of the radial distribution function.

19.18 LIQUID CRYSTALS

In certain liquids new phases, which resemble both liquid and solid phases, appear when the liquid is cooled. These phases have a translucent or cloudy appearance and are called liquid crystals.

In a liquid of asymmetric molecules the molecular axes are arranged at random. But in liquid crystals there is some kind of alignment. As shown in Fig. 19.17 there are three types of liquid crystals. In *nematic* liquid crystals the long axes of the molecules are lined up. The molecular axes are parallel to each other, but the molecules are not arranged in layers. The word nematic was coined from the Greek root for thread to describe the appearance of this particular type of liquid crystal under a microscope. Nematic liquid crystals have a translucent appearance because they scatter light strongly.

In *cholesteric* liquid crystals the molecular axes are aligned, and the molecules are arranged in layers in which the orientation of the axes shifts in a regular way in going from one layer to the next, as shown in Fig. 19.17. The distance measured perpendicular to the layers through which the direction of alignments shifts 360° is of the order of the wavelength of visible light. As a result of the strong Bragg reflection of light cholesteric liquid crystals have vivid iridescent colors. The pitch of the spiral and the reflected color depends sensitively on the temperature, and so these liquid crystals have been used to measure skin and other surface temperatures. The name cholesteric comes from the fact that many derivatives of cholesterol (but not cholesterol itself) form this type of liquid crystal.

The third type of liquid crystals, *smectic*, are formed by certain molecules with chemically dissimilar parts. The chemically similar parts attract each other, and there is a tendency to form layers as well as to have the molecules aligned in one direction, as illustrated in Fig. 19.17. Smectic phases are soaplike in feel and structure and may have some relationships with cell membranes.

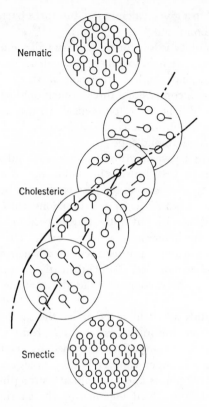

Nematic

Cholesteric

Smectic

Fig. 19.17 Structures of liquid crystals.

19.19 BINDING FORCES IN CRYSTALS

A number of different types of binding forces are involved in holding crystals together. The physical properties of a crystal are very dependent on the type of bonding.

Ionic crystals are held together by the electrostatic forces between ions. The lattice energy determined from heat of formation and heat of vaporization measurements agrees with that calculated on the assumption that the units of the crystal are ions held together by electrostatic forces.

In ionic crystals there is no fixed directed force of attraction. Although the ionic crystals are strong, they are likely to be brittle. They have very little elasticity and cannot easily be bent or worked. The melting points of ionic crystals are generally high (NaCl, 800°; KCl, 790°). In ionic crystals some of the atoms may be held together by covalent bonds to form ions having definite positions and orientations in the crystal lattice. For example, in calcium carbonate a carbonate ion does not "belong" to a given calcium ion, but three particular oxygen atoms do belong to a given carbon atom.

The electric conductivity is low for reasons we will see in the next section.

Covalent crystals, which are held together by covalent bonds in three dimensions, are strong and hard and have high melting points. An example of this type is the diamond structure which consists of a three-dimensional network of carbon atoms each bonded to four others in a tetrahedral arrangement. The C—C bond distance is the same as in aliphatic compounds (1.54 Å). Silicon and zinc sulfide also form crystals of the diamond type.

The great difference between graphite and diamond can be understood in terms of the crystal lattice. Graphite has hexagonal networks in sheets like benzene rings. The distance between atoms in the plane is 1.42 Å, but the distance between these atomic layer planes is 3.35 Å. In two directions, then, the carbon atoms are tightly held as in the diamond, but in the third direction the force of attraction is much less. As a result one layer can slip over another. The crystals are flaky, and yet the material is not wholly disintegrated by a shearing action. This planar structure is part of the explanation of the lubricating action of graphite, but this action also depends on absorbed gases, and the coefficient of friction is much higher in a vacuum.

Van der Waals' crystals are held together by the same forces (Section 14.15) that cause deviation from the ideal gas law and produce condensation at sufficiently low temperatures. Examples are provided by crystals of neutral organic compounds and rare gases. Since van der Waals' forces are weak such molecular crystals have low melting points and low cohesive strengths.

Hydrogen-bonded crystals are held together by the sharing of protons between electronegative atoms (Section 14.8). Hydrogen bonds are involved in many organic and inorganic crystals and in the structure of ice and water. They are comparatively weak bonds but play an extremely important role in determining the atomic arrangement in such hydrogen-bonded substances as proteins and polynucleotides.

Metallic bonds exist only between large aggregates of atoms. This type of bonding gives metals their characteristic properties, opaque, lustrous, malleable, and good conduction of electricity and heat. Metallic bonding is due to the outer, or valence electrons. The wave functions for these electrons are sufficiently distributed over space that they have appreciable probability densities at distances equal to the interatomic distances in metals. The over-lapping of the wave functions for the valence electrons in metals results in orbitals that extend over the entire crystal. The electrons pass throughout the volume of the crystal and for certain purposes we may consider that there is an electron gas—except that we shall see that it is fundamentally different from other gases.

There is a gradual transition between metallic and nonmetallic properties. Atoms with fewer and more loosely held electrons form metals with the most prominent metallic properties. Examples are sodium, copper, and gold. As the number of valence electrons increases and they are held more tightly, there is a transition to covalent properties.

The close-packed structures are often found in metals because the binding energy per unit volume is maximized. In both of the close-packed structures for atoms of equal size, 12 spheres touch the central sphere, 6 in the midplane, 3 above, and 3 below.

19.20 ELECTRONIC STRUCTURE OF SOLIDS

Theories of the electronic structures of solids have the task of explaining why the properties of solids extend over such a wide range. For example, the electric conductivities range from about 10^8 to 10^{-18} Ω^{-1} m^{-1}.

We have seen in Section 14.2 that as atoms are brought together their wave functions combine to form bonding and antibonding orbitals. As more atoms are brought together this process continues. For example, Fig. 19.18 illustrates the splitting that occurs when six hydrogen atoms are brought together in a linear array. The combination of six $1s$ wave functions produces six orbitals, three bonding and three antibonding. As electrons are fed into this energy level scheme they will first fill the lower energy bonding orbitals, 2 at a time. As the number of interacting atoms is increased the number of energy levels increases and they become more and more closely spaced within a band, but the width of the band at a given internuclear separation does not increase. We use the term *band* to distinguish the groups of levels arising from different atomic orbitals. In Fig. 19.18 there is the $1s$ band and the $2s$ band. Thus, in contrast to molecules, we find in a solid bands of energy levels containing very large numbers of discrete levels, but with relatively large separations in energy between bands. These separations between allowed bands are called energy gaps.

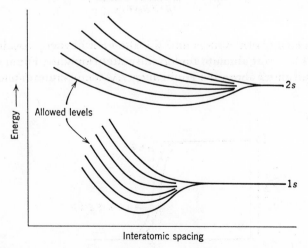

Fig. 19.18 Energy levels for a linear array of six hydrogen atoms as a function of internuclear distance.

We have been discussing the so-called "tight-binding approximation" for the electronic energy levels of solids. This theory works better for the inner electrons rather than the valence electrons. In order to explain the electrical conductivity of solids we go back to the free electron approximation of Drude, Lorentz, and others. According to this theory electrons within metals have relatively large mean-free paths and are free to move when an electric field is applied. The electric conductivity κ is given by

$$\kappa = Nqu \tag{19.11}$$

where N is the number of current carriers per unit volume, q is the charge carried by each carrier, and u is its mobility (Section 11.5). Although relatively successful in explaining electric and thermal conductivities, this theory does not lead to correct contributions by the conduction electrons to the heat capacity of a metal. Electrons contribute a very small amount in comparison with the classical expectation (Section 9.5) of $\frac{3}{2}R$ per mole.

In order to understand the small contribution of conduction electrons to the heat capacity, we need to remember that electrons are put into energy levels in accord with the Pauli principle, beginning at the lowest energies. At absolute zero the energy levels are filled to a sharply defined energy, referred to as the Fermi energy E_F. In contrast to the translational energies of gas molecules, where Maxwell-Boltzmann statistics allow any number of particles to have exactly the same energy, electrons follow Fermi-Dirac statistics, which means that only one particle is allowed in each state of the system. According to Fermi-Dirac statistics the probability $P(E)$ of a state of energy E is

$$P(E) = \frac{1}{e^{(E-E_F)/kT} + 1} \tag{19.12}$$

where E is the energy of the state, and E_F is the Fermi energy. As illustrated in Fig. 19.19, $P(E) = 1$ at absolute zero for all states below the Fermi energy, and $P(E) = 0$ for all states above the Fermi energy. At temperatures above absolute

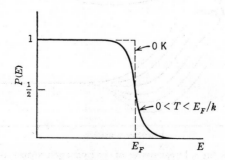

Fig. 19.19   Fermi-Dirac distribution of electron energies in a metal.

zero some electrons just below the Fermi energy will be promoted to energies above E_F. When $(E - E_F) \gg kT$ the Fermi-Dirac distribution function reduces to

$$P(E) = e^{-(E-E_F)/kT} \tag{19.13}$$

which is the classical Boltzmann distribution. We can see from Fig. 19.19 why valence electrons make such a small contribution to the heat capacity. Only a small fraction of the electrons near the Fermi energy pick up more energy when the temperature is raised one degree.

The Fermi energy is the electrochemical potential of the electrons and determines their tendency to move at an interface, just like the chemical potential does for a substance.

The differences between insulators, metals, and semiconductors may be understood in terms of the extent of filling of energy bands. In an insulator all the bands that contain electrons are completely filled, and the gap to the next band is large. In a metal the highest occupied band is approximately half filled. As the temperature is reduced, metals become better conductors of electricity because the thermal vibrations of the atoms of the lattice are diminished, and so the atoms interfere less with the motion of the conduction electrons.

There are two situations where a material is a semiconductor. If there are only a few electrons in the highest occupied level, the solid has a low electric conductivity because there are so few current carriers. If the highest occupied band is almost completely filled, the solid has a low electric conductivity because there are so few levels into which the electrons can move. The vacant position left behind in a covalent bond when an electron is released and is free to wander through the crystal is referred to as "positive hole" or simply as a "hole." When an electric field is applied, holes move in the direction of the field as a result of the shift of valence bond electrons in the opposite direction.

The motions of electrons in an insulator, metal, and semiconductor are analogous to the possible movements of cars on a floor of a parking garage. If a floor is completely filled, there is no possibility of movement. If there are only a few vacancies on a floor, there are few opportunities for movement. If a floor is half filled, there is the maximum possibility for movement. If there are only a few cars on a floor, they can move easily, but the total amount of movement is small because of the small number of cars.

Semiconductors with only a few electrons in the upper band are referred to as n-type because the charge carriers determining the conductivity are $negative$ electrons. Semiconductors with only a few empty states, or holes, are referred to as p-type because the charge carriers are $positive$. Pure germanium and pure silicon are weakly ionized solids, just as water is a weakly ionized liquid. At room temperature the intrinsic carrier concentrations $((e^-) = (h^+))$ in Si and Ge are of the same order of magnitude as the ion concentration $((H^+) = (OH^-))$ for water.

We refer here to intrinsic semiconductors, that is, pure substances that are semiconductors. In Section 19.23 we will discuss extrinsic semiconductors.

The electric conductivity of semiconductors increases with T, in contrast with metals because more electrons are excited to the conduction band.

19.21 POINT IMPERFECTIONS

Real crystals have imperfections or defects. These imperfections can be characterized as to whether the defect is at a point, along a line, or over a surface. Imperfections are the key to many of the physical and chemical properties of crystals.

A point imperfection may result from the absence of an atom (vacancy), presence of an impurity atom at a lattice site (substitutional impurity atom), presence of an impurity atom at an interstice (interstitial impurity atom), or displacement of an atom to an interstitial site (self interstitial).

Various point imperfections are shown in Fig. 19.20. The two basic types of imperfections are interstitial atoms and vacancies. The capacity of a crystal to accommodate interstitial atoms is very dependent on how open the crystal lattice is and the size of the interstitial atom. If the repeating units in the crystal lattice are electrically neutral, vacancies produce no particular problem with respect to the overall balance of electric charge. But in an ionic crystal vacancies must be balanced so that the crystal as a whole is electrically neutral. In a *Frenkel defect* the vacancies are compensated for by an interstitial atom of the same type. In a *Schottky defect* anion and cation vacancies occur in equal numbers. These defects lower the density of the crystal, but Frenkel defects do not change the density significantly.

A certain number of defects are expected in pure crystals at temperatures above absolute zero because of thermal agitation. Indeed we can regard electrons in any but the lowest lying energy levels and the accompanying "holes" as imperfections.

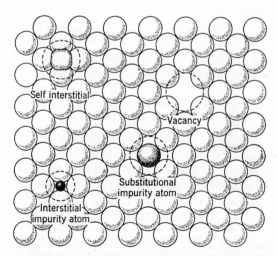

Fig. 19.20 Point imperfections in crystals.

19.22 LINE DEFECTS AND PLANE DEFECTS

Pure single crystals deform permanently at shear forces orders of magnitude lower than predicted theoretically for perfect crystals. In 1934 line defects were postulated to explain the ease with which single crystals deform permanently. The two principal-line defects are called *edge dislocations* and *screw dislocations*. An edge dislocation corresponds to the edge of an atomic plane that terminates within the crystal rather than passing all the way through. The way in which an edge dislocation facilitates shear in a crystal is illustrated in Fig. 19.21. During shear the dislocation moves across the crystal with the net effect that the top half of the crystal is displaced one lattice distance with respect to the lower half of the crystal. The number of dislocation lines passing through a unit area within an ordinary crystal is of the order of 10^6 cm^{-2} or more.

Near an edge dislocation, the atoms are pushed together above the edge and pulled apart below the edge. Thus impurity atoms with larger diameters than the solvent atoms tend to concentrate below the edge, and impurity atoms with smaller diameters tend to concentrate above the edge. This binding of impurity at the dislocation tends to make it more difficult to move a dislocation in an impure material than in a pure material. Therefore alloys require greater shear forces for permanent deformation than do pure crystals.

A screw dislocation forms a continuous helical ramp of one set of atomic planes about the dislocation line, as shown in Fig. 19.22. A screw dislocation provides for easy crystal growth because atoms can be added at the step. Screw dislocations can move in a crystal subjected to appropriate shear forces, and they give rise to the same type of permanent deformation as an edge dislocation. Real dislocations are often mixtures of edge dislocations and screw dislocations.

Plane defects include the surfaces around the tiny grains that make up crystalline metals. Neighboring grains have unrelated crystallographic orientations, and the boundaries between grains are regions of strain in which impurities tend to concentrate. Grain boundaries may be detected by etching a polished metal surface.

Another type of plane defect has already been referred to in connection with the close packing of spheres, that is, irregular stacking of atomic planes (Section

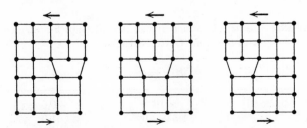

Fig. 19.21 Motion of an edge dislocation under shear. (N. B. Hannay, *Solid-State Chemistry*, (c) 1967. Reprinted by permission of Prentice-Hall, Inc. Englewood Cliffs, New Jersey.)

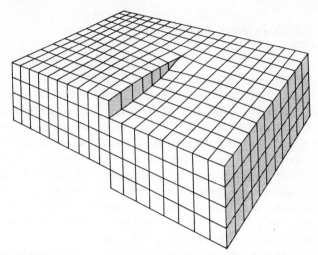

Fig. 19.22   Screw dislocation. (N. B. Hannay, *Solid-State Chemistry*, (c) 1967. Reprinted by permission of Prentice-Hall, Inc. Englewood Cliffs, New Jersey.)

19.13). For example, in face-centered cubic packing the normal stacking sequence is *ABCABC* . . . . If a *B* layer accidentally forms on top of a *C* layer and the reverse stacking sequence proceeds, the sequence is *ABCBACBA* . . . . The two parts of the crystal on either side of the plane defect are mirror images and are referred to as twins.

19.23   INTRINSIC AND EXTRINSIC SEMICONDUCTORS

Intrinsic semiconductors are pure substances that have conductivities between that of conductors and insulators. The conductivities of four semiconductors are summarized in Table 19.2. These solids all have the same crystal structure; the atoms are bonded covalently with four adjacent atoms in a tetrahedral arrangement. At absolute zero the valence band is completely filled, and the conduction

Table 19.2   Electric Conductivity and Band Gaps for Elements with the Same Crystal Structure

	$\kappa, \Omega^{-1}\,\mathrm{m}^{-1}$	eV
C (diamond)	$<10^{-4}$	6.0
Si	1.5×10^{-3}	1.0
Ge	2	0.7
Sn (gray)	>100	0.08

band is empty. These pure solids have the electrical conductivity that they do because a few electrons are excited to the conduction band.

Extrinsic semiconductors are materials whose semiconducting properties are due to the addition of small amounts of other elements. For example, let us consider the addition of phosphorous and aluminum to silicon. Since each atom of Si forms four covalent bonds with its neighbors, we can represent pure silicon as shown in Fig. 19.23a. When an atom of P replaces an atom of Si there is an extra valence electron from P that is not required to form covalent bonds. The presence of phosphorous atoms produces extra energy levels in the band gap below the conduction band that are called *donor* levels. In this particular instance the extra energy levels are only 0.012 eV below the conduction band. Electrons promoted from there to the conduction band move under the influence of an applied electric field. Thus silicon "doped" with phosphorous is an *n*-type semiconductor.

If silicon is doped with aluminum a *p*-type semiconductor is formed, as illustrated in Fig. 19.23b. Aluminum has one less valence electron than silicon, and so each aluminum atom creates an electron hole. In this situation aluminum is referred to as an *acceptor* because it can accept electrons. This produces extra energy levels in the band gap just above the valence band. These levels are referred to as acceptor levels since electrons can jump from the valence band to the acceptor levels. This leaves a positive hole in the valence band that can conduct current.

Added atoms are not necessarily ionized. If the energy required to produce this ionization is quite small, the ionization is essentially complete at room temperature. However, as the temperature is progressively lowered, charge carriers are progressively "frozen out" on the impurity atoms. Thus the ionization of added atoms can be treated as a problem in chemical equilibrium. The donor D and acceptor A ionizations are

$$D = D^+ + e^- \qquad (19.14)$$

$$A = A^- + h^+ \qquad (19.15)$$

The concentrations of the species on the right-hand side increase with increasing temperature. The slope of the plot of the logarithm of the concentration of e^- or

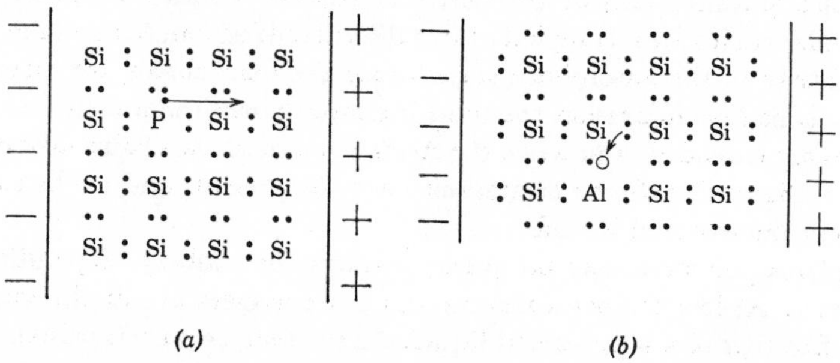

(a) (b)

Fig. 19.23 (a) Silicon crystal containing substitutional aluminum atoms. (b) Silicon crystal containing substitutional phosphorus atoms.

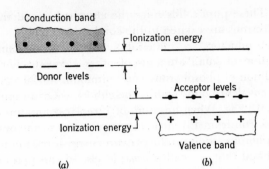

Fig. 19.24 Electron energy levels in the gap between the valence and conduction bands. (a) Ionized donors. (b) Ionized acceptors.

h^+ versus $1/T$ gives information about the ionization energy of the donor or acceptor.

The energy levels for donors and acceptors lie in the forbidden energy gap, as shown in Fig. 19.24. The donor levels are close to the conduction band because electrons are promoted from there to the conduction band. The acceptor levels are close to the valence band because the acceptor can pick up electrons from the valence band and produce holes there that contribute to the electric conductivity.

Semiconductors are of tremendous practical importance in rectifiers, amplifiers, electronic circuits, and computers.

19.24 MECHANICAL PROPERTIES OF POLYMERS

Polymeric materials display a very wide range of physical properties: they are hard or soft, leathery or rubbery, brittle or tough, and meltable and nonmeltable. These properties depend on the molecular structure of the polymer, and the various properties are obtained by the choice of the polymer and its treatment. The physical properties depend to quite an extent on whether the polymer is (a) completely amorphous or (b) partly crystalline. In an amorphous polymer the molecular chains are all tangled up in a disordered fashion. Long-chain polymers exist primarily in the amorphous state. Since the long chains are mostly in the randomly coiled configuration the solid is elastic; when the material is stretched the chains are extended, and when the stress is removed the chains take on a more random configuration. Some crystallinity may be present, and to the extent that it is present the material is stiffer.

Amorphous polymers may be glassy, leathery, or rubbery, depending on the temperature. At low temperatures amorphous polymers are in the glassy state, which is like that of a supercooled liquid. As the temperature is raised, there is a transition at the "glass-transition temperature" from the glassy to the leathery state. There is a marked change in physical properties but no discontinuity in the density. Below the glass-transition temperature even amorphous polymers are rigid and brittle. The atoms and small groups of atoms vibrate about mean positions, but it is not possible for molecular segments to slide over each other. Above the glass-transition temperature an amorphous polymer becomes leathery or elastic and a crystalline polymer becomes more flexible and less brittle. In amorphous polymers large molecular segments are able to slide over each other, and the characteristic plastic properties are obtained. For both amorphous and crystalline polymers the rate of change of density with temperature is greater above T_g than below it because of the greater increase in molecular motion. The transition from the glassy to the rubberlike state generally extends over a range of about 50°, but the temperature range in which it occurs depends on the type of polymer. If there are rather long sections of the molecular chains between cross links and centers of entanglement, long segments will be in Brownian motion, and rubberlike properties will result.

The glass-transition temperature is raised by vulcanization, which introduces cross linkages, that is, bonds are formed between adjacent threadlike molecules, thus restricting the molecular motions.

Many polymeric solids have properties that are intermediate between those of ideal solids and ideal liquids. In a perfectly elastic solid the stress is directly proportional to the strain but independent of the rate of strain. In a perfectly viscous fluid the stress is directly proportional to the rate of strain but independent of the strain itself. If the stress depends on both the strain and the rate of strain the material is said to be *viscoelastic*. Polymers in particular show such behavior because of the complicated way in which such long chainlike molecules interact with each other.*

The mechanical properties of polymers are profoundly dependent on the stereoregularity of their chains. Polymers of monomers like styrene have asymmetric carbon atoms. By using certain solid catalysts it is possible to prepare polymers in which all the adjacent asymmetric atoms, at least for long periods of the chain, have the same steric configuration.

There are two types of stereoregular vinyl polymers, *isotactic* and *syndiotactic*, depending on whether successive pseudoasymmetric carbon atoms have the same or opposite enantiomorphic configurations. These two types of structures are illustrated in Fig. 19.25, along with a structure without order. Structures with a lesser degree of order than isotactic or syndiotactic are referred to as *atactic*. Some polymers (stereoblock polymers) have alternating isotactic and syndiotactic sections. K. Ziegler and G. Natta, who first prepared such *isotactic* polymers, shared the 1963 Nobel prize for chemistry. The regularity in structure of isotactic polymers makes it possible for the polymer chains to fit into a crystal lattice. This increased orderliness increases the melting point, rigidity, and toughness above those of corresponding atactic polymers.

Some polymers contain regions in which the chains are arranged in an orderly, three-dimensional array. These crystalline regions typically have dimensions of

---

* J. D. Ferry, *Viscoelastic Properties of Polymers*, Wiley, New York, 1961.

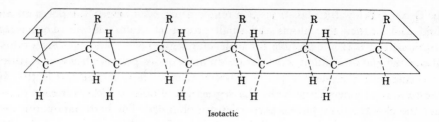

Isotactic

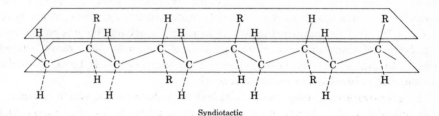

Syndiotactic

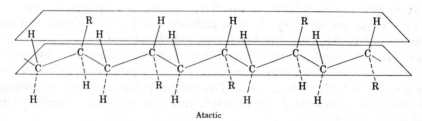

Atactic

Fig. 19.25   Steric configuration of polymer chains.

the order of 100 Å, and they have an important influence on the physical properties of the polymer. The extent of crystallinity increases when the polymer is stretched because this draws the chains together and reduces the random thermal motions. For example, natural rubber is ordinarily amorphous at room temperature but becomes oriented and crystalline when stretched. The amount of crystallinity may be investigated by X-ray diffraction and by studying the volume of the polymer as a function of temperature.

Crystallites occur most frequently in polymers having a comparatively simple and symmetrical structure or polymer chains that tend to associate by the formation of hydrogen bonds. There are crystalline regions in nylon, polyethylene, and other polymers in which the chains have sufficient chemical regularity. The individual crystals in a polymer sample are usually too small to be seen by visible-light microscopy. Larger crystals of long chain polymers may be obtained from solution.

References

H. J. M. Bowen (et al.) and E. L. Sutton (et al.) suppl. ed., *Tables of Interatomic Distances in Molecules and Ions*, Chemical Society, London, 1965.

M. J. Buerger, *Elementary Crystallography*, Wiley, New York, 1956, repr. 1963.

M. J. Buerger, *Introduction to Crystal Geometry*, McGraw-Hill Book Co., New York, 1971.

A. K. Galwey, *Chemistry of Solids*, Chapman and Hall, London, 1967.

J. P. Glusker and K. N. Trueblood, *Crystal Structure Analysis: A Primer*, Oxford University Press, London, 1972.

B. Gruber, *Theory of Crystal Defects*, Academic Press, New York, 1966.

N. B. Hannay, *Solid State Chemistry*, Prentice-Hall, Englewood Cliffs, N.J., 1967.

International Tables for X-ray Crystallography; Vol. 1, *Symmetry Groups*, 1952; Vol. 2, *Mathematical Tables*, 1959; Vol. 3, *Physical and Chemical Tables*, 1962. The Kynoch Press, Birmingham, England.

C. Kittel, *Introduction to Solid State Physics*, Wiley, New York, 1966.

L. Mandelkern, *Crystallization in Polymers*, McGraw-Hill Book Co., New York, 1964.

W. J. Moore, *Seven Solid States*, W. A. Benjamin, Inc., New York 1967.

L. Pauling, *The Nature of the Chemical Bond*, Cornell University Press, Ithaca, 1960.

D. E. Sands, *Introduction to Crystallography*, W. A. Benjamin, Inc., New York, 1969.

G. A. Somorjai, *The Structure and Chemistry of Solid Surfaces*, Wiley-Interscience, New York, 1969.

G. H. Stout and L. H. Jensen, *X-Ray Structure Determination. A Practical Guide*, The Macmillan Co., New York, 1968.

M. M. Woolfson, *An Introduction to X-Ray Crystallography*, Cambridge University Press, Cambridge, 1970.

R. W. G. Wycoff, *Crystal Structure*, (in five sections with supplements), Wiley-Interscience, New York, 1959.

Problems

19.1 The X-ray diffraction pattern of a crystal is obtained using X rays from a copper target, and a certain reflection is found at $10°\ 27'$. With a molybdenum target the same reflection was found at $4°\ 48'$. Given the wavelength of the X rays from the copper target ($\lambda = 1.540$ Å), calculate the wavelength of the X rays from the molybdenum target.

Ans. 0.710 Å.

19.2 Copper forms cubic crystals. When an X-ray powder pattern of crystalline copper is taken using X rays from a copper target (the wavelength of the $K\alpha$ line is 1.5405 Å), reflections are found at $\theta = 21.65°, 25.21°, 37.06°, 44.96°, 47.58°$, and other larger angles. (*a*) What type of lattice is formed by copper? (*b*) What is the length of a side of the unit cell at this temperature? (*c*) What is the density of copper?

Ans. (*a*) Face centered, (*b*) 3.616 Å, (*c*) 8.93 g cm^{-3}.

19.3 The density of potassium chloride at 18° is 1.9893 g cm^{-3}, and the length of a side of the unit cell is 6.29082 Å, as determined by X-ray diffraction. Calculate the Avogadro constant using the values of the atomic weights given in the front cover.

Ans. 6.0213×10^{23} mol^{-1}.

19.4 Calculate the density of diamond from the fact that it has a face-centered cubic structure with two atoms per lattice point and a unit cell edge of 3.569 Å.

Ans. 3.509 g cm^{-3}.

19.5 A solution of carbon in face-centered cubic iron has a density of 8.105 g cm^{-3} and a unit cell edge of 3.583 Å. Are the carbon atoms interstitial, or do they substitute for iron atoms in the lattice? What is the weight percent carbon?   *Ans.* Interstitial, 0.51%.

19.6   The crystal unit cell of magnesium oxide is a cube 4.20 Å on an edge. The structure is interpenetrating face centered. What is the density of crystalline MgO?

Ans. 3.62 g cm^{-3}.

19.7   Tungsten forms body-centered cubic crystals. From the fact that the density of tungsten is 19.3 g cm^{-3} calculate (*a*) the length of the side of this unit cell and (*b*) d_{200}, d_{110}, and d_{222}.   *Ans.* (*a*) 3.16 Å, (*b*) $d_{200} = 1.58$, $d_{110} = 2.23$, $d_{222} = 0.912$ Å.

19.8   Molybdenum forms body-centered cubic crystals and at 20° the density is 10.3 g cm^{-3}. Calculate the distance between the centers of the nearest molybdenum atoms.

Ans. 2.725 Å.

19.9   Cesium chloride, bromide, and iodide form interpenetrating simple cubic crystals rather than interpenetrating face-centered cubic crystals like the other alkali halides. The length of the side of the unit cell of CsCl is 4.121 ± 0.003 Å. (*a*) What is the density? (*b*) Calculate the ion radius of Cs$^+$, assuming that the ions touch along a diagonal through the unit cell and that the ion radius of Cl$^-$ is 1.81 Å.   *Ans.* (*a*) 3.99 g cm^{-3}, (*b*) 1.77 Å.

19.10   A close-packed structure of uniform spheres has a cubic unit cell with a side of 8 Å. What is the radius of the spherical molecule?   *Ans.* 2.82 Å.

19.11   If spherical molecules of 5 Å radius are packed in cubic close packing, and body-centered cubic, what are the lengths of the sides of the cubic unit cells in the two cases?

Ans. 14.142 Å, 11.547 Å

19.12   At 550° C the conductivity of solid NaCl is $2 \times 10^{-4}\ \Omega^{-1}\ m^{-1}$. Since the sodium ions are smaller than the chloride ions (see Table 19.1), they are responsible for most of the electric conductivity. What is the ionic mobility of Na$^+$ under these conditions?

Ans. $5.5 \times 10^{-14}\ m^2\ V^{-1}\ s^{-1}$.

19.13   Calculate the relation of the Fermi energy E_F for a semiconductor to the energy gap E_g between the top of the valence band and the bottom of the conduction band.

Ans. $E_F = E_g/2$.

19.14   A diamond has an electric conductivity of $5.1 \times 10^{-5}\ \Omega^{-1}\ m^{-1}$ at 25°. Assuming that the mobilities of the carriers is independent of temperature calculate its conductivity at 35° if the energy gap is 6 V.   *Ans.* $2.2 \times 10^{-3}\ \Omega^{-1}\ m^{-1}$.

19.15   When rubber is allowed to contract, the mechanical work obtained is equal to the force times the displacement in the direction of the force and is given by $dw = -f\,dL$, where f is force and L is the length of the piece of rubber. If the contraction is carried out reversibly, the first law may be written

$$dU = \left(\frac{\partial U}{\partial T}\right)_L dT + \left(\frac{\partial U}{\partial L}\right)_T dL = T\,dS + f\,dL$$

if pressure-volume work is neglected. (*a*) Show that

$$\left(\frac{\partial S}{\partial T}\right)_L = \frac{1}{T}\left(\frac{\partial U}{\partial T}\right)_L \qquad \left(\frac{\partial S}{\partial L}\right)_T = \frac{1}{T}\left[\left(\frac{\partial U}{\partial L}\right)_T - f\right]$$

(*b*) By using the fact that the order of differentiation used to obtain $\partial^2 S/\partial L\ \partial T$ is immaterial, show that

$$\left(\frac{\partial U}{\partial L}\right)_T = f - T\left(\frac{\partial f}{\partial T}\right)_L$$

19.16   Calculate the angles at which reflections would be obtained in problem 2 if an iron target was used in the X-ray tube rather than copper. Given: λ for the $K\alpha$ line of iron is 1.937 Å.

19.17 Calculate the highest-order diffraction line that can be observed for the 100 planes of NaCl using an X-ray tube with a copper target ($\lambda = 1.54$ Å).

19.18 What is the number of nearest neighbors in atomic crystals of the (a) primitive, (b) body-centered, and (c) face-centered types?

19.19 The X-ray powder pattern for molybdenum has reflections at $\theta = 20.25°$, $29.30°$, $36.82°$, $43.81°$, $50.69°$, $58.00°$, $66.30°$, and other larger angles when Cu $K\alpha$ X rays are used ($\lambda = 1.5405$ Å). (a) What type of cubic crystal is formed by molybdenum? (b) What is the length of a side of the unit cell at this temperature? (c) What is the density of molybdenum?

19.20 Using X rays with a wavelength of 1.54 Å from a copper target, it is found that the first reflection from a face of a crystal of potassium chloride at 25° is at $\theta = 14°\ 12'$. Calculate (a) the length of the side of the unit cell for this interpenetrating face-centered lattice and (b) the density of the crystal.

19.21 The density of platinum is 21.45 g cm^{-3} at 20°. Given the fact that the crystal is face-centered cubic, calculate the length of the side of the unit cell.

19.22 Diamond is face-centered cubic with a unit cell edge of 3.569 Å. There are two atoms per lattice point. What is the carbon-carbon bond length in diamond?

19.23 A substance forms face-centered cubic crystals. Its density is 1.984 g cm^{-3}, and the length of the edge of the unit is 6.30 Å. Calculate the molecular weight.

19.24 The density of calcium oxide is 3.32 g cm^{-3}. If the length of a side of the cubic unit cell is 4.81 Å, how many molecules of CaO are there per unit cell?

19.25 From the fact that the length of the side of the unit cell for lithium is 3.51 Å calculate the atomic radius of Li. Lithium forms body-centered cubic crystals.

19.26 A crystal has a body-centered structure with a unit cell side of 8 Å. If the lattice points are occupied by spheres, what is their radius?

19.27 Silicon is doped with aluminum to the extent of 10^{-8} of the atoms. The conductivity κ is $4.0\ \Omega^{-1}\ m^{-1}$. What is the mobility of holes? (The density of silicon is 2.4 g cm^{-3}.)

19.28 What concentration of donor atoms in germanium is required to produce an electric conductivity of $10^2\ \Omega^{-1}\ m^{-1}$? Assume that all donor atoms are ionized and calculate the fraction of the atoms that must be donor atoms. The mobility of electrons in germanium is $0.39\ m^2\ V^{-1}\ s^{-1}$.

19.29 The experiment described in Section 19.11 using sodium chloride and X rays from a palladium target is repeated with X rays from a copper target for which the weighted average wavelength of $K_{\alpha 1}$ and $K_{\alpha 2}$ is 1.5418 Å. At what angle will the 200 reflection be obtained?

19.30 A face-centered cubic unit cell may be represented as another crystal system. What is this crystal system and what are the angles? Why is it more convenient to use the face-centered cubic unit cell when it applies, rather than this other crystal system?

19.31 Calculate the angles at which the first-, second-, and third-order reflections are obtained from planes 5 Å apart, using X rays with a wavelength of 1 Å.

19.32 By means of the Bragg method, a cubic crystal may be oriented in different directions to obtain d_{100}, d_{110}, and d_{111}. Show that the ratios of the distances between the three different sets of planes are:

Primitive cubic $d_{100}:d_{110}:d_{111} = 1:0.707:0.578$
Face-centered cubic $d_{200}:d_{220}:d_{111} = 1:0.707:1.155$
Body-centered cubic $d_{200}:d_{110}:d_{222} = 1:1.414:0.578$

19.33   The diamond has a face-centered cubic crystal lattice, and there are eight atoms in a unit cell. Its density is 3.51 g cm^{-3}. Calculate the first six angles at which reflections would be obtained using an X-ray beam of wavelength 0.712 Å.

19.34   (a) Metallic iron at 20° is studied by the Bragg method, in which the crystal is oriented so that a reflection is obtained from the planes parallel to the sides of the cubic crystal, then from planes cutting diagonally through opposite edges, and finally from planes cutting diagonally through opposite corners. Reflections are first obtained at $\theta = 11°\ 36'$, $8°\ 3'$, and $20°\ 26'$, respectively. What type of cubic lattice does iron have at 20°? (b) Metallic iron also forms cubic crystals at 1100°, but the reflections determined as described in (a) occur at $\theta = 9°\ 8'$, $12°\ 57'$, and $7°\ 55'$, respectively. What type of cubic lattice does iron have at 1100°? (c) The density of iron at 20° is 7.86 g cm^{-3}. What is the length of a side of the unit cell at 20°? (d) What is the wavelength of the X rays used? (e) What is the density of iron at 1100°?

19.35   Potassium bromide has a face-centred cubic lattice, and the edge of the unit cell is 6.54 Å. What is the density of the crystal?

19.36   Insulin forms crystals of the orthorhombic type with unit-cell dimensions of $130 \times 74.8 \times 30.9$ Å. If the density of the crystal is 1.315 g cm^{-3} and there are 6 insulin molecules per unit cell, what is the molecular weight of the protein insulin?

19.37   Aluminum forms face-centered cubic crystals, and the length of the side of the unit cell is 4.050 Å at 25°. Calculate (a) the density of aluminum at this temperature and (b) the distances between (300), (220), and (111) planes.

19.38   Tantalum crystallizes with a body-centered cubic lattice. Its density is 17.00 g cm^{-3}. (a) How many atoms of tantalum are there in a unit cell? (b) What is the length of a unit cell? (c) What is the distance between (200) planes? (d) What is the distance between (110) planes? (e) What is the distance between (222) planes?

19.39   Aluminum forms face-centered cubic crystals, and at 20° the closest interatomic distance is 2.862 Å. Calculate the density of the crystal.

19.40   Calculate the length of the side of the unit cell of potassium iodide (face-centered cubic) from the ionic radii in Table 19.1. Calculate the density of the crystal.

19.41   The ions Na$^+$, Mg^{2+}, and Al^{3+} have the same number of extranuclear electrons. Would you expect the ion radius to increase or decrease with increasing atomic number? Why?

19.42   For uniform spheres of radius r calculate the length of the side of the unit cell for (a) hexagonal close packing, (b) face-centered cubic, and (c) primitive cubic.

19.43   The common form of ice has a tetrahedral structure with protons located on the lines between oxygen atoms. A given proton is closer to one oxygen atom than the other and is said to belong to the closer oxygen atom. How many different orientations of a water molecule in space are possible in this lattice?

CHAPTER 20

MACROMOLECULES

This term is generally applied to molecules with molecular weights greater than 10,000. Macromolecules in the form of proteins, polynucleotides, and polysaccharides are necessary for life, and their structures make possible complex functions. Macromolecules in the form of synthetic high polymers are the basis for synthetic fibers, plastics, and synthetic rubber. The relation between the physical properties of these materials and their molecular structures is, of course, of utmost importance.

In this chapter we will consider proteins and synthetic high polymers. Information about molecular weight, molecular weight distribution, and shape may be obtained from studies of viscosity, ultracentrifugation, diffusion, osmotic pressure, and light scattering.

20.1 PROTEIN STRUCTURE AND FUNCTION

Proteins are polymers of amino acids ($H_2NCHRCO_2H$) that have molecular weights greater than about 15,000. About 22 amino acids occur in proteins, and they are all of the L configuration. In proteins and polypeptides, amino-acid residues are connected through amide bonds.

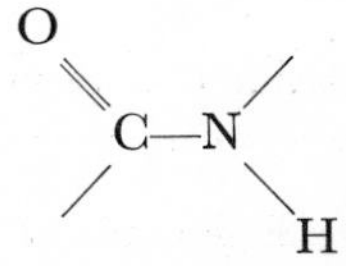

$$
\begin{array}{c}
O \\
\diagdown \\
C-N \\
\diagup \quad \diagdown \\
\qquad\quad H
\end{array}
$$

The nuclei of H, N, C, and O lie in a plane because of resonance, and the connecting bonds are trans. Smaller polymers of amino acids called oligopeptides form random coils in solution, but proteins have a more or less fixed three-dimensional structure held together by hydrogen bonds (Section 14.8), disulfide bonds (—S—S—) between cystine residues, and ionic and van der Waals interactions. The amino-acid sequences of many proteins have been determined, and the complete three-dimensional structures of a couple of dozen proteins have been determined by X-ray diffraction (Chapter 19). One small protein (ribonuclease) has been synthesized in the laboratory by two different methods. In this case the polypeptide with the amino-acid residues in the correct order coils up to yield the same three-dimensional structure as the native protein.

Proteins play many roles in living things; enzymes, antibodies, and certain structural materials are proteins. Viruses are nucleoproteins—that is, combinations of protein and nucleic acid.

The remarkable catalytic activity of enzymes discussed in Section 10.26 is due to the formation of catalytic sites by bringing together the necessary side chain groups of amino acids, chelating metal ions, and in some cases attaching other groups, like heme.

The number of possible proteins is essentially limitless. Let us consider only the number of possible proteins in the smallest size group with $M = 15,000$ having about 150 amino-acid residues. In making up the chain, any of 22 amino acids may follow any of 22 amino acids so there are 22 choices for the first residue, 22 for the second, . . . , or $(22)^{150} = 10^{202}$ different sequences possible.

20.2 ACID-BASE PROPERTIES OF PROTEINS

The fact that proteins are polyvalent acids and bases is an important feature of their structure. The pK's of titratable groups in proteins are given in Table 20.1. The pK of a given group in a protein varies over a range because of the effect of the neighboring parts of the protein and because of the electrostatic effect of the charges on the rest of the protein molecule. If the net charge on the protein molecule is positive, as it is in sufficiently acidic solutions, it is easier for a proton to get away from an acidic group and pK's are lowered. If the net charge on the protein molecule is negative, as it is in sufficiently alkaline solutions, it is harder for a proton to get away from an acidic group, and pK's are raised. As a result of these effects, parts of the titration curve of a protein may be steeper than the titration curves of the constituent amino-acid side chains. For each end of a polypeptide chain there will be an α-carboxyl group or an α-amino group. Additional electric

Table 20.1   pK's of Acidic Groups in Proteins (25°)

α—CO_2H	3.0–3.6
—CO_2H in aspartic and glutamic acids	3.0–4.7
imidazole in histidine	5.6–7.0

$$-C\!\!=\!\!=\!\!CH$$
$$\begin{array}{cc} | & | \\ N & N \\ \end{array}$$
$$\begin{array}{c} \diagdown \quad \diagup\!\!\diagup \\ C \\ H \end{array}$$

α—NH_2	7.5–8.4
—SH in cysteine	9.1–10.8
ε—NH_2 in lysine	9.4–10.6
phenolic hydroxyl in tyrosine	9.8–10.4
guanidine in arginine	12.0–13.0

$$\begin{array}{cc} H & NH \\ | & \| \\ -(CH_2)_3-N-C-NH_2 \end{array}$$

charges result from the binding of ions by the protein. At the isoelectric point the number of positive and negative charges is equal so that the protein does not move in an applied electric field.

The acidic and basic groups of a protein are concentrated on the outside of the protein. They are hydrophylic (water loving) groups as opposed to hydrophobic groups like aliphatic side chains and tend to increase the solubility of the protein.

At extreme pH values a protein molecule may become unwound and is said to be denatured. This process may be either reversible or irreversible, depending on the protein and the treatment. Denaturation is often irreversible because it is accompanied by a great decrease in solubility so that the protein precipitates from solution. Proteins may also be denatured by heating, even at neutral pH values, or by the addition of urea, $(H_2N)_2CO$, or other substances that weaken the bonding that holds the protein in its globular form.

20.3 ELECTROPHORESIS

Since protein molecules have net electric charges away from the isoelectric point they move when an electric field is applied to a solution; this process is referred to as electrophoresis when colloids or macromolecules are involved. The ionic mobility u, which is calculated in the same way as for other ions (see Section 11.5), depends on the net charge and the frictional coefficient.

One of the most important applications of electrophoresis is in the analysis of naturally occurring mixtures of colloids, such as proteins, polysaccharides, and nucleic acids, and of the products obtained in the course of fractionations to obtain purified components. In the electrophoresis experiment sharp boundaries between protein solution and buffer are formed in a special U-tube provided with electrodes, and the movement of boundaries is followed with a schlieren optical system (see Section 11.10). These experiments are usually carried out at 4° C, where water has its maximum density, so that the temperature gradient in the electrophoresis cell caused by the heating effect of the current is accompanied by the smallest possible density gradient. Density gradients horizontally across the cell tend to produce convection. Figure 20.1* gives an example of an electrophoresis pattern for normal human blood plasma in 0.10 ionic strength sodium diethylbarbiturate buffer of pH 8.6 after 150 min at 6.0 V cm^{-1} at 1°. The refractive-index gradient is plotted versus the distance in the cell, which is plotted horizontally. One pattern is obtained in the limb of the cell in which the proteins are moving downwards, and the other pattern in the limb in which the proteins are moving upwards. The positions of the initial boundaries are indicated by the rear ends of the arrows. The various proteins are represented by albumin, α_1, α_2, β, and γ globulins, and fibrinogen, ϕ. The area under a given peak is very nearly proportional to the concentration of the protein producing that boundary. Thus the percentage of albumin, for example, may be approximated by dividing

---

* R. A. Alberty, "Introduction to Electrophoresis," *J. Chem. Educ.*, **25**, 426, 619 (1948).

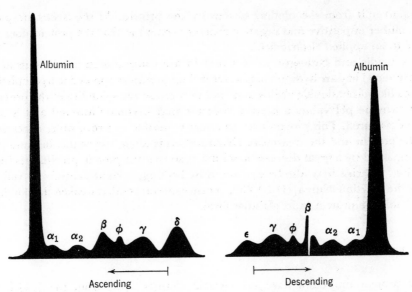

Fig. 20.1 Electrophoretic schlieren patterns for normal human blood plasma.

the area of the albumin peak by the total of the areas of all the protein peaks. The ϵ boundary in the descending pattern and the δ boundary in the ascending pattern are not caused by protein components, but are related to the changes in salt concentration which occur in ordinary transference experiments near the initial boundary position.

All the components of human plasma, as indicated by Fig. 20.1, have been separated by taking advantage of the differences in the solubilities of the various proteins in ethyl alcohol solutions of various pH values containing low concentrations of salts at low temperatures. Albumin, which represents about 60% of the plasma proteins by weight, is largely responsible for the osmotic pressure of plasma, since it has the lowest molecular weight of the plasma proteins.

If an electric field is applied across a membrane or a porous plug, it may be found that the liquid (usually water) moves through the pores. The relative motion of solvent and solid phase is the same as if the solid particles could move; that is, the solvent moves toward the cathode if the plug or membrane is negatively charged. In this phenomenon, which is referred to as *electroösmosis*, it may be considered that the solvent is carried along by the ions in the neighborhood of the solid surface which have a sign opposite to that of the surface.

20.4 HELICAL STRUCTURES IN POLYPEPTIDES AND PROTEINS

In the amide group the bond lengths and angles can be considered to be fixed. Only two angles have to be given per alpha carbon to specify the configuration of the chain. These two angles, ϕ and ψ, are defined in Fig. 20.2. Certain combinations of these angles are impossible, as shown by the two smaller figures on the right. In the example at the top the electron clouds of the nonbonded carboxyl oxygens would overlap, and in the example at the bottom the electron clouds of the nonbonded hydrogen atoms would overlap. Actually many possible combinations of angles ϕ and ψ are eliminated by contacts with other parts of the chain. When R groups are larger, there is greater steric hindrance and greater restrictions on possible values of ϕ and ψ.

If all the ψ values are the same in a chain, and all the θ values are the same, then a helical structure results. A peptide helix is characterized by the number n of amino-acid residues per turn of helix and by the distance d traversed parallel to the helix axis per amino-acid residue.

The α helix is sufficiently important to show the structures of the right-handed and left-handed varieties in Fig. 20.3. It is primarily the right-handed helix that

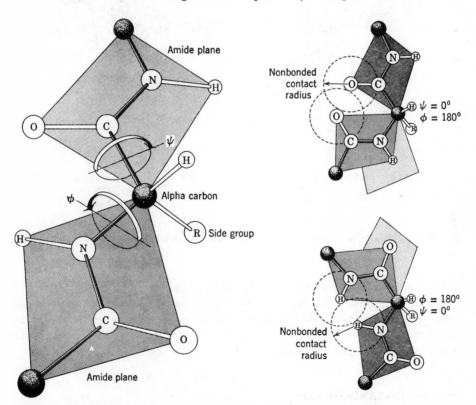

Fig. 20.2 Orientations of peptide planes in the polypeptide chain. The relative orientations of two successive planes are defined by the angles ϕ (C_α—N bond) and ψ (C_α—C bond). The two orientations shown on the right are impossible because of the overlap of carbonyl oxygens (top) and overlap of hydrogens (bottom). (From R. E. Dickerson and I. Geis, *The Structure and Action of Proteins*, W. A. Benjamin, Inc., Menlo Park, California, copyright (c) 1969 by Dickerson and Geis.)

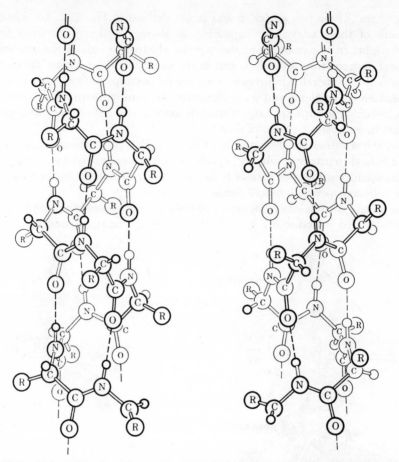

Fig. 20.3 Alpha helix of a polypeptide. A left-hand helix is shown on the left and a right-hand helix on the right.

exists in proteins. The α helix is held together by hydrogen bonds between a carbonyl oxygen and the N—H of the fourth residue along the chain. This produces an especially stable structure with $n = 3.6$ residues per turn. In the direction of the axis of the helix there is a residue every 1.5 Å. There is the minimum steric hindrance in polyglycine.

20.5 HELIX-RANDOM COIL TRANSITIONS IN POLYPEPTIDES

A number of synthetic polypeptides show structural transitions in solution when there is a change in temperature or pH. A single polypeptide chain can exist in three states in solution, as illustrated in Fig. 20.4.

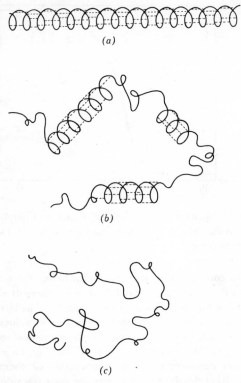

Fig. 20.4 Configurational states of a polypeptide. (*a*) helix, (*b*) intermediate form, (*c*) random coil (from L. Peller, *J. Phys. Chem.*, **63**, 1194 (1959)).

If it were possible to have a polypeptide chain in a vacuum the helix (*a*) would be the stable form at low temperature and the random coil form *c* would be the stable form at high temperatures. For the helix–random coil reaction with ΔH positive and ΔS positive, ΔG would be positive at low temperatures and negative at a sufficiently high temperature. However, the polypeptide interacts strongly with solvent, and as a result the random coil may have a smaller entropy than the helix in some solvents and a lower enthalpy. In such a solvent raising the temperature would cause the polypeptide to undergo a transition from random coil to helix. Such a transition for poly-γ-benzyl-*L*-glutamate in a mixture of dichloroacetic acid and dichloroethane is shown in Fig. 20.5.

At an intermediate temperature the configurational state would be that represented in Fig. 20.4*b*, in which there are alternating random coil and helical regions. The transition from helix to random coil may be much sharper than for a simple chemical equilibrium. In other words large changes in molecular properties may occur as a result of small changes in temperature, pressure, pH, or other environmental conditions. The change takes place over a narrow range of the environmental variable because of its cooperative nature. In a polypeptide the formation of the first helical section is the least probable. Once the first loop has formed further helix forms more easily. This may be understood in molecular terms as follows: to form the first loop of a helix the carbonyl oxygen of the *i*th residue must hydrogen bond with the NH of the $(i + 4)$th residue. This requires that six ϕ, and six ψ

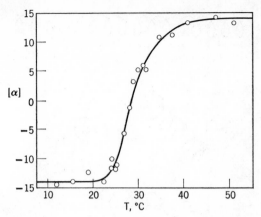

Fig. 20.5   Temperature dependence of the optical rotation of poly-γ-benzyl-L-glutamate ($M = 350,000$) in dichloroacetic acid-dichloroethane (80:20) (P. Doty and J. T. Yang, *J. Am. Chem. Soc.*, **78**, 499 (1956)).

angles have the values corresponding with an alpha helix. Thus the formation of the first loop is unfavorable from an entropy standpoint and is energetically favorable because a single hydrogen bond is formed. However, the addition of helix to a helical sequence requires that only one ϕ and one ψ be twisted to the needed values per hydrogen bond formed. Thus an additional helix tends to form on existing helical sections rather than in the midst of a random coil. As a consequence transitions from, for example, helix to coil configuration occur over narrower ranges of temperature or compositions than would otherwise be the case. Ising developed the theory for such transitions and applied it to ferromagnetism.

20.6   MOLECULAR WEIGHT DISTRIBUTIONS OF CONDENSATION POLYMERS

The simplest type of polymerization is linear polymerization in which monomer units are hooked end-to-end to form a chain without branches. For example, a hydroxyacid, HO—R—CO_2H, can be polymerized to form a polyester

$$
\begin{array}{ccccccc}
 & O & & O & & & O \\
 & \| & & \| & & & \| \\
\text{HO—R—C—O—R—C—} & \cdots & \text{—O—R—C—OH}
\end{array}
$$

with the elimination of water. The number of monomer units in a chain is called the *degree of polymerization*, i. Thus $i = M/M_0$, where M is the molecular weight of the polymer, and M_0 is the residue weight of the monomer.

A given sample of polyester contains a mixture of molecules, and we can describe this distribution by specifying the fraction of molecules of each different degree of polymerization. The molecular weight distribution may be calculated* for the idealized case that the

---

* P. J. Flory, *J. Am. Chem. Soc.*, **58**, 1877 (1936). P. J. Flory, *Principles of Polymer Chemistry*, Cornell University Press, Ithaca, 1953.

reactivity of the functional groups (carboxyl and hydroxyl for a polyester) is independent of the size of the molecule.

Suppose we start with n_0 monomer molecules AB, for example $HORCO_2H$, and imagine that the condensation polymerization continues until the number of unreacted A (or B) groups remaining is n. The extent of reaction p is the fraction of A (or B) groups which have reacted

$$p = \frac{n_0 - n}{n_0} = 1 - \frac{n}{n_0} \qquad (20.1)$$

The average number of monomer units in the polymer molecules, $\bar{X}_n$ (number-average degree of polymerization), is equal to the initial number of molecules n_0 divided by the number of molecules remaining. Since each independent molecule has a free A (or B) group,

$$\bar{X}_n = \frac{n_0}{n} = \frac{1}{1 - p} \qquad (20.2)$$

where the second form has been obtained by use of equation 20.1. For the average number of monomer units in the polymer molecules to be 100, it is evident that the extent of reaction p will have to be equal to 0.99.

To obtain the distribution of molecular weights, consider the probabilities π_i of finding molecules of various degrees of polymerization i in a partially polymerized sample. The probability π_1 of finding an unreacted A (or B) group in the mixture is $1 - p$. The probability π_2 that an AB has reacted with another AB to form ABAB is $p(1 - p)$ since the probability of two independent events is the product of the two independent probabilities for having a bond p and not having a bond $1 - p$. The probability π_3 of finding ABABAB, that is, two successive bonds, is $p^2(1 - p)$. To generalize, consider

$$\begin{aligned}
\pi_1 &= 1 - p \\
\pi_2 &= p(1 - p) \\
\pi_3 &= p^2(1 - p) \\
\pi_i &= p^{i-1}(1 - p)
\end{aligned} \qquad (20.3)$$

One test of these probabilities is that the sum of the probabilities of all of the different chain lengths in a sample ($i = 1$ to ∞) must equal unity

$$\begin{aligned}
\sum_{i=1}^{\infty} \pi_i = \sum_{i=1}^{\infty} p^{i-1}(1 - p) &= (1 - p) \sum_{i=1}^{\infty} p^{i-1} \\
&= (1 - p)(1 + p + p^2 + \cdots) \\
&= \frac{1 - p}{1 - p} = 1
\end{aligned} \qquad (20.4)$$

Since $(1 - p)^{-1} = 1 + p + p^2 + p^3 + \cdots$. The quantity π_i may be interpreted as the mole fraction of i-mer. As may be seen from Fig. 20.6a the mole fraction decreases steadily with increase in degree of polymerization i.

It is of perhaps greater interest to know the fraction of monomer units that have been incorporated into polymer of a particular degree of polymerization. The probability W_i of finding a monomer unit in an i-mer is equal to the fraction of monomer units in i-mers.

$$W_i = \frac{in_i}{n_0} \qquad (20.5)$$

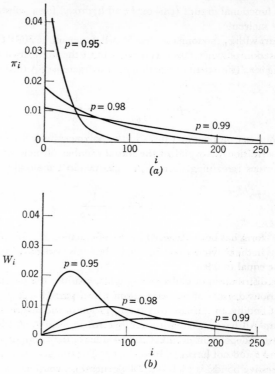

Fig. 20.6 (*a*) Mole fraction distribution of condensation polymer for extents of reaction *p* of 0.95, 0.98, and 0.99. (*b*) Weight fraction distribution of condensation polymer for extents of reaction *p* of 0.95, 0.98, 0.99.

where n_i is the number of *i*-mer molecules in the sample and n_0 is the number of monomer units initially present. The weight contributed by each monomer unit is the same, excepting the units that form the head and tail of the chain, and so W_i represents the weight fraction of *i*-mer. Since $n_i/n = p^{i-1}(1-p)$; then

$$W_i = \frac{ip^{i-1}(1-p)n}{n_0} \tag{20.6}$$

Introducing $n/n_0 = 1 - p$ from equation 20.1, we have

$$W_i = ip^{i-1}(1-p)^2 \tag{20.7}$$

In contrast with the mole fraction of *i*-mer, π_i, the weight fraction *i*-mer, W_i, goes through a maximum. The weight fractions are plotted for several extents of reaction in Fig. 20.6*b*. (These plots correspond with those in Fig. 20.6*a*.) The essential correctness of these distributions for condensation polymers has been confirmed by fractionating polymer samples by solubility methods into narrow ranges of degree of polymerization and determining the amounts and molecular weights of the various fractions.

Synthetic high polymers, and some naturally occurring polymers, have a distribution of molecular weights which may be represented by a graph such as Fig. 20.7. The molecular

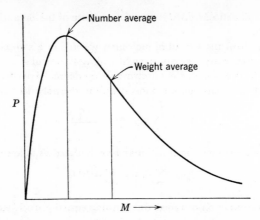

Fig. 20.7 Distribution of molecular weights and types of average molecular weight. The ordinate gives the probability of occurrence of molecules with a molecular weight of M.

weight distribution is, of course, not continuous, but if the molecular weights are high the distribution can be considered to be continuous. The ordinate of Fig. 20.7 is the *probability density P* for molecular weight. The quantity $P\,dM$ is the probability of finding molecules having molecular weights in the range M to $M + dM$ (cf., the discussion of the probability density function for molecular velocities). The probability of finding molecules in the range M to $M + dM$ is simply the fraction of the molecules having molecular weights in this range.

When there is a range of molecular weights, different experimental methods yield different types of average molecular weights. The *number average molecular weight M_n* is equal to the weight of the whole sample divided by the number of molecules in it.

$$M_n = \frac{\sum_i n_i M_i}{\sum_i n_i} \tag{20.8}$$

Here n_i is the number of molecules of molecular weight M_i per gram of dry polymer. This type of average is obtained from the measurement of the osmotic pressure of solutions containing the polymer, since this property is dependent on the number of molecules of solute per unit volume.

The *weight average molecular weight* weights molecules proportionally to their molecular weight in the averaging process; that is, the molecular weight M_i is multiplied by the weight $n_i M_i$ of material of that molecular weight rather than by the number of molecules. The weight average molecular weight M_w is defined by

$$M_w = \frac{\sum_i n_i M_i^2}{\sum_i n_i M_i} \tag{20.9}$$

This type of average is obtained by the study of light scattering (Section 20.13) because the amount of light scattered depends on the size as well as the number of particles.

The weight average molecular weight must always be greater than the number average molecular weight, except for a sample in which all the molecules have the same weight, so that $M_n = M_w$. The ratio M_w/M_n is a useful measure of the breadth of the molecular-weight distribution.

Some information about the spread of molecular weights in a sample may be obtained by comparing the number average and weight average molecular weights. The number average molecular weight is simply the number average degree of polymerization (equation 20.2) times the molecular weight of a monomer unit in the polymer M_0.

$$M_n = \sum \pi_i i M_0 = \frac{M_0}{1 - p} \tag{20.10}$$

The weight average molecular weight M_w may be calculated as follows:

$$M_w = \sum i M_i = \sum_i i W_i M_0 \tag{20.11}$$

where W_i is the weight fraction of i-mer. Substituting equation 20.7, we see that

$$M_w = M_0 (1 - p)^2 \sum i^2 p^{i-1} \tag{20.12}$$

$$= M_0 \frac{1 + p}{1 - p}$$

since

$$\sum_i i^2 p^{i-1} = \frac{1 + p}{(1 - p)^3} \tag{20.13}$$

The ratio of these average molecular weights is $M_w/M_n = 1 + p$. As we have seen from Fig. 20.6a and b it is necessary for p to be in the range 0.99 to 1 for high polymer to be formed; thus the weight average molecular weight M_w is twice the number average molecular weight M_n, for a high polymer produced by condensation. To obtain high molecular weight material it is important to eliminate traces of monofunctional reactants (acid or alcohol) which cause termination at one end of the chain or the other.

20.7 CONFIGURATIONS OF POLYMER CHAINS

A long polymer chain may assume an enormous number of configurations of essentially identical energy because of the possibility of rotation about single bonds. The average configuration may be calculated for various models of the polymer chain (i.e., certain bond lengths and bond angles) using statistical methods. These theoretical results are of great importance for an understanding of rubberlike elasticity (Section 19.24) and of hydrodynamic and thermodynamic properties of dilute polymer solutions. The simple theory of polymer configurations is very much like the theory of the random flight of a gas molecule with a constant free path, with the free path corresponding to the bond distance in the polymer.

To discuss the probabilities of various configurations it is convenient to imagine a vector r from one end of a polymer chain to the other. It can be shown* that for a freely jointed chain in three dimensions the probability $W(r)\, dr$ that this vector will have a length r is given by

$$W(r)\, dr = (\beta/\pi^{1/2})^3 e^{-\beta^2 r^2} 4\pi r^2\, dr \tag{20.14}$$

---

* P. Flory, *Principles of Polymer Chemistry*, Cornell University Press, Ithaca, N.Y., 1953, p. 404.

where

$$\beta = \frac{\sqrt{\frac{3}{2}}}{n^{\frac{1}{2}}l} \tag{20.15}$$

where l is the bond length and n is the number bonds. The full extension length of the molecule would be nl. A plot of $W(r)$ versus r goes through a maximum just like the plot for probability density of molecular speed versus speed (Section 9.3). The most convenient measure of the average end-to-end distance is the root-mean-square length

$$\langle r^2 \rangle^{\frac{1}{2}} = \left[\int_0^\infty r^2 W(r) \, dr \right]^{\frac{1}{2}} \tag{20.16}$$

$$= n^{\frac{1}{2}}l$$

Thus it is seen that the root-mean-square end-to-end distance for this simple model of a polymer molecule is proportional to the square root of the number of segments and the proportionality factor is the bond length.

Actually the extension of a real polymer chain depends very much on the solvent. When a polymer is dissolved in a good solvent, for example when polystyrene is dissolved in toluene, the segment-solvent contacts are energetically more favorable than segment-segment contacts. This causes the molecule to stretch out in solution so that segment-segment contacts are minimized, and the Gibbs free energy is a minimum. On the other hand, when a polymer is dissolved in a poor solvent, the segment-segment contacts are favored over segment-solvent contacts. This causes the molecule to tend to ball up in solution in opposition to the thermal motions that tend to extend it. If the solvent is too poor, the polymer will not dissolve.

Intermediate between good and poor solvents is the theta (θ) solvent in which the dimensional contraction resulting from the poorer solvency exactly cancels the increase in size due to the finite volumes of the segments. Cyclohexane is a theta solvent for polystyrene at $34°$ C. For a theta solvent the value of B in the osmotic pressure equation 20.31 is equal to zero.

20.8 INTRINSIC VISCOSITY

The experimental determination of the coefficient of viscosity has been discussed earlier (Section 11.1) in connection with other irreversible processes. Large molecules make considerable contributions to the viscosities of solutions, and since viscosity is readily measured this provides an important method for studying macromolecules.

The theoretical background for the use of viscosity measurements to study macromolecules starts with Einstein's derivation, in 1906, of an equation relating the viscosity η of a dilute suspension of small rigid spheres to the volume fraction ϕ occupied by the spherical particles. Einstein derived

$$\lim_{\phi \to 0} \left[\frac{(\eta/\eta_0) - 1}{\phi} \right] = \frac{\eta_{sp}}{\phi} = \frac{5}{2} \tag{20.17}$$

where η_0 is the viscosity of the solvent. The quantity $(\eta/\eta_0) - 1$ is of frequent occurrence in the theory of viscosity and is known as the specific viscosity, η_{sp}.

It is interesting to note that the viscosity (for the dilute suspensions covered by this equation) is independent of the size of the spheres and is dependent only on the volume fraction ϕ that they occupy.

Since most synthetic high-polymer molecules consist of chains of monomer units and are threadlike and flexible in solution, they cannot be considered to be solid spheres. Because of the difficulty of knowing the fraction of the volume occupied by high-polymer molecules in solution and the fact that the *reduced viscosity*, η_{sp}/c, is not independent of concentration, it is necessary to utilize the intrinsic viscosity $[\eta]$, defined by

$$[\eta] = \lim_{c \to 0} \frac{\eta_{sp}}{c} \tag{20.18}$$

where c is the concentration of the high polymer in grams per deciliter. The intrinsic viscosity is obtained by plotting the ratio of specific viscosity to concentration against concentration for a series of solutions and extrapolating to zero concentration.

In solution certain polymer molecules have a randomly kinked and coiled configuration that is constantly changing; in other cases the molecules are not so flexible. The "size" of the molecule, measured for example by the average end-to-end distance of the chain, is not directly proportional to the molecular weight. The intrinsic viscosity depends, however, on the molecular weight. For a series of samples of the same polymer in a given solvent and at a constant temperature the following empirical relation is obeyed quite well:

$$[\eta] = KM^a \tag{20.19}$$

Here K and a are constants that may be determined by measuring the intrinsic viscosities of a series of samples of a polymer for which the molecular weights have been measured by another method, say by use of light scattering. Since the values of K and a are known for a large number of solvents and temperatures, this method for determining molecular weights is widely used because of the simple apparatus required.

The value of a for a theta solvent (Section 20.7) is about 0.5. For better solvents a is larger, and for a good solvent the polymer molecule behaves like a random coil, and a is about 0.8.

The intrinsic viscosity of DNA depends somewhat on the ionic strength; at high ionic strengths a lower value of $[\eta]$ is obtained because the electrostatic shielding of the negatively charged backbone phosphate groups reduces the rigidity of the molecule observed at low ionic strengths. However, this effect is much less with DNA than with flexible polyelectrolytes. High molecular weight DNA may be degraded (reduced in molecular weight) by the shear gradients encountered in pipetting solutions. In order to measure the viscosities of DNA solutions at very low shear it has been necessary to use rotating cylinder, or Couette, viscometers.

20.9   VELOCITY SEDIMENTATION

Particles of finely divided solid sediment in a fluid if their density is higher than that of the suspension medium. Their equivalent radius can be calculated from the rate of sedimentation using the expression for the frictional coefficient, $f = 6\pi\eta r$, given in Section 11.1.

Dissolved molecules tend to sediment in the earth's gravitational field or to float upwards, depending on their density relative to that of the solvent, but this tendency is counteracted by the translational kinetic energy of the molecules. However, sufficiently powerful ultracentrifuges have been built to cause even molecules as small as those of sucrose to sediment at measurable rates. Svedberg[*] was the leader in the development of ultracentrifuges, which he defined as centrifuges adapted for quantitative measurements of convection-free and vibration-free sedimentation. There are two distinct types of ultracentrifuge experiments: (1) those where the velocity of sedimentation of a component of the solution is measured (sedimentation velocity), and (2) those where the redistribution of molecules is determined at equilibrium (sedimentation equilibrium).

The acceleration of a particle in a centrifugal field is equal to $\omega^2 r$, where ω is the velocity of the centrifuge in radians per second (that is, 2π times the number of revolutions per second) and r is the distance of the particle from the axis of rotation. Velocity ultracentrifuges in which r is about 6 cm are commonly operated at 60,000 rpm or 1000 rps, and so the acceleration is

$$\omega^2 r = (2\pi 1000 \text{ s}^{-1})^2 (0.06 \text{ m}) = 2.36 \times 10^6 \text{ m s}^{-2}$$

Since the acceleration of the earth's field is 9.80 m s^{-2}, the acceleration is 240,000 times greater than in the earth's field.

A solution to be studied in the velocity ultracentrifuge is placed in a cell with thick quartz windows. The cell has a sector shape when viewed at right angles to the plane of rotation of the centrifuge rotor, since the sedimentation takes place radially. As the high-molecular-weight component throughout the solution sediments, a moving boundary is formed, behind which there is only solvent. The movement of such boundaries in the cell may be followed with the schlieren optical system, as was mentioned in connection with diffusion (Section 11.10).

Figure 20.8 shows the schlieren patterns for an ultracentrifuge experiment with the enzyme fumarase at 50,400 rpm. The second and third photographs were taken 35 and 70 min later than the top photograph. If additional components with different rates of sedimentation had been present, additional peaks would be evident in the schlieren photograph. Thus, the ultracentrifuge is useful in analyzing complex mixtures such as blood plasma.

The velocity of sedimentation, dr/dt, divided by the centrifugal acceleration $\omega^2 r$, is called the *sedimentation coefficient* S.

$$S = \frac{dr/dt}{\omega^2 r} \tag{20.20}$$

---

[*] T. Svedberg and K. O. Pedersen, *The Ultracentrifuge*, Oxford University Press, Oxford, 1940.

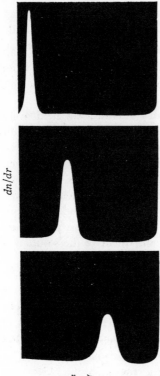

Fig. 20.8 Schlieren patterns for an ultracentrifuge experiment with fumarase at 50,400 rpm. The lower photographs were taken 35 and 70 min later than the top photograph. The protein is dissolved in pH 6.8 phosphate buffer.

The sedimentation coefficients of proteins fall in the range 10^{-13} s to 200×10^{-13} s, and the unit 10^{-13} s is called a *svedberg*.

If a boundary is r_1 centimeters from the axis of the centrifuge at time t_1 and r_2 centimeters from the axis at time t_2, the sedimentation coefficient may be calculated from

$$S = \frac{1}{\omega^2(t_2 - t_1)} \ln \frac{r_2}{r_1} \tag{20.21}$$

which is obtained by integrating equation 20.20.

Example 20.1   In the ultracentrifuge experiment illustrated in Fig. 20.8 the distance from the axis of the ultracentrifuge to the boundary was 5.949 cm in the top photograph and 6.731 cm in the bottom photograph taken 70 min later. Since the speed of the rotor was 50,400 rpm, $\omega^2 = 2.79 \times 10^7$. Using equation 20.21, we see that

$$S = \frac{1}{\omega^2(t_2 - t_1)} \ln \frac{r_2}{r_1} = \frac{2.303 \log{(6.731/5.949)}}{(2.79 \times 10^7)(60)(70)}$$

$$= 10.5 \times 10^{-13} \text{ s}$$

This is the sedimentation coefficient at 28.2°, the temperature of the experiment. Making a correction to 20° in water, taking into account the change of viscosity and density, a value of 8.90 svedbergs is obtained.

The sedimentation coefficient by itself cannot be used to determine the molecular weight of the sedimenting component unless the molecules are spherical. If both the sedimentation and diffusion coefficients are measured, however, the molecular weight may be calculated without any assumptions whatsoever about the shape. The equation on which this calculation is based may be derived by setting the force of the centrifugal field on the particle equal to the frictional force, $f(dr/dt)$, where f is the frictional coefficient of the molecule and dr/dt is the speed of sedimentation. The force of the field on a particle of mass m and partial specific volume $\bar{v}$ suspended in a medium of density ρ is

$$m(1 - \bar{v}\rho)\omega^2 r \doteq \frac{M}{N_A}(1 - \bar{v}\rho)\omega^2 r \tag{20.22}$$

where $(1 - \bar{v}\rho)$ is the buoyancy factor.

The sedimenting molecule or particle will be accelerated by the field until its velocity is such that the frictional force is equal to the force of the field

$$f\frac{dr}{dt} = \frac{M}{N_A}(1 - \bar{v}\rho)\omega^2 r \tag{20.23}$$

or

$$\frac{dr/dt}{\omega^2 r} = S = \frac{M(1 - \bar{v}\rho)}{N_A f} \tag{20.24}$$

Since the velocity of sedimentation is so low that there is no appreciable orientation of the molecules, the frictional coefficient involved in sedimentation is taken to be the same as that involved in diffusion. Introduction of equation 11.34 into equation 20.24 yields

$$M = \frac{RTS}{D(1 - \bar{v}\rho)} \tag{20.25}$$

To calculate the molecular weight from measured values of S and D it is necessary to correct sedimentation and diffusion coefficients to the same temperature, usually 20°, and if S and D depend appreciably on concentration, to zero concentration. Equation 20.25 has probably been the most widely used in the calculation of molecular weights of proteins, and the wide range of molecular weights which can be obtained by this method is indicated by Table 20.2.

Table 20.2   Physical Constants of Proteins at 20° in Water and Molecular Weights

Protein	$S \times 10^{13}$, s	$D \times 10^{11}$, m^2 s^{-1}	$\bar{v}$, cm^3 g^{-1}	M, g mol^{-1}
Beef insulin	1.7	15	0.72	12,000
Lactalbumin	1.9	10.6	0.75	17,400
Ovalbumin	3.6	7.8	0.75	44,000
Serum albumin	4.3	6.15	0.735	64,000
Serum globulin	7.1	4.0	0.75	167,000
Urease	18.6	3.4	0.73	490,000
Tobacco mosaic virus	185	0.53	0.72	40,000,000

Example 20.2 Using the data of Table 20.2 the molecular weight of serum albumin may be calculated as follows (the density of water at $20°$ is 0.9982 g cm^{-2}).

$$M = \frac{RTS}{D(1 - \bar{v}\rho)}$$

$$M = \frac{(8.31 \text{ J K}^{-1} \text{ mol}^{-1})(293 \text{ K})(4.3 \times 10^{-13} \text{ s})}{(6.15 \times 10^{-11} \text{ m}^2 \text{ s}^{-1})[1 - (0.735 \times 10^{-3} \text{ m}^3 \text{ kg}^{-1})(0.9982 \times 10^3 \text{ kg m}^{-3})]}$$

$$= 64.0 \text{ kg mol}^{-1}$$

$$= 64,000 \text{ g mol}^{-1}$$

In measuring the sedimentation coefficient of DNA significant concentration effects are encountered at concentrations as low as 10 mg per liter. Taking advantage of the strong nucleotide absorption around 260 nm the velocity of the sedimenting boundary may be measured even at this low concentration using an ultraviolet absorption optical system. In another method for studying such dilute solutions a thin layer of DNA solution may be placed on a preformed density gradient of CsCl solution created in a ultracentrifuge cell by the centrifugal field causing the sedimentation. In these experiments the density gradient stabilizes the system from convection.

20.10 EQUILIBRIUM ULTRA-CENTRIFUGATION

In contrast with velocity sedimentation this is a thermodynamic measurement that can yield a value for M all by itself. In an equilibrium ultracentrifuge experiment the centrifugation is continued until the tendency of the molecules to sediment is balanced by the opposing tendency to diffuse into the region of lower concentration. Since this redistribution in the solution is an equilibrium phenomenon, the relation between the concentrations at two levels and the molecular weight may be derived by the methods of thermodynamics. For ideal solutions the molecular weight may be calculated from

$$M = \frac{2RT \ln (c_2/c_1)}{(1 - \bar{v}\rho)\omega^2(r_2^2 - r_1^2)} \tag{20.26}$$

where c_1 is the concentration at distance r_1 from the axis, c_2 is the concentration at distance r_2, ω is the angular velocity of the rotor (radians per second), and ρ is the density of the solution. Optical methods are used to determine c_2 and c_1.

Sedimentation equilibrium may also be used to set up a density gradient due to a low molecular weight substance (for example CsCl) so that a mixture of macromolecules may be separated into bands according to their density. In a density gradient a substance will tend to come to rest at a level where its density is the same as the solution. In 1957 Meselson, Stahl, and Vinograd used CsCl gradients (formed from about 7 M solutions by sedimentation equilibrium) to separate different types of DNA. By this method it is also possible to separate

DNA labeled with ^{15}N from unlabeled DNA. By use of this technique and isotopic nitrogen Meselson and Stahl were able to show in 1958 that when DNA is replicated each daughter double helix consists of one strand from the parent DNA and one newly synthesized strand.

20.11 DIFFUSION

Methods for measuring the diffusion coefficient of gases (Chapter 9) and solutes in liquids (Chapter 11) were discussed earlier. The diffusion coefficient of a homogeneous high polymer is of special importance because it may be combined with the sedimentation coefficient to obtain the molecular weight. The diffusion coefficient D is related to the frictional coefficient f of a molecule by equation 11.34, which is

$$D = \frac{RT}{N_A f} \tag{20.27}$$

For the special case of spherical particles, molecular weights may be estimated from diffusion coefficients alone. Combining equations 11.1 and 11.34 yields

$$D = \frac{RT}{N_A 6\pi\eta r} \tag{20.28}$$

Thus, for spherical particles the radius may be calculated from the measured diffusion coefficient. Since it is more familiar to think of size in terms of molecular weight, the relation between D and the molecular weight of the spherical particle may be derived from equation 20.28 by introducing

$$\frac{M\bar{v}}{N_A} = \frac{4}{3}\pi r^3 \tag{20.29}$$

where $\bar{v}$ is the partial specific volume (that is, $\bar{V}$, divided by M).

$$D = \frac{RT}{N_A 6\pi\eta}\left(\frac{4\pi N_A}{3M\bar{v}}\right)^{1/3} \tag{20.30}$$

For spherical particles, then, the diffusion coefficient is inversely proportional to the cube root of the molecular weight. Of course if the particles or molecules are not spherical, the value of the molecular weight calculated from equation 20.30 will not be correct. This equation, however, does give the maximum molecular weight that is consistent with a given D and $\bar{v}$. For a nonspherical particle the molecular weight would be smaller.

Example 20.3 Using the diffusion coefficient and $\bar{v}$ for ovalbumin given in Table 20.2, calculate the maximum molecular weight this protein might have. The coefficient of viscosity of water at $20°$ is 0.001005 Pa s. Using equation 20.30 we see that

$$D = 7.8 \times 10^{-11}\,\mathrm{m^2\,s^{-1}}$$
$$= \frac{(8.31\,\mathrm{J\,K^{-1}\,mol^{-1}})(293\,\mathrm{K})}{(6.02\times10^{23}\,\mathrm{mol^{-1}})(6\pi)(0.001005\,\mathrm{J\,m^{-3}\,s})}\left[\frac{4\pi(6.02\times10^{23}\,\mathrm{mol^{-1}})}{3\,M(0.75\times10^{-3}\,\mathrm{m^3\,mol^{-1}})}\right]^{\frac{1}{3}}$$
$$M = 69.0\,\mathrm{kg\,mol^{-1}}$$
$$= 69{,}000\,\mathrm{g\,mol^{-1}}$$

The ovalbumin molecule is not spherical, as shown by the fact that the actual molecular weight calculated from the sedimentation and diffusion coefficients is 44,000.

20.12   OSMOTIC PRESSURE

Measurements of osmotic pressure are useful in the molecular weight range up to about 500,000 g mol^{-1}, provided good enough semipermeable membranes may be obtained. The theory of this method has been described in Section 4.7. If the solute is heterogeneous, the number average molecular weight is obtained.

In solutions of proteins or other colloidal electrolytes it is necessary to distinguish between the *total osmotic pressure*, which would be obtained with a membrane impermeable to both salt and protein, and the *colloid osmotic pressure*, which is obtained with a membrane permeable to salt ions but not to protein. The latter type of membrane is always used when it is desired to obtain the molecular weight of the protein or other colloidal electrolyte.

For colloidal electrolytes in solutions with low concentrations of electrolytes, the measured osmotic pressure is greater than that expected for the colloidal ions alone. This is a result of the fact that, although the salt ions may pass through the membrane, they will not be distributed equally at equilibrium. Donnan showed that because of the high molecular weight ion on one side of the membrane, the concentration of the small ion of the same sign as the macroion is lower on that side of the membrane than in the salt solution and that this is compensated by an increased concentration of the small ion of opposite charge. The Donnan effect may be reduced by increasing the salt concentration and, if possible, adjusting the pH to the isoelectric point (Section 20.1) of the colloidal electrolyte.

It is convenient to add a term to the simple osmotic pressure equation so that it can be used to represent data over a wider range of concentration:

$$\frac{\Pi}{c} = \frac{RT}{M} + Bc \qquad (20.31)$$

The value of B is dependent upon polymer-polymer interactions, and for particular solvents and temperatures may be equal to zero. In using equation 20.31 a plot of Π/c versus c is extrapolated to zero concentration. Then M is calculated from

$$\lim_{c\to 0}\frac{\Pi}{c} = \frac{RT}{M} \qquad (20.32)$$

20.13 LIGHT SCATTERING*

Pure fluids scatter light because of local fluctuations in density. When a solute is added the intensity of scattered light increases because there are in addition local fluctuations in concentration of solute that produce even larger fluctuations in refractive index. These fluctuations are related to the molecular weight, and so the measurement of light scattering offers a means of measuring M. The study of the angular dependence of light scattering yields information on axial ratio if one of the dimensions is comparable to the wavelength of the light used.

A simple apparatus for the measurement of the intensity of light scattered from a solution is shown in Fig. 20.9.

When a molecule is irradiated with plane polarized light an alternating electric dipole moment is induced in the molecule by the electric field of the radiation. If the molecule is isotropic (has the same properties in all directions), the induced dipole moment will be in the same direction as the electric field and will be proportional to the polarizability α (Section 14.12) and the electric field strength. The oscillating dipole that results in this way will radiate electromagnetic waves. The intensity of the scattered light is directly proportional to the square of the induced dipole moment, since it is proportional to the square of the scattered amplitude, and is therefore directly proportional to the square of the polarizability α.

Rayleigh showed that the intensity i (energy per cm^2 per sec) of unpolarized light scattered through an angle θ by a single molecule is given by

$$\frac{i}{I_0} = \frac{4\pi^4\alpha^2(1 + \cos^2\theta)}{\lambda^4 r^2} \tag{20.33}$$

where I_0 is the intensity of incident unpolarized light. The scattered intensity i is measured at distance r from the scattering molecule and at an angle θ measured

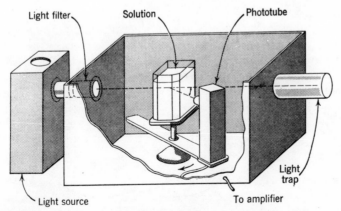

Fig. 20.9 Apparatus for measuring light scattered by a solution.

---

* K. A. Stacey, *Light Scattering in Physical Chemistry*, Academic Press, New York, 1956.

from the emergent beam, as indicated in Fig. 20.9. The scattered intensity is inversely proportional to the fourth power of the wave length λ of the light used. Thus, blue light is scattered to a greater extent than red light, and therefore the light transmitted through a suspension of particles is reddish. We are familiar with these effects in the blue sky and red sunset.

The dependence of intensity i on the scattering angle θ is shown in Fig. 20.10, which shows separately the scattered intensities from the two polarized components of the light from the source, adding up to give the scattered intensity given by equation 20.33.

The polarizability α of a molecule may be calculated from the rate of change of the refractive index with concentration. The refractive index n of a solution, of the solvent n_0, and the polarizability α of a molecule are related by equation 14.52, which may be written

$$n^2 - n_0^2 = \frac{4\pi N \alpha}{V} \tag{20.34}$$

where N is the number of molecules in volume V. The polarizability may be expressed in terms of the refractive index gradient by writing equation 20.34 as

$$\alpha = \frac{(n + n_0)(n - n_0)V}{4\pi N} = \frac{n_0}{2\pi}\left(\frac{dn}{dc}\right)\frac{M}{N_A} \tag{20.35}$$

where the second form is obtained by substituting $dn/dc = (n - n_0)/c$, concentration $c = MN/N_A V$, and using the approximation $2n_0$ for $n + n_0$, since we are considering dilute solutions. The refractive index increment dn/dc may be obtained from measurements of n as a function of c, and so it will be considered to be a known in the following equations. Inserting equation 20.35 in equation 20.33 and multiplying by the number of molecules per cm³ (that is, cN_A/M) yields the following equation for the intensity i scattered from solution in excess of that scattered by solvent.

$$\frac{i}{I_0} = \frac{2\pi^2 n_0^2 (1 + \cos^2\theta)(dn/dc)^2 Mc}{N_A \lambda^4 r^2} \tag{20.36}$$

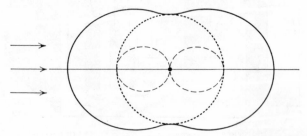

Fig. 20.10   Scattered intensities from light polarized in the plane of the source, sample, and observer (dashed line) and from light polarized perpendicular to this plane (dotted line). The outer solid line gives the sum of the two intensities that is, it gives the angular distribution that is characteristic of unpolarized light.

If this expression is integrated over all angles to obtain the total scattered intensity, it is found that the *transmitted* intensity I is given by

$$I = I_0 e^{-\tau x} \tag{20.37}$$

where τ is the turbidity, and x is the thickness of the cell in the direction of the incident beam. This is of the form of Beers' law for light absorption (Section 15.12). The result of this calculation is that

$$\tau = \frac{32\pi^3 n_0^2 (dn/dc)^2 Mc}{3 N_A \lambda^4} = HMc \tag{20.38}$$

where

$$H = \frac{32\pi^3 n_0^2 (dn/dc)^2}{3 N_A \lambda^4} \tag{20.39}$$

is a constant for a particular solvent, solute, and wavelength. Since the turbidity τ is proportional to molecular weight this method is more useful as the molecular weight increases, in contrast to osmotic pressure, which is inversely proportional to molecular weight.

The molecular weight of the scattering solute can be calculated from equation 20.38, but it is much more sensitive experimentally to determine the scattered intensity rather than the transmitted intensity. However, experimental results are usually expressed in terms of τ.

The above theory is for ideal solutions, and in practice it is necessary to plot Hc/τ versus c and extrapolate to zero concentration to obtain the value of Hc/τ that will yield the correct molecular weight from equation 20.38. In fact it can be shown that for a nonideal polymer solution for which the osmotic pressure (Section 20.12) is given by

$$\frac{\Pi}{cRT} = \frac{1}{M} + Bc + \cdots \tag{20.40}$$

the turbidity is given by

$$\frac{Hc}{\tau} = \frac{1}{M} + 2Bc + \cdots \tag{20.41}$$

If there are particles of different sizes having the same composition, the polarizability α is proportional to the mass of the particle. The intensity of the light scattered by a single particle is proportional to the square of its polarizability and therefore to the square of its mass. If the sample contains a distribution of molecular weights, the larger particles make a proportionately larger contribution to the scattering than an equal weight of smaller ones. It can be shown that for a heterogeneous polymer the weight average molecular weight (equation 20.9) is obtained from equation 20.41. The osmotic pressure yields the number average molecular weight.

If one of the dimensions of a molecule is comparable to the wavelength of light, the molecule no longer acts as a single point in scattering light, and there is interference between light waves scattered from different parts of the molecule.

It is found that large molecules scatter more light forward, in the direction of the beam, than the small molecules. Thus for larger molecules or particles the angular dependence of the light scattering gives information as to shape. The development of lasers (Section 18.11) has made it possible to study the line broadening of light scattered by solutions of macromolecules. Measurement of the spectrum of scattered light permits the determination of the diffusion coefficient.*

20.14 GEL PERMEATION CHROMATOGRAPHY

In gel permeation chromatography a solution of a polymer having a distribution of molecular weights is passed through a column packed with a rigid porous gel. The polymer network is swollen by the solvent and has pores of various sizes. As the liquid phase containing the polymer passes through the column, the polymer molecules diffuse into all parts of the gel not mechanically barred to them. The smaller molecules diffuse further into the gel and therefore spend more time in the solid phase of the column. The smaller polymer molecules therefore travel through the column more slowly than the larger polymer molecules that cannot diffuse into the gel because of their size. Thus a chromatographic separation by size is obtained. Information about molecular weight distribution is important because the properties of a polymeric material depend on the range of molecular weights as well as on the average molecular weight. Molecular weight distributions may also be obtained by use of the ultracentrifuge.

References

F. W. Billmeyer, *Textbook of Polymer Chemistry*, Wiley-Interscience, New York, 1957.

T. M. Birshtein and O. B. Ptitsyn, translated by S. N. Timasheff and M. J. Timasheff, *Conformations of Macromolecules*, Wiley-Interscience, New York, 1966.

V. A. Bloomfield, D. A. Crothers, and I. Tinoco, Jr., *Physical Chemistry of Nucleic Acids*, Harper & Row, New York, 1974.

R. E. Dickerson and I. Geis, *The Structure and Action of Proteins*, Harper and Row, New York, 1969.

P. J. Flory, *Principles of Polymer Chemistry*, Cornell University Press, Ithaca, 1953.

M. Kaufman, *Giant Molecules*, Doubleday and Co., Garden City, New York, 1968.

H. Morawetz, *Macromolecules in Solution*, Wiley-Interscience, New York, 1965.

K. J. Mysels, *Introduction to Colloid Chemistry*, Wiley-Interscience, New York, 1959.

D. Poland and H. A. Scheraga, *Theory of Helix-Coil Transitions in Biopolymers*, Academic Press, New York, 1970.

C. Tanford, *Physical Chemistry of Macromolecules*, Wiley, New York, 1961.

J. D. Watson, *Molecular Biology of the Gene*, W. A. Benjamin, Inc., New York, 1965.

J. W. Williams, *Ultracentrifugation of Macromolecules: Modern Topics*, Academic Press, New York, 1973.

---

* S. B. Dubin, J. H. Lunacek, and G. B. Benedek, *Proc. Natl. Acad. of Sciences*, **57**, 1164 (1967).

Problems

20.1 A turn of α helix (3.6 residues) is 5.41 Å in length measured parallel to the helix axis. If your hair grows 6 in. per year, how many amino-acid residues must be added to each α helix in the keratin fiber per second? *Ans. 32 residues per sec.*

20.2 A sample of polymer contains 0.50 mole fraction with molecular weight 100,000 and 0.50 mole fraction with molecular weight 200,000. Calculate (a) M_n and (b) M_w. *Ans. (a) 150,000, (b) 167,000 g mol^{-1}.*

20.3 For a condensation polymerization of a hydroxyacid in which 99% of the acid groups are used up, calculate (a) the average number of monomer units in the polymer molecules, (b) the probability that a given molecule will have the number of residues given by this value, and (c) the weight fraction having this particular number of monomer units. *Ans. (a) 100, (b) 3.7 × 10^{-3}, (c) 3.7 × 10^{-3}.*

20.4 In the condensation polymerization of a hydroxyacid with a residue weight of 200 it is found that 99% of the acid groups are used up. Calculate (a) the number average molecular weight and (b) the weight average molecular weight. *Ans. (a) 20,000, (b) 39,800.*

20.5 It is found that the root-mean-square end-to-end distance for a polymethylene polymer is 75 Å. Assuming the polymer is homogeneous with respect to chain length and that the chain is freely joined, what is the molecular weight? The C—C bond distance is 1.54 Å. *Ans. 33,200 g mol^{-1}.*

20.6 The relative viscosities of a series of solutions of a sample of polystyrene in toluene were determined with an Ostwald viscometer at 25°.

Concentration, g/100 cm³	0.249	0.499	0.999	1.998
η/η_0	1.355	1.782	2.879	6.090

The ratio η_{sp}/c is plotted against c and extrapolated to zero concentration to obtain the intrinsic viscosity. If the constants in equation 20.19 are $K = 3.7 \times 10^{-4}$ and $a = 0.62$ for this polymer, calculate the molecular weight. *Ans. 500,000 g mol^{-1}.*

20.7 The intrinsic viscosity of a sample of polystyrene at 34° C is 0.84 dl g^{-1} (deciliters per gram) in toluene and 0.40 dl g^{-1} in cyclohexane, a theta (θ) solvent. Calculate the viscosity average molecular weight using the following empirical relation for polystyrene in toluene at 34°:

$$[\eta] = 1.15 \times 10^{-4} M^{0.72}$$

where the intrinsic viscosity is in dl g^{-1}. *Ans. 230,000 g mol^{-1}.*

20.8 Calculate the sedimentation coefficient of tobacco mosaic virus from the fact that the boundary moves with a velocity of 0.454 cm hr^{-1} in an ultracentrifuge at a speed of 10,000 rpm at a distance of 6.5 cm from the axis of the centrifuge rotor. *Ans. 185 × 10^{-13} s.*

20.9 The sedimentation and diffusion coefficients for hemoglobin corrected to 20° in water are 4.41 × 10^{-13} s and 6.3 × 10^{-11} m² s^{-1}, respectively. If $\bar{v} = 0.749$ cm³ g^{-1} and $\rho_{H_2O} = 0.998$ g cm^{-3} at this temperature, calculate the molecular weight of the protein. If there is 1 g atom of iron per 17,000 g of protein, how many atoms of iron are there per hemoglobin molecule? *Ans. 68,000 g mol^{-1}, 4.*

20.10 The diffusion coefficient for serum globulin at 20° in a dilute aqueous salt solution is 4.0 × 10^{-11} m² s^{-1}. If the molecules are assumed to be spherical, calculate their molecular weight. Given: $\eta_{H_2O} = 0.001005$ Pa s at 20° and $\bar{v} = 0.75$ cm³ g^{-1} for the protein. *Ans. 512,000 g mol^{-1}.*

20.11   Estimate the diffusion coefficient in water at 25° of a spherical molecule with a molecular weight of 10^5 and $\bar{v} = 0.73$ cm^3 g^{-1}.   *Ans.* $D = 7.96 \times 10^{-11}$ m^2 s^{-1}.
20.12   Calculate the diffusion coefficient in water at 25° for spherical particles of 5 Å radius.   *Ans.* 4.88×10^{-10} m^2 s^{-1}.
20.13   The protein human plasma albumin has a molecular weight of 69,000. Calculate the osmotic pressure of a solution of this protein containing 2 g per 100 cm^3 at 25° in (*a*) Torr and (*b*) millimeters of water. The experiment is carried out using a salt solution for solvent and a membrane permeable to salt.   *Ans.* (*a*) 5.39 Torr, (*b*) 73.2 mm H$_2$O.
20.14   The following osmotic pressures were measured for solutions of a sample of polyisobutylene in benzene at 25°:

c, g/100 cm^3	0.500	1.00	1.50	2.00
Π, g/cm^2	0.505	1.03	1.58	2.15

Calculate the number average molecular weight from the value of Π/c extrapolated to zero concentration of the polymer. [The pressures may be converted into atmospheres dividing by (76 cm atm^{-1})(13.53 g cm^{-3}) = 1028 g cm^{-2} atm^{-1}.]   *Ans.* 256,000 g mol^{-1}.
20.15   A beam of sodium *D* light (589 nm) is passed through 100 cm of an aqueous solution of sucrose containing 10 g sucrose per 100 cm^3. Calculate I/I_0, where I_0 is the intensity that would have been obtained with pure water, given that $M = 342.30$ and $dn/dc = 0.15$ g^{-1} cm^3 for sucrose. The refractive index of water at 20° is 1.333 for the sodium *D* line.   *Ans.* 0.9938.

20.16   Calculate the (*a*) number and (*b*) weight average molecular weights for the following mixture of high-polymer fractions: 1 g of $M = 20,000$, 2 g of $M = 50,000$, and 0.5 g of $M = 100,000$.
20.17   For a condensation polymerization of a hydroxyacid in which 95% of the acid groups is used up, calculate (*a*) the average number of monomer units in the polymer molecules, (*b*) the probability that a molecule chosen at random will have this number of residues, and (*c*) the weight fraction having this particular number of monomer units.
20.18   For the polymer described in problem 20.17 calculate the number average and weight average molecular weights.
20.19   For a polymethylene chain, H(CH$_2$)$_n$H, of molecular weight 100,000, (*a*) calculate the end-to-end distance assuming an exactly linear chain and a C—C distance of 1.54 Å. (*b*) Calculate the root-mean-square end-to-end distance assuming chain is freely joined. (*c*) Calculate the root-mean-square end-to-end distance assuming that the bond angle is 109° 28′. If the valence angles are fixed at θ, the root mean-square-end-to-end distance is given by

$$\langle r^2 \rangle^{\frac{1}{2}} = n^{\frac{1}{2}} l \left(\frac{1 + \cos\theta}{1 - \cos\theta} \right)^{\frac{1}{2}}$$

20.20   A sample of polystyrene was dissolved in toluene, and the following flow times in an Ostwald viscometer at 25° were obtained for different concentrations:

Concentration, g/100 cm^3	0	0.1	0.3	0.6	0.9
Time, s	86.0	99.5	132	194	301

If the constants in equation 20.19 are $K = 3.7 \times 10^{-4}$ and $a = 0.62$ for this polymer, calculate the molecular weight.

20.21 Show that if the effective hydrodynamic radius of a random-coil polymer molecule in solution is proportional to the radius of gyration, the constant a in $[\eta] = KM^a$ is expected to be $\frac{1}{2}$.

20.22 The sedimentation coefficient of gamma globulin at $20°$ is 7.1×10^{-13} s. Calculate how far the protein boundary will sediment in $\frac{1}{2}$ hr if the speed of the centrifuge is 60,000 rpm and the initial boundary is 6.50 cm from the axis of rotation.

20.23 Assuming that DNA with molecular weight 10^7 is a rigid rod calculate the sedimentation coefficient S that would be expected in dilute aqueous solution at $25°$. The rod would be about 51,400 Å long and 22 Å in diameter.

The translational frictional coefficient f for a cigar-shaped ellipsoid of revolution of semimajor axis a and semiminor axis b is

$$f = 6\pi\eta_0 a/\ln (2a/b)$$

where η_0 is the viscosity of the solvent.

The sedimentation coefficient S can be calculated from the molecular weight M, partial specific volume $\bar{v}$, solvent density ρ and frictional coefficient f using

$$S = M(1 - \bar{v}\rho)/N_A f$$

The partial specific volume of the sodium salt of DNA is 0.556 cm^3 g^{-1}.

The experimental value of S is 22.5×10^{-13} s. How do you interpret the difference from your calculated value?

20.24 Using data in Table 20.2, verify the molecular weight given for urease.

20.25 Assuming that a protein of molecular weight 16,000 behaves in aqueous solution as a sphere, calculate
(a) diffusion coefficient D,
(b) standard deviation σ of a diffusing boundary after 10 hr,
(c) the sedimentation coefficient,
(d) the distance a molecule sediments in 1 hr at 60,000 rpm starting at 6.0 cm from the axis of rotation.
At $20°$ C

$$\eta = 0.001005 \text{ Pa s}$$
$$\bar{v} = 0.75 \text{ cm}^3 \text{ g}^{-1}$$
$$\rho_{H_2O} = 0.9982 \text{ g cm}^{-3}$$

20.26 Using the diffusion coefficient of lactalbumin in Table 20.2, calculate the maximum molecular weight this substance could have. The highest molecular weight is obtained by assuming that the molecules are spherical. The absolute viscosity of water at $20°$ is 0.001005 Pa s.

20.27 For spherical particles with a radius of 20 Å in water at $25°$ ($\eta = 0.001$ Pa s) calculate the (a) frictional coefficient, (b) the diffusion coefficient, and (c) the ionic mobility at zero ionic strength, assuming one charge per particle (see Section 11.9).

20.28 The following osmotic pressures of polyvinyl acetate in dioxane were measured by G. V. Browning and J. D. Ferry at $25°$:

c, g/100 cm^3	0.292	0.579	0.810	1.140
Π, cm of solvent	0.73	1.76	2.73	4.68

Calculate the number average molecular weight. The density of dioxane is 1.035 g cm^{-3}.

20.29 Human blood plasma contains approximately 40 g of albumin ($M = 69{,}000$) and 20 g of globulin ($M = 160{,}000$) per liter. Calculate the colloid osmotic pressure at 37°, ignoring the Donnan effect.

20.30 A solution of high polymer in benzene has a concentration of 1 g/100 cm³ and a refractive index for the sodium D line (5890 Å) of 1.5021. The refractive index of benzene under these conditions is 1.5011. The turbidity τ is 2×10^{-4} cm⁻¹. What is the molecular weight of the polymer? What is I/I_0 for a 10-cm cell?

20.31 Calculate number average and weight average molecular weights for the following mixture: 1 g of polymer of $M = 10^4$ and 1 g of polymer of $M = 10^6$.

20.32 Construct a plot of the probability of finding a monomer unit in an i-mer versus i for $p = 0.97$ for a linear polymerization.

20.33 For a free jointed polymethylene chain of molecular weight 100,000, plot the probability density for end-to-end distance r versus r.

20.34 At low polymer concentrations plots of η_{sp}/c versus c are linear and may be represented by

$$\frac{\eta_{sp}}{c} = [\eta] + k_1[\eta]^2 c$$

where $[\eta]$ is the intrinsic viscosity and k_1 is a constant. The intrinsic viscosity may also be obtained by plotting $(1/c) \ln (1 + \eta_{sp})$ versus c. Show that the initial slope of this latter plot is given by $[\eta]^2(k_1 - \frac{1}{2})$.

20.35 The values of K and a for polystyrene dissolved in toluene at 25° are 3.7×10^{-4} and 0.62, respectively. Calculate the molecular weight of a sample of polystyrene having an intrinsic viscosity of 0.74.

20.36 In an ultracentrifuge experiment with egg albumin in a buffered aqueous solution at a pH of 6, the boundary moved 6.3 mm in 105 min, as determined by measurement of the change in position of the refractive-index gradient with time. The speed of the centrifuge was 57,000 rpm. The distance from the center of rotation was 6.43 cm. Show that the sedimentation coefficient for this protein is 3.5×10^{-13} s after multiplying by 0.81 to correct for density and viscosity.

20.37 Given the diffusion coefficient for sucrose at 20° C in water ($D = 45.5 \times 10^{-11}$ m² s⁻¹), calculate its sedimentation coefficient. The partial specific volume $\bar{v}$ is 0.630 cm³ g⁻¹.

20.38 Calculate the translational diffusion coefficient of the DNA of $M = 10^7$ discussed in problem 20.23. What would be the half-width of a boundary after 24-hr diffusion?

20.39 A sharp boundary is formed between a solution of hemoglobin in a buffer and the buffer solution at 25°. After 10 hr the half-width of the concentration-gradient curve at the inflection point is 0.226 cm. What is the diffusion coefficient of hemoglobin under these conditions?

20.40 The diffusion coefficient of a certain virus having spherical particles is 0.50×10^{-11} m² s⁻¹ at 0° in a solution with a viscosity of 0.00180 Pa s. Calculate the molecular weight of this virus, assuming that the density of the virus is 1 g cm⁻³.

20.41 Assuming that serum albumin molecules are spherical calculate the radius from the fact that the diffusion coefficient at 20° is 6.15×10^{-7} cm² s⁻¹. The coefficient of viscosity of water at this temperature is 0.001 Pa s.

20.42 Estimate the diffusion coefficients in water at 25° of spherical molecules with molecular weights of 10^3 and 10^5. ($\bar{v} = 0.730$ cm³ g⁻¹.)

20.43 A protein solution containing 6 g of protein per 100 cm³ of solution has a colloid osmotic pressure of 22 mm of water at 25° at the isoelectric point. What is the molecular weight of the protein?

20.44 Egg albumin has a molecular weight of 44,000. What is the osmotic pressure at 25° of a solution containing 5 g per liter?

20.45 Calculate the velocity of an ultracentrifuge in revolutions per minute required to sediment bushy stunt virus $(M = 10,700,000)$ so that its concentration at the bottom of cell $(r_2 = 6.5$ cm$)$ is $5 \times$ greater than its concentration at the meniscus $(r_1 = 6.2$ cm$)$ at equilibrium at 25°. The partial specific volume of the virus is 0.74 cm^3 g^{-1} and the density of the aqueous salt solution used was 1.00 g cm^{-3}.

APPENDIX

UNITS AND GENERAL PHYSICAL CONSTANTS

SI Units

The International System of units was defined and given official status by the General Conference on Weights and Measures in 1960.*

The seven SI base units are defined as follows:

Physical Quantity	Name of Unit	Symbol	Definition
Length	meter	m	1,650,763.73 wavelengths in vacuum of the orange-red line of the spectrum of krypton-86
Mass	kilogram	kg	A cylinder of platinum-iridium alloy kept by the International Bureau of Weights and Measures in Paris
Time	second	s	The duration of 9,192,631,770 cycles of the radiation associated with a specific transition of the cesium atom
Electric current	ampere	A	The magnitude of the current that, when flowing through each of two long parallel wires separated by 1 m in free space, results in a force between the two wires of 2×10^{-7} N for each meter of length
Thermodynamic temperature	kelvin	K	Origin is at absolute zero and the triple point of water is 273.16 K
Luminous intensity	candela	cd	The luminous intensity, in the perpendicular direction, of a surface of 1/600,000 sq m of a black body at the temperature of freezing platinum under a pressure of 101,325 N m^{-2}
Amount of substance	mole	mol	Amount of substance that contains as many elementary entities as there are carbon atoms in 0.012 kg of carbon-12

* See *J. Chem. Ed.*, **48**, 569 (1971). M. A. Paul, *J. Chem. Documentation*, **11**, 3 (1971). M. L. McGlashan, *Pure and Applied Chem.*, **21**, No. 1 (1970). M. L. McGlashan, *Physiochemical Quantities and Units*, The Royal Institute of Chemistry, London, 1971. M. L. McGlashan, *Annual Review of Physical Chemistry*, **24**, 51 (1974).

It is important to be able to convert from cgs or other units to SI. Some of these conversion factors are as follows:

Physical Quantity	Name of Unit	Symbol	Definition in SI Units
Length	Ångstrom	Å	10^{-10} m (10^{-1} nm)
Energy	electronvolt	eV	1.6022×10^{-19} J
	wave number	cm^{-1}	1.986×10^{-23} J
	calorie (I.T.)		4.1868 J
	calorie (15° C)		4.1855 J
	calorie (thermochemical)		4.184 J
	erg		10^{-7} J
Force	dyne		10^{-5} N
Pressure	atmosphere		101.325 kN m^{-2}
	Torr		133.322 N m^{-2}
Electric charge	e.s.u.		3.334×10^{-10} C
dipole moment	debye (10^{-18} e.s.u. cm)		3.334×10^{-30} C m
Magnetic field strength	oersted		79.6 A m^{-1}
magnetic flux density	gauss	G	10^{-4} T

Values of Fundamental Constants*

Constant	Symbol	Value (with uncertainty)
Speed of light in vacuum	c	$2.99792458(1) \times 10^8$ m s^{-1}
Permittivity of vacuum	ϵ_0	$8.85418782(5) \times 10^{-12}$ C^2 N^{-1} m^{-2}
Elementary charge	e	$1.6021892(46) \times 10^{-19}$ C
Planck constant	h	$6.626176(36) \times 10^{-34}$ J s
Avogadro constant	N_A	$6.022045(31) \times 10^{23}$ mol^{-1}
Atomic mass unit	$u = 10^{-3}$ kg mol$^{-1}/N_A$	$1.6605655(86) \times 10^{-27}$ kg
Rest mass of electron	m_e	$9.109534(47) \times 10^{-31}$ kg
Rest mass of proton	m_p	$1.6726485(86) \times 10^{-27}$ kg
Rest mass of neutron	m_n	$1.6749543(86) \times 10^{-27}$ kg
Faraday constant	$F = N_A e$	$9.648456(27) \times 10^4$ C mol^{-1}
Rydberg constant	R_∞	$1.097373177(83) \times 10^7$ m^{-1}
Hartree energy	H	$4.359814(24) \times 10^{-18}$ J
Bohr radius	a_0	$5.2917706(44) \times 10^{-11}$ m
Bohr magneton	μ_B	$9.274078(36) \times 10^{-24}$ J T^{-1}
Nuclear magneton	μ_N	$5.050824(20) \times 10^{-27}$ J T^{-1}
Gas constant	R	$8.31441(26)$ J K^{-1} mol^{-1}
Boltzmann constant	$k = R/N_A$	$1.380662(44) \times 10^{-23}$ J K^{-1}

* *Manual of Symbols and Terminology for Physicochemical Quantities and Units, International Union of Pure and Applied Chemistry*, Butterworths, London, 1973.

The digits in parentheses following a numerical value represent the standard deviation of that value in terms of the final listed digits.

Greek Alphabet

A	α	Alpha	N	ν	Nu
B	β	Beta	Ξ	ξ	Xi
Γ	γ	Gamma	O	o	Omicron
Δ	δ	Delta	Π	π	Pi
E	ϵ	Epsilon	P	ρ	Rho
Z	ζ	Zeta	Σ	σ	Sigma
H	η	Eta	T	τ	Tau
Θ	θ	Theta	Υ	υ	Upsilon
I	ι	Iota	Φ	ϕ	Phi
K	κ	Kappa	X	χ	Chi
Λ	λ	Lambda	Ψ	ψ	Psi
M	μ	Mu	Ω	ω	Omega

INDEX

LOGARITHMS

Natural Numbers	0	1	2	3	4	5	6	7	8	9	Proportional Parts								
											1	2	3	4	5	6	7	8	9
10	0000	0043	0086	0128	0170	0212	0253	0294	0334	0374	4	8	12	17	21	25	29	33	37
11	0414	0453	0492	0531	0569	0607	0645	0682	0719	0755	4	8	11	15	19	23	26	30	34
12	0792	0828	0864	0899	0934	0969	1004	1038	1072	1106	3	7	10	14	17	21	24	28	31
13	1139	1173	1206	1239	1271	1303	1335	1367	1399	1430	3	6	10	13	16	19	23	26	29
14	1461	1492	1523	1553	1584	1614	1644	1673	1703	1732	3	6	9	12	15	18	21	24	27
15	1761	1790	1818	1847	1875	1903	1931	1959	1987	2014	3	6	8	11	14	17	20	22	25
16	2041	2068	2095	2122	2148	2175	2201	2227	2253	2279	3	5	8	11	13	16	18	21	24
17	2304	2330	2355	2380	2405	2430	2455	2480	2504	2529	2	5	7	10	12	15	17	20	22
18	2553	2577	2601	2625	2648	2672	2695	2718	2742	2765	2	5	7	9	12	14	16	19	21
19	2788	2810	2833	2856	2878	2900	2923	2945	2967	2989	2	4	7	9	11	13	16	18	20
20	3010	3032	3054	3075	3096	3118	3139	3160	3181	3201	2	4	6	8	11	13	15	17	19
21	3222	3243	3263	3284	3304	3324	3345	3365	3385	3404	2	4	6	8	10	12	14	16	18
22	3424	3444	3464	3483	3502	3522	3541	3560	3579	3598	2	4	6	8	10	12	14	15	17
23	3617	3636	3655	3674	3692	3711	3729	3747	3766	3784	2	4	6	7	9	11	13	15	17
24	3802	3820	3838	3856	3874	3892	3909	3927	3945	3962	2	4	5	7	9	11	12	14	16
25	3979	3997	4014	4031	4048	4065	4082	4099	4116	4133	2	3	5	7	9	10	12	14	15
26	4150	4166	4183	4200	4216	4232	4249	4265	4281	4298	2	3	5	7	8	10	11	13	15
27	4314	4330	4346	4362	4378	4393	4409	4425	4440	4456	2	3	5	6	8	9	11	13	14
28	4472	4487	4502	4518	4533	4548	4564	4579	4594	4609	2	3	5	6	8	9	11	12	14
29	4624	4639	4654	4669	4683	4698	4713	4728	4742	4757	1	3	4	6	7	9	10	12	13
30	4771	4786	4800	4814	4829	4843	4857	4871	4886	4900	1	3	4	6	7	9	10	11	13
31	4914	4928	4942	4955	4969	4983	4997	5011	5024	5038	1	3	4	6	7	8	10	11	12
32	5051	5065	5079	5092	5105	5119	5132	5145	5159	5172	1	3	4	5	7	8	9	11	12
33	5185	5198	5211	5224	5237	5250	5263	5276	5289	5302	1	3	4	5	6	8	9	10	12
34	5315	5328	5340	5353	5366	5378	5391	5403	5416	5428	1	3	4	5	6	8	9	10	11
35	5441	5453	5465	5478	5490	5502	5514	5527	5539	5551	1	2	4	5	6	7	9	10	11
36	5563	5575	5587	5599	5611	5623	5635	5647	5658	5670	1	2	4	5	6	7	8	10	11
37	5682	5694	5705	5717	5729	5740	5752	5763	5775	5786	1	2	3	5	6	7	8	9	10
38	5798	5809	5821	5832	5843	5855	5866	5877	5888	5899	1	2	3	5	6	7	8	9	10
39	5911	5922	5933	5944	5955	5966	5977	5988	5999	6010	1	2	3	4	5	7	8	9	10
40	6021	6031	6042	6053	6064	6075	6085	6096	6107	6117	1	2	3	4	5	6	8	9	10
41	6128	6138	6149	6160	6170	6180	6191	6201	6212	6222	1	2	3	4	5	6	7	8	9
42	6232	6243	6253	6263	6274	6284	6294	6304	6314	6325	1	2	3	4	5	6	7	8	9
43	6335	6345	6355	6365	6375	6385	6395	6405	6415	6425	1	2	3	4	5	6	7	8	9
44	6435	6444	6454	6464	6474	6484	6493	6503	6513	6522	1	2	3	4	5	6	7	8	9
45	6532	6542	6551	6561	6571	6580	6590	6599	6609	6618	1	2	3	4	5	6	7	8	9
46	6628	6637	6646	6656	6665	6675	6684	6693	6702	6712	1	2	3	4	5	6	7	7	8
47	6721	6730	6739	6749	6758	6767	6776	6785	6794	6803	1	2	3	4	5	5	6	7	8
48	6812	6821	6830	6839	6848	6857	6866	6875	6884	6893	1	2	3	4	4	5	6	7	8
49	6902	6911	6920	6928	6937	6946	6955	6964	6972	6981	1	2	3	4	4	5	6	7	8
50	6990	6998	7007	7016	7024	7033	7042	7050	7059	7067	1	2	3	3	4	5	6	7	8
51	7076	7084	7093	7101	7110	7118	7126	7135	7143	7152	1	2	3	3	4	5	6	7	8
52	7160	7168	7177	7185	7193	7202	7210	7218	7226	7235	1	2	2	3	4	5	6	7	7
53	7243	7251	7259	7267	7275	7284	7292	7300	7308	7316	1	2	2	3	4	5	6	6	7
54	7324	7332	7340	7348	7356	7364	7372	7380	7388	7396	1	2	2	3	4	5	6	6	7